International Series in Operations Research & Management Science

Volume 228

Series Editor

Camille C. Price
Stephen F. Austin State University, TX, USA

Associate Series Editor

Joe Zhu
Worcester Polytechnic Institute, MA, USA

Founding Series Editor

Frederick S. Hillier
Stanford University, CA, USA

More information about this series at http://www.springer.com/series/6161

David G. Luenberger • Yinyu Ye

Linear and Nonlinear Programming

Fourth Edition

Springer

David G. Luenberger
Department of Management Science
and Engineering
Stanford University
Stanford, CA, USA

Yinyu Ye
Department of Management Science
and Engineering
Stanford University
Stanford, CA, USA

ISSN 0884-8289 ISSN 2214-7934 (electronic)
International Series in Operations Research & Management Science
ISBN 978-3-319-18841-6 ISBN 978-3-319-18842-3 (eBook)
DOI 10.1007/978-3-319-18842-3

Library of Congress Control Number: 2015942692

Springer Cham Heidelberg New York Dordrecht London

Printed on acid-free paper

Springer International Publishing AG Switzerland is part of Springer Science+Business Media (www.springer.com)

To Susan, Robert, Jill, and Jenna;
Daisun, Fei, Tim, and Kaylee

Preface

This book is intended as a text covering the central concepts of practical optimization techniques. It is designed for either self-study by professionals or classroom work at the undergraduate or graduate level for students who have a technical background in engineering, mathematics, or science. Like the field of optimization itself, which involves many classical disciplines, the book should be useful to system analysts, operations researchers, numerical analysts, management scientists, and other specialists from the host of disciplines from which practical optimization applications are drawn. The prerequisites for convenient use of the book are relatively modest; the prime requirement being some familiarity with introductory elements of linear algebra. Certain sections and developments do assume some knowledge of more advanced concepts of linear algebra, such as eigenvector analysis, or some background in sets of real numbers, but the text is structured so that the mainstream of the development can be faithfully pursued without reliance on this more advanced background material.

Although the book covers primarily material that is now fairly standard, this edition emphasizes methods that are both state-of-the-art and popular. One major insight is the connection between the purely analytical character of an optimization problem, expressed perhaps by properties of the necessary conditions, and the behavior of algorithms used to solve a problem. This was a major theme of the first edition of this book and the fourth edition expands and further illustrates this relationship.

As in the earlier editions, the material in this fourth edition is organized into three separate parts. Part I is a self-contained introduction to linear programming, a key component of optimization theory. The presentation in this part is fairly conventional, covering the main elements of the underlying theory of linear programming, many of the most effective numerical algorithms, and many of its important special applications. Part II, which is independent of Part I, covers the theory of unconstrained optimization, including both derivations of the appropriate optimality conditions and an introduction to basic algorithms. This part of the book explores the general properties of algorithms and defines various notions of convergence. Part III extends the concepts developed in the second part to constrained optimization problems. Except for a few isolated sections, this part is also independent of Part I. It is possible to go directly into Parts II and III omitting Part I, and, in fact, the book has been used in this way in many universities. Each part of the book contains enough material to form the basis of a one-quarter course. In either classroom use or for self-study, it is important not to overlook the suggested exercises at the end of each chapter. The selections generally include exercises of a computational variety designed to test one's understanding of a particular algorithm, a theoretical variety designed to test one's understanding of a given theoretical development, or of the variety that extends the presentation of the chapter to new applications or theoretical areas. One should attempt at least four or five exercises from each chapter. In progressing through the book it would be unusual to read straight through from cover to cover. Generally, one will wish to skip around. In order to facilitate this mode, we have indicated sections of a specialized or digressive nature with an asterisk*.

New to this edition is a special Chap. 6 devoted to Conic Linear Programming, a powerful generalization of Linear Programming. While the constraint set in a normal linear program is defined by a finite number of linear inequalities of finite-dimensional vector variables, the constraint set in conic linear programming may be defined, for example, as a linear combination of symmetric positive semi-definite matrices of a given dimension. Indeed, many conic structures are possible and useful in a variety of applications. It must be recognized, however, that conic linear programming is an advanced topic, requiring special study.

Another important topic is an accelerated steepest descent method that exhibits superior convergence properties, and for this reason, has become quite popular. The proof of the convergence property for both standard and accelerated steepest descent methods are presented in Chap. 8.

As the field of optimization advances, addressing greater complexity, treating problems with ever more variables (as in Big Data situations), ranging over diverse applications. The field responds yo these challenges, developing new algorithms, building effective software, and expanding overall theory. An example of a valuable new development is the work on big data problems. Surprisingly, coordinate descent, with randomly selected coordinates at each step, is quite effective as explained in Chap. 8. As another example some problems are formulated so that the unknowns can be split into two sub groups, there are linear constraints and the objective function is separable with respect to the two groups of variables. The augmented Lagrangian can be computed and it is natural to use an alternating series method. We discuss the alternating direction method with multipliers as a dual method in Chap. 14. Interestingly, this method is convergent for when the number of partition groups is two, but not for finer partitions.

We wish to thank the many students and researchers who over the years have given us comments concerning the book and those who encouraged us to carry out this revision.

Stanford, CA, USA D.G. Luenberger
Stanford, CA, USA Y. Ye
January 2015

Contents

Chapter 1
Introduction

1.1 Optimization

The concept of optimization is now well rooted as a principle underlying the analysis of many complex decision or allocation problems. It offers a certain degree of philosophical elegance that is hard to dispute, and it often offers an indispensable degree of operational simplicity. Using this optimization philosophy, one approaches a complex decision problem, involving the selection of values for a number of interrelated variables, by focusing attention on a single objective designed to quantify performance and measure the quality of the decision. This one objective is maximized (or minimized, depending on the formulation) subject to the constraints that may limit the selection of decision variable values. If a suitable single aspect of a problem can be isolated and characterized by an objective, be it profit or loss in a business setting, speed or distance in a physical problem, expected return in the environment of risky investments, or social welfare in the context of government planning, optimization may provide a suitable framework for analysis.

It is, of course, a rare situation in which it is possible to fully represent all the complexities of variable interactions, constraints, and appropriate objectives when faced with a complex decision problem. Thus, as with all quantitative techniques of analysis, a particular optimization formulation should be regarded only as an approximation. Skill in modeling, to capture the essential elements of a problem, and good judgment in the interpretation of results are required to obtain meaningful conclusions. Optimization, then, should be regarded as a tool of conceptualization and analysis rather than as a principle yielding the philosophically correct solution.

Skill and good judgment, with respect to problem formulation and interpretation of results, is enhanced through concrete practical experience and a thorough understanding of relevant theory. Problem formulation itself always involves a tradeoff between the conflicting objectives of building a mathematical model sufficiently complex to accurately capture the problem description and building a model that is

D.G. Luenberger, Y. Ye, *Linear and Nonlinear Programming*, International Series in Operations Research & Management Science 228,
DOI 10.1007/978-3-319-18842-3_1

tractable. The expert model builder is facile with both aspects of this tradeoff. One aspiring to become such an expert must learn to identify and capture the important issues of a problem mainly through example and experience; one must learn to distinguish tractable models from nontractable ones through a study of available technique and theory and by nurturing the capability to extend existing theory to new situations.

This book is centered around a certain optimization structure—that characteristic of linear and nonlinear programming. Examples of situations leading to this structure are sprinkled throughout the book, and these examples should help to indicate how practical problems can be often fruitfully structured in this form. The book mainly, however, is concerned with the development, analysis, and comparison of algorithms for solving general subclasses of optimization problems. This is valuable not only for the algorithms themselves, which enable one to solve given problems, but also because identification of the collection of structures they most effectively solve can enhance one's ability to formulate problems.

1.2 Types of Problems

The content of this book is divided into three major parts: Linear Programming, Unconstrained Problems, and Constrained Problems. The last two parts together comprise the subject of nonlinear programming.

Linear Programming

Linear programming is without doubt the most natural mechanism for formulating a vast array of problems with modest effort. A linear programming problem is characterized, as the name implies, by linear functions of the unknowns; the objective is linear in the unknowns, and the constraints are linear equalities or linear inequalities in the unknowns. One familiar with other branches of linear mathematics might suspect, initially, that linear programming formulations are popular because the mathematics is nicer, the theory is richer, and the computation simpler for linear problems than for nonlinear ones. But, in fact, these are *not* the primary reasons. In terms of mathematical and computational properties, there are much broader classes of optimization problems than linear programming problems that have elegant and potent theories and for which effective algorithms are available. It seems that the popularity of linear programming lies primarily with the formulation phase of analysis rather than the solution phase—and for good cause. For one thing, a great number of constraints and objectives that arise in practice *are* indisputably linear. Thus, for example, if one formulates a problem with a budget constraint restricting the total amount of money to be allocated among two different commodities, the budget constraint takes the form $x_1 + x_2 \leq B$, where x_i, $i = 1, 2$, is the amount allocated to activity i, and B is the budget. Similarly, if the objective is, for example, maximum weight, then it can be expressed as $w_1x_1 + w_2x_2$, where w_j, $i = 1, 2$, is the unit weight of the commodity i. The overall problem would be expressed as

$$\begin{aligned} \text{maximize } & w_1x_1 + w_2x_2 \\ \text{subject to } & x_1 + x_2 \leq B, \\ & x_1 \geq 0, \ x_2 \geq 0, \end{aligned}$$

which is an elementary linear program. The linearity of the budget constraint is extremely natural in this case and does not represent simply an approximation to a more general functional form.

Another reason that linear forms for constraints and objectives are so popular in problem formulation is that they are often the least difficult to define. Thus, even if an objective function is not purely linear by virtue of its inherent definition (as in the above example), it is often far easier to define it as being linear than to decide on some other functional form and convince others that the more complex form is the best possible choice. Linearity, therefore, by virtue of its simplicity, often is selected as the easy way out or, when seeking generality, as the only functional form that will be equally applicable (or nonapplicable) in a class of similar problems.

Of course, the theoretical and computational aspects do take on a somewhat special character for linear programming problems—the most significant development being the simplex method. This algorithm is developed in Chaps. 2 and 3. More recent interior point methods are nonlinear in character and these are developed in Chap. 5.

Unconstrained Problems

It may seem that unconstrained optimization problems are so devoid of structural properties as to preclude their applicability as useful models of meaningful problems. Quite the contrary is true for two reasons. First, it can be argued, quite convincingly, that if the scope of a problem is broadened to the consideration of all relevant decision variables, there may then be no constraints—or put another way, constraints represent artificial delimitations of scope, and when the scope is broadened the constraints vanish. Thus, for example, it may be argued that a budget constraint is not characteristic of a meaningful problem formulation; since by borrowing at some interest rate it is always possible to obtain additional funds, and hence rather than introducing a budget constraint, a term reflecting the cost of funds should be incorporated into the objective. A similar argument applies to constraints describing the availability of other resources which at some cost (however great) could be supplemented.

The second reason that many important problems can be regarded as having no constraints is that constrained problems are sometimes easily converted to unconstrained problems. For instance, the sole effect of equality constraints is simply to limit the degrees of freedom, by essentially making some variables functions of others. These dependencies can sometimes be explicitly characterized, and a new problem having its number of variables equal to the true degree of freedom can be determined. As a simple specific example, a constraint of the form $x_1 + x_2 = B$ can be eliminated by substituting $x_2 = B - x_1$ everywhere else that x_2 appears in the problem.

Aside from representing a significant class of practical problems, the study of unconstrained problems, of course, provides a stepping stone toward the more general case of constrained problems. Many aspects of both theory and algorithms are most naturally motivated and verified for the unconstrained case before progressing to the constrained case.

Constrained Problems

In spite of the arguments given above, many problems met in practice are formulated as constrained problems. This is because in most instances a complex problem such as, for example, the detailed production policy of a giant corporation, the planning of a large government agency, or even the design of a complex device cannot be directly treated in its entirety accounting for all possible choices, but instead must be decomposed into separate subproblems—each subproblem having constraints that are imposed to restrict its scope. Thus, in a planning problem, budget constraints are commonly imposed in order to decouple that one problem from a more global one. Therefore, one frequently encounters general nonlinear constrained mathematical programming problems.

The general mathematical programming problem can be stated as

$$\begin{aligned} \text{minimize} \quad & f(\mathbf{x}) \\ \text{subject to} \quad & h_j(\mathbf{x}) = 0, \ i = 1, 2, \ \ldots, \ m \\ & g_j(\mathbf{x}) \leq 0, \ j = 1, 2, \ p \\ & \mathbf{x} \in S. \end{aligned}$$

In this formulation, $\mathbf{x}$ is an n-dimensional vector of unknowns, $\mathbf{x} = (x_1, x_2, \ldots, x_n)$, and f, h_i, $i = 1, 2, \ldots, m$, and g_j, $j = 1, 2, \ldots, p$, are real-valued functions of the variables $x_1, x_2, \ldots, x_n$. The set S is a subset of n-dimensional space. The function f is the *objective function* of the problem and the equations, inequalities, and set restrictions are *constraints*.

Generally, in this book, additional assumptions are introduced in order to make the problem smooth in some suitable sense. For example, the functions in the problem are usually required to be continuous, or perhaps to have continuous derivatives. This ensures that small changes in $\mathbf{x}$ lead to small changes in other values associated with the problem. Also, the set S is not allowed to be arbitrary but usually is required to be a connected region of n-dimensional space, rather than, for example, a set of distinct isolated points. This ensures that small changes in $\mathbf{x}$ can be made. Indeed, in a majority of problems treated, the set S is taken to be the entire space; there is no set restriction.

In view of these smoothness assumptions, one might characterize the problems treated in this book as *continuous variable programming*, since we generally discuss problems where all variables and function values can be varied continuously. In fact, this assumption forms the basis of many of the algorithms discussed, which operate essentially by making a series of small movements in the unknown $\mathbf{x}$ vector.

1.3 Size of Problems

One obvious measure of the complexity of a programming problem is its size, measured in terms of the number of unknown variables or the number of constraints. As might be expected, the size of problems that can be effectively solved has been increasing with advancing computing technology and with advancing theory. Today, with present computing capabilities, however, it is reasonable to distinguish three classes of problems: *small-scale problems* having about five or fewer unknowns and constraints; *intermediate-scale problems* having from about five to a hundred or a thousand variables; and *large-scale problems* having perhaps thousands or even millions of variables and constraints. This classification is not entirely rigid, but it reflects at least roughly not only size but the basic differences in approach that accompany different size problems. As a rough rule, small-scale problems can be solved by hand or by a small computer. Intermediate-scale problems can be solved on a personal computer with general purpose mathematical programming codes. Large-scale problems require sophisticated codes that exploit special structure and usually require large computers.

Much of the basic theory associated with optimization, particularly in nonlinear programming, is directed at obtaining necessary and sufficient conditions satisfied by a solution point, rather than at questions of computation. This theory involves mainly the study of Lagrange multipliers, including the Karush-Kuhn-Tucker Theorem and its extensions. It tremendously enhances insight into the philosophy of constrained optimization and provides satisfactory basic foundations for other important disciplines, such as the theory of the firm, consumer economics, and optimal control theory. The interpretation of Lagrange multipliers that accompanies this theory is valuable in virtually every optimization setting. As a basis for computing numerical solutions to optimization, however, this theory is far from adequate, since it does not consider the difficulties associated with solving the equations resulting from the necessary conditions.

If it is acknowledged from the outset that a given problem is too large and too complex to be efficiently solved by hand (and hence it is acknowledged that a computer solution is desirable), then one's theory should be directed toward development of procedures that exploit the efficiencies of computers. In most cases this leads to the abandonment of the idea of solving the set of necessary conditions in favor of the more direct procedure of searching through the space (in an intelligent manner) for ever-improving points.

Today, search techniques can be effectively applied to more or less general nonlinear programming problems. Problems of great size, *large-scale programming* problems, can be solved if they possess special structural characteristics, especially sparsity, that can be exploited by a solution method. Today linear programming software packages are capable of automatically identifying sparse structure within the input data and taking advantage of this sparsity in numerical computation. It is now not uncommon to solve linear programs of up to a million variables and constraints, as long as the structure is sparse. Problem-dependent methods, where the structure is not automatically identified, are largely directed to transportation and network flow problems as discussed in the book.

This book focuses on the aspects of general theory that are most fruitful for computation in the widest class of problems. While necessary and sufficient conditions are examined and their application to small-scale problems is illustrated, our primary interest in such conditions is in their role as the core of a broader theory applicable to the solution of larger problems. At the other extreme, although some instances of structure exploitation are discussed, we focus primarily on the general continuous variable programming problem rather than on special techniques for special structures.

1.4 Iterative Algorithms and Convergence

The most important characteristic of a high-speed computer is its ability to perform repetitive operations efficiently, and in order to exploit this basic characteristic, most algorithms designed to solve large optimization problems are iterative in nature. Typically, in seeking a vector that solves the programming problem, an initial vector $\mathbf{x}_0$ is selected and the algorithm generates an improved vector $\mathbf{x}_1$. The process is repeated and a still better solution $\mathbf{x}_2$ is found. Continuing in this fashion, a sequence of ever-improving points $\mathbf{x}_0, \mathbf{x}_1, \ldots, \mathbf{x}_k, \ldots$, is found that approaches a solution point $\mathbf{x}^*$. For linear programming problems solved by the simplex method, the generated sequence is of finite length, reaching the solution point exactly after a finite (although initially unspecified) number of steps. For nonlinear programming problems or interior-point methods, the sequence generally does not ever exactly reach the solution point, but converges toward it. In operation, the process is terminated when a point sufficiently close to the solution point, for practical purposes, is obtained.

The theory of iterative algorithms can be divided into three (somewhat overlapping) aspects. The first is concerned with the creation of the algorithms themselves. Algorithms are not conceived arbitrarily, but are based on a creative examination of the programming problem, its inherent structure, and the efficiencies of digital computers. The second aspect is the verification that a given algorithm will in fact generate a sequence that converges to a solution point. This aspect is referred to as *global convergence analysis*, since it addresses the important question of whether the algorithm, when initiated far from the solution point, will eventually converge to it. The third aspect is referred to as *local convergence analysis* or *complexity analysis* and is concerned with the rate at which the generated sequence of points converges to the solution. One cannot regard a problem as solved simply because an algorithm is known which will converge to the solution, since it may require an exorbitant amount of time to reduce the error to an acceptable tolerance. It is essential when prescribing algorithms that some estimate of the time required be available. It is the convergence-rate aspect of the theory that allows some quantitative evaluation and comparison of different algorithms, and at least crudely, assigns a measure of tractability to a problem, as discussed in Sect. 1.1.

A modern-day technical version of Confucius' most famous saying, and one which represents an underlying philosophy of this book, might be, "One good theory is worth a thousand computer runs." Thus, the convergence properties of an iterative algorithm can be estimated with confidence either by performing numerous computer experiments on different problems or by a simple well-directed theoretical analysis. A simple theory, of course, provides invaluable insight as well as the desired estimate.

For linear programming using the simplex method, solid theoretical statements on the speed of convergence were elusive, because the method actually converges to an exact solution in a finite number of steps. The question is how many steps might be required. This question was finally resolved when it was shown that it was possible for the number of steps to be exponential in the size of the program. The situation is different for interior point algorithms, which essentially treat the problem by introducing nonlinear terms, and which therefore do not generally obtain a solution in a finite number of steps but instead converge toward a solution.

For nonlinear programs, including interior point methods applied to linear programs, it is meaningful to consider the speed of convergence. There are many different classes of nonlinear programming algorithms, each with its own convergence characteristics. However, in many cases the convergence properties can be deduced analytically by fairly simple means, and this analysis is substantiated by computational experience. Presentation of convergence analysis, which seems to be the natural focal point of a theory directed at obtaining specific answers, is a unique feature of this book.

There are in fact two aspects of convergence-rate theory. The first is generally known as *complexity analysis* and focuses on how fast the method converges overall, distinguishing between polynomial-time algorithms and non-polynomial-time algorithms. The second aspect provides more detailed analysis of how fast the method converges in the final stages, and can provide comparisons between different algorithms. Both of these are treated in this book.

The convergence-rate theory presented has two somewhat surprising but definitely pleasing aspects. First, the theory is, for the most part, extremely simple in nature. Although initially one might fear that a theory aimed at predicting the speed of convergence of a complex algorithm might itself be doubly complex, in fact the associated convergence analysis often turns out to be exceedingly elementary, requiring only a line or two of calculation. Second, a large class of seemingly distinct algorithms turns out to have a common convergence rate. Indeed, as emphasized in the later chapters of the book, there is a *canonical rate* associated with a given programming problem that seems to govern the speed of convergence of many algorithms when applied to that problem. It is this fact that underlies the potency of the theory, allowing definitive comparisons among algorithms to be made even without detailed knowledge of the problems to which they will be applied. Together these two properties, simplicity and potency, assure convergence analysis a permanent position of major importance in mathematical programming theory.

Part I
Linear Programming

Chapter 2
Basic Properties of Linear Programs

2.1 Introduction

A linear program (LP) is an optimization problem in which the objective function is linear in the unknowns and the constraints consist of linear equalities and linear inequalities. The exact form of these constraints may differ from one problem to another, but as shown below, any linear program can be transformed into the following *standard form*:

$$\begin{array}{ll} \text{minimize} & c_1x_1 + c_2x_2 + \ldots + c_nx_n \\ \text{subject to} & a_{11}x_1 + a_{12}x_2 + \ldots + a_{1n}x_n \;= b_1 \\ & a_{21}x_1 + a_{22}x_2 + \ldots + a_{2n}x_n \;= b_2 \\ & \vdots \qquad\qquad\qquad\qquad\qquad\quad \vdots \\ & a_{m1}x_1 + a_{m2}x_2 + \cdots + a_{mn}x_n = b_m \\ \text{and} & x_1 \geqslant 0,\ x_2 \geqslant 0,\ \ldots,\ x_n \geqslant 0, \end{array} \tag{2.1}$$

where the b_i's, c_i's and a_{ij}'s are fixed real constants, and the x_i's are real numbers to be determined. We always assume that each equation has been multiplied by minus unity, if necessary, so that each $b_i \geqslant 0$.

In more compact vector notation,[1] this standard problem becomes

$$\begin{array}{ll} \text{minimize} & \mathbf{c}^T\mathbf{x} \\ \text{subject to} & \mathbf{Ax} = \mathbf{b} \quad \text{and} \quad \mathbf{x} \geqslant \mathbf{0}. \end{array} \tag{2.2}$$

Here $\mathbf{x}$ is an n-dimensional column vector, $\mathbf{c}^T$ is an n-dimensional row vector, $\mathbf{A}$ is an $m \times n$ matrix, and $\mathbf{b}$ is an m-dimensional column vector. The vector inequality $\mathbf{x} \geqslant \mathbf{0}$ means that each component of $\mathbf{x}$ is nonnegative.

[1] See Appendix A for a description of the vector notation used throughout this book.

D.G. Luenberger, Y. Ye, *Linear and Nonlinear Programming*, International Series in Operations Research & Management Science 228,
DOI 10.1007/978-3-319-18842-3_2

Before giving some examples of areas in which linear programming problems arise naturally, we indicate how various other forms of linear programs can be converted to the standard form.

Example 1 (Slack Variables). Consider the problem

$$\begin{array}{ll} \text{minimize} & c_1x_1 + c_2x_2 + \cdots + c_nx_n \\ \text{subject to} & a_{11}x_1 + a_{12}x_2 + \cdots + a_{1n}x_n \leqslant b_1 \\ & a_{21}x_1 + a_{22}x_2 + \cdots + a_{2n}x_n \leqslant b_2 \\ & \vdots \qquad\qquad\qquad\qquad\qquad\qquad \vdots \\ & a_{m1}x_1 + a_{m2}x_2 + \cdots + a_{mn}x_n \leqslant b_m \\ \text{and} & x_1 \geqslant 0,\ x_2 \geqslant 0,\ \ldots,\ x_n \geqslant 0, \end{array}$$

In this case the constraint set is determined entirely by linear inequalities. The problem may be alternatively expressed as

$$\begin{array}{ll} \text{minimize} & c_1x_1 + c_2x_2 + \cdots + c_nx_n \\ \text{subject to} & a_{11}x_1 + a_{12}x_2 + \cdots + a_{1n}x_n + y_1 \qquad\qquad = b_1 \\ & a_{21}x_1 + a_{22}x_2 + \cdots + a_{2n}x_n \quad + y_2 \qquad = b_2 \\ & \vdots \qquad\qquad\qquad\qquad\qquad\qquad\qquad\qquad \vdots \\ & a_{m1}x_1 + a_{m2}x_2 + \cdots + a_{mn}x_n \qquad\quad + y_m = b_m \\ \text{and} & x_1 \geqslant 0,\ x_2 \geqslant 0,\ \ldots,\ x_n \geqslant 0, \\ \text{and} & y_1 \geqslant 0,\ y_2 \geqslant 0,\ \ldots,\ y_m \geqslant 0. \end{array}$$

The new positive variables y_i introduced to convert the inequalities to equalities are called *slack variables* (or more loosely, *slacks*). By considering the problem as one having $n + m$ unknowns $x_1, x_2, \ldots, x_n, y_1, y_2, \ldots, y_m$, the problem takes the standard form. The $m \times (n + m)$ matrix that now describes the linear equality constraints is of the special form $[\mathbf{A}, \mathbf{I}]$ (that is, its columns can be partitioned into two sets; the first n columns make up the original A matrix and the last m columns make up an $m \times m$ identity matrix).

Example 2 (Surplus Variables). If the linear inequalities of Example 1 are reversed so that a typical inequality is

$$a_{i1}x_1 + a_{i2}x_2 + \cdots + a_{in}x_n \geqslant b_i,$$

it is clear that this is equivalent to

$$a_{i1}x_1 + a_{i2}x_2 + \cdots + a_{in}x_n - y_i = b_i$$

with $y_i \geqslant 0$. Variables, such as y_i, adjoined in this fashion to convert a "greater than or equal to" inequality to equality are called *surplus variables*.

It should be clear that by suitably multiplying by minus unity, and adjoining slack and surplus variables, any set of linear inequalities can be converted to standard form if the unknown variables are restricted to be nonnegative.

Example 3 (Free Variables—First Method). If a linear program is given in standard form except that one or more of the unknown variables is not required to be nonnegative, the problem can be transformed to standard form by either of two simple techniques.

To describe the first technique, suppose in (2.1), for example, that the restriction $x_1 \geqslant 0$ is not present and hence x_1 is free to take on either positive or negative values. We then write

$$x_1 = u_1 - v_1, \tag{2.3}$$

where we require $u_1 \geqslant 0$ and $v_1 \geqslant 0$. If we substitute $u_1 - v_1$ for x_1 everywhere in (2.1), the linearity of the constraints is preserved and all variables are now required to be nonnegative. The problem is then expressed in terms of the $n+1$ variables $u_1, v_1, x_2, x_3, \ldots, x_n$.

There is obviously a certain degree of redundancy introduced by this technique, however, since a constant added to u_1 and v_1 does not change x_1 (that is, the representation of a given value x_1 is not unique). Nevertheless, this does not hinder the simplex method of solution.

Example 4 (Free Variables—Second Method). A second approach for converting to standard form when x_1 is unconstrained in sign is to eliminate x_1 together with one of the constraint equations. Take any one of the m equations in (2.1) which has a nonzero coefficient for x_1. Say, for example,

$$a_{i1}x_1 + a_{i2}x_2 + \cdots + a_{in}x_n = b_i, \tag{2.4}$$

where $a_{i1} \neq 0$. Then x_1 can be expressed as a linear combination of the other variables plus a constant. If this expression is substituted for x_1 everywhere in (2.1), we are led to a new problem of exactly the same form but expressed in terms of the variables $x_2, x_3, \ldots, x_n$ only. Furthermore, the ith equation, used to determine x_1, is now identically zero and it too can be eliminated. This substitution scheme is valid since any combination of nonnegative variables $x_2, x_3, \ldots, x_n$ leads to a feasible x_1 from (2.4), since the sign of x_1 is unrestricted. As a result of this simplification, we obtain a standard linear program having $n-1$ variables and $m-1$ constraint equations. The value of the variable x_1 can be determined after solution through (2.4).

Example 5 (Specific Case). As a specific instance of the above technique consider the problem

$$\begin{aligned} \text{minimize}\quad & x_1 + 3x_2 + 4x_3 \\ \text{subject to}\quad & x_1 + 2x_2 + x_3 = 5 \\ & 2x_1 + 3x_2 + x_3 = 6 \\ & x_2 \geqslant 0, \qquad x_3 \geqslant 0. \end{aligned}$$

Since x_1 is free, we solve for it from the first constraint, obtaining

$$x_1 = 5 - 2x_2 - x_3. \tag{2.5}$$

Substituting this into the objective and the second constraint, we obtain the equivalent problem (subtracting five from the objective)

$$\begin{aligned} \text{minimize} \quad & x_2 + 3x_3 \\ \text{subject to} \quad & x_2 + x_3 = 4 \\ & x_2 \geqslant 0, \; x_3 \geqslant 0, \end{aligned}$$

which is a problem in standard form. After the smaller problem is solved (the answer is $x_2 = 4$, $x_3 = 0$) the value for $x_1 (x_1 = -3)$ can be found from (2.5).

2.2 Examples of Linear Programming Problems

Linear programming has long proved its merit as a significant model of numerous allocation problems and economic phenomena. The continuously expanding literature of applications repeatedly demonstrates the importance of linear programming as a general framework for problem formulation. In this section we present some classic examples of situations that have natural formulations.

Example 1 (The Diet Problem). How can we determine the most economical diet that satisfies the basic minimum nutritional requirements for good health? Such a problem might, for example, be faced by the dietitian of a large army. We assume that there are available at the market n different foods and that the jth food sells at a price c_j per unit. In addition there are m basic nutritional ingredients and, to achieve a balanced diet, each individual must receive at least b_i units of the ith nutrient per day. Finally, we assume that each unit of food j contains a_{ij} units of the ith nutrient.

If we denote by x_j the number of units of food j in the diet, the problem then is to select the x_j's to minimize the total cost

$$c_1x_1 + c_2x_2 + \cdots + c_nx_n$$

subject to the nutritional constraints

$$a_{i1}x_1 + a_{i2}x_2 + \cdots + a_{in}x_n \geqslant b_i, \; i = 1, \ldots, m,$$

and the nonnegativity constraints

$$x_1 \geqslant 0, \; x_2 \geqslant 0, \; \ldots, \; x_n \geqslant 0$$

on the food quantities.

This problem can be converted to standard form by subtracting a nonnegative surplus variable from the left side of each of the m linear inequalities. The diet problem is discussed further in Chap. 4.

Example 2 (Manufacturing Problem). Suppose we own a facility that is capable of manufacturing n different products, each of which may require various amounts of m different resources. Each product can be produced at any level $x_j \geqslant 0$, $j = 1, 2, \ldots, n$, and each unit of the jth product can sell for p_j dollars and needs a_{ij} units of the ith resource, $i = 1, 2, \ldots, m$. Assuming linearity of the production facility, if we are given a set of m numbers $b_1, b_2, \ldots, b_m$ describing the available quantities of the m resources, and we wish to manufacture products at maximum revenue, ours decision problem is a linear program to maximize

$$p_1x_1 + p_2x_2 + \cdots + p_nx_n$$

subject to the resource constraints

$$a_{i1}x_1 + a_{i2}x_2 + \cdots + a_{in}x_n \leqslant b_i, \; i = 1, \ldots, m$$

and the nonnegativity constraints on all production variables.

Example 3 (The Transportation Problem). Quantities $a_1, a_2, \ldots, a_m$, respectively, of a certain product are to be shipped from each of m locations and received in amounts $b_1, b_2, \ldots, b_n$, respectively, at each of n destinations. Associated with the shipping of a unit of product from origin i to destination j is a shipping cost c_{ij}. It is desired to determine the amounts x_{ij} to be shipped between each origin–destination pair $i = 1, 2, \ldots, m$; $j = 1, 2, \ldots, n$; so as to satisfy the shipping requirements and minimize the total cost of transportation.

To formulate this problem as a linear programming problem, we set up the array shown below:

$$\begin{array}{|cccc|c} x_{11} & x_{12} & \cdots & x_{1n} & a_1 \\ x_{21} & x_{22} & \cdots & x_{2n} & a_2 \\ \cdot & & & \cdot & \cdot \\ \cdot & & & \cdot & \cdot \\ \cdot & & & \cdot & \cdot \\ x_{m1} & x_{m2} & \cdots & x_{mn} & a_m \\ \hline \multicolumn{4}{c}{} & \\ b_1 & b_2 & \cdots & b_n & \end{array}$$

The ith row in this array defines the variables associated with the ith origin, while the jth column in this array defines the variables associated with the jth destination. The problem is to place nonnegative variables x_{ij} in this array so that the sum across the ith row is a_j, the sum down the jth column is b_j, and the weighted sum $\sum_{j=1}^{n}\sum_{i=1}^{m} c_{ij}x_{ij}$, representing the transportation cost, is minimized.

Thus, we have the linear programming problem:

$$\begin{aligned} \text{minimize} \quad & \sum_{ij} c_{ij}x_{ij} \\ \text{subject to} \quad & \sum_{j=1}^{n} x_{ij} = a_j \quad \text{for} \quad i = 1, 2, \ldots, m \qquad (2.6) \\ & \sum_{i=1}^{m} x_{ij} = b_j \quad \text{for} \quad j = 1, 2, \ldots, n \qquad (2.7) \end{aligned}$$

$$x_{ij} \geqslant 0 \text{ for } i = 1, 2, \ldots, m; \; j = 1, 2, \ldots, n.$$

In order that the constraints (2.6) and (2.7) be consistent, we must, of course, assume that $\sum_{i=1}^{m} a_i = \sum_{j=1}^{n} b_j$ which corresponds to assuming that the total amount shipped is equal to the total amount received.

The transportation problem is now clearly seen to be a linear programming problem in mn variables. The equations (2.6) and (2.7) can be combined and expressed in matrix form in the usual manner and this results in an $(m+n) \times (mn)$ coefficient matrix consisting of zeros and ones only.

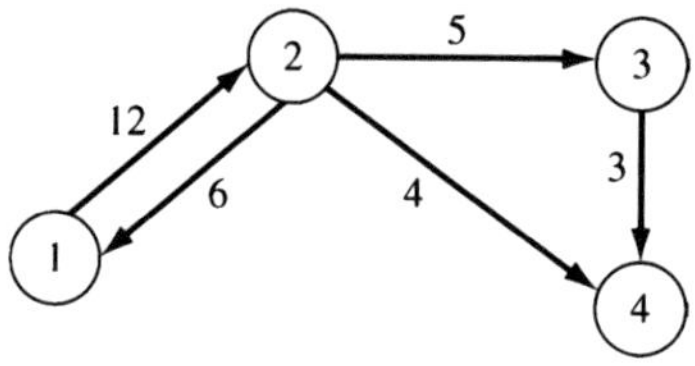

Fig. 2.1 A network with capacities

Example 4 (The Maximal Flow Problem). Consider a capacitated network (see Fig. 2.1, and Appendix D) in which two special nodes, called the *source* and the *sink*, are distinguished. Say they are nodes 1 and m, respectively. All other nodes must satisfy the strict conservation requirement; that is, the net flow into these nodes must be zero. However, the source may have a net outflow and the sink a net inflow. The outflow f of the source will equal the inflow of the sink as a consequence of the conservation at all other nodes. A set of arc flows satisfying these conditions is said to be a *flow* in the network of value f. The maximal flow problem is that of determining the maximal flow that can be established in such a network. When written out, it takes the form

$$\begin{aligned}
&\text{minimize } f \\
&\text{subject to } \sum_{j=1}^{n} x_{1j} - \sum_{j=1}^{n} x_{j1} - f = 0 \\
&\sum_{j=1}^{n} x_{ij} - \sum_{j=1}^{n} x_{ji} \qquad = 0, \qquad i \neq 1, m \qquad (2.8) \\
&\sum_{j=1}^{n} x_{mj} - \sum_{j=1}^{n} x_{jm} + f = 0 \\
&0 \leq x_{ij} \leq k_{ij}, \qquad \text{forall } i, \; j,
\end{aligned}$$

where $k_{ij} = 0$ for those no-arc pairs (i, j).

Example 5 (A Warehousing Problem). Consider the problem of operating a warehouse, by buying and selling the stock of a certain commodity, in order to maximize profit over a certain length of time. The warehouse has a fixed capacity C, and there is a cost r per unit for holding stock for one period. The price, p_i, of the commodity is known to fluctuate over a number of time periods—say months, indexed by i. In any period the same price holds for both purchase or sale. The warehouse is originally empty and is required to be empty at the end of the last period.

To formulate this problem, variables are introduced for each time period. In particular, let x_i denote the level of stock in the warehouse at the beginning of period i. Let u_i denote the amount bought during period i, and let s_i denote the amount sold during period i. If there are n periods, the problem is

$$\begin{array}{lllll}
\text{maximize} & \sum_{i=1}^{n}(p_i(s_i - u_i) - rx_i) & & & \\
\text{subject to} & x_{i+1} & = x_i + u_i - s_i & & i = 1, 2, \ldots, n-1 \\
& 0 & = x_n + u_n - s_n & & \\
& x_i + z_i & = C & & i = 2, \ldots, n \\
& x_1 = 0, \; x_i \geqslant 0, \; u_i \geqslant 0, & & & s_i \geqslant 0, \; z_i \geqslant 0,
\end{array}$$

where z_i is a slack variable. If the constraints are written out explicitly for the case $n = 3$, they take the form

$-u_1 + s_1$	$+x_2$		$=0$
	$-x_2 - u_2 + s_2$	$+x_3$	$=0$
	$x_2 \qquad + z_2$		$=C$
		$-x_3 - u_3 + s_3$	$=0$
		$x_3 \qquad + z_3$	$=C$

Note that the coefficient matrix can be partitioned into blocks corresponding to the variables of the different time periods. The only blocks that have nonzero entries are the diagonal ones and the ones immediately above the diagonal. This structure is typical of problems involving time.

Example 6 (Linear Classifier and Support Vector Machine). Suppose several d-dimensional data points are classified into two distinct classes. For example, two-dimensional data points may be grade averages in science and humanities for different students. We also know the academic major of each student, as being in science or humanities, which serves as the classification. In general we have vectors $\mathbf{a}_i \in E^d$ for $i = 1, 2, \ldots, n_1$ and vectors $\mathbf{b}_j \in E^d$ for $j = 1, 2, \ldots, n_2$. We wish to find a hyperplane that separates the $\mathbf{a}_i$'s from the $\mathbf{b}_j$'s. Mathematically we wish to find $\mathbf{y} \in E^d$ and a number β such that

$$\begin{array}{ll}
\mathbf{a}_i^T\mathbf{y} + \beta \geqslant 1 & \text{for all } i \\
\mathbf{b}_j^T\mathbf{y} + \beta \leqslant -1 & \text{for all } j,
\end{array}$$

where $\{\mathbf{x} : \mathbf{x}^T\mathbf{y} + \beta = 0\}$ is the desired hyperplane, and the separation is defined by the $+1$ and -1. This is a linear program. See Fig. 2.2.

Example 7 (Combinatorial Auction). Suppose there are m mutually exclusive potential states and only one of them will be true at maturity. For example, the states may correspond to the winning horse in a race of m horses, or the value of a stock index, falling within m intervals. An auction organizer who establishes a *parimutuel* auction is prepared to issue contracts specifying subsets of the m possibilities that pay \$1 if the final state is one of those designated by the contract, and zero otherwise. There are n participants who may place orders with the organizer for the purchase of such contracts. An order by the jth participant consists of an m-vector $\mathbf{a}_j = (a_{1j}, a_{2j}, \ldots, a_{mj})^T$ where each component is either 0 or 1, a one indicating a desire to be paid if the corresponding state occurs.

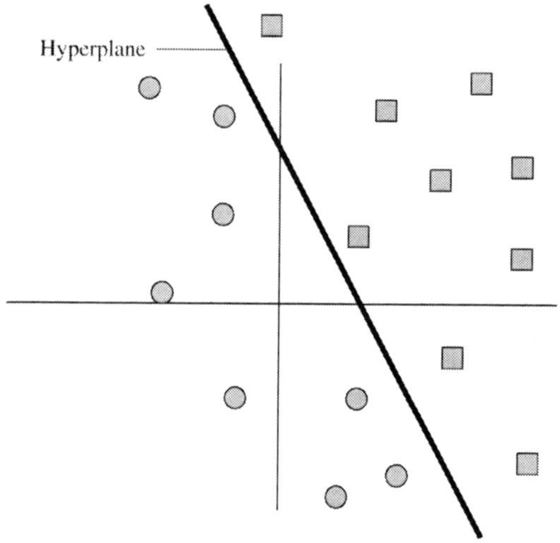

Fig. 2.2 Support vector for data classification

Accompanying the order is a number π_j which is the price limit the participant is willing to pay for one unit of the order. Finally, the participant also declares the maximum number q_j of units he or she is willing to accept under these terms.

The auction organizer, after receiving these various orders, must decide how many contracts to fill. Let x_j be the (real) number of units awarded to the jth order. Then the jth participant will pay $\pi_j x_j$. The total amount paid by all participants is $\boldsymbol{\pi}^T\mathbf{x}$, where $\mathbf{x}$ is the vector of x_j's and π is the vector of prices.

If the outcome is the ith state, the auction organizer must pay out a total of $\sum_{j=1}^{n} a_{ij}x_j = (\mathbf{A}\mathbf{x})_j$. The organizer would like to maximize profit in the worst possible case, and does this by solving the problem

$$\begin{aligned} \text{maximize} \;\; & \boldsymbol{\pi}^T\mathbf{x} - \max_i(\mathbf{A}\mathbf{x})_i \\ \text{subject to} \;\; & \mathbf{0} \leqslant \mathbf{x} \leqslant \mathbf{q}. \end{aligned}$$

This problem can be expressed alternatively as selecting $\mathbf{x}$ and scalar s to

$$\begin{aligned} \text{maximize } & \boldsymbol{\pi}^T \mathbf{x} - s \\ \text{subject to } & \mathbf{A}\mathbf{x} - \mathbf{1}s \leqslant \mathbf{0} \\ & \mathbf{0} \leqslant \mathbf{x} \leqslant \mathbf{q} \end{aligned}$$

where $\mathbf{1}$ is the vector of all 1's. Notice that the profit will always be nonnegative, since $\mathbf{x} = \mathbf{0}$ is feasible.

2.3 Basic Solutions

Consider the system of equalities

$$\mathbf{A}\mathbf{x} = \mathbf{b}, \tag{2.9}$$

where $\mathbf{x}$ is an n-vector, $\mathbf{b}$ an m-vector, and $\mathbf{A}$ is an $m \times n$ matrix. Suppose that from the n columns of $\mathbf{A}$ we select a set of m linearly independent columns (such a set exists if the rank of $\mathbf{A}$ is m). For notational simplicity assume that we select the first m columns of $\mathbf{A}$ and denote the $m \times m$ matrix determined by these columns by $\mathbf{B}$. The matrix $\mathbf{B}$ is then nonsingular and we may uniquely solve the equation.

$$\mathbf{B}\mathbf{x}_\mathbf{B} = \mathbf{b} \tag{2.10}$$

for the m-vector $\mathbf{x}_\mathbf{B}$. By putting $\mathbf{x} = (\mathbf{x}_\mathbf{B}, \mathbf{0})$ (that is, setting the first m components of $\mathbf{x}$ equal to those of $\mathbf{x}_\mathbf{B}$ and the remaining components equal to zero), we obtain a solution to $\mathbf{A}\mathbf{x} = \mathbf{b}$. This leads to the following definition.

Definition. Given the set of m simultaneous linear equations in n unknowns (2.9), let $\mathbf{B}$ be any nonsingular $m \times m$ submatrix made up of columns of $\mathbf{A}$. Then, if all $n-m$ components of $\mathbf{x}$ not associated with columns of $\mathbf{B}$ are set equal to zero, the solution to the resulting set of equations is said to be a *basic solution* to (2.9) with respect to the basis B. The components of $\mathbf{x}$ associated with columns of $\mathbf{B}$ are called *basic variables*.

In the above definition we refer to $\mathbf{B}$ as a basis, since $\mathbf{B}$ consists of m linearly independent columns that can be regarded as a basis for the space E^m. The basic solution corresponds to an expression for the vector $\mathbf{b}$ as a linear combination of these basis vectors. This interpretation is discussed further in the next section.

In general, of course, Eq. (2.9) may have no basic solutions. However, we may avoid trivialities and difficulties of a nonessential nature by making certain elementary assumptions regarding the structure of the matrix $\mathbf{A}$. First, we usually assume that $n > m$, that is, the number of variables x_j exceeds the number of equality constraints. Second, we usually assume that the rows of $\mathbf{A}$ are linearly independent, corresponding to linear independence of the m equations. $\mathbf{A}$ linear dependency among the rows of $\mathbf{A}$ would lead either to contradictory constraints and hence no solutions to (2.9), or to a redundancy that could be eliminated. Formally, we explicitly make the following assumption in our development, unless noted otherwise.

***Full Rank Assumption**. The $m \times n$ matrix $\mathbf{A}$ has $m < n$, and the m rows of $\mathbf{A}$ are linearly independent.*

Under the above assumption, the system (2.9) will always have a solution and, in fact, it will always have at least one basic solution.

The basic variables in a basic solution are not necessarily all nonzero. This is noted by the following definition.

Definition. If one or more of the basic variables in a basic solution has value zero, that solution is said to be a *degenerate basic solution*.

We note that in a nondegenerate basic solution the basic variables, and hence the basis $\mathbf{B}$, can be immediately identified from the positive components of the solution. There is ambiguity associated with a degenerate basic solution, however, since the zero-valued basic and some of nonbasic variables can be interchanged.

So far in the discussion of basic solutions we have treated only the equality constraint (2.9) and have made no reference to positivity constraints on the variables. Similar definitions apply when these constraints are also considered. Thus, consider now the system of constraints

$$\mathbf{Ax} = \mathbf{b}, \ \mathbf{x} \geqslant \mathbf{0}, \tag{2.11}$$

which represent the constraints of a linear program in standard form.

Definition. A vector $\mathbf{x}$ satisfying (2.11) is said to be *feasible* for these constraints. A feasible solution to the constraints (2.11) that is also basic is said to be a *basic feasible solution*; if this solution is also a degenerate basic solution, it is called a *degenerate basic feasible solution*.

2.4 The Fundamental Theorem of Linear Programming

In this section, through the fundamental theorem of linear programming, we establish the primary importance of basic feasible solutions in solving linear programs. The method of proof of the theorem is in many respects as important as the result itself, since it represents the beginning of the development of the simplex method. The theorem (due to Carathéodory) itself shows that it is necessary only to consider basic feasible solutions when seeking an optimal solution to a linear program because the optimal value is always achieved at such a solution.

Corresponding to a linear program in standard form

$$\begin{aligned} &\text{minimize} \ \ \mathbf{c}^T\mathbf{x} \\ &\text{subject to} \ \ \mathbf{Ax} = \mathbf{b}, \ \mathbf{x} \geqslant \mathbf{0} \end{aligned} \tag{2.12}$$

a feasible solution to the constraints that achieves the minimum value of the objective function subject to those constraints is said to be an *optimal feasible solution*. If this solution is basic, it is an *optimal basic feasible solution*.

Fundamental Theorem of Linear Programming. *Given a linear program in standard form (2.12) where* $\mathbf{A}$ *is an* $m \times n$ *matrix of rank* m,

i) *if there is a feasible solution, there is a basic feasible solution;*
ii) *if there is an optimal feasible solution, there is an optimal basic feasible solution.*

Proof of (i). Denote the columns of $\mathbf{A}$ by $\mathbf{a}_1, \mathbf{a}_2, \ldots, \mathbf{a}_n$. Suppose $\mathbf{x} = (x_1, x_2, \ldots, x_n)$ is a feasible solution. Then, in terms of the columns of $\mathbf{A}$, this solution satisfies:

$$x_1\mathbf{a}_1 + x_2\mathbf{a}_2 + \cdots + x_n\mathbf{a}_n = \mathbf{b}.$$

Assume that exactly p of the variables x_i are greater than zero, and for convenience, that they are the first p variables. Thus

$$x_1\mathbf{a}_1 + x_2\mathbf{a}_2 + \cdots + x_p\mathbf{a}_p = \mathbf{b}. \tag{2.13}$$

There are now two cases, corresponding as to whether the set $\mathbf{a}_1, \mathbf{a}_2, \ldots, \mathbf{a}_p$ is linearly independent or linearly dependent.

Case 1: Assume $\mathbf{a}_1, \mathbf{a}_2, \ldots, \mathbf{a}_p$ are linearly independent. Then clearly, $p \leqslant m$. If $p = m$, the solution is basic and the proof is complete. If $p < m$, then, since A has rank m, $m - p$ vectors can be found from the remaining $n - p$ vectors so that the resulting set of m vectors is linearly independent. (See Exercise 12.) Assigning the value zero to the corresponding $m - p$ variables yields a (degenerate) basic feasible solution.

Case 2: Assume $\mathbf{a}_1, \mathbf{a}_2, \ldots, \mathbf{a}_p$ are linearly dependent. Then there is a nontrivial linear combination of these vectors that is zero. Thus there are constants $y_1, y_2, \ldots, y_p$, at least one of which can be assumed to be positive, such that

$$y_1\mathbf{a}_1 + y_2\mathbf{a}_2 + \cdots + y_p\mathbf{a}_p = \mathbf{0}. \tag{2.14}$$

Multiplying this equation by a scalar ε and subtracting it from (2.13), we obtain

$$(x_1 - \varepsilon y_1)\mathbf{a}_1 + (x_2 - \varepsilon y_2)\mathbf{a}_2 + \cdots + (x_p - \varepsilon y_p)\mathbf{a}_p = \mathbf{b}. \tag{2.15}$$

This equation holds for every ε, and for each ε the components $x_j - \varepsilon y_j$ correspond to a solution of the linear equalities—although they may violate $x_i - \varepsilon y_i \geqslant 0$. Denoting $\mathbf{y} = (y_1, y_2, \ldots, y_p, 0, 0, \ldots, 0)$, we see that for any ε

$$\mathbf{x} - \varepsilon\mathbf{y} \tag{2.16}$$

is a solution to the equalities. For $\varepsilon = 0$, this reduces to the original feasible solution. As ε is increased from zero, the various components increase, decrease, or remain constant, depending upon whether the corresponding y_i is negative, positive, or zero. Since we assume at least one y_i is positive, at least one component will decrease as ε is increased. We increase ε to the first point where one or more components become zero. Specifically, we set

$$\varepsilon = \min\{x_i/y_i : y_i > 0\}.$$

For this value of ε the solution given by (2.16) is feasible and has at most $p - 1$ positive variables. Repeating this process if necessary, we can eliminate positive variables until we have a feasible solution with corresponding columns that are linearly independent. At that point Case 1 applies. ∎

Proof of (ii). Let $\mathbf{x} = (x_1, x_2, \ldots, x_n)$ be an optimal feasible solution and, as in the proof of (i) above, suppose there are exactly p positive variables $x_1, x_2, \ldots, x_p$. Again there are two cases; and Case 1, corresponding to linear independence, is exactly the same as before.

Case 2 also goes exactly the same as before, but it must be shown that for any ε the solution (2.16) is optimal. To show this, note that the value of the solution $\mathbf{x} - \varepsilon\mathbf{y}$ is

$$\mathbf{c}^T\mathbf{x} - \varepsilon\mathbf{c}^T\mathbf{y}. \tag{2.17}$$

For ε sufficiently small in magnitude, $\mathbf{x} - \varepsilon\mathbf{y}$ is a feasible solution for positive or negative values of ε. Thus we conclude that $\mathbf{c}^T\mathbf{y} = 0$. For, if $\mathbf{c}^T\mathbf{y} \neq 0$, an ε of small magnitude and proper sign could be determined so as to render (2.17) smaller than $\mathbf{c}^T\mathbf{x}$ while maintaining feasibility. This would violate the assumption of optimality of $\mathbf{x}$ and hence we must have $\mathbf{c}^T\mathbf{y} = 0$.

Having established that the new feasible solution with fewer positive components is also optimal, the remainder of the proof may be completed exactly as in part (i). ∎

This theorem reduces the task of solving a linear program to that of searching over basic feasible solutions. Since for a problem having n variables and m constraints there are at most

$$\binom{n}{m} = \frac{n!}{m!(n-m)!}$$

basic solutions (corresponding to the number of ways of selecting m of n columns), there are only a finite number of possibilities. Thus the fundamental theorem yields an obvious, but terribly inefficient, finite search technique. By expanding upon the technique of proof as well as the statement of the fundamental theorem, the efficient simplex procedure is derived.

It should be noted that the proof of the fundamental theorem given above is of a simple algebraic character. In the next section the geometric interpretation of this theorem is explored in terms of the general theory of convex sets. Although the geometric interpretation is aesthetically pleasing and theoretically important, the reader should bear in mind, lest one be diverted by the somewhat more advanced arguments employed, the underlying elementary level of the fundamental theorem.

2.5 Relations to Convexity

Our development to this point, including the above proof of the fundamental theorem, has been based only on elementary properties of systems of linear equations. These results, however, have interesting interpretations in terms of the theory of convex sets that can lead not only to an alternative derivation of the fundamental theorem, but also to a clearer geometric understanding of the result. The main link between the algebraic and geometric theories is the formal relation between basic feasible solutions of linear inequalities in standard form and extreme points of polytopes. We establish this correspondence as follows. The reader is referred to Appendix B for a more complete summary of concepts related to convexity, but the definition of an extreme point is stated here.

Definition. A point $\mathbf{x}$ in a convex set C is said to be an *extreme point* of C if there are no two distinct points $\mathbf{x}_1$ and $\mathbf{x}_2$ in C such that $\mathbf{x} = \alpha\mathbf{x}_1 + (1-\alpha)\mathbf{x}_2$ for some α, $0 < \alpha < 1$.

An extreme point is thus a point that does not lie strictly within a line segment connecting two other points of the set. The extreme points of a triangle, for example, are its three vertices.

Theorem (Equivalence of Extreme Points and Basic Solutions). *Let $\mathbf{A}$ be an $m \times n$ matrix of rank m and $\mathbf{b}$ an m-vector. Let K be the convex polytope consisting of all n-vectors $\mathbf{x}$ satisfying*

$$\mathbf{A}\mathbf{x} = \mathbf{b}, \ \mathbf{x} \geqslant \mathbf{0}. \tag{2.18}$$

A vector $\mathbf{x}$ is an extreme point of K if and only if $\mathbf{x}$ is a basic feasible solution to (2.18).

Proof. Suppose first that $\mathbf{x} = (x_1, x_2, \ldots, x_m, 0, 0, \ldots, 0)$ is a basic feasible solution to (2.18). Then

$$x_1\mathbf{a}_1 + x_2\mathbf{a}_2 + \cdots + x_m\mathbf{a}_m = \mathbf{b},$$

where $\mathbf{a}_1, \mathbf{a}_2, \ldots, \mathbf{a}_m$, the first m columns of $\mathbf{A}$, are linearly independent. Suppose that $\mathbf{x}$ could be expressed as a convex combination of two other points in K; say, $\mathbf{x} = \alpha\mathbf{y} + (1-\alpha)\mathbf{z}$, $0 < \alpha < 1$, $\mathbf{y} \neq \mathbf{z}$. Since all components of $\mathbf{x}$, $\mathbf{y}$, $\mathbf{z}$ are nonnegative and since $0 < \alpha < 1$, it follows immediately that the last $n - m$ components of $\mathbf{y}$ and $\mathbf{z}$ are zero. Thus, in particular, we have

$$y_1\mathbf{a}_1 + y_2\mathbf{a}_2 + \cdots + y_m\mathbf{a}_m = \mathbf{b}$$

and

$$z_1\mathbf{a}_1 + z_2\mathbf{a}_2 + \cdots + z_m\mathbf{a}_m = \mathbf{b}.$$

Since the vectors $\mathbf{a}_1, \mathbf{a}_2, \ldots, \mathbf{a}_m$ are linearly independent, however, it follows that $\mathbf{x} = \mathbf{y} = \mathbf{z}$ and hence $\mathbf{x}$ is an extreme point of K.

Conversely, assume that $\mathbf{x}$ is an extreme point of K. Let us assume that the nonzero components of $\mathbf{x}$ are the first k components. Then

$$x_1\mathbf{a}_1 + x_2\mathbf{a}_2 + \cdots + x_k\mathbf{a}_k = \mathbf{b},$$

with $x_i > 0,\ i = 1, 2,\ \ldots,\ k$. To show that $\mathbf{x}$ is a basic feasible solution it must be shown that the vectors $\mathbf{a}_1,\ \mathbf{a}_2,\ \ldots,\ \mathbf{a}_k$ are linearly independent. We do this by contradiction. Suppose $\mathbf{a}_1,\ \mathbf{a}_2,\ \ldots,\ \mathbf{a}_k$ are linearly dependent. Then there is a nontrivial linear combination that is zero:

$$y_1\mathbf{a}_1 + y_2\mathbf{a}_2 + \cdots + y_k\mathbf{a}_k = 0.$$

Define the n-vector $\mathbf{y} = (y_1,\ y_2,\ \ldots,\ y_k,\ 0, 0,\ \ldots,\ 0)$. Since $x_i > 0,\ 1 \leqslant i \leqslant k$, it is possible to select ε such that

$$\mathbf{x} + \varepsilon\mathbf{y} \geqslant 0, \quad \mathbf{x} - \varepsilon\mathbf{y} \geqslant 0.$$

We then have $\mathbf{x} = \frac{1}{2}(\mathbf{x}+\varepsilon\mathbf{y})+\frac{1}{2}(\mathbf{x}-\varepsilon\mathbf{y})$ which expresses $\mathbf{x}$ as a convex combination of two distinct vectors in K. This cannot occur, since $\mathbf{x}$ is an extreme point of K. Thus $\mathbf{a}_1,\ \mathbf{a}_2,\ \ldots,\ \mathbf{a}_k$ are linearly independent and $\mathbf{x}$ is a basic feasible solution. (Although if $k < m$, it is a degenerate basic feasible solution.) ∎

This correspondence between extreme points and basic feasible solutions enables us to prove certain geometric properties of the convex polytope K defining the constraint set of a linear programming problem.

Corollary 1. *If the convex set K corresponding to (2.18) is nonempty, it has at least one extreme point.*

Proof. This follows from the first part of the Fundamental Theorem and the Equivalence Theorem above. ∎

Corollary 2. *If there is a finite optimal solution to a linear programming problem, there is a finite optimal solution which is an extreme point of the constraint set.*

Corollary 3. *The constraint set K corresponding to (2.18) possesses at most a finite number of extreme points.*

Proof. There are obviously only a finite number of basic solutions obtained by selecting m basis vectors from the n columns of $\mathbf{A}$. The extreme points of K are a subset of these basic solutions. ∎

Finally, we come to the special case which occurs most frequently in practice and which in some sense is characteristic of well-formulated linear programs—the case where the constraint set K is nonempty and bounded. In this case we combine the results of the Equivalence Theorem and Corollary 3 above to obtain the following corollary.

Corollary 4. *If the convex polytope K corresponding to (2.18) is bounded, then K is a convex polyhedron, that is, K consists of points that are convex combinations of a finite number of points.*

Some of these results are illustrated by the following examples:

Example 1. Consider the constraint set in E^3 defined by

$$x_1 + x_2 + x_3 = 1$$
$$x_1 \geqslant 0,\ x_2 \geqslant 0,\ x_3 \geqslant 0.$$

This set is illustrated in Fig. 2.3. It has three extreme points, corresponding to the three basic solutions to $x_1 + x_2 + x_3 = 1$.

Example 2. Consider the constraint set in E^3 defined by

$$\begin{aligned} x_1 + x_2 + x_3 &= 1 \\ 2x_1 + 3x_2 \quad &= 1 \\ x_1 \geqslant 0,\ x_2 \geqslant 0,\ x_3 &\geqslant 0. \end{aligned}$$

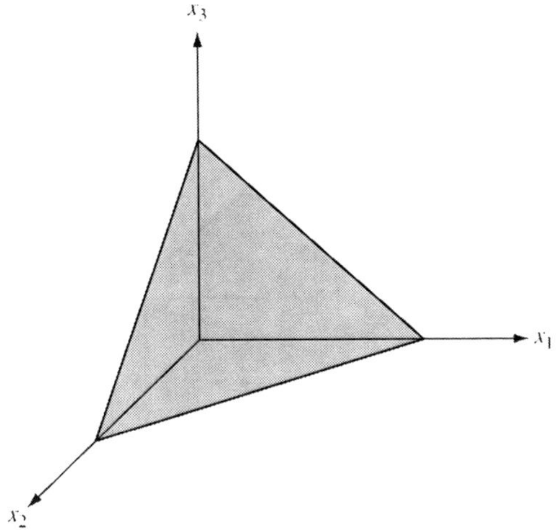

Fig. 2.3 Feasible set for Example 1

This set is illustrated in Fig. 2.4. It has two extreme points, corresponding to the two basic feasible solutions. Note that the system of equations itself has three basic solutions, (2, −1, 0), (1/2, 0, 1/2), (0, 1/3, 2/3), the first of which is not feasible.

Example 3. Consider the constraint set in E^2 defined in terms of the inequalities

$$\begin{aligned} x_1 + \frac{8}{3}x_2 &\leqslant 4 \\ x_1 + \quad x_2 &\leqslant 2 \\ 2x_1 \qquad &\leqslant 3 \\ x_1 \geqslant 0, \quad x_2 &\geqslant 0. \end{aligned}$$

This set is illustrated in Fig. 2.5. We see by inspection that this set has five extreme points. In order to compare this example with our general results we must introduce slack variables to yield the equivalent set in E^5:

$$\begin{aligned} x_1 + \frac{8}{3}x_2 + x_3 \qquad\qquad &= 4 \\ x_1 + x_2 \qquad + x_4 \qquad &= 2 \\ 2x_1 \qquad\qquad + x_5 &= 3 \\ x_1 \geqslant 0,\ x_2 \geqslant 0,\ x_3 \geqslant 0,\ x_4 \geqslant 0,\ x_5 &\geqslant 0. \end{aligned}$$

A basic solution for this system is obtained by setting any two variables to zero and solving for the remaining three. As indicated in Fig. 2.5, each edge of the figure corresponds to one variable being zero, and the extreme points are the points where two variables are zero.

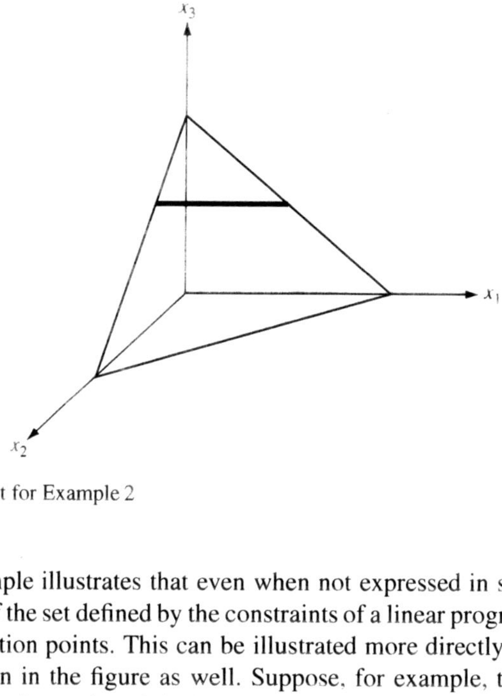

Fig. 2.4 Feasible set for Example 2

The last example illustrates that even when not expressed in standard form the extreme points of the set defined by the constraints of a linear program correspond to the possible solution points. This can be illustrated more directly by including the objective function in the figure as well. Suppose, for example, that in Example 3 the objective function to be minimized is $-2x_1 - x_2$. The set of points satisfying $-2x_1 - x_2 = z$ for fixed z is a line. As z varies, different parallel lines are obtained as shown in Fig. 2.6. The optimal value of the linear program is the smallest value of z for which the corresponding line has a point in common with the feasible set. It should be reasonably clear, at least in two dimensions, that the points of solution will always include an extreme point. In the figure this occurs at the point (3/2, 1/2) with $z = -7/2$.

2.6 Exercises

1. Convert the following problems to standard form:

$$\text{(a) minimize } x + 2y + 3z$$
$$\text{subject to } 2 \leqslant x + y \leqslant 3$$
$$4 \leqslant x + z \leqslant 5$$
$$x \geqslant 0,\ y \geqslant 0,\ z \geqslant 0.$$

$$\text{(b) minimize } x + y + z$$
$$\text{subject to } x + 2y + 3z = 10$$
$$x \geqslant 1,\ y \geqslant 2,\ z \geqslant 1.$$

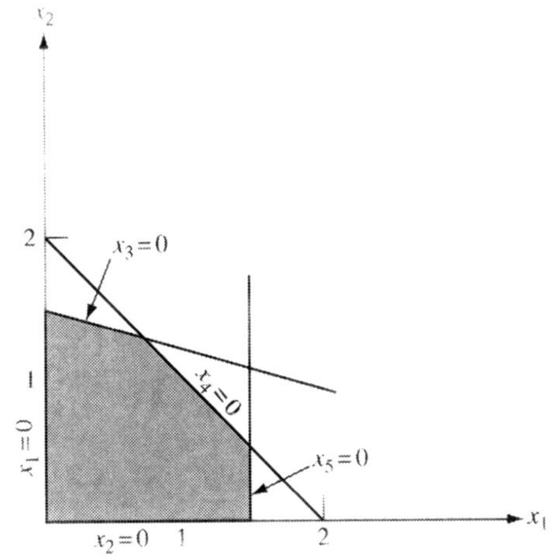

Fig. 2.5 Feasible set for Example 3

2. A manufacturer wishes to produce an alloy that is, by weight, 30 % metal A and 70 % metal B. Five alloys are available at various prices as indicated below:

Alloy	1	2	3	4	5
%A	10	25	50	75	95
% B	90	75	50	25	5
Price/lb	\$ 5	\$ 4	\$ 3	\$ 2	\$ 1.50

The desired alloy will be produced by combining some of the other alloys. The manufacturer wishes to find the amounts of the various alloys needed and to determine the least expensive combination. Formulate this problem as a linear program.

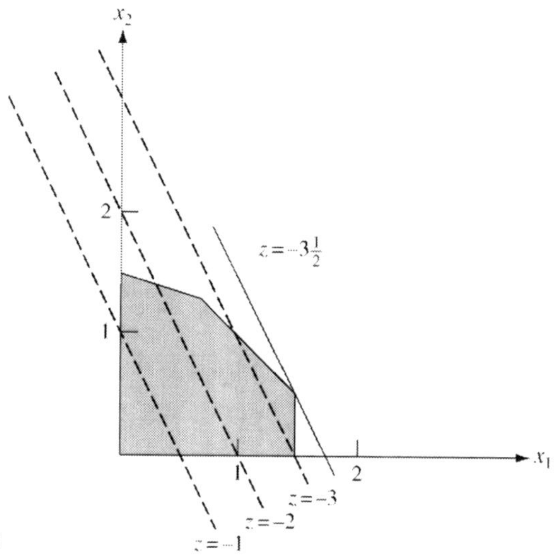

5

Fig. 2.6 Illustration of extreme point solution

3. An oil refinery has two sources of crude oil: a light crude that costs \$35/barrel and a heavy crude that costs \$30/barrel. The refinery produces gasoline, heating oil, and jet fuel from crude in the amounts per barrel indicated in the following table:

	Gasoline	Heating oil	Jet fuel
Light crude	0.3	0.2	0.3
Heavy crude	0.3	0.4	0.2

The refinery has contracted to supply 900,000 barrels of gasoline, 800,000 barrels of heating oil, and 500,000 barrels of jet fuel. The refinery wishes to find the amounts of light and heavy crude to purchase so as to be able to meet its obligations at minimum cost. Formulate this problem as a linear program.

4. A small firm specializes in making five types of spare automobile parts. Each part is first cast from iron in the casting shop and then sent to the finishing shop where holes are drilled, surfaces are turned, and edges are ground. The required worker-hours (per 100 units) for each of the parts of the two shops are shown below:

Part	1	2	3	4	5
Casting	2	1	3	3	1
Finishing	3	2	2	1	1

The profits from the parts are \$30, \$20, \$40, \$25, and \$10 (per 100 units), respectively. The capacities of the casting and finishing shops over the next month are 700 and 1,000 worker-hours, respectively. Formulate the problem of determining the quantities of each spare part to be made during the month so as to maximize profit.

5. Convert the following problem to standard form and solve:

$$\begin{aligned} \text{maximize} \quad & x_1 + 4x_2 + x_3 \\ \text{subject to} \quad & 2x_1 - 2x_2 + x_3 = 4 \\ & x_1 \qquad\qquad - x_3 = 1 \\ & x_2 \geqslant 0, \qquad x_3 \geqslant 0. \end{aligned}$$

6. A large textile firm has two manufacturing plants, two sources of raw material, and three market centers. The transportation costs between the sources and the plants and between the plants and the markets are as follows:

		Plant A	Plant B
Source	1	\$1/ton	\$1.50/ton
	2	\$2/ton	\$1.50/ton

		Market 1	Market 2	Market 3
Plant	A	\$4/ton	\$2/ton	\$1/ton
	B	\$3/ton	\$4/ton	\$2/ton

Ten tons are available from source 1 and 15 tons from source 2. The three market centers require 8 tons, 14 tons, and 3 tons. The plants have unlimited processing capacity.

(a) Formulate the problem of finding the shipping patterns from sources to plants to markets that minimizes the total transportation cost.

(b) Reduce the problem to a single standard transportation problem with two sources and three destinations. (*Hint*: Find minimum cost paths from sources to markets.)

(c) Suppose that plant A has a processing capacity of 8 tons, and plant B has a processing capacity of 7 tons. Show how to reduce the problem to two separate standard transportation problems.

7. A businessman is considering an investment project. The project has a lifetime of 4 years, with cash flows of −\$100,000, +\$50,000, +\$70,000, and +\$30,000 in each of the 4 years, respectively. At any time he may borrow funds at the rates of 12 %, 22 %, and 34 % (total) for 1, 2, or 3 periods, respectively. He may loan funds at 10 % per period. He calculates the *present value* of a project as the maximum amount of money he would pay now, to another party, for the project, assuming that he has no cash on hand and must borrow and lend to pay the other party and operate the project while maintaining a nonnegative cash

balance after all debts are paid. Formulate the project valuation problem in a linear programming framework.

8. Convert the following problem to a linear program in standard form:

$$\begin{aligned} &\text{minimize } |x| + |y| + |z| \\ &\text{subject to } x + y \leqslant 1 \\ &\qquad\qquad 2x + z = 3. \end{aligned}$$

9. A class of piecewise linear functions can be represented as $f(\mathbf{x}) = \text{Maximum}$ $(\mathbf{c}_1^T\mathbf{x}+d_1,\ \mathbf{c}_2^T\mathbf{x}+d_2,\ \ldots,\ \mathbf{c}_p^T\mathbf{x}+d_p)$. For such a function f, consider the problem

$$\begin{aligned} &\text{minimize } f(\mathbf{x}) \\ &\text{subject to } \mathbf{Ax} = \mathbf{b},\ \mathbf{x} \geqslant \mathbf{0}. \end{aligned}$$

Show how to convert this problem to a linear programming problem.

10. A small computer manufacturing company forecasts the demand over the next n months to be d_i, $i = 1, 2, \ldots, n$. In any month it can produce r units, using *regular* production, at a cost of b dollars per unit. By using *overtime*, it can produce additional units at c dollars per unit, where $c > b$. The firm can store units from month to month at a cost of s dollars per unit per month. Formulate the problem of determining the production schedule that minimizes cost. (*Hint*: See Exercise 9.)

11. Discuss the situation of a linear program that has one or more columns of the $\mathbf{A}$ matrix equal to zero. Consider both the case where the corresponding variables are required to be nonnegative and the case where some are free.

12. Suppose that the matrix $\mathbf{A} = (\mathbf{a}_1,\ \mathbf{a}_2,\ \ldots,\ \mathbf{a}_n)$ has rank m, and that for some $p < m$, $\mathbf{a}_1,\ \mathbf{a}_2,\ \ldots,\ \mathbf{a}_p$ are linearly independent. Show that $m - p$ vectors from the remaining $n - p$ vectors can be adjoined to form a set of m linearly independent vectors.

13. Suppose that $\mathbf{x}$ is a feasible solution to the linear program (2.12), with $\mathbf{A}$ an $m \times n$ matrix of rank m. Show that there is a feasible solution $\mathbf{y}$ having the same value (that is, $\mathbf{c}^T\mathbf{y} = \mathbf{c}^T\mathbf{x}$) and having at most $m + 1$ positive components.

14. What are the basic solutions of Example 3, Sect. 2.5?

15. Let S be a convex set in E^n and S^* a convex set in E^m. Suppose $\mathbf{T}$ is an $m \times n$ matrix that establishes a one-to-one correspondence between S and S^*, i.e., for every $\mathbf{s} \in S$ there is $\mathbf{s}^* \in S^*$ such that $\mathbf{Ts} = \mathbf{s}^*$, and for every $\mathbf{s}^* \in S^*$ there is a single $\mathbf{s} \in S$ such that $\mathbf{Ts} = \mathbf{s}^*$. Show that there is a one-to-one correspondence between extreme points of S and S^*.

16. Consider the two linear programming problems in Example 1, Sect. 2.1, one in E^n and the other in E^{n+m}. Show that there is a one-to-one correspondence between extreme points of these two problems.

References

2.1–2.4 The approach taken in this chapter, which is continued in the next, is the more or less standard approach to linear programming as presented in, for example, Dantzig [D6], Hadley [H1], Gass [G4], Simonnard [S6], Murty [M11], and Gale [G2]. Also see Bazaraa, Jarvis, and H. F. Sherali [B6], Bertsimas and Tsitsiklis [B13], Cottle, [C6], Dantzig and Thapa [D9, D10], Nash and Sofer [N1], Saigal [S1], and Vanderbei [V3].

2.5 An excellent discussion of this type can be found in Simonnard [S6].

Chapter 3
The Simplex Method

The idea of the simplex method is to proceed from one basic feasible solution (that is, one extreme point) of the constraint set of a problem in standard form to another, in such a way as to continually decrease the value of the objective function until a minimum is reached. The results of Chap. 2 assure us that it is sufficient to consider only basic feasible solutions in our search for an optimal feasible solution. This chapter demonstrates that an efficient method for moving among basic solutions to the minimum can be constructed.

In the first five sections of this chapter the simplex machinery is developed from a careful examination of the system of linear equations that defines the constraints and the basic feasible solutions of the system. This approach, which focuses on individual variables and their relation to the system, is probably the simplest, but unfortunately is not easily expressed in compact form. In the last few sections of the chapter, the simplex method is viewed from a matrix theoretic approach, which focuses on all variables together. This more sophisticated viewpoint leads to a compact notational representation, increased insight into the simplex process, and to alternative methods for implementation.

3.1 Pivots

To obtain a firm grasp of the simplex procedure, it is essential that one first understand the process of pivoting in a set of simultaneous linear equations. There are two dual interpretations of the pivot procedure.

D.G. Luenberger, Y. Ye, *Linear and Nonlinear Programming*, International Series in Operations Research & Management Science 228,
DOI 10.1007/978-3-319-18842-3_3

First Interpretation

Consider the set of simultaneous linear equations

$$
\begin{aligned}
a_{11}x_1 + a_{12}x_2 + \ldots + a_{1n}x_n &= b_1\\
a_{21}x_1 + a_{22}x_2 + \ldots + a_{2n}x_n &= b_2\\
\vdots \qquad\qquad\qquad & \quad \vdots\\
a_{m1}x_1 + a_{m2}x_2 + \ldots + a_{mn}x_n &= b_m,
\end{aligned}
\tag{3.1}
$$

where $m \leqslant n$. In matrix form we write this as

$$\mathbf{Ax} = \mathbf{b}. \tag{3.2}$$

In the space E^n we interpret this as a collection of m linear relations that must be satisfied by a vector $\mathbf{x}$. Thus denoting by $\mathbf{a}^j$ the ith row of $\mathbf{A}$ we may express (3.1) as:

$$
\begin{aligned}
\mathbf{a}^1\mathbf{x} &= b_1\\
\mathbf{a}^2\mathbf{x} &= b_2\\
&\vdots\\
\mathbf{a}^{\mathbf{m}}\mathbf{x} &= b_m.
\end{aligned}
\tag{3.3}
$$

This corresponds to the most natural interpretation of (3.1) as a set of m equations.

If $m < n$ and the equations are linearly independent, then there is not a unique solution but a whole linear variety of solutions (see Appendix B). A unique solution results, however, if $n - m$ additional independent linear equations are adjoined. For example, we might specify $n - m$ equations of the form $\mathbf{e}^k\mathbf{x} = 0$, where $\mathbf{e}^k$ is the kth unit vector (which is equivalent to $x_k = 0$), in which case we obtain a basic solution to (3.1). Different basic solutions are obtained by imposing different additional equations of this special form.

If Eq. (3.3) are linearly independent, we may replace a given equation by any nonzero multiple of itself plus any linear combination of the other equations in the system. This leads to the well-known Gaussian reduction schemes, whereby multiples of equations are systematically subtracted from one another to yield either a triangular or canonical form. It is well known, and easily proved, that if the first m columns of A are linearly independent, the system (3.1) can, by a sequence of such multiplications and subtractions, be converted to the following *canonical form*:

$$
\begin{aligned}
x_1 \qquad & +\bar{a}_{1(m+1)}x_{m+1} + \bar{a}_{1(m+2)}x_{m+2} + \cdots + \bar{a}_{1n}x_n = \bar{a}_{10}\\
x_2 \qquad & +\bar{a}_{2(m+1)}x_{m+1} + \bar{a}_{2(m+2)}x_{m+2} + \cdots + \bar{a}_{2n}x_n = \bar{a}_{20}\\
& \qquad\qquad \vdots \qquad\qquad\qquad\qquad \vdots\\
x_m & +\bar{a}_{m(m+1)}x_{m+1} + \bar{a}_{m(m+2)}x_{m+2} + \cdots + \bar{a}_{mn}x_n = \bar{a}_{m0}.
\end{aligned}
\tag{3.4}
$$

Corresponding to this canonical representation of the system, the variables x_1, $x_2, \ldots, x_m$ are called *basic* and the other variables are *nonbasic*. The corresponding basic solution is then:

$$x_1 = \bar{a}_{10},\ x_2 = \bar{a}_{20},\ \ldots,\ x_m = \bar{a}_{m0},\ x_{m+1} = 0,\ \ldots,\ x_n = 0,$$

or in vector form: $\mathbf{x} = (\bar{\mathbf{a}}_0, \mathbf{0})$ where $\bar{\mathbf{a}}_0$ is m-dimensional and $\mathbf{0}$ is the $(n-m)$-dimensional zero vector.

Actually, we relax our definition somewhat and consider a system to be in *canonical form* if, among the n variables, there are m basic ones with the property that each appears in only one equation, its coefficient in that equation is unity, and no two of these m variables appear in any one equation. This is equivalent to saying that a system is in canonical form if by some reordering of the equations and the variables it takes the form (3.4).

Also it is customary, from the dictates of economy, to represent the system (3.4) by its corresponding array of coefficients or *tableau*:

$$\begin{array}{cccccccccc}
x_1 & x_2 & x_3 & \cdots & x_m & x_{m+1} & x_{m+2} & \cdots & x_n & \\
1 & 0 & 0 & \cdots & 0 & \bar{a}_{1(m+1)} & \bar{a}_{1(m+2)} & \cdots & \bar{a}_{1n} & \bar{a}_{10} \\
0 & 1 & 0 & \cdots & 0 & \bar{a}_{2(m+1)} & \bar{a}_{2(m+2)} & \cdots & \cdot & \bar{a}_{20} \\
0 & 0 & 1 & \cdots & \cdot & \cdot & \cdot & & \cdot & \cdot \\
\cdot & \cdot & \cdot & & \cdot & \cdot & \cdot & & \cdot & \cdot \\
\cdot & \cdot & \cdot & & \cdot & \cdot & \cdot & & \cdot & \cdot \\
\cdot & \cdot & \cdot & & \cdot & \cdot & \cdot & & \cdot & \cdot \\
0 & 0 & 0 & \cdots & 1 & \bar{a}_{m(m+1)} & \bar{a}_{m(m+2)} & \cdots & \bar{a}_{mn} & \bar{a}_{m0}
\end{array} \tag{3.5}$$

The question solved by pivoting is this: given a system in canonical form, suppose a basic variable is to be made nonbasic and a nonbasic variable is to be made basic; what is the new canonical form corresponding to the new set of basic variables? The procedure is quite simple. Suppose in the canonical system (3.4) we wish to replace the basic variable x_p, $1 \leqslant p \leqslant m$, by the nonbasic variable x_q. This can be done if and only if $\bar{a}_{pq}$ is nonzero; it is accomplished by dividing row p by $\bar{a}_{pq}$ to get a unit coefficient for x_q in the pth equation, and then subtracting suitable multiples of row p from each of the other rows in order to get a zero coefficient for x_q in all other equations. This transforms the qth column of the tableau so that it is zero except in its pth entry (which is unity) and does not affect the columns of the other basic variables. Denoting the coefficients of the new system in canonical form by $\bar{a}'_{ij}$, we have explicitly

$$\begin{cases} \bar{a}'_{ij} = \bar{a}_{ij} - \frac{\bar{a}_{iq}}{\bar{a}_{pq}}\bar{a}_{pj}, & i \neq p \\ \bar{a}'_{pj} = \frac{\bar{a}_{pj}}{\bar{a}_{pq}}. \end{cases} \tag{3.6}$$

Equation (3.6) are the pivot equations that arise frequently in linear programming. The element $\bar{a}_{pq}$ in the original system is said to be the *pivot element*.

Example 1. Consider the system in canonical form:

$$\begin{aligned} x_1 \qquad\quad + x_4 + x_5 - x_6 &= 5 \\ x_2 \quad + 2x_4 - 3x_5 + x_6 &= 3 \\ x_3 - \quad x_4 + 2x_5 - x_6 &= -1. \end{aligned}$$

Let us find the basic solution having basic variables $x_4,\ x_5,\ x_6$. We set up the coefficient array below:

x_1	x_2	x_3	x_4	x_5	x_6	
1	0	0	(1)	1	−1	5
0	1	0	2	−3	1	3
0	0	1	−1	2	−1	−1

The circle indicated is our first pivot element and corresponds to the replacement of x_1 by x_4 as a basic variable. After pivoting we obtain the array

x_1	x_2	x_3	x_4	x_5	x_6	
1	0	0	1	1	−1	5
−2	1	0	0	(−5)	3	−7
1	0	1	0	3	−2	4

and again we have circled the next pivot element indicating our intention to replace x_2 by x_5. We then obtain

x_1	x_2	x_3	x_4	x_5	x_6	
3/5	1/5	0	1	0	−2/5	18/5
2/5	−1/5	0	0	1	−3/5	7/5
−1/5	3/5	1	0	0	(−1/5)	−1/5

Continuing, there results

x_1	x_2	x_3	x_4	x_5	x_6	
1	−1	−2	1	0	0	4
1	−2	−3	0	1	0	2
1	−3	−5	0	0	1	1

From this last canonical form we obtain the new basic solution

$$x_4 = 4, \quad x_5 = 2, \quad x_6 = 1.$$

Second Interpretation

The set of simultaneous equations represented by (3.1) and (3.2) can be interpreted in E^m as a vector equation. Denoting the columns of A by $\mathbf{a}_1,\ \mathbf{a}_2,\ \ldots,\ \mathbf{a}_n$ we write (3.1) as

$$x_1\mathbf{a}_1 + x_2\mathbf{a}_2 + \cdots + x_n\mathbf{a}_n = \mathbf{b}. \tag{3.7}$$

In this interpretation we seek to express $\mathbf{b}$ as a linear combination of the $\mathbf{a}_i$'s.

If $m < n$ and the vectors $\mathbf{a}_j$ span E^m then there is not a unique solution but a whole family of solutions. The vector $\mathbf{b}$ has a unique representation, however, as a linear combination of a given linearly independent subset of these vectors. The corresponding solution with $(n - m)$ x_j variables set equal to zero is a basic solution to (3.1).

Suppose now that we start again with a system in the canonical form corresponding to the tableau:

$$\begin{array}{cccccccccc}
\mathbf{a}_1 & \mathbf{a}_2 & \mathbf{a}_3 & \cdots & \mathbf{a}_m & \mathbf{a}_{m+1} & \mathbf{a}_{m+2} & \cdots & \mathbf{a}_n & \mathbf{b} \\
1 & 0 & 0 & \cdots & 0 & \bar{a}_{1(m+1)} & \bar{a}_{1(m+2)} & \cdots & \bar{a}_{1n} & \bar{a}_{10} \\
0 & 1 & 0 & \cdots & 0 & \bar{a}_{2(m+1)} & \bar{a}_{2(m+2)} & \cdots & \cdot & \bar{a}_{20} \\
0 & 0 & 1 & \cdots & \cdot & \cdot & \cdot & & \cdot & \cdot \\
\cdot & \cdot & \cdot & & \cdot & \cdot & \cdot & & \cdot & \cdot \\
\cdot & \cdot & \cdot & & \cdot & \cdot & \cdot & & \cdot & \cdot \\
\cdot & \cdot & \cdot & & \cdot & \cdot & \cdot & & \cdot & \cdot \\
0 & 0 & 0 & \cdots & 1 & \bar{a}_{m(m+1)} & \bar{a}_{m(m+2)} & \cdots & \bar{a}_{mn} & \bar{a}_{m0}
\end{array} \tag{3.8}$$

In this case the first m vectors form a basis. Furthermore, every other vector represented in the tableau can be expressed as a linear combination of these basis vectors by simply reading the coefficients down the corresponding column. Thus

$$\mathbf{a}_j = \bar{a}_{1j}\mathbf{a}_1 + \bar{a}_{2j}\mathbf{a}_2 + \cdots + \bar{a}_{mj}\mathbf{a}_m. \tag{3.9}$$

The tableau can be interpreted as giving the representations of the vectors $\mathbf{a}_j$ in terms of the basis; the jth column of the tableau is the representation for the vector $\mathbf{a}_j$. In particular, the expression for $\mathbf{b}$ in terms of the basis is given in the last column.

We now consider the operation of replacing one member of the basis by another vector not already in the basis. Suppose for example we wish to replace the basis vector $\mathbf{a}_p$, $1 \leqslant p \leqslant m$, by the vector $\mathbf{a}_q$. Provided that the first m vectors with $\mathbf{a}_p$ replaced by $\mathbf{a}_q$ are linearly independent these vectors constitute a basis and every vector can be expressed as a linear combination of this new basis. To find the new representations of the vectors we must update the tableau. The linear independence condition holds if and only if $\bar{a}_{pq} \neq 0$.

Any vector $\mathbf{a}_j$ can be expressed in terms of the old array through (3.9). For $\mathbf{a}_q$ we have

$$\mathbf{a}_q = \sum_{\substack{i=1 \\ i \neq p}}^{m} \bar{a}_{iq}\mathbf{a}_i + \bar{a}_{pq}\mathbf{a}_p$$

from which we may solve for $\mathbf{a}_p$,

$$\mathbf{a}_p = \frac{1}{\bar{a}_{pq}}\mathbf{a}_q - \sum_{\substack{i=1 \\ i \neq p}}^{m} \frac{\bar{a}_{iq}}{\bar{a}_{pq}}\mathbf{a}_i. \tag{3.10}$$

Substituting (3.10) into (3.9) we obtain:

$$\mathbf{a}_j = \sum_{\substack{i=1 \\ i \neq p}}^{m} \left(\bar{a}_{ij} - \frac{\bar{a}_{iq}}{\bar{a}_{pq}} \bar{a}_{pj} \right) \mathbf{a}_i + \frac{\bar{a}_{pj}}{\bar{a}_{pq}} \mathbf{a}_q. \tag{3.11}$$

Denoting the coefficients of the new tableau, which give the linear combinations, by $\bar{a}'_{ij}$ we obtain immediately from (3.11)

$$\begin{cases} \bar{a}'_{ij} = \bar{a}_{ij} - \frac{\bar{a}_{iq}}{\bar{a}_{pq}} \bar{a}_{pj}, i \neq p \\ \bar{a}'_{pj} = \frac{\bar{a}_{pj}}{\bar{a}_{pq}}. \end{cases} \tag{3.12}$$

These formulas are identical to (3.6).

If a system of equations is not originally given in canonical form, we may put it into canonical form by adjoining the m unit vectors to the tableau and, starting with these vectors as the basis, successively replace each of them with columns of A using the pivot operation.

Example 2. Suppose we wish to solve the simultaneous equations

$$\begin{aligned} x_1 + x_2 - x_3 &= 5 \\ 2x_1 - 3x_2 + x_3 &= 3 \\ -x_1 + 2x_2 - x_3 &= -1. \end{aligned}$$

To obtain an original basis, we form the augmented tableau

$\mathbf{e}_1$	$\mathbf{e}_2$	$\mathbf{e}_3$	$\mathbf{a}_1$	$\mathbf{a}_2$	$\mathbf{a}_3$	$\mathbf{b}$
1	0	0	1	1	−1	5
0	1	0	2	−3	1	3
0	0	1	−1	2	−1	−1

and replace $\mathbf{e}_1$ by $\mathbf{a}_1$, $\mathbf{e}_2$ by $\mathbf{a}_2$, and $\mathbf{e}_3$ by $\mathbf{a}_3$. The required operations are identical to those of Example 1.

3.2 Adjacent Extreme Points

In Chap. 2 it was discovered that it is only necessary to consider basic feasible solutions to the system

$$\mathbf{Ax} = \mathbf{b}, \ \ \mathbf{x} \geqslant \mathbf{0} \tag{3.13}$$

when solving a linear program, and in the previous section it was demonstrated that the pivot operation can generate a new basic solution from an old one by replacing one basic variable by a nonbasic variable. It is clear, however, that although the pivot operation takes one basic solution into another, the nonnegativity of the solution will not in general be preserved. Special conditions must be satisfied in order that a pivot operation maintain feasibility. In this section we show how it is possible to select pivots so that we may transfer from one basic feasible solution to another.

We show that although it is not possible to arbitrarily specify the pair of variables whose roles are to be interchanged and expect to maintain the nonnegativity condition, it is possible to arbitrarily specify which nonbasic variable is to become basic and then determine which basic variable should become nonbasic. As is conventional, we base our derivation on the vector interpretation of the linear equations although the dual interpretation could alternatively be used.

Nondegeneracy Assumption

Many arguments in linear programming are substantially simplified upon the introduction of the following.

Nondegeneracy Assumption: Every basic feasible solution of (3.13) is a nondegenerate basic feasible solution.

This assumption is invoked throughout our development of the simplex method, since when it does not hold the simplex method can break down if it is not suitably amended. The assumption, however, should be regarded as one made primarily for convenience, since all arguments can be extended to include degeneracy, and the simplex method itself can be easily modified to account for it.

Determination of Vector to Leave Basis

Suppose we have the basic feasible solution $\mathbf{x} = (x_1, x_2, \ldots, x_m, 0, 0, \ldots, 0)$ or, equivalently, the representation

$$x_1\mathbf{a}_1 + x_2\mathbf{a}_2 + \cdots + x_m\mathbf{a}_m = \mathbf{b}. \tag{3.14}$$

Under the nondegeneracy assumption, $x_j > 0$, $i = 1, 2, \ldots, m$. Suppose also that we have decided to bring into the representation the vector $\mathbf{a}_q$, $q > m$. We have available a representation of $\mathbf{a}_q$ in terms of the current basis

$$\mathbf{a}_q = \bar{a}_{1q}\mathbf{a}_1 + \bar{a}_{2q}\mathbf{a}_2 + \cdots + \bar{a}_{mq}\mathbf{a}_m. \tag{3.15}$$

Multiplying (3.15) by a variable $\varepsilon \geqslant 0$ and subtracting from (3.14), we have

$$(x_1 - \varepsilon\bar{a}_{1q})\mathbf{a}_1 + (x_2 - \varepsilon\bar{a}_{2q})\mathbf{a}_2 + \cdots + (x_m - \varepsilon\bar{a}_{mq})\mathbf{a}_m + \varepsilon\mathbf{a}_q = \mathbf{b}. \tag{3.16}$$

Thus, for any $\varepsilon \geqslant 0$ (3.16) gives $\mathbf{b}$ as a linear combination of at most $m+1$ vectors. For $\varepsilon = 0$ we have the old basic feasible solution. As ε is increased from zero, the coefficient of $\mathbf{a}_q$ increases, and it is clear that for small enough ε, (3.16) gives a feasible but nonbasic solution. The coefficients of the other vectors will either increase or decrease linearly as ε is increased. If any decrease, we may set ε equal to the value corresponding to the first place where one (or more) of the coefficients vanishes. That is

$$\varepsilon = \min_i \{x_j/\bar{a}_{iq} : \bar{a}_{iq} > 0\}. \tag{3.17}$$

In this case we have a new basic feasible solution, with the vector $\mathbf{a}_q$ replacing the vector $\mathbf{a}_p$, where p corresponds to the minimizing index in (3.17). If the minimum in (3.17) is achieved by more than a single index i, then the new solution is degenerate and any of the vectors with zero component can be regarded as the one that left the basis.

If none of the $\bar{a}_{iq}$'s are positive, then all coefficients in the representation (3.16) increase (or remain constant) as ε is increased, and no new basic feasible solution is obtained. We observe, however, that in this case, where none of the $\bar{a}_{iq}$'s are positive, there are feasible solutions to (3.13) having arbitrarily large coefficients. This means that the set K of feasible solutions to (3.13) is unbounded, and this special case, as we shall see, is of special significance in the simplex procedure.

In summary, we have deduced that, given a basic feasible solution and an arbitrary vector $\mathbf{a}_q$, there is either a new basic feasible solution having $\mathbf{a}_q$ in its basis and one of the original vectors removed, or the set of feasible solutions is unbounded.

Let us consider how the calculation of this section can be displayed in our tableau. We assume that corresponding to the constraints

$$\mathbf{Ax} = \mathbf{b}, \;\; \mathbf{x} \geqslant \mathbf{0},$$

we have a tableau of the form (3.8). Note that the tableau may be the result of several pivot operations applied to the original tableau, but in any event, it represents a solution with basis $\mathbf{a}_1, \mathbf{a}_2, \ldots, \mathbf{a}_m$. We assume that $\bar{a}_{10}, \bar{a}_{20}, \ldots, \bar{a}_{m0}$ are nonnegative, so that the corresponding basic solution $x_1 = \bar{a}_{10}, x_2 = \bar{a}_{20}, \ldots, x_m = \bar{a}_{m0}$ is feasible. We wish to bring into the basis the vector $\mathbf{a}_q$, $q > m$, and maintain feasibility. In order to determine which element in the qth column to use as the pivot (and hence which vector in the basis will leave), we use (3.17) and compute the ratios $x_i/\bar{a}_{iq} = \bar{a}_{i0}/\bar{a}_{iq}$, $i = 1, 2, \ldots, m$, select the smallest nonnegative ratio, and pivot on the corresponding $\bar{a}_{iq}$.

Example 3. Consider the system

$\mathbf{a}_1$	$\mathbf{a}_2$	$\mathbf{a}_3$	$\mathbf{a}_4$	$\mathbf{a}_5$	$\mathbf{a}_6$	$\mathbf{b}$
1	0	0	2	4	6	4
0	1	0	1	2	3	3
0	0	1	−1	2	1	1

which has basis $\mathbf{a}_1$, $\mathbf{a}_2$, $\mathbf{a}_3$ yielding a basic feasible solution $\mathbf{x} = (4, 3, 1, 0, 0, 0)$. Suppose we elect to bring $\mathbf{a}_4$ into the basis. To determine which element in the fourth column is the appropriate pivot, we compute the three ratios:

$$4/2 = 2, \ 3/1 = 3, \ 1/-1 = -1$$

and select the smallest nonnegative one. This gives 2 as the pivot element. The new tableau is

$\mathbf{a}_1$	$\mathbf{a}_2$	$\mathbf{a}_3$	$\mathbf{a}_4$	$\mathbf{a}_5$	$\mathbf{a}_6$	$\mathbf{b}$
1/2	0	0	1	2	3	2
−1/2	1	0	0	0	0	1
1/2	0	1	0	4	4	3

with corresponding basic feasible solution $\mathbf{x} = (0, 1, 3, 2, 0, 0)$.

Our derivation of the method for selecting the pivot in a given column that will yield a new feasible solution has been based on the vector interpretation of the equation $\mathbf{Ax} = \mathbf{b}$. An alternative derivation can be constructed by considering the dual approach that is based on the rows of the tableau rather than the columns. Briefly, the argument runs like this: if we decide to pivot on $\bar{a}_{pq}$, then we first divide the pth row by the pivot element $\bar{a}_{pq}$ to change it to unity. In order that the new $\bar{a}_{p0}$ remain positive, it is clear that we must have $\bar{a}_{pq} > 0$. Next we subtract multiples of the pth row from each other row in order to obtain zeros in the qth column. In this process the new elements in the last column must remain nonnegative—if the pivot was properly selected. The full operation is to subtract, from the ith row, $\bar{a}_{iq}/\bar{a}_{pq}$ times the pth row. This yields a new solution obtained directly from the last column:

$$x_i' = x_i - \frac{\bar{a}_{iq}}{\bar{a}_{pq}} x_p.$$

For this to remain nonnegative, it follows that $x_p/\bar{a}_{pq} \leqslant x_i/\bar{a}_{iq}$, and hence again we are led to the conclusion that we select p as the index i minimizing $x_i/\bar{a}_{iq}$.

Geometrical Interpretations

Corresponding to the two interpretations of pivoting and extreme points developed algebraically, are two geometrical interpretations. The first is in *activity space*, the space where $\mathbf{x}$ is represented. This is perhaps the most natural space to consider, and it was used in Sect. 2.5. Here the feasible region is shown directly as a convex set, and basic feasible solutions are extreme points. Adjacent extreme points are points that lie on a common edge.

The second geometrical interpretation is in *requirements space*, the space where the columns of $\mathbf{A}$ and $\mathbf{b}$ are represented. The fundamental relation is

$$\mathbf{a}_1 x_1 + \mathbf{a}_2 x_2 + \cdots + \mathbf{a}_n x_n = \mathbf{b}.$$

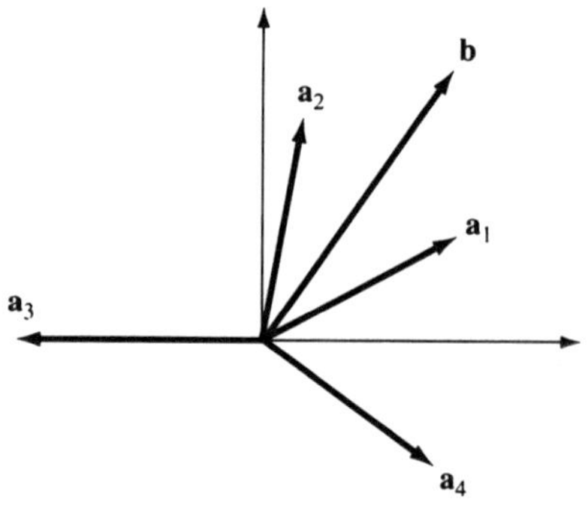

Fig. 3.1 Constraint representation in requirements space

An example for $m = 2$, $n = 4$ is shown in Fig. 3.1. A feasible solution defines a representation of $\mathbf{b}$ as a positive combination of the $\mathbf{a}_i$'s. A basic feasible solution will use only m positive weights. In the figure a basic feasible solution can be constructed with positive weights on $\mathbf{a}_1$ and $\mathbf{a}_2$ because $\mathbf{b}$ lies between them. A basic feasible solution cannot be constructed with positive weights on $\mathbf{a}_1$ and $\mathbf{a}_4$. Suppose we start with $\mathbf{a}_1$ and $\mathbf{a}_2$ as the initial basis. Then an adjacent basis is found by bringing in some other vector. If $\mathbf{a}_3$ is brought in, then clearly $\mathbf{a}_2$ must go out. On the other hand, if $\mathbf{a}_4$ is brought in, $\mathbf{a}_1$ must go out.

3.3 Determining a Minimum Feasible Solution

In the last section we showed how it is possible to pivot from one basic feasible solution to another (or determine that the solution set is unbounded) by arbitrarily selecting a column to pivot on and then appropriately selecting the pivot in that column. The idea of the simplex method is to select the column so that the resulting new basic feasible solution will yield a lower value to the objective function than the previous one. This then provides the final link in the simplex procedure. By an elementary calculation, which is derived below, it is possible to determine which vector should enter the basis so that the objective value is reduced, and by another simple calculation, derived in the previous section, it is possible to then determine which vector should leave in order to maintain feasibility.

Suppose we have a basic feasible solution

$$(\mathbf{x_B},\ \mathbf{0}) = (\bar{a}_{10},\ \bar{a}_{20},\ \ldots,\ \bar{a}_{m0}, 0, 0,\ \ldots,\ 0)$$

together with a tableau having an identity matrix appearing in the first m columns as shown in tableau (3.8). The value of the objective function corresponding to any solution $\mathbf{x}$ is

$$z = c_1x_1 + c_2x_2 + \cdots + c_nx_n, \tag{3.18}$$

and hence for the basic solution, the corresponding value is

$$z_0 = \mathbf{c}_{\mathbf{B}}^T \mathbf{x}_{\mathbf{B}}, \tag{3.19}$$

where $\mathbf{c}_{\mathbf{B}}^T = [c_1, c_2, \ldots, c_m]$.

Although it is natural to use the basic solution $(\mathbf{x}_{\mathbf{B}}, \mathbf{0})$ when we have the tableau (3.8), it is clear that if arbitrary values are assigned to $x_{m+1}, x_{m+2}, \ldots, x_n$, we can easily solve for the remaining variables as

$$\begin{aligned} x_1 &= \bar{a}_{10} - \sum_{j=m+1}^{n} \bar{a}_{1j} x_j \\ x_2 &= \bar{a}_{20} - \sum_{j=m+1}^{n} \bar{a}_{2j} x_j \\ &\vdots \\ x_m &= \bar{a}_{m0} - \sum_{j=m+1}^{n} \bar{a}_{mj} x_j. \end{aligned} \tag{3.20}$$

Using (3.20) we may eliminate $x_1, x_2, \ldots, x_m$ from the general formula (3.18). Doing this we obtain

$$\begin{aligned} z = \mathbf{c}^T \mathbf{x} = z_0 &+ (c_{m+1} - z_{m+1}) x_{m+1} \\ &+ (c_{m+2} - z_{m+2}) x_{m+2} + \cdots + (c_n - z_n) x_n \end{aligned} \tag{3.21}$$

where

$$z_j = \bar{a}_{1j} c_1 + \bar{a}_{2j} c_2 + \cdots + \bar{a}_{mj} c_m, \; m+1 \leqslant j \leqslant n, \tag{3.22}$$

which is the fundamental relation required to determine the pivot column. The important point is that this equation gives the values of the objective function z for any solution of $\mathbf{A}\mathbf{x} = \mathbf{b}$ in terms of the variables $x_{m+1}, \ldots, x_n$. From it we can determine if there is any advantage in changing the basic solution by introducing one of the nonbasic variables. For example, if $c_j - z_j$ is negative for some j, $m+1 \leqslant j \leqslant n$, then increasing x_j from zero to some positive value would decrease the total cost, and therefore would yield a better solution. The formula (3.21) and (3.22) automatically take into account the changes that would be required in the values of the basic variables $x_1, x_2, \ldots, x_m$ to accommodate the change in x_j.

Let us derive these relations from a different viewpoint. Let $\bar{\mathbf{a}}_j$ be the jth column of the tableau. Then any solution satisfies

$$x_1 \mathbf{e}_1 + x_2 \mathbf{e}_2 + \cdots + x_m \mathbf{e}_m = \bar{\mathbf{a}}_0 - x_{m+1} \bar{\mathbf{a}}_{m+1} - x_{m+2} \bar{\mathbf{a}}_{m+2} - \cdots - x_n \bar{\mathbf{a}}_n.$$

Taking the inner product of this vector equation with $\mathbf{c}_{\mathbf{B}}^T$, we have

$$\sum_{i=1}^{m} c_j x_j = \mathbf{c}_{\mathbf{B}}^T \bar{\mathbf{a}}_0 - \sum_{j=m+1}^{n} z_j x_j,$$

where $z_j = \mathbf{c}_{\mathbf{B}}^T \bar{\mathbf{a}}_j$. Thus, adding $\sum_{j=m+1}^{n} c_j x_j$ to both sides,

$$\mathbf{c}^T \mathbf{x} = z_0 + \sum_{j=m+1}^{n} (c_j - z_j) x_j \tag{3.23}$$

as before.

We now state the condition for improvement, which follows easily from the above observation, as a theorem.

Theorem (Improvement of Basic Feasible Solution). *Given a nondegenerate basic feasible solution with corresponding objective value z_0, suppose that for some j there holds $c_j - z_j < 0$. Then there is a feasible solution with objectivevalue $z < z_0$. If the column $\mathbf{a}_j$ can be substituted for some vector in the originalbasis to yield a new basic feasible solution, this new solution will have $z < z_0$. If $\mathbf{a}_j$ cannot be substituted to yield a basic feasible solution, then the solutionset K is unbounded and the objective function can be made arbitrarily small (toward minus infinity).*

Proof. The result is an immediate consequence of the previous discussion. Let $(x_1, x_2, \ldots, x_m, 0, 0, \ldots, 0)$ be the basic feasible solution with objective value z_0 and suppose $c_{m+1} - z_{m+1} < 0$. Then, in any case, new feasible solutions can be constructed of the form $(x'_1, x'_2, \ldots, x'_m, x'_{m+1}, 0, 0, \ldots, 0)$ with $x'_{m+1} > 0$. Substituting this solution in (3.21) we have

$$z - z_0 = (c_{m+1} - z_{m+1}) x'_{m+1} < 0,$$

and hence $z < z_0$ for any such solution. It is clear that we desire to make x'_{m+1} as large as possible. As x'_{m+1} is increased, the other components increase, remain constant, or decrease. Thus x'_{m+1} can be increased until one $x'_i = 0$, $i \leqslant m$, in which case we obtain a new basic feasible solution, or if none of the x'_i's decrease, x'_{m+1} can be increased without bound indicating an unbounded solution set and an objective value without lower bound. ∎

We see that if at any stage $c_j - z_j < 0$ for some j, it is possible to make x_j positive and decrease the objective function. The final question remaining is whether $c_j - z_j \geqslant 0$ for all j implies optimality.

Optimality Condition Theorem. *If for some basic feasible solution $c_j - z_j \geqslant 0$ for all j, then that solution is optimal.*

Proof. This follows immediately from (3.21), since any other feasible solution must have $x_i \geqslant 0$ for all i, and hence the value z of the objective will satisfy $z - z_0 \geqslant 0$. ∎

Since the constants $c_j - z_j$ play such a central role in the development of the simplex method, it is convenient to introduce the somewhat abbreviated notation $r_j = c_j - z_j$ and refer to the r_j's as the *relative cost coefficients* or, alternatively, the *reduced cost coefficients* (both terms occur in common usage). These coefficients measure the cost of a variable relative to a given basis. (For notational convenience we extend the definition of relative cost coefficients to basic variables as well; the relative cost coefficient of a basic variable is zero.)

We conclude this section by giving an economic interpretation of the relative cost coefficients. Let us agree to interpret the linear program

$$\begin{aligned} &\text{minimize} \quad \mathbf{c}^T\mathbf{x} \\ &\text{subject to} \quad \mathbf{A}\mathbf{x} = \mathbf{b}, \ \mathbf{x} \geqslant \mathbf{0} \end{aligned}$$

as a diet problem (see Sect. 2.2) where the nutritional requirements must be met exactly. A column of **A** gives the nutritional equivalent of a unit of a particular food. With a given basis consisting of, say, the first m columns of **A**, the corresponding simplex tableau shows how any food (or more precisely, the nutritional content of any food) can be constructed as a combination of foods in the basis. For instance, if carrots are not in the basis we can, using the description given by the tableau, construct a *synthetic* carrot which is nutritionally equivalent to a carrot, by an appropriate combination of the foods in the basis.

In considering whether or not the solution represented by the current basis is optimal, we consider a certain food not in the basis—say carrots—and determine if it would be advantageous to bring it into the basis. This is very easily determined by examining the cost of carrots as compared with the cost of synthetic carrots. If carrots are food j, then the unit cost of carrots is c_j. The cost of a unit of synthetic carrots is, on the other hand,

$$z_j = \sum_{i=1}^{m} c_j \bar{a}_{ij}.$$

If $r_j = c_j - z_j < 0$, it is advantageous to use real carrots in place of synthetic carrots, and carrots should be brought into the basis.

In general each z_j can be thought of as the price of a unit of the column $\mathbf{a}_j$ when constructed from the current basis. The difference between this synthetic price and the direct price of that column determines whether that column should enter the basis.

3.4 Computational Procedure: Simplex Method

In previous sections the theory, and indeed much of the technique, necessary for the detailed development of the simplex method has been established. It is only necessary to put it all together and illustrate it with examples.

In this section we assume that we begin with a basic feasible solution and that the tableau corresponding to $\mathbf{A}\mathbf{x} = \mathbf{b}$ is in the canonical form for this solution. Methods for obtaining this first basic feasible solution, when one is not obvious, are described in the next section.

In addition to beginning with the array $\mathbf{A}\mathbf{x} = \mathbf{b}$ expressed in canonical form corresponding to a basic feasible solution, we append a row at the bottom consisting of the relative cost coefficients and the negative of the current cost. The result is a *simplex tableau*.

Thus, if we assume the basic variables are (in order) $x_1, \ x_2, \ldots, \ x_m$, the simplex tableau takes the initial form shown in Fig. 3.2.

The basic solution corresponding to this tableau is

$$x_j = \begin{cases} \bar{a}_{i0} & 0 \leqslant i \leqslant m \\ 0 & m+1 \leqslant i \leqslant n \end{cases}$$

which we have assumed is feasible, that is, $\bar{a}_{i0} \geqslant 0, \ i = 1, 2, \ \ldots, \ m$. The corresponding value of the objective function is z_0.

$\mathbf{a}_1$	$\mathbf{a}_2$	$\cdots$	$\mathbf{a}_m$	$\mathbf{a}_{m+1}$	$\mathbf{a}_{m+2}$	$\cdots$	$\mathbf{a}_j$	$\cdots$	$\mathbf{a}_n$	$\mathbf{b}$
1	0	$\cdots$	0	$\bar{a}_{1(m+1)}$	$\bar{a}_{1(m+2)}$	$\cdots$	$\bar{a}_{1j}$	$\cdots$	$\bar{a}_{1n}$	$\bar{a}_{10}$
0	1		0	$\bar{a}_{2(m+1)}$	$\bar{a}_{2(m+2)}$	$\cdots$	$\bar{a}_{2j}$	$\cdots$	$\bar{a}_{2n}$	$\bar{a}_{10}$
·	·		·	·	·		·		·	·
·	·		·	·	·		·		·	·
0	0		0	$\bar{a}_{i(m+1)}$	$\bar{a}_{i(m+2)}$	$\cdots$	$\bar{a}_{ij}$	$\cdots$	$\bar{a}_{in}$	$\bar{a}_{i0}$
·	·		·	·	·		·		·	·
·	·		·	·	·		·		·	·
0	0		1	$\bar{a}_{m(m+1)}$	$\bar{a}_{m(m+2)}$	$\cdots$	$\bar{a}_{mj}$	$\cdots$	$\bar{a}_{mn}$	$\bar{a}_{m0}$
0	0		0	r_{m+1}	r_{m+2}	$\cdots$	r_j	$\cdots$	r_n	$-z_0$

Fig. 3.2 Canonical simplex tableau

The relative cost coefficients r_j indicate whether the value of the objective will increase or decrease if x_j is pivoted into the solution. If these coefficients are all nonnegative, then the indicated solution is optimal. If some of them are negative, an improvement can be made (assuming nondegeneracy) by bringing the corresponding component into the solution. When more than one of the relative cost coefficients is negative, any one of them may be selected to determine in which column to pivot. Common practice is to select the most negative value. (See Exercise 13 for further discussion of this point.)

Some more discussion of the relative cost coefficients and the last row of the tableau is warranted. We may regard z as an additional variable and

$$c_1x_1 + c_2x_2 + \cdots + c_nx_n - z = 0$$

as another equation. A basic solution to the augmented system will have $m + 1$ basic variables, but we can require that z be one of them. For this reason it is not necessary to add a column corresponding to z, since it would always be $(0, 0, \ \ldots, \ 0, 1)$. Thus, initially, a last row consisting of the c_j's and a right-hand side of zero can be appended to the standard array to represent this additional equation. Using standard pivot operations, the elements in this row corresponding to basic variables can be reduced to zero. This is equivalent to transforming the additional equation to the form

$$r_{m+1}x_{m+1} + r_{m+2}x_{m+2} + \cdots + r_nx_n - z = -z_0. \tag{3.24}$$

This must be equivalent to (3.23), and hence the r_j's obtained are the relative cost coefficients. Thus, the last row can be treated operationally like any other row: just start with c_j's and reduce the terms corresponding to basic variables to zero by row operations.

After a column q is selected in which to pivot, the final selection of the pivot element is made by computing the ratio $\bar{a}_{i0}/\bar{a}_{iq}$ for the positive elements $\bar{a}_{iq}$, $i = 1, 2, \ldots, m$, of the qth column and selecting the element p yielding the minimum ratio. Pivoting on this element will maintain feasibility as well as (assuming nondegeneracy) decrease the value of the objective function. If there are ties, any element yielding the minimum can be used. If there are no nonnegative elements in the column, the problem is unbounded. After updating the entire tableau with $\bar{a}_{pq}$ as pivot and transforming the last row in the same manner as all other rows (except row q), we obtain a new tableau in canonical form. The new value of the objective function again appears in the lower right-hand corner of the tableau.

The simplex algorithm can be summarized by the following steps:

Step 0. Form a tableau as in Fig. 3.2 corresponding to a basic feasible solution. The relative cost coefficients can be found by row reduction.
Step 1. If each $r_j \geqslant 0$, stop; the current basic feasible solution is optimal.
Step 2. Select q such that $r_q < 0$ to determine which nonbasic variable is to become basic.
Step 3. Calculate the ratios $\bar{a}_{i0}/\bar{a}_{iq}$ for $\bar{a}_{iq} > 0$, $i = 1, 2, \ldots, m$. If no $\bar{a}_{iq} > 0$, stop; the problem is unbounded. Otherwise, select p as the index i corresponding to the minimum ratio.
Step 4. Pivot on the pqth element, updating all rows including the last. Return to Step 1.

Proof that the algorithm solves the problem (again assuming nondegeneracy) is essentially established by our previous development. The process terminates only if optimality is achieved or unboundedness is discovered. If neither condition is discovered at a given basic solution, then the objective is strictly decreased. Since there are only a finite number of possible basic feasible solutions, and no basis repeats because of the strictly decreasing objective, the algorithm must reach a basis satisfying one of the two terminating conditions.

Example 1. Maximize $3x_1 + x_2 + 3x_3$ subject to

$$\begin{aligned} 2x_1 + x_2 + x_3 &\leqslant 2 \\ x_1 + 2x_2 + 3x_3 &\leqslant 5 \\ 2x_1 + 2x_2 + x_3 &\leqslant 6 \\ x_1 \geqslant 0, \ x_2 \geqslant 0, \ x_3 &\geqslant 0. \end{aligned}$$

To transform the problem into standard form so that the simplex procedure can be applied, we change the maximization to minimization by multiplying the objective function by minus one, and introduce three nonnegative slack variables x_4, x_5, x_6. We then have the initial tableau

	$\mathbf{a}_1$	$\mathbf{a}_2$	$\mathbf{a}_3$	$\mathbf{a}_4$	$\mathbf{a}_5$	$\mathbf{a}_6$	$\mathbf{b}$
	②	①	1	1	0	0	2
	1	2	③	0	1	0	5
	2	2	1	0	0	1	6
$\mathbf{r}^T$	−3	−1	−3	0	0	0	0

First tableau

The problem is already in canonical form with the three slack variables serving as the basic variables. We have at this point $r_j = c_j - z_j = c_j$, since the costs of the slacks are zero. Application of the criterion for selecting a column in which to pivot shows that any of the first three columns would yield an improved solution. In each of these columns the appropriate pivot element is determined by computing the ratios $\bar{a}_{i0}/\bar{a}_{ij}$ and selecting the smallest positive one. The three allowable pivots are all circled on the tableau. It is only necessary to determine one allowable pivot, and normally we would not bother to calculate them all. For hand calculation on problems of this size, however, we may wish to examine the allowable pivots and select one that will minimize (at least in the short run) the amount of division required. Thus for this example we select the second column and result in:

2	1	1	1	0	0	2
−3	0	①	−2	1	0	1
−2	0	−1	−2	0	1	2
−1	0	−2	1	0	0	2

Second tableau

We note that the objective function—we are using the negative of the original one—has decreased from zero to minus two. We now pivot on ①.

⑤	1	0	3	−1	0	1
−3	0	1	−2	1	0	1
−5	0	0	−4	1	1	3
−7	0	0	−3	2	0	4

Third tableau

The value of the objective function has now decreased to minus four and we may pivot in either the first or fourth column. We select ⑤.

1	1/5	0	3/5	−1/5	0	1/5
0	3/5	1	−1/5	2/5	0	8/5
0	1	0	−1	0	1	4
0	7/5	0	6/5	3/5	0	27/5

Fourth tableau

Since the last row has no negative elements, we conclude that the solution corresponding to the fourth tableau is optimal. Thus $x_1 = 1/5$, $x_2 = 0$, $x_3 = 8/5$, $x_4 = 0$, $x_5 = 0$, $x_6 = 4$ is the optimal solution with a corresponding value of the (negative) objective of $-(27/5)$.

Degeneracy

It is possible that in the course of the simplex procedure, degenerate basic feasible solutions may occur. Often they can be handled as a nondegenerate basic feasible solution. However, it is possible that after a new column q is selected to enter the basis, the minimum of the ratios $\bar{a}_{i0}/\bar{a}_{iq}$ may be zero, implying that the zero-valued basic variable is the one to go out. This means that the new variable x_q will come in at zero value, the objective will not decrease, and the new basic feasible solution will also be degenerate. Conceivably, this process could continue for a series of steps until, finally, the original degenerate solution is again obtained. The result is a *cycle* that could be repeated indefinitely.

Methods have been developed to avoid such cycles (see Exercises 15–17 for a full discussion of one of them, which is based on perturbing the problem slightly so that zero-valued variables are actually small positive values, and Exercise 32 for Bland's rule, which is simpler). In practice, however, such procedures are found to be unnecessary. When degenerate solutions are encountered, the simplex procedure generally does not enter a cycle. However, anticycling procedures are simple, and many codes incorporate such a procedure for the sake of safety.

3.5 Finding a Basic Feasible Solution

A basic feasible solution is sometimes immediately available for linear programs. For example, in problems with constraints of the form

$$\mathbf{Ax} \leqslant \mathbf{b}, \quad \mathbf{x} \geqslant \mathbf{0} \tag{3.25}$$

with $\mathbf{b} \geqslant \mathbf{0}$, a basic feasible solution to the corresponding standard form of the problem is provided by the slack variables. This provides a means for initiating the simplex procedure. The example in the last section was of this type. An initial basic feasible solution is not always apparent for other types of linear programs, however, and it is necessary to develop a means for determining one so that the simplex method can be initiated. Interestingly (and fortunately), an auxiliary linear program and corresponding application of the simplex method can be used to determine the required initial solution.

By elementary straightforward operations the constraints of a linear programming problem can always be expressed in the form

$$\mathbf{Ax} = \mathbf{b}, \quad \mathbf{x} \geqslant \mathbf{0} \tag{3.26}$$

with $\mathbf{b} \geqslant \mathbf{0}$. In order to find a solution to (3.26) consider the artificial minimization problem

$$\begin{aligned} \text{minimize} \quad & \sum_{i=1}^{m} u_j \\ \text{subject to} \quad & \mathbf{Ax} + \mathbf{u} = \mathbf{b} \\ & \mathbf{x} \geqslant \mathbf{0}, \ \mathbf{u} \geqslant \mathbf{0} \end{aligned} \tag{3.27}$$

where $\mathbf{u} = (u_1, u_2, \ldots, u_m)$ is a vector of artificial variables. If there is a feasible solution to (3.26), then it is clear that (3.27) has a minimum value of zero with $\mathbf{u} = \mathbf{0}$. If (3.26) has no feasible solution, then the minimum value of (3.27) is greater than zero.

Now (3.27) is itself a linear program in the variables $\mathbf{x}$, $\mathbf{u}$, and the system is already in canonical form with basic feasible solution $\mathbf{u} = \mathbf{b}$. If (3.27) is solved using the simplex technique, a basic feasible solution is obtained at each step. If the minimum value of (3.27) is zero, then the final basic solution will have all $u_j = 0$, and hence barring degeneracy, the final solution will have no u_j variables basic. If in the final solution some u_j are both zero and basic, indicating a degenerate solution, these basic variables can be exchanged for nonbasic x_j variables (again at zero level) to yield a basic feasible solution involving x variables only. (However, the situation is more complex if A is not of full rank. See Exercise 21.)

Example 1. Find a basic feasible solution to

$$\begin{aligned} 2x_1 + x_2 + 2x_3 &= 4 \\ 3x_1 + 3x_2 + x_3 &= 3 \\ x_1 \geqslant 0, \ x_2 \geqslant 0, \ x_3 &\geqslant 0. \end{aligned}$$

We introduce artificial variables $x_4 \geqslant 0$, $x_5 \geqslant 0$ and an objective function $x_4 + x_5$. The initial tableau is

	x_1	x_2	x_3	x_4	x_5	$\mathbf{b}$
	2	1	2	1	0	4
	3	3	1	0	1	3
$\mathbf{c}^T$	0	0	0	1	1	0

Initial tableau

A basic feasible solution to the expanded system is given by the artificial variables. To initiate the simplex procedure we must update the last row so that it has zero components under the basic variables. This yields:

	2	1	2	1	0	4
	③	3	1	0	1	3
$\mathbf{r}^T$	−5	−4	−3	0	0	−7

First tableau

Pivoting in the column having the most negative bottom row component as indicated, we obtain:

0	−1	(4/3)	1	−2/3	2
1	1	1/3	0	1/3	1
0	1	−4/3	0	5/3	−2

Second tableau

In the second tableau there is only one choice for pivot, and it leads to the final tableau shown.

0	−3/4	1	3/4	−1/2	3/2
1	5/4	0	−1/4	1/2	1/2
0	0	0	1	1	0

Final tableau

Both of the artificial variables have been driven out of the basis, thus reducing the value of the objective function to zero and leading to the basic feasible solution to the original problem

$$x_1 = 1/2, \quad x_2 = 0, \quad x_3 = 3/2.$$

Using artificial variables, we attack a general linear programming problem by use of the *two-phase method*. This method consists simply of a *phase I* in which artificial variables are introduced as above and a basic feasible solution is found (or it is determined that no feasible solutions exist); and a *phase II* in which, using the basic feasible solution resulting from phase I, the original objective function is minimized. During phase II the artificial variables and the objective function of phase I are omitted. Of course, in phase I artificial variables need be introduced only in those equations that do not contain slack variables.

Example 2. Consider the problem

$$\begin{aligned} \text{minimize}\quad & 4x_1 + x_2 + x_3 \\ \text{subject to}\quad & 2x_1 + x_2 + 2x_3 = 4 \\ & 3x_1 + 3x_2 + x_3 = 3 \\ & x_1 \geqslant 0, \quad x_2 \geqslant 0, \quad x_3 \geqslant 0. \end{aligned}$$

There is no basic feasible solution apparent, so we use the two-phase method. The first phase was done in Example 1 for these constraints, so we shall not repeat it here. We give only the final tableau with the columns corresponding to the artificial variables deleted, since they are not used in phase II. We use the new cost function in place of the old one. Temporarily writing $\mathbf{c}^T$ in the bottom row we have

	x_1	x_2	x_3	$\mathbf{b}$
	0	−3/4	1	3/2
	1	5/4	0	1/2
$\mathbf{c}^T$	4	1	1	0

Initial tableau

Transforming the last row so that zeros appear in the basic columns, we have

0	−3/4	1	3/2
1	(5/4)	0	1/2
0	−13/4	0	−7/2

First tableau

3/5	0	1	9/5
4/5	1	0	2/5
13/5	0	0	−11/5

Second tableau

and hence the optimal solution is $x_1 = 0,\ x_2 = 2/5,\ x_3 = 9/5$.

Example 3 (A Free Variable Problem).

$$\begin{array}{ll} \text{minimize} & -2x_1 + 4x_2 + 7x_3 + x_4 + 5x_5 \\ \text{subject to} & -x_1 + x_2 + 2x_3 + x_4 + 2x_5 = 7 \\ & -x_1 + 2x_2 + 3x_3 + x_4 + x_5 = 6 \\ & -x_1 + x_2 + x_3 + 2x_4 + x_5 = 4 \\ & x_1 \text{ free},\ x_2 \geqslant 0,\ x_3 \geqslant 0,\ x_4 \geqslant 0,\ x_5 \geqslant 0. \end{array}$$

Since x_1 is free, it can be eliminated, as described in Chap. 2, by solving for x_1 in terms of the other variables from the first equation and substituting everywhere else. This can all be done with the simplex tableau as follows:

	x_1	x_2	x_3	x_4	x_5	$\mathbf{b}$
	−(1)	1	2	1	2	7
	−1	2	3	1	1	6
	−1	1	1	2	1	4
$\mathbf{c}^T$	−2	4	7	1	5	0

Initial tableau

We select any nonzero element in the first column to pivot on—this will eliminate x_1.

1	−1	−2	−1	−2	−7
0	1	1	0	−1	−1
0	0	−1	1	−1	−3
0	2	3	−1	1	−14

Equivalent problem

We now save the first row for future reference, but our linear program only involves the sub-tableau indicated. There is no obvious basic feasible solution for this problem, so we introduce artificial variables x_6 and x_7.

	x_2	x_3	x_4,	x_5	x_6	x_7	$\mathbf{b}$
	−1	−1	0	1	1	0	1
	0	1	−1	1	0	1	3
$\mathbf{c}^T$	0	0	0	0	1	1	0

Initial tableau for phase I

Transforming the last row appropriately we obtain

	x_2	x_3	x_4	x_5	x_6	x_7	$\mathbf{b}$
	−1	−1	0	①	1	0	1
	0	1	−1	1	0	1	3
$\mathbf{r}^T$	1	0	1	−2	0	0	−4

First tableau—phase I

x_2	x_3	x_4	x_5	x_6	x_7	$\mathbf{b}$
−1	−1	0	1	1	0	1
①	2	−1	0	−1	1	2
−1	−2	1	0	2	0	−2

Second tableau—phase I

0	1	−1	1	0	1	3
1	2	−1	0	−1	1	2
0	0	0	0	1	1	0

Final tableau—phase I

Now we go back to the equivalent reduced problem

	x_2	x_3	x_4	x_5	$\mathbf{b}$
	0	1	−1	1	3
	1	2	−1	0	2
$\mathbf{c}^T$	2	3	−1	1	−14

Initial tableau—phase II

Transforming the last row appropriately we proceed with:

0	1	−1	1	3
0	②	−1	0	2
0	−2	2	0	−21

First tableau—phase II

−1/2	0	−1/2	1	2
1/2	1	−1/2	0	1
1	0	1	0	−19

Final tableau—phase II

The solution $x_3 = 1,\ x_5 = 2$ can be inserted in the expression for x_1 giving

$$x_1 = -7 + 2 \cdot 1 + 2 \cdot 2 = -1;$$

thus the final solution is

$$x_1 = -1,\ x_2 = 0,\ x_3 = 1,\ x_4 = 0,\ x_5 = 2.$$

3.6 Matrix Form of the Simplex Method

Although the elementary pivot transformations associated with the simplex method are in many respects most easily discernible in the tableau format, with attention focused on the individual elements, there is much insight to be gained by studying a matrix interpretation of the procedure. The vector-matrix relationships that exist between the various rows and columns of the tableau lead, however, not only to increased understanding but also, in a rather direct way, to the *revised simplex procedure* which in many cases can result in considerable computational advantage. The matrix formulation is also a natural setting for the discussion of dual linear programs and other topics related to linear programming.

A preliminary observation in the development is that the tableau at any point in the simplex procedure can be determined solely by a knowledge of which variables are basic. As before we denote by $\mathbf{B}$ the submatrix of the original A matrix consisting of the m columns of A corresponding to the basic variables. These columns are linearly independent and hence the columns of $\mathbf{B}$ form a basis for E^m. We refer to $\mathbf{B}$ as the basis matrix.

As usual, let us assume that $\mathbf{B}$ consists of the first m columns of A. Then by partitioning $\mathbf{A}$, $\mathbf{x}$, and $\mathbf{c}^T$ as

$$\mathbf{A} = [\mathbf{B},\ \mathbf{D}]$$
$$\mathbf{x} = (\mathbf{x_B},\ \mathbf{x_D}), \quad \mathbf{c}^T = \left[\mathbf{c}_\mathbf{B}^T,\ \mathbf{c}_\mathbf{D}^T\right],$$

the standard linear program becomes

$$\begin{aligned} \text{minimize} \quad & \mathbf{c}_\mathbf{B}^T\mathbf{x_B} + \mathbf{c}_\mathbf{D}^T\mathbf{x_D} \\ \text{subject to} \quad & \mathbf{B}\mathbf{x_B} + \mathbf{D}\mathbf{x_D} = \mathbf{b} \\ & \mathbf{x_B} \geqslant \mathbf{0},\ \mathbf{x_D} \geqslant \mathbf{0}. \end{aligned} \tag{3.28}$$

The basic solution, which we assume is also feasible, corresponding to the basis $\mathbf{B}$ is $\mathbf{x} = (\mathbf{x_B},\ 0)$ where $\mathbf{x_B} = \mathbf{B}^{-1}\mathbf{b}$. The basic solution results from setting $\mathbf{x_D} = 0$. However, for any value of $\mathbf{x_D}$ the necessary value of $\mathbf{x_B}$ can be computed from (3.28) as

$$\mathbf{x_B} = \mathbf{B}^{-1}\mathbf{b} - \mathbf{B}^{-1}\mathbf{D}\mathbf{x_D}, \tag{3.29}$$

and this general expression when substituted in the cost function yields

$$\begin{aligned} z &= \mathbf{c}_\mathbf{B}^T(\mathbf{B}^{-1}\mathbf{b} - \mathbf{B}^{-1}\mathbf{D}\mathbf{x_D}) + \mathbf{c}_\mathbf{D}^T\mathbf{x_D} \\ &= \mathbf{c}_\mathbf{B}^T\mathbf{B}^{-1}\mathbf{b} + (\mathbf{c}_\mathbf{D}^T - \mathbf{c}_\mathbf{B}^T\mathbf{B}^{-1}\mathbf{D})\mathbf{x_D}, \end{aligned} \tag{3.30}$$

which expresses the cost of any solution to (3.28) in terms of $\mathbf{x_D}$. Thus

$$\mathbf{r}_\mathbf{D}^T = \mathbf{c}_\mathbf{D}^T - \mathbf{c}_\mathbf{B}^T\mathbf{B}^{-1}\mathbf{D} \tag{3.31}$$

is the relative cost vector (for nonbasic variables). It is the components of this vector that are used to determine which vector to bring into the basis.

Having derived the vector expression for the relative cost it is now possible to write the simplex tableau in matrix form. The initial tableau takes the form

$$\left[\begin{array}{c|c} \mathbf{A} & \mathbf{b} \\ \hline \mathbf{c}^T & 0 \end{array}\right] = \left[\begin{array}{c|c|c} \mathbf{B} & \mathbf{D} & \mathbf{b} \\ \hline \mathbf{c}_{\mathbf{B}}^T & \mathbf{c}_{\mathbf{D}}^T & 0 \end{array}\right], \tag{3.32}$$

which is not in general in canonical form and does not correspond to a point in the simplex procedure. If the matrix $\mathbf{B}$ is used as a basis, then the corresponding tableau becomes

$$\mathbf{T} = \left[\begin{array}{c|c|c} \mathbf{I} & \mathbf{B}^{-1}\mathbf{D} & \mathbf{B}^{-1}\mathbf{b} \\ \hline \mathbf{0} & \mathbf{c}_{\mathbf{D}}^T - \mathbf{C}_{\mathbf{B}}^T\mathbf{B}^{-1}\mathbf{D} & -\mathbf{c}_{\mathbf{B}}^T\mathbf{B}^{-1}\mathbf{b} \end{array}\right], \tag{3.33}$$

which is the matrix form we desire.

*The Revised Simplex Method and LU Decomposition

Extensive experience with the simplex procedure applied to problems from various fields, and having various values of n and m, has indicated that the method can be expected to converge to an optimum solution in about m, or perhaps $3m/2$, pivot operations. (Except in the worst case. See Chap. 5.) Thus, particularly if m is much smaller than n, that is, if the matrix $\mathbf{A}$ has far fewer rows than columns, pivots will occur in only a small fraction of the columns during the course of optimization.

Since the other columns are not explicitly used, it appears that the work expended in calculating the elements in these columns after each pivot is, in some sense, wasted effort. The revised simplex method is a scheme for ordering the computations required of the simplex method so that unnecessary calculations are avoided. In fact, even if pivoting is eventually required in all columns, but m is small compared to n, the revised simplex method can frequently save computational effort.

The revised form of the simplex method is this: Given the inverse $\mathbf{B}^{-1}$ of a current basis, and the current solution $\mathbf{x_B} = \bar{\mathbf{a}}_0 = \mathbf{B}^{-1}\mathbf{b}$,

Step 1. Calculate the current relative cost coefficients $\mathbf{r}_{\mathbf{D}}^T = \mathbf{c}_{\mathbf{D}}^T - \mathbf{c}_{\mathbf{B}}^T\mathbf{B}^{-1}\mathbf{D}$. This can best be done by first calculating $\mathbf{y}^T = \mathbf{c}_{\mathbf{B}}^T\mathbf{B}^{-1}$ and then the relative cost vector $\mathbf{r}_{\mathbf{D}}^T = \mathbf{c}_{\mathbf{D}}^T - \mathbf{y}^T\mathbf{D}$. If $\mathbf{r_D} \geqslant \mathbf{0}$ stop; the current solution is optimal.

Step 2. Determine which vector $\mathbf{a}_q$ is to enter the basis by selecting the most negative cost coefficient; and calculate $\bar{\mathbf{a}}_q = \mathbf{B}^{-1}\mathbf{a}_q$ which gives the vector $\mathbf{a}_q$ expressed in terms of the current basis.

Step 3. If no $\bar{a}_{iq} > 0$, stop; the problem is unbounded. Otherwise, calculate the ratios $\bar{a}_{i0}/\bar{a}_{iq}$ for $\bar{a}_{iq} > 0$ to determine which vector is to leave the basis.

Step 4. Update $\mathbf{B}^{-1}$ and the current solution $\mathbf{B}^{-1}\mathbf{b}$. Return to Step 1.

Updating of $\mathbf{B}^{-1}$ is accomplished by the usual pivot operations applied to an array consisting of $\mathbf{B}^{-1}$ and $\bar{\mathbf{a}}_q$, where the pivot is the appropriate element in $\bar{\mathbf{a}}_q$. Of course $\mathbf{B}^{-1}\mathbf{b}$ may be updated at the same time by adjoining it as another column.

One may go one step further in the matrix interpretation of the simplex method and note that execution of a single simplex cycle is not explicitly dependent on having $\mathbf{B}^{-1}$ but rather on the ability to solve linear systems with $\mathbf{B}$ as the coefficient matrix. A decomposition of $\mathbf{B} = \mathbf{LU}$ can be updated where $\mathbf{L}$ is a lower triangular matrix and $\mathbf{U}$ is an upper triangular matrix. Then each of the linear systems can be solved by solving two triangular systems.

3.7 Simplex Method for Transportation Problems

The transportation problem was stated briefly in Chap. 2. We restate it here. There are m origins that contain various amounts of a commodity that must be shipped to n destinations to meet demand requirements. Specifically, origin i contains an amount a_i, and destination j has a requirement of amount b_j. It is assumed that the system is *balanced* in the sense that total supply equals total demand. That is,

$$\sum_{i=1}^{m} a_i = \sum_{j=1}^{n} b_j. \tag{3.34}$$

The numbers a_i and b_j, $i = 1, 2, \ldots, m$; $j = 1, 2, \ldots, n$, are assumed to be nonnegative, and in many applications they are in fact nonnegative integers. There is a unit cost c_{ij} associated with the shipping of the commodity from origin i to destination j. The problem is to find the shipping pattern between origins and destinations that satisfies all the requirements and minimizes the total shipping cost.

In mathematical terms the above problem can be expressed as finding a set of x_{ij}'s, $i = 1, 2, \ldots, m$; $j = 1, 2, \ldots, n$, to

$$\begin{aligned} \text{minimize} \quad & \sum_{i=1}^{m} \sum_{j=1}^{n} c_{ij} x_{ij} \\ \text{subject to} \quad & \sum_{j=1}^{n} x_{ij} = a_i \quad \text{for} \quad i = 1, 2, \ldots, m \\ & \sum_{i=1}^{m} x_{ij} = b_j \quad \text{for} \quad j = 1, 2, \ldots, n \\ & x_{ij} \geqslant 0 \quad \text{for all} \quad i \text{ and } j. \end{aligned} \tag{3.35}$$

This mathematical problem, together with the assumption (3.34), is the general transportation problem. In the shipping context, the variables x_{ij} represent the amounts of the commodity shipped from origin i to destination j.

The structure of the problem can be seen more clearly by writing the constraint equations in standard form:

$$
\begin{array}{lllll}
x_{11} + x_{12} + \cdots + x_{1n} & & & & = a_1 \\
& x_{21} + x_{22} + \cdots + x_{2n} & & & = a_2 \\
& & & & \vdots \\
& & & x_{m1} + x_{m2} + \cdots + x_{mn} & = a_m \\
\hline
x_{11} & + x_{21} & & x_{m1} & = b_1 \\
\quad x_{12} & \quad + x_{22} & & \quad + x_{m2} & = b_2 \\
& & & & \vdots \\
\qquad x_{1n} & \qquad + x_{2n} & & \qquad + x_{mn} & = b_n
\end{array}
\tag{3.36}
$$

The structure is perhaps even more evident when the coefficient matrix $\mathbf{A}$ of the system of equations above is expressed in vector-matrix notation as

$$
\mathbf{A} = \begin{bmatrix} \mathbf{1}^T & & & \\ & \mathbf{1}^T & & \\ & & \ddots & \\ & & & \mathbf{1}^T \\ \mathbf{I} & \mathbf{I} & \cdots & \mathbf{I} \end{bmatrix}, \tag{3.37}
$$

where $\mathbf{1} = (1, 1, \ldots, 1)$ is n-dimensional, and where each $\mathbf{I}$ is an $n \times n$ identity matrix.

In practice it is usually unnecessary to write out the constraint equations of the transportation problem in the explicit form (3.36). A specific transportation problem is generally defined by simply presenting the data in compact form, such as:

$$
\mathbf{a} = (a_1, a_2, \ldots, a_m)
$$

$$
\mathbf{C} = \begin{bmatrix} c_{11} & c_{12} & \cdots & c_{1n} \\ c_{21} & c_{22} & \cdots & c_{2n} \\ c_{m1} & c_{m2} & \cdots & c_{mn} \end{bmatrix}.
$$

$$
\mathbf{b} = (b_1, b_2, \ldots, b_n)
$$

The solution can also be represented by an $m \times n$ array, and as we shall see, all computations can be made on arrays of a similar dimension.

Example 1. As an example, which will be solved completely in a later section, a specific transportation problem with four origins and five destinations is defined by

$$
\mathbf{a} = (30, 80, 10, 60)
$$

$$
\mathbf{C} = \begin{bmatrix} 3 & 4 & 6 & 8 & 9 \\ 2 & 2 & 4 & 5 & 5 \\ 2 & 2 & 2 & 3 & 2 \\ 3 & 3 & 2 & 4 & 2 \end{bmatrix}.
$$

$$
\mathbf{b} = (10, 50, 20, 80, 20)
$$

Note that the balance requirement is satisfied, since the sum of the supply and the demand are both 180.

Finding a Basic Feasible Solution

A first step in the study of the structure of the transportation problem is to show that there is always a feasible solution, thus establishing that the problem is well defined. A feasible solution can be found by allocating shipments from origins to destinations in proportion to supply and demand requirements. Specifically, let S be equal to the total supply (which is also equal to the total demand). Then let $x_{ij} = a_i b_j / S$ for $i = 1, 2, \ldots, m$; $j = 1, 2, \ldots, n$. The reader can easily verify that this is a feasible solution. We also note that the solutions are bounded, since each x_{ij} is bounded by a_i (and by b_j). A bounded program with a feasible solution has an optimal solution. Thus, a transportation problem always has an optimal solution.

A second step in the study of the structure of the transportation problem is based on a simple examination of the constraint equations. Clearly there are m equations corresponding to origin constraints and n equations corresponding to destination constraints—a total of $n + m$. However, it is easily noted that the sum of the origin equations is

$$\sum_{i=1}^{m} \sum_{j=1}^{n} x_{ij} = \sum_{i=1}^{m} a_i, \tag{3.38}$$

and the sum of the destination equations is

$$\sum_{j=1}^{n} \sum_{i=1}^{m} x_{ij} = \sum_{j=1}^{n} b_j. \tag{3.39}$$

The left-hand sides of these equations are equal. Since they were formed by two distinct linear combinations of the original equations, it follows that the equations in the original system are not independent. The right-hand sides of (3.38) and (3.39) are equal by the assumption that the system is balanced, and therefore the two equations are, in fact, consistent. However, it is clear that the original system of equations is redundant. This means that one of the constraints can be eliminated without changing the set of feasible solutions. Indeed, *any* one of the constraints can be chosen as the one to be eliminated, for it can be reconstructed from those remaining. It follows that a basis for the transportation problem consists of $m + n - 1$ vectors, and a nondegenerate basic feasible solution consists of $m + n - 1$ variables. The simple solution found earlier in this section is clearly not a basic solution.

There is a straightforward way to compute an initial basic feasible solution to a transportation problem. The method is worth studying at this stage because it introduces the computational process that is the foundation for the general solution technique based on the simplex method. It also begins to illustrate the fundamental property of the structure of transportation problems.

The Northwest Corner Rule

This procedure is conducted on the *solution array* shown below:

x_{11}	x_{12}	x_{13}	$\cdots$	x_{1n}	a_1
x_{21}	x_{22}	x_{23}	$\cdots$	x_{2n}	a_2
$\vdots$					$\vdots$
x_{m1}	x_{m2}	x_{m3}	$\cdots$	x_{mn}	a_m
b_1	b_2	b_3	$\cdots$	b_n	

(3.40)

The individual elements of the array appear in *cells* and represent a solution. An empty cell denotes a value of zero.

Beginning with all empty cells, the procedure is given by the following steps:

Step 1. Start with the cell in the upper left-hand corner.
Step 2. Allocate the maximum feasible amount consistent with row and column sum requirements involving that cell. (At least one of these requirements will then be met.)
Step 3. Move one cell to the right if there is any remaining row requirement (supply). Otherwise move one cell down. If all requirements are met, stop; otherwise go to Step 2.

The procedure is called the *Northwest Corner Rule* because at each step it selects the cell in the upper left-hand corner of the subarray consisting of current nonzero row and column requirements.

Example 1. A basic feasible solution constructed by the Northwest corner Rule is shown below for Example 1 of the last section.

10	20				30
	30	20	30		80
			10		10
			40	20	60
10	50	20	80	20	

(3.41)

In the first step, at the upper left-hand corner, a maximum of 10 units could be allocated, since that is all that was required by column 1. This left $30 - 10 = 20$ units required in the first row. Next, moving to the second cell in the top row, the remaining 20 units were allocated. At this point the row 1 requirement is met, and it is necessary to move down to the second row. The reader should be able to follow the remaining steps easily.

There is the possibility that at some point both the row and column requirements corresponding to a cell may be met. The next entry will then be a zero, indicating a degenerate basic solution. In such a case there is a choice as to where to place the zero. One can either move right or move down to enter the zero. Two examples of degenerate solutions to a problem are shown below:

30				30
20	20			40
	0	20		20
		20	40	60
50	20	40	40	

30				30
20	20	0		40
		20		20
		20	40	60
50	20	40	40	

It should be clear that the Northwest Corner Rule can be used to obtain different basic feasible solutions by first permuting the rows and columns of the array before the procedure is applied. Or equivalently, one can do this indirectly by starting the procedure at an arbitrary cell and then considering successive rows and columns in an arbitrary order.

Basis Triangularity

We now establish the most important structural property of the transportation problem: the triangularity of all bases. This property simplifies the process of solution of a system of equations whose coefficient matrix corresponds to a basis, and thus leads to efficient implementation of the simplex method.

The concept of upper and lower triangular matrices was introduced in connection with Gaussian elimination methods, see Appendix C. It is useful at this point to generalize slightly the notion of upper and lower triangularity.

Definition. A nonsingular square matrix $\mathbf{M}$ is said to be *triangular* if by a permutation of its rows and columns it can be put in the form of a lower triangular matrix.

There is a simple and useful procedure for determining whether a given matrix $\mathbf{M}$ is triangular:

Step 1. Find a row with exactly one nonzero entry.
Step 2. Form a submatrix of the matrix used in Step 1 by crossing out the row found in Step 1 and the column corresponding to the nonzero entry in that row. Return to Step 1 with this submatrix.
If this procedure can be continued until all rows have been eliminated, then the matrix is triangular. It can be put in lower triangular form explicitly by arranging the rows and columns in the order that was determined by the procedure.

Example 1. Shown below on the left is a matrix before the above procedure is applied to it. Indicated along the edges of this matrix is the order in which the rows and columns are indexed according to the procedure. Shown at the right is the same matrix when its rows and columns are permuted according to the order found.

$$\begin{bmatrix} 1 & 2 & 0 & 1 & 0 & 2 \\ 4 & 1 & 0 & 5 & 0 & 0 \\ 0 & 0 & 0 & 4 & 0 & 0 \\ 2 & 1 & 7 & 2 & 1 & 3 \\ 2 & 3 & 2 & 0 & 0 & 3 \\ 0 & 2 & 0 & 1 & 0 & 0 \end{bmatrix} \quad \begin{matrix} 3 \\ 6 \\ 2 \\ 1 \\ 5 \\ 4 \end{matrix} \begin{bmatrix} 4 & 0 & 0 & 0 & 0 & 0 \\ 1 & 2 & 0 & 0 & 0 & 0 \\ 5 & 1 & 4 & 0 & 0 & 0 \\ 1 & 2 & 1 & 2 & 0 & 0 \\ 0 & 3 & 2 & 3 & 2 & 0 \\ 2 & 1 & 2 & 3 & 7 & 1 \end{bmatrix}$$
$$\begin{matrix} 4 & 2 & 1 & 6 & 3 & 5 \end{matrix}$$

Triangularization

We are now prepared to derive the most important structural property of the transportation problem.

Basis Triangularity Theorem. *Every basis of the transportation problem is triangular.*

Proof. Refer to the system of constraints (3.36). Let us change the sign of the top half of the system; then the coefficient matrix of the system consists of entries that are either $+1$, -1, or 0. Following the result of the theorem in Sect. 3.7, delete any one of the equations to eliminate the redundancy. From the resulting coefficient matrix, form a basis **B** by selecting a nonsingular subset of $m + n - 1$ columns.

Each column of **B** contains at most two nonzero entries, a $+ 1$ and a $- 1$. Thus there are at most $2(m+n-1)$ nonzero entries in the basis. However, if every column contained two nonzero entries, then the sum of all rows would be zero, contradicting the nonsingularity of **B**. Thus at least one column of **B** must contain only one nonzero entry. This means that the total number of nonzero entries in **B** is less than $2(m + n - 1)$. It then follows that there must be a row with only one nonzero entry; for if every row had two or more nonzero entries, the total number would be at least $2(m+n-1)$. This means that the first step of the procedure for verifying triangularity is satisfied. A similar argument can be applied to the submatrix of **B** obtained by crossing out the row with the single nonzero entry and the column corresponding to that entry; that submatrix must also contain a row with a single nonzero entry. This argument can be continued, establishing that the basis **B** is triangular. ∎

Example 2. As an illustration of the Basis Triangularity Theorem, consider the basis selected by the Northwest Corner Rule in Example 1. This basis is represented below, except that only the basic variables are indicated, not their values.

x_{11}	x_{12}				30
	x_{22}	x_{23}	x_{24}		80
			x_{34}		10
			x_{44}	x_{45}	60
10	50	20	80	20	

A row in a basis matrix corresponds to an equation in the original system and is associated with a constraint either on a row or column sum in the solution array. In this example the equation corresponding to the first column sum contains only one basis variable, x_{11}. The value of this variable can be found immediately to be 10. The next equation corresponds to the first row sum. The corresponding variable is x_{12}, which can be found to be 20, since x_{11} is known. Progression in this manner through the basis variables is equivalent to back substitution.

The importance of triangularity is, of course, the associated method of *back substitution* for the solution of a triangular system of equations, as discussed in Appendix C. Moreover, since any basis matrix is triangular and all nonzero elements are equal to one (or minus one if the signs of some equations are changed), it follows that the process of back substitution will simply involve repeated additions and subtractions of the given row and column sums. No multiplication is required. It therefore follows that if the original row and column totals are integers, the values of all basic variables will be integers. This is an important result, which we summarize by a corollary to the Basis Triangularity Theorem.

Corollary. *If the row and column sums of a transportation problem are integers, then the basic variables in any basic solution are integers.*

The Transportation Simplex Method

Now that the structural properties of the transportation problem have been developed, it is a relatively straightforward task to work out the details of the simplex method for the transportation problem. A major objective is to exploit fully the triangularity property of bases in order to achieve both computational efficiency and a compact representation of the method. The method used is actually a direct adaptation of the version of the revised simplex method presented in the first part of Sect. 3.6. The basis is never inverted; instead, its triangular form is used directly to solve for all required variables.

Simplex Multipliers

Simplex multipliers are associated with the constraint equations. In this case we partition the vector of multipliers as $\mathbf{y} = (\mathbf{u}, \mathbf{v})$. Here, u_i represents the multiplier associated with the ith row sum constraint, and v_j represents the multiplier associated with the jth column sum constraint. Since one of the constraints is redundant, an arbitrary value may be assigned to any one of the multipliers (see Exercise 4, Chap. 4). For notational simplicity we shall at this point set $v_n = 0$.

Given a basis $\mathbf{B}$, the simplex multipliers are found to be the solution to the equation $\mathbf{y}^T\mathbf{B} = \mathbf{c}_{\mathbf{B}}^T$. To determine the explicit form of these equations, we again refer to the original system of constraints (3.36). If x_{ij} is basic, then the corresponding column from $\mathbf{A}$ will be included in $\mathbf{B}$. This column has exactly two +1 entries: one in the ith position of the top portion and one in the jth position of the bottom portion. This column thus generates the simplex multiplier equation $u_i + v_j = c_{ij}$, since u_i and v_j are the corresponding components of the multiplier vector. Overall, the simplex multiplier equations are

$$u_i + v_j = c_{ij}, \tag{3.42}$$

for all i, j for which x_{ij} is basic. The coefficient matrix of this system is the transpose of the basis matrix and hence it is triangular. Thus, this system can be solved by back substitution. This is similar to the procedure for finding the values of basic variables and, accordingly, as another corollary of the Triangular Basis Theorem, an integer property holds for simplex multipliers.

Corollary. *If the unit costs c_{ij} of a transportation problem are all integers, then (assuming one simplex multiplier is set arbitrarily equal to an integer) the simplex multipliers associated with any basis are integers.*

Once the simplex multipliers are known, the relative cost coefficients for nonbasic variables can be found in the usual manner as $\mathbf{r}_{\mathbf{D}}^T = \mathbf{c}_{\mathbf{D}}^T - \mathbf{y}^T\mathbf{D}$. In this case the relative cost coefficients are

$$r_{ij} = c_{ij} - u_i - v_j \quad \text{for} \quad \begin{array}{l} i = 1, 2, \ldots, m \\ j = 1, 2, \ldots, n. \end{array} \tag{3.43}$$

This relation is valid for basic variables as well if we define relative cost coefficients for them—having value zero.

Given a basis, computation of the simplex multipliers is quite similar to the calculation of the values of the basic variables. The calculation is easily carried out on an array of the form shown below, where the circled elements correspond to the positions of the basic variables in the current basis.

c_{11}	c_{12}	c_{13}	$\cdots$	c_{1n}	u_1
c_{21}	c_{22}	c_{23}	$\cdots$	c_{2n}	u_2
$\vdots$				$\vdots$	$\vdots$
c_{m1}		$\cdots$		c_{mn}	u_m
v_1	v_2		$\cdots$	v_n	

In this case the main part of the array, with the coefficients c_{ij}, remains fixed, and we calculate the extra column and row corresponding to $\mathbf{u}$ and $\mathbf{v}$.

The procedure for calculating the simplex multipliers is this:

Step 1. Assign an arbitrary value to any one of the multipliers.
Step 2. Scan the rows and columns of the array until a circled element c_{ij} is found such that either u_i or v_j (but not both) has already been determined.
Step 3. Compute the undetermined u_i or v_j from the equation $c_{ij} = u_i + v_j$. If all multipliers are determined, stop. Otherwise, return to Step 2.
The triangularity of the basis guarantees that this procedure can be carried through to determine all the simplex multipliers.

Example 1. Consider the cost array of Example 1 of Sect. 5.1, which is shown below with the circled elements corresponding to a basic feasible solution (found by the Northwest Corner Rule). Only these numbers are used in the calculation of the multipliers.

$$\begin{bmatrix} ③ & ④ & 6 & 8 & 9 \\ 2 & ② & ④ & ⑤ & 5 \\ 2 & 2 & 2 & ③ & 2 \\ 3 & 3 & 2 & ④ & ② \end{bmatrix}.$$

We first arbitrarily set $v_5 = 0$. We then scan the cells, searching for a circled element for which only one multiplier must be determined. This is the bottom right corner element, and it gives $u_4 = 2$. Then, from the equation $4 = 2 + v_4$, v_4 is found to be 2. Next, u_3 and u_2 are determined, then v_3 and v_2, and finally u_1 and v_1. The result is shown below:

						u
	③	④	6	8	9	5
	2	②	④	⑤	5	3
	2	2	2	③	2	1
	3	3	2	④	②	2
v	−2	−1	1	2	0	

Cycle of Change

In accordance with the general simplex procedure, if a nonbasic variable has an associated relative cost coefficient that is negative, then that variable is a candidate for entry into the basis. As the value of this variable is gradually increased, the values of the current basic variables will change continuously in order to maintain feasibility. Then, as usual, the value of the new variable is increased precisely to the point where one of the old basic variables is driven to zero.

We must work out the details of how the values of the current basic variables change as a new variable is entered. If the new basic vector is $\mathbf{d}$, then the change in the other variables is given by $-\mathbf{B}^{-1}\mathbf{d}$, where $\mathbf{B}$ is the current basis. Hence, once again we are faced with a problem of solving a system associated with the triangular basis, and once again the solution has special properties. In the next theorem recall that $\mathbf{A}$ is defined by (3.37).

Theorem. *Let* $\mathbf{B}$ *be a basis from* $\mathbf{A}$ *(ignoring one row), and let* $\mathbf{d}$ *be another column. Then the components of the vector* $\mathbf{w} = \mathbf{B}^{-1}\mathbf{d}$ *are either* 0, +1, *or* −1.

Proof. Let $\mathbf{w}$ be the solution to the equation $\mathbf{B}\mathbf{w} = \mathbf{d}$. Then $\mathbf{w}$ is the representation of $\mathbf{d}$ in terms of the basis. This equation can be solved by Cramer's rule as

$$w_k = \frac{\det \mathbf{B}_k}{\det \mathbf{B}},$$

where $\mathbf{B}_k$ is the matrix obtained by replacing the kth column of $\mathbf{B}$ by $\mathbf{d}$. Both $\mathbf{B}$ and $\mathbf{B}_k$ are submatrices of the original constraint matrix $\mathbf{A}$. The matrix $\mathbf{B}$ may be put in triangular form with all diagonal elements equal to $+1$. Hence, accounting for the sign change that may result from the combined row and column interchanges, $\det \mathbf{B} = +1$ or -1. Likewise, it can be shown (see Exercise 3) that $\det \mathbf{B}_k = 0, +1$, or -1. We conclude that each component of $\mathbf{w}$ is either 0, $+1$, or -1. ∎

The implication of the above result is that when a new variable is added to the solution at a unit level, the current basic variables will each change by $+1$, -1, or 0. If the new variable has a value θ, then, correspondingly, the basic variables change by $+\theta$, $-\theta$, or 0. It is therefore only necessary to determine the signs of change for each basic variable.

The determination of these signs is again accomplished by row and column scanning. Operationally, one assigns a + to the cell of the entering variable to represent a change of $+\theta$, where θ is yet to be determined. Then +'s, −'s, and 0's are assigned, one by one, to the cells of some basic variables, indicating changes of $+\theta$, $-\theta$, or 0 to maintain a solution. As usual, after each step there will always be an equation that uniquely determines the sign to be assigned to another basic variable. The result will be a sequence of pluses and minuses assigned to cells that form a cycle leading from the cell of the entering variable back to that cell. In essence, the new change is part of a cycle of redistribution of the commodity flow in the transportation system.

Once the sequence of +'s, −'s, and 0's is determined, the new basic feasible solution is found by setting the level of the change θ. This is set so as to drive one of the old basic variables to zero. One must simply examine those basic variables for which a minus sign has been assigned, for these are the ones that will decrease as the new variable is introduced. Then θ is set equal to the smallest magnitude of these variables. This value is added to all cells that have a + assigned to them and subtracted from all cells that have a − assigned. The result will be the new basic feasible solution.

The procedure is illustrated by the following example.

Example 2. A completed solution array is shown below:

		10^0			10
		20^-		10^+	30
20^+	10^0			30^-	60
10^0					10
10^-		+	40^0		50
40	10	30	40	40	

In this example x_{53} is the entering variable, so a plus sign is assigned there. The signs of the other cells were determined in the order x_{13}, x_{23}, x_{25}, x_{35}, x_{32}, x_{31}, x_{41}, x_{51}, x_{54}. The smallest variable with a minus assigned to it is $x_{51} = 10$. Thus we set $\theta = 10$.

The Transportation Simplex Algorithm

It is now possible to put together the components developed to this point in the form of a complete revised simplex procedure for the transportation problem. The steps are:

Step 1. Compute an initial basic feasible solution using the Northwest Corner Rule or some other method.
Step 2. Compute the simplex multipliers and the relative cost coefficients. If all relative cost coefficients are nonnegative, stop; the solution is optimal. Otherwise, go to Step 3.
Step 3. Select a nonbasic variable corresponding to a negative cost coefficient to enter the basis (usually the one corresponding to the most negative cost coefficient). Compute the cycle of change and set θ equal to the smallest basic variable with a minus assigned to it. Update the solution. Go to Step 2.

Example 3. We can now completely solve the problem that was introduced in Example 1 of the first section. The requirements and a first basic feasible solution obtained by the Northwest Corner Rule are shown below. The plus and minus signs indicated on the array should be ignored at this point, since they cannot be computed until the next step is completed.

10	20				30
	30	20^-	30^+		80
			10^0		10
		+	40^-	20^0	60
10	50	20	80	20	

The cost coefficients of the problem are shown in the array below, with the circled cells corresponding to the current basic variables. The simplex multipliers, computed by row and column scanning, are shown as well.

③	④	6	8	9	5
2	②	④	⑤	5	3
2	2	2	③	2	1
3	3	2	④	②	2
−2	−1	1	2	0	

The relative cost coefficients are found by subtracting $u_j + v_j$ from c_{ij}. In this case the only negative result is in cell 4,3; so variable x_{43} will be brought into the basis. Thus a + is entered into this cell in the original array, and the cycle of zeros and plus and minus signs is determined as shown in that array. (It is not necessary to continue scanning once a complete cycle is determined.)

The smallest basic variable with a minus sign is 20 and, accordingly, 20 is added or subtracted from elements of the cycle as indicated by the signs. This leads to the new basic feasible solution shown in the array below:

10	20				30
	30		50		80
			10		10
		20	20	20	60
10	50	20	80	20	

The new simplex multipliers corresponding to the new basis are computed, and the cost array is revised as shown below. In this case all relative cost coefficients are positive, indicating that the current solution is optimal.

③	④	6	8	9	5
2	②	4	⑤	5	3
2	2	2	③	2	1
3	3	②	④	②	2
−2	−1	0	2	0	

Degeneracy

As in all linear programming problems, degeneracy, corresponding to a basic variable having the value zero, can occur in the transportation problem. If degeneracy is encountered in the simplex procedure, it can be handled quite easily by introduction of the standard perturbation method (see Exercise 15, Chap. 3). In this method a zero-valued basic variable is assigned the value ε and is then treated in the usual way. If it later leaves the basis, then the ε can be dropped.

Example 4. To illustrate the method of dealing with degeneracy, consider a modification of Example 3, with the fourth row sum changed from 60 to 20 and the fourth column sum changed from 80 to 40. Then the initial basic feasible solution found by the Northwest Corner Rule is degenerate. An ε is placed in the array for the zero-valued basic variable as shown below:

10	20				30
	30	20^-	30^+		80
			10^0		10
		+	ε^-	20^0	20
10	50	20	40	20	

The relative cost coefficients will be the same as in Example 3, and hence again x_{43} should be chosen to enter, and the cycle of change is the same as before. In this case, however, the change is only ε, and variable x_{44} leaves the basis. The new relative cost coefficients are all positive, indicating that the new solution is optimal. Now the ε can be dropped to yield the final solution (which is, itself, degenerate in this case).

10	20				30
	30	20	30		80
			10		10
		ε		20	20
10	50	20	40	20	

*3.8 Decomposition

Large linear programming problems usually have some special structural form that can (and should) be exploited to develop efficient computational procedures. One common structure is where there are a number of separate activity areas that are linked through common resource constraints. An example is provided by a multidivisional firm attempting to minimize the total cost of its operations. The divisions of the firm must each meet internal requirements that do not interact with the constraints of other divisions; but in addition there are common resources that must be shared among divisions and thereby represent linking constraints.

A problem of this form can be solved by the Dantzig-Wolfe decomposition method described in this section. The method is an iterative process where at each step a number of separate subproblems are solved. The subproblems are themselves linear programs within the separate areas (or within divisions in the example of the firm). The objective functions of these subproblems are varied from iteration to iteration and are determined by a separate calculation based on the results of the previous iteration. This action coordinates the individual subproblems so that, ultimately, the solution to the overall problem is solved. The method can be derived as a special version of the revised simplex method, where the subproblems correspond to evaluation of reduced cost coefficients for the main problem.

To describe the method we consider the linear program in standard form

$$\begin{aligned} &\text{minimize} \quad \mathbf{c}^T\mathbf{x} \\ &\text{subject to} \quad \mathbf{A}\mathbf{x} = \mathbf{b}, \ \mathbf{x} \geqslant \mathbf{0}. \end{aligned} \tag{3.44}$$

Suppose, for purposes of this entire section, that the $\mathbf{A}$ matrix has the special "block-angular" structure:

$$\mathbf{A} = \begin{bmatrix} \mathbf{L}_1 & \mathbf{L}_2 & \cdots & \mathbf{L}_N \\ \mathbf{A}_1 & & & \\ & \mathbf{A}_2 & & \\ & & \ddots & \\ & & & \mathbf{A}_N \end{bmatrix} \tag{3.45}$$

By partitioning the vectors $\mathbf{x}$, $\mathbf{c}^T$, and $\mathbf{b}$ consistent with this partition of $\mathbf{A}$, the problem can be rewritten as

$$\begin{aligned} \text{minimize} \quad & \sum_{i=1}^{N} \mathbf{c}_i^T \mathbf{x}_i \\ \text{subject to} \quad & \sum_{i=1}^{N} \mathbf{L}_i \mathbf{x}_i = \mathbf{b}_0 \\ & \mathbf{A}_i \mathbf{x}_i = \mathbf{b}_i \\ & \mathbf{x}_i \geqslant \mathbf{0}, \; i = 1, \ldots, N. \end{aligned} \tag{3.46}$$

This may be viewed as a problem of minimizing the total cost of N different linear programs that are independent except for the first constraint, which is a linking constraint of, say, dimension m.

Each of the subproblems is of the form

$$\begin{aligned} \text{minimize} \quad & \mathbf{c}_i^T \mathbf{x}_i \\ \text{subject to} \quad & \mathbf{A}_i \mathbf{x}_i = \mathbf{b}_i, \; \mathbf{x}_i \geqslant \mathbf{0}. \end{aligned}$$

The constraint set for the ith subproblem is $S_i = \{\mathbf{x}_i : \mathbf{A}_i\mathbf{x}_i = \mathbf{b}_i, \; \mathbf{x}_i \geqslant \mathbf{0}\}$. As for any linear program, this constraint set S_i is a polytope and can be expressed as the intersection of a finite number of closed half-spaces. There is no guarantee that each S_i is bounded, even if the original linear program (3.44) has a bounded constraint set. We shall assume for simplicity, however, that each of the polytopes S_i, $i = 1, \ldots, N$ is indeed bounded and hence is a polyhedron. One may guarantee that this assumption is satisfied by placing artificial (large) upper bounds on each $\mathbf{x}_i$.

Under the boundedness assumption, each polyhedron S_i consists entirely of points that are convex combinations of its extreme points. Thus, if the extreme points of S_i are $\{\mathbf{x}_{i1}, \mathbf{x}_{i2}, \ldots, \mathbf{x}_{iK_i}\}$, then any point $\mathbf{x}_i \in S_i$ can be expressed in the form

$$\begin{aligned} \mathbf{x}_i &= \sum_{j=1}^{K_i} \alpha_{ij}\mathbf{x}_{ij}, \\ \text{where } & \sum_{j=1}^{K_i} \alpha_{ij} = 1 \\ \text{and } & \alpha_{ij} \geqslant 0, \quad j = 1, \ldots, K_i. \end{aligned} \tag{3.47}$$

The α_{ij}'s are the weighting coefficients of the extreme points.

We now convert the original linear program to an equivalent *master problem*, of which the objective is to find the optimal weighting coefficients for each polyhedron, S_i. Corresponding to each extreme point $\mathbf{x}_{ij}$ in S_i, define $p_{ij} = \mathbf{c}_i^T \mathbf{x}_{ij}$ and $\mathbf{q}_{ij} = \mathbf{L}_i\mathbf{x}_{ij}$. Clearly p_{ij} is the equivalent cost of the extreme point $\mathbf{x}_{ij}$, and $\mathbf{q}_{ij}$ is its equivalent activity vector in the linking constraints.

Then the original linear program (3.44) is equivalent, using (3.47), to the *master problem*:

$$\begin{aligned} &\text{minimize} \quad \sum_{i=1}^{N}\sum_{j=1}^{K_i} p_{ij}\alpha_{ij} \\ &\text{subject to} \quad \sum_{i=1}^{N}\sum_{j=1}^{K_i} \mathbf{q}_{ij}\alpha_{ij} = \mathbf{b}_0 \\ &\left.\begin{aligned} &\sum_{j=1}^{K_i} \alpha_{ij} = 1 \\ &\alpha_{ij} \geqslant 0, \quad j = 1, \ldots, K_i \end{aligned}\right\} i = 1, \ldots, N. \end{aligned} \tag{3.48}$$

This master problem has variables

$$\boldsymbol{\alpha}^T = (\alpha_{11}, \ldots, \alpha_{1K_1}, \alpha_{21}, \ldots, \alpha_{2K_2}, \ldots, \alpha_{N1}, \ldots, \alpha_{NK_N})$$

and can be expressed more compactly as

$$\begin{aligned} &\text{minimize} \quad \mathbf{p}^T\boldsymbol{\alpha} \\ &\text{subject to} \quad \mathbf{Q}\boldsymbol{\alpha} = \mathbf{g}, \boldsymbol{\alpha} \geqslant \mathbf{0} \end{aligned} \tag{3.49}$$

where $\mathbf{g}^T = [\mathbf{b}_0^T, 1, 1, \ldots, 1]$; the element of $\mathbf{p}$ associated with α_{ij} is p_{ij}; and the column of $\mathbf{Q}$ associated with α_{ij} is

$$\begin{bmatrix} \mathbf{q}_{ij} \\ \mathbf{e}_i \end{bmatrix},$$

with $\mathbf{e}_i$ denoting the ith unit vector in E^N.

Suppose that at some stage of the revised simplex method for the master problem we know the basis $\mathbf{B}$ and corresponding simplex multipliers $\mathbf{y}^T = \mathbf{p}_{\mathbf{B}}^T\mathbf{B}^{-1}$. The corresponding relative cost vector is $\mathbf{r}_{\mathbf{D}}^T = \mathbf{c}_{\mathbf{D}}^T - \mathbf{y}^T\mathbf{D}$, having components

$$r_{ij} = p_{ij} - \mathbf{y}^T \begin{bmatrix} \mathbf{q}_{ij} \\ \mathbf{e}_i \end{bmatrix}. \tag{3.50}$$

It is not necessary to calculate all the r_{ij}'s; it is only necessary to determine the minimal r_{ij}. If the minimal value is nonnegative, the current solution is optimal and the process terminates. If, on the other hand, the minimal element is negative, the corresponding column should enter the basis.

The search for the minimal element in (3.50) is normally made with respect to nonbasic columns only. The search can be formally extended to include basic columns as well, however, since for basic elements

$$p_{ij} - \mathbf{y}^T \begin{bmatrix} \mathbf{q}_{ij} \\ \mathbf{e}_i \end{bmatrix} = 0.$$

The extra zero values do not influence the subsequent procedure, since a new column will enter only if the minimal value is less than zero.

We therefore define r^* as the minimum relative cost coefficient for *all* possible basis vectors. That is,

$$r^* = \underset{i\in\{1,\ldots,N\}}{\text{minimum}} \left\{ r_i^* = \underset{j\in\{1,\ldots,K_i\}}{\text{minimum}} \{ p_{ij} - \mathbf{y}^T \begin{bmatrix} \mathbf{q}_{ij} \\ \mathbf{e}_i \end{bmatrix} \} \right\}.$$

Using the definitions of p_{ij} and $\mathbf{q}_{ij}$, this becomes

$$r_i^* = \underset{j\in\{1,\ldots,K_i\}}{\text{minimum}} \left\{ \mathbf{c}_i^T \mathbf{x}_{ij} - \mathbf{y}_0^T \mathbf{L}_j \mathbf{x}_{ij} - y_{m+i} \right\}, \tag{3.51}$$

where $\mathbf{y}_0$ is the vector made up of the first m elements of $\mathbf{y}$, m being the number of rows of $\mathbf{L}_j$ [the number of linking constraints in (3.47)].

The minimization problem in (3.51) is actually solved by the ith *subproblem*:

$$\begin{aligned} &\text{minimize} \quad (\mathbf{c}_i^T - \mathbf{y}_0^T \mathbf{L}_j)\mathbf{x}_j \\ &\text{subject to} \quad \mathbf{A}_j \mathbf{x}_j = \mathbf{b}_j, \ \mathbf{x}_j \geqslant \mathbf{0} \end{aligned} \tag{3.52}$$

This follows from the fact that y_{m+i} is independent of the extreme point index j (since $\mathbf{y}$ is fixed during the determination of the r_j's), and that the solution of (3.52) must be that extreme point of S_i, say $\mathbf{x}_{ik}$, of minimum cost, using the adjusted cost coefficients $\mathbf{c}_i^T - \mathbf{y}_0^T \mathbf{L}_j$.

Thus, an algorithm for this special version of the revised simplex method applied to the master problem is the following: Given a basis $\mathbf{B}$

Step 1. Calculate the current basic solution $\mathbf{x}_\mathbf{B}$, and solve $\mathbf{y}^T\mathbf{B} = \mathbf{c}_\mathbf{B}^T$ for $\mathbf{y}$.
Step 2. For each $i = 1, 2, \ldots, N$, determine the optimal solution $\mathbf{x}_i^*$ of the ith subproblem (3.52) and calculate

$$r_i^* = (\mathbf{c}_i^T - \mathbf{y}_0^T \mathbf{L}_j)\mathbf{x}_i^* - \mathbf{y}_{m+i}. \tag{3.53}$$

If all $r_i^* > 0$, stop; the current solution is optimal.
Step 3. Determine which column is to enter the basis by selecting the minimal r_i^*.
Step 4. Update the basis of the master problem as usual.

This algorithm has an interesting economic interpretation in the context of a multidivisional firm minimizing its total cost of operations as described earlier. Division i's activities are internally constrained by $\mathbf{A}\mathbf{x}_i = \mathbf{b}_j$, and the common resources $\mathbf{b}_0$ impose linking constraints. At Step 1 of the algorithm, the firm's central management formulates its current master plan, which is perhaps suboptimal, and announces a new set of prices that each division must use to revise its recommended strategy at Step 2. In particular, $-\mathbf{y}_0$ reflects the new prices that higher management has placed on the common resources. The division that reports the greatest rate of potential cost improvement has its recommendations incorporated in the new master plan at Step 3, and the process is repeated. If no cost improvement is possible, central management settles on the current master plan.

3.9 Summary

The simplex method is founded on the fact that the optimal value of a linear program, if finite, is always attained at a basic feasible solution. Using this foundation there are two ways in which to visualize the simplex process. The first is to view the process as one of continuous change. One starts with a basic feasible solution and imagines that some nonbasic variable is increased slowly from zero. As the value of this variable is increased, the values of the current basic variables are continuously adjusted so that the overall vector continues to satisfy the system of linear equality constraints. The change in the objective function due to a unit change in this nonbasic variable, taking into account the corresponding required changes in the values of the basic variables, is the relative cost coefficient associated with the nonbasic variable. If this coefficient is negative, then the objective value will be continuously improved as the value of this nonbasic variable is increased, and therefore one increases the variable as far as possible, to the point where further increase would violate feasibility. At this point the value of one of the basic variables is zero, and that variable is declared nonbasic, while the nonbasic variable that was increased is declared basic.

The other viewpoint is more discrete in nature. Realizing that only basic feasible solutions need be considered, various bases are selected and the corresponding basic solutions are calculated by solving the associated set of linear equations. The logic for the systematic selection of new bases again involves the relative cost coefficients and, of course, is derived largely from the first, continuous, viewpoint.

Problems of special structure are important both for applications and for theory. The transportation problem represents an important class of linear programs with structural properties that lead to an efficient implementation of the simplex method. The most important property of the transportation problem is that any basis is triangular. This means that the basic variables can be found, one by one, directly by back substitution, and the basis need never be inverted. Likewise, the simplex multipliers can be found by back substitution, since they solve a set of equations involving the transpose of the basis. Moreover, when any basis matrix is triangular and all nonzero elements are equal to one (or minus one if the signs of some equations are changed), it follows that the process of back substitution will simply involve repeated additions and subtractions of the given row and column sums. No multiplication or division is required. It therefore follows that if the original right-hand side are integers, the values of all basic variables will be integers. Hence, an optimal basic solution, where each entry is integral, always exists; that is, there is no gap between continuous linear program and integer linear program (or the integrality gap is zero). The transportation problem can be generalized to a minimum cost flow problem in a network. This leads to the interpretation of a simplex basis as corresponding to a spanning tree in the network; see Appendix D.

Many linear programming methods have implemented a *Presolver* procedure to eliminate redundant or duplicate constraints and/or value fixed variables, and to check possible constraint inconsistency and unboundedness. This typically results in problem size reduction and possible infeasibility detection.

3.10 Exercises

1. Using pivoting, solve the simultaneous equations

$$3x_1 + 2x_2 = 5$$
$$5x_1 + x_2 = 9.$$

2. Using pivoting, solve the simultaneous equations

$$x_1 + 2x_2 + x_3 = 7$$
$$2x_1 - x_2 + 2x_3 = 6$$
$$x_1 + x_2 + 3x_3 = 12.$$

3. Solve the equations in Exercise 2 by Gaussian elimination as described in Appendix C.
4. Suppose **B** is an $m \times m$ square nonsingular matrix, and let the tableau **T** be constructed, $\mathbf{T} = [\mathbf{I}, \mathbf{B}]$ where I is the $m \times m$ identity matrix. Suppose that pivot operations are performed on this tableau so that it takes the form $[\mathbf{C}, \mathbf{I}]$. Show that $\mathbf{C} = \mathbf{B}^{-1}$.
5. Show that if the vectors $\mathbf{a}_1, \mathbf{a}_2, \ldots, \mathbf{a}_m$ are a basis in E^m, the vectors $\mathbf{a}_1$, $\mathbf{a}_2, \ldots, \mathbf{a}_{p-1}$,
$\mathbf{a}_q, \mathbf{a}_{p+1}, \ldots, \mathbf{a}_m$ also are a basis if and only if $\bar{a}_{pq} \neq 0$, where $\bar{a}_{pq}$ is defined by the tableau (3.5).
6. If $r_j > 0$ for every j corresponding to a variable x_j that is not basic, show that the corresponding basic feasible solution is the unique optimal solution.
7. Show that a degenerate basic feasible solution may be optimal without satisfying $r_j \geqslant 0$ for all j.
8.

(a) Using the simplex procedure, solve

$$\begin{array}{ll} \text{maximize} & -x_1 + x_2 \\ \text{subject to} & x_1 - x_2 \leqslant 2 \\ & x_1 + x_2 \leqslant 6 \\ & x_1 \geqslant 0, \quad x_2 \geqslant 0. \end{array}$$

(b) Draw a graphical representation of the problem in x_1, x_2 space and indicate the path of the simplex steps.

(c) Repeat for the problem

$$\begin{array}{ll} \text{maximize} & x_1 + x_2 \\ \text{subject to} & -2x_1 + x_2 \leqslant 1 \\ & x_1 - x_2 \leqslant 1 \\ & x_1 \geqslant 0, \quad x_2 \geqslant 0. \end{array}$$

9. Using the simplex procedure, solve the spare-parts manufacturer's problem (Exercise 4, Chap. 2).
10. Using the simplex procedure, solve

$$\begin{array}{lrcrcrcrl} \text{minimize} & 2x_1 & + & 4x_2 & + & x_3 & + & x_4 & \\ \text{subject to} & x_1 & + & 3x_2 & & & + & x_4 & \leqslant 4 \\ & 2x_1 & + & x_2 & & & & & \leqslant 3 \\ & & & x_2 & + & 4x_3 & + & x_4 & \leqslant 3 \\ & x_i \geqslant 0 & & & & i = 1, 2, 3, 4. & & & \end{array}$$

11. For the linear program of Exercise 10
(a) How much can the elements of $\mathbf{b} = (4, 3, 3)$ be changed without changing the optimal basis?
(b) How much can the elements of $\mathbf{c} = (2, 4, 1, 1)$ be changed without changing the optimal basis?
(c) What happens to the optimal cost for small changes in $\mathbf{b}$?
(d) What happens to the optimal cost for small changes in $\mathbf{c}$?
12. Consider the problem

$$\begin{array}{lrcrcrl} \text{minimize} & x_1 & - & 3x_2 & - & 0.4x_3 & \\ \text{subject to} & 3x_1 & - & x_2 & + & 2x_3 & \leqslant 7 \\ & -2x_1 & + & 4x_2 & & & \leqslant 12 \\ & -4x_1 & + & 3x_2 & + & 3x_3 & \leqslant 14 \\ & x_1 \geqslant 0, & & x_2 \geqslant 0, & & x_3 \geqslant 0. & \end{array}$$

(a) Find an optimal solution.
(b) How many optimal basic feasible solutions are there?
(c) Show that if $c_4 + \frac{1}{3}a_{14} + \frac{4}{5}a_{24} \geqslant 0$, then another activity x_4 can be introduced with cost coefficient c_1 and activity vector (a_{14}, a_{24}, a_{34}) without changing the optimal solution.
13. Rather than select the variable corresponding to the most negative relative cost coefficient as the variable to enter the basis, it has been suggested that a better criterion would be to select that variable which, when pivoted in, will produce the greatest improvement in the objective function. Show that this criterion leads to selecting the variable x_k corresponding to the index k minimizing $\max\limits_{i, \bar{a}_{ik} > 0} r_k \bar{a}_{i0} / \bar{a}_{ik}$.
14. In the ordinary simplex method one new vector is brought into the basis and one removed at every step. Consider the possibility of bringing two new vectors into the basis and removing two at each stage. Develop a complete procedure that operates in this fashion.
15. *Degeneracy*. If a basic feasible solution is degenerate, it is then theoretically possible that a sequence of degenerate basic feasible solutions will be generated that endlessly cycles without making progress. It is the purpose of this exercise and the next two to develop a technique that can be applied to the simplex method to avoid this *cycling*.

Corresponding to the linear system $\mathbf{Ax} = \mathbf{b}$ where $\mathbf{A} = [\mathbf{a}_1, \mathbf{a}_2, \ldots, \mathbf{a}_n]$ define the perturbed system $\mathbf{Ax} = \mathbf{b}(\varepsilon)$ where $\mathbf{b}(\varepsilon) = \mathbf{b} + \varepsilon\mathbf{a}_1 + \varepsilon^2\mathbf{a}_2 + \cdots + \varepsilon^n\mathbf{a}_n$, $\varepsilon > 0$. Show that if there is a basic feasible solution (possibly degenerate) to the unperturbed system with basis $\mathbf{B} = [\mathbf{a}_1, \mathbf{a}_2, \ldots, \mathbf{a}_m]$, then corresponding to the same basis, there is a nondegenerate basic feasible solution to the perturbed system for some range of $\varepsilon > 0$.

16. Show that corresponding to any basic feasible solution to the perturbed system of Exercise 15, which is nondegenerate for some range of $\varepsilon > 0$, and to a vector $\mathbf{a}_k$ not in the basis, there is a unique vector $\mathbf{a}_j$ in the basis which when replaced by $\mathbf{a}_k$ leads to a basic feasible solution; and that solution is nondegenerate for a range of $\varepsilon > 0$.
17. Show that the tableau associated with a basic feasible solution of the perturbed system of Exercise 15, and which is nondegenerate for a range of $\varepsilon > 0$, is identical with that of the unperturbed system except in the column under $\mathbf{b}(\varepsilon)$. Show how the proper pivot in a given column to preserve feasibility of the perturbed system can be determined from the tableau of the unperturbed system. Conclude that the simplex method will avoid cycling if whenever there is a choice in the pivot element of a column k, arising from a tie in the minimum of $\bar{a}_{i0}/\bar{a}_{ik}$ among the elements $i \in I_0$, the tie is resolved by finding the minimum of $\bar{a}_{i1}/\bar{a}_{ik}$, $i \in I_0$. If there still remainties among elements $i \in I$, the process is repeated with $\bar{a}_{i2}/\bar{a}_{ik}$, etc., until there is a unique element.
18. Using the two-phase simplex procedure solve

(a)

$$\begin{aligned}
\text{minimize} \quad & -3x_1 + x_2 + 3x_3 - x_4 \\
\text{subject to} \quad & x_1 + 2x_2 - x_3 + x_4 = 0 \\
& 2x_1 - 2x_2 + 3x_3 + 3x_4 = 9 \\
& x_1 - x_2 + 2x_3 - x_4 = 6 \\
& x_1 \geqslant 0,\ i = 1, 2, 3, 4.
\end{aligned}$$

(b)

$$\begin{aligned}
\text{minimize} \quad & x_1 + 6x_2 - 7x_3 + x_4 + 5x_5 \\
\text{subject to} \quad & 5x_1 - 4x_2 + 13x_3 - 2x_4 + x_5 = 20 \\
& x_1 - x_2 + 5x_3 - x_4 + x_5 = 8 \\
& x_1 \geqslant 0,\ i = 1, 2, 3.4, 5.
\end{aligned}$$

19. Solve the oil refinery problem (Exercise 3, Chap. 2).
20. Show that in the phase I procedure of a problem that has feasible solutions, if an artificial variable becomes nonbasic, it need never again be made basic. Thus, when an artificial variable becomes nonbasic its column can be eliminated from future tableaus.
21. Suppose the phase I procedure is applied to the system $\mathbf{Ax} = \mathbf{b}$, $\mathbf{x} \geqslant \mathbf{0}$, and that the resulting tableau (ignoring the cost row) has the form

$x_1\ x_2 \cdots x_k$	$x_{k+1} \cdots x_n$	$y_1\ y_2 \cdots y_k$	$y_{k+1} \cdots y_m$	
1			$0 \ \cdots \ 0$	$\bar{b}_1$
1	$\mathbf{R}_1$	$\mathbf{S}_1$	$0 \ \cdots \ 0$	
			$\vdots$	$\vdots$
1			$0 \ \cdots \ 0$	$\bar{b}_k$
$0 \ 0 \cdots \ 0$			1	0
$\vdots$	$\mathbf{R}_2$	$\mathbf{S}_2$	1	$\vdots$
$0 \ \cdots \ 0$			1	0

This corresponds to having $m - k$ basic artificial variables at zero level.

(a) Show that any nonzero element in $\mathbf{R}_2$ can be used as a pivot to eliminate a basic artificial variable, thus yielding a similar tableau but with k increased by one.

(b) Suppose that the process in (a) has been repeated to the point where $\mathbf{R}_2 = \mathbf{0}$. Show that the original system is redundant, and show how phase II may proceed by eliminating the bottom rows.

(c) Use the above method to solve the linear program

$$
\begin{array}{ll}
\text{minimize} & 2x_1 + 6x_2 + x_3 + x_4 \\
\text{subject to} & x_1 + 2x_2 \qquad\quad + x_4 = 6 \\
 & x_1 + 2x_2 + x_3 + x_4 = 7 \\
 & x_1 + 3x_2 - x_3 + 2x_4 = 7 \\
 & x_1 + x_2 + x_3 \qquad\qquad = 5 \\
 & x_1 \geqslant 0, \quad x_2 \geqslant 0, \quad x_3 \geqslant 0, \quad x_4 \geqslant 0.
\end{array}
$$

22. Find a basic feasible solution to

$$
\begin{aligned}
x_1 + 2x_2 - x_3 + x_4 &= 3 \\
2x_1 + 4x_2 + x_3 + 2x_4 &= 12 \\
x_1 + 4x_2 + 2x_3 + x_4 &= 9 \\
x_1 \geqslant 0, \quad i &= 1, 2, 3, 4.
\end{aligned}
$$

23. Consider the system of linear inequalities $\mathbf{Ax} \geqslant \mathbf{b}$, $\mathbf{x} \geqslant \mathbf{0}$ with $\mathbf{b} \geqslant \mathbf{0}$. This system can be transformed to standard form by the introduction of m surplus variables so that it becomes $\mathbf{Ax}-\mathbf{y} = \mathbf{b}$, $\mathbf{x} \geqslant \mathbf{0}$, $\mathbf{y} \geqslant \mathbf{0}$. Let $b_k = \max_i b_i$ and consider the new system in standard form obtained by adding the kth row to the negative of every other row. Show that the new system requires the addition of only a single artificial variable to obtain an initial basic feasible solution.
Use this technique to find a basic feasible solution to the system.

$$
\begin{aligned}
x_1 + 2x_2 + x_3 &\geqslant 4 \\
2x_1 + x_2 + x_3 &\geqslant 5 \\
2x_1 + 3x_2 + 2x_3 &\geqslant 6 \\
x_j \geqslant 0, \ i = 1, 2, 3.&
\end{aligned}
$$

24. It is possible to combine the two phases of the two-phase method into a single procedure by the *big-M method*. Given the linear program in standard form

$$\begin{aligned} \text{minimize} \quad & \mathbf{c}^T\mathbf{x} \\ \text{subject to} \quad & \mathbf{A}\mathbf{x} = \mathbf{b}, \ \mathbf{x} \geqslant \mathbf{0}, \end{aligned}$$

one forms the approximating problem

$$\begin{aligned} \text{minimize} \quad & \mathbf{c}^T\mathbf{x} + M\sum_{i=1}^{m} u_i \\ \text{subject to} \quad & \mathbf{A}\mathbf{x} + \mathbf{u} = \mathbf{b} \\ & \mathbf{x} \geqslant \mathbf{0}, \ \mathbf{u} \geqslant \mathbf{0}. \end{aligned}$$

In this problem $\mathbf{u} = (u_1, u_2, \dots, u_m)$ is a vector of artificial variables and M is a large constant. The term $M\sum_{i=1}^{m} u_j$ serves as a penalty term for nonzero u_i's.

If this problem is solved by the simplex method, show the following:

(a) If an optimal solution is found with $\mathbf{y} = \mathbf{0}$, then the corresponding $\mathbf{x}$ is an optimal basic feasible solution to the original problem.

(b) If for every $M > 0$ an optimal solution is found with $\mathbf{y} \neq \mathbf{0}$, then the original problem is infeasible.

(c) If for every $M > 0$ the approximating problem is unbounded, then the original problem is either unbounded or infeasible.

(d) Suppose now that the original problem has a finite optimal value $V(\infty)$. Let $V(M)$ be the optimal value of the approximating problem. Show that $V(M) \leqslant V(\infty)$.

(e) Show that for $M_1 \leqslant M_2$ we have $V(M_1) \leqslant V(M_2)$.

(f) Show that there is a value M_0 such that for $M \geqslant M_0$, $V(M) = V(\infty)$, and hence conclude that the big-M method will produce the right solution for large enough values of M.

25. Using the revised simplex method find a basic feasible solution to

$$\begin{aligned} x_1 \ +2x_2 - \ x_3 + x_4 &= 3 \\ 2x_1 \ +4x_2 + x_3 + 2x_4 &= 12 \\ x_1 \ +4x_2 + 2x_3 + x_4 &= 9 \\ x_1 \ \geqslant 0, \ i &= 1, 2, 3, 4. \end{aligned}$$

26. The following tableau is an intermediate stage in the solution of a minimization problem:

	$\mathbf{y}_1$	$\mathbf{y}_2$	$\mathbf{y}_3$	$\mathbf{y}_4$	$\mathbf{y}_5$	$\mathbf{y}_6$	$\mathbf{y}_0$
	1	2/3	0	0	4/3	0	4
	0	−7/3	3	1	−2/3	0	2
	0	−2/3	−2	0	2/3	1	2
$\mathbf{r}^T$	0	8/3	−11	0	4/3	0	−8

(a) Determine the next pivot element.
(b) Given that the inverse of the current basis is

$$\mathbf{B}^{-1} = [\mathbf{a}_1,\ \mathbf{a}_4,\ \mathbf{a}_6]^{-1} = \frac{1}{3}\begin{bmatrix} 1 & 1 & -1 \\ 1 & -2 & 2 \\ -1 & 2 & 1 \end{bmatrix}$$

and the corresponding cost coefficients are

$$\mathbf{c}_{\mathbf{B}}^T = (c_1,\ c_4,\ c_6) = (-1,\ -3, 1),$$

find the original problem.

27. In many applications of linear programming it may be sufficient, for practical purposes, to obtain a solution for which the value of the objective function is within a predetermined tolerance ε from the minimum value z^*. Stopping the simplex algorithm at such a solution rather than searching for the true minimum may considerably reduce the computations.
(a) Consider a linear programming problem for which the sum of the variables is known to be bounded above by s. Let z_0 denote the current value of the objective function at some stage of the simplex algorithm, $(c_j - z_j)$ the corresponding relative cost coefficients, and

$$M = \max(z_j - c_j)j.$$

Show that if $M \leqslant \varepsilon/s$, then $z_0 - z^* \leq \varepsilon$.
(b) Consider the transportation problem described in Sect. 2.2 (Example 3). Assuming this problem is solved by the simplex method and it is sufficient to obtain a solution within ε tolerance from the optimal value of the objective function, specify a stopping criterion for the algorithm in terms of ε and the parameters of the problem.

28. A matrix $\mathbf{A}$ is said to be *totally unimodular* if the determinant of every square submatrix formed from it has value 0, +1, or −1

    (a) Show that the matrix $\mathbf{A}$ defining the equality constraints of a transportation problem is totally unimodular.
    (b) In the system of equations $\mathbf{Ax} = \mathbf{b}$, assume that $\mathbf{A}$ is totally unimodular and that all elements of $\mathbf{A}$ and $\mathbf{b}$ are integers. Show that all basic solutions have integer components.

29. For the arrays below:

    (a) Compute the basic solutions indicated. (*Note*: They may be infeasible.)
    (b) Write the equations for the basic variables, corresponding to the indicated basic solutions, in lower triangular form.

	x	x	10
	x		20
x		x	30
20	20	20	

x		x	10
	x		20
	x	x	30
20	20	20	

30. For the arrays of cost coefficients below, the circled positions indicate basic variables.

    (a) Compute the simplex multipliers.
    (b) Write the equations for the simplex multipliers in upper triangular form, and compare with Part(b) of Exercise 4.

3	⑥	⑦
2	④	3
①	5	②

③	6	⑦
2	④	3
1	⑤	②

31. Consider the modified transportation problem where there is more available at origins than is required at destinations (i.e., $\sum_{i=1}^{m} a_i > \sum_{j=1}^{n} b_j$).

$$
\begin{aligned}
\text{minimize} \quad & \sum_{j=1}^{m}\sum_{i=1}^{n} c_{ij}x_{ij} \\
\text{subject to} \quad & \sum_{j=1}^{n} x_{ij} \leqslant a_i, \quad i = 1, 2, \ldots, m \\
& \sum_{i=1}^{n} x_{ij} = b_j, \quad j = 1, 2, \ldots, n \\
& x_{ij} \geqslant 0, \text{ for all } i,\ j.
\end{aligned}
$$

    (a) Show how to convert it to an ordinary transportation problem.
    (b) Suppose there is a storage cost of s_i per unit at origin i for goods not transported to a destination. Repeat Part(a) with this assumption.

32. Solve the following transportation problem, which is an original example of Hitchcock.

$$
\mathbf{a} = \begin{pmatrix} 25 & 25 & 50 \end{pmatrix} \qquad \mathbf{C} = \begin{bmatrix} 10 & 5 & 6 & 7 \\ 8 & 2 & 7 & 6 \\ 9 & 3 & 4 & 8 \end{bmatrix}
$$
$$
\mathbf{b} = \begin{pmatrix} 15 & 20 & 30 & 35 \end{pmatrix}
$$

33. In a transportation problem, suppose that two rows or two columns of the cost coefficient array differ by a constant. Show that the problem can be reduced by combining those rows or columns.
34. The transportation problem is often solved more quickly by carefully selecting the starting basic feasible solution. The *matrix minimum* technique for finding a starting solution is: (3.34) Find the lowest cost unallocated cell in the array, and allocate the maximum possible to it, (3.35) Reduce the corresponding row and column requirements, and drop the row or column having zero remaining requirement. Go back to Step 1 unless all remaining requirements are zero.

    (a) Show that this procedure yields a basic feasible solution.
    (b) Apply the method to Exercise 7.

35. *The caterer problem.* A caterer is booked to cater a banquet each evening for the next T days. He requires r_t clean napkins on the tth day for $t = 1, 2, \ldots, T$. He may send dirty napkins to the laundry, which has two speeds of service—fast and slow. The napkins sent to the fast service will be ready for the next day's banquet; those sent to the slow service will be ready for the banquet 2 days later. Fast and slow service cost c_1 and c_2 per napkin, respectively, with $c_1 > c_2$. The caterer may also purchase new napkins at any time at cost c_0. He has an initial stock of s napkins and wishes to minimize the total cost of supplying fresh napkins.

    (a) Formulate the problem as a transportation problem. (*Hint*: Use $T+1$ sources and T destinations.)
    (b) Using the values $T = 4$, $s = 200$, $r_1 = 100$, $r_2 = 130$, $r_3 = 150$, $r_4 = 140$, $c_1 = 6$, $c_2 = 4$, $c_0 = 12$, solve the problem.

36. *The marriage assignment problem.* A group of n men and n women live on an island. The amount of happiness that the ith man and the jth woman derive by spending a fraction x_{ij} of their lives together is $c_{ij}x_{ij}$. What is the nature of the living arrangements that maximizes the total happiness of the islanders?
37. *Anticycling Rule.* A remarkably simple procedure for avoiding cycling was developed by Bland, and we discuss it here.
    Bland's Rule. *In the simplex method*:
    (a) *Select the column to enter the basis by* $j = \min\{j : r_j < 0\}$; *that is, select the lowest indexed favorable column.*
    (b) *In case ties occur in the criterion for determining which column is to leave the basis, select the one with lowest index.*

    We can prove by contradiction that the use of Bland's rule prohibits cycling. Suppose that cycling occurs. During the cycle a finite number of columns enter and leave the basis. Each of these columns enters at level zero, and the cost function does not change.

    Delete all rows and columns that do not contain pivots during a cycle, obtaining a new linear program that also cycles. Assume that this reduced linear program has m rows and n columns. Consider the solution stage where column n is about to leave the basis, being replaced by column p. The corresponding tableau is as follows (where the entries shown are explained below):

$$
\begin{array}{c|ccccc}
 & \mathbf{a}_1 \cdots & \mathbf{a}_p & \cdots & \mathbf{a}_n & \mathbf{b} \\
 & & \leqslant 0 & & 0 & 0 \\
 & & \leqslant 0 & & 0 & 0 \\
 & & \vdots & & \vdots & \vdots \\
 & & > 0 & & 1 & 0 \\
\hline
\mathbf{c}^T & & < 0 & & 0 & 0
\end{array}
$$

    Without loss of generality, we assume that the current basis consists of the last m columns. In fact, we may define the reduced linear program in terms of this tableau, calling the current coefficient array $\mathbf{A}$ and the current relative cost vector $\mathbf{c}$. In this tableau we pivot on a_{mp}, so $a_{mp} > 0$. By Part(b) of Bland's rule,

$\mathbf{a}_n$ can leave the basis only if there are no ties in the ratio test, and since $\mathbf{b} = \mathbf{0}$ because all rows are in the cycle, it follows that $a_{ip} \leqslant 0$ for all $i \neq m$.
Now consider the situation when column n is about to reenter the basis. Part(a) of Bland's rule ensures that $r_n < 0$ and $r_j \geqslant 0$ for all $i \neq n$. Apply the formula $r_i = c_i - \mathbf{y}^T\mathbf{a}_i$ to the last m columns to show that each component of $\mathbf{y}$ except y_m is nonpositive; and $y_m > 0$. Then use this to show that $r_p = c_p - \mathbf{y}^T\mathbf{a}_p < c_p < 0$, contradicting $r_p \geqslant 0$.

38. Use the Dantzig-Wolfe decomposition method to solve

$$\begin{array}{ll} \text{minimize} & -4x_1 - x_2 - 3x_3 - 2x_4 \\ \text{subject to} & 2x_1 + 2x_2 + x_3 + 2x_4 \leqslant 6 \\ & \qquad x_2 + 2x_3 + 3x_4 \leqslant 4 \\ & \qquad 2x_1 + x_2 \qquad\quad \leqslant 5 \\ & x_2 \qquad\qquad\qquad\quad \leqslant 1 \\ & \qquad\quad -x_3 + 2x_4 \leqslant 2 \\ & \qquad\qquad x_3 + 2x_4 \leqslant 6 \\ & x_1 \geqslant 0,\ x_2 \geqslant 0,\ x_3 \geqslant 0,\ x_4 \geqslant 0. \end{array}$$

References

3.1–3.5 All of this is now standard material contained in most courses in linear programming. See the references cited at the end of Chap. 2. For the original work in this area, see Dantzig [D2] for development of the simplex method; Orden [O2] for the artificial basis technique; Dantzig, Orden and Wolfe [D8], Orchard-Hays [O1], and Dantzig [D4] for the revised simplex method; and Charnes and Lemke [C3] and Dantzig [D5] for upper bounds. The synthetic carrot interpretation is due to Gale [G2].

3.6 The idea of using **LU** decomposition for the simplex method is due to Bartels and Golub [B2]. See also Bartels [B1]. For a nice simple introduction to Gaussian elimination, see Forsythe and Moler [F15]. For an expository treatment of modern computer implementation issues of linear programming, see Murtagh [M9].

3.7 The transportation problem in its present form was first formulated by Hitchcock [H11]. Koopmans [K8] also contributed significantly to the early development of the problem. The simplex method for the transportation problem was developed by Dantzig [D3]. Most textbooks on linear programming include a discussion of the transportation problem. See especially Simonnard [S6], Murty [M11], and Bazaraa and Jarvis [B5]. The method of changing basis is often called the *stepping stone method*. The assignment problem has a long and interesting history. The important fact that the integer problem is solved by a standard linear programming problem follows from a theorem of Birkhoff [B16], which states that the extreme points of the set of feasible assignments are permutation matrices.

3.8 For a more comprehensive description of the Dantzig and Wolfe [D11] decomposition method, see Dantzig [D6].
The degeneracy technique discussed in Exercises 15–17 is due to Charnes [C2]. The anticycling method of Exercise 35 is due to Bland [B19]. For the state of the art in Simplex solvers see Bixby [B18].

Chapter 4
Duality and Complementarity

Associated with every linear program, and intimately related to it, is a corresponding dual linear program. Both programs are constructed from the same underlying cost and constraint coefficients but in such a way that if one of these problems is one of minimization the other is one of maximization, and the optimal values of the corresponding objective functions, if finite, are equal. The variables of the dual problem can be interpreted as prices associated with the constraints of the original (primal) problem, and through this association it is possible to give an economically meaningful characterization to the dual whenever there is such a characterization for the primal.

The variables of the dual problem are also intimately related to the calculation of the relative cost coefficients in the simplex method. Thus, a study of duality sharpens our understanding of the simplex procedure and motivates certain alternative solution methods. Indeed, the simultaneous consideration of a problem from both the primal and dual viewpoints often provides significant computational advantage as well as economic insight.

4.1 Dual Linear Programs

In this section we define the dual program that is associated with a given linear program. Initially, we depart from our usual strategy of considering programs in standard form, since the duality relationship is most symmetric for programs expressed solely in terms of inequalities. Specifically then, we define duality through the pair of programs displayed below.

D.G. Luenberger, Y. Ye, *Linear and Nonlinear Programming*, International Series in Operations Research & Management Science 228,
DOI 10.1007/978-3-319-18842-3_4

$$\begin{array}{llll} & \text{Primal} & & \text{Dual} \\ \text{minimize} & \mathbf{c}^T\mathbf{x} & \text{maximize} & \mathbf{y}^T\mathbf{b} \\ \text{subject to} & \mathbf{A}\mathbf{x} \geqslant \mathbf{b} & \text{subject to} & \mathbf{y}^T\mathbf{A} \leqslant \mathbf{c}^T \\ & \mathbf{x} \geqslant \mathbf{0} & & \mathbf{y} \geqslant \mathbf{0} \end{array} \tag{4.1}$$

If $\mathbf{A}$ is an $m \times n$ matrix, then $\mathbf{x}$ is an m-dimensional column vector, $\mathbf{b}$ is an n-dimensional column vector, $\mathbf{c}^T$ is an n-dimensional row vector, and $\mathbf{y}^T$ is an m-dimensional row vector. The vector $\mathbf{x}$ is the variable of the primal program, and $\mathbf{y}$ is the variable of the dual program.

The pair of programs (4.1) is called the *symmetric form* of duality and, as explained below, can be used to define the dual of any linear program. It is important to note that the role of primal and dual can be reversed. Thus, studying in detail the process by which the dual is obtained from the primal: interchange of cost and constraint vectors, transposition of coefficient matrix, reversal of constraint inequalities, and change of minimization to maximization; we see that this same process applied to the dual yields the primal. Put another way, if the dual is transformed, by multiplying the objective and the constraints by minus unity, so that it has the structure of the primal (but is still expressed in terms of $\mathbf{y}$), its corresponding dual will be equivalent to the original primal.

The dual of any linear program can be found by converting the program to the form of the primal shown above. For example, given a linear program in standard form

$$\begin{array}{ll} \text{minimize} & \mathbf{c}^T\mathbf{x} \\ \text{subject to} & \mathbf{A}\mathbf{x} = \mathbf{b},\ \mathbf{x} \geqslant \mathbf{0}, \end{array}$$

we write it in the equivalent form

$$\begin{array}{ll} \text{minimize} & \mathbf{c}^T\mathbf{x} \\ \text{subject to} & \mathbf{A}\mathbf{x} \geqslant \mathbf{b} \\ & -\mathbf{A}\mathbf{x} \geqslant -\mathbf{b} \\ & \mathbf{x} \geqslant \mathbf{0}, \end{array}$$

which is in the form of the primal of (4.1) but with coefficient matrix $\begin{bmatrix} \mathbf{A} \\ -\mathbf{A} \end{bmatrix}$. Using a dual vector partitioned as $(\mathbf{u}, \mathbf{v})$, the corresponding dual is

$$\begin{array}{ll} \text{maximize} & \mathbf{u}^T\mathbf{b} - \mathbf{v}^T\mathbf{b} \\ \text{subject to} & \mathbf{u}^T\mathbf{A} - \mathbf{v}^T\mathbf{A} \leqslant \mathbf{c}^T \\ & \mathbf{u} \geqslant \mathbf{0},\ \mathbf{v} \geqslant \mathbf{0}. \end{array}$$

Letting $\mathbf{y} = \mathbf{u} - \mathbf{v}$ we may simplify the representation of the dual program so that we obtain the pair of problems displayed below:

$$\begin{array}{llll} & \text{Primal} & & \text{Dual} \\ \text{minimize} & \mathbf{c}^T\mathbf{x} & \text{maximize} & \mathbf{y}^T\mathbf{b} \\ \text{subject to} & \mathbf{A}\mathbf{x} = \mathbf{b},\ \mathbf{x} \geqslant \mathbf{0} & \text{subject to} & \mathbf{y}^T\mathbf{A} \leqslant \mathbf{c}^T. \end{array} \tag{4.2}$$

This is the *asymmetric form* of the duality relation. In this form the dual vector $\mathbf{y}$ (which is really a composite of $\mathbf{u}$ and $\mathbf{v}$) is not restricted to be nonnegative.

Similar transformations can be worked out for any linear program to first get the primal in the form (4.1), calculate the dual, and then simplify the dual to account for special structure.

In general, if some of the linear inequalities in the primal (4.1) are changed to equality, the corresponding components of $\mathbf{y}$ in the dual become free variables. If some of the components of $\mathbf{x}$ in the primal are free variables, then the corresponding inequalities in $\mathbf{y}^T\mathbf{A} \leqslant \mathbf{c}^T$ are changed to equality in the dual. We mention again that these are not arbitrary rules but are direct consequences of the original definition and the equivalence of various forms of linear programs.

Example 1 (Dual of the Diet Problem). The diet problem, Example 1, Sect. 2.2, was the problem faced by a dietitian trying to select a combination of foods to meet certain nutritional requirements at minimum cost. This problem has the form

$$\begin{aligned} &\text{minimize} \quad \mathbf{c}^T\mathbf{x} \\ &\text{subject to} \quad \mathbf{A}\mathbf{x} \geqslant \mathbf{b}, \ \mathbf{x} \geqslant \mathbf{0} \end{aligned}$$

and hence can be regarded as the primal program of the symmetric pair above. We describe an interpretation of the dual problem.

Imagine a pharmaceutical company that produces in pill form each of the nutrients considered important by the dietitian. The pharmaceutical company tries to convince the dietitian to buy pills, and thereby supply the nutrients directly rather than through purchase of various foods. The problem faced by the drug company is that of determining positive unit prices $\lambda_1, \lambda_2, \ldots, \lambda_m$ for the nutrients so as to maximize revenue while at the same time being competitive with real food. To be competitive with real food, the cost of a unit of food i made synthetically from pure nutrients bought from the druggist must be no greater than c_i, the market price of the food. Thus, denoting by $\mathbf{a}_i$ the ith food, the company must satisfy $\mathbf{y}^T\mathbf{a}_i \leqslant \mathbf{c}_i$ for each i. In matrix form this is equivalent to $\mathbf{y}^T\mathrm{A} \leqslant \mathbf{c}^T$. Since b_j units of the jth nutrient will be purchased, the problem of the druggist is

$$\begin{aligned} &\text{maximize} \quad \mathbf{y}^T\mathbf{b} \\ &\text{subject to} \quad \mathbf{y}^T\mathbf{A} \leqslant \mathbf{c}^T, \ \mathbf{y} \geqslant \mathbf{0}, \end{aligned}$$

which is the dual problem.

Example 2 (Dual of the Transportation Problem). The transportation problem, Example 3, Sect. 2.2, is the problem, faced by a manufacturer, of selecting the pattern of product shipments between several fixed origins and destinations so as to minimize transportation cost while satisfying demand. Referring to (4.6) and (4.7) of Chap. 2, the problem is in standard form, and hence the asymmetric version of the duality relation applies. There is a dual variable for each constraint. In this case we denote the variables $u_i, i = 1, 2, \ldots, m$ for (4.6) and $v_j, \ j = 1, 2, \ \ldots, \ n$ for (4.7). Accordingly, the dual is

$$\begin{array}{lll} \text{maximize} & \sum_{i=1}^{m} a_i u_i + \sum_{j=1}^{n} b_j v_j & \\ \text{subject to} & u_i + v_j \leqslant c_{ij}, & i = 1, 2, \ldots, m, \\ & & j = 1, 2, \ldots, n. \end{array}$$

To interpret the dual problem, we imagine an entrepreneur who, feeling that he can ship more efficiently, comes to the manufacturer with the offer to buy his product at the plant sites (origins) and sell it at the warehouses (destinations). The product price that is to be used in these transactions varies from point to point, and is determined by the entrepreneur in advance. He must choose these prices, of course, so that his offer will be attractive to the manufacturer.

The entrepreneur, then, must select prices $-u_1, -u_2, \ldots, -u_m$ for the m origins and $v_1, v_2, \ldots, v_n$ for the n destinations. To be competitive with usual transportation modes, his prices must satisfy $u_i + v_j \leqslant c_{ij}$ for all i, j, since $u_i + v_j$ represents the net amount the manufacturer must pay to sell a unit of product at origin i and buy it back again at destination j. Subject to this constraint, the entrepreneur will adjust his prices to maximize his revenue. Thus, his problem is as given above.

4.2 The Duality Theorem

To this point the relation between the primal and dual programs has been simply a formal one based on what might appear as an arbitrary definition. In this section, however, the deeper connection between a program and its dual, as expressed by the Duality Theorem, is derived.

The proof of the Duality Theorem given in this section relies on the Separating Hyperplane Theorem (Appendix B) and is therefore somewhat more advanced than previous arguments. It is given here so that the most general form of the Duality Theorem is established directly. An alternative approach is to use the theory of the simplex method to derive the duality result. A simplified version of this alternative approach is given in the next section.

Throughout this section we consider the primal program in standard form

$$\begin{array}{ll} \text{minimize} & \mathbf{c}^T\mathbf{x} \\ \text{subject to} & \mathbf{A}\mathbf{x} = \mathbf{b}, \ \mathbf{x} \geqslant \mathbf{0} \end{array} \tag{4.3}$$

and its corresponding dual

$$\begin{array}{ll} \text{maximize} & \mathbf{y}^T\mathbf{b} \\ \text{subject to} & \mathbf{y}^T\mathbf{A} \leqslant \mathbf{c}^T. \end{array} \tag{4.4}$$

In this section it is *not* assumed that $\mathbf{A}$ is necessarily of full rank. The following lemma is easily established and gives us an important relation between the two problems.

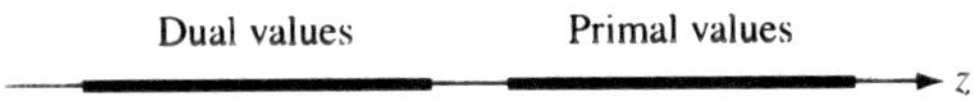

Fig. 4.1 Relation of primal and dual values

Lemma 1 (Weak Duality Lemma). *If* $\mathbf{x}$ *and* $\mathbf{y}$ *are feasible for (4.3) and (4.4), respectively, then* $\mathbf{c}^T\boldsymbol{x} \geqslant \mathbf{y}^T\boldsymbol{b}$.

Proof. We have
$$\mathbf{y}^T\mathbf{b} = \mathbf{y}^T\mathbf{A}\mathbf{x} \leqslant \mathbf{c}^T\mathbf{x},$$
the last inequality being valid since $\mathbf{x} \geqslant \mathbf{0}$ and $\mathbf{y}^T\mathbf{A} \leqslant \mathbf{c}^T$. ∎

This lemma shows that a feasible vector to either problem yields a bound on the value of the other problem. The values associated with the primal are all larger than the values associated with the dual as illustrated in Fig. 4.1. Since the primal seeks a minimum and the dual seeks a maximum, each seeks to reach the other. From this we have an important corollary.

Corollary. *If* $\mathbf{x}_0$ *and* $\mathbf{y}_0$ *are feasible for (4.3) and (4.4), respectively, and if* $\mathbf{c}^T\mathbf{x}_0 = \mathbf{y}_0^T\mathbf{b}$, *then* $\mathbf{x}_0$ *and* $\mathbf{y}_0$ *are optimal for their respective problems.*

The above corollary shows that if a pair of feasible vectors can be found to the primal and dual programs with equal objective values, then these are both optimal. The Duality Theorem of linear programming states that the converse is also true, and that, in fact, the two regions in Fig. 4.1 actually have a common point; there is no "gap."

Duality Theorem of Linear Programming. *If either of the problems* (4.3) *or* (4.4) *has a finite optimal solution, so does the other, and the corresponding values of the objective functions are equal. If either problem has an unbounded objective, the other problem has no feasible solution.*

Proof. We note first that the second statement is an immediate consequence of Lemma 1. For if the primal is unbounded and $\mathbf{y}$ is feasible for the dual, we must have $\mathbf{y}^T\mathbf{b} \leqslant -M$ for arbitrarily large M, which is clearly impossible.

Second we note that although the primal and dual are not stated in symmetric form it is sufficient, in proving the first statement, to assume that the primal has a finite optimal solution and then show that the dual has a solution with the same value. This follows because either problem can be converted to standard form and because the roles of primal and dual are reversible.

Suppose (4.3) has a finite optimal solution with value z_0. In the space E^{m+1} define the convex set
$$C = \{(r, \mathbf{w}) : r = tz_0 - \mathbf{c}^T\mathbf{x},\ \mathbf{w} = t\mathbf{b} - \mathbf{A}\mathbf{x},\ \mathbf{x} \geqslant \mathbf{0},\ t \geqslant 0\}.$$

It is easily verified that C is in fact a closed convex cone. We show that the point (1, $\mathbf{0}$) is not in C. If $\mathbf{w} = t_0\mathbf{b} - \mathbf{A}\mathbf{x}_0 = \mathbf{0}$ with $t_0 > 0$, $\mathbf{x}_0 \geqslant \mathbf{0}$, then $\mathbf{x} = \mathbf{x}_0/t_0$ is feasible for (4.3) and hence $r/t_0 = z_0 - \mathbf{c}^T\mathbf{x} \leqslant \mathbf{0}$; which means $r \leqslant \mathbf{0}$. If $\mathbf{w} = -\mathbf{A}\mathbf{x}_0 = \mathbf{0}$

with $\mathbf{x}_0 \geqslant \mathbf{0}$ and $\mathbf{c}^T\mathbf{x}_0 = -1$, and if $\mathbf{x}$ is any feasible solution to (4.3), then $\mathbf{x} + \alpha\mathbf{x}_0$ is feasible for any $\alpha \geqslant 0$ and gives arbitrarily small objective values as α is increased. This contradicts our assumption on the existence of a finite optimum and thus we conclude that no such $\mathbf{x}_0$ exists. Hence $(1, \mathbf{0}) \notin C$.

Now since C is a closed convex set, there is by Theorem 4.4, Sect. B.3, a hyperplane separating $(1, \mathbf{0})$ and C. Thus there is a nonzero vector $[s, \mathbf{y}] \in E^{m+1}$ and a constant c such that

$$s < c = \inf\{sr + \mathbf{y}^T\mathbf{w} : (r,\ \mathbf{w}) \in C\}.$$

Now since C is a cone, it follows that $c \geqslant 0$. For if there were $(r,\ \mathbf{w}) \in C$ such that $sr + \mathbf{y}^T\mathbf{w} < 0$, then $\alpha(r,\ \mathbf{w})$ for large α would violate the hyperplane inequality. On the other hand, since $(0,\ \mathbf{0}) \in C$ we must have $c \leqslant 0$. Thus $c = 0$. As a consequence $s < 0$, and without loss of generality we may assume $s = -1$.

We have to this point established the existence of $\mathbf{y} \in E^m$ such that

$$-r + \mathbf{y}^T\mathbf{w} \geqslant 0$$

for all $(r,\ \mathbf{w}) \in C$. Equivalently, using the definition of C,

$$(\mathrm{c} - \mathbf{y}^T\mathbf{A})\mathbf{x} - tz_0 + t\mathbf{y}^T\mathbf{b} \geqslant 0$$

for all $\mathbf{x} \geqslant \mathbf{0}, t \geqslant 0$. Setting $t = 0$ yields $\mathbf{y}^T\mathbf{A} \leqslant \mathbf{c}^T$, which says $\mathbf{y}$ is feasible for the dual. Setting $\mathbf{x} = \mathbf{0}$ and $t = 1$ yields $\mathbf{y}^T\mathbf{b} \geqslant z_0$, which in view of Lemma 1 and its corollary shows that $\mathbf{y}$ is optimal for the dual. ∎

4.3 Relations to the Simplex Procedure

In this section the Duality Theorem is proved by making explicit use of the characteristics of the simplex procedure. As a result of this proof it becomes clear that once the primal is solved by the simplex procedure a solution to the dual is readily obtainable.

Suppose that for the linear program

$$\begin{aligned} &\text{minimize} && \mathbf{c}^T\mathbf{x} \\ &\text{subject to} && \mathbf{Ax} = \mathbf{b},\ \mathbf{x} \geqslant \mathbf{0}, \end{aligned} \tag{4.5}$$

we have the optimal basic feasible solution $\mathbf{x} = (\mathbf{x}_\mathrm{B},\ \mathbf{0})$ with corresponding basis $\mathbf{B}$. We shall determine a solution of the dual program

$$\begin{aligned} &\text{maximize} && \mathbf{y}^T\mathbf{b} \\ &\text{subject to} && \mathbf{y}^T\mathbf{A} \leqslant \mathbf{c}^T \end{aligned} \tag{4.6}$$

in terms of $\mathbf{B}$.

We partition $\mathbf{A}$ as $\mathbf{A} = [\mathbf{B}, \mathbf{D}]$. Since the basic feasible solution $\mathbf{x_B} = \mathbf{B}^{-1}\mathbf{b}$ is optimal, the relative cost vector $\mathbf{r}$ must be nonnegative in each component. From Sect. 3.6 we have

$$\mathbf{r}_\mathbf{D}^T = \mathbf{c}_\mathbf{D}^T - \mathbf{c}_\mathbf{B}^T\mathbf{B}^{-1}\mathbf{D},$$

and since $\mathbf{r_D}$ is nonnegative in each component we have $\mathbf{c}_\mathbf{B}^T\mathbf{B}^{-1}\mathbf{D} \leqslant \mathbf{c}_\mathbf{D}^T$.

Now define $\mathbf{y}^T = \mathbf{c}_\mathbf{B}^T\mathbf{B}^{-1}$. We show that this choice of $\mathbf{y}$ solves the dual problem. We have

$$\mathbf{y}^T\mathbf{A} = [\mathbf{y}^T\mathbf{B}, \mathbf{y}^T\mathbf{D}] = [\mathbf{c}_\mathbf{B}^T, \mathbf{c}_\mathbf{B}^T\mathbf{B}^{-1}\mathbf{D}] \leqslant [\mathbf{c}_\mathbf{B}^T, \mathbf{c}_\mathbf{D}^T] = \mathbf{c}^T.$$

Thus since $\mathbf{y}^T\mathbf{A} \leqslant \mathbf{c}^T$, $\mathbf{y}$ is feasible for the dual. On the other hand,

$$\mathbf{y}^T\mathbf{b} = \mathbf{c}_\mathbf{B}^T\mathbf{B}^{-1}\mathbf{b} = \mathbf{c}_\mathbf{B}^T\mathbf{x_B},$$

and thus the value of the dual objective function for this $\mathbf{y}$ is equal to the value of the primal problem. This, in view of Lemma 1, Sect. 4.2, establishes the optimality of $\mathbf{y}$ for the dual. The above discussion yields an alternative derivation of the main portion of the Duality Theorem.

Theorem. *Let the linear program (4.5) have an optimal basic feasible solution corresponding to the basis $\mathbf{B}$. Then the vector $\mathbf{y}$ satisfying $\mathbf{y}^T = \mathbf{c}_\mathbf{B}^T\mathbf{B}^{-1}$ is an optimal solution to the dual program (4.6). The optimal values of both problems are equal.*

We turn now to a discussion of how the solution of the dual can be obtained directly from the final simplex tableau of the primal. Suppose that embedded in the original matrix $\mathbf{A}$ is an $m \times m$ identity matrix. This will be the case if, for example, m slack variables are employed to convert inequalities to equalities. Then in the final tableau the matrix $\mathbf{B}^{-1}$ appears where the identity appeared in the beginning. Furthermore, in the last row the components corresponding to this identity matrix will be $\mathbf{c}_\mathbf{I}^T - \mathbf{c}_\mathbf{B}^T\mathbf{B}^{-1}$, where $\mathbf{c_I}$ is the m-vector representing the cost coefficients of the variables corresponding to the columns of the original identity matrix. Thus by subtracting these cost coefficients from the corresponding elements in the last row, the negative of the solution $\mathbf{y}^T = \mathbf{c}_\mathbf{B}^T\mathbf{B}^{-1}$ to the dual is obtained. In particular, if, as is the case with slack variables, $\mathbf{c_I} = \mathbf{0}$, then the elements in the last row under $\mathbf{B}^{-1}$ are equal to the negative of components of the solution to the dual.

Example. Consider the primal program

$$\begin{aligned} \text{minimize} \quad & -x_1 - 4x_2 - 3x_3 \\ \text{subject to} \quad & 2x_1 + 2x_2 + x_3 \leqslant 4 \\ & x_1 + 2x_2 + 2x_3 \leqslant 6 \\ & x_1 \geqslant 0,\ x_2 \geqslant 0,\ x_3 \geqslant 0. \end{aligned}$$

This can be solved by introducing slack variables and using the simplex procedure. The appropriate sequence of tableaus is given below without explanation.

2	②	1	1	0	4
1	2	2	0	1	6
−1	−4	−3	0	0	0
1	1	1/2	1/2	0	2
−1	0	①	−1	1	2
3	0	−1	2	0	8
3/2	1	0	1	−1/2	1
−1	0	1	−1	1	2
2	0	0	1	1	10

The optimal solution is $x_1 = 0,\ x_2 = 1,\ x_3 = 2$. The corresponding dual program is

$$\begin{array}{ll} \text{maximize} & 4\lambda_1 + 6\lambda_2 \\ \text{subject to} & 2\lambda_1 + \lambda_2 \leqslant -1 \\ & 2\lambda_1 + 2\lambda_2 \leqslant -4 \\ & \lambda_1 + 2\lambda_2 \leqslant -3 \\ & \lambda_1 \leqslant 0,\ \lambda_2 \leqslant 0. \end{array}$$

The optimal solution to the dual is obtained directly from the last row of the simplex tableau under the columns where the identity appeared in the first tableau: $\lambda_1 = -1,\ \lambda_2 = -1$.

Geometric Interpretation

The duality relations can be viewed in terms of the dual interpretations of linear constraints emphasized in Chap. 3. Consider a linear program in standard form. For sake of concreteness we consider the problem

$$\begin{array}{ll} \text{minimize} & 18x_1 + 12x_2 + 2x_3 + 6x_4 \\ \text{subject to} & 3x_1 + x_2 - 2x_3 + x_4 = 2 \\ & x_1 + 3x_2 - x_4 = 2 \\ & x_1 \geqslant 0,\ x_2 \geqslant 0,\ x_3 \geqslant 0,\ x_4 \geqslant 0. \end{array}$$

The columns of the constraints are represented in requirements space in Fig. 4.2. A basic solution represents construction of $\mathbf{b}$ with positive weights on two of the $\mathbf{a}_i$'s. The dual problem is

$$\begin{array}{ll} \text{maximize} & 2\lambda_1 + 2\lambda_2 \\ \text{subject to} & 3\lambda_1 + \lambda_2 \leqslant 18 \\ & \lambda_1 + 3\lambda_2 \leqslant 12 \\ & -2\lambda_1 \leqslant 2 \\ & \lambda_1 - \lambda_2 \leqslant 6. \end{array}$$

The dual problem is shown geometrically in Fig. 4.3. Each column $\mathbf{a}_i$ of the primal defines a constraint of the dual as a half-space whose boundary is orthogonal

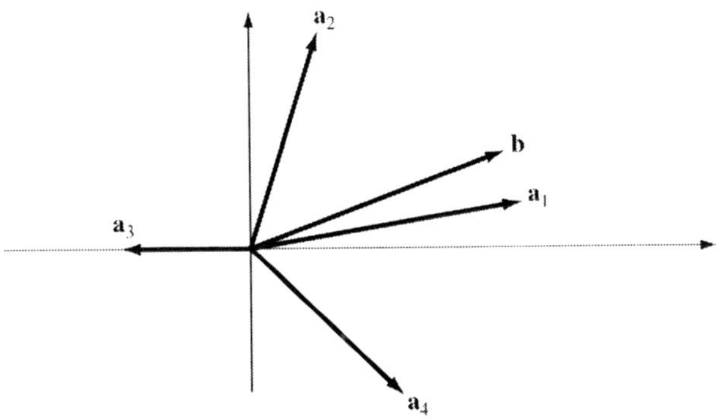

Fig. 4.2 The primal requirements space

to that column vector and is located at a point determined by c_i. The dual objective is maximized at an extreme point of the dual feasible region. At this point exactly two dual constraints are active. These active constraints correspond to an optimal basis of the primal. In fact, the vector defining the dual objective is a positive linear combination of the vectors. In the specific example, **b** is a positive combination of $\mathbf{a}_1$ and $\mathbf{a}_2$. The weights in this combination are the x_i's in the solution of the primal.

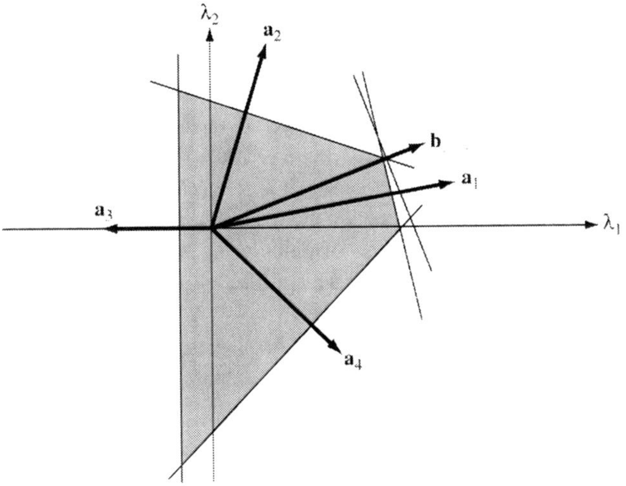

Fig. 4.3 The dual in activity space

Simplex Multipliers

We conclude this section by giving an economic interpretation of the relation between the simplex basis and the vector $\mathbf{y}$. At any point in the simplex procedure we may form the vector $\mathbf{y}$ satisfying $\mathbf{y}^T = \mathbf{c}_\mathbf{B}^T\mathbf{B}^{-1}$. This vector is not a solution to the dual unless $\mathbf{B}$ is an optimal basis for the primal, but nevertheless, it has an economic interpretation. Furthermore, as we have seen in the development of the revised simplex method, this $\mathbf{y}$ vector can be used at every step to calculate the relative cost coefficients. For this reason $\mathbf{y}^T = \mathbf{c}_\mathbf{B}^T\mathbf{B}^{-1}$, corresponding to any basis, is often called the vector of *simplex multipliers*.

Let us pursue the economic interpretation of these simplex multipliers. As usual, denote the columns of $\mathbf{A}$ by $\mathbf{a}_1, \mathbf{a}_2, \ldots, \mathbf{a}_n$ and denote by $\mathbf{e}_1, \mathbf{e}_2, \ldots, \mathbf{e}_m$ the m unit vectors in E^m. The components of the $\mathbf{a}_i$'s and $\mathbf{b}$ tell how to construct these vectors from the $\mathbf{e}_i$'s.

Given any basis $\mathbf{B}$, however, consisting of m columns of $\mathbf{A}$, any other vector can be constructed (synthetically) as a linear combination of these basis vectors. If there is a unit cost c_i associated with each basis vector $\mathbf{a}_i$, then the cost of a (synthetic) vector constructed from the basis can be calculated as the corresponding linear combination of the c_i's associated with the basis. In particular, the cost of the jth unit vector, $\mathbf{e}_j$, when constructed from the basis $\mathbf{B}$, is λ_j, the jth component of $\mathbf{y}^T = \mathbf{c}_\mathbf{B}^T\mathbf{B}^{-1}$. Thus the λ_j's can be interpreted as synthetic prices of the unit vectors.

Now, any vector can be expressed in terms of the basis $\mathbf{B}$ in two steps: (1) express the unit vectors in terms of the basis, and then (2) express the desired vector as a linear combination of unit vectors. The corresponding synthetic cost of a vector constructed from the basis $\mathbf{B}$ can correspondingly be computed directly by: (1) finding the synthetic price of the unit vectors, and then (2) using these prices to evaluate the cost of the linear combination of unit vectors. Thus, the simplex multipliers can be used to quickly evaluate the synthetic cost of any vector that is expressed in terms of the unit vectors. The difference between the true cost of this vector and the synthetic cost is the relative cost. The process of calculating the synthetic cost of a vector, with respect to a given basis, by using the simplex multipliers is sometimes referred to as *pricing out* the vector.

Optimality of the primal corresponds to the situation where every vector $\mathbf{a}_1, \mathbf{a}_2, \ldots, \mathbf{a}_n$ is cheaper when constructed from the basis than when purchased directly at its own price. Thus we have $\mathbf{y}^T\mathbf{a}_i \leqslant c_i$ for $i = 1, 2, \ldots, n$ or equivalently $\mathbf{y}^T\mathbf{A} \leqslant \mathbf{c}^T$.

4.4 Sensitivity and Complementary Slackness

The optimal values of the dual variables in a linear program can, as we have seen, be interpreted as prices. In this section this interpretation is explored in further detail.

Sensitivity

Suppose in the linear program

$$\begin{aligned}&\text{minimize} \quad \mathbf{c}^T\mathbf{x}\\&\text{subject to} \quad \mathbf{Ax} = \mathbf{b},\ \mathbf{x} \geqslant \mathbf{0},\end{aligned} \tag{4.7}$$

the optimal basis is $\mathbf{B}$ with corresponding solution $(\mathbf{x_B}, \mathbf{0})$, where $\mathbf{x_B} = \mathbf{B}^{-1}\mathbf{b}$. A solution to the corresponding dual is $\mathbf{y}^T = \mathbf{c}_\mathbf{B}^T\mathbf{B}^{-1}$.

Now, assuming nondegeneracy, small changes in the vector $\mathbf{b}$ will not cause the optimal basis to change. Thus for $\mathbf{b} + \Delta\mathbf{b}$ the optimal solution is

$$\mathbf{x} = (\mathbf{x_B} + \Delta\mathbf{x_B}, \mathbf{0}),$$

where $\Delta\mathbf{x_B} = \mathbf{B}^{-1}\Delta\mathbf{b}$. Thus the corresponding increment in the cost function is

$$\Delta z = \mathbf{c}_\mathbf{B}^T\Delta\mathbf{x_B} = \mathbf{y}^T\Delta\mathbf{b}. \tag{4.8}$$

This equation shows that $\mathbf{y}$ gives the sensitivity of the optimal cost with respect to small changes in the vector $\mathbf{b}$. In other words, if a new program were solved with $\mathbf{b}$ changed to $\mathbf{b} + \Delta\mathbf{b}$, the change in the optimal value of the objective function would be $\mathbf{y}^T\Delta\mathbf{b}$.

This interpretation of the dual vector $\mathbf{y}$ is intimately related to its interpretation as a vector of simplex multipliers. Since λ_j is the price of the unit vector $\mathbf{e}_j$ when constructed from the basis $\mathbf{B}$, it directly measures the change in cost due to a change in the jth component of the vector $\mathbf{b}$. Thus, λ_j may equivalently be considered as the *marginal price* of the component b_j, since if b_j is changed to $b_j + \Delta b_j$ the value of the optimal solution changes by $\lambda_j \Delta b_j$.

If the linear program is interpreted as a diet problem, for instance, then λ_j is the maximum price per unit that the dietitian would be willing to pay for a small amount of the jth nutrient, because decreasing the amount of nutrient that must be supplied by food will reduce the food bill by λ_j dollars per unit. If, as another example, the linear program is interpreted as the problem faced by a manufacturer who must select levels $x_1, x_2, \ldots, x_n$ of n production activities in order to meet certain required levels of output $b_1, b_2, \ldots, b_m$ while minimizing production costs, the λ_i's are the marginal prices of the outputs. They show directly how much the production cost varies if a small change is made in the output levels.

Complementary Slackness

The optimal solutions to primal and dual programs satisfy an additional relation that has an economic interpretation. This relation can be stated for any pair of dual linear programs, but we state it here only for the asymmetric and the symmetric pairs defined in Sect. 4.1.

Theorem. (Complementary slackness—asymmetric form). *Let* **x** *and* **y** *be feasible solutions for the primal and dual programs, respectively, in the pair (4.2). A necessary and sufficient condition that they both be optimal solutions is that*[†] *for all i*

i) $x_i > 0 \Rightarrow \mathbf{y}^T \boldsymbol{a}_i = c_i$
ii) $x_i = 0 \Leftarrow \mathbf{y}^T \boldsymbol{a}_j < c_j$.

Proof. If the stated conditions hold, then clearly $(\mathbf{y}^T\mathbf{A} - \mathbf{c}^T)\mathbf{x} = 0$. Thus $\mathbf{y}^T\mathbf{b} = \mathbf{c}^T\mathbf{x}$, and by the corollary to Lemma 1, Sect. 4.2, the two solutions are optimal. Conversely, if the two solutions are optimal, it must hold, by the Duality Theorem, that $\mathbf{y}^T\mathbf{b} = \mathbf{c}^T\mathbf{x}$ and hence that $(\mathbf{y}^T\mathbf{A} - \mathbf{c}^T)\mathbf{x} = 0$. Since each component of **x** is nonnegative and each component of $\mathbf{y}^T\mathbf{A} - \mathbf{c}^T$ is nonpositive, the conditions (i) and (ii) must hold. ∎

Theorem. (Complementary slackness—symmetric form). *Let* **x** *and* **y** *be feasible solutions for the primal and dual programs, respectively, in the pair (4.1). A necessary and sufficient condition that they both be optimal solutions is that for all i and j*

i) $x_i > 0 \Rightarrow \mathbf{y}^T \mathbf{a}_i = c_i$
ii) $x_i = 0 \Leftarrow \mathbf{y}^T \mathbf{a}_i < c_i$
iii) $\lambda_j > 0 \Rightarrow \mathbf{a}^j\mathbf{x} = b_j$
iv) $\lambda_j = 0 \Leftarrow \mathbf{a}^j\mathbf{x} > b_j$,

(where $\mathbf{a}^j$ *is the jth row of* **A***).*

Proof. This follows by transforming the previous theorem. ∎

The complementary slackness conditions have a rather obvious economic interpretation. Thinking in terms of the diet problem, for example, which is the primal part of a symmetric pair of dual problems, suppose that the optimal diet supplies more than b_j units of the jth nutrient. This means that the dietitian would be unwilling to pay anything for small quantities of that nutrient, since availability of it would not reduce the cost of the optimal diet. This, in view of our previous interpretation of λ_j as a marginal price, implies $\lambda_j = 0$ which is (iv) of Theorem 4.4. The other conditions have similar interpretations which the reader can work out.

4.5 Max Flow–Min Cut Theorem

One of the most exemplary pairs of linear primal and dual problems is the max-flow and min-cut theorem, which we describe in this section. The maximal flow problem described in Chap. 2 can be expressed more compactly in terms of the node–arc incidence matrix (see Appendix D). Let **x** be the vector of arc flows x_{ij} (ordered in any way). Let **A** be the corresponding node-arc incidence matrix. Finally, let **e** be a

[†] The symbol ⇒ means "implies" and ⇐ means "is implied by."

vector with dimension equal to the number of nodes and having a + 1 component on node 1, a − 1 on node m, and all other components zero. The maximal flow problem is then

$$\begin{aligned} &\text{maximize} \ f \\ &\text{subject to} \ \mathbf{Ax} - f\mathbf{e} = \mathbf{0} \\ &\qquad\qquad\quad \mathbf{x} \leqslant \mathbf{k}. \end{aligned} \tag{4.9}$$

The coefficient matrix of this problem is equal to the node–arc incidence matrix with an additional column for the flow variable f. Any basis of this matrix is triangular, and hence as indicated by the theory in the transportation problem in Chap. 3, the simplex method can be effectively employed to solve this problem. However, instead of the simplex method, a simple algorithm based on the tree algorithm (also see Appendix D) can be used.

Max Flow Augmenting Algorithm

The basic strategy of the algorithm is quite simple. First we recognize that it is possible to send nonzero flow from node 1 to node m only if node m is reachable from node 1. The tree procedure can be used to determine if m is in fact reachable; and if it is reachable, the algorithm will produce a path from 1 to m. By examining the arcs along this path, we can determine the one with minimum capacity. We may then construct a flow equal to this capacity from 1 to m by using this path. This gives us a strictly positive (and integer-valued) initial flow.

Next consider the nature of the network at this point in terms of additional flows that might be assigned. If there is already flow x_{ij} in the arc (i, j), then the effective capacity of that arc is reduced by x_{ij}(to $k_{ij} - x_{ij}$), since that is the maximal amount of additional flow that can be assigned to that arc. On the other hand, the effective reverse capacity, on the arc (j, i), is increased by x_{ij}(to $k_{ji} + x_{ij}$), since a small incremental backward flow is actually realized as a reduction in the forward flow through that arc. Once these changes in capacities have been made, the tree procedure can again be used to find a path from node 1 to node m on which to assign additional flow. (Such a path is termed an *augmenting path.*) Finally, if m is not reachable from 1, no additional flow can be assigned, and the procedure is complete.

It is seen that the method outlined above is based on repeated application of the tree procedure, which is implemented by labeling and scanning. By including slightly more information in the labels than in the basic tree algorithm, the minimum arc capacity of the augmenting path can be determined during the initial scanning, instead of by reexamining the arcs after the path is found. A typical label at a node i has the form (k, c_i), where k denotes a precursor node and c_i is the maximal flow that can be sent from the source to node i through the path created by the previous labeling and scanning. The complete procedure is this:

Step 0. Set all $x_{ij} = 0$ and $f = 0$.
Step 1. Label node 1 $(-, \infty)$. All other nodes are unlabeled.
Step 2. Select any labeled node i for scanning. Say it has label (k, c_i). For all unlabeled nodes j such that (i, j) is an arc with $x_{ij} < k_{ij}$, assign the label (i, c_j), where $c_j = \min\{c_i, k_{ij} - x_{ij}\}$. For all unlabeled nodes j such that (j, i) is an arc with $x_{ji} > 0$, assign the label (i, c_j), where $c_j = \min\{c_i, x_{ji}\}$.
Step 3. Repeat Step 2 until either node m is labeled or until no more labels can be assigned. In this latter case, the current solution is optimal.
Step 4. (Augmentation.) If the node m is labeled (i, c_m), then increase f and the flow on arc (i, m) by c_m. Continue to work backward along the augmenting path determined by the nodes, increasing the flow on each arc of the path by c_m. Return to Step 1.
The validity of the algorithm should be fairly apparent, that is, the finite termination of the algorithm. However, a complete proof is deferred until we consider the max flow-min cut theorem below.

Example. An example of the above procedure is shown in Fig. 4.4. Node 1 is the source, and node 6 is the sink. The original network with capacities indicated on the arcs is shown in Fig. 4.4a. Also shown in that figure are the initial labels obtained by the procedure. In this case the sink node is labeled, indicating that a flow of 1 unit can be achieved. The augmenting path of this flow is shown in Fig. 4.4b. Numbers in square boxes indicate the total flow in an arc. The new labels are then found and added to that figure. Note that node 2 cannot be labeled from node 1 because there is no unused capacity in that direction. Node 2 can, however, be labeled from node 4, since the existing flow provides a reverse capacity of 1 unit. Again the sink is labeled, and 1 unit more flow can be constructed. The augmenting path is shown in Fig. 4.4c. A new labeling is appended to that figure. Again the sink is labeled, and an additional 1 unit of flow can be sent from source to sink. The path of this 1 unit is shown in Fig. 4.4d. Note that it includes a flow from node 4 to node 2, even though flow was not allowed in this direction in the original network. This flow is allowable now, however, because there is already flow in the opposite direction. The total flow at this point is shown in Fig. 4.4e. The flow levels are again in square boxes. This flow is maximal, since only the source node can be labeled.

Max Flow–Min Cut Theorem

A great deal of insight and some further results can be obtained through the introduction of the notion of *cuts* in a network. Given a network with source node 1 and sink node m, divide the nodes arbitrarily into two sets S and $\bar{S}$ such that the source node is in S and the sink is in $\bar{S}$. The set of arcs from S to $\bar{S}$ is a *cut* and is denoted $(S, \bar{S})$. The *capacity* of the cut is the sum of the capacities of the arcs in the cut.

An example of a cut is shown in Fig. 4.5. The set S consists of nodes 1 and 2, while $\bar{S}$ consists of 3, 4, 5, 6. The capacity of this cut is 4.

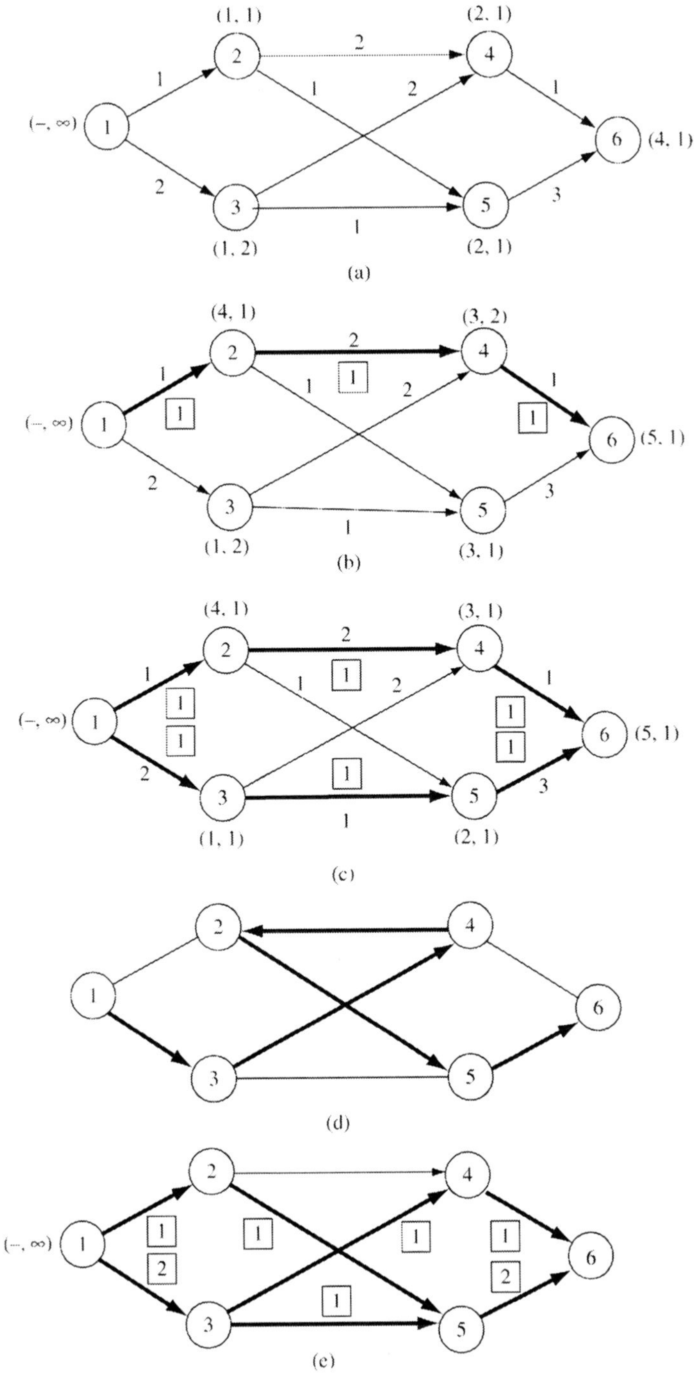

Fig. 4.4 Illustration of algorithmic steps of the maximal flow example

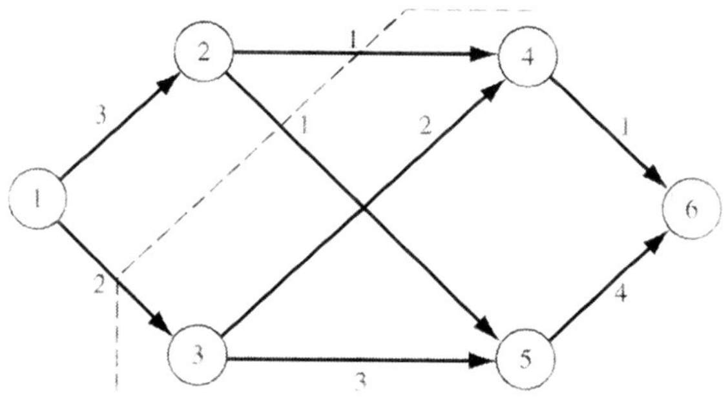

Fig. 4.5 A cut

It should be clear that a path from node 1 to node m must include at least one arc in any cut, for the path must have an arc from the set S to the set $\bar{S}$. Furthermore, it is clear that the maximal amount of flow that can be sent through a cut is equal to its capacity. Thus each cut gives an upper bound on the value of the maximal flow problem. The max flow-min cut theorem states that equality is actually achieved for some cut. That is, the maximal flow is equal to the minimal cut capacity. It should be noted that the proof of the theorem also establishes the maximality of the flow obtained by the maximal flow algorithm.

Max Flow–Min Cut Theorem. *In a network the maximal flow between a source and a sink is equal to the minimal cut capacity of all cuts separating the source and sink.*

Proof. Since any cut capacity must be greater than or equal to the maximal flow, it is only necessary to exhibit a flow and a cut for which equality is achieved. Begin with a flow in the network that cannot be augmented by the maximal flow algorithm. For this flow find the effective arc capacities of all arcs for incremental flow changes as described earlier and apply the labeling procedure of the maximal flow algorithm. Since no augmenting path exists, the algorithm must terminate before the sink is labeled.

Let S and $\bar{S}$ consist of all labeled and unlabeled nodes, respectively. This defines a cut separating the source from the sink. All arcs originating in S and terminating in $\bar{S}$ have zero incremental capacity, or else a node in $\bar{S}$ could have been labeled. This means that each arc in the cut is saturated by the original flow; that is, the flow is equal to the capacity. Any arc originating in $\bar{S}$ and terminating in S, on the other hand, must have zero flow; otherwise, this would imply a positive incremental capacity in the reverse direction, and the originating node in $\bar{S}$ would be labeled. Thus, there is a total flow from S to $\bar{S}$ equal to the cut capacity, and zero flow from $\bar{S}$ to S. This means that the flow from source to sink is equal to the cut capacity. Thus the cut capacity must be minimal, and the flow must be maximal. ∎

In the network of Fig. 4.4, the minimal cut corresponds to the S consisting only of the source. That cut capacity is 3. Note that in accordance with the max flow-min cut theorem, this is equal to the value of the maximal flow, and the minimal cut is determined by the final labeling in Fig. 4.4e. In Fig. 4.5 the cut shown is also minimal, and the reader should easily be able to determine the pattern of maximal flow.

Relation to Duality

The character of the max flow–min cut theorem suggests a connection with the Duality Theorem. We conclude this section by exploring this connection.

The maximal flow problem is a linear program, which is expressed formally by (4.9). The dual problem is found to be

$$\begin{aligned} \text{minimize} \quad & \mathbf{w}^T\mathbf{k} \\ \text{subject to} \quad & \mathbf{u}^T\mathbf{A} = \mathbf{w}^T \\ & \mathbf{u}^T\mathbf{e} = 1 \\ & \mathbf{w} \geq \mathbf{0}. \end{aligned} \tag{4.10}$$

When written out in detail, the dual is

$$\begin{aligned} \text{minimize} \quad & \sum_{ij} w_{ij}k_{ij} \\ \text{subject to} \quad & u_i - u_j = w_{ij} \\ & u_1 - u_m = 1 \\ & w_{ij} \geq 0. \end{aligned} \tag{4.11}$$

A pair $i,\ j$ is included in the above only if $(i,\ j)$ is an arc of the network.

A feasible solution to this dual problem can be found in terms of any cut set $(S, \bar{S})$. In particular, it is easily seen that

$$u_i = \begin{cases} 1 & \text{if} \quad i \in S \\ 0 & \text{if} \quad i \in \bar{S} \end{cases} \tag{4.12}$$

$$w_{ij} = \begin{cases} 1 & \text{if} \quad (i,\ j) \in (S, \bar{S}) \\ 0 & \text{otherwise} \end{cases}$$

is a feasible solution. The value of the dual problem corresponding to this solution is the cut capacity. If we take the cut set to be the one determined by the labeling procedure of the maximal flow algorithm as described in the proof of the theorem above, it can be seen to be optimal by verifying the complementary slackness conditions (a task we leave to the reader). The minimum value of the dual is therefore equal to the minimum cut capacity.

4.6 The Dual Simplex Method

Often there is available a basic solution to a linear program which is not feasible but which prices out optimally; that is, the simplex multipliers are feasible for the dual problem. In the simplex tableau this situation corresponds to having no negative elements in the bottom row but an infeasible basic solution. Such a situation may arise, for example, if a solution to a certain linear programming problem is calculated and then a new problem is constructed by changing the vector $\mathbf{b}$. In such situations a basic feasible solution to the dual is available and hence it is desirable to pivot in such a way as to optimize the dual.

Rather than constructing a tableau for the dual problem (which, if the primal is in standard form; involves m free variables and n nonnegative slack variables), it is more efficient to work on the dual from the primal tableau. The complete technique based on this idea is the dual simplex method. In terms of the primal problem, it operates by maintaining the optimality condition of the last row while working toward feasibility. In terms of the dual problem, however, it maintains feasibility while working toward optimality.

Given the linear program

$$\begin{aligned} &\text{minimize} \quad \mathbf{c}^T\mathbf{x} \\ &\text{subject to} \quad \mathbf{A}\mathbf{x} = \mathbf{b}, \ \mathbf{x} \geqslant \mathbf{0}, \end{aligned} \tag{4.13}$$

suppose a basis $\mathbf{B}$ is known such that $\mathbf{y}$ defined by $\mathbf{y}^T = \mathbf{c}_{\mathbf{B}}^T\mathbf{B}^{-1}$ is feasible for the dual. In this case we say that the corresponding basic solution to the primal, $\mathbf{x}_{\mathbf{B}} = \mathbf{B}^{-1}\mathbf{b}$, is *dual feasible*. If $\mathbf{x}_B \geqslant \mathbf{0}$ then this solution is also primal feasible and hence optimal.

The given vector $\mathbf{y}$ is feasible for the dual and thus satisfies $\mathbf{y}^T\mathbf{a}_j \leqslant c_j$, for $j = 1, 2, \ldots, n$. Indeed, assuming as usual that the basis is the first m columns of $\mathbf{A}$, there is equality

$$\mathbf{y}^T\mathbf{a}_j = c_j, \text{ for } j = 1, 2, \ldots, m, \tag{4.14a}$$

and (barring degeneracy in the dual) there is inequality

$$\mathbf{y}^T\mathbf{a}_j < c_j, \text{ for } j = m+1, \ldots, n. \tag{4.14b}$$

To develop one cycle of the dual simplex method, we find a new vector $\bar{\mathbf{y}}$ such that one of the equalities becomes an inequality and one of the inequalities becomes equality, while at the same time increasing the value of the dual objective function. The m equalities in the new solution then determine a new basis.

Denote the ith row of $\mathbf{B}^{-1}$ by $\mathbf{u}^i$. Then for

$$\bar{\mathbf{y}}^T = \mathbf{y}^T - \varepsilon\mathbf{u}^i, \tag{4.15}$$

we have $\bar{\mathbf{y}}^T\mathbf{a}_j = \mathbf{y}^T\mathbf{a}_j - \varepsilon\mathbf{u}^j\mathbf{a}_j$. Thus, recalling that $z_j = \mathbf{y}^T\mathbf{a}_j$ and noting that $\mathbf{u}^i\mathbf{a}_j = y_{ij}$, the ijth element of the tableau, we have

$$\bar{\mathbf{y}}^T\mathbf{a}_j = c_j, \qquad j = 1, 2, \ldots, m, \quad i \neq j \tag{4.16a}$$

$$\bar{\mathbf{y}}^T\mathbf{a}_i = c_i - \varepsilon \tag{4.16b}$$

$$\bar{\mathbf{y}}^T\mathbf{a}_j = z_j - \varepsilon y_{ij}, \quad j = m + 1, \quad m + 2, \ldots, n. \tag{4.16c}$$

Also,

$$\bar{\mathbf{y}}^T\mathbf{b} = \mathbf{y}^T\mathbf{b} - \varepsilon \mathbf{x}_{\mathbf{B}i}. \tag{4.17}$$

These last equations lead directly to the algorithm:

Step 1. Given a dual feasible basic solution $\mathbf{x_B}$, if $\mathbf{x_B} \geqslant \mathbf{0}$ the solution is optimal. If $\mathbf{x_B}$ is not nonnegative, select an index i such that the ith component of $\mathbf{x_B}$, $\mathbf{x}_{\mathbf{B}i} < 0$.

Step 2. If all $y_{ij} \geqslant 0$, $j = 1, 2, \ldots, n$, then the dual has no maximum (this follows since by (4.16) $\bar{\boldsymbol{\lambda}}$ is feasible for all $\varepsilon > 0$). If $y_{ij} < 0$ for some j, then let

$$\varepsilon_0 = \frac{z_k - c_k}{y_{ik}} = \min_j \left\{ \frac{z_j - c_j}{y_{ij}} : y_{ij} < 0 \right\}. \tag{4.18}$$

Step 3. Form a new basis $\mathbf{B}$ by replacing $\mathbf{a}_i$ by $\mathbf{a}_k$. Using this basis determine the corresponding basic dual feasible solution $\mathbf{x_B}$ and return to Step 1.

The proof that the algorithm converges to the optimal solution is similar in its details to the proof for the primal simplex procedure. The essential observations are: (a) from the choice of k in (4.18) and from (4.16a), (4.16b), 4.16c) the new solution will again be dual feasible; (b) by (4.17) and the choice $\mathbf{x}_{\mathbf{B}_i} < 0$, the value of the dual objective will increase; (c) the procedure cannot terminate at a nonoptimum point; and (d) since there are only a finite number of bases, the optimum must be achieved in a finite number of steps.

Example. A form of problem arising frequently is that of minimizing a positive combination of positive variables subject to a series of "greater than" type inequalities having positive coefficients. Such problems are natural candidates for application of the dual simplex procedure. The classical diet problem is of this type as is the simple example below.

$$\begin{array}{ll} \text{minimize} & 3x_1 + 4x_2 + 5x_3 \\ \text{subject to} & x_i + 2x_2 + 3x_3 \geqslant 5 \\ & 2x_1 + 2x_2 + x_3 \geqslant 6 \\ x_1 \geqslant 0, & x_2 \geqslant 0, \quad x_3 \geqslant 0. \end{array}$$

By introducing surplus variables and by changing the sign of the inequalities we obtain the initial tableau

−1	−2	−3	1	0	−5
−②	−2	−1	0	1	−6
3	4	5	0	0	0

Initial tableau

The basis corresponds to a dual feasible solution since all of the $c_j - z_j$'s are nonnegative. We select any $\mathbf{x}_{\mathbf{B}_i} < 0$, say $x_5 = -6$, to remove from the set of basic variables. To find the appropriate pivot element in the second row we compute the ratios $(z_j - c_j)/y_{2j}$ and select the minimum positive ratio. This yields the pivot indicated. Continuing, the remaining tableaus are

0	−①	−5/2	1	−1/2	−2
1	1	1/2	0	−1/2	3
0	1	7/2	0	3/2	9

Second tableau

0	1	5/2	−1	1/2	2
1	0	−2	1	−1	1
0	0	1	1	1	11

Final tableau

The third tableau yields a feasible solution to the primal which must be optimal. Thus the solution is $x_1 = 1,\ x_2 = 2,\ x_3 = 0$.

*4.7 *The Primal-Dual Algorithm

In this section a procedure is described for solving linear programming problems by working simultaneously on the primal and the dual problems. The procedure begins with a feasible solution to the dual that is improved at each step by optimizing an *associated restricted primal* problem. As the method progresses it can be regarded as striving to achieve the complementary slackness conditions for optimality. Originally, the primal-dual method was developed for solving a special kind of linear program arising in network flow problems, and it continues to be the most efficient procedure for these problems. (For general linear programs the dual simplex method is most frequently used). In this section we describe the generalized version of the algorithm and point out an interesting economic interpretation of it. We consider the program

$$\begin{aligned} &\text{minimize} \quad \mathbf{c}^T\mathbf{x} \\ &\text{subject to} \quad \mathbf{A}\mathbf{x} = \mathbf{b}, \mathbf{x} \geqslant \mathbf{0} \end{aligned} \tag{4.19}$$

and the corresponding dual program

$$\begin{aligned} &\text{maximize} \quad \mathbf{y}^T\mathbf{b} \\ &\text{subject to} \quad \mathbf{y}^T\mathbf{A} \leqslant \mathbf{c}^T. \end{aligned} \tag{4.20}$$

Given a feasible solution $\mathbf{y}$ to the dual, define the subset P of $1, 2, \ldots, n$ by $i \in P$ if $\mathbf{y}^T\mathbf{a}_i = c_i$ where $\mathbf{a}_i$ is the ith column of $\mathbf{A}$. Thus, since $\mathbf{y}$ is dual feasible, it follows that $i \notin P$ implies $\mathbf{y}^T\mathbf{a}_i < c_i$. Now corresponding to $\mathbf{y}$ and P, we define the *associated restricted primal* problem

$$\begin{aligned} &\text{minimize} \quad \mathbf{1}^T\mathbf{y} \\ &\text{subject to} \quad \mathbf{A}\mathbf{x} + \mathbf{y} = \mathbf{b} \\ &\qquad\qquad \mathbf{x} \geqslant \mathbf{0}, \qquad x_i = 0 \text{ for } i \notin P \\ &\qquad\qquad \mathbf{y} \geqslant \mathbf{0}, \end{aligned} \tag{4.21}$$

where $\mathbf{1}$ denotes the m-vector $(1, 1, \ldots, 1)$.

The dual of this associated restricted primal is called the *associated restricted dual*. It is

$$\begin{aligned} &\text{maximize} \quad \mathbf{u}^T\mathbf{b} \\ &\text{subject to} \quad \mathbf{u}^T\mathbf{a}_i \leqslant \mathbf{0}, \quad i \notin P \\ &\qquad\qquad \mathbf{u} \leqslant \mathbf{1}. \end{aligned} \tag{4.22}$$

The condition for optimality of the primal-dual method is expressed in the following theorem.

Primal-Dual Optimality Theorem. *Suppose that* $\mathbf{y}$ *is feasible for the dual and that* $\mathbf{x}$ *and* $\mathbf{y} = 0$ *is feasible (and of course optimal) for the associated restricted primal. Then* $\mathbf{x}$ *and* $\mathbf{y}$ *are optimal for the original primal and dual programs, respectively.*

Proof. Clearly $\mathbf{x}$ is feasible for the primal. Also we have $\mathbf{c}^T\mathbf{x} = \mathbf{y}^T\mathbf{A}\mathbf{x}$, because $\mathbf{y}^T\mathbf{A}$ is identical to $\mathbf{c}^T$ on the components corresponding to nonzero elements of $\mathbf{x}$. Thus $\mathbf{c}^T\mathbf{x} = \mathbf{y}^T\mathbf{A}\mathbf{x} = \mathbf{y}^T\mathbf{b}$ and optimality follows from Lemma 1, Sect. 4.2. ∎

The primal–dual method starts with a feasible solution to the dual and then optimizes the associated restricted primal. If the optimal solution to this associated restricted primal is not feasible for the primal, the feasible solution to the dual is improved and a new associated restricted primal is determined. Here are the details:

Step 1. Given a feasible solution $\mathbf{y}_0$ to the dual program (4.20), determine the associated restricted primal according to (4.21).

Step 2. Optimize the associated restricted primal. If the minimal-value of this problem is zero, the corresponding solution is optimal for the original primal by the Primal-Dual Optimality Theorem.

Step 3. If the minimal value of the associated restricted primal is strictly positive, obtain from the final simplex tableau of the restricted primal, the solution $\mathbf{u}_0$ of the associated restricted dual (4.22). If there is no j for which $\mathbf{u}_0^T\mathbf{a}_j > 0$ conclude the primal has no feasible solutions. If, on the other hand, for at least one j, $\mathbf{u}_0^T\mathbf{a}_j > 0$, define the new dual feasible vector

$$\mathbf{y} = \mathbf{y}_0 + \varepsilon_0\mathbf{u}_0$$

where

$$\varepsilon_0 = \frac{c_k - \mathbf{y}_0^T\mathbf{a}_k}{\mathbf{u}_0^T\mathbf{a}_k} = \min_j \left\{ \frac{c_j - \mathbf{y}_0^T\mathbf{a}_j}{\mathbf{u}_0^T\mathbf{a}_j} : \mathbf{u}_0^T\mathbf{a}_j > 0 \right\}.$$

Now go back to Step 1 using this $\mathbf{y}$.

To prove convergence of this method a few simple observations and explanations must be made. First we verify the statement made in Step 3 that $\mathbf{u}_0^T\mathbf{a}_j \leqslant 0$ for all j

implies that the primal has no feasible solution. The vector $\mathbf{y}_\varepsilon = \mathbf{y}_0 + \varepsilon\mathbf{u}_0$ is feasible for the dual problem for all positive ε, since $\mathbf{u}_0^T\mathbf{A} \leqslant 0$. In addition, $\mathbf{y}_\varepsilon^T\mathbf{b} = \mathbf{y}_0^T\mathbf{b} + \varepsilon\mathbf{u}_0^T\mathbf{b}$ and, since $\mathbf{u}_0^T\mathbf{b} = \mathbf{1}^T y > 0$, we see that as ε is increased we obtain an unbounded solution to the dual. In view of the Duality Theorem, this implies that there is no feasible solution to the primal.

Next suppose that in Step 3, for at least one j, $\mathbf{u}_0^T\mathbf{a}_j > 0$. Again we define the family of vectors $\mathbf{y}_\varepsilon = \mathbf{y}_0 + \varepsilon\mathbf{u}_0$. Since $\mathbf{u}_0$ is a solution to (4.22) we have $\mathbf{u}_0^T\mathbf{a}_i \leqslant 0$ for $i \in P$, and hence for small positive ε the vector $\mathbf{y}_\varepsilon$ is feasible for the dual. We increase ε to the first point where one of inequalities $\mathbf{y}_\varepsilon^T\mathbf{a}_j < c_j$, $j \notin P$ becomes an equality. This determines $\varepsilon_0 > 0$ and k. The new $\mathbf{y}$ vector corresponds to an increased value of the dual objective $\mathbf{y}^T\mathbf{b} = \mathbf{y}_0^T\mathbf{b} + \varepsilon\mathbf{u}_0^T\mathbf{b}$. In addition, the corresponding new set P now includes the index k. Any other index i that corresponded to a positive value of x_i in the associated restricted primal is in the new set P, because by complementary slackness $\mathbf{u}_0^T\mathbf{a}_i = 0$ for such an i and thus $\mathbf{y}^T\mathbf{a}_i = \mathbf{y}_0^T\mathbf{a}_i + \varepsilon_0\mathbf{u}_0^T\mathbf{a}_i = c_i$. This means that the old optimal solution is feasible for the new associated restricted primal and that $\mathbf{a}_k$ can be pivoted into the basis. Since $\mathbf{u}_0^T\mathbf{a}_k > 0$, pivoting in $\mathbf{a}_k$ will decrease the value of the associated restricted primal.

In summary, it has been shown that at each step either an improvement in the associated primal is made or an infeasibility condition is detected. Assuming nondegeneracy, this implies that no basis of the associated primal is repeated—and since there are only a finite number of possible bases, the solution is reached in a finite number of steps.

The primal-dual algorithm can be given an interesting interpretation in terms of the manufacturing problem in Example 2, Sect. 2.2. Suppose we own a facility that is capable of engaging in n different production activities each of which produces various amounts of m commodities. Each activity i can be operated at any level $x_i \geqslant 0$, but when operated at the unity level the ith activity costs c_i dollars and yields the m commodities in the amounts specified by the m-vector $\mathbf{a}_i$. Assuming linearity of the production facility, if we are given a vector $\mathbf{b}$ describing output requirements of the m commodities, and we wish to produce these at minimum cost, ours is the primal problem.

Imagine that an entrepreneur *not knowing* the value of our requirements vector $\mathbf{b}$ decides to sell us these requirements directly. He assigns a price vector $\mathbf{y}_0$ to these requirements such that $\mathbf{y}_0^T\mathbf{A} \leqslant \mathbf{c}$. In this way his prices are competitive with our production activities, and he can assure us that purchasing directly from him is no more costly than engaging activities. As owner of the production facilities we are reluctant to abandon our production enterprise but, on the other hand, we deem it not frugal to engage an activity whose output can be duplicated by direct purchase for lower cost. Therefore, we decide to engage only activities that cannot be duplicated cheaper, and at the same time we attempt to minimize the total business volume given the entrepreneur. Ours is the associated restricted primal problem.

Upon receiving our order, the greedy entrepreneur decides to modify his prices in such a manner as to keep them competitive with our activities but increase the cost of our order. As a reasonable and simple approach he seeks new prices of the form

$$\mathbf{y} = \mathbf{y}_0 + \varepsilon \mathbf{u}_0,$$

where he selects $\mathbf{u}_0$ as the solution to

$$\begin{aligned} \text{maximize} \quad & \mathbf{u}^T \mathbf{y} \\ \text{subject to} \quad & \mathbf{u}^T \mathbf{a}_i \leqslant \mathbf{0}, \quad i \in P \\ & \mathbf{u} \leqslant \mathbf{1}. \end{aligned}$$

The first set of constraints is to maintain competitiveness of his new price vector for small ε, while the second set is an arbitrary bound imposed to keep this subproblem bounded. It is easily shown that the solution $\mathbf{u}_0$ to this problem is identical to the solution of the associated dual (4.22). After determining the maximum ε to maintain feasibility, he announces his new prices.

At this point, rather than concede to the price adjustment, we recalculate the new minimum volume order based on the new prices. As the greedy (and shortsighted) entrepreneur continues to change his prices in an attempt to maximize profit he eventually finds he has reduced his business to zero! At that point we have, with his help, solved the original primal problem.

Example. To illustrate the primal-dual method and indicate how it can be implemented through use of the tableau format consider the following problem:

$$\begin{aligned} \text{minimize} \quad & 2x_1 + x_2 + 4x_3 \\ \text{subject to} \quad & x_1 + x_2 + 2x_3 = 3 \\ & 2x_1 + x_2 + 3x_3 = 5 \\ & x_1 \geqslant 0, \quad x_2 \geqslant 0, \quad x_3 \geqslant 0. \end{aligned}$$

Because all of the coefficients in the objective function are nonnegative, $\mathbf{y} = (0, 0)$ is a feasible vector for the dual. We lay out the simplex tableau shown below

	$\mathbf{a}_1$	$\mathbf{a}_2$	$\mathbf{a}_3$	·	·	$\mathbf{b}$
	1	1	2	1	0	3
	2	1	3	0	1	5
	−3	−2	−5	0	0	−8
$c_i - \mathbf{y}^T \mathbf{a}_i \rightarrow$	2	1	4	·	·	·

First tableau

To form this tableau we have adjoined artificial variables in the usual manner. The third row gives the relative cost coefficients of the associated primal problem—the same as the row that would be used in a phase I procedure. In the fourth row are listed the $c_i - \mathbf{y}^T \mathbf{a}_i$'s for the current $\mathbf{y}$. The allowable columns in the associated restricted primal are determined by the zeros in this last row.

Since there are no zeros in the last row, no progress can be made in the associated restricted primal and hence the original solution $x_1 = x_2 = x_3 = 0$, $y_1 = 3$, $y_2 = 5$ is optimal for this $\mathbf{y}$. The solution $\mathbf{u}_0$ to the associated restricted dual is $\mathbf{u}_0 = (1, 1)$, and the numbers $-\mathbf{u}_0^T \mathbf{a}_i$, $i = 1, 2, 3$ are equal to the first three elements in the third row.

Thus, we compute the three ratios $\frac{2}{3}, \frac{1}{2}, \frac{4}{5}$ from which we find $\varepsilon_0 = \frac{1}{2}$. The new values for the fourth row are now found by adding ε_0 times the (first three) elements of the third row to the fourth row.

$\mathbf{a}_1$	$\mathbf{a}_2$	$\mathbf{a}_3$	·	·	$\mathbf{b}$
1	①	2	1	0	3
2	1	3	0	1	5
−3	−2	−5	0	0	−8
1/2	0	3/2	·	·	·

Second tableau

Minimizing the new associated restricted primal by pivoting as indicated we obtain

$\mathbf{a}_1$	$\mathbf{a}_2$	$\mathbf{a}_3$	·	·	$\mathbf{b}$
1	1	2	1	0	3
1	0	1	−1	1	2
−1	0	−1	2	0	−2
−1/2	0	3/2	·	·	·

Now we again calculate the ratios $\frac{1}{2}, \frac{3}{2}$ obtaining $\varepsilon_0 = \frac{1}{2}$, and add this multiple of the third row to the fourth row to obtain the next tableau.

$\mathbf{a}_1$	$\mathbf{a}_2$	$\mathbf{a}_3$	·	·	$\mathbf{b}$
1	1	2	1	0	3
①	0	1	−1	1	2
−1	0	−1	2	0	−2
0	0	1	·	·	·

Third tableau

optimizing the new restricted primal we obtain the tableau:

$\mathbf{a}_1$	$\mathbf{a}_2$	$\mathbf{a}_3$	·	·	$\mathbf{b}$
0	1	1	2	−1	1
1	0	1	−1	1	2
0	0	0	1	1	0
0	0	1	·	·	·

Final tableau

Having obtained feasibility in the primal, we conclude that the solution is also optimal: $x_1 = 2,\ x_2 = 1,\ x_3 = 0$.

4.8 Summary

There is a corresponding dual linear program associated with every (primal) linear program. Both programs share the same underlying cost and constraint coefficients. We have demonstrated rich theorems to relate the pair. The variables of the dual problem can be interpreted as prices associated with the constraints of the original (primal) problem, and through this association it is possible to give an economically meaningful characterization to the dual whenever there is such a characterization for the primal.

Mathematically, the pair also establish an optimality certificate to each other: one cannot claim an optimal objective value unless you find an solution for the dual to achieve the same value of the dual objective. This also leads to the set of optimality conditions, including the complementarity conditions, that we would see many times in the rest of the book.

4.9 Exercises

1. Verify in detail that the dual of a linear program is the original problem.
2. Show that if a linear inequality in a linear program is changed to equality, the corresponding dual variable becomes free.
3. Find the dual of
$$\begin{array}{ll} \text{minimize} & \mathbf{c}^T\mathbf{x} \\ \text{subject to} & \mathbf{Ax} = \mathbf{b},\ \mathbf{x} \geqslant \mathbf{a} \\ \text{where} & \mathbf{a} \geqslant \mathbf{0}. \end{array}$$
4. Show that in the transportation problem the linear equality constraints are not linearly independent, and that in an optimal solution to the dual problem the dual variables are not unique. Generalize this observation to any linear program having redundant equality constraints.
5. Construct an example of a primal problem that has no feasible solutions and whose corresponding dual also has no feasible solutions.
6. Let $\mathbf{A}$ be an $m \times n$ matrix and $\mathbf{b}$ be an n-vector. Prove that $\mathbf{Ax} \leqslant \mathbf{0}$ implies $\mathbf{c}^T\mathbf{x} \leqslant \mathbf{0}$ if and only if $\mathbf{c}^T = \mathbf{y}^T\mathbf{A}$ for some $\mathbf{y} \geqslant \mathbf{0}$. Give a geometric interpretation of the result.
7. There is in general a strong connection between the theories of optimization and free competition, which is illustrated by an idealized model of activity location. Suppose there are n economic activities (various factories, homes, stores, etc.) that are to be individually located on n distinct parcels of land. If activity i is located on parcel j that activity can yield s_{ij} units (dollars) of value.
If the assignment of activities to land parcels is made by a central authority, it might be made in such a way as to maximize the total value generated. In other words, the assignment would be made so as to maximize $\sum_i \sum_j s_{ij}x_{ij}$ where
$$x_{ij} = \begin{cases} 1 & \text{if activity } i \text{ is assigned to parcel } j \\ 0 & \text{otherwise} \end{cases}$$

More explicitly this approach leads to the optimization problem

$$\begin{aligned}
\text{maximize } & \sum_i \sum_j s_{ij} x_{ij} \\
\text{subject to } & \sum_j x_{ij} = 1, \; i = 1, 2, \ldots, n \\
& \sum_i x_{ij} = 1, \; j = 1, 2, \ldots, n \\
& x_{ij} \geqslant 0, \quad x_{ij} = 0 \text{ or } 1.
\end{aligned}$$

Actually, it can be shown that the final requirement ($x_{ij} = 0$ or 1) is automatically satisfied at any extreme point of the set defined by the other constraints, so that in fact the optimal assignment can be found by using the simplex method of linear programming.

If one considers the problem from the viewpoint of free competition, it is assumed that, rather than a central authority determining the assignment, the individual activities bid for the land and thereby establish prices.

(a) Show that there exists a set of activity prices p_i, $i = 1, 2, \ldots, n$ and land prices q_j, $j = 1, 2, \ldots, n$ such that

$$p_i + q_j \geqslant s_{ij}, \quad i = 1, 2, \ldots, n, \quad j = 1, 2, \ldots, n$$

with equality holding if in an optimal assignment activity i is assigned to parcel j.

(b) Show that Part (a) implies that if activity i is optimally assigned to parcel j and if j' is any other parcel

$$s_{ij} - q_j \geqslant s_{ij'} - q_{j'}.$$

Give an economic interpretation of this result and explain the relation between free competition and optimality in this context.

(c) Assuming that each s_{ij} is positive, show that the prices can all be assumed to be nonnegative.

8. Construct the dual of the combinatorial auction problem of Example 7 of Chap. 2, and give an economical interpretation for each type of the dual variables.
9. *Game theory* is in part related to linear programming theory. Consider the game in which player X may select any one of m moves, and player Y may select any one of n moves. If X selects i and Y selects j, then X wins an amount a_{ij} from Y. The game is repeated many times. Player X develops a *mixed* strategy where the various moves are played according to probabilities represented by the components of the vector $\mathbf{x} = (x_1, x_2, \ldots, x_m)$, where $x_1 \geqslant 0$, $i = 1, 2, \ldots, m$ and $\sum_{i=1}^{m} x_i = 1$. Likewise Y develops a mixed strategy $\mathbf{y} = (y_1, y_2, \ldots, y_n)$, where $y_i \geqslant 0$, $i = 1, 2, \ldots, n$ and $\sum_{i=1}^{n} y_i = 1$. The average payoff to X is then $\mathbf{P}(\mathbf{x}, \mathbf{y}) = \mathbf{x}^T \mathbf{A} \mathbf{y}$.

(a) Suppose X selects $\mathbf{x}$ as the solution to the linear program

$$\begin{aligned} &\text{maximize } \mathbf{A} \\ &\text{subject to } \sum_{i=1}^{m} x_i = 1 \\ &\qquad\qquad \sum_{i=1}^{m} x_i a_{ij} \geqslant A, \; j = 1, 2, \ldots, n \\ &\qquad\qquad x_i \geqslant 0, \qquad i = 1, 2, \ldots, m. \end{aligned}$$

Show that X is guaranteed a payoff of at least A no matter what $\mathbf{y}$ is chosen by Y.

(b) Show that the dual of the problem above is

$$\begin{aligned} &\text{minimize } B \\ &\text{subject to } \sum_{j=1}^{n} y_j = 1 \\ &\qquad\qquad \sum_{j=1}^{n} a_{ij} y_j \leqslant B, \; i = 1, 2, \ldots, m \\ &\qquad\qquad y_j \geqslant 0, \qquad j = 1, 2, \ldots, n. \end{aligned}$$

(c) Prove that $\max A = \min B$. (The common value is called the *value* of the game.)

(d) Consider the "matching" game. Each player selects heads or tails. If the choices match, X wins \$1 from Y; if they do not match, Y wins \$1 from X. Find the value of this game and the optimal mixed strategies.

(e) Repeat Part (d) for the game where each player selects either 1, 2, or 3. The player with the highest number wins \$1 unless that number is exactly 1 higher than the other player's number, in which case he loses \$3. When the numbers are equal there is no payoff.

10. Consider the primal linear program in the standard form. Suppose that this program and its dual are feasible. Let $\mathbf{y}$ be a known optimal solution to the dual.

(a) If the kth equation of the primal is multiplied by $\mu \neq 0$, determine an optimal solution $\mathbf{w}$ to the dual of this new problem.

(b) Suppose that, in the original primal, we add μ times the kth equation to the rth equation. What is an optimal solution $\mathbf{w}$ to the corresponding dual problem?

(c) Suppose, in the original primal, we add μ times the kth row of $\mathbf{A}$ to $\mathbf{c}$. What is an optimal solution to the corresponding dual problem?

11. Consider the linear program (P) of the form

$$\begin{aligned} &\text{minimize } \mathbf{q}^T \mathbf{z} \\ &\text{subject to } \mathbf{M}\mathbf{z} \geq -\mathbf{q}, \; \mathbf{z} \geq \mathbf{0} \end{aligned}$$

in which the matrix $\mathbf{M}$ is *skew symmetric*; that is, $\mathbf{M} = -\mathbf{M}^T$.

(a) Show that problem (P) and its dual are the same.

(b) A problem of the kind in part (a) is said to be *self-dual*. An example of a self-dual problem has

$$\mathbf{M} = \begin{bmatrix} \mathbf{0} & -\mathbf{A}^T \\ \mathbf{A} & \mathbf{0} \end{bmatrix}, \quad \mathbf{q} = \begin{bmatrix} \mathbf{c} \\ -\mathbf{b} \end{bmatrix}, \quad \mathbf{z} = \begin{bmatrix} \mathbf{x} \\ \mathbf{y} \end{bmatrix}.$$

Give an interpretation of the problem with this data.

(c) Show that a self-dual linear program has an optimal solution if and only if it is feasible.

12. A company may manufacture n different products, each of which uses various amounts of m limited resources. Each unit of product i yields a profit of c_i dollars and uses a_{ji} units of the jth resource. The available amount of the jth resource is b_j. To maximize profit the company selects the quantities x_i to be manufactured of each product by solving

$$\begin{array}{ll} \text{maximize } \mathbf{c}^T\mathbf{x} \\ \text{subject to } \mathbf{Ax} = \mathbf{b},\ \mathbf{x} \geqslant \mathbf{0}. \end{array}$$

The unit profits c_i already take into account the variable cost associated with manufacturing each unit. In addition to that cost, the company incurs a fixed overhead H, and for accounting purposes it wants to allocate this overhead to each of its products. In other words, it wants to adjust the unit profits so as to account for the overhead. Such an overhead allocation scheme must satisfy two conditions: (4.1) Since H is fixed regardless of the product mix, the overhead allocation scheme must not alter the optimal solution, (4.2) All the overhead must be allocated; that is, the optimal value of the objective with the modified cost coefficients must be H dollars lower than z—the original optimal value of the objective.

(a) Consider the allocation scheme in which the unit profits are modified according to $\hat{\mathbf{c}}^T = \mathbf{c}^T - r\mathbf{y}_0^T\mathbf{A}$, where $\mathbf{y}_0$ is the optimal solution to the original dual and $r = H/z_0$ (assume $H \leqslant z_0$).

  (i) Show that the optimal $\mathbf{x}$ for the modified problem is the same as that for the original problem, and the new dual solution is $\hat{\mathbf{y}}_0 = (1 - r)\mathbf{y}_0$.

  (ii) Show that this approach fully allocates H.

(b) Suppose that the overhead can be traced to each of the resource constraints. Let $H_i \geqslant 0$ be the amount of overhead associated with the ith resource, where $\sum_{i=1}^{m} H_i \leqslant z_0$ and $r_i = H_i/b_i \leqslant \lambda_i^0$ for $i = 1, \ldots, m$. Based on this information, an allocation scheme has been proposed where the unit profits are modified such that $\hat{\mathbf{c}}^T = \mathbf{c}^T - \mathbf{r}^T\mathbf{A}$.

  (i) Show that the optimal $\mathbf{x}$ for this modified problem is the same as that for the original problem, and the corresponding dual solution is $\hat{\mathbf{y}}_0 = \mathbf{y}_0 - \mathbf{r}$.

  (ii) Show that this scheme fully allocates H.

13. Solve the linear inequalities

$$\begin{aligned} -2x_1 + 2x_2 &\leqslant -1 \\ 2x_1 - x_2 &\leqslant 2 \\ -4x_2 &\leqslant 3 \\ -15x_1 - 12x_2 &\leqslant -2 \\ 12x_1 + 20x_2 &\leqslant -1. \end{aligned}$$

Note that x_1 and x_2 are *not* restricted to be positive. Solve this problem by considering the problem of maximizing $0 \cdot x_1 + 0 \cdot x_2$ subject to these constraints, taking the dual and using the simplex method.

14. (a) Using the simplex method solve

$$\begin{aligned} \text{minimize} \quad & 2x_1 - x_2 \\ \text{subject to} \quad & 2x_1 - x_2 - x_3 \geqslant 3 \\ & x_1 - x_2 + x_3 \geqslant 2 \\ & x_i \geqslant 0, \; i = 1, 2, 3. \end{aligned}$$

(*Hint*: Note that $x_1 = 2$ gives a feasible solution.)

(b) What is the dual problem and its optimal solution?

15. (a) Using the simplex method solve

$$\begin{aligned} \text{minimize} \quad & 2x_1 + 3x_2 + 2x_3 + 2x_4 \\ \text{subject to} \quad & x_1 + 2x_2 + x_3 + 2x_4 = 3 \\ & x_1 + x_2 + 2x_3 + 4x_4 = 5 \\ & x_i \geqslant 0, \; i = 1, 2, 3, 4. \end{aligned}$$

(b) Using the work done in Part (a) and the dual simplex method, solve the same problem but with the right-hand sides of the equations changed to 8 and 7 respectively.

16. For the problem

$$\begin{aligned} \text{minimize} \quad & 5x_1 + 3x_2 \\ \text{subject to} \quad & 2x_1 - x_2 + 4x_3 \leqslant 4 \\ & x_1 + x_2 + 2x_3 \leqslant 5 \\ & 2x_1 - x_2 + x_3 \geqslant 1 \\ & x_1 \geqslant 0, \quad x_2 \geqslant 0, \quad x_3 \geqslant 0; \end{aligned}$$

(a) Using a single pivot operation with pivot element 1, find a feasible solution.
(b) Using the simplex method, solve the problem.
(c) What is the dual problem?
(d) What is the solution to the dual?

17. Solve the following problem by the dual simplex method:

$$\begin{array}{rl} \text{minimize} & -7x_1 + 7x_2 - 2x_3 - x_4 - 6x_5 \\ \text{subject to} & 3x_1 - x_2 + x_3 - 2x_4 = -3 \\ & 2x_1 + x_2 + x_4 + x_5 = 4 \\ & -x_1 + 3x_2 - 3x_4 + x_6 = 12 \\ \text{and} & x_i \geqslant 0, i = 1, \ldots, 6. \end{array}$$

18. Given the linear programming problem in standard form (4.3) suppose a basis $\mathbf{B}$ and the corresponding (not necessarily feasible) primal and dual basic solutions $\mathbf{x}$ and $\mathbf{y}$ are known. Assume that at least one relative cost coefficient $c_i - \mathbf{y}^T\mathbf{a}_i$ is negative. Consider the auxiliary problem

$$\begin{array}{rl} \text{minimize} & \mathbf{c}^T\mathbf{x} \\ \text{subject to} & \mathbf{A}\mathbf{x} = \mathbf{b} \\ & \sum_{i \in T} x_i + y = M \\ & \mathbf{x} \geqslant \mathbf{0}, \; y \geqslant 0, \end{array}$$

where $T = \{i : c_i - \mathbf{y}^T\mathbf{a}_i < 0\}$, y is a slack variable, and M is a large positive constant. Show that if k is the index corresponding to the most negative relative cost coefficient in the original solution, then $(\mathbf{y}, \; c_k - \mathbf{y}^T\mathbf{a}_k)$ is dual feasible for the auxiliary problem. Based on this observation, develop a big-M artificial constraint method for the dual simplex method. (Refer to Exercise 24, Chap. 3.)

19. A textile firm is capable of producing three products—x_1, x_2, x_3. Its production plan for next month must satisfy the constraints

$$\begin{array}{c} x_1 + 2x_2 + 2x_3 \leqslant 12 \\ 2x_1 + 4x_2 + x_3 \leqslant f \\ x_1 \geqslant 0, \quad x_2 \geqslant 0, \quad x_3 \geqslant 0. \end{array}$$

The first constraint is determined by equipment availability and is fixed. The second constraint is determined by the availability of cotton. The net profits of the products are 2, 3, and 3, respectively, exclusive of the cost of cotton and fixed costs.

(a) Find the shadow price λ_2 of the cotton input as a function of f. (*Hint*: Use the dual simplex method.) Plot $\lambda_2(f)$ and the net profit $z(f)$ exclusive of the cost for cotton.
(b) The firm may purchase cotton on the open market at a price of 1/6. However, it may acquire a limited amount at a price of 1/12 from a major supplier that it purchases from frequently. Determine the net profit of the firm $\pi(s)$ as a function of s.

20. A certain telephone company would like to determine the maximum number of long-distance calls from Westburgh to Eastville that it can handle at any one time. The company has cables linking these cities via several intermediary cities as follows:

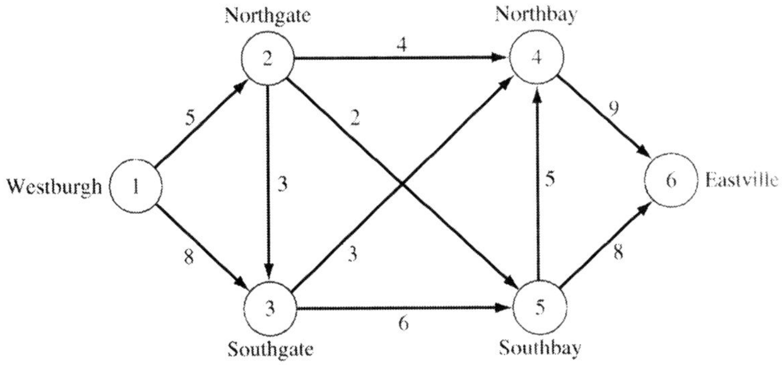

Each cable can handle a maximum number of calls simultaneously as indicated in the figure. For example, the number of calls routed from Westburgh to Northgate cannot exceed five at any one time. A call from Westburgh to Eastville can be routed through any other city, as long as there is a cable available that is not currently being used to its capacity. In addition to determining the maximum number of calls from Westburgh to Eastville, the company would, of course, like to know the optimal routing of these calls. Assume calls can be routed only in the directions indicated by the arrows.

(a) Formulate the above problem as a linear programming problem with upper bounds.
(*Hint*: Denote by x_{ij} the number of calls routed from city i to city j.)
(b) Find the solution by inspection of the graph.

21. Apply the maximal flow algorithm to the network below. All arcs have capacity 1 unless otherwise indicated.

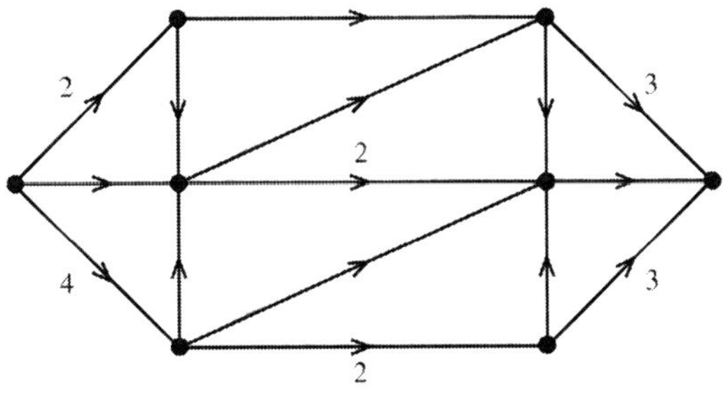

22. Consider the problem

$$\begin{array}{ll} \text{minimize} & 2x_1 + x_2 + 4x_3 \\ \text{subject to} & x_1 + x_2 + 2x_3 = 3 \\ & 2x_1 + x_2 + 3x_3 = 5 \\ & x_i \geqslant 0, \quad x_2 \geqslant 0, \quad x_3 \geqslant 0. \end{array}$$

(a) What is the dual problem?
(b) Note that $\mathbf{y} = (1, 0)$ is feasible for the dual. Starting with this $\mathbf{y}$, solve the primal using the primal-dual algorithm.

23. Show that in the associated restricted dual of the primal-dual method the objective $\mathbf{y}^T\mathbf{b}$ can be replaced by $\mathbf{y}^T\mathbf{y}$.
24. Consider the primal feasible region in standard form $\mathbf{Ax} = \mathbf{b}$, $\mathbf{x} \geqslant \mathbf{0}$, where $\mathbf{A}$ is an $m \times n$ matrix, $\mathbf{b}$ is a constant nonzero m-vector, and $\mathbf{x}$ is a variable n-vector.

(a) A variable x_i is said to be a *null variable* if $x_i = 0$ in every feasible solution. Prove that, if the feasible region is non-empty, x_i is a null variable if and only if there is a nonzero vector $\mathbf{y} \in E^m$ such that $\mathbf{y}^T\mathbf{A} \geq \mathbf{0}$, $\mathbf{y}^T\mathbf{b} = 0$. and the *ith* component of $\mathbf{y}^T\mathbf{A}$ is strictly positive.
(b) *Strict complementarity* Let the feasible region be nonempty. Then there is a feasible $\mathbf{x}$ and vector $\mathbf{y} \in E^m$ such that

$$\mathbf{y}^T\mathbf{A} \geq \mathbf{0},\ \mathbf{y}^T\mathbf{b} = 0,\ \mathbf{y}^T\mathbf{b} + \mathbf{x} > \mathbf{0}.$$

(c) A variable x_i is a *nonextremal variable* if $x_i > 0$ in every feasible solution. Prove that, if the feasible region is non-empty, x_i is a nonextremal variable if and only if there is $\mathbf{y} \in E^m$ and $\mathbf{d} \in E^n$ such that $\mathbf{y}^T\mathbf{A} = \mathbf{d}^T$, where $d_i = -1$, $d_j \geqslant 0$ for $j \neq i$; and such that $\mathbf{y}^T\mathbf{b} < 0$.

References

4.1–4.4 Again most of the material in this chapter is now quite standard. See the references of Chap. 2. A particularly careful discussion of duality can be found in Simonnard [S6].

4.5 Koopmans [K8] was the first to discover the relationship between bases and tree structures in a network. The classic reference for network flow theory is Ford and Fulkerson [F13]. For discussion of even more efficient versions of the maximal flow algorithm, see Lawler [L2] and Papadimitriou and Steiglitz [P2]. The Hungarian method for the assignment problem was designed by Kuhn [K10]. It is called the Hungarian method because it was based on work by the Hungarian mathematicians Egerváry and König. Ultimately, this led to the general primal–dual algorithm for linear programming.

4.6 The dual simplex method is due to Lemke [L4].

4.7 The general primal-dual algorithm is due to Dantzig, Ford and Fulkerson [D7]. See also Ford and Fulkerson [F13]. The economic interpretation given in this section is apparently novel.
The concepts of reduction are due to Shefi [S5], who has developed a complete theory in this area. For more details along the lines presented here, see Luenberger [L15].

Chapter 5
Interior-Point Methods

Linear programs can be viewed in two somewhat complementary ways. They are, in one view, a class of continuous optimization problems each with continuous variables defined on a convex feasible region and with a continuous objective function. They are, therefore, a special case of the general form of problem considered in this text. However, linearity implies a certain degree of degeneracy, since for example the derivatives of all functions are constants and hence the differential methods of general optimization theory cannot be directly used. From an alternative view, linear programs can be considered as a class of combinatorial problems because it is known that solutions can be found by restricting attention to the vertices of the convex polyhedron defined by the constraints. Indeed, this view is natural when considering network problems such as those of early chapters. However, the number of vertices may be large, up to $n!/m!(n-m)!$, making direct search impossible for even modest size problems.

The simplex method embodies both of these viewpoints, for it restricts attention to vertices, but exploits the continuous nature of the variables to govern the progress from one vertex to another, defining a sequence of adjacent vertices with improving values of the objective as the process reaches an optimal point. The simplex method, with ever-evolving improvements, has for five decades provided an efficient general method for solving linear programs.

Although it performs well in practice, visiting only a small fraction of the total number of vertices, a definitive theory of the simplex method's performance was unavailable. However, in 1972, Klee and Minty showed by examples that for certain linear programs the simplex method will examine every vertex f. These examples proved that in the worst case, the simplex method requires a number of steps that is exponential in the size of the problem.

D.G. Luenberger, Y. Ye, *Linear and Nonlinear Programming*, International Series in Operations Research & Management Science 228,
DOI 10.1007/978-3-319-18842-3_5

In view of this result, many researchers believed that a good algorithm, different than the simplex method, might be devised whose number of steps would be polynomial rather than exponential in the program's size—that is, the time required to compute the solution would be bounded above by a polynomial in the size of the problem.[1]

Indeed, in 1979, a new approach to linear programming, Khachiyan's ellipsoid method was announced with great acclaim. The method is quite different in structure than the simplex method, for it constructs a sequence of shrinking ellipsoids each of which contains the optimal solution set and each member of the sequence is smaller in volume than its predecessor by at least a certain fixed factor. Therefore, the solution set can be found to any desired degree of approximation by continuing the process. Khachiyan proved that the ellipsoid method, developed during the 1970s by other mathematicians, is a polynomial-time algorithm for linear programming.

Practical experience, however, was disappointing. In almost all cases, the simplex method was much faster than the ellipsoid method. However, Khachiyan's ellipsoid method showed that polynomial time algorithms for linear programming do exist. It left open the question of whether one could be found that, in practice, was faster than the simplex method.

It is then perhaps not surprising that the announcement by Karmarkar in 1984 of a new polynomial time algorithm, an interior-point method, with the potential to improve the practical effectiveness of the simplex method made front-page news in major newspapers and magazines throughout the world. It is this interior-point approach that is the subject of this chapter and the next.

This chapter begins with a brief introduction to complexity theory, which is the basis for a way to quantify the performance of iterative algorithms, distinguishing polynomial-time algorithms from others.

Next the example of Klee and Minty showing that the simplex method is not a polynomial-time algorithm in the worst case is presented. Following that the ellipsoid algorithm is defined and shown to be a polynomial-time algorithm. These two sections provide a deeper understanding of how the modern theory of linear programming evolved, and help make clear how complexity theory impacts linear programming. However, the reader may wish to consider them optional and omit them at first reading.

The development of the basics of interior-point theory begins with Sect. 5.4 which introduces the concept of barrier functions and the analytic center. Section 5.5 introduces the central path which underlies interior-point algorithms. The relations between primal and dual in this context are examined. An overview of the details of specific interior-point algorithms based on the theory are presented in Sects. 5.6 and 5.7

---

[1] We will be more precise about complexity notions such as "polynomial algorithm" in Sect. 5.1 below.

5.1 Elements of Complexity Theory

Complexity theory is arguably the foundation for analysis of computer algorithms. The goal of the theory is twofold: to develop criteria for measuring the effectiveness of various algorithms (and thus, be able to compare algorithms using these criteria), and to assess the inherent difficulty of various problems.

The term *complexity* refers to the amount of resources required by a computation. In this chapter we focus on a particular resource, namely, computing time. In complexity theory, however, one is not interested in the execution time of a program implemented in a particular programming language, running on a particular computer over a particular input. This involves too many contingent factors. Instead, one wishes to associate to an algorithm more intrinsic measures of its time requirements.

Roughly speaking, to do so one needs to define:

- a notion of *input size*,
- a set of *basic operations*, and
- a *cost* for each basic operation.

The last two allow one to associate a *cost of a computation*. If x is any input, the cost $C(x)$ of the computation with input x is the sum of the costs of all the basic operations performed during this computation.

Let $\mathcal{A}$ be an algorithm and $\mathcal{J}_n$ be the set of all its inputs having size n. The *worst-case cost function* of $\mathcal{A}$ is the function $T^w_{\mathcal{A}}$ defined by

$$T^w_{\mathcal{A}}(n) = \sup_{x \in \mathcal{J}_n} C(x).$$

If there is a probability structure on $\mathcal{J}_n$ it is possible to define the *average-case cost function* $T^a_{\mathcal{A}}$ given by

$$T^a_{\mathcal{A}}(n) = \mathrm{E}_n(C(x)).$$

where E_n is the expectation over $\mathcal{J}_n$. However, the average is usually more difficult to find, and there is of course the issue of what probabilities to assign.

We now discuss how the objects in the three items above are selected. The selection of a set of basic operations is generally easy. For the algorithms we consider in this chapter, the obvious choice is the set $\{+, -, \times, /, \leq\}$ of the four arithmetic operations and the comparison. Selecting a notion of input size and a cost for the basic operations depends on the kind of data dealt with by the algorithm. Some kinds can be represented within a fixed amount of computer memory; others require a variable amount.

Examples of the first are fixed-precision floating-point numbers, stored in a fixed amount of memory (usually 32 or 64 bits). For this kind of data the size of an element is usually taken to be 1 and consequently to have *unit size* per number.

Examples of the second are integer numbers which require a number of bits approximately equal to the logarithm of their absolute value. This (base 2) logarithm is usually referred to as the *bit size* of the integer. Similar ideas apply for rational numbers.

Let A be some kind of data and $\mathbf{x} = (x_1, \ldots, x_n) \in A^n$. If A is of the first kind above then we define size($\mathbf{x}$) = n. Otherwise, we define size($\mathbf{x}$) = $\sum_{i=1}^{n}$ bit-size (x_i).

The cost of operating on two unit-size numbers is taken to be 1 and is called the *unit cost*. In the bit-size case, the cost of operating on two numbers is the product of their bit-sizes (for multiplications and divisions) or their maximum (for additions, subtractions, and comparisons).

The consideration of integer or rational data with their associated bit size and bit cost for the arithmetic operations is usually referred to as the *Turing model of computation*. The consideration of idealized reals with unit size and unit cost is today referred as the *real number arithmetic model*. When comparing algorithms, one should make clear which model of computation is used to derive complexity bounds.

A basic concept related to both models of computation is that of *polynomial time*. An algorithm $\mathcal{A}$ is said to be a polynomial time algorithm if $T_{\mathcal{A}}^{w}(n)$ is bounded above by a polynomial. A problem can be solved in polynomial time if there is a polynomial time algorithm solving the problem. The notion of *average polynomial time* is defined similarly, replacing $T_{\mathcal{A}}^{w}$ by $T_{\mathcal{A}}^{a}$.

The notion of polynomial time is usually taken as the formalization of efficiency in complexity theory.

*5.2 *The Simplex Method Is Not Polynomial-Time

When the simplex method is used to solve a linear program in standard form with coefficient matrix $\mathbf{A} \in E^{m \times n}$, $\mathbf{b} \in E^m$ and $\mathbf{c} \in E^n$, the number of pivot steps to solve the problem starting from a basic feasible solution is typically a small multiple of m: usually between $2m$ and $3m$. In fact, Dantzig observed that for problems with $m \leq 50$ and $n \leq 200$ the number of iterations is ordinarily less than $1.5m$.

At one time researchers believed—and attempted to prove—that the simplex algorithm (or some variant thereof) always requires a number of iterations that is bounded by a polynomial expression in the problem size. That was until Victor Klee and George Minty exhibited a class of linear programs each of which requires an exponential number of iterations when solved by the conventional simplex method.

One form of the Klee–Minty example is

$$
\begin{aligned}
&\text{maximize } \sum_{j=1}^{n} 10^{n-j} x_j \\
&\text{subject to } 2\sum_{j=1}^{i-1} 10^{i-j} x_j + x_i \leq 100^{i-1} i = 1, \ldots, n \qquad (5.1) \\
&\qquad\qquad\qquad x_j \geq 0 \quad j = 1, \ldots, n
\end{aligned}
$$

The problem above is easily cast as a linear program in standard form.

A specific case is that for $n = 3$, giving

$$\begin{array}{lllll} \text{maximize} & 100x_1 + 10x_2 + x_3 & & & \\ \text{subject to} & x_1 & & \leq & 1 \\ & 20x_1 + x_2 & & \leq & 100 \\ & 200x_1 + 20x_2 + x_3 & & \leq & 10,000 \\ & x_1 \geqslant 0,\ x_2 \geqslant 0,\ x_3 \geqslant 0. & & & \end{array}$$

In this case, we have three constraints and three variables (along with their nonnegativity constraints). After adding slack variables, the problem is in standard form. The system has $m = 3$ equations and $n = 6$ nonnegative variables. It can be verified that it takes $2^3 - 1 = 7$ pivot steps to solve the problem with the simplex method when at each step the pivot column is chosen to be the one with the largest (because this a maximization problem) reduced cost. (See Exercise 1.)

The general problem of the class (1) takes $2^n - 1$ pivot steps and this is in fact the number of vertices minus one (which is the starting vertex). To get an idea of how bad this can be, consider the case where $n = 50$. We have $2^{50} - 1 \approx 10^{15}$. In a year with 365 days, there are approximately 3×10^7 s. If a computer ran continuously, performing a million pivots of the simplex algorithm per second, it would take approximately

$$\frac{10^{15}}{3 \times 10^7 \times 10^6} \approx 33 \text{ years}$$

to solve a problem of this class using the greedy pivot selection rule.

Although it is not polynomial in the worst case, the simplex method remains one of major solvers for linear programming. In fact, the method has been recently proved to be (strongly) polynomial for solving the Markov Decision Process with any fixed discount rate.

*5.3 *The Ellipsoid Method

The basic ideas of the ellipsoid method stem from research done in the 1960s and 1970s mainly in the Soviet Union (as it was then called) by others who preceded Khachiyan. In essence, the idea is to enclose the region of interest in ever smaller ellipsoids.

The significant contribution of Khachiyan was to demonstrate in that under certain assumptions, the ellipsoid method constitutes a polynomially bounded algorithm for linear programming.

The version of the method discussed here is really aimed at finding a point of a polyhedral set Ω given by a system of linear inequalities.

$$\Omega = \{\mathbf{y} \in E^m : \mathbf{y}^T\mathbf{a}_j \leq c_j,\ j = 1, \ldots n\}$$

Finding a point of Ω can be thought of as equivalent to solving a linear programming problem.

Two important assumptions are made regarding this problem:

(A1) There is a vector $\mathbf{y}_0 \in E^m$ and a scalar $R > 0$ such that the closed ball $S(\mathbf{y}_0, R)$ with center $\mathbf{y}_0$ and radius R, that is

$$\{\mathbf{y} \in E^m : |\mathbf{y} - \mathbf{y}_0| \leq R\},$$

contains Ω.

(A2) If Ω is nonempty, there is a known scalar $r > 0$ such that Ω contains a ball of the form $S(\mathbf{y}^*, r)$ with center at $\mathbf{y}^*$ and radius r. (This assumption implies that if Ω is nonempty, then it has a nonempty interior and its volume is at least $\text{vol}(S(\mathbf{0}, r))$.)[2]

Definition. An *ellipsoid* in E^m is a set of the form

$$E = \{\mathbf{y} \in E^m : (\mathbf{y} - \mathbf{z})^T \mathbf{Q}(\mathbf{y} - \mathbf{z}) \leq 1\}$$

where $\mathbf{z} \in E^m$ is a given point (called the *center*) and $\mathbf{Q}$ is a positive definite matrix (see Sect. A.4 of Appendix A) of dimension $m \times m$. This ellipsoid is denoted $E(\mathbf{z}, \mathbf{Q})$.

The unit sphere $S(\mathbf{0}, 1)$ centered at the origin $\mathbf{0}$ is a special ellipsoid with $\mathbf{Q} = \mathbf{I}$, the identity matrix.

The axes of a general ellipsoid are the eigenvectors of $\mathbf{Q}$ and the lengths of the axes are $\lambda_1^{-1/2}, \lambda_2^{-1/2}, \ldots, \lambda_m^{-1/2}$, where the λ_i's are the corresponding eigenvalues. It can be shown that the volume of an ellipsoid is

$$\text{vol}(E) = \text{vol}(S(\mathbf{0}, 1))\Pi_{i=1}^m \lambda_i^{-1/2} = \text{vol}(S(\mathbf{0}, 1)) \det(\mathbf{Q}^{-1/2}).$$

Cutting Plane and New Containing Ellipsoid

In the ellipsoid method, a series of ellipsoids E_k is defined, with centers $\mathbf{y}_k$ and with the defining $\mathbf{Q} = \mathbf{B}_k^{-1}$, where $\mathbf{B}_k$ is symmetric and positive definite.

At each iteration of the algorithm, we have $\Omega \subset E_k$. It is then possible to check whether $\mathbf{y}_k \in \Omega$. If so, we have found an element of Ω as required. If not, there is at least one constraint that is violated. Suppose $\mathbf{a}_j^T \mathbf{y}_k > c_j$. Then

$$\Omega \subset \frac{1}{2} E_k = \{\mathbf{y} \in E_k : \mathbf{a}_j^T \mathbf{y} \leq \mathbf{a}_j^T \mathbf{y}_k\}$$

This set is half of the ellipsoid, obtained by cutting the ellipsoid in half through its center (Fig. 5.1).

---

[2] The (topological) interior of any set Ω is the set of points in Ω which are the centers of some balls contained in Ω.

The successor ellipsoid E_{k+1} is defined to be the minimal-volume ellipsoid containing $(1/2)E_k$. It is constructed as follows. Define

$$\tau = \frac{1}{m+1},\ \delta = \frac{m^2}{m^2-1},\ \sigma = 2\tau.$$

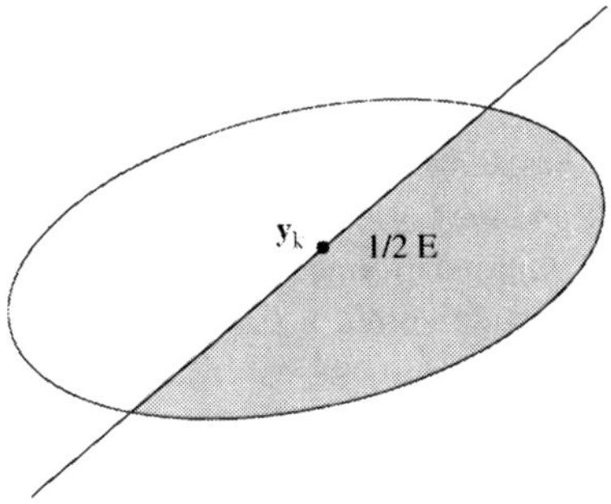

Fig. 5.1 A half-ellipsoid

Then put

$$\begin{aligned} \mathbf{y}_{k+1} &= \mathbf{y}_k - \frac{\tau}{(\mathbf{a}_j^T\mathbf{B}_k\mathbf{a}_j)^{1/2}}\mathbf{B}_k\mathbf{a}_j \\ \mathbf{B}_{k+1} &= \delta\left(\mathbf{B}_k - \sigma\frac{\mathbf{B}_k\mathbf{a}_j\mathbf{a}_j^T\mathbf{B}_k}{\mathbf{a}_j^T\mathbf{B}_k\mathbf{a}_j}\right) \end{aligned} \tag{5.2}$$

Theorem 1. *The ellipsoid $E_{k+1} = E(\mathbf{y}_{k+1}, \mathbf{B}_{k+1}^{-1})$ defined as above is the ellipsoid of least volume containing $(1/2)E_k$. Moreover,*

$$\frac{\text{vol}(E_{k+1})}{\text{vol}(E_k)} = \left(\frac{m^2}{m^2-1}\right)^{(m-1)/2}\frac{m}{m+1} < \exp\left(-\frac{1}{2(m+1)}\right) < 1.$$

Proof. We shall not prove the statement about the new ellipsoid being of least volume, since that is not necessary for the results that follow. To prove the remainder of the statement, we have

$$\frac{\text{vol}(E_{k+1})}{\text{vol}(E_k)} = \frac{\det(\mathbf{B}_{k+1}^{1/2})}{\det(\mathbf{B}_k^{1/2})}$$

For simplicity, by a change of coordinates, we may take $\mathbf{B}_k = \mathbf{I}$. Then $\mathbf{B}_{k+1}$ has $m-1$ eigenvalues equal to $\delta = \frac{m^2}{m^2-1}$ and one eigenvalue equal to $\delta - 2\delta\tau = \frac{m^2}{m^2-1}(1-\frac{2}{m+1}) = (\frac{m}{m+1})^2$. The reduction in volume is the product of the square roots of these, giving the equality in the theorem.

Then using $(1+x)^p \leqslant e^{xp}$, we have

$$\left(\frac{m^2}{m^2-1}\right)^{(m-1)/2} \frac{m}{m+1} = \left(1+\frac{1}{m^2-1}\right)^{(m-1)/2} \left(1-\frac{1}{m+1}\right)$$
$$< \exp\left(\frac{1}{2(m+1)} - \frac{1}{(m+1)}\right) = \exp\left(-\frac{1}{2(m+1)}\right). \blacksquare$$

Convergence

The ellipsoid method is initiated by selecting $\mathbf{y}_0$ and R such that condition (A1) is satisfied. Then $\mathbf{B}_0 = R^2\mathbf{I}$, and the corresponding E_0 contains Ω. The updating of the E_k's is continued until a solution is found.

Under the assumptions stated above, a single repetition of the ellipsoid method reduces the volume of an ellipsoid to one-half of its initial value in $O(m)$ iterations. (See Appendix A for O notation.) Hence it can reduce the volume to less than that of a sphere of radius r in $O(m^2 \log(R/r))$ iterations, since its volume is bounded from below by $\mathrm{vol}(S(\mathbf{0}, 1))r^m$ and the initial volume is $\mathrm{vol}(S(\mathbf{0}, 1))R^m$. Generally a single iteration requires $O(m^2)$ arithmetic operations. Hence the entire process requires $O(m^4 \log(R/r))$ arithmetic operations.[3]

Ellipsoid Method for Usual Form of LP

Now consider the linear program (where $\mathbf{A}$ is $m \times n$)

$$\text{(P)} \quad \begin{array}{ll} \text{maximize} & \mathbf{c}^T\mathbf{x} \\ \text{subject to} & \mathbf{Ax} \leq \mathbf{b},\ \mathbf{x} \geq \mathbf{0} \end{array}$$

and its dual

$$\text{(D)} \quad \begin{array}{ll} \text{minimize} & \mathbf{y}^T\mathbf{b} \\ \text{subject to} & \mathbf{y}^T\mathbf{A} \geq \mathbf{c}^T,\ \mathbf{y} \geq \mathbf{0}. \end{array}$$

Note that both problems can be solved by finding a feasible point to inequalities

$$\begin{aligned} -\mathbf{c}^T\mathbf{x} + \mathbf{b}^T\mathbf{y} &\leq 0 \\ \mathbf{Ax} &\leq \mathbf{b} \\ -\mathbf{A}^T\mathbf{y} &\leq -\mathbf{c} \\ \mathbf{x},\ \mathbf{y} &\geq \mathbf{0}, \end{aligned} \tag{5.3}$$

where both $\mathbf{x}$ and $\mathbf{y}$ are variables. Thus, the total number of arithmetic operations for solving a linear program is bounded by $O((m+n)^4 \log(R/r))$.

[3] Assumption (A2) is sometimes too strong. It has been shown, however, that when the data consists of integers, it is possible to perturb the problem so that (A2) is satisfied and if the perturbed problem has a feasible solution, so does the original Ω.

5.4 The Analytic Center

The new interior-point algorithms introduced by Karmarkar move by successive steps inside the feasible region. It is the interior of the feasible set rather than the vertices and edges that plays a dominant role in this type of algorithm. In fact, these algorithms purposely avoid the edges of the set, only eventually converging to one as a solution.

Our study of these algorithms begins in the next section, but it is useful at this point to introduce a concept that definitely focuses on the interior of a set, termed the set's analytic center. As the name implies, the center is away from the edge.

In addition, the study of the analytic center introduces a special structure, termed a *barrier* or *potential* that is fundamental to interior-point methods.

Consider a set $\mathcal{S}$ in a subset of $\mathcal{X}$ of E^n defined by a group of inequalities as

$$\mathcal{S} = \{\mathbf{x} \in \mathcal{X} : g_j(\mathbf{x}) \geqslant 0, \ j = 1, 2, \ldots, m\},$$

and assume that the functions g_j are continuous. $\mathcal{S}$ has a nonempty interior $\mathring{\mathcal{S}} = \{\mathbf{x} \in \mathcal{X} : g_j(\mathbf{x}) > 0, \text{ all } j\}$. Associated with this definition of the set is the *potential function*

$$\psi(\mathbf{x}) = -\sum_{j=1}^{m} \log g_j(\mathbf{x})$$

defined on $\mathring{\mathcal{S}}$.

The *analytic center* of $\mathcal{S}$ is the vector (or set of vectors) that minimizes the potential; that is, the vector (or vectors) that solve

$$\min \psi(\mathbf{x}) = \min \left\{ -\sum_{j=1}^{m} \log g_j(\mathbf{x}) : \mathbf{x} \in \mathcal{X}, \ g_j(\mathbf{x}) > 0 \text{ for each } j \right\}.$$

Example 1 (A Cube). Consider the set $\mathcal{S}$ defined by $x_i \geqslant 0$, $(1 - x_i) \geqslant 0$, for $i = 1, 2, \ldots, n$. This is $\mathcal{S} = [0, 1]^n$, the unit cube in E^n. The analytic center can be found by differentiation to be $x_i = 1/2$, for all i. Hence, the analytic center is identical to what one would normally call the center of the unit cube.

In general, the analytic center depends on how the set is defined—on the particular inequalities used in the definition. For instance, the unit cube is also defined by the inequalities $x_i \geqslant 0$, $(1 - x_i)^d \geqslant 0$ with odd $d > 1$. In this case the solution is $x_i = 1/(d+1)$ for all i. For large d this point is near the inner corner of the unit cube.

Also, the addition of redundant inequalities can change the location of the analytic center. For example, repeating a given inequality will change the center's location.

There are several sets associated with linear programs for which the analytic center is of particular interest. One such set is the feasible region itself. Another is the set of optimal solutions. There are also sets associated with dual and primal-dual formulations. All of these are related in important ways.

Let us illustrate by considering the analytic center associated with a bounded polytope Ω in E^m represented by $n(> m)$ linear inequalities; that is,

$$\Omega = \{\mathbf{y} \in E^m : \mathbf{c}^T - \mathbf{y}^T\mathbf{A} \geqslant \mathbf{0}\},$$

where $\mathbf{A} \in E^{m \times n}$ and $\mathbf{c} \in E^n$ are given and $\mathbf{A}$ has rank m. Denote the interior of Ω by

$$\mathring{\Omega} = \{\mathbf{y} \in E^m : \mathbf{c}^T - \mathbf{y}^T\mathbf{A} > \mathbf{0}\}.$$

The potential function for this set is

$$\psi_\Omega(\mathbf{y}) \equiv -\sum_{j=1}^{n} \log(c_j - \mathbf{y}^T\mathbf{a}_j) = -\sum_{j=1}^{n} \log s_j, \tag{5.4}$$

where $\mathbf{s} \equiv \mathbf{c} - \mathbf{A}^T\mathbf{y}$ is a *slack vector*. Hence the potential function is the negative sum of the logarithms of the slack variables.

The analytic center of Ω is the interior point of Ω that minimizes the potential function. This point is denoted by $\mathbf{y}^a$ and has the associated $\mathbf{s}^a = \mathbf{c} - \mathbf{A}^T\mathbf{y}^a$. The pair $(\mathbf{y}^a, \mathbf{s}^a)$ is uniquely defined, since the potential function is strictly convex (see Sect. 7.4) in the bounded convex set Ω.

Setting to zero the derivatives of $\psi(\mathbf{y})$ with respect to each y_i gives

$$\sum_{j=1}^{n} \frac{a_{ij}}{c_j - \mathbf{y}^T\mathbf{a}_j} = 0, \text{ for all } i.$$

which can be written

$$\sum_{j=1}^{n} \frac{a_{ij}}{s_j} = 0, \text{ for all } i.$$

Now define $x_j = 1/s_j$ for each j. We introduce the notation

$$\mathbf{x} \circ \mathbf{s} \equiv (x_1 s_1,\ x_2 s_2,\ \ldots,\ x_n s_n)^T,$$

which is *component multiplication*. Then the analytic center is defined by the conditions

$$\begin{aligned} \mathbf{x} \circ \mathbf{s} &= \mathbf{1} \\ \mathbf{A}\mathbf{x} &= \mathbf{0} \\ \mathbf{A}^T\mathbf{y} + \mathbf{s} &= \mathbf{c}. \end{aligned}$$

The analytic center can be defined when the interior is empty or equalities are present, such as

$$\Omega = \{\mathbf{y} \in E^m : \mathbf{c}^T - \mathbf{y}^T\mathbf{A} \geqslant \mathbf{0},\ \mathbf{B}\mathbf{y} = \mathbf{b}\}.$$

In this case the analytic center is chosen on the linear surface $\{\mathbf{y} : \mathbf{B}\mathbf{y} = \mathbf{b}\}$ to maximize the product of the slack variables $\mathbf{s} = \mathbf{c} - \mathbf{A}^T\mathbf{y}$. Thus, in this context the interior of Ω refers to the interior of the positive orthant of slack variables: $R^n_+ \equiv \{\mathbf{s} : \mathbf{s} \geqslant \mathbf{0}\}$. This definition of interior depends only on the region of the slack variables. Even if there is only a single point in Ω with $\mathbf{s} = \mathbf{c} - \mathbf{A}^T\mathbf{y}$ for some $\mathbf{y}$ where $\mathbf{By} = \mathbf{b}$ with $\mathbf{s} > \mathbf{0}$, we still say that $\mathring{\Omega}$ is not empty.

5.5 The Central Path

The concept underlying interior-point methods for linear programming is to use nonlinear programming techniques of analysis and methodology. The analysis is often based on differentiation of the functions defining the problem. Traditional linear programming does not require these techniques since the defining functions are linear. Duality in general nonlinear programs is typically manifested through Lagrange multipliers (which are called dual variables in linear programming). The analysis and algorithms of the remaining sections of the chapter use these nonlinear techniques. These techniques are discussed systematically in later chapters, so rather than treat them in detail at this point, these current sections provide only minimal detail in their application to linear programming. It is expected that most readers are already familiar with the basic method for minimizing a function by setting its derivative to zero, and for incorporating constraints by introducing Lagrange multipliers. These methods are discussed in detail in Chaps. 11–15.

The computational algorithms of nonlinear programming are typically iterative in nature, often characterized as search algorithms. At any step with a given point, a direction for search is established and then a move in that direction is made to define the next point. There are many varieties of such search algorithms and they are systematically presented throughout the text. In this chapter, we use versions of Newton's method as the search algorithm, but we postpone a detailed study of the method until later chapters.

Not only have nonlinear methods improved linear programming, but interior-point methods for linear programming have been extended to provide new approaches to nonlinear programming. This chapter is intended to show how this merger of linear and nonlinear programming produces elegant and effective methods. These ideas take an especially pleasing form when applied to linear programming. Study of them here, even without all the detailed analysis, should provide good intuitive background for the more general manifestations.

Consider a primal linear program in standard form

$$\begin{aligned} \text{(LP) minimize} \quad & \mathbf{c}^T\mathbf{x} \\ \text{subject to} \quad & \mathbf{Ax} = \mathbf{b},\ \mathbf{x} \geqslant \mathbf{0}. \end{aligned} \tag{5.5}$$

We denote the feasible region of this program by $\mathcal{F}_p$. We assume that $\mathring{\mathcal{F}}_p = \{\mathbf{x} : \mathbf{Ax} = \mathbf{b},\ \mathbf{x} > \mathbf{0}\}$ is nonempty and the optimal solution set of the problem is bounded.

Associated with this problem, we define for $\mu \geqslant 0$ the *barrier problem*

$$\text{(BP)} \quad \text{minimize } \mathbf{c}^T\mathbf{x} - \mu \sum_{j=1}^{n} \log x_j \tag{5.6}$$
$$\text{subject to } \mathbf{Ax} = \mathbf{b}, \ \mathbf{x} > \mathbf{0}.$$

It is clear that $\mu = 0$ corresponds to the original problem (5.5). As $\mu \to \infty$, the solution approaches the analytic center of the feasible region (when it is bounded), since the barrier term swamps out $\mathbf{c}^T\mathbf{x}$ in the objective. As μ is varied continuously toward 0, there is a path $\mathbf{x}(\mu)$ defined by the solution to (BP). This path $\mathbf{x}(\mu)$ is termed the *primal central path*. As $\mu \to 0$ this path converges to the analytic center of the optimal face $\{\mathbf{x} : \mathbf{c}^T\mathbf{x} = z^*, \ \mathbf{Ax} = \mathbf{b}, \ \mathbf{x} \geqslant \mathbf{0}\}$, where z^* is the optimal value of (LP).

A strategy for solving (LP) is to solve (BP) for smaller and smaller values of μ and thereby approach a solution to (LP). This is indeed the basic idea of interior-point methods.

At any $\mu > 0$, under the assumptions that we have made for problem (5.5), the necessary and sufficient conditions for a unique and bounded solution are obtained by introducing a *Lagrange multiplier* vector $\mathbf{y}$ for the linear equality constraints to form the *Lagrangian* (see Chap. 11)

$$\mathbf{c}^T\mathbf{x} - \mu \sum_{j=1}^{n} \log x_j - \mathbf{y}^T(\mathbf{Ax} - \mathbf{b}).$$

The derivatives with respect to the x_j's are set to zero, leading to the conditions

$$c_j - \mu/x_j - \mathbf{y}^T\mathbf{a}_j = 0, \ \text{for each } j$$

or equivalently

$$\mu\mathbf{X}^{-1}\mathbf{1} + \mathbf{A}^T\mathbf{y} = \mathbf{c} \tag{5.7}$$

where as before $\mathbf{a}_j$ is the jth column of $\mathbf{A}$, $\mathbf{1}$ is the vector of 1's, and $\mathbf{X}$ is the diagonal matrix whose diagonal entries are the components of $\mathbf{x} > \mathbf{0}$. Setting $s_j = \mu/x_j$ the complete set of conditions can be rewritten

$$\begin{aligned} \mathbf{x} \circ \mathbf{s} &= \mu\mathbf{1} \\ \mathbf{Ax} &= \mathbf{b} \\ \mathbf{A}^T\mathbf{y} + \mathbf{s} &= \mathbf{c}. \end{aligned} \tag{5.8}$$

Note that $\mathbf{y}$ is a dual feasible solution and $\mathbf{c} - \mathbf{A}^T\mathbf{y} > \mathbf{0}$ (see Exercise 4).

Example 2 (A Square Primal). Consider the problem of maximizing x_1 within the unit square $\mathcal{S} = [0, \ 1]^2$. The problem is formulated as

$$\begin{array}{lll} \min & -x_1 & \\ \text{s.t.} & x_1 + x_3 = 1 & \\ & & x_2 + x_4 = 1 \\ & x_1 \geqslant 0,\ x_2 \geqslant 0,\ x_3 \geqslant 0,\ x_4 \geqslant 0. & \end{array}$$

Here x_3 and x_4 are slack variables for the original problem to put it in standard form. The optimality conditions for $\mathbf{x}(\mu)$ consist of the original two linear constraint equations and the four equations

$$y_1 + s_1 = -1,\ y_2 + s_2 = 0,\ y_1 + s_3 = 0,\ y_2 + s_4 = 0$$

together with the relations $s_i = \mu/x_i$ for $i = 1, 2\ldots, 4$. These equations are readily solved with a series of elementary variable eliminations to find

$$x_1(\mu) = \frac{1 - 2\mu \pm \sqrt{1 + 4\mu^2}}{2}$$
$$x_2(\mu) = 1/2.$$

Using the "+" solution, it is seen that as $\mu \to 0$ the solution goes to $\mathbf{x} \to (1, 1/2)$. Note that this solution is not a corner of the cube. Instead it is at the analytic center of the optimal face $\{\mathbf{x} : x_1 = 1,\ 0 \leqslant x_2 \leqslant 1\}$. See Fig. 5.2. The limit of $\mathbf{x}(\mu)$ as $\mu \to \infty$ can be seen to be the point (1/2, 1/2). Hence, the central path in this case is a straight line progressing from the analytic center of the square (at $\mu \to \infty$) to the analytic center of the optimal face (at $\mu \to 0$).

Dual Central Path

Now consider the dual problem

$$\begin{aligned} \text{(LD) maximize } & \mathbf{y}^T\mathbf{b} \\ \text{subject to } & \mathbf{y}^T\mathbf{A} + \mathbf{s}^T = \mathbf{c}^T,\ \mathbf{s} \geqslant \mathbf{0}. \end{aligned}$$

We may apply the barrier approach to this problem by formulating the problem

$$\begin{aligned} \text{(BD) maximize } & \mathbf{y}^T\mathbf{b} + \mu \sum_{j=1}^{n} \log s_j \\ \text{subject to } & \mathbf{y}^T\mathbf{A} + \mathbf{s}^T = \mathbf{c}^T, \qquad \mathbf{s} > \mathbf{0}. \end{aligned}$$

We assume that the dual feasible set $\mathcal{F}_d$ has an interior $\overset{\circ}{\mathcal{F}}_d = \{(\mathbf{y},\ \mathbf{s}) : \mathbf{y}^T\mathbf{A} + \mathbf{s}^T = \mathbf{c}^T,\ \mathbf{s} > \mathbf{0}\}$ is nonempty and the optimal solution set of (LD) is bounded. Then, as μ is varied continuously toward 0, there is a path $(\mathbf{y}(\mu),\ \mathbf{s}(\mu))$ defined by the solution to (BD). This path is termed the *dual central path*.

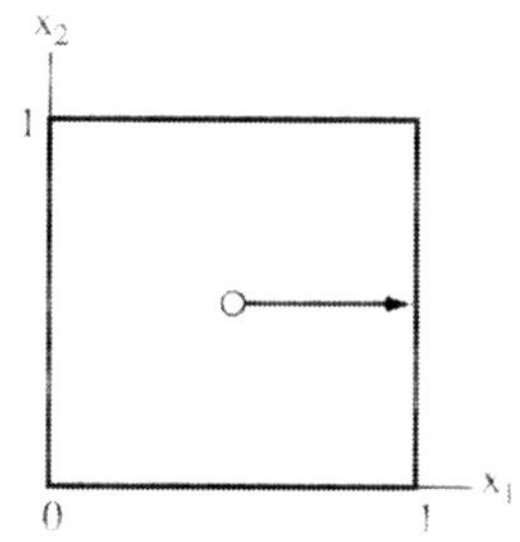

Fig. 5.2 The analytic path for the square

To work out the necessary and sufficient conditions we introduce $\mathbf{x}$ as a Lagrange multiplier and form the Lagrangian

$$\mathbf{y}^T\mathbf{b} + \mu \sum_{j=1}^{n} \log s_j - (\mathbf{y}^T\mathbf{A} + \mathbf{s}^T - \mathbf{c}^T)\mathbf{x}.$$

Setting to zero the derivative with respect to y_i leads to

$$b_i - \mathbf{a}^i\mathbf{x} = 0, \text{ for all } i$$

where $\mathbf{a}^i$ is the ith row of A. Setting to zero the derivative with respect to s_j leads to

$$\mu/s_j - x_j = 0, \text{ or } 1 - x_j s_j = 0, \text{ for all } j.$$

Combining these equations and including the original constraint yields the complete set of conditions which are identical to the optimality conditions for the primal central path (5.8). Note that $\mathbf{x}$ is indeed a primal feasible solution and $\mathbf{x} > \mathbf{0}$.

To see the geometric representation of the dual central path, consider the dual level set

$$\Omega(z) = \{\mathbf{y} : \mathbf{c}^T - \mathbf{y}^T\mathbf{A} \geqslant \mathbf{0},\ \mathbf{y}^T\mathbf{b} \geqslant z\}$$

for any $z < z^*$ where z^* is the optimal value of (LD). Then, the analytic center $(\mathbf{y}(z),\ \mathbf{s}(z))$ of $\Omega(z)$ coincides with the dual central path as z tends to the optimal value z^* from below. This is illustrated in Fig. 5.3, where the feasible region of the dual set (not the primal) is shown. The level sets $\Omega(z)$ are shown for various values of z. The analytic centers of these level sets correspond to the dual central path.

Example 3 (The Square Dual). Consider the dual of Example 2. This is

$$\begin{array}{ll} \max & y_1 + y_2 \\ \text{subject to} & y_1 \leqslant -1 \\ & y_2 \leqslant 0. \end{array}$$

(The values of s_1 and s_2 are the slack variables of the inequalities.) The solution to the dual barrier problem is easily found from the solution of the primal barrier problem to be

$$y_1(\mu) = -1 - \mu/x_1(\mu), \ y_2 = -2\mu.$$

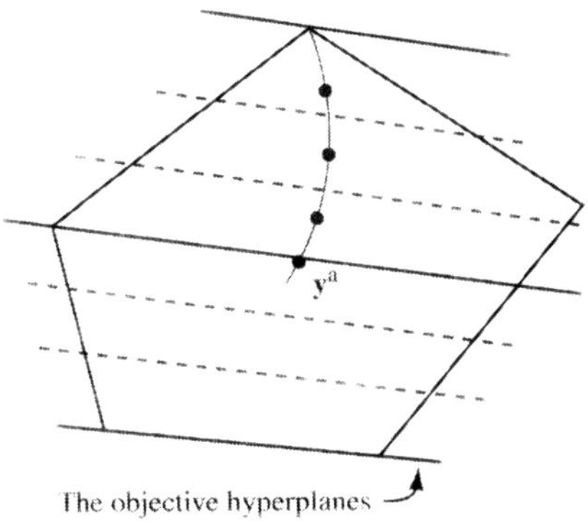

Fig. 5.3 The central path as analytic centers in the dual feasible region

As $\mu \to 0$, we have $y_1 \to -1, \ y_2 \to 0$, which is the unique solution to the dual LP. However, as $\mu \to \infty$, the vector y is unbounded, for in this case the dual feasible set is itself unbounded.

Primal–Dual Central Path

Suppose the feasible region of the primal (LP) has interior points and its optimal solution set is bounded. Then, the dual also has interior points (see Exercise 4). The primal–dual path is defined to be the set of vectors ($\mathbf{x}(\mu) > \mathbf{0}, \ \mathbf{y}(\mu), \ \mathbf{s}(\mu) > \mathbf{0}$) that satisfy the conditions

$$\begin{aligned} \mathbf{x} \circ \mathbf{s} &= \mu \mathbf{1} \\ \mathbf{A}\mathbf{x} &= \mathbf{b} \\ \mathbf{A}^T\mathbf{y} + \mathbf{s} &= \mathbf{c} \end{aligned}$$

for $0 \leqslant \mu \leqslant \infty$. Hence the central path is defined without explicit reference to an optimization problem. It is simply defined in terms of the set of equality and inequality conditions.

Since conditions (5.8) and (5.9) are identical, the primal–dual central path can be split into two components by projecting onto the relevant space, as described in the following proposition.

Proposition 1. *Suppose the feasible sets of the primal and dual programs contain interior points. Then the primal–dual central path* ($\mathbf{x}(\mu), \ \mathbf{y}(\mu), \ \mathbf{s}(\mu)$) *exists for all* $\mu, \ 0 \leqslant \mu < \infty$. *Furthermore,* $\mathbf{x}(\mu)$ *is the primal central path, and* ($\mathbf{y}(\mu), \ \mathbf{s}(\mu)$) *is the dual central path. More-*

over, $\mathbf{x}(\mu)$ *and* $(\mathbf{y}(\mu), \mathbf{s}(\mu))$ *converge to the analytic centers of the optimal primal solution and dual solution faces, respectively, as* $\mu \to 0$.

Duality Gap

Let $(\mathbf{x}(\mu), \mathbf{y}(\mu), \mathbf{s}(\mu))$ be on the primal-dual central path. Then from (5.9) it follows that

$$\mathbf{c}^T\mathbf{x} - \mathbf{y}^T\mathbf{b} = \mathbf{y}^T\mathbf{A}\mathbf{x} + \mathbf{s}^T\mathbf{x} - \mathbf{y}^T\mathbf{b} = \mathbf{s}^T\mathbf{x} = n\mu.$$

The value $\mathbf{c}^T\mathbf{x} - \mathbf{y}^T\mathbf{b} = \mathbf{s}^T\mathbf{x}$ is the difference between the primal objective value and the dual objective value. This value is always nonnegative (see the weak duality lemma in Sect. 4.2) and is termed the *duality gap*. At any point on the primal–dual central path, the duality gap is equal to $n\mu$. It is clear that as $\mu \to 0$ the duality gap goes to zero, and hence both $\mathbf{x}(\mu)$ and $(\mathbf{y}(\mu), \mathbf{s}(\mu))$ approach optimality for the primal and dual, respectively.

The duality gap provides a measure of closeness to optimality. For any primal feasible $\mathbf{x}$, the value $\mathbf{c}^T\mathbf{x}$ gives an upper bound as $\mathbf{c}^T\mathbf{x} \geqslant z^*$ where z^* is the optimal value of the primal. Likewise, for any dual feasible pair $(\mathbf{y}, \mathbf{s})$, the value $\mathbf{y}^T\mathbf{b}$ gives a lower bound as $\mathbf{y}^T\mathbf{b} \leqslant z^*$. The difference, the duality gap $g = \mathbf{c}^T\mathbf{x} - \mathbf{y}^T\mathbf{b}$, provides a bound on z^* as $z^* \geqslant \mathbf{c}^T\mathbf{x} - g$. Hence if at a feasible point $\mathbf{x}$, a dual feasible $(\mathbf{y}, \mathbf{s})$ is available, the quality of $\mathbf{x}$ can be measured as $\mathbf{c}^T\mathbf{x} - z^* \leqslant g$.

5.6 Solution Strategies

The various definitions of the central path directly suggest corresponding strategies for solution of a linear program. We outline three general approaches here: the primal barrier or path-following method, the primal-dual path-following method and the primal-dual potential-reduction method, although the details of their implementation and analysis must be deferred to later chapters after study of general nonlinear methods. Table 5.1 depicts these solution strategies and the simplex methods described in Chaps. 3 and 4 with respect to how they meet the three optimality conditions: Primal Feasibility, Dual Feasibility, and Zero-Duality during the iterative process.

Table 5.1 Properties of algorithms

	P-F	D-F	0-Duality
Primal simplex	√		√
Dual simplex		√	√
Primal barrier	√		
Primal-dual path-following	√	√	
Primal-dual potential-reduction	√	√	

For example, the primal simplex method keeps improving a primal feasible solution, maintains the zero-duality gap (complementarity slackness condition) and moves toward dual feasibility; while the dual simplex method keeps improving a dual feasible solution, maintains the zero-duality gap (complementarity condition) and moves toward primal feasibility (see Sect. 4.3). The primal barrier method keeps improving a primal feasible solution and moves toward dual feasibility and complementarity; and the primal-dual interior-point methods keep improving a primal and dual feasible solution pair and move toward complementarity.

Primal Barrier Method

A direct approach is to use the barrier construction and solve the the problem

$$\begin{array}{ll} \text{minimize} & \mathbf{c}^T\mathbf{x} - \mu \sum_{j=1}^{n} \log x_j \\ \text{subject to} & \mathbf{A}\mathbf{x} = \mathbf{b},\ \mathbf{x} > \mathbf{0}, \end{array} \tag{5.9}$$

for a very small value of μ. In fact, if we desire to reduce the duality gap to ε it is only necessary to solve the problem for $\mu = \varepsilon/n$. Unfortunately, when μ is small, the problem (5.9) could be highly ill-conditioned in the sense that the necessary conditions are nearly singular. This makes it difficult to directly solve the problem for small μ.

An overall strategy, therefore, is to start with a moderately large μ (say $\mu = 100$) and solve that problem approximately. The corresponding solution is a point approximately on the primal central path, but it is likely to be quite distant from the point corresponding to the limit of $\mu \to 0$. However this solution point at $\mu = 100$ can be used as the starting point for the problem with a slightly smaller μ, for this point is likely to be close to the solution of the new problem. The value of μ might be reduced at each stage by a specific factor, giving $\mu_{k+1} = \gamma\mu_k$, where γ is a fixed positive parameter less than one and k is the stage count.

If the strategy is begun with a value μ_0, then at the kth stage we have $\mu_k = \gamma^k\mu_0$. Hence to reduce μ_k/μ_0 to below ε, requires

$$k = \frac{\log \varepsilon}{\log \gamma}$$

stages.

Often a version of Newton's method for minimization is used to solve each of the problems. For the current strategy, Newton's method works on problem (5.9) with fixed μ by considering the central path equations (5.8)

$$\begin{aligned} \mathbf{x} \circ \mathbf{s} &= \mu\mathbf{1} \\ \mathbf{A}\mathbf{x} &= \mathbf{b} \\ \mathbf{A}^T\mathbf{y} + \mathbf{s} &= \mathbf{c}. \end{aligned} \tag{5.10}$$

From a given point $\mathbf{x} \in \mathring{\mathcal{F}}_p$, Newton's method moves to a closer point $\mathbf{x}^+ \in \mathring{\mathcal{F}}_p$ by moving in the directions $\mathbf{d}_\mathbf{x}$, $\mathbf{d}_\mathbf{y}$ and $\mathbf{d}_\mathbf{s}$ determined from the linearized version of (5.10)

$$\begin{aligned} \mu\mathbf{X}^{-2}\mathbf{d}_\mathbf{x} + \mathbf{d}_\mathbf{s} &= \mu\mathbf{X}^{-1}1 - \mathbf{c}, \\ \mathbf{A}\mathbf{d}_\mathbf{x} &= \mathbf{0}, \\ -\mathbf{A}^T\mathbf{d}_\mathbf{y} - \mathbf{d}_\mathbf{s} &= \mathbf{0}. \end{aligned} \tag{5.11}$$

(Recall that $\mathbf{X}$ is the diagonal matrix whose diagonal entries are components of $\mathbf{x} > \mathbf{0}$.) The new point is then updated by taking a step in the direction of $\mathbf{d}_x$, as $\mathbf{x}^+ = \mathbf{x} + \mathbf{d}_x$.

Notice that if $\mathbf{x} \circ \mathbf{s} = \mu\mathbf{1}$ for some $\mathbf{s} = \mathbf{c} - \mathbf{A}^T\mathbf{y}$, then $\mathbf{d} \equiv (\mathbf{d}_\mathbf{x},\ \mathbf{d}_\mathbf{y},\ \mathbf{d}_\mathbf{s}) = \mathbf{0}$ because the current point satisfies $\mathbf{A}\mathbf{x} = \mathbf{b}$ and hence is already the central path solution for μ. If some component of $\mathbf{x} \circ \mathbf{s}$ is less than μ, then $\mathbf{d}$ will tend to increment the solution so as to increase that component. The converse will occur for components of $\mathbf{x} \circ \mathbf{s}$ greater than μ.

This process may be repeated several times until a point close enough to the proper solution to the barrier problem for the given value of μ is obtained. That is, until the necessary and sufficient conditions (5.7) are (approximately) satisfied.

There are several details involved in a complete implementation and analysis of Newton's method. These items are discussed in later chapters of the text. However, the method works well if either μ is moderately large, or if the algorithm is initiated at a point very close to the solution, exactly as needed for the barrier strategy discussed in this subsection.

To solve (5.11), premultiplying both sides by $\mathbf{X}^2$ we have

$$\mu\mathbf{d}_\mathbf{x} + \mathbf{X}^2\mathbf{d}_\mathbf{s} = \mu\mathbf{X}1 \ - \mathbf{X}^2\mathbf{c}.$$

Then, premultiplying by $\mathbf{A}$ and using $\mathbf{A}\mathbf{d}_\mathbf{x} = \mathbf{0}$, we have

$$\mathbf{A}\mathbf{X}^2\mathbf{d}_\mathbf{s} = \mu\mathbf{A}\mathbf{X}\mathbf{1} - \mathbf{A}\mathbf{X}^2\mathbf{c}.$$

Using $\mathbf{d}_\mathbf{s} = -\mathbf{A}^T\mathbf{d}_\mathbf{y}$ we have

$$(\mathbf{A}\mathbf{X}^2\mathbf{A}^T)\mathbf{d}_\mathbf{y} = -\mu\mathbf{A}\mathbf{X}\mathbf{1} + \mathbf{A}\mathbf{X}^2\mathbf{c}.$$

Thus, $\mathbf{d}_\mathbf{y}$ can be computed by solving the above linear system of equations. Then $\mathbf{d}_\mathbf{s}$ can be found from the third equation in (5.11) and finally $\mathbf{d}_\mathbf{x}$ can be found from the first equation in (5.11), together this amounts to $O(nm^2 + m^3)$ arithmetic operations for each Newton step.

Primal-Dual Path-Following

Another strategy for solving a linear program is to follow the central path from a given initial primal-dual solution pair. Consider a linear program in standard form

Primal	Dual
minimize $\mathbf{c}^T\mathbf{x}$	maximize $\mathbf{y}^T\mathbf{b}$
subject to $\mathbf{A}\mathbf{x} = \mathbf{b},\ \mathbf{x} \geqslant \mathbf{0}$	subject to $\mathbf{y}^T\mathbf{A} \leqslant \mathbf{c}^T$.

Assume that the interior of both primal and dual feasible regions $\mathring{\mathcal{F}} \neq \emptyset$; that is, both[4]

$$\mathring{\mathcal{F}}_p = \{\mathbf{x} : \mathbf{A}\mathbf{x} = \mathbf{b},\ \mathbf{x} > \mathbf{0}\} \neq \emptyset \quad \text{and} \quad \mathring{\mathcal{F}}_d = \{(\mathbf{y},\ \mathbf{s}) : \mathbf{s} = \mathbf{c} - \mathbf{A}^T\mathbf{y} > \mathbf{0}\} \neq \emptyset;$$

and denote by z^* the optimal objective value.

The central path can be expressed as

$$C = \left\{(\mathbf{x},\ \mathbf{y},\ \mathbf{s}) \in \mathring{\mathcal{F}} : \mathbf{x} \circ \mathbf{s} = \frac{\mathbf{x}^T\mathbf{s}}{n}\mathbf{1}\right\}$$

in the primal-dual form. On the path we have $\mathbf{x} \circ \mathbf{s} = \mu\mathbf{1}$ and hence $\mathbf{s}^T\mathbf{x} = n\mu$. A *neighborhood* of the central path C is of the form

$$\mathcal{N}(\eta) = \{(\mathbf{x}, \mathbf{y},\ \mathbf{s}) \in \mathring{\mathcal{F}} : |\mathbf{s} \circ \mathbf{x} - \mu\mathbf{1}| < \eta\mu,\ \text{where}\ \mu = \mathbf{s}^T\mathbf{x}/n\} \tag{5.12}$$

for some $\eta \in (0,\ 1)$, say $\eta = 1/4$. This can be thought of as a tube whose center is the central path.

The idea of the path-following method is to move within a tubular neighborhood of the central path toward the solution point. A suitable initial point $(\mathbf{x}^0,\ \mathbf{y}^0,\ \mathbf{s}^0) \in \mathcal{N}(\eta)$ can be found by solving the barrier problem for some fixed μ_0 or from an initialization phase proposed later. After that, step by step moves are made, alternating between a predictor step and a corrector step. After each pair of steps, the point achieved is again in the fixed given neighborhood of the central path, but closer to the linear program's solution set.

The predictor step is designed to move essentially parallel to the true central path. The step $\mathbf{d} \equiv (\mathbf{d}_\mathbf{x},\ \mathbf{d}_\mathbf{y},\ \mathbf{d}_\mathbf{s})$ is determined from the linearized version of the primal–dual central path equations of (5.9), as

$$\begin{aligned} \mathbf{s} \circ \mathbf{d}_\mathbf{x} + \mathbf{x} \circ \mathbf{d}_\mathbf{s} &= \gamma\mu\mathbf{1} - \mathbf{x} \circ \mathbf{s}, \\ \mathbf{A}\mathbf{d}_\mathbf{x} &= \mathbf{0}, \\ -\mathbf{A}^T\mathbf{d}_\mathbf{y} - \mathbf{d}_\mathbf{s} &= \mathbf{0}, \end{aligned} \tag{5.13}$$

where here one selects $\gamma = 0$. (To show the dependence of $\mathbf{d}$ on the current pair $(\mathbf{x},\ \mathbf{s})$ and the parameter γ, we write $\mathbf{d} = \mathbf{d}(\mathbf{x},\ \mathbf{s},\ \gamma)$.)

---

[4] The symbol $\emptyset$ denotes the empty set.

The new point is then found by taking a step in the direction of $\mathbf{d}$, as $(\mathbf{x}^+, \mathbf{y}^+, \mathbf{s}^+) = (\mathbf{x}, \mathbf{y}, \mathbf{s}) + \alpha(\mathbf{d}_\mathbf{x}, \mathbf{d}_\mathbf{y}, \mathbf{d}_\mathbf{s})$, where α is the step-size. Note that $\mathbf{d}_\mathbf{x}^T\mathbf{d}_\mathbf{s} = -\mathbf{d}_\mathbf{x}^T\mathbf{A}^T\mathbf{d}_\mathbf{y} = 0$ here. Then

$$(\mathbf{x}^+)^T\mathbf{s}^+ = (\mathbf{x} + \alpha\mathbf{d}_\mathbf{x})^T(\mathbf{s} + \alpha\mathbf{d}_\mathbf{s}) = \mathbf{x}^T\mathbf{s} + \alpha(\mathbf{d}_\mathbf{x}^T\mathbf{s} + \mathbf{x}^T\mathbf{d}_\mathbf{s}) = (1 - \alpha)\mathbf{x}^T\mathbf{s},$$

where the last step follows by multiplying the first equation in (5.13) by $\mathbf{1}^T$. Thus, the predictor step reduces the duality gap by a factor $1 - \alpha$. The maximum possible step-size α in that direction is made in that parallel direction without going outside of the neighborhood $\mathcal{N}(2\eta)$.

The corrector step essentially moves perpendicular to the central path in order to get closer to it. This step moves the solution back to within the neighborhood $\mathcal{N}(\eta)$, and the step is determined by selecting $\gamma = 1$ in (5.13) with $\mu = \mathbf{x}^T\mathbf{s}/n$. Notice that if $\mathbf{x} \circ \mathbf{s} = \mu\mathbf{1}$, then $\mathbf{d} = \mathbf{0}$ because the current point is already a central path solution.

This corrector step is identical to one step of the barrier method. Note, however, that the predictor–corrector method requires only one sequence of steps, each consisting of a single predictor and corrector. This contrasts with the barrier method which requires a complete sequence for each μ to get back to the central path, and then an outer sequence to reduce the μ's.

One can prove that for any $(\mathbf{x}, \mathbf{y}, \mathbf{s}) \in \mathcal{N}(\eta)$ with $\mu = \mathbf{x}^T\mathbf{s}/n$, the step-size in the predictor stop satisfies

$$\alpha \geqslant \frac{1}{2\sqrt{n}}.$$

Thus, the iteration complexity of the method is $O(\sqrt{n})\log(1/\varepsilon))$ to achieve $\mu/\mu_0 \leqslant \varepsilon$ where $n\mu_0$ is the initial duality gap. Moreover, one can prove that the step-size $\alpha \to 1$ as $\mathbf{x}^T\mathbf{s} \to 0$, that is, the duality reduction speed is accelerated as the gap becomes smaller.

Primal-Dual Potential Reduction Algorithm

In this method a *primal-dual potential function* is used to measure the solution's progress. The potential is reduced at each iteration. There is no restriction on either neighborhood or step-size during the iterative process as long as the potential is reduced. The greater the reduction of the potential function, the faster the convergence of the algorithm. Thus, from a practical point of view, potential-reduction algorithms may have an advantage over *path-following algorithms* where iterates are confined to lie in certain neighborhoods of the central path.

For $\mathbf{x} \in \mathring{\mathcal{F}}_p$ and $(\mathbf{y}, \mathbf{s}) \in \mathring{\mathcal{F}}_d$ the primal-dual potential function is defined by

$$\psi_{n+\rho}(\mathbf{x}, \mathbf{s}) \equiv (n + \rho)\log(\mathbf{x}^T\mathbf{s}) - \sum_{j=1}^{n} \log(x_j s_j), \qquad (5.14)$$

where $\rho \geqslant 0$.

From the arithmetic and geometric mean inequality (also see Exercise 10) we can derive that

$$n\log(\mathbf{x}^T\mathbf{s}) - \sum_{j=1}^{n}\log(x_j s_j) \geqslant n\log n.$$

Then

$$\psi_{n+\rho}(\mathbf{x},\ \mathbf{s}) = \rho\log(\mathbf{x}^T\mathbf{s}) + n\log(\mathbf{x}^T\mathbf{s}) - \sum_{j=1}^{n}\log(x_j s_j) \geqslant \rho\log(\mathbf{x}^T\mathbf{s}) + n\log n. \quad (5.15)$$

Thus, for $\rho > 0$, $\psi_{n+\rho}(\mathbf{x},\ \mathbf{s}) \to -\infty$ implies that $\mathbf{x}^T\mathbf{s} \to 0$. More precisely, we have from (5.15)

$$\mathbf{x}^T\mathbf{s} \leqslant \exp\left(\frac{\psi_{n+\rho}(\mathbf{x},\mathbf{s}) - n\log n}{\rho}\right).$$

Hence the primal–dual potential function gives an explicit bound on the magnitude of the duality gap.

The objective of this method is to drive the potential function down toward minus infinity. The method of reduction is a version of Newton's method (5.13). In this case we select $\gamma = n/(n+\rho)$ in (5.13). Notice that is a combination of a predictor and corrector choice. The predictor uses $\gamma = 0$ and the corrector uses $\gamma = 1$. The primal–dual potential method uses something in between. This seems logical, for the predictor moves parallel to the central path toward a lower duality gap, and the corrector moves perpendicular to get close to the central path. This new method does both at once. Of course, this intuitive notion must be made precise.

For $\rho \geqslant \sqrt{n}$, there is in fact a guaranteed decrease in the potential function by a fixed amount δ (see Exercises 12 and 13). Specifically,

$$\psi_{n+\rho}(\mathbf{x}^+,\ \mathbf{s}^+) - \psi_{n+\rho}(\mathbf{x},\ \mathbf{s}) \leqslant -\delta \quad (5.16)$$

for a constant $\delta \geqslant 0.2$. This result provides a theoretical bound on the number of required iterations and the bound is competitive with other methods. However, a faster algorithm may be achieved by conducting a line search along direction $\mathbf{d}$ to achieve the greatest reduction in the primal-dual potential function at each iteration.

We outline the algorithm here:

Step 1. Start at a point $(\mathbf{x}_0,\ \mathbf{y}_0,\ \mathbf{s}_0) \in \mathring{\mathcal{F}}$ with $\psi_{n+\rho}(\mathbf{x}_0,\ \mathbf{s}_0) \leq \rho\log((\mathbf{s}_0)^T\mathbf{x}_0) + n\log n + O(\sqrt{n}\log n)$ which is determined by an initiation procedure, as discussed in Sect. 5.7. Set $\rho \geq \sqrt{n}$. Set $k = 0$ and $\gamma = n/(n+\rho)$. Select an accuracy parameter $\varepsilon > 0$.

Step 2. Set $(\mathbf{x},\ \mathbf{s}) = (\mathbf{x}_k,\ \mathbf{s}_k)$ and compute $(\mathbf{d}_\mathbf{x},\ \mathbf{d}_\mathbf{y},\ \mathbf{d}_\mathbf{s})$ from (5.13).

Step 3. Let $\mathbf{x}_{k+1} = \mathbf{x}_k + \bar{\alpha}\mathbf{d}_\mathbf{x}$, $\mathbf{y}_{k+1} = \mathbf{y}_k + \bar{\alpha}\mathbf{d}_\mathbf{y}$, and $\mathbf{s}_{k+1} = \mathbf{s}_k + \bar{\alpha}\mathbf{d}_\mathbf{s}$ where

$$\bar{\alpha} = \arg\min_{\alpha\geq 0}\psi_{n+\rho}(\mathbf{x}_k + \alpha\mathbf{d}_\mathbf{x},\ \mathbf{s}_k + \alpha\mathbf{d}_\mathbf{s}).$$

Step 4. Let $k = k+1$. If $\frac{\mathbf{s}_k^T\mathbf{x}_k}{\mathbf{s}_0^T\mathbf{x}_0} \leq \varepsilon$, Stop. Otherwise return to Step 2.

Theorem 2. *The algorithm above terminates in at most $O(\rho\log(n/\varepsilon))$ iterations with*

$$\frac{(\mathbf{s}_k)^T\mathbf{x}_k}{(\mathbf{s}_0)^T\mathbf{x}_0} \le \varepsilon.$$

Proof. Note that after k iterations, we have from (5.16)

$$\psi_{n+\rho}(\mathbf{x}_k, \mathbf{s}_k) \le \psi_{n+\rho}(\mathbf{x}_0, \mathbf{s}_0) - k\cdot\delta \le \rho\log((\mathbf{s}_0)^T\mathbf{x}_0) + n\log n + O(\sqrt{n}\log n) - k\cdot\delta.$$

Thus, from the inequality (5.15),

$$\rho\log(\mathbf{s}_k^T\mathbf{x}_k) + n\log n \le \rho\log(\mathbf{s}_0^T\mathbf{x}_0) + n\log n + O(\sqrt{n}\log n) - k\cdot\delta,$$

or

$$\rho(\log(\mathbf{s}_k^T\mathbf{x}_k) - \log(\mathbf{s}_0^T\mathbf{x}_0)) \le -k\cdot\delta + O(\sqrt{n}\log n).$$

Therefore, as soon as $k \ge O(\rho\log(n/\varepsilon))$, we must have

$$\rho(\log(\mathbf{s}_k^T\mathbf{x}_k) - \log(\mathbf{s}_0^T\mathbf{x}_0)) \le -\rho\log(1/\varepsilon),$$

or

$$\frac{\mathbf{s}_k^T\mathbf{x}_k}{\mathbf{s}_0^T\mathbf{x}_0} \le \varepsilon. \blacksquare$$

Theorem 2 holds for any $\rho \ge \sqrt{n}$. Thus, by choosing $\rho = \sqrt{n}$, the iteration complexity bound becomes $O(\sqrt{n}\log(n/\varepsilon))$.

Iteration Complexity

The computation of each iteration basically requires solving (5.13) for **d**. Note that the first equation of (5.13) can be written as

$$\mathbf{S}\mathbf{d}_\mathbf{x} + \mathbf{X}\mathbf{d}_\mathbf{s} = \gamma\mu\mathbf{1} - \mathbf{X}\mathbf{S}\mathbf{1}$$

where **X** and **S** are two diagonal matrices whose diagonal entries are components of $\mathbf{x} > \mathbf{0}$ and $\mathbf{s} > \mathbf{0}$, respectively. Premultiplying both sides by $\mathbf{S}^{-1}$ we have

$$\mathbf{d}_\mathbf{x} + \mathbf{S}^{-1}\mathbf{X}\mathbf{d}_\mathbf{s} = \gamma\mu\mathbf{S}^{-1}\mathbf{1} - \mathbf{x}.$$

Then, premultiplying by **A** and using $\mathbf{A}\mathbf{d}_\mathbf{x} = \mathbf{0}$, we have

$$\mathbf{A}\mathbf{S}^{-1}\mathbf{X}\mathbf{d}_\mathbf{s} = \gamma\mu\mathbf{A}\mathbf{S}^{-1}\mathbf{1} - \mathbf{A}\mathbf{x} = \gamma\mu\mathbf{A}\mathbf{S}^{-1}\mathbf{1} - \mathbf{b}.$$

Using $\mathbf{d}_\mathbf{s} = -\mathbf{A}^T\mathbf{d}_\mathbf{y}$ we have

$$(\mathbf{A}\mathbf{S}^{-1}\mathbf{X}\mathbf{A}^T)\mathbf{d}_\mathbf{y} = \mathbf{b} - \gamma\mu\mathbf{A}\mathbf{S}^{-1}\mathbf{1}.$$

Thus, the primary computational cost of each iteration of the interior-point algorithm discussed in this section is to form and invert the normal matrix $\mathbf{AXS}^{-1}\mathbf{A}^T$, which typically requires $O(nm^2 + m^3)$ arithmetic operations. However, an approximation of this matrix can be updated and inverted using far fewer arithmetic operations. In fact, using a rank-one technique (see Chap. 10) to update the approximate inverse of the normal matrix during the iterative progress, one can reduce the average number of arithmetic operations per iteration to $O(\sqrt{n}m^2)$. Thus, if the relative tolerance ε is viewed as a variable, we have the following total arithmetic operation complexity bound to solve a linear program:

Corollary. *Let* $\rho = \sqrt{n}$. *Then, the algorithm above Theorem 2 terminates in at most* $O(nm^2 \log(n/\varepsilon))$ *arithmetic operations.*

5.7 Termination and Initialization

There are several remaining important issues concerning interior-point algorithms for linear programs. The first issue involves termination. Unlike the simplex method which terminates with an exact solution, interior-point algorithms are continuous optimization algorithms that generate an infinite solution sequence converging to an optimal solution. If the data of a particular problem are integral or rational, an argument is made that, after the worst-case time bound, an exact solution can be rounded from the latest approximate solution. Several questions arise. First, under the real number computation model (that is, the data consists of real numbers), how can we terminate at an exact solution? Second, regardless of the data's status, is there a practical test, which can be computed cost-effectively during the iterative process, to identify an exact solution so that the algorithm can be terminated before the worse-case time bound? Here, by exact solution we mean one that could be found using exact arithmetic, such as the solution of a system of linear equations, which can be computed in a number of arithmetic operations bounded by a polynomial in n.

The second issue involves initialization. Almost all interior-point algorithms require the regularity assumption that $\mathring{\mathcal{F}} \neq \emptyset$. What is to be done if this is not true? A related issue is that interior-point algorithms have to start at a strictly feasible point near the central path.

*Termination

Complexity bounds for interior-point algorithms generally depend on an ε which must be zero in order to obtain an exact optimal solution. Sometimes it is advantageous to employ an early termination or rounding method while ε is still moderately large. There are five basic approaches.

- A "purification" procedure finds a feasible corner whose objective value is at least as good as the current interior point. This can be accomplished in strongly polynomial time (that is, the complexity bound is a polynomial only in the dimensions m and n). One difficulty is that there may be many non-optimal vertices close to the optimal face, and the procedure might require many pivot steps for difficult problems.
- A second method seeks to identify an optimal basis. It has been shown that if the linear program is nondegenerate, the unique optimal basis may be identified early. The procedure seems to work well for some problems but it has difficulty if the problem is degenerate. Unfortunately, most real linear programs are degenerate.
- The third approach is to slightly perturb the data such that the new program is nondegenerate and its optimal basis remains one of the optimal bases of the original program. There are questions about how and when to perturb the data during the iterative process, decisions which can significantly affect the success of the effort.
- The fourth approach is to guess the optimal face and find a feasible solution on that face. It consists of two phases: the first phase uses interior point algorithms to identify the complementarity partition (P^*, Z^*) (see Exercise 6), and the second phase adapts the simplex method to find an optimal primal (or dual) basic solution and one can use (P^*, Z^*) as a starting base for the second phase. This method is often called the cross-over method. It is guaranteed to work in finite time and is implemented in several popular linear programming software packages.
- The fifth approach is to guess the optimal face and project the current interior point onto the interior of the optimal face. See Fig. 5.4. The termination criterion is guaranteed to work in finite time.

The fourth and fifth methods above are based on the fact that (as observed in practice and subsequently proved) many interior-point algorithms for linear programming generate solution sequences that converge to a strictly complementary solution or an interior solution on the optimal face; see Exercise 8.

Initialization

Most interior-point algorithms must be initiated at a strictly feasible point. The complexity of obtaining such an initial point is the same as that of solving the linear program itself. More importantly, a complete algorithm should accomplish two tasks: (1) detect the infeasibility or unboundedness status of the problem, then (2) generate an optimal solution if the problem is neither infeasible nor unbounded.

Several approaches have been proposed to accomplish these goals:

- The primal and dual can be combined into a single linear feasibility problem, and a feasible point found. Theoretically, this approach achieves the currently best iteration complexity bound, that is, $O(\sqrt{n}\log(1/\varepsilon))$. Practically, a significant disadvantage of this approach is the doubled dimension of the system of equations that must be solved at each iteration.

- The big-M method can be used by adding one or more artificial column(s) and/or row(s) and a huge penalty parameter M to force solutions to become feasible during the algorithm. A major disadvantage of this approach is the numerical problems caused by the addition of coefficients of large magnitude.
- Phase I-then-Phase II methods are effective. A major disadvantage of this approach is that the two (or three) related linear programs must be solved sequentially.

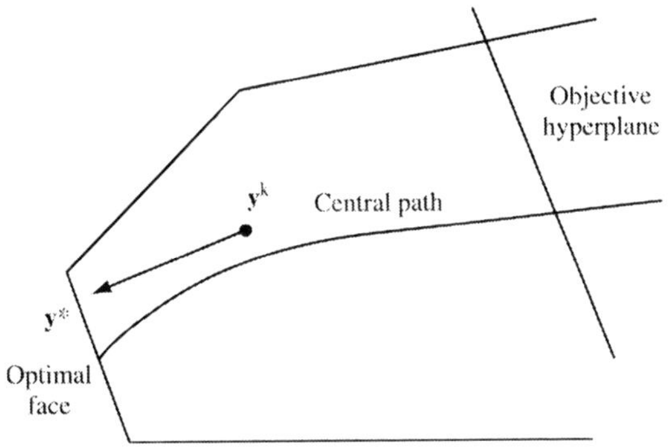

Fig. 5.4 Illustration of the projection of an interior point onto the optimal face

- A modified Phase I-Phase II method approaches feasibility and optimality simultaneously. To our knowledge, the currently best iteration complexity bound of this approach is $O(n\log(1/\varepsilon))$, as compared to $O(\sqrt{n}\log(1/\varepsilon))$ of the three above. Other disadvantages of the method include the assumption of non-empty interior and the need of an objective lower bound.

The HSD Algorithm

There is an algorithm, termed the *Homogeneous Self-Dual Algorithm* that overcomes the difficulties mentioned above. The algorithm achieves the theoretically best $O(\sqrt{n}\log(1/\varepsilon))$ complexity bound and is often used in linear programming software packages.

The algorithm is based on the construction of a homogeneous and self-dual linear program related to (LP) and (LD) (see Sect. 5.5). We now briefly explain the two major concepts, homogeneity and self-duality, used in the construction.

In general, a system of linear equations of inequalities is *homogeneous* if the right hand side components are all zero. Then if a solution is found, any positive multiple of that solution is also a solution. In the construction used below, we allow a single inhomogeneous constraint, often called a *normalizing constraint*. Karmarkar's original canonical form is a homogeneous linear program.

A linear program is termed *self-dual* if the dual of the problem is equivalent to the primal. The advantage of self-duality is that we can apply a primal-dual interior-point algorithm to solve the self-dual problem *without* doubling the dimension of the linear system solved at each iteration.

The homogeneous and self-dual linear program (HSDP) is constructed from (LP) and (LD) in such a way that the point $\mathbf{x} = \mathbf{1}$, $\mathbf{y} = \mathbf{0}$, $\tau = 1$, $z = 1$, $\theta = 1$ is feasible. The primal program is

$$
\begin{array}{lllllll}
(HSDP) & \text{minimize} & & & & (n+1)\theta & \\
& \text{subject to} & & \mathbf{A}\mathbf{x} & -\mathbf{b}\tau & +\bar{\mathbf{b}}\theta & = \mathbf{0}, \\
& & -\mathbf{A}^T\mathbf{y} & & +\mathbf{c}\tau & -\bar{\mathbf{c}}\theta & \geq \mathbf{0}, \\
& & \mathbf{b}^T\mathbf{y} & -\mathbf{c}^T\mathbf{x} & & +\bar{z}\theta & \geq 0, \\
& & -\bar{\mathbf{b}}^T\mathbf{y} & +\bar{\mathbf{c}}^T\mathbf{x} & -\bar{z}\tau & & = -(n+1), \\
& & \mathbf{y}\ \text{free}, & \mathbf{x} \geq \mathbf{0}, & \tau \geq 0, & \theta\ \text{free}; &
\end{array}
$$

where

$$\bar{\mathbf{b}} = \mathbf{b} - \mathbf{A}\mathbf{1},\ \bar{\mathbf{c}} = \mathbf{c} - \mathbf{1},\ \bar{z} = \mathbf{c}^T\mathbf{1} + 1. \tag{5.17}$$

Notice that $\bar{\mathbf{b}}$, $\bar{\mathbf{c}}$, and $\bar{\mathbf{z}}$ represent the "infeasibility" of the initial primal point, dual point, and primal-dual "gap," respectively. They are chosen so that the system is feasible. For example, for the point $\mathbf{x} = \mathbf{1}$, $\mathbf{y} = \mathbf{0}$, $\tau = 1$, $\theta = 1$, the last equation becomes

$$0 + \mathbf{c}^T\mathbf{x} - \mathbf{1}^T\mathbf{x} - (\mathbf{c}^T\mathbf{x} + 1) = -n - 1.$$

Note also that the top two constraints in (HSDP), with $\tau = 1$ and $\theta = 0$, represent primal and dual feasibility (with $\mathbf{x} \geq \mathbf{0}$). The third equation represents reversed weak duality (with $\mathbf{b}^T\mathbf{y} \geq \mathbf{c}^T\mathbf{x}$) rather than the reverse. So if these three equations are satisfied with $\tau = 1$ and $\theta = 0$ they define primal and dual optimal solutions. Then, to achieve primal and dual feasibility for $\mathbf{x} = \mathbf{1}$, $(\mathbf{y}, \mathbf{s}) = (\mathbf{0}, \mathbf{1})$, we add the artificial variable θ. The fourth constraint is added to achieve self-duality.

The problem is self-dual because its overall coefficient matrix has the property that its transpose is equal to its negative. It is *skew-symmetric*.

Denote by $\mathbf{s}$ the slack vector for the second constraint and by κ the slack scalar for the third constraint. Denote by $\mathcal{F}_h$ the set of all points $(\mathbf{y}, \mathbf{x}, \tau, \theta, \mathbf{s}, \kappa)$ that are feasible for (HSDP). Denote by $\mathcal{F}_h^0$ the set of strictly feasible points with $(\mathbf{x}, \tau, \mathbf{s}, \kappa) > \mathbf{0}$ in $\mathcal{F}_h$. By combining the constraints (Exercise 14) we can write the last (equality) constraint as

$$\mathbf{1}^T\mathbf{x} + \mathbf{1}^T\mathbf{s} + \tau + \kappa - (n+1)\theta = (n+1), \tag{5.18}$$

which serves as a normalizing constraint for (HSDP). This implies that for $0 \leq \theta \leq 1$ the variables in this equation are bounded.

We state without proof the following basic result.

Theorem 1. *Consider problems (HSDP).*

(i) (HSDP) has an optimal solution and its optimal solution set is bounded.

(ii) The optimal value of (HSDP) is zero, and

$$(\mathbf{y}, \mathbf{x}, \tau, \theta, \mathbf{s}, \kappa) \in \mathcal{F}_h \textit{ implies that } (n+1)\theta = \mathbf{x}^T\mathbf{s} + \tau\kappa.$$

(iii) There is an optimal solution $(\mathbf{y}^*, \mathbf{x}^*, \tau^*, \theta^* = 0, \mathbf{s}^*, \kappa^*) \in \mathcal{F}_h$ *such that*

$$\begin{pmatrix} \mathbf{x}^* + \mathbf{s}^* \\ \tau^* + \kappa^* \end{pmatrix} > \mathbf{0},$$

which we call a strictly self-complementary solution.

Part (ii) of the theorem shows that as θ goes to zero, the solution tends toward satisfying complementary slackness between $\mathbf{x}$ and $\mathbf{s}$ and between τ and κ. Part (iii) shows that at a solution with $\theta = 0$, the complementary slackness is strict in the sense that at least one member of a complementary pair must be positive. For example, $x_1 s_1 = 0$ is required by complementary slackness, but in this case $x_1 = 0$, $s_1 = 0$ will not occur; exactly one of them must be positive.

We now relate optimal solutions to (HSDP) to those for (LP) and (LD).

Theorem 2. *Let* $(\mathbf{y}^*, \mathbf{x}^*, \tau^*, \theta^* = 0, \mathbf{s}^*, \kappa^*)$ *be a strictly-self complementary solution for (HSDP).*

(i) (LP) has a solution (feasible and bounded) if and only if $\tau^* > 0$. *In this case,* $\mathbf{x}^*/\tau^*$ *is an optimal solution for (LP) and* $\mathbf{y}^*/\tau^*$, $\mathbf{s}^*/\tau^*$ *is an optimal solution for (LD).*

(ii) (LP) has no solution if and only if $\kappa^* > 0$. *In this case,* $\mathbf{x}^*/\kappa^*$ *or* $\mathbf{y}^*/\kappa^*$ *or both are certificates for proving infeasibility: if* $\mathbf{c}^T\mathbf{x}^* < 0$ *then (LD) is infeasible; if* $-\mathbf{b}^T\mathbf{y}^* < 0$ *then (LP) is infeasible; and if both* $\mathbf{c}^T\mathbf{x}^* < 0$ *and* $-\mathbf{b}^T\mathbf{y}^* < 0$ *then both (LP) and (LD) are infeasible.*

Proof. We prove the second statement. We first assume that one of (LP) and (LD) is infeasible, say (LD) is infeasible. Then there is some certificate $\bar{\mathbf{x}} \geq \mathbf{0}$ such that $\mathbf{A}\bar{\mathbf{x}} = \mathbf{0}$ and $\mathbf{c}^T\bar{\mathbf{x}} = -1$. Let $(\bar{\mathbf{y}} = \mathbf{0}, \bar{\mathbf{s}} = \mathbf{0})$ and

$$\alpha = \frac{n+1}{\mathbf{1}^T\bar{\mathbf{x}} + \mathbf{1}^T\bar{\mathbf{s}} + 1} > 0.$$

Then one can verify that

$$\tilde{\mathbf{y}}^* = \alpha\bar{\mathbf{y}}, \ \tilde{\mathbf{x}}^* = \alpha\bar{\mathbf{x}}, \ \tilde{\tau}^* = 0, \ \tilde{\theta}^* = 0, \ \tilde{\mathbf{s}}^* = \alpha\bar{\mathbf{s}}, \ \tilde{\kappa}^* = \alpha$$

is a self-complementary solution for (HSDP). Since the supporting set (the set of positive entries) of a strictly complementary solution for (HSDP) is unique (see Exercise 6), $\kappa^* > 0$ at any strictly complementary solution for (HSDP).

Conversely, if $\tau^* = 0$, then $\kappa^* > 0$, which implies that $\mathbf{c}^T\mathbf{x}^* - \mathbf{b}^T\mathbf{y}^* < 0$, i.e., at least one of $\mathbf{c}^T\mathbf{x}^*$ and $-\mathbf{b}^T\mathbf{y}^*$ is strictly less than zero. Let us say $\mathbf{c}^T\mathbf{x}^* < 0$. In addition, we have

$$\mathbf{A}\mathbf{x}^* = \mathbf{0}, \ \mathbf{A}^T\mathbf{y}^* + \mathbf{s}^* = \mathbf{0}, \ (\mathbf{x}^*)^T\mathbf{s}^* = \mathbf{0} \text{ and } \mathbf{x}^* + \mathbf{s}^* > \mathbf{0}.$$

From Farkas' lemma (Exercise 5), $\mathbf{x}^*/\kappa^*$ is a certificate for proving dual infeasibility. The other cases hold similarly. ∎

To solve (HSDP), we have the following theorem that resembles the the central path analyzed for (LP) and (LD).

Theorem 3. *Consider problem (HSDP). For any* $\mu > 0$*, there is a unique* $(\mathbf{y}, \mathbf{x}, \tau, \theta, \mathbf{s}, \kappa)$ *in* $\overset{\circ}{\mathcal{F}}_h$*, such that*

$$\begin{pmatrix} \mathbf{x} \circ \mathbf{s} \\ \tau\kappa \end{pmatrix} = \mu \mathbf{1}.$$

Moreover, $(\mathbf{x}, \tau) = (\mathbf{1}, 1)$, $(\mathbf{y}, \mathbf{s}, \kappa) = (\mathbf{0}, \mathbf{0}, 1)$ *and* $\theta = 1$ *is the solution with* $\mu = 1$.

Theorem 3 defines an endogenous path associated with (HSDP):

$$C = \left\{ (\mathbf{y}, \mathbf{x}, \tau, \theta, \mathbf{s}, \kappa) \in \mathcal{F}_h^0 : \begin{pmatrix} \mathbf{x} \circ \mathbf{s} \\ \tau\kappa \end{pmatrix} = \frac{\mathbf{x}^T \mathbf{s} + \tau\kappa}{n+1} \mathbf{1} \right\}.$$

Furthermore, the potential function for (HSDP) can be defined as

$$\psi_{n+1+\rho}(\mathbf{x}, \tau, \mathbf{s}, \kappa) = (n + 1 + \rho) \log(\mathbf{x}^T \mathbf{s} + \tau\kappa) - \sum_{j=1}^{n} \log(x_j s_j) - \log(\tau\kappa), \quad (5.19)$$

where $\rho \geq 0$. One can then apply the interior-point algorithms described earlier to solve (HSDP) from the initial point $(\mathbf{x}, \tau) = (\mathbf{1}, 1)$, $(\mathbf{y}, \mathbf{s}, \kappa) = (\mathbf{0}, \mathbf{1}, 1)$ and $\theta = 1$ with $\mu = (\mathbf{x}^T \mathbf{s} + \tau\kappa)/(n + 1) = 1$.

The HSDP method outlined above enjoys the following properties:

- It does not require regularity assumptions concerning the existence of optimal, feasible, or interior feasible solutions.
- It can be initiated at $\mathbf{x} = \mathbf{1}$, $\mathbf{y} = \mathbf{0}$ and $\mathbf{s} = \mathbf{1}$, feasible or infeasible, on the central ray of the positive orthant (cone), and it does not require a big-M penalty parameter or lower bound.
- Each iteration solves a system of linear equations whose dimension is almost the same as that used in the standard (primal-dual) interior-point algorithms.
- If the linear program has a solution, the algorithm generates a sequence that approaches feasibility and optimality simultaneously; if the problem is infeasible or unbounded, the algorithm produces an infeasibility certificate for at least one of the primal and dual problems; see Exercise 5.

5.8 Summary

The simplex method has for decades been an efficient method for solving linear programs, despite the fact that there are no theoretical results to support its efficiency. Indeed, it was shown that in the worst case, the method may visit every vertex of the feasible region and this can be exponential in the number of variables and constraints. If on practical problems the simplex method behaved according to the worst case, even modest problems would require years of computer time to solve. The ellipsoid method was the first method that was proved to converge in time proportional to a polynomial in the size of the program, rather than to an exponential in the size. However, in practice, it was disappointingly less fast than the simplex method. Later, the interior-point method of Karmarkar significantly advanced the field of linear programming, for it not only was proved to be a polynomial-time method, but it was found in practice to be faster than the simplex method when applied to general linear programs.

The interior-point method is based on introducing a logarithmic barrier function with a weighting parameter μ; and now there is a general theoretical structure defining the analytic center, the central path of solutions as $\mu \to 0$, and the duals of these concepts. This structure is useful for specifying and analyzing various versions of interior point methods.

Most methods employ a step of Newton's method to find a point near the central path when moving from one value of μ to another. One approach is the predictor-corrector method, which first takes a step in the direction of decreasing μ and then a corrector step to get closer to the central path. Another method employs a potential function whose value can be decreased at each step, which guarantees convergence and assures that intermediate points simultaneously make progress toward the solution while remaining close to the central path.

Complete algorithms based on these approaches require a number of other features and details. For example, once systematic movement toward the solution is terminated, a final phase may move to a nearby vertex or to a non-vertex point on a face of the constraint set. Also, an initial phase must be employed to obtain an feasible point that is close to the central path from which the steps of the search algorithm can be started. These features are incorporated into several commercial software packages, and generally they perform well, able to solve very large linear programs in reasonable time.

5.9 Exercises

1. Using the simplex method, solve the program (5.1) and count the number of pivots required.
2. Prove the volume reduction rate in Theorem 1 for the ellipsoid method.
3. Develop a cutting plane method, based on the ellipsoid method, to find a point satisfying convex inequalities

$$f_i(\mathbf{x}) \leqslant 0,\ i = 1,\ \ldots,\ m,\ |\mathbf{x}|^2 \leqslant E^2,$$

   where f_i's are convex functions of $\mathbf{x}$ in C^1.
4. Consider the linear program (5.5) and assume that $\mathring{\mathcal{F}}_p = \{\mathbf{x} : \mathbf{A}\mathbf{x} = \mathbf{b},\ \mathbf{x} > \mathbf{0}\}$ is nonempty and its optimal solution set is bounded. Show that the dual of the problem has a nonempty interior.
5. (Farkas' lemma) Prove: Exactly one of the feasible sets $\{\mathbf{x} : \ \mathbf{A}\mathbf{x} = \mathbf{b},\ \mathbf{x} \geqslant \mathbf{0}\}$ and $\{\mathbf{y} : \mathbf{y}^T\mathbf{A} \leqslant \mathbf{0},\ \mathbf{y}^T\mathbf{b} = 1\}$ is nonempty. A vector $\mathbf{y}$ in the latter set is called an infeasibility certificate for the former.

6. (Strict complementarity) Consider any linear program in standard form and its dual and let both of them be feasible. Then, there always exists a strictly complementary solution pair, $(\mathbf{x}^*, \mathbf{y}^*, \mathbf{s}^*)$, such that

$$x_j^* s_j^* = 0 \text{ and } x_j^* + s_j^* > 0 \text{ for all } j.$$

Moreover, the supports of $\mathbf{x}^*$ and $\mathbf{s}^*$, $P^* = \{j : x_j^* > 0\}$ and $Z^* = \{j : x_j^* > 0\}$, are invariant among all strictly complementary solution pairs.

7. (Central path theorem) Let $(\mathbf{x}(\mu), \mathbf{y}(\mu), \mathbf{s}(\mu))$ be the central path of (5.9). Then prove

   (a) The central path point $(\mathbf{x}(\mu), \mathbf{y}(\mu), \mathbf{s}(\mu))$ is bounded for $0 < \mu \leqslant \mu^0$ and any given $0 < \mu^0 < \infty$.

   (b) For $0 < \mu' < \mu$,

$$\mathbf{c}^T\mathbf{x}(\mu') \leqslant \mathbf{c}^T\mathbf{x}(\mu) \text{ and } \mathbf{b}^T\mathbf{y}(\mu') \geqslant \mathbf{b}^T\mathbf{y}(\mu).$$

   Furthermore, if $\mathbf{x}(\mu') \neq \mathbf{x}(\mu)$ and $\mathbf{y}(\mu') \neq \mathbf{y}(\mu)$,

$$\mathbf{c}^T\mathbf{x}(\mu') < \mathbf{c}^T\mathbf{x}(\mu) \text{ and } \mathbf{b}^T\mathbf{y}(\mu') > \mathbf{b}^T\mathbf{y}(\mu).$$

   (c) $(\mathbf{x}(\mu), \mathbf{y}(\mu), \mathbf{s}(\mu))$ converges to an optimal solution pair for (LP) and (LD). Moreover, the limit point $\mathbf{x}(0)_{P^*}$ is the analytic center on the primal optimal face, and the limit point $\mathbf{s}(0)_{Z^*}$ is the analytic center on the dual optimal face, where (P^*, Z^*) is the strict complementarity partition of the index set $\{1, 2, \ldots, n\}$.

8. Consider a primal-dual interior point $(\mathbf{x}, \mathbf{y}, \mathbf{s}) \in \mathcal{N}(\eta)$ where $\eta < 1$. Prove that there is a fixed quantity $\delta > 0$ such that

$$x_j \geqslant \delta, \text{ for all } j \in P^*$$

and

$$s_j \geqslant \delta, \text{ for all } j \in Z^*,$$

where (P^*, Z^*) is defined in Exercise 6.

9. (Potential level theorem) Define the potential level set

$$\Psi(\delta) := \{(\mathbf{x}, \mathbf{y}, \mathbf{s}) \in \mathring{\mathcal{F}} : \psi_{n+\rho}(\mathbf{x}, \mathbf{s}) \leqslant \delta\}.$$

   Prove

   (a)

$$\Psi(\delta^1) \subset \Psi(\delta^2) \text{ if } \delta^1 \leqslant \delta^2.$$

   (b) For every δ, $\Psi(\delta)$ is bounded and its closure $\overline{\Psi}(\delta)$ has non-empty intersection with the solution set.

10. Given $\mathbf{0} < \mathbf{x},\ \mathbf{0} < \mathbf{s} \in E^n$, show that

$$n \log(\mathbf{x}^T\mathbf{s}) - \sum_{j=1}^{n} \log(x_j s_j) \geqslant n \log n$$

and

$$\mathbf{x}^T\mathbf{s} \leqslant \exp\left[\frac{\psi_{n+p}(\mathbf{x}, \mathbf{s}) - n \log n}{p}\right].$$

11. (Logarithmic approximation) If $\mathbf{d} \in E^n$ such that $|\mathbf{d}|_\infty < 1$ then

$$\mathbf{1}^T\mathbf{d} \geqslant \sum_{i=1}^{n} \log(1 + d_i) \geqslant \mathbf{1}^T\mathbf{d} - \frac{|\mathbf{d}|^2}{2(1 - |\mathbf{d}|_\infty)}.$$

[Note:If $\mathbf{d} = (d_1,\ d_2,\ \dots d_n)$ then $|\mathbf{d}|_\infty \equiv \max_i\{d_j\}$.]

12. Let the direction $(\mathbf{d_x},\ \mathbf{d_y},\ \mathbf{d_s})$ be generated by system (5.13) with $\gamma = n/(n+\rho)$ and $\mu = \mathbf{x}^T\mathbf{s}/n$, and let the step size be

$$\alpha = \frac{\theta \sqrt{\min(\mathbf{Xs})}}{|(\mathbf{XS})^{-1/2}(\frac{\mathbf{x}^T\mathbf{s}}{(n+\rho)}\mathbf{1} - \mathbf{Xs})|}, \tag{5.20}$$

where θ is a positive constant less than 1. Let

$$\mathbf{x}^+ = \mathbf{x} + \alpha\mathbf{d_x},\ \mathbf{y}^+ = \mathbf{y} + \alpha\mathbf{d_y},\ \text{and}\ \mathbf{s}^+ = \mathbf{s} + \alpha\mathbf{d_s}.$$

Then, using Exercise 11 and the concavity of the logarithmic function show $(\mathbf{x}^+,\ \mathbf{y}^+,\ \mathbf{s}^+) \in \overset{\circ}{\mathcal{F}}$ and

$$\begin{aligned}&\psi_{n+\rho}(\mathbf{x}^+,\ \mathbf{s}^+) - \psi_{n+\rho}(\mathbf{x},\ \mathbf{s})\\ &\quad\leqslant -\theta\sqrt{\min(\mathbf{Xs})}\,|(\mathbf{Xs})^{-1/2}(1 - \frac{(n+\rho)}{\mathbf{x}^T\mathbf{s}}\mathbf{Xs})| + \frac{\theta^2}{2(1-\theta)}.\end{aligned}$$

13. Let $\mathbf{v} = \mathbf{Xs}$ in Exercise 12. Prove

$$\sqrt{\min(\mathbf{v})}|\mathbf{V}^{-1/2}(\mathbf{1} - \frac{(n+\rho)}{\mathbf{1}^T\mathbf{V}}\mathbf{v})| \geqslant \sqrt{3/4},$$

where $\mathbf{V}$ is the diagonal matrix of $\mathbf{v}$. Thus, the two exercises imply

$$\psi_{n+\rho}(\mathbf{x}^+,\ \mathbf{s}^+) - \psi_{n+\rho}(\mathbf{x},\ \mathbf{s}) \leqslant -\theta\sqrt{3/4} + \frac{\theta^2}{2(1-\theta)} = -\delta$$

for a constant δ. One can verify that $\delta > 0.2$ when $\theta = 0.4$.

14. Prove property (5.18) for (HDSP).
15. Prove Theorem 1

References

5.1 Computation and complexity models were developed by a number of scientists; see, e.g., Cook [C5], Hartmanis and Stearns [H5] and Papadimitriou and Steiglitz [P2] for the bit complexity models and Blum et al. [B21] for the real number arithmetic model. For a general discussion of complexity see Vavasis [V4]. For a comprehensive treatment which served as the basis for much of this chapter, see Ye [Y3].

5.2 The Klee Minty example is presented in [K5]. Much of this material is based on a teaching note of Cottle on Linear Programming taught at Stanford [C6]. Practical performances of the simplex method can be seen in Bixby [B18]. The simplex method efficiency for the Markov Decision Process is due to Ye [269].

5.3 The ellipsoid method was developed by Khachiyan [K4]; more developments of the ellipsoid method can be found in Bland, Goldfarb and Todd [B20].

5.3 The analytic center for a convex polyhedron given by linear inequalities was introduced by Huard [H12], and later by Sonnevend [S8]. The barrier function was introduced by Frisch [F19]. The central path was analyzed in McLinden [M3], Megiddo [M4], and Bayer and Lagarias [B3, B4], Gill et al. [G5].

5.5 Path-following algorithms were first developed by Renegar [R1]. A primal barrier or path-following algorithm was independently analyzed by Gonzaga [G13]. Both Gonzaga [G13] and Vaidya [V1] extended the rank-one updating technique [K2] for solving the Newton equation of each iteration, and proved that each iteration uses $O(n^{2.5})$ arithmetic operations on average. Kojima, Mizuno and Yoshise [K6] and Monteiro and Adler [M7] developed a symmetric primal-dual path-following algorithm with the same iteration and arithmetic operation bounds.

5.6–5.7 Predictor-corrector algorithms were developed by Mizuno et al. [M6]. A more practical predictor-corrector algorithm was proposed by Mehrotra [M5] (also see Lustig et al. [L19] and Zhang and Zhang [Z3]). Mehrotra's technique has been used in almost all linear programming interior-point implementations. A primal potential reduction algorithm was initially proposed by Karmarkar [K2]. The primal-dual potential function was proposed by Tanabe [T2] and Todd and Ye [T5]. The primal-dual potential reduction algorithm was developed by Ye [Y1], Freund [F18], Kojima, Mizuno and Yoshise [K7], Goldfarb and Xiao [G11], Gonzaga and Todd [G14], Todd [T4], Tunçel [T10], Tutuncu [T11], and others. The homogeneous and self-dual embedding method can be found in Ye et al. [Y2], Luo et al. [L18], Andersen and Ye [A5], and many others. It is also implemented in most linear programming software packages such as SEDUMI of Sturm [S11].

5.1–5.7 There are several comprehensive text books which cover interior-point linear programming algorithms. They include Bazaraa, Jarvis and Sherali [B6], Bertsekas [B12], Bertsimas and Tsitsiklis [B13], Cottle [C6], Cottle, Pang and Stone [C7], Dantzig and Thapa [D9, D10], Fang and Puthenpura [F2], den Hertog [H6], Murty [M12], Nash and Sofer [N1], Nesterov [N2], Roos et al. [R4], Renegar [R2], Saigal [S1], Vanderebei [V3], and Wright [W8].

Chapter 6
Conic Linear Programming

6.1 Convex Cones

Conic Linear Programming, hereafter CLP, is a natural extension of Linear programming (LP). In LP, the variables form a vector which is required to be componentwise nonnegative, while in CLP they are points in a pointed convex cone (see Appendix B.1) of an Euclidean space, such as vectors as well as matrices of finite dimensions. For example, Semidefinite programming (SDP) is a kind of CLP, where the variable points are symmetric matrices constrained to be positive semidefinite. Both types of problems may have linear equality constraints as well. Although CLPs have long been known to be convex optimization problems, no efficient solution algorithm was known until about two decades ago, when it was discovered that interior-point algorithms for LP discussed in Chap. 5, can be adapted to solve certain CLPs with both theoretical and practical efficiency. During the same period, it was discovered that CLP, especially SDP, is representative of a wide assortment of applications, including combinatorial optimization, statistical computation, robust optimization, Euclidean distance geometry, quantum computing, optimal control, etc. CLP is now widely recognized as a powerful mathematical computation model of general importance.

First, we illustrate several convex cones popularly used in conic linear optimization.

Example 1. The followings are all (closed) convex cones.

- The n-dimensional non-negative orthant, $E^n_+ = \{\mathbf{x} \in E^n : \mathbf{x} \geq \mathbf{0}\}$, is a convex cone.
- The set of all n-dimensional symmetric positive semidefinite matrices, denoted by $\mathcal{S}^n_+$, is a convex cone, called the *positive semidefinite matrix cone*. When $\mathbf{X}$ is positive semidefinite (positive definite), we often write the property as $\mathbf{X} \succeq (\succ)\mathbf{0}$.

D.G. Luenberger, Y. Ye, *Linear and Nonlinear Programming*, International Series in Operations Research & Management Science 228,
DOI 10.1007/978-3-319-18842-3_6

- The set $\{(u;\mathbf{x}) \in E^{n+1} : u \geq |\mathbf{x}|_p\}$ is a convex cone in E^{n+1}, called the *p-order cone* where $1 \leq p < \infty$. When $p = 2$, the cone is called second-order cone or "Ice-cream" cone.

Sometimes, we use the notion of conic inequalities $\mathbf{P} \succeq_K \mathbf{Q}$ or $\mathbf{Q} \preceq_K \mathbf{P}$, in which cases we simply mean $\mathbf{P} - \mathbf{Q} \in K$.

Suppose $\mathbf{A}$ and $\mathbf{B}$ are $k \times n$ matrices. We define the inner product

$$\mathbf{A} \bullet \mathbf{B} = \text{trace}(\mathbf{A}^T\mathbf{B}) = \sum_{i,j} a_{ij}b_{ij}.$$

When $k = 1$, they become n-dimensional vectors and the inner product is the standard dot product of two vectors. In SDP, this definition is almost always used for the case where the matrices are both square and symmetric. The matrix norm associated with the inner product is called *Frobenius norm*:

$$|\mathbf{X}|_f = \sqrt{\mathbf{X} \bullet \mathbf{X}}.$$

For a cone K, the dual of K is the cone

$$K^* := \{\mathbf{Y} : \mathbf{X} \bullet \mathbf{Y} \geq 0 \quad \text{for all} \quad \mathbf{X} \in K\}.$$

It is not difficult to see that the dual cones of the first two cones in Example 1 are all them self, respectively; while the dual cone of the p-order cone is the q-order cone where

$$\frac{1}{p} + \frac{1}{q} = 1.$$

One can see that when $p = 2$, $q = 2$ as well; that is, they are both 2-order cones. For a closed convex cone K, the dual of the dual cone is itself.

6.2 Conic Linear Programming Problem

Now let $\mathbf{C}$ and $\mathbf{A}_i$, $i = 1, 2, \ldots, m$, be given matrices of $E^{k\times n}$, $\mathbf{b} \in E^m$, and $\mathcal{K}$ be a closed convex cone in $E^{k\times n}$. And let $\mathbf{X}$ be an unknown matrix of $E^{k\times n}$. Then, the standard form (primal) conic linear programming problem is

$$\begin{array}{ll}\text{(CLP)} & \text{minimize } \mathbf{C} \bullet \mathbf{X} \\ & \text{subject to } \mathbf{A}_i \bullet \mathbf{X} = b_i,\ i = 1, 2, \ldots, m,\ \mathbf{X} \in K.\end{array} \tag{6.1}$$

Note that in CLP we minimize a linear function of the decision matrix constrained in cone K and subject to linear equality constraints.

For convenience, we define an operator from a symmetric matrix to a vector:

$$\mathcal{A}\mathbf{X} = \begin{pmatrix} \mathbf{A}_1 \bullet \mathbf{X} \\ \mathbf{A}_2 \bullet \mathbf{X} \\ \cdots \\ \mathbf{A}_m \bullet \mathbf{X} \end{pmatrix}. \tag{6.2}$$

Then, CLP can be written in a compact form:

$$\begin{array}{ll} \text{(CLP)} & \text{minimize } \mathbf{C} \bullet \mathbf{X} \\ & \text{subject to } \mathcal{A}\mathbf{X} = \mathbf{b}, \ \mathbf{X} \in K. \end{array}$$

When cone K is the non-negative orthant E^n_+, CLP reduces to linear programming (LP) in the standard form, where $\mathcal{A}$ becomes the constraint matrix $\mathbf{A}$. When K is the positive semidefinite cone $\mathcal{S}^n_+$, CLP is called semidefinite programming (SDP); and when K is the p-order cone, it is called p-order cone programming. In particular, when $p = 2$, the model is called second-order cone programming (SOCP). Frequently, we write variable $\mathbf{X}$ in (CLP) as $\mathbf{x}$ if it is indeed a vector, such as when K is the nonnegative orthant or p-order cone.

One can see that the problem (SDP) (that is, (6.1) with the semidefinite cone) generalizes classical linear programming in standard form:

$$\begin{array}{l} \text{minimize } \mathbf{c}^T\mathbf{x}, \\ \text{subject to } \mathbf{A}\mathbf{x} = \mathbf{b}, \ \mathbf{x} \geq \mathbf{0}. \end{array}$$

Define $\mathbf{C} = \text{Diag}[c_1,\ c_2,\ \ldots,\ c_n]$, and let $\mathbf{A}_i = \text{Diag}[a_{i1},\ a_{i2},\ \ldots,\ a_{in}]$ for $i = 1, 2, \ldots m$. The unknown is the $n \times n$ symmetric matrix $\mathbf{X}$ which is constrained by $\mathbf{X} \succeq \mathbf{0}$. Since the trace of $\mathbf{C} \bullet \mathbf{X}$ and $\mathbf{A}_i \bullet \mathbf{X}$ depend only on the diagonal elements of $\mathbf{X}$, we may restrict the solutions $\mathbf{X}$ to diagonal matrices. It follows that in this case the SDP problem is equivalent to a linear program, since a diagonal matrix is positive semidefinite is and only if its all diagonal elements are nonnegative.

One can further see the role of cones in the following examples.

Example 1. Consider the following optimization problems with three variables.

- This is a linear programming problem in standard form:

$$\begin{array}{l} \text{minimize } 2x_1 + x_2 + x_3 \\ \text{subject to } x_1 + x_2 + x_3 = 1, \\ \qquad\qquad (x_1; x_2; x_3) \geq \mathbf{0}. \end{array}$$

- This is a semidefinite programming problem where the dimension of the matrix is two:

$$\begin{array}{l} \text{minimize } 2x_1 + x_2 + x_3 \\ \text{subject to } x_1 + x_2 + x_3 = 1, \\ \qquad\qquad \begin{bmatrix} x_1 & x_2 \\ x_2 & x_3 \end{bmatrix} \succeq \mathbf{0}, \end{array}$$

Let

$$\mathbf{C} = \begin{bmatrix} 2 & .5 \\ .5 & 1 \end{bmatrix} \quad \text{and} \quad \mathbf{A}_1 = \begin{bmatrix} 1 & .5 \\ .5 & 1 \end{bmatrix}.$$

Then, the problem can be written in a standard SDP form

$$\begin{array}{l} \text{minimize } \mathbf{C} \bullet \mathbf{X} \\ \text{subject to } \mathbf{A}_1 \bullet \mathbf{X} = 1, \ \mathbf{X} \in \mathcal{S}^2_+. \end{array}$$

- This is a second-order cone programming problem:

$$\begin{array}{l} \text{minimize } 2x_1 + x_2 + x_3 \\ \text{subject to } x_1 + x_2 + x_3 = 1, \\ \qquad\qquad \sqrt{x_2^2 + x_3^2} \le x_1. \end{array}$$

We present several application examples to illustrate the flexibility of this formulation.

Example 2 (Binary Quadratic Optimization). Consider a binary quadratic maximization problem

$$\begin{array}{l} \text{maximize } \mathbf{x}^T\mathbf{Q}\mathbf{x} + 2\mathbf{c}^T\mathbf{x} \\ \text{subject to } x_j = \{1, -1\}, \text{ for all } j = 1, \ldots, n, \end{array}$$

which is a difficult nonconvex optimization problem. The problem can be rewritten as

$$\begin{array}{l} z^* \equiv \text{maximize} \begin{bmatrix} \mathbf{x} \\ 1 \end{bmatrix}^T \begin{bmatrix} \mathbf{Q} & \mathbf{c} \\ \mathbf{c}^T & 0 \end{bmatrix} \begin{bmatrix} \mathbf{x} \\ 1 \end{bmatrix} \\ \qquad \text{subject to } (x_j)^2 = 1, \text{ for all } j = 1, \ldots, n, \end{array}$$

which can be also written as a homogeneous quadratic binary problem

$$\begin{array}{l} z^* \equiv \text{maximize} \begin{bmatrix} \mathbf{Q} & \mathbf{c} \\ \mathbf{c}^T & \mathbf{0} \end{bmatrix} \bullet \begin{bmatrix} \mathbf{x} \\ x_{n+1} \end{bmatrix} \begin{bmatrix} \mathbf{x} \\ x_{n+1} \end{bmatrix}^T \\ \text{subject to } \mathbf{I}_j \bullet \begin{bmatrix} \mathbf{x} \\ x_{n+1} \end{bmatrix} \begin{bmatrix} \mathbf{x} \\ x_{n+1} \end{bmatrix}^T = 1, \text{ for all } j = 1, \ldots, n+1, \end{array}$$

where $\mathbf{I}_j$ is the $(n+1) \times (n+1)$ matrix whose components are all zero except at the jth position on the main diagonal where it is 1. Let $(\mathbf{x}^*; \ x^*_{n+1})$ be an optimal solution for the homogeneous problem. Then, one can see that $\mathbf{x}^*/x^*_{n+1}$ would be an optimal solution to the original problem.

Since $\begin{bmatrix} \mathbf{x} \\ x_{n+1} \end{bmatrix} \begin{bmatrix} \mathbf{x} \\ x_{n+1} \end{bmatrix}^T$ forms a positive-semidefinite matrix (with rank equal to 1), a *semidefinite relaxation* of the problem is defined as

$$
\begin{aligned}
z^{SDP} \equiv \text{maximize} & \begin{bmatrix} \mathbf{Q} & \mathbf{c} \\ \mathbf{c}^T & 0 \end{bmatrix} \bullet \mathbf{Y} \\
\text{subject to } & \mathbf{I}_j \bullet \mathbf{Y} = 1, \text{ for all } j = 1, \ldots, n+1, \\
& \mathbf{Y} \in \mathcal{S}_+^{n+1},
\end{aligned} \tag{6.3}
$$

where the symmetric matrix $\mathbf{Y}$ has dimension $n+1$. Obviously, z^{SDP} is a upper bound of z^*, since the rank-1 requirement is not enforced in the relaxation.

Let's see how to use the relaxation. For simplicity, assuming $z^{SDP} > 0$, it has been shown that in many cases of this problem an optimal SDP solution either constitutes an exact solution or can be rounded to a good approximate solution of the original problem. In the former case, one can show that a rank-1 optimal solution matrix $\mathbf{Y}$ exists for the semidefinite relaxation and it can be found by using a rank-reduction procedure. For the latter case, one can, using a randomized rank-reduction procedure or the principle components of $\mathbf{Y}$, find a rank-1 feasible solution matrix $\hat{\mathbf{Y}}$ such that

$$
\begin{bmatrix} \mathbf{Q} & \mathbf{c} \\ \mathbf{c}^T & \mathbf{0} \end{bmatrix} \bullet \hat{\mathbf{Y}} \geq \alpha \cdot Z^{SDP} \geq \alpha \cdot Z^*
$$

for a provable factor $0 < \alpha \leq 1$. Thus, one can find a feasible solution to the original problem whose objective value is no less than a factor α of the true maximal objective cost.

Example 3 (Sensor Localization). This problem is that of determining the location of sensors (for example, several cell phones scattered in a building) when measurements of some of their separation Euclidean distances can be determined, but their specific locations are not known. In general, suppose there are n unknown points $\mathbf{x}_j \in E^d$, $j = 1, \ldots, n$. We consider an edge to be a path between two points, say, i and j. There is a known subset N_e of pairs (edges) ij for which the separation distance d_{ij} is known. For example, this distance might be determined by the signal strength or delay time between the points. Typically, in the cell phone example, N_e contains those edges whose lengths are small so that there is a strong radio signal. Then, the localization problem is to find locations $\mathbf{x}_j$, $j = 1, \ldots, n$, such that

$$
|\mathbf{x}_i - \mathbf{x}_j|^2 = (d_{ij})^2, \text{ for all } (i, j) \in N_e,
$$

subject to possible rotation and translation. (If the locations of some of the sensors are known, these may be sufficient to determine the rotation and translation as well.)

Let $\mathbf{X} = [\mathbf{x}_1 \ \mathbf{x}_2 \ \ldots \ \mathbf{x}_n]$ be the $d \times n$ matrix to be determined. Then

$$
|\mathbf{x}_i - \mathbf{x}_j|^2 = (\mathbf{e}_i - \mathbf{e}_j)^T \mathbf{X}^T \mathbf{X} (\mathbf{e}_i - \mathbf{e}_j),
$$

where $\mathbf{e}_i \in E^n$ is the vector with 1 at the i**th** position and zero everywhere else. Let $\mathbf{Y} = \mathbf{X}^T\mathbf{X}$. Then the semidefinite relaxation of the localization problem is to find $\mathbf{Y}$ such that

$$(\mathbf{e}_i - \mathbf{e}_j)(\mathbf{e}_i - \mathbf{e}_j)^T \bullet \mathbf{Y} = (d_{ij})^2, \text{ for all } (i,\ j) \in N_e,$$
$$\mathbf{Y} \succeq \mathbf{0}.$$

This problem is one of finding a feasible solution; the objective function is null. But if the distance measurements have noise, one can add additional variables and an error objective to minimize. For example,

$$\begin{array}{ll} \text{minimize} & \sum_{(i,j)\in N_e} |z_{ij}| \\ \text{subject to} & (\mathbf{e}_i - \mathbf{e}_j)(\mathbf{e}_i - \mathbf{e}_j)^T \bullet \mathbf{Y} + z_{ij} = (d_{ij})^2, \text{ for all } (i,\ j) \in N_e, \\ & \mathbf{Y} \succeq \mathbf{0}. \end{array}$$

This problem can be converted into a conic linear program with mixed nonnegative orthant and semidefinite cones.

Under certain graph structure, an optimal SDP solution $\mathbf{Y}$ of the formulation would be guaranteed rank-d so that it constitutes an exact solution of the original problem. Also, in general $\mathbf{Y}$ can be rounded to a good approximate solution of the original problem. For example, one can, using a randomized rank-reduction procedure or the d principle components of $\mathbf{Y}$, find a rank-d solution matrix $\hat{\mathbf{Y}}$.

6.3 Farkas' Lemma for Conic Linear Programming

We first introduce the notion of "interior" of cones.

Definition 1. We call $\mathbf{X}$ an interior point of cone K if and only if, for any point $\mathbf{Y} \in K^*$, $\mathbf{Y} \bullet \mathbf{X} = 0$ implies $\mathbf{Y} = \mathbf{0}$.

The set of interior points of K is denoted by $\overset{\circ}{K}$.

Theorem 1. *The interior of the followings convex cones are given as:*

- *The interior of the non-negative orthant cone is the set of all vectors where every entry is positive.*
- *The interior of the positive semidefinite cone is the set of all positive definite matrices.*
- *The interior of p-order cone is the set of* $\{(u; \mathbf{x}) \in E^{n+1} : u > |\mathbf{x}|_p\}$.

We give a sketch of the proof for the second order cone, i.e., $p = 2$. Let $(\bar{u}; \bar{\mathbf{x}}) \neq \mathbf{0}$ be any second-order cone point but $\bar{u} = |\bar{\mathbf{x}}|$. Then, we can choose a dual cone (also the second-order cone) point $(v; \mathbf{y})$ such that

$$v = \alpha\bar{u},\ \mathbf{y} = -\alpha\bar{\mathbf{x}},$$

for a positive α. Note that

$$(\bar{u}; \bar{\mathbf{x}}) \bullet (v; \mathbf{y}) = \alpha\bar{v}^2 - \alpha|\bar{\mathbf{x}}|^2 = 0.$$

Then, one can let $\alpha \to \infty$ so that $(v; \mathbf{y})$ cannot be bounded.

Now let $(\bar{u}; \bar{\mathbf{x}})$ be any given second-order cone point with $\bar{u} > |\bar{\mathbf{x}}|$. We like to prove that, for any dual cone (also the second-order cone) point $(v; \mathbf{y})$,

$$(\bar{u}; \bar{\mathbf{x}}) \bullet (v; \mathbf{y}) = 0$$

implies that $(v; \mathbf{y})$ is bounded. Note that

$$0 = (\bar{u}; \bar{\mathbf{x}}) \bullet (v; \mathbf{y}) = \bar{u}v + \bar{\mathbf{x}} \bullet \mathbf{y}$$

or

$$\bar{u}v \leq -\bar{\mathbf{x}} \bullet \mathbf{y} \leq |\bar{\mathbf{x}}||\mathbf{y}|.$$

If $v = 0$, we must have $\mathbf{y} = \mathbf{0}$; otherwise,

$$\bar{u} \leq |\bar{\mathbf{x}}||\mathbf{y}|/v \leq |\mathbf{x}|,$$

which contradicts $\bar{u} > |\bar{\mathbf{x}}|$.

We leave the proof of the following proposition as an exercise.

Proposition 1. *Let* $\mathbf{X} \in \overset{\circ}{K}$ *and* $\mathbf{Y} \in K^*$. *Then For any nonnegative constant* κ, $Y \bullet X \leq \kappa$ *implies that* $\mathbf{Y}$ *is bounded.*

Let us now consider the feasible region of (CLP) (6.1):

$$\mathcal{F} := \{\mathbf{X} : \ \mathcal{A}\mathbf{X} = \mathbf{b}, \ \mathbf{X} \in K\};$$

where the interior of the feasible region is

$$\overset{\circ}{\mathcal{F}} := \{\mathbf{X} : \ \mathcal{A}\mathbf{X} = \mathbf{b}, \ \mathbf{X} \in \overset{\circ}{K}\}.$$

If $\mathcal{F}$ is empty with $K = E^n_+$, from Farkas' lemma for linear programming, a vector $\mathbf{y} \in E^m$, with $\mathbf{y}^T A \leq \mathbf{0}$ and $\mathbf{y}^T\mathbf{b} > 0$, always exists and is called an infeasibility certificate for the system $\{\mathbf{x} : \ A\mathbf{x} = \mathbf{b}, \ \mathbf{x} \geq \mathbf{0}\}$.

Does this alternative relations hold for K being a general closed convex one? Let us rigorousize the question. Let us define the reverse operator of (6.2) from a vector to a matrix:

$$\mathbf{y}^T\mathcal{A} = \sum_{i=1}^{m} \mathbf{A}_i y_i. \tag{6.4}$$

Note that, by the definition, for any matrix $\mathbf{X} \in E^{k \times n}$

$$\mathbf{y}^T\mathcal{A} \bullet \mathbf{X} = \mathbf{y}^T(\mathcal{A}\mathbf{X}),$$

that is, the association property holds. Also, $(\mathbf{y}^T\mathcal{A})^T = \mathcal{A}^T\mathbf{y}$, that is, the transpose operation applies here as well.

Then, the question becomes: when $\mathcal{F}$ is empty, does there exist a vector $\mathbf{y} \in E^m$ such that $-\mathbf{y}^T\mathcal{A} \in K^*$ and $\mathbf{y}^T\mathbf{b} > 0$? Similarly, one can ask: when set

$\{\mathbf{y} : \mathbf{C}^T - \mathbf{y}^T \mathcal{A} \in K\}$ is empty, does there exist a matrix $\mathbf{X} \in K^*$ such that $\mathcal{A}\mathbf{X} = \mathbf{0}$ and $\mathbf{C} \bullet \mathbf{X} < 0$? Note that the answer to the second question is also "yes" when $K = E^n_+$.

Example 1. The answer to either question is "not necessarily"; see example below.

- For the first question, consider $K = \mathcal{S}^2_+$ and

$$\mathbf{A}_1 = \begin{bmatrix} 1 & 0 \\ 0 & 0 \end{bmatrix}, \quad \mathbf{A}_2 = \begin{bmatrix} 0 & 1 \\ 1 & 0 \end{bmatrix}$$

  and

$$\mathbf{b} = \begin{bmatrix} 0 \\ 2 \end{bmatrix}$$

- For the second question, consider $K = \mathcal{S}^2_+$ and

$$C = \begin{bmatrix} 0 & 1 \\ 1 & 0 \end{bmatrix} \quad \text{and} \quad A_1 = \begin{bmatrix} 1 & 0 \\ 0 & 0 \end{bmatrix}.$$

However, if the data set $\mathcal{A}$ satisfies additional conditions, the answer would be "yes"; see theorem below.

Theorem 2 (Farkas' Lemma for CLP). *We have*

- *Consider set*

$$\mathcal{F}_p := \{\mathbf{X} : \mathcal{A}\mathbf{X} = \mathbf{b}, \ \mathbf{X} \in K\}.$$

  Suppose that there exists a vector $\mathring{\mathbf{y}}$ *such that* $-\mathring{\mathbf{y}}^T \mathcal{A} \in \mathring{K^*}$. *Then,*

  1. *Set* $C := \{\mathcal{A}\mathbf{X} \in E^m : \mathbf{X} \in K\}$ *is a closed convex set;*
  2. $\mathcal{F}_p$ *has a (feasible) solution if and only if set* $\{\mathbf{y} : -\mathbf{y}^T \mathcal{A} \in K^*, \ \mathbf{y}^T \mathbf{b} > 0\}$ *has no feasible solution.*

- *Consider set*

$$\mathcal{F}_d := \{\mathbf{y} : \mathbf{C}^T - \mathbf{y}^T \mathcal{A} \in K\}.$$

  Suppose that there exists a vector $\mathring{\mathbf{X}} \in \mathring{K^*}$ *such that* $\mathcal{A}\mathring{\mathbf{X}} = \mathbf{0}$. *Then,*

  1. *Set* $C := \{\mathbf{S} - \mathbf{y}^T \mathcal{A} : \mathbf{S} \in K\}$ *is a closed convex set;*
  2. $\mathcal{F}_d$ *has a (feasible) solution if and only if set* $\{\mathbf{X} : \mathcal{A}\mathbf{X} = \mathbf{0}, \ \mathbf{X} \in K^*, \ \mathbf{C} \bullet \mathbf{X} < 0\}$ *has no feasible solution.*

Proof. We prove the first statement of the theorem. We prove the first part. It is clear that C is a convex set. To prove that C is a closed set, we need to show that if $\mathbf{y}^k := \mathcal{A}\mathbf{X}^k \in E^m$ for $\mathbf{X}^k \in K$, $k = 1, \ldots,$ converges to a vector $\bar{\mathbf{y}}$, then $\bar{\mathbf{y}} \in C$ or there is $\bar{\mathbf{X}} \in K$ such that $\bar{\mathbf{y}} := \mathcal{A}\bar{\mathbf{X}}$. Without loss of generality, we assume that $\mathbf{y}^k$ is a bounded sequence. Then, we have, for a positive constant c,

$$c \geq -(\mathring{\mathbf{y}})^T \mathbf{y}^k = -(\mathring{\mathbf{y}})^T (\mathcal{A}\mathbf{X}^k) = -(\mathring{\mathbf{y}})^T \mathcal{A} \bullet \mathbf{X}^k, \forall k.$$

Since $-(\overset{\circ}{\mathbf{y}})^T\mathcal{A} \in \overset{\circ}{K^*}$, by definition, the sequence of $\mathbf{X}^k$ is also bounded. Then there is at least an accumulate point $\bar{\mathbf{X}} \in K$ because K is a closed cone. Thus, we must have $\bar{\mathbf{y}} := \mathcal{A}\bar{\mathbf{X}}$.

We now prove the second part. If $\mathcal{F}_p$ has a feasible solution $\bar{\mathbf{X}}$. Then, let $\mathbf{y}$ make $-\mathbf{y}^T\mathcal{A} \in K^*$

$$-\mathbf{y}^T\mathbf{b} = -\mathbf{y}^T(\mathcal{A}\bar{\mathbf{X}}) = -\mathbf{y}^T\mathcal{A} \bullet \bar{\mathbf{X}} \geq 0.$$

Thus, it must be true $\mathbf{y}^T\mathbf{b} \leq 0$, that is, $\{\mathbf{y} : \ -\mathbf{y}^T\mathcal{A} \in K^*, \ \mathbf{y}^T\mathbf{b} > 0\}$ must be empty.

On the other hand, let $\mathcal{F}_p$ has no feasible solution, or equivalently, $\mathbf{b} \notin C$. We now show that $\{\mathbf{y} : \ -\mathbf{y}^T\mathcal{A} \in K^*, \ \mathbf{y}^T\mathbf{b} > 0\}$ must be nonempty.

Since C is a closed convex set, from the separating hyperplane theorem, there must exist a $\bar{\mathbf{y}} \in E^m$ such that

$$\bar{\mathbf{y}}^T\mathbf{b} > \bar{\mathbf{y}}^T\mathbf{y}, \ \forall \mathbf{y} \in C,$$

or, from $\mathbf{y} = \mathcal{A}\mathbf{X}, \ \mathbf{X} \in K$, we have

$$\bar{\mathbf{y}}^T\mathbf{b} > \bar{\mathbf{y}}^T(\mathcal{A}\mathbf{X}) = \bar{\mathbf{y}}^T\mathcal{A} \bullet \mathbf{X}, \ \forall \mathbf{X} \in K.$$

That is, $\bar{\mathbf{y}}^T\mathcal{A} \bullet \mathbf{X}$ is bounded above for all $\mathbf{X} \in K$.

Immediately, we see $\bar{\mathbf{y}}^T\mathbf{b} > 0$ since $\mathbf{0} \in K$. Next, it must be true $-\bar{\mathbf{y}}^T\mathcal{A} \in K^*$. Otherwise, we must be able to find an $\bar{\mathbf{X}} \in K$ such that $-\bar{\mathbf{y}}^T\mathcal{A} \bullet \bar{\mathbf{X}} < 0$ by the definition of K and its dual K^*. For any positive constant α we maintain $\alpha\bar{\mathbf{X}} \in K$ and let α go to ∞. Then, $\bar{\mathbf{y}}^T\mathcal{A} \bullet (\alpha\bar{\mathbf{X}})$ goes to ∞, contradicting the fact that $\bar{\mathbf{y}}^T\mathcal{A} \bullet \mathbf{X}$ is bounded above for all $\mathbf{X} \in K$. Thus, $\bar{\mathbf{y}}$ is a feasible solution in $\{\mathbf{y} : \ -\mathbf{y}^T\mathcal{A} \in K^*, \ \mathbf{y}^T\mathbf{b} > 0\}$. ∎

Note that C may not be a closed set if the interior condition of Theorem 2 is not met. Consider $\mathbf{A}_1$, $\mathbf{A}_2$ and $\mathbf{b}$ in Example 1, and we have

$$C = \left\{ \mathcal{A}\mathbf{X} = \begin{bmatrix} \mathbf{A}_1 \bullet \mathbf{X} \\ \mathbf{A}_2 \bullet \mathbf{X} \end{bmatrix} : \ \mathbf{X} \in \mathcal{S}^2_+ \right\}.$$

Let

$$\mathbf{X}^k = \begin{bmatrix} \frac{1}{k} & 1 \\ 1 & k \end{bmatrix} \in \mathcal{S}^2_+, \ \forall k = 1, \ldots.$$

Then we see

$$\mathbf{y}^k = \mathcal{A}\mathbf{X}^k = \begin{bmatrix} \frac{1}{k} \\ 2 \end{bmatrix}.$$

As $k \to \infty$ we see $\mathbf{y}^k$ converges $\mathbf{b}$, but $\mathbf{b}$ is *not* in C.

6.4 Conic Linear Programming Duality

Because conic linear programming is an extension of classical linear programming, it would seem that there is a natural dual to the primal problem, and that this dual is itself a conic linear program. This is indeed the case, and it is related to the primal in much the same way as primal and dual linear programs are related. Furthermore, the primal and dual together lead to the formation a primal-dual solution method, which is discussed later in this chapter.

The dual of the (primal) CLP (6.1) is

$$\begin{array}{ll}\text{(CLD)} & \text{maximize } \mathbf{y}^T\mathbf{b}\\ & \text{subject to } \sum_i^m y_i\mathbf{A}_i + \mathbf{S} = \mathbf{C}^T,\ \mathbf{S} \in K^*.\end{array} \tag{6.5}$$

On written in a compact form:

$$\begin{array}{ll}\text{(CLD)} & \text{maximize } \mathbf{y}^T\mathbf{b}\\ & \text{subject to } \mathbf{y}^T\mathcal{A} + \mathbf{S} = \mathbf{C}^T,\ \mathbf{S} \in K^*.\end{array}$$

Notice that $\mathbf{S}$ represents a slack matrix, and hence the problem can alternatively be expressed as

$$\begin{array}{l}\text{maximize } \mathbf{y}^T\mathbf{b}\\ \text{subject to } \sum_i^m y_i\mathbf{A}_i \preceq_{K^*} \mathbf{C}^T.\end{array} \tag{6.6}$$

Recall that conic inequality $\mathbf{Q} \preceq_K \mathbf{P}$ means $\mathbf{P} - \mathbf{Q} \in K$.

Again, just like linear programming, the dual of (CLD) will be (CLP), and they form a primal and dual pair. Whichever is the primal, then the other will be the dual. We would see more primal and dual relations later.

Example 1. Here are dual problems to the three instances in Example 1 where $\mathbf{y}$ is just a scalar.

- The dual to the linear programming instance:

$$\begin{array}{ll}\text{maximize} & y\\ \text{subject to} & y(1,\ 1,\ 1) + (s_1,\ s_2,\ s_3) = (2,\ 1,\ 1),\\ & \mathbf{s} = (s_1,\ s_2,\ s_3) \in K^* = E^3_+.\end{array}$$

- The dual to semidefinite programming instance:

$$\begin{array}{ll}\text{maximize} & y\\ \text{subject to} & y\mathbf{A}_1 + \mathbf{S} = \mathbf{C},\\ & \mathbf{S} \in K^* = \mathcal{S}^2_+,\end{array}$$

where recall

$$\mathbf{C} = \begin{bmatrix} 2 & .5 \\ .5 & 1 \end{bmatrix} \quad \text{and} \quad \mathbf{A}_1 = \begin{bmatrix} 1 & .5 \\ .5 & 1 \end{bmatrix}.$$

- The dual to the second-order cone instance:

$$\begin{array}{ll} \text{maximize} & y \\ \text{subject to} & y(1,\ 1,\ 1) + (s_1,\ s_2,\ s_3) = (2,\ 1,\ 1), \\ & \sqrt{s_2^2 + s_3^2} \le s_1, \text{ or } \mathbf{s} = (s_1,\ s_2,\ s_3) \text{ in second-order cone.} \end{array}$$

Let us consider a couple of more dual examples of the problems we posted earlier.

Example 2 (The Dual of Binary Quadratic Maximization). Consider the semidefinite relaxation (6.3) for the binary quadratic maximization problem. It's dual is

$$\begin{array}{ll} \text{minimize} & \sum_{j=1}^{n+1} y_j \\ \text{subject to} & \sum_{j=1}^{n+1} y_j \mathbf{I}_j - \mathbf{S} = \begin{bmatrix} \mathbf{Q} & \mathbf{c} \\ \mathbf{c}^T & 0 \end{bmatrix}, \ \mathbf{S} \succeq \mathbf{0}. \end{array}$$

Note that

$$\sum_{j=1}^{n+1} y_j \mathbf{I}_j - \begin{bmatrix} \mathbf{Q} & \mathbf{c} \\ \mathbf{c}^T & 0 \end{bmatrix}$$

is exactly the Hessian matrix of the Lagrange function of the quadratic maximization problem; see Chap. 11. Therefore, there is a close connection between the Lagrange and conic dualities. The problems is to find a diagonal matrix $\text{Diag}[(y_1;\ldots;y_{n+1})]$ such that the Lagrange Hessian is positive semidefinite and its sum of diagonal elements is minimized.

Example 3 (The Dual of Sensor Localization). Consider the semidefinite programming relaxation for the sensor localization problem (with no noises). It's dual is

$$\begin{array}{ll} \text{maximize} & \sum_{(i,j)\in N_e} y_{ij} \\ \text{subject to} & \sum_{(i,j)\in N_e} y_{ij}(\mathbf{e}_i - \mathbf{e}_j)(\mathbf{e}_i - \mathbf{e}_j)^T + \mathbf{S} = \mathbf{0},\ \mathbf{S} \succeq \mathbf{0}. \end{array}$$

Here, y_{ij} represents an internal force or tension on edge $(i,\ j)$. Obviously, $y_{ij} = 0$ for all $(i,\ j) \in N_e$ is a feasible solution for the dual. However, finding non-trivial internal forces is a fundamental problem in network and structure design, and the maximization of the dual would help to achieve the goal.

Many optimization problems can be directly cast in the CLD form.

Example 4 (Euclidean Facility Location). This problem is to determine the location of a facility serving n clients placed in a Euclidean space, whose known locations are denoted by $\mathbf{a}_j \in E^d$, $j = 1, \ldots, n$. The location of the facility would minimize the sum of the Euclidean distances from the facility to each of the clients. Let the location decision be vector $\mathbf{f} \in E^d$. Then the problem is

$$\text{minimize } \sum_{j=1}^{n} |\mathbf{f} - \mathbf{a}_j| .$$

The problem can be reformulated as

$$\begin{array}{ll} \text{minimize} & \sum_{j=1}^{n} \delta_j \\ \text{subject to} & \mathbf{s}_j + \mathbf{f} = \mathbf{a}_j, \ \forall j = 1, \ldots, n, \\ & |\mathbf{s}_j| \le \delta_j, \quad \forall j = 1, \ldots, n. \end{array}$$

This is a conic formulation in the (CLD) form. To see it clearly, let $d = 2$ and $n = 3$ in the example, and let

$$\mathbf{A}^T = \begin{bmatrix} 1 & 0 & 0 & 0 & 0 \\ 0 & 0 & 0 & -1 & 0 \\ 0 & 0 & 0 & 0 & -1 \\ 0 & 1 & 0 & 0 & 0 \\ 0 & 0 & 0 & -1 & 0 \\ 0 & 0 & 0 & 0 & -1 \\ 0 & 0 & 1 & 0 & 0 \\ 0 & 0 & 0 & -1 & 0 \\ 0 & 0 & 0 & 0 & -1 \end{bmatrix} \in E^{9\times 5}, \ \mathbf{b} = \begin{bmatrix} 1 \\ 1 \\ 1 \\ 0 \\ 0 \end{bmatrix} \in E^5, \ \mathbf{c} = \begin{bmatrix} 0 \\ 0 \\ 0 \\ \mathbf{a}_1 \\ \mathbf{a}_2 \\ \mathbf{a}_3 \end{bmatrix} \in E^9,$$

and variable vector

$$\mathbf{y} = [\delta_1; \ \delta_2; \ \delta_3; \ \mathbf{f}] \in E^5.$$

Then, the facility location problem becomes

$$\begin{array}{ll} \text{minimize} & \mathbf{y}^T\mathbf{b} \\ \text{subject to} & \mathbf{y}^T\mathbf{A} + \mathbf{s}^T = \mathbf{c}^T, \ \mathbf{s} \in K; \end{array}$$

where K is the product of three second-order cones each of which has dimension 3. More precisely, the first three elements of $\mathbf{s} \in E^9$ are in the 3-dimensional second-order cone; and so are the second three elements and the third three elements of $\mathbf{s}$. In general, the product of (possibly mixed) cones, say K_1, K_2 and K_3, is denoted by $K_1 \oplus K_2 \oplus K_3$, and $\mathbf{X} \in K_1 \oplus K_2 \oplus K_3$ means that $\mathbf{X}$ is divided into three components such that

$$\mathbf{X} = (\mathbf{X}_1; \ \mathbf{X}_2; \ \mathbf{X}_3), \text{ where } \mathbf{X}_1 \in K_1, \ \mathbf{X}_2 \in K_2, \text{ and } \mathbf{X}_3 \in K_3.$$

The dual of the facility location problem would be in the (CLP) form:

$$\begin{array}{ll} \text{minimize} & \mathbf{c}^T\mathbf{x} \\ \text{subject to} & \mathbf{A}\mathbf{x} = \mathbf{b}, \ \mathbf{x} \in K^*; \end{array}$$

where

$$K^* = (K_1 \oplus K_2 \oplus K_3)^* = K_1^* \oplus K_2^* \oplus K_3^*.$$

That is, in this particular problem, the first three elements of $\mathbf{x} \in E^9$ are in the 3-dimensional second-order cone; and so are the second three elements and the third three elements of $\mathbf{x}$.

Consider further the equality constraints, the dual can be simplified as

$$\begin{array}{lll} \text{maximize} & \sum_{j=1}^{3} \mathbf{a}_j^T \mathbf{x}_j & \\ \text{subject to} & \sum_{j=1}^{3} \mathbf{x}_j & = \mathbf{0} \in E^2, \\ & |\mathbf{x}_j| & \leq 1, \ \forall\, j = 1, 2, 3. \end{array}$$

Example 5 (Quadratic Constraints). Quadratic constraints can be transformed to linear semidefinite form by using the concept of *Schur complements*. Let $\mathbf{A}$ be a (symmetric) m-dimension positive definite matrix, $\mathbf{C}$ be a symmetric n-dimension matrix, and $\mathbf{B}$ be an $m \times n$ matrix. Then, matrix

$$\mathbf{S} = \mathbf{C} - \mathbf{B}^T \mathbf{A}^{-1} \mathbf{B}$$

is called the Schur complement of $\mathbf{A}$ in the matrix

$$\mathbf{Z} = \begin{bmatrix} \mathbf{A} & \mathbf{B} \\ \mathbf{B}^T & \mathbf{C} \end{bmatrix}.$$

Moreover, $\mathbf{Z}$ is positive semidefinite if and only if $\mathbf{S}$ is positive semidefinite.

Now consider a general quadratic constraint of the form

$$\mathbf{y}^T \mathbf{B}^T \mathbf{B} \mathbf{y} - \mathbf{c}^T \mathbf{y} - d \leq 0. \tag{6.7}$$

This is equivalent to

$$\begin{bmatrix} \mathbf{I} & \mathbf{B}\mathbf{y} \\ \mathbf{y}^T \mathbf{B}^T & \mathbf{c}^T \mathbf{y} + d \end{bmatrix} \succeq \mathbf{0} \tag{6.8}$$

because the Schur complement of this matrix with respect to $\mathbf{I}$ is the negative of the left side of the original constraint (6.7). Note that in this larger matrix, the variable $\mathbf{y}$ appears only affinely, not quadratically.

Indeed, (6.8) can be written as

$$\mathbf{P}(\mathbf{y}) = \mathbf{P}_0 + y_1 \mathbf{P}_1 + y_2 \mathbf{P}_2 + \cdots y_n \mathbf{P}_n \succeq \mathbf{0}, \tag{6.9}$$

where

$$\mathbf{P}_0 = \begin{bmatrix} \mathbf{I} & \mathbf{0} \\ \mathbf{0} & d \end{bmatrix}, \ \mathbf{P}_i = \begin{bmatrix} \mathbf{0} & \mathbf{b}_i \\ \mathbf{b}_i^T & c_i \end{bmatrix} \text{ for } i = 1, 2, \ldots n$$

with $\mathbf{b}_i$ being the **ith** column of $\mathbf{B}$ and c_i being the ith component of $\mathbf{c}$. The constraint (6.9) is of the form that appears in the dual form of a semidefinite program.

There is a more efficient mixed semidefinite and second-order cone formulation of the inequality (6.7) to reduce the dimension of semidefinite cone. We first introduce slack variable **s** and s_0 by linear constraints:

$$\mathbf{B}\mathbf{y} - \mathbf{s} = \mathbf{0}$$

Then, we let $|\mathbf{s}| \le s_0$ (or $(s_0; \mathbf{s})$ in the second-order cone) and

$$\begin{bmatrix} 1 & s_0 \\ s_0 & \mathbf{c}^T\mathbf{y} + d \end{bmatrix} \succeq \mathbf{0}.$$

Again, the matrix constraint is of the dual form of a semidefinite cone, but its dimension is fixed at 2.

Suppose the original optimization problem has a quadratic objective: minimize $q(\mathbf{x})$. The objective can be written instead as: minimize t subject to $q(\mathbf{x}) \le t$, and then this constraint as well as any number of other quadratic constraints can be transformed to semidefinite constraints, and hence the entire problem converted to a mixed second-order cone and semidefinite program. This approach is useful in many applications, especially in various problems of financial engineering and control theory.

The duality is manifested by the relation between the optimal values of the primal and dual programs. The weak form of this relation is spelled out in the following lemma, the proof of which, like the weak form of other duality relations we have studied, is essentially an accounting issue.

Weak Duality in CLP. *Let* $\mathbf{X}$ *be feasible for (CLP) and* $(\mathbf{y}, \mathbf{S})$ *feasible for (CLD). Then,*

$$\mathbf{C} \bullet \mathbf{X} \ge \mathbf{y}^T\mathbf{b}.$$

Proof. By direct calculation

$$\begin{aligned} \mathbf{C} \bullet \mathbf{X} - \mathbf{y}^T\mathbf{b} &= (\sum_{i=1}^{m} y_i\mathbf{A}_i + \mathbf{S}) \bullet \mathbf{X} - \mathbf{y}^T\mathbf{b} \\ &= \sum_{i=1}^{m} y_i(\mathbf{A}_i \bullet \mathbf{X}) + \mathbf{S} \bullet \mathbf{X} - \mathbf{y}^T\mathbf{b} \\ &= \sum_{i=1}^{m} y_i b_i + \mathbf{S} \bullet \mathbf{X} - \mathbf{y}^T\mathbf{b} \\ &= \mathbf{S} \bullet \mathbf{X} \ge 0, \end{aligned}$$

where the last inequality comes from $\mathbf{X} \in K$ and $\mathbf{S} \in K^*$. ∎

As in other instances of duality, the strong duality of conic linear programming is weak unless other conditions hold. For example, the duality gap may not be zero at optimality in the following SDP instance.

Example 6. The following semidefinite program has a duality gap:

$$\mathbf{C} = \begin{bmatrix} 0 & 1 & 0 \\ 1 & 0 & 0 \\ 0 & 0 & 0 \end{bmatrix}, \mathbf{A}_1 = \begin{bmatrix} 0 & 0 & 0 \\ 0 & 1 & 0 \\ 0 & 0 & 0 \end{bmatrix}, \mathbf{A}_2 = \begin{bmatrix} 0 & -1 & 0 \\ -1 & 0 & 0 \\ 0 & 0 & 2 \end{bmatrix}$$

and

$$\mathbf{b} = \begin{bmatrix} 0 \\ 2 \end{bmatrix}.$$

The primal minimal objective value is 0 achieved by

$$\mathbf{X} = \begin{bmatrix} 0 & 0 & 0 \\ 0 & 0 & 0 \\ 0 & 0 & 1 \end{bmatrix}$$

and the dual maximal objective value is -2 achieved by $\mathbf{y} = [0, \ -1]$; so the duality gap is 2.

However, under certain technical conditions, there would be no duality gap. One condition is related to weather or not the primal feasible region $\mathcal{F}_p$ or dual feasible region has an interior feasible solution. We say $\mathcal{F}_p$ has an interior (feasible solution) if and only if

$$\overset{\circ}{\mathcal{F}}_p := \{\mathbf{X} : \ \mathcal{A}\mathbf{X} = \mathbf{b}, \ \mathbf{X} \in \overset{\circ}{K}\}$$

is non-empty, and $\mathcal{F}_d$ has an interior feasible solution if and only if

$$\overset{\circ}{\mathcal{F}}_d := \{(\mathbf{y}, \mathbf{S}) : \ \mathbf{y}^T\mathcal{A} + \mathbf{S} = \mathbf{C}, \ \mathbf{S} \in \overset{\circ}{K}{}^*\}$$

is non-empty. We state here a version of the strong duality theorem.

Strong Duality in (CLP).

i) *Let (CLP) or (CLD) be infeasible, and furthermore the other be feasible and has an interior. Then the other is unbounded.*

ii) *Let (CLP) and (CLD) be both feasible, and furthermore one of them has an interior. Then there is no duality gap between (CLP) and (CLD).*

iii) *Let (CLP) and (CLD) be both feasible and have interior. Then, both have optimal solutions with no duality gap.*

Proof. We let cone $H = K \oplus E^1_+$ in the following proof.

i) Suppose $\mathcal{F}_d$ is empty and $\mathcal{F}_p$ is feasible and has an interior feasible solution. Then, we have an $\bar{\mathbf{X}} \in \overset{\circ}{K}$ and $\bar{\tau} = 1$ that is an interior feasible solution to (homogeneous) conic system:

$$\mathcal{A}\bar{\mathbf{X}} - \mathbf{b}\bar{\tau} = \mathbf{0}, \ (\bar{\mathbf{X}}, \bar{\tau}) \in \overset{\circ}{H}.$$

Now, for any z^*, we form an alternative system pair based on Farkas' Lemma (Theorem 2):

$$\{(\mathbf{X}, \tau) : \mathcal{A}\mathbf{X} - \mathbf{b}\tau = \mathbf{0},\ \mathbf{C} \bullet \mathbf{X} - z^*\tau < 0,\ (\mathbf{X}, \tau) \in H\},$$

and

$$\{(\mathbf{y}; \mathbf{S}, \kappa) : \mathcal{A}^T\mathbf{y} + \mathbf{S} = \mathbf{C},\ -\mathbf{b}^T\mathbf{y} + \kappa = -z^*,\ (\mathbf{S}, \kappa) \in H^*\}.$$

But the latter is infeasible, so that the former has a feasible solution $(\mathbf{X}, \tau)$. At such a feasible solution, if $\tau > 0$, we have $\mathbf{C} \bullet (\mathbf{X}/\tau) < z^*$ for any z^*. Otherwise, $\tau = 0$ implies that a new solution $\bar{\mathbf{X}} + \alpha\mathbf{X}$ is feasible for (CLP) for any positive α; and, as $\alpha \to \infty$, the objective value of the new solution goes to $-\infty$. Hence, either way we have a feasible solution for (CLP) whose objective value is unbounded from below.

ii) Let $\mathcal{F}_p$ be feasible and have an interior feasible solution, and let z^* be its objective infimum. Again, we have an alternative system pair as listed in the proof of i). But now the former is infeasible, so that we have a solution for the latter. From the Weak Duality theorem $\mathbf{b}^T\mathbf{y} \le z^*$, thus we must have $\kappa = 0$, that is, we have a solution $(\mathbf{y}, \mathbf{S})$ such that

$$\mathcal{A}^T\mathbf{y} + \mathbf{S} = \mathbf{C},\ \mathbf{b}^T\mathbf{y} = z^*,\ \mathbf{S} \in K^*.$$

iii) We only need to prove that there exist a solution $\mathbf{X} \in \mathcal{F}_p$ such that $\mathbf{C} \bullet \mathbf{X} = z^*$, that is, the infimum of (CLP) is attainable. But this is just the other side of the proof given that $\mathcal{F}_d$ is feasible and has an interior feasible solution, and z^* is also the supremum of (CLD). ∎

Again, if one of (CLP) and (CLD) has no interior feasible solution, the common objective value may not be attainable. For example,

$$C = \begin{bmatrix} 1 & 0 \\ 0 & 0 \end{bmatrix},\ A_1 = \begin{bmatrix} 0 & 1 \\ 1 & 0 \end{bmatrix}, \quad \text{and} \quad b_1 = 2.$$

The dual is feasible but has no interior, while the primal has an interior. The common objective value equals 0, but no primal solution attaining the infimum value.

Most of these examples that make the strong duality failed are superficial, and a small perturbation would overcome the failure. Thus, in real applications and in the rest of the chapter, we may assume that both (CLP) and (CLD) have interior when they are feasible. Consequently, any primal and dual optimal solution pair must satisfy the optimality conditions:

$$\begin{aligned} \mathbf{C} \bullet \mathbf{X} - \mathbf{y}^T\mathbf{b} &= 0 \\ \mathcal{A}\mathbf{X} &= \mathbf{b} \\ \mathbf{y}^T\mathcal{A} + \mathbf{S} &= \mathbf{C}^T \\ \mathbf{X} \in K, \quad \mathbf{S} &\in K^* \end{aligned} ; \tag{6.10}$$

or

$$\begin{aligned} \mathbf{X} \bullet \mathbf{S} &= 0 \\ \mathcal{A}\mathbf{X} &= \mathbf{b} \\ \mathbf{y}^T \mathcal{A} + \mathbf{S} &= \mathbf{C}^T \\ \mathbf{X} \in K, &\quad \mathbf{S} \in K^* \end{aligned} \tag{6.11}$$

We now present an application of the strong duality theorem.

Example 7 (Robust Portfolio Design). The Markowitz portfolio design model (also see 5) is

$$\begin{aligned} &\text{minimize } \mathbf{x}^T \boldsymbol{\Sigma} \mathbf{x} \\ &\text{subject to } \mathbf{1}^T \mathbf{x} = 1, \ \boldsymbol{\pi}^T \mathbf{x} \geq \pi, \end{aligned}$$

where $\boldsymbol{\Sigma}$ is the covariance matrix and $\boldsymbol{\pi}$ is the expect return rate vector of a set of stocks, and π is the desired return rate of the portfolio. The problem can be equivalently written as a mixed conic problem

$$\begin{aligned} &\text{minimize } \boldsymbol{\Sigma} \bullet \mathbf{X} \\ &\text{subject to } \mathbf{1}^T \mathbf{x} = 1, \ \boldsymbol{\pi}^T \mathbf{x} \geq \pi, \\ &\qquad\qquad \mathbf{X} - \mathbf{x}\mathbf{x}^T \succeq \mathbf{0}. \end{aligned}$$

Now suppose $\boldsymbol{\Sigma}$ is incomplete and/or uncertain, and it is expressed by

$$\boldsymbol{\Sigma}_0 + \sum_{i=1}^m y_i \boldsymbol{\Sigma}_i (\succeq \mathbf{0}),$$

for some variables y_i's. Then, we like to solve a robust model

$$\begin{aligned} &\text{minimize} \left\{ \begin{array}{l} \max_{\mathbf{y}} \left(\boldsymbol{\Sigma}_0 + \sum_{i=1}^m y_i \boldsymbol{\Sigma}_i \right) \bullet \mathbf{X} \\ \quad \text{s.t. } \boldsymbol{\Sigma}_0 + \sum_{i=1}^m y_i \boldsymbol{\Sigma}_i \succeq \mathbf{0} \end{array} \right\} \\ &\text{subject to } \mathbf{1}^T \mathbf{x} = 1, \ \boldsymbol{\pi}^T \mathbf{x} \geq \pi, \\ &\qquad\qquad \mathbf{X} - \mathbf{x}\mathbf{x}^T \succeq \mathbf{0}. \end{aligned}$$

The inner problem is an SDP problem. Assuming strong duality holds, we replace it by its dual, and have

$$\begin{aligned} &\text{minimize} \left\{ \begin{array}{l} \min_{\mathbf{Y}} \boldsymbol{\Sigma}_0 \bullet (\mathbf{Y} + \mathbf{X}) \\ \quad \text{s.t. } \boldsymbol{\Sigma}_i \bullet (\mathbf{Y} + \mathbf{X}) = 0, \ \forall i = 1, \ldots, m, \\ \qquad\quad \mathbf{Y} \succeq \mathbf{0} \end{array} \right\} \\ &\text{subject to } \mathbf{1}^T \mathbf{x} = 1, \ \boldsymbol{\pi}^T \mathbf{x} \geq \pi, \\ &\qquad\qquad \mathbf{X} - \mathbf{x}\mathbf{x}^T \succeq \mathbf{0}. \end{aligned}$$

Then, we can integrate the two minimization problems together and form

$$\begin{aligned} &\text{minimize } \Sigma_0 \bullet (\mathbf{Y} + \mathbf{X}) \\ &\text{subject to } \mathbf{1}^T\mathbf{x} = 1, \ \boldsymbol{\pi}^T\mathbf{x} \geq \pi, \\ &\qquad \Sigma_i \bullet (\mathbf{Y} + \mathbf{X}) = 0, \ \forall i = 1, \ldots, m, \\ &\qquad \mathbf{Y} \succeq \mathbf{0}, \ \mathbf{X} - \mathbf{x}\mathbf{x}^T \succeq \mathbf{0}. \end{aligned}$$

6.5 Complementarity and Solution Rank of SDP

In linear programming, since $\mathbf{x} \geq \mathbf{0}$ and $\mathbf{s} \geq \mathbf{0}$,

$$0 = \mathbf{x} \bullet \mathbf{s} = \mathbf{x}^T\mathbf{s} = \sum_{j=1}^{n} x_j s_j$$

implies that $x_j s_j = 0$ for all $j = 1, \ldots, n$. This property is often called complementarity. Thus, besides feasibility, and optimal linear programming solution pair must satisfy complementarity.

Now consider semidefinite cone $\mathcal{S}^n_+$. Since $\mathbf{X} \succeq \mathbf{0}$ and $\mathbf{S} \succeq \mathbf{0}$, $0 = \mathbf{X} \bullet \mathbf{S}$ implies $\mathbf{XS} = \mathbf{0}$, that is, the regular matrix product of the two is a zero matrix. In other words, every column (or row) of $\mathbf{X}$ is orthogonal to every column (or row) of $\mathbf{X}$. We also call such property complementarity. Thus, besides feasibility, an optimal semidefinite programming solution pair must satisfy complementarity.

Proposition 1. *Let* $\mathbf{X}^*$ *and* $(\mathbf{y}^*, \mathbf{S}^*)$ *be any optimal SDP solution pair with zero duality gap. Then complementarity of* $\mathbf{X}^*$ *and* $\mathbf{S}^*$ *implies*

$$\text{rank}(\mathbf{X}^*) + \text{rank}(\mathbf{S}^*) \leq n.$$

Furthermore, is there an optimal (dual) $\mathbf{S}^*$ *such that* rank$\mathbf{S}^* \geq d$, *then the rank of any optimal (primal)* $\mathbf{X}^*$ *is bounded above by* $n - d$, *where integer* $0 \leq d \leq n$; *and the converse is also true.*

In certain SDP problems, one may be interested in finding an optimal solution whose rank is minimal, while the interior-point algorithm for SDP (developed later) typically generates solution whose rank is maximal for primal and dual, respectively. Thus, a rank reduction method sometimes is necessary to achieve this goal. For linear programming in the standard form, it is known that if there is an optimal solution, then there is an optimal *basic* solution $\mathbf{x}^*$ whose positive entries have at most m many. Is there a similar structural fact for semidefinite programming? In deed, we have

Proposition 2. *If there is an optimal solution for SDP, then there is an optimal solution of SDP whose rank* r *satisfies* $\frac{r(r+1)}{2} \leq m$.

The proposition resembles the linear programming fundamental theorem of Carathéodory in Sect. 2.4. We now give a sketch of similar constructive proof, as well as several other rank-reduction methods.

Null-Space Rank Reduction

Let $\mathbf{X}^*$ be an optimal solution of SDP with rank r. If $r(r+1)/2 > m$, we orthonormally factorize $\mathbf{X}^*$

$$\mathbf{X}^* = (\mathbf{V}^*)^T\mathbf{V}^*, \quad \mathbf{V}^* \in E^{r\times n}.$$

Then we consider a related SDP problem

$$\begin{aligned} &\text{minimize } \mathbf{V}^*\mathbf{C}(\mathbf{V}^*)^T \bullet \mathbf{U} \\ &\text{subject to } \mathbf{V}^*\mathbf{A}_i(\mathbf{V}^*)^T \bullet \mathbf{U} = b_i,\ i = 1,\ldots,m \\ &\qquad\qquad \mathbf{U} \in \mathcal{S}^r_+. \end{aligned} \tag{6.12}$$

Note that, for any feasible solution of (6.12) one can construct a feasible solution for original SDP using

$$\mathbf{X}(\mathbf{U}) = (\mathbf{V}^*)^T\mathbf{U}\mathbf{V}^* \quad \text{and} \quad \mathbf{C} \bullet \mathbf{X}(\mathbf{U}) = \mathbf{V}^*\mathbf{C}(\mathbf{V}^*)^T \bullet \mathbf{U}.$$

Thus, the minimal value of (6.12) is also z^*, and in particular $\mathbf{U} = \mathbf{I}$ (the identity matrix) is an minimizer of (6.12), since

$$\mathbf{V}^*\mathbf{C}(\mathbf{V}^*)^T \bullet \mathbf{I} = \mathbf{C} \bullet (\mathbf{V}^*)^T\mathbf{V}^* = \mathbf{C} \bullet \mathbf{X}^* = z^*.$$

Also, one can show that any feasible solution $\mathbf{U}$ of (6.12) is its minimizer, so that $\mathbf{X}(\mathbf{U})$ is a minimizer of original SDP.

Consider the system of homogeneous linear equations:

$$\mathbf{V}^*\mathbf{A}_i(\mathbf{V}^*)^T \bullet \mathbf{W} = 0,\ i = 1,\ldots,m.$$

where $\mathbf{W} \in \mathcal{S}^r$ (i.e., a $r \times r$ symmetric matrices that does not need to be semidefinite). This system has $r(r+1)/2$ real variables and m equations. Thus, as long as $r(r+1)/2 > m$, we must be able to find a symmetric matrix $\mathbf{W} \neq \mathbf{0}$ to satisfy all the m equations. Without loss of generality, let $\mathbf{W}$ be either indefinite or negative semidefinite (if it is positive semidefinite, we take $-\mathbf{W}$ as $\mathbf{W}$), that is, $\mathbf{W}$ have at least one negative eigenvalue. Then we consider

$$\mathbf{U}(\alpha) = \mathbf{I} + \alpha\mathbf{W}.$$

Choosing a α^* sufficiently large such that $\mathbf{U}(\alpha^*) \succeq \mathbf{0}$ and it has at least one 0 eigenvalue (or rank$\mathbf{U}(\alpha^*) < r$). Note that

$$\mathbf{V}^*\mathbf{A}_i(\mathbf{V}^*)^T \bullet \mathbf{U}(\alpha^*) = \mathbf{V}^*\mathbf{A}_i(\mathbf{V}^*)^T \bullet (\mathbf{I} + \alpha^*\mathbf{W}) = \mathbf{V}^*\mathbf{A}_i(\mathbf{V}^*)^T \bullet I = b_i,\ i = 1,\ldots,m.$$

That is, $\mathbf{U}(\alpha^*)$ is feasible and also optimal for (6.12). Thus, $\mathbf{X}(\mathbf{U}(\alpha^*))$ is a new minimizer for the original SDP, and its rank is strictly less than r. This process can be repeated till the system of homogeneous linear equations has only all-zero solution, which is necessary when $r(r+1)/2 \le m$. Such a solution rank reduction procedure is called the Null-space reduction, which is deterministic.

To see an application of Proposition 2, consider a general quadratic minimization with sphere constraint

$$\begin{aligned} z^* \equiv \ & \text{minimize } \mathbf{x}^T\mathbf{Q}\mathbf{x} + 2\mathbf{c}^T\mathbf{x} \\ & \text{subject to } |\mathbf{x}|^2 = 1, \ \mathbf{x} \in E^n, \end{aligned}$$

where Q is general. The problem has an SDP relaxation:

$$\begin{aligned} z^{SDP} \equiv \ & \text{maximize } \begin{bmatrix} \mathbf{Q} & \mathbf{c} \\ \mathbf{c}^T & 0 \end{bmatrix} \bullet \mathbf{Y} \\ & \text{subject to } \begin{bmatrix} \mathbf{I} & \mathbf{0} \\ \mathbf{0}^T & 0 \end{bmatrix} \bullet \mathbf{Y} = 1, \\ & \qquad \begin{bmatrix} \mathbf{0} & \mathbf{0} \\ \mathbf{0}^T & 1 \end{bmatrix} \bullet \mathbf{Y} = 1, \\ & \qquad \mathbf{Y} \in \mathcal{S}_+^{n+1}. \end{aligned}$$

Note that the relaxation and its dual both have interior so that the strong duality theorem holds, and it must have a rank-1 optimal SDP solution because $m = 2$. But a rank-1 optimal SDP solution would be optimal to the original quadratic minimization with sphere constraint. Thus, we must have $z^* = z^{SDP}$.

Gaussian Projection Reduction

There is also a randomized procedure to produce an approximate SDP solution with a desired low rank d. Again, let $\mathbf{X}^*$ be an optimal solution of SDP with rank $r > d$ and we factorize $\mathbf{X}^*$ as

$$\mathbf{X}^* = (\mathbf{V}^*)^T\mathbf{V}^*, \quad \mathbf{V}^* \in E^{r\times n}.$$

We then generate i.i.d. Gaussian random variables ξ_i^j with mean 0 and variance $1/d$, $i = 1, \ldots, r;\ j = 1, \ldots, d$, and form random vectors $\xi^j = (\xi_1^j; \ldots; \xi_r^j)$, $j = 1, \ldots, d$. Finally, we let

$$\hat{\mathbf{X}} = (\mathbf{V}^*)^T \left[\sum_{j=1}^{d} \xi^j (\xi^j)^T \right] \mathbf{V}^*.$$

Note that the rank of $\hat{\mathbf{X}}$ is d and

$$\mathrm{E}(\hat{\mathbf{X}}) = (\mathbf{V}^*)^T \mathrm{E} \left[\sum_{j=1}^{d} \xi^j (\xi^j)^T \right] \mathbf{V}^* = (\mathbf{V}^*)^T \mathbf{I} \mathbf{V}^* = \mathbf{X}^*.$$

One can further show that $\hat{\mathbf{X}}$ would be a good rank-d approximate SDP solution in many cases.

Randomized Binary Reduction

As discussed in the binary QP optimization, we like to produce a vector $\mathbf{x}$ where each entry is either 1 or -1. A procedure to achieve this is as follows. Let $\mathbf{X}^*$ be any optimal solution of SDP and we factorize $\mathbf{X}^*$ as

$$\mathbf{X}^* = (\mathbf{V}^*)^T\mathbf{V}^*, \quad \mathbf{V}^* \in E^{n\times n}.$$

Then, we generate a random n-dimensional vector ξ where each entry is a i.i.d. Gaussian random variable with mean 0 and variance 1. Then we let

$$\hat{\mathbf{x}} = \text{sign}((\mathbf{V}^*)^T\xi)$$

where

$$\text{sign}(x) = \begin{cases} 1 & \text{if } x \geq 0 \\ -1 & \text{otherwise.} \end{cases}$$

It was proved by Sheppard [228]:

$$\text{E}[\hat{x}_i\hat{x}_j] = \frac{2}{\pi}\arcsin(\mathbf{X}^*_{ij}), \quad i, j = 1, 2, \ldots, n.$$

Obviously, each entry of $\hat{\mathbf{x}}$ is either 1 or -1.

One can further show $\hat{\mathbf{x}}$ would be a good approximate solution to the original binary QP. Let us consider the (homogeneous) binary quadratic maximization problem

$$\begin{aligned} z^* := \ &\text{maximize } \mathbf{x}^T\mathbf{Q}\mathbf{x} \\ &\text{subject to } x_j = \{1, \ -1\}, \text{ for all } j = 1, \ \ldots, \ n, \end{aligned}$$

where we assume Q is positive semidefinite. Then, the SDP relaxation would be

$$\begin{aligned} z^{SDP} := \ &\text{maximize } \mathbf{Q} \bullet \mathbf{X} \\ &\text{subject to } \mathbf{I}_j \bullet \mathbf{X} = 1, \text{ for all } j = 1, \ \ldots, \ n, \\ &\qquad \mathbf{X} \in \mathcal{S}^n_+; \end{aligned}$$

and let $\mathbf{X}^*$ be any optimal solution, from which we produced a random binary vector $\hat{\mathbf{x}}$. Let us evaluate the expected objective value

$$\text{E}(\hat{\mathbf{x}}^T\mathbf{Q}\hat{\mathbf{x}}) = \text{E}(\mathbf{Q} \bullet \hat{\mathbf{x}}\hat{\mathbf{x}}^T) = \mathbf{Q} \bullet \text{E}(\hat{\mathbf{x}}\hat{\mathbf{x}}^T) = Q \bullet \frac{2}{\pi}\arcsin[\mathbf{X}^*] = \frac{2}{\pi}(Q \bullet \arcsin[\mathbf{X}^*]),$$

where $\arcsin[\mathbf{X}^*] \in \mathcal{S}^n$ whose $(i, \ j)$ the entry equals $\arcsin(\mathbf{X}^*_{ij})$. One can further show

$$\arcsin[\mathbf{X}^*] - \mathbf{X}^* \geq \mathbf{0}$$

so that (from $Q \succeq \mathbf{0}$)

$$Q \bullet \arcsin[\mathbf{X}^*] \geq Q \bullet \mathbf{X}^* = z^{SDP} \geq z^*,$$

that is, the expected objective value of $\hat{\mathbf{x}}$ is no less than factor $\frac{2}{\pi}$ of the maximal value of the binary QP.

The randomized binary reduction can be extended to quadratic optimization with simple bound constraints such as $x_j^2 \leq 1$.

6.6 Interior-Point Algorithms for Conic Linear Programming

Since (CLP) is a convex minimization problem, many optimization algorithms are applicable for solving it. However, the most natural conic linear programming algorithm seems to be an extension of the interior-point linear programming algorithm described in Chap. 5. We describe what it is now.

To develop efficient interior-point algorithms, the key is to find a suitable barrier or potential function. There is a general theory on selection of barrier functions for (CLP), depending on the convex cone involved. We present few for the convex cones listed in Example 1.

Example 1. The following are barrier function for each of the convex cones.

- The n-dimensional non-negative orthant E^n_+:

$$B(\mathbf{x}) = -\sum_{j=1}^{n} \log(x_j).$$

- The n-dimensional semidefinite cone $\mathcal{S}^n_+$:

$$B(\mathbf{X}) = -\log(\det \mathbf{X}).$$

- The $(n+1)$-dimensional second-order cone $\{(u; \mathbf{x}) : u \geq |\mathbf{x}|\}$:

$$B(u; \mathbf{x}) = -\log(u^2 - |\mathbf{x}|^2).$$

In the rest of the section, we devote our discussion on solving (SDP). Similar to LP, we consider (SDP) with the barrier function added in the objective:

$$\begin{aligned} (SDPB)\ \text{minimize}\ & \mathbf{C} \bullet \mathbf{X} - \mu \log\det(\mathbf{X}) \\ \text{subject to}\ & \mathbf{X} \in \overset{\circ}{\mathcal{F}}_p, \end{aligned}$$

or (SDD) with the barrier function added in the objective:

$$\begin{aligned} (SDDB)\ \text{maximize}\ & \mathbf{y}^T\mathbf{b} + \mu \log\det(\mathbf{S}) \\ \text{subject to}\ & (\mathbf{y}, \mathbf{S}) \in \overset{\circ}{\mathcal{F}}_d, \end{aligned}$$

where again $\mu > 0$ is called the barrier weight parameter. For a given μ, the minimizers of (SDPB) and (SDDB) satisfy conditions:

$$\begin{aligned} \mathbf{XS} &= \mu\mathbf{I} \\ \mathcal{A}\mathbf{X} &= \mathbf{b} \\ \mathcal{A}^T\mathbf{y} + \mathbf{S} &= \mathbf{C} \\ \mathbf{X} \succ \mathbf{0}, &\quad \mathbf{S} \succ \mathbf{0} \end{aligned} \tag{6.13}$$

Since

$$\mu = \frac{\text{trace}(\mathbf{XS})}{n} = \frac{X \bullet S}{n} = \frac{\mathbf{C} \bullet \mathbf{X} - \mathbf{y}^T\mathbf{b}}{n},$$

so that μ equals the average of complementarity or duality gap. And, these minimizers, denoted by $(\mathbf{X}(\mu), \mathbf{y}(\mu), \mathbf{S}(\mu))$, form the central path of SDP for $mu \in (0, \infty)$. It is known that when $\mu \to 0$, $(\mathbf{X}(\mu), \mathbf{y}(\mu), \mathbf{S}(\mu))$ tends to an optimal solution pair whose rank is maximal (Exercise 11).

We can also extend the primal-dual potential function from LP to SDP as a descent merit function:

$$\psi_{n+\rho}(\mathbf{X}, \mathbf{S}) = (n + \rho)\log(\mathbf{X} \bullet \mathbf{S}) - \log(\det(\mathbf{X}) \cdot \det(\mathbf{S}))$$

where $\rho \geq 0$. Note that if $\mathbf{X}$ and $\mathbf{S}$ are diagonal matrices, these definitions reduce to those for linear programming.

Once we have an interior feasible point $(\mathbf{X}, \mathbf{y}, \mathbf{S})$, we can generate a new iterate $(\mathbf{X}^+, \mathbf{y}^+, \mathbf{S}^+)$ by solving for $(\mathbf{D}_x, \mathbf{d}_y, \mathbf{D}_s)$ from the primal-dual system of linear equations

$$\begin{aligned} \mathbf{D}^{-1}\mathbf{D}_x\mathbf{D}^{-1} + \mathbf{D}_s &= \frac{n}{n+\rho}\mu\mathbf{X}^{-1} - \mathbf{S}, \\ \mathbf{A}_i \bullet \mathbf{D}_x &= 0, \text{ for all } i, \\ \sum\nolimits_i^m (\mathbf{d}_y)_i\mathbf{A}_i + \mathbf{D}_s &= \mathbf{0}, \end{aligned} \tag{6.14}$$

where $\mathbf{D}$ is the (scaling) matrix

$$\mathbf{D} = \mathbf{X}^{\frac{1}{2}}(\mathbf{X}^{\frac{1}{2}}\mathbf{S}\mathbf{X}^{\frac{1}{2}})^{-\frac{1}{2}}\mathbf{X}^{\frac{1}{2}}$$

and $\mu = \mathbf{X} \bullet \mathbf{S}/n$. Then one assigns $\mathbf{X}^+ = \mathbf{X} + \alpha\mathbf{D}_x$, $\mathbf{y}^+ = \mathbf{y} + \alpha\mathbf{d}_y$, and $\mathbf{S}^+ = \mathbf{s} + \alpha\mathbf{D}_s$ for a step size $\alpha > 0$. Furthermore, it can be shown that there exists a step size $\alpha = \bar{\alpha}$ such that

$$\psi_{n+\rho}(\mathbf{X}^+, \mathbf{S}^+) - \psi_{n+\rho}(\mathbf{X}, \mathbf{S}) \leq -\delta$$

for a constant $\delta > 0.2$.

We outline the algorithm here

Step 1. Given $(\mathbf{X}^0, \mathbf{y}^0, \mathbf{S}^0) \in \overset{\circ}{\mathcal{F}}$. Set $\rho \geq \sqrt{n}$ and $k := 0$.
Step 2. Set $(\mathbf{X}, \mathbf{S}) = (\mathbf{X}^k, \mathbf{S}^k)$ and compute $(\mathbf{D}_x, \mathbf{d}_y, \mathbf{D}_s)$ from (6.14).

Step 3. Let $\mathbf{X}^{k+1} = \mathbf{X}^k + \bar{\alpha}\mathbf{D_x}$, $\mathbf{y}^{k+1} = \mathbf{y}^k + \bar{\alpha}\mathbf{d_y}$, and $\mathbf{S}^{k+1} = \mathbf{S}^k + \bar{\alpha}\mathbf{D_s}$, where

$$\bar{\alpha} = \arg\min_{\alpha \geq 0} \psi_{n+\rho}(\mathbf{X}^k + \alpha\mathbf{D_x}, \mathbf{S}^k + \alpha\mathbf{D_s}).$$

Step 4. Let $k := k + 1$. If $\frac{\mathbf{X}^k \bullet \mathbf{S}^k}{\mathbf{X}^0 \bullet \mathbf{S}^0} \leq \epsilon$, Stop. Otherwise return to Step 2.

Theorem 3. *Let $\psi_{n+\rho}(\mathbf{X}^0, \mathbf{S}^0) \leq \rho \log(\mathbf{X}^0 \bullet \mathbf{S}^0) + n \log n$. Then, the algorithm terminates in at most $O(\rho \log(n/\epsilon))$ iterations.*

Initialization: The HSD Algorithm

The linear programming Homogeneous Self-Dual Algorithm is also extendable to conic linear programming. Consider the minimization problem Homogeneous self-dual algorithm! for conic linear programming

$$\begin{array}{llllll}
(HSDCLP) \min & & & & (n+1)\theta & \\
\text{s.t.} & & \mathcal{A}\mathbf{X} & -\mathbf{b}\tau & +\bar{\mathbf{b}}\theta & = \mathbf{0}, \\
-\mathcal{A}^T\mathbf{y} & & & +\mathbf{C}\tau & -\bar{\mathbf{C}}\theta & = \mathbf{S} \in K^*, \\
\mathbf{b}^T\mathbf{y} & -\mathbf{C} \bullet \mathbf{X} & & & +\bar{z}\theta & = \kappa \geq 0, \\
-\bar{\mathbf{b}}^T\mathbf{y} & +\bar{\mathbf{C}} \bullet \mathbf{X} & & -\bar{z}\tau & & = -(n+1), \\
\mathbf{y} \text{ free}, & \mathbf{X} \in K, & \tau \geq 0, & \theta \text{ free}, & &
\end{array}$$

where

$$\bar{\mathbf{b}} = \mathbf{b} - \mathcal{A}\mathbf{X}^0, \quad \bar{\mathbf{C}} = \mathbf{C} - \mathbf{S}^0, \quad \bar{z} = \mathbf{C} \bullet \mathbf{X}^0 + 1$$

Here $\mathbf{X}^0$ and $\mathbf{S}^0$ are any pair of interior points in the interior of K and K^* such that they form a central path point with $\mu = 1$. Note that $\mathbf{X}^0$ and $\mathbf{S}^0$ don't need to satisfy other equality constraint, so that they can be easily identified. For examples, $\mathbf{x}^0 = \mathbf{y}^0 = \mathbf{1}$ for the nonnegative orthant cone; $\mathbf{x}^0 = \mathbf{y}^0 = (1; \mathbf{0})$ for the p-order cone; and $\mathbf{X}^0 = \mathbf{X}^0 = \mathbf{I}$ for the semidefinite cone.

Let $\mathcal{F}$ be the set of all feasible points $(\mathbf{y}, \mathbf{X} \in K, \tau \geq 0, \theta, \mathbf{S} \in K^*, \kappa \geq 0)$. Then $\overset{\circ}{\mathcal{F}}$ is the set of interior feasible points $(\mathbf{y}, \mathbf{X} \in \overset{\circ}{K}, \tau > 0, \theta, \mathbf{S} \in \overset{\circ}{K}{}^*, \kappa > 0)$.

Theorem 4. *Consider the conic optimization (HSDCLP).*

i) *(HSDCLP) is self-dual, that is, its dual has an identical form of (HSDCLP).*
ii) *(HSDCLP) has an optimal solution and its optimal solution set is bounded.*
iii) *(HSDCLP) has an interior feasible point*

$$\mathbf{y} = \mathbf{0}, \quad \mathbf{X} = \mathbf{X}^0, \quad \tau = 1, \quad \theta = 1, \quad \mathbf{S} = \mathbf{S}^0, \quad \kappa = 1.$$

iv) *For any feasible point $(\mathbf{y}, \mathbf{X}, \tau, \theta, \mathbf{S}, \kappa) \in \mathcal{F}$*

$$\mathbf{S}^0 \bullet \mathbf{X} + \mathbf{X}^0 \bullet \mathbf{S} + \tau + \kappa - (n+1)\theta = (n+1),$$

and

$$\mathbf{X} \bullet \mathbf{S} + \tau\kappa = (n+1)\theta.$$

v) *The optimal objective value of (HSDCLP) is zero, that is, any optimal solution of (HSDCLP) has*

$$\mathbf{X}^* \bullet \mathbf{S}^* + \tau^*\kappa^* = (n+1)\theta^* = 0.$$

Now we are ready to apply the interior-point algorithm, starting from a available initial interior-point feasible solution, to solve (HSDCLP). The question is: how is an optimal solution of (HSDCLP) related to optimal solutions of original (CLP) and (CLD)? We present the next theorem, and leave this proof as an exercise.

Theorem 5. *Let* $(\mathbf{y}^*, \mathbf{X}^*, \tau^*, \theta^* = 0, \mathbf{S}^*, \kappa^*)$ *be a (maximal rank) optimal solution of (HSD-CLP) (as it is typically computed by interior-point algorithms).*

i) (CLP) and (CLD) have an optimal solution pair if and only if $\tau^* > 0$. *In this case,* $\mathbf{X}^*/\tau^*$ *is an optimal solution for (CLP) and* $(\mathbf{y}^*/\tau^*, \mathbf{S}^*/\tau^*)$ *is an optimal solution for (CLD).*

ii) (CLP) or (CLD) has an infeasibility certificate if and only if $\kappa^* > 0$. *In this case,* $\mathbf{X}^*/\kappa^*$ *or* S^*/κ^* *or both are certificates for proving infeasibility; see Farkas' lemma for CLP.*

iii) For all other cases, $\tau^* = \kappa^* = 0$.

6.7 Summary

A relatively new class of mathematical programming problems, Conic linear programming (hereafter CLP), is a natural extension of Linear programming that is a central decision model in Management Science and Operations Research. In CLP, the unknown is a vector or matrix in a closed convex cone while its entries are also restricted by some linear equalities and/or inequalities.

One of cones is the semidefinite cone, that is, the set of all symmetric positive semidefinite matrices in a given dimension. There is a variety of interesting and important practical problems that can be naturally cast in this form. Because many problems which appear nonlinear (such as quadratic problems) become essentially linear in semidefinite form. We have described some of these applications and selected results in Combinatory Optimization, Robust Optimization, and Engineering Sensor Network. We have also illustrated some analyses to show why CLP is an effective model to tackle these difficult optimization problems.

We present fundamental theorems underlying conic linear programming. These theorems include Farkas' lemma, weak and strong dualities, and solution rank structure. We show the common features and differences of these theorems between LP and CLP.

The efficient interior-point algorithms for linear programming can be extended to solving these problems as well. We describe these extensions applied to general conic programming problems. These algorithms closely parallel those for linear programming. There is again a central path and potential functions, and Newton's method is a good way to follow the path or reduce the potential function. The homogeneous and self-dual algorithm, which is popularly used for linear programming, is also extended to CLP.

6.8 Exercises

1. Prove that

   i) The dual cone of E^n_+ is itself.
   ii) The dual cone of $\mathcal{S}^n_+$ is itself.
   iii) The dual cone of p-order cone is the q-order cone where $\frac{1}{p} + \frac{1}{q} = 1$ and $1 \le p \le \infty$.

2. When both K_1 and K_2 are closed convex cones. Show

   i) $(K_1^*)^* = K_1$.
   ii) $K_1 \subset K_2 \Longrightarrow K_2^* \subset K_1^*$.
   iii) $(K_1 \oplus K_2)^* = K_1^* \oplus K_2^*$.
   iv) $(K_1 + K_2)^* = K_1^* \cap K_2^*$.
   v) $(K_1 \cap K_2)^* = K_1^* + K_2^*$.

   Note: by definition $S + T = \{\mathbf{s} + \mathbf{t} : \mathbf{s} \in S,\ \mathbf{t} \in T\}$.

3. Prove the following:

   i) Theorem 1.
   ii) Proposition 1.
   iii) Let $\mathbf{X} \in \overset{\circ}{K}$ and $\mathbf{Y} \in \overset{\circ}{K^*}$. Then $\mathbf{X} \bullet \mathbf{Y} > 0$.

4. Guess an optimal solution and the optimal objective value of each instance of Example 1.
5. Prove the second statement of Theorem 2.
6. Verify the weak duality theorem of the three CLP instances in Example 1 in Sect. 6.2 and Example 1 in Sect. 6.4.
7. Consider the SDP relaxation of the sensor network localization problem with four sensors:

$$\begin{aligned}(\mathbf{e}_i - \mathbf{e}_j)(\mathbf{e}_i - \mathbf{e}_j)^T \bullet \mathbf{X} = 1,\ \forall i < j = 1, 2, 3, 4,\\ \mathbf{X} \in \mathcal{S}^4_+,\end{aligned}$$

   in which $m = 6$. Show that the SDP problem has the solution with rank 3, which reaches the bound of Proposition 2.
8. Let $\mathbf{A}$ and $\mathbf{B}$ be two symmetric and positive semidefinite matrices. Prove that $\mathbf{A} \bullet \mathbf{B} \ge 0$, and $\mathbf{A} \bullet \mathbf{B} = 0$ implies $\mathbf{AB} = \mathbf{0}$.
9. Let $\mathbf{X}$ and $\mathbf{S}$ both be positive definite. Prove that

$$n \log(\mathbf{X} \bullet \mathbf{S}) - \log(\det(\mathbf{X}) \cdot \det(\mathbf{S})) \ge n \log n.$$

10. Consider a SDP and the potential level set

$$\mathbf{\Psi}(\delta) = \{(\mathbf{X},\ \mathbf{y},\ \mathbf{S}) \in \overset{\circ}{\mathcal{F}} : \psi_{n+\rho}(\mathbf{X},\ \mathbf{S}) \le \delta\}.$$

Prove that

$$\Psi(\delta^1) \subset \Psi(\delta^2) \quad \text{if} \quad \delta^1 \leq \delta^2,$$

and for every δ, $\boldsymbol{\Psi}(\delta)$ is bounded and its closure $\overline{\boldsymbol{\Psi}}(\delta)$ has non-empty intersection with the SDP solution set.

11. Let both (SDP) and (SDD) have interior feasible points. Then for any $0 < \mu < \infty$, the central path point $(\mathbf{X}(\mu),\ \mathbf{y}(\mu),\ \mathbf{S}(\mu))$ exists and is unique. Moreover,

    i) the central path point $(\mathbf{X}(\mu),\ \mathbf{y}(\mu),\ \mathbf{S}(\mu))$ is bounded where $0 < \mu \leq \mu^0$ for any given $0 < \mu^0 < \infty$.
    ii) For $0 < \mu' < \mu$,

    $$\mathbf{C} \bullet \mathbf{X}(\mu') < \mathbf{C} \bullet \mathbf{X}(\mu) \quad \text{and} \quad \mathbf{b}^T\mathbf{y}(\mu') > \mathbf{b}^T\mathbf{y}(\mu)$$

    if $\mathbf{X}(\mu) \neq \mathbf{X}(\mu')$ and $\mathbf{y}(\mu) \neq \mathbf{y}(\mu')$.
    iii) $(\mathbf{X}(\mu),\ \mathbf{y}(\mu),\ \mathbf{S}(\mu))$ converges to an optimal solution pair for (SDP) and (SDD), and the rank of the limit of $\mathbf{X}(\mu)$ is maximal among all optimal solutions of (SDP) and the rank of the limit $\mathbf{S}(\mu)$ is maximal among all optimal solutions of (SDD).

12. Prove the logarithmic approximation lemma for SDP. Let $\mathbf{D} \in \mathcal{S}^n$ and $|\mathbf{D}|_\infty < 1$. Then,

$$\text{trace}(\mathbf{D}) \geq \log\det(\mathbf{I} + \mathbf{D}) \geq \text{trace}(\mathbf{D}) - \frac{|\mathbf{D}|^2}{2(1 - |\mathbf{D}|_\infty)}.$$

13. Let $\mathbf{V} \in \overset{\circ}{\mathcal{S}}{}^n_+$ and $\rho \geq \sqrt{n}$. Then,

$$\frac{|\mathbf{V}^{-1/2} - \frac{n+\rho}{\mathbf{I}\bullet\mathbf{V}}\mathbf{V}^{1/2}|}{|\mathbf{V}^{-1/2}|_\infty} \geq \sqrt{3/4}.$$

14. Prove both Theorems 4 and 5.

References

6.1 Most of the materials presented can be found from convex analysis, such as Rockeafellar [219].

6.2 Semidefinite relaxations have appeared in relation to relaxations discrete optimization problems. In Lovasz and Shrijver [159], a "lifting" procedure is presented to obtain a problem in $\Re^{n^2}$; and then the problem is projected back to obtain tighter inequalities; see also Balas et al. [12]. Then, there have been several remarkable results of SDP relaxations for combinatorial optimization. The binary QP, a generalized Max-Cut problem, was studied by Goemans and Williamson [G8] and Nesterov [189]. Other SDP relaxations can be found in the survey by Luo et al. [171] and references therein. More CLP applications can be found in Boyd et al [B22], Vandenberghe and Boyd [V2], and Lobo, Vandenberghe and Boyd [156], Lasserre [150], Parrilo [204], etc.

The sensor localization problem described here is due to Biswas and Ye [B17]. Note that we can view the Sensor Network Localization problem as a Graph Realization or Embedding problem in Euclidean spaces, see So and Ye [231] and references therein; and it is related to the Euclidean Distance Matrix Completion Problems, see Alfakih et al. [3] and Laurent [151].

6.3 Farkas' lemma for conic linear constraints are closely linked to convex analysis (i.e, Rockeafellar [219]) and the CLP duality theorems commented next.

6.4 The conic formulation of the Euclidean facility location problem was due to Xue and Ye [264]. For discussion of Schur complements see Boyd and Vanderberghe [B23]. Robust optimization models using SDP can be found in Ben-Tal and Nemirovski [26] and Goldfarb and Iyengar [112], and etc. The SDP duality theory was studied by Barvinok [16], Nesterov and Nemirovskii [N2], Ramana [214], Ramana e al. [215], etc. The SDP example with a duality gap was constructed by R. Freund (private communication).

6.5 Complementarity and rank. The exact rank theorem described here is due to Pataki [205], also see Barvinok [15]. A analysis of the Gaussian projection was presented by So et al. [232] which can be sees as a generalization of the Johnson and Lindenstrauss theorem [137]. The expectation of the randomized binary reduction is due to Sheppard [228] in 1900, and it was extensively used in Goemans and Williamson [G8] and Nesterov [189], Ye [265], and Bertsimas and Ye, [31].

6.6 In interior-point algorithms, the search direction $(\mathbf{D_x}, \mathbf{d_y}, \mathbf{D_s})$ can be determined by Newton's method with three different scalings: primal, dual and primal-dual. A primal-scaling (potential reduction) algorithm for semidefinite programming is due to Alizadeh [A4, A3] where Yinyu Ye "suggested studying the primal-dual potential function for this problem" and "looking at symmetric preserving scalings of the form $X_0^{-1/2}XX_0^{-1/2}$", and to Nesterov and Nemirovskii [N2]. A dual-scaling algorithm was developed by Benson et al. [25] which exploits the sparse structure of the dual SDP. The primal-dual SDP algorithm described here is due to Nesterov and Todd [N3] and references therein.

Efficient interior-point algorithms are also developed for optimization over the second-order cone; see Nesterov and Nemirovskii [N2] and Xue and Ye [264]. These algorithms have established the best approximation complexity results for certain combinatorial location problems.

The homogeneous and self-dual initialization model was originally developed by Ye, Todd and Mizuno for LP [Y2], and for SDP by de Klerk et al. [72], Luo et al. [L18], and Nesterov et al. [191], and it became the foundational algorithm implemented in Sturm [S11] and Andersen [6].

Part II
Unconstrained Problems

Chapter 7
Basic Properties of Solutions and Algorithms

In this chapter we consider optimization problems of the form

$$\begin{aligned} &\text{minimize} \quad f(\mathbf{x}) \\ &\text{subject to} \quad \mathbf{x} \in \Omega, \end{aligned} \tag{7.1}$$

where f is a real-valued function and Ω, the feasible set, is a subset of E^n. Throughout most of the chapter attention is restricted to the case where $\Omega = E^n$, corresponding to the completely unconstrained case, but sometimes we consider cases where Ω is some particularly simple subset of E^n.

The first and third sections of the chapter characterize the first- and second-order conditions that must hold at a solution point of (7.1). These conditions are simply extensions to E^n of the well-known derivative conditions for a function of a single variable that hold at a maximum or a minimum point. The fourth and fifth sections of the chapter introduce the important classes of convex and concave functions that provide zeroth-order conditions as well as a natural formulation for a global theory of optimization and provide geometric interpretations of the derivative conditions derived in the first two sections.

The final sections of the chapter are devoted to basic convergence characteristics of algorithms. Although this material is not exclusively applicable to optimization problems but applies to general iterative algorithms for solving other problems as well, it can be regarded as a fundamental prerequisite for a modern treatment of optimization techniques. Two essential questions are addressed concerning iterative algorithms. The first question, which is qualitative in nature, is whether a given algorithm in some sense yields, at least in the limit, a solution to the original problem. This question is treated in Sect. 7.6, and conditions sufficient to guarantee appropriate convergence are established. The second question, the more quantitative one, is related to how fast the algorithm converges to a solution. This question is defined more precisely in Sect. 7.7. Several special types of convergence, which arise frequently in the development of algorithms for optimization, are explored.

D.G. Luenberger, Y. Ye, *Linear and Nonlinear Programming*, International Series in Operations Research & Management Science 228,
DOI 10.1007/978-3-319-18842-3_7

7.1 First-Order Necessary Conditions

Perhaps the first question that arises in the study of the minimization problem (7.1) is whether a solution exists. The main result that can be used to address this issue is the theorem of Weierstras, which states that if f is continuous and Ω is compact, a solution exists (see Appendix A.6). This is a valuable result that should be kept in mind throughout our development; however, our primary concern is with characterizing solution points and devising effective methods for finding them.

In an investigation of the general problem (7.1) we distinguish two kinds of solution points: *local minimum points*, and *global minimum points*.

Definition. A point $\mathbf{x}^* \in \Omega$ is said to be a *relative minimum point* or a *local minimum point* of f over Ω if there is an $\varepsilon > 0$ such that $f(\mathbf{x}) \geqslant f(\mathbf{x}^*)$ for all $\mathbf{x} \in \Omega$ within a distance ε of $\mathbf{x}^*$ (that is, $\mathbf{x} \in \Omega$ and $|\mathbf{x} - \mathbf{x}^*| < \varepsilon$). If $f(\mathbf{x}) > f(\mathbf{x}^*)$ for all $\mathbf{x} \in \Omega$, $\mathbf{x} \neq \mathbf{x}^*$, within a distance ε of $\mathbf{x}^*$, then $\mathbf{x}^*$ is said to be a *strict relative minimum point* of f over Ω.

Definition. A point $\mathbf{x}^* \in \Omega$ is said to be a *global minimum point* of f over Ω if $f(\mathbf{x}) \geqslant f(\mathbf{x}^*)$ for all $\mathbf{x} \in \Omega$. If $f(\mathbf{x}) > f(\mathbf{x}^*)$ for all $\mathbf{x} \in \Omega$, $\mathbf{x} \neq \mathbf{x}^*$, then $\mathbf{x}^*$ is said to be a *strict global minimum point* of f over Ω.

In formulating and attacking problem (7.1) we are, by definition, explicitly asking for a global minimum point of f over the set Ω. Practical reality, however, both from the theoretical and computational viewpoint, dictates that we must in many circumstances be content with a relative minimum point. In deriving necessary conditions based on the differential calculus, for instance, or when searching for the minimum point by a convergent stepwise procedure, comparisons of the values of nearby points is all that is possible and attention focuses on relative minimum points. Global conditions and global solutions can, as a rule, only be found if the problem possesses certain convexity properties that essentially guarantee that any relative minimum is a global minimum. Thus, in formulating and attacking problem (7.1) we shall, by the dictates of practicality, usually consider, implicitly, that we are asking for a relative minimum point. If appropriate conditions hold, this will also be a global minimum point.

Feasible Directions

To derive necessary conditions satisfied by a relative minimum point $\mathbf{x}^*$, the basic idea is to consider movement away from the point in some given direction. Along any given direction the objective function can be regarded as a function of a single variable, the parameter defining movement in this direction, and hence the ordinary calculus of a single variable is applicable. Thus given $\mathbf{x} \in \Omega$ we are motivated to say that a vector $\mathbf{d}$ is a *feasible direction at* $\mathbf{x}$ if there is an $\bar{\alpha} > 0$ such that $\mathbf{x} + \alpha\mathbf{d} \in \Omega$ for all α, $0 \leqslant \alpha \leqslant \bar{\alpha}$. With this simple concept we can state some simple conditions satisfied by relative minimum points.

Proposition 1 (First-Order Necessary Conditions). *Let Ω be a subset of E^n and let $f \in C^1$ be a function on Ω. If $\mathbf{x}^*$ is a relative minimum point of f over Ω, then for any $\mathbf{d} \in E^n$ that is a feasible direction at $\mathbf{x}^*$, we have $\nabla f(\mathbf{x}^*)\mathbf{d} \geqslant 0$.*

Proof. For any α, $0 \leqslant \alpha \leqslant \bar{\alpha}$, the point $\mathbf{x}(\alpha) = \mathbf{x}^* + \alpha\mathbf{d} \in \Omega$. For $0 \leqslant \alpha \leqslant \bar{\alpha}$ define the function $g(\alpha) = f(\mathbf{x}(\alpha))$. Then g has a relative minimum at $\alpha = 0$. A typical g is shown in Fig. 7.1. By the ordinary calculus we have

$$g(\alpha) - g(0) = g'(0)\alpha + o(\alpha), \tag{7.2}$$

where $o(\alpha)$ denotes terms that go to zero faster than α (see Appendix A). If $g'(0) < 0$ then, for sufficiently small values of $\alpha > 0$, the right side of (7.2) will be negative, and hence $g(\alpha) - g(0) < 0$, which contradicts the minimal nature of $g(0)$. Thus $g'(0) = \nabla f(\mathbf{x}^*)\mathbf{d} \geqslant 0$. ∎

A very important special case is where $\mathbf{x}^*$ is in the interior of Ω (as would be the case if $\Omega = E^n$). In this case there are feasible directions emanating in every direction from $\mathbf{x}^*$, and hence $\nabla f(\mathbf{x}^*)\mathbf{d} \geqslant 0$ for all $\mathbf{d} \in E^n$. This implies $\nabla f(\mathbf{x}^*) = 0$. We state this important result as a corollary.

***Corollary** (Unconstrained Case). Let Ω be a subset of E^n, and let $f \in C^1$ be function' on Ω. If $\mathbf{x}^*$ is a relative minimum point of f over Ω and if $\mathbf{x}^*$ is an interior point of Ω, then $\nabla f(\mathbf{x}^*) = 0$.*

The necessary conditions in the pure unconstrained case lead to n equations (one for each component of ∇f) in n unknowns (the components of $\mathbf{x}^*$), which in many cases can be solved to determine the solution. In practice, however, as demonstrated in the following chapters, an optimization problem is solved directly without explicitly attempting to solve the equations arising from the necessary conditions. Nevertheless, these conditions form a foundation for the theory.

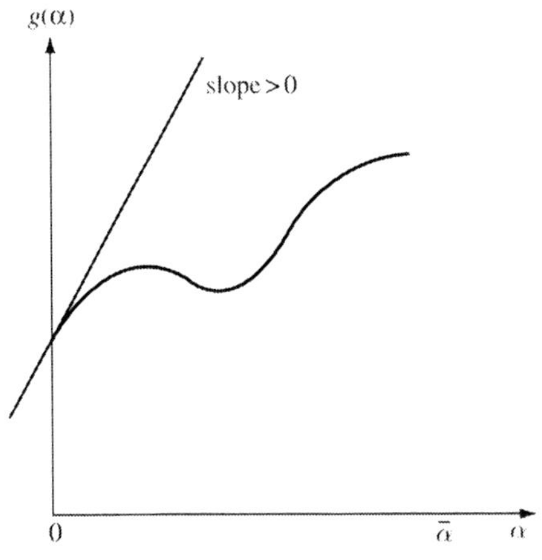

Fig. 7.1 Construction for proof

Example 1. Consider the problem

$$\text{minimize } f(x_1,\ x_2) = x_1^2 - x_1x_2 + x_2^2 - 3x_2.$$

There are no constraints, so $\Omega = E^2$. Setting the partial derivatives of f equal to zero yields the two equations

$$\begin{aligned} 2x_1 - x_2 &= 0 \\ -x_1 + 2x_2 &= 3. \end{aligned}$$

These have the unique solution $x_1 = 1,\ x_2 = 2$, which is a global minimum point of f.

Example 2. Consider the problem

$$\begin{aligned} &\text{minimize} \quad f(x_1,\ x_2) = x_1^2 - x_1 + x_2 + x_1x_2 \\ &\text{subject to} \quad x_1 \geqslant 0, \qquad x_2 \geqslant 0. \end{aligned}$$

This problem has a global minimum at $x_1 = \frac{1}{2},\ x_2 = 0$. At this point

$$\begin{aligned} \frac{\partial f}{\partial x_1} &= 2x_1 - 1 + x_2 = 0 \\ \frac{\partial f}{\partial x_2} &= 1 + x_1 = \frac{3}{2}. \end{aligned}$$

Thus, the partial derivatives do not both vanish at the solution, but since any feasible direction must have an x_2 component greater than or equal to zero, we have $\nabla f(\mathbf{x}^*)\mathbf{d} \geqslant 0$ for all $\mathbf{d} \in E^2$ such that $\mathbf{d}$ is a feasible direction at the point $(1/2, 0)$.

7.2 Examples of Unconstrained Problems

Unconstrained optimization problems occur in a variety of contexts, but most frequently when the problem formulation is simple. More complex formulations often involve explicit functional constraints. However, many problems with constraints are frequently converted to unconstrained problems, such as using the barrier functions, e.g., the analytic center problem for (dual) linear programs. We present a few more examples here that should begin to indicate the wide scope to which the theory applies.

Example 1 (Logistic Regression). Recall the classification problem where we have vectors $\mathbf{a}_i \in E^d$ for $i = 1, 2, \ldots, n_1$ in a class, and vectors $\mathbf{b}_j \in E^d$ for $j = 1, 2, \ldots, n_2$ not. Then we wish to find $\mathbf{y} \in E^d$ and a number β such that

$$\frac{\exp(\mathbf{a}_i^T\mathbf{y} + \beta)}{1 + \exp(\mathbf{a}_i^T\mathbf{y} + \beta)}$$

is close to 1 for all i, and

$$\frac{\exp(\mathbf{b}_j^T\mathbf{y}+\beta)}{1+\exp(\mathbf{b}_j^T\mathbf{y}+\beta)}$$

is close to 0 for all j. The problem can be cast as a unconstrained optimization problem, called the max-likelihood,

$$\text{maximize}_{\mathbf{y},\beta}\left(\prod_i \frac{\exp(\mathbf{a}_i^T\mathbf{y}+\beta)}{1+\exp(\mathbf{a}_i^T\mathbf{y}+\beta)}\right)\left(\prod_j\left(1-\frac{\exp(\mathbf{b}_j^T\mathbf{y}+\beta)}{1+\exp(\mathbf{b}_j^T\mathbf{y}+\beta)}\right)\right),$$

which can be also equivalently, using a logarithmic transformation, written as

$$\text{minimize}_{\mathbf{y},\beta}\ \sum_i \log\left(1+\exp(-\mathbf{a}_i^T\mathbf{y}-\beta)\right)+\sum_j \log\left(1+\exp(\mathbf{b}_j^T\mathbf{y}+\beta)\right).$$

Example 2 (Utility Maximization). A common problem in economic theory is the determination of the best way to combine various inputs in order to maximize a utility function $f(x_1, x_2, \ldots, x_n)$ (in the monetary unit) of the amounts x_j of the inputs, $i = 1, 2, \ldots, n$. The unit prices of the inputs are $p_1, p_2, \ldots, p_n$. The producer wishing to maximize profit must solve the problem

$$\text{maximize } f(x_1, x_2, \ldots, x_n) - p_1x_1 - p_2x_2 \ldots - p_nx_n.$$

The first-order necessary conditions are that the partial derivatives with respect to the x_i's each vanish. This leads directly to the n equations

$$\frac{\partial f}{\partial x_i}(x_1, x_2, \ldots, x_n) = p_i, \quad i = 1, 2, \ldots, n.$$

These equations can be interpreted as stating that, at the solution, the marginal value due to a small increase in the ith input must be equal to the price p_i.

Example 3 (Parametric Estimation). A common use of optimization is for the purpose of function approximation. Suppose, for example, that through an experiment the value of a function g is observed at m points, $x_1, x_2, \ldots, x_m$. Thus, values $g(x_1), g(x_2), \ldots, g(x_m)$ are known. We wish to approximate the function by a polynomial

$$h(x) = a_nx^n + a_{n-1}x^{n-1} + \ldots + a_0$$

of degree n (or less), where $n < m$. Corresponding to any choice of the approximating polynomial, there will be a set of errors $\varepsilon_k = g(x_k) - h(x_k)$. We define the best approximation as the polynomial that minimizes the sum of the squares of these errors; that is, minimizes

$$\sum_{k=1}^{m}(\varepsilon_k)^2.$$

This in turn means that we minimize

$$f(\mathbf{a}) = \sum_{k=1}^{m} \left[g(x_k) - \left(a_n x_k^n + a_{n-1} x_k^{n-1} + \ldots + a_0\right)\right]^2$$

with respect to $\mathbf{a} = (a_0, a_1, \ldots, a_n)$ to find the best coefficients. This is a quadratic expression in the coefficients $\mathbf{a}$. To find a compact representation for this objective we define $q_{ij} = \sum_{k=1}^{m} (x_k)^{i+j}$, $b_j = \sum_{k=1}^{m} g(x_k)(x_k)^j$ and $c = \sum_{k=1}^{m} g(x_k)^2$. Then after a bit of algebra it can be shown that

$$f(\mathbf{a}) = \mathbf{a}^T \mathbf{Q} \mathbf{a} - 2\mathbf{b}^T \mathbf{a} + c$$

where $\mathbf{Q} = [q_{ij}]$, $\mathbf{b} = (b_1, b_2, \ldots, b_{n+1})$.

The first-order necessary conditions state that the gradient of f must vanish. This leads directly to the system of $n + 1$ equations

$$\mathbf{Q}\mathbf{a} = \mathbf{b}.$$

These can be solved to determine $\mathbf{a}$.

Example 4 (Selection Problem). It is often necessary to select an assortment of factors to meet a given set of requirements. An example is the problem faced by an electric utility when selecting its power-generating facilities. The level of power that the company must supply varies by time of the day, by day of the week, and by season. Its power-generating requirements are summarized by a curve, $h(x)$, as shown in Fig. 7.2a, which shows the total hours in a year that a power level of at least x is required for each x. For convenience the curve is normalized so that the upper limit is unity.

The power company may meet these requirements by installing generating equipment, such as (7.1) nuclear or (7.2) coal-fired, or by purchasing power from a central energy grid. Associated with type $i(i = 1, 2)$ of generating equipment is a yearly unit capital cost b_i and a unit operating cost c_i. The unit price of power purchased from the grid is c_3.

Nuclear plants have a high capital cost and low operating cost, so they are used to supply a base load. Coal-fired plants are used for the intermediate level, and power is purchased directly only for peak demand periods. The requirements are satisfied as shown in Fig. 7.2b, where x_1 and x_2 denote the capacities of the nuclear and coal-fired plants, respectively. (For example, the nuclear power plant can be visualized as consisting of x_1/Δ small generators of capacity Δ, where Δ is small. The first such generator is on for about $h(\Delta)$ hours, supplying $\Delta h(\Delta)$ units of energy; the next supplies $\Delta h(2\Delta)$ units, and so forth. The total energy supplied by the nuclear plant is thus the area shown.)

The total cost is

$$f(x_1, x_2) = b_1 x_1 + b_2 x_2 + c_1 \int_0^{x_1} h(x)dx + c_2 \int_{x_1}^{x_1+x_2} h(x)dx + c_3 \int_{x_1+x_2}^{1} h(x)dx,$$

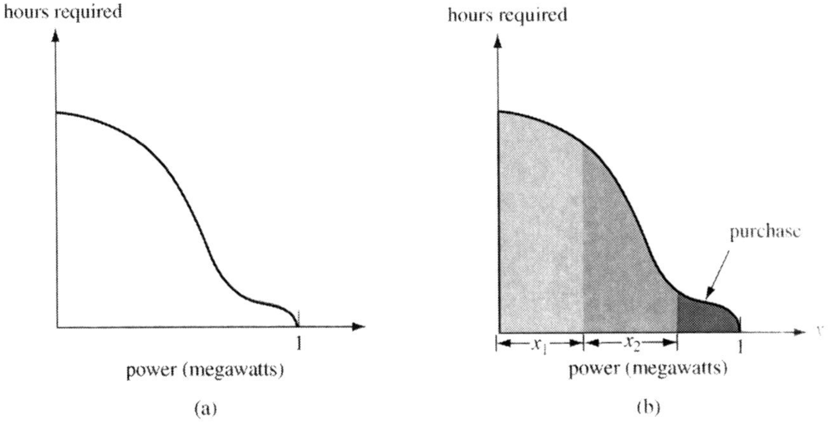

Fig. 7.2 (**a**) Power requirement curve; (**b**) x_1 and x_2 denote the capacities of the nuclear and coal-fired plants, respectively

and the company wishes to minimize this over the set defined by

$$x_1 \geqslant 0, \quad x_2 \geqslant 0, \quad x_1 + x_2 \leqslant 1.$$

Assuming that the solution is interior to the constraints, by setting the partial derivatives equal to zero, we obtain the two equations

$$\begin{aligned} b_1 + (c_1 - c_2)h(x_1) + (c_2 - c_3)h(x_1 + x_2) &= 0 \\ b_2 + (c_2 - c_3)h(x_1 + x_2) &= 0, \end{aligned}$$

which represent the necessary conditions.

If $x_1 = 0$, then the general necessary condition theorem shows that the first equality could relax to $\geqslant 0$. Likewise, if $x_2 = 0$, then the second equality could relax to $\geqslant 0$. The case $x_1 + x_2 = 1$ requires a bit more analysis (see Exercise 2).

7.3 Second-Order Conditions

The proof of Proposition 1 in Sect. 7.1 is based on making a first-order approximation to the function f in the neighborhood of the relative minimum point. Additional conditions can be obtained by considering higher-order approximations.

The second-order conditions, which are defined in terms of the Hessian matrix $\nabla^2 f$ of second partial derivatives of f (see Appendix A), are of extreme theoretical importance and dominate much of the analysis presented in later chapters.

Proposition 1 (Second-Order Necessary Conditions). *Let Ω be a subset of E^n and let $f \in C^2$ be a function on Ω. If $\mathbf{x}^*$ is a relative minimum point of f over Ω, then for any $\mathbf{d} \in E^n$ that is a feasible direction at $\mathbf{x}^*$ we have*

$$\text{i) } \nabla f(\mathbf{x}^*)\mathbf{d} \geqslant 0 \tag{7.3}$$

$$\text{ii) if } \nabla f(\mathbf{x}^*)\mathbf{d} = 0, \text{ then } \mathbf{d}^T \nabla^2 f(\mathbf{x}^*)\mathbf{d} \geqslant 0. \tag{7.4}$$

Proof. The first condition is just Proposition 1, and the second applies only if $\nabla f(\mathbf{x}^*)\mathbf{d} = 0$. In this case, introducing $\mathbf{x}(\alpha) = \mathbf{x}^* + \alpha\mathbf{d}$ and $g(\alpha) = f(\mathbf{x}(\alpha))$ as before, we have, in view of $g'(0) = 0$,

$$g(\alpha) - g(0) = \frac{1}{2}g''(0)\alpha^2 + o(\alpha^2).$$

If $g''(0) < 0$ the right side of the above equation is negative for sufficiently small α which contradicts the relative minimum nature of $g(0)$. Thus

$$g''(0) = \mathbf{d}^T \nabla^2 f(\mathbf{x}^*)\mathbf{d} \geqslant 0. \blacksquare$$

Example 1. For the same problem as Example 2 of Sect. 7.1, we have for $\mathbf{d} = (d_1, d_2)$

$$\nabla f(\mathbf{x}^*)\mathbf{d} = \frac{3}{2}d_2.$$

Thus condition (ii) of Proposition 1 applies only if $d_2 = 0$. In that case we have $\mathbf{d}^T \nabla^2 f(\mathbf{x}^*)\mathbf{d} = 2d_1^2 \geqslant 0$, so condition (ii) is satisfied.

Again of special interest is the case where the minimizing point is an interior point of Ω, as, for example, in the case of completely unconstrained problems. We then obtain the following classical result.

Proposition 2 (Second-Order Necessary Conditions—Unconstrained Case). *Let $\mathbf{x}^*$ be an interior point of the set Ω, and suppose $\mathbf{x}^*$ is a relative minimum point over Ω of the function $f \in C^2$. Then*

$$\text{i) } \nabla f(\mathbf{x}^*) = 0 \tag{7.5}$$

$$\text{ii) for all } \mathbf{d},\ \mathbf{d}^T \nabla^2 f(\mathbf{x}^*)\mathbf{d} \geqslant 0. \tag{7.6}$$

For notational simplicity we often denote $\nabla^2 f(\mathbf{x})$, the $n \times n$ matrix of the second partial derivatives of f, the Hessian of f, by the alternative notation $\mathbf{F}(\mathbf{x})$. Condition (ii) is equivalent to stating that the matrix $\mathbf{F}(\mathbf{x}^*)$ is positive semidefinite. As we shall see, the matrix $\mathbf{F}(\mathbf{x}^*)$, which arises here quite naturally in a discussion of necessary conditions, plays a fundamental role in the analysis of iterative methods for solving unconstrained optimization problems. The structure of this matrix is the primary determinant of the rate of convergence of algorithms designed to minimize the function f.

Example 2. Consider the problem

$$\begin{aligned} &\text{minimize} \quad f(x_1, x_2) = x_1^3 - x_1^2 x_2 + 2x_2^2 \\ &\text{subject to} \quad x_1 \geqslant 0, \quad x_2 \geqslant 0. \end{aligned}$$

If we assume that the solution is in the interior of the feasible set, that is, if $x_1 > 0$, $x_2 > 0$, then the first-order necessary conditions are

$$3x_1^2 - 2x_1 x_2 = 0, \quad -x_1^2 + 4x_2 = 0.$$

There is a solution to these at $x_1 = x_2 = 0$ which is a boundary point, but there is also a solution at $x_1 = 6$, $x_2 = 9$. We note that for x_1 fixed at $x_1 = 6$, the objective attains a relative minimum with respect to x_2 at $x_2 = 9$. Conversely, with x_2 fixed at $x_2 = 9$, the objective attains a relative minimum with respect to x_1 at $x_1 = 6$. Despite this fact, the point $x_1 = 6$, $x_2 = 9$ is not a relative minimum point, because the Hessian matrix is

$$\mathbf{F} = \begin{bmatrix} 6x_1 - 2x_2 & -2x_1 \\ -2x_1 & 4 \end{bmatrix},$$

which, evaluated at the proposed solution $x_1 = 6$, $x_2 = 9$, is

$$\mathbf{F} = \begin{bmatrix} 18 & -12 \\ -12 & 4 \end{bmatrix}.$$

This matrix is not positive semidefinite, since its determinant is negative. Thus the proposed solution is not a relative minimum point.

Sufficient Conditions for a Relative Minimum

By slightly strengthening the second condition of Proposition 2 above, we obtain a set of conditions that imply that the point $\mathbf{x}^*$ is a relative minimum. We give here the conditions that apply only to unconstrained problems, or to problems where the minimum point is interior to the feasible region, since the corresponding conditions for problems where the minimum is achieved on a boundary point of the feasible set are a good deal more difficult and of marginal practical or theoretical value. A more general result, applicable to problems with functional constraints, is given in Chap. 11.

Proposition 3 (Second-Order Sufficient Conditions—Unconstrained Case). *Let $f \in C^2$ be function defined on a region in which the point $\mathbf{x}^*$ is an interior point. Suppose in addition that*

$$\text{i) } \nabla f(\mathbf{x}^*) = \mathbf{0} \tag{7.7}$$

$$\text{ii) } \mathbf{F}(\mathbf{x}^*) \text{ is positive definite} \tag{7.8}$$

Then $\mathbf{x}^*$ is a strict relative minimum point of f.

Proof. Since $\mathbf{F}(\mathbf{x}^*)$ is positive definite, there is an $a > 0$ such that for all $\mathbf{d}$, $\mathbf{d}^T\mathbf{F}(\mathbf{x}^*)\mathbf{d} \geqslant a|\mathbf{d}|^2$. Thus by the Taylor's Theorem (with remainder)

$$\begin{aligned} f(\mathbf{x}^* + \mathbf{d}) - f(\mathbf{x}^*) &= \frac{1}{2}\mathbf{d}^T\mathbf{F}(\mathbf{x}^*)\mathbf{d} + o(|\mathbf{d}|^2) \\ &\geqslant (a/2)|\mathbf{d}|^2 + o(|\mathbf{d}|^2) \end{aligned}$$

For small $|\mathbf{d}|$ the first term on the right dominates the second, implying that both sides are positive for small $\mathbf{d}$. ∎

7.4 Convex and Concave Functions

In order to develop a theory directed toward characterizing global, rather than local, minimum points, it is necessary to introduce some sort of convexity assumptions. This results not only in a more potent, although more restrictive, theory but also provides an interesting geometric interpretation of the second-order sufficiency result derived above.

Definition. A function f defined on a convex set Ω is said to be *convex* if, for every $\mathbf{x}_1, \mathbf{x}_2 \in \Omega$ and every α, $0 \leqslant \alpha \leqslant 1$, there holds

$$f(\alpha\mathbf{x}_1 + (1-\alpha)\mathbf{x}_2) \leqslant \alpha f(\mathbf{x}_1) + (1-\alpha)f(\mathbf{x}_2).$$

If, for every α, $0 < \alpha < 1$, and $\mathbf{x}_1 \neq \mathbf{x}_2$, there holds

$$f(\alpha\mathbf{x}_1 + (1-\alpha)\mathbf{x}_2) < \alpha f(\mathbf{x}_1) + (1-\alpha)f(\mathbf{x}_2),$$

then f is said to be *strictly convex*.

Several examples of convex or nonconvex functions are shown in Fig. 7.3. Geometrically, a function is convex if the line joining two points on its graph lies nowhere below the graph, as shown in Fig. 7.3a, or, thinking of a function in two dimensions, it is convex if its graph is bowl shaped.

Next we turn to the definition of a concave function.

Definition. A function g defined on a convex set Ω is said to be *concave* if the function $f = -g$ is convex. The function g is *strictly concave* if $-g$ is strictly convex.

Combinations of Convex Functions

We show that convex functions can be combined to yield new convex functions and that convex functions when used as constraints yield convex constraint sets.

Proposition 1. *Let f_1 and f_2 be convex functions on the convex set Ω. Then the function $f_1 + f_2$ is convex on Ω.*

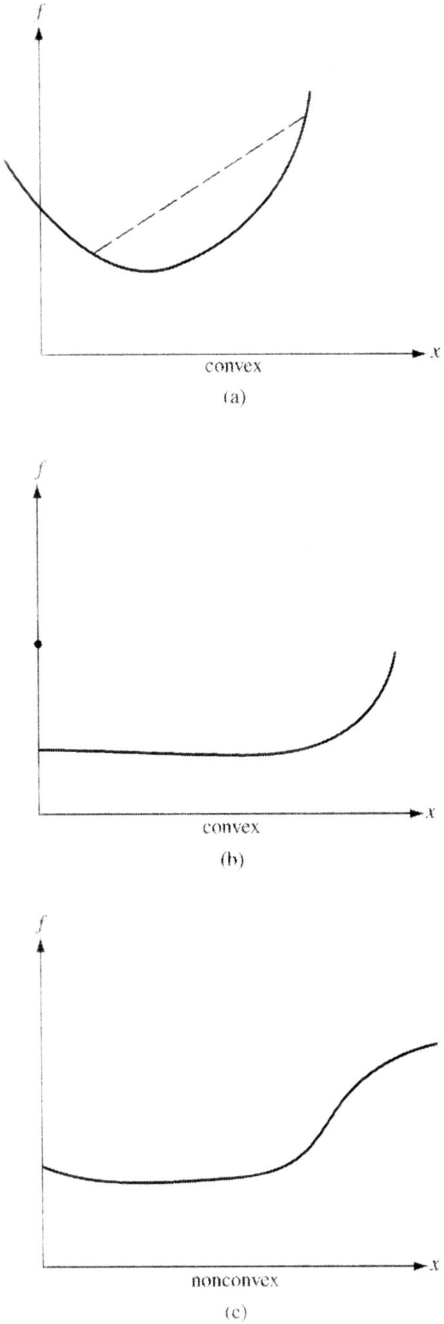

Fig. 7.3 Convex and nonconvex functions

Proof. Let $\mathbf{x}_1,\ \mathbf{x}_2 \in \Omega$, and $0 < \alpha < 1$. Then

$$f_1(\alpha\mathbf{x}_1 + (1-\alpha)\mathbf{x}_2) + f_2(\alpha\mathbf{x}_1) + (1-\alpha)\mathbf{x}_2)$$
$$\leqslant \alpha[f_1(\mathbf{x}_1) + f_2(\mathbf{x}_1)] + (1-\alpha)[f_1(\mathbf{x}_2) + f_2(\mathbf{x}_2)]. \blacksquare$$

Proposition 2. *Let f be a convex function over the convex set Ω. Then the function af is convex for any $a \geqslant 0$.*

Proof. Immediate. ∎

Note that through repeated application of the above two propositions it follows that a positive combination $a_1 f_1 + a_2 f_2 + \ldots + a_m f_m$ of convex functions is again convex.

Finally, we consider sets defined by convex inequality constraints.

Proposition 3. *Let f be a convex function on a convex set Ω. The set $\Gamma_c = \{\mathbf{x} : \mathbf{x} \in \Omega,\ f(\mathbf{x}) \leqslant c\}$ is convex for every real number c.*

Proof. Let $\mathbf{x}_1,\ \mathbf{x}_2 \in \Gamma_c$. Then $f(\mathbf{x}_1) \leqslant c,\ f(\mathbf{x}_2) \leqslant c$ and for $0 < \alpha < 1$,

$$f(\alpha\mathbf{x}_1 + (1-\alpha)\mathbf{x}_2) \leqslant \alpha f(\mathbf{x}_1) + (1-\alpha) f(\mathbf{x}_2) \leqslant c.$$

Thus $\alpha\mathbf{x}_1 + (1-\alpha)\mathbf{x}_2 \in \Gamma_c$. ∎

We note that, since the intersection of convex sets is also convex, the set of points simultaneously satisfying

$$f_1(\mathbf{x}) \leqslant c_1,\ f_2(\mathbf{x}) \leqslant c_2,\ \ldots,\ f_m(\mathbf{x}) \leqslant c_m,$$

where each f_i is a convex function, defines a convex set. This is important in mathematical programming, since the constraint set is often defined this way.

Properties of Differentiable Convex Functions

If a function f is differentiable, then there are alternative characterizations of convexity.

Proposition 4. *Let $f \in C^1$. Then f is convex over a convex set Ω if and only if*

$$f(\mathbf{y}) \geqslant f(\mathbf{x}) + \nabla f(\mathbf{x})(\mathbf{y} - \mathbf{x}) \tag{7.9}$$

for all $\mathbf{x},\ \mathbf{y} \in \Omega$.

Proof. First suppose f is convex. Then for all $\alpha,\ 0 \leqslant \alpha \leqslant 1$,

$$f(\alpha\mathbf{y} + (1-\alpha)\mathbf{x}) \leqslant \alpha f(\mathbf{y}) + (1-\alpha) f(\mathbf{x}).$$

Thus for $0 < \alpha \leqslant 1$

$$\frac{f(\mathbf{x} + \alpha(\mathbf{y} - \mathbf{x})) - f(\mathbf{x})}{\alpha} \leqslant f(\mathbf{y}) - f(\mathbf{x}).$$

Letting $\alpha \to 0$ we obtain

$$\nabla f(\mathbf{x})\,(\mathbf{y} - \mathbf{x}) \leqslant f(\mathbf{y}) - f(\mathbf{x}).$$

This proves the "only if" part.

Now assume

$$f(\mathbf{y}) \geqslant f(\mathbf{x}) + \nabla f(\mathbf{x})\,(\mathbf{y} - \mathbf{x})$$

for all $\mathbf{x},\ \mathbf{y} \in \Omega$. Fix $\mathbf{x}_1,\ \mathbf{x}_2 \in \Omega$ and $\alpha,\ 0 \leqslant \alpha \leqslant 1$. Setting $\mathbf{x} = \alpha\mathbf{x}_1 + (1 - \alpha)\mathbf{x}_2$ and alternatively $\mathbf{y} = \mathbf{x}_1$ or $\mathbf{y} = \mathbf{x}_2$, we have

$$f(\mathbf{x}_1) \geqslant f(\mathbf{x}) + \nabla f(\mathbf{x})(\mathbf{x}_1 - \mathbf{x}) \tag{7.10}$$

$$f(\mathbf{x}_2) \geqslant f(\mathbf{x}) + \nabla f(\mathbf{x})(\mathbf{x}_2 - \mathbf{x}). \tag{7.11}$$

Multiplying (7.10) by α and (7.11) by $(1 - \alpha)$ and adding, we obtain

$$\alpha f(\mathbf{x}_1) + (1 - \alpha) f(\mathbf{x}_2) \geqslant f(\mathbf{x}) + \nabla f(\mathbf{x})[\alpha\mathbf{x}_1 + (1 - \alpha)\mathbf{x}_2 - \mathbf{x}].$$

But substituting $\mathbf{x} = \alpha\mathbf{x}_1 + (1 - \alpha)\mathbf{x}_2$, we obtain

$$\alpha f(\mathbf{x}_1) + (1 - \alpha) f(\mathbf{x}_2) \geqslant f(\alpha\mathbf{x}_1 + (1 - \alpha)\mathbf{x}_2). \blacksquare$$

The statement of the above proposition is illustrated in Fig. 7.4. It can be regarded as a sort of dual characterization of the original definition illustrated in Fig. 7.3. The original definition essentially states that linear interpolation between two points overestimates the function, while the above proposition states that linear approximation based on the local derivative underestimates the function.

For twice continuously differentiable functions, there is another characterization of convexity.

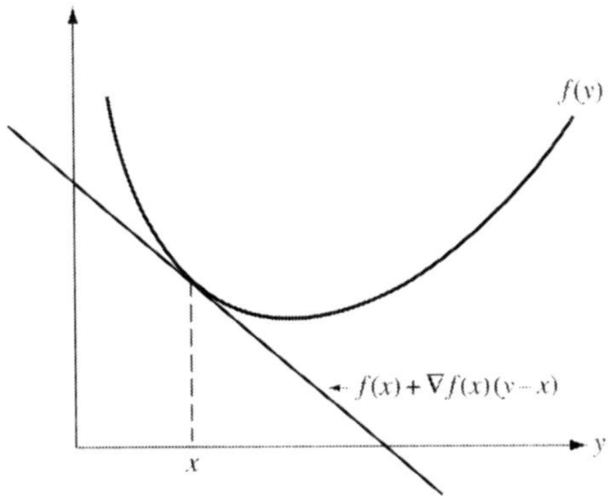

Fig. 7.4 Illustration of Proposition 4

Proposition 5. *Let $f \in C^2$. Then f is convex over a convex set Ω containing an interior point if and only if the Hessian matrix $\mathbf{F}$ of f is positive semidefinite throughout Ω.*

Proof. By Taylor's theorem we have

$$f(\mathbf{y}) = f(\mathbf{x}) = \nabla f(\mathbf{x})(\mathbf{y} - \mathbf{x}) + \frac{1}{2}(\mathbf{y} - \mathbf{x})^T \mathbf{F}(\mathbf{x} + \alpha(\mathbf{y} - \mathbf{x}))(\mathbf{y} - \mathbf{x}) \tag{7.12}$$

for some α, $0 \leqslant \alpha \leqslant 1$. Clearly, if the Hessian is everywhere positive semidefinite, we have

$$f(\mathbf{y}) \geqslant f(\mathbf{x}) + \nabla f(\mathbf{x})(\mathbf{y} - \mathbf{x}), \tag{7.13}$$

which in view of Proposition 4 implies that f is convex.

Now suppose the Hessian is not positive semidefinite at some point $\mathbf{x} \in \Omega$. By continuity of the Hessian it can be assumed, without loss of generality, that $\mathbf{x}$ is an interior point of Ω. There is a $\mathbf{y} \in \Omega$ such that $(\mathbf{y} - \mathbf{x})^T \mathbf{F}(\mathbf{x})(\mathbf{y} - \mathbf{x}) < 0$. Again by the continuity of the Hessian, $\mathbf{y}$ may be selected so that for all α, $0 \leqslant \alpha \leqslant 1$,

$$(\mathbf{y} - \mathbf{x})^T \mathbf{F}(\mathbf{x} + \alpha(\mathbf{y} - \mathbf{x}))\,(\mathbf{y} - \mathbf{x}) < 0.$$

This in view of (7.12) implies that (7.13) does not hold; which in view of Proposition 4 implies that f is not convex. ∎

The Hessian matrix is the generalization to E^n of the concept of the curvature of a function, and correspondingly, positive definiteness of the Hessian is the generalization of positive curvature. Convex functions have positive (or at least nonnegative) curvature in every direction. Motivated by these observations, we sometimes refer to a function as being *locally convex* if its Hessian matrix is positive semidefinite in a small region, and *locally strictly convex* if the Hessian is positive definite in the region. In these terms we see that the second-order sufficiency result of the last section requires that the function be locally strictly convex at the point $\mathbf{x}^*$. Thus, even the local theory, derived solely in terms of the elementary calculus, is actually intimately related to convexity—at least locally. For this reason we can view the two theories, local and global, not as disjoint parallel developments but as complementary and interactive. Results that are based on convexity apply even to nonconvex problems in a region near the solution, and conversely, local results apply to a global minimum point.

7.5 Minimization and Maximization of Convex Functions

We turn now to the three classic results concerning minimization or maximization of convex functions.

Theorem 1. *Let f be a convex function defined on the convex set Ω. Then the set Γ where f achieves its minimum is convex, and any relative minimum of f is a global minimum.*

Proof. If f has no relative minima the theorem is valid by default. Assume now that c_0 is the minimum of f. Then clearly $\Gamma = \{\mathbf{x} : f(\mathbf{x}) \leqslant c_0, \ \mathbf{x} \in \Omega\}$ and this is convex by Proposition 3 of the last section.

Suppose now that $\mathbf{x}^* \in \Omega$ is a relative minimum point of f, but that there is another point $\mathbf{y} \in \Omega$ with $f(\mathbf{y}) < f(\mathbf{x}^*)$. On the line $\alpha\mathbf{y} + (1-\alpha)\mathbf{x}^*$, $0 < \alpha < 1$ we have

$$f(\alpha\mathbf{y} + (1-\alpha)\mathbf{x}^*) \leqslant \alpha f(\mathbf{y}) + (1-\alpha)f(\mathbf{x}^*) < f(\mathbf{x}^*),$$

contradicting the fact that $\mathbf{x}^*$ is a relative minimum point. ∎

We might paraphrase the above theorem as saying that for convex functions, all minimum points are located together (in a convex set) and all relative minima are global minima. The next theorem says that if f is continuously differentiable and convex, *then* satisfaction of the first-order necessary conditions are both necessary and sufficient for a point to be a global minimizing point.

Theorem 2. *Let $f \in C^1$ be convex on the convex set Ω. If there is a point $\mathbf{x}^* \in \Omega$ such that, for all $\mathbf{y} \in \Omega$, $\nabla f(\mathbf{x}^*)(\mathbf{y} - \mathbf{x}^*) \geqslant 0$, then $\mathbf{x}^*$ is a global minimum point of f over Ω.*

Proof. We note parenthetically that since $\mathbf{y}-\mathbf{x}^*$ is a feasible direction at $\mathbf{x}^*$, the given condition is equivalent to the first-order necessary condition stated in Sect. 7.1. The proof of the proposition is immediate, since by Proposition 4 of the last section

$$f(\mathbf{y}) \geqslant f(\mathbf{x}^*) + \nabla f(\mathbf{x}^*)(\mathbf{y} - \mathbf{x}^*) \geqslant f(\mathbf{x}^*). \ ∎$$

Next we turn to the question of maximizing a convex function over a convex set. There is, however, no analog of Theorem 1 for maximization; indeed, the tendency is for the occurrence of numerous nonglobal relative maximum points. Nevertheless, it is possible to prove one important result. It is not used in subsequent chapters, but it is useful for some areas of optimization.

Theorem 3. *Let f be a convex function defined on the bounded, closed convex set Ω. If f has a maximum over Ω it is achieved at an extreme point of Ω.*

Proof. Suppose f achieves a global maximum at $\mathbf{x}^* \in \Omega$. We show first that this maximum is achieved at some boundary point of Ω. If $\mathbf{x}^*$ is itself a boundary point, then there is nothing to prove, so assume $\mathbf{x}^*$ is not a boundary point. Let L be any line passing through the point $\mathbf{x}^*$. The intersection of this line with Ω is an interval of the line L having end points $\mathbf{y}_1$, $\mathbf{y}_2$ which are boundary points of Ω, and we have $\mathbf{x}^* = \alpha\mathbf{y}_1 + (1-\alpha)\mathbf{y}_2$ for some α, $0 < \alpha < 1$. By convexity of f

$$f(\mathbf{x}^*) \leqslant \alpha f(\mathbf{y}_1) + (1-\alpha)f(\mathbf{y}_2) \leqslant \max\{f(\mathbf{y}_1), \ f(\mathbf{y}_2)\}.$$

Thus either $f(\mathbf{y}_1)$ or $f(\mathbf{y}_2)$ must be at least as great as $f(\mathbf{x}^*)$. Since $\mathbf{x}^*$ is a maximum point, so is either $\mathbf{y}_1$ or $\mathbf{y}_2$.

We have shown that the maximum, if achieved, must be achieved at a boundary point of Ω. If this boundary point, $\mathbf{x}^*$, is an extreme point of Ω there is nothing more to prove. If it is not an extreme point, consider the intersection of Ω with a supporting hyperplane H at $\mathbf{x}^*$. This intersection, T_1, is of dimension $n-1$ or less and the global maximum of f over T_1 is equal to $f(\mathbf{x}^*)$ and must be achieved at a boundary point $\mathbf{x}_1$ of T_1. If this boundary point is an extreme point of T_1, it is also an extreme point of Ω by Lemma 1, Sect. B.4, and hence the theorem is proved. If $\mathbf{x}_1$ is not an extreme point of T_1, we form T_2, the intersection of T_1 with a hyperplane in E^{n-1} supporting T_1 at $\mathbf{x}_1$. This process can continue at most a total of n times when a set T_n of dimension zero, consisting of a single point, is obtained. This single point is an extreme point of T_n and also, by repeated application of Lemma 1, Sect. B.4, an extreme point of Ω. ∎

*7.6 *Zero-Order Conditions

We have considered the problem

$$\begin{aligned} &\text{minimize} && f(\mathbf{x}) \\ &\text{subject to} && \mathbf{x} \in \Omega \end{aligned} \tag{7.14}$$

to be unconstrained because there are no functional constraints of the form $g(\mathbf{x}) \leqslant b$ or $h(\mathbf{x}) = c$. However, the problem is of course constrained by the set Ω. This constraint influences the first- and second-order necessary and sufficient conditions through the relation between feasible directions and derivatives of the function f. Nevertheless, there is a way to treat this constraint without reference to derivatives. The resulting conditions are then of zero order. These necessary conditions require that the problem be convex is a certain way, while the sufficient conditions require no assumptions at all. The simplest assumptions for the necessary conditions are that Ω is a convex set and that f is a convex function on all of E^n.

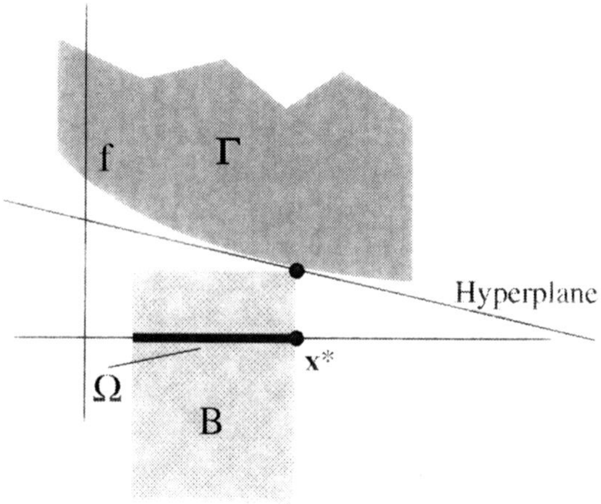

Fig. 7.5 The epigraph, the tubular region, and the hyperplane

To derive the necessary conditions under these assumptions consider the set $\Gamma \subset E^{n+1} = \{(r, \mathbf{x}) : r \geqslant f(\mathbf{x}), \mathbf{x} \in E^n\}$. In a figure of the graph of f, the set Γ is the region above the graph, shown in the upper part of Fig. 7.5. This set is called the *epigraph* of f. It is easy to verify that the set Γ is convex if f is a convex function.

Suppose that $\mathbf{x}^* \in \Omega$ is the minimizing point with value $f^* = f(\mathbf{x}^*)$. We construct a tubular region with cross section Ω and extending vertically from $-\infty$ up to f^*, shown as B in the upper part of Fig. 7.5. This is also a convex set, and it overlaps the set Γ only at the boundary point $(f^*, \mathbf{b}^*)$ above $\mathbf{x}^*$(or possibly many boundary points if f is flat near $\mathbf{x}^*$).

According to the separating hyperplane theorem (Appendix B), there is a hyperplane separating these two sets. This hyperplane can be represented by a nonzero vector of the form $(s, \lambda) \in E^{n+1}$ with s a scalar and $\lambda \in E^n$, and a separation constant c. The separation conditions are

$$sr + \lambda^T \mathbf{x} \geqslant c \text{ for all } \mathbf{x} \in E^n \text{ and } r \geqslant f(\mathbf{x}) \tag{7.15}$$

$$sr + \lambda^T \mathbf{x} \leqslant c \text{ for all } \mathbf{x} \in \Omega \text{ and } r \leqslant f^*. \tag{7.16}$$

It follows that $s \neq 0$; for otherwise $\lambda \neq \mathbf{0}$ and then (7.15) would be violated for some $\mathbf{x} \in E^n$. It also follows that $s \geqslant 0$ since otherwise (7.16) would be violated by very negative values of r. Hence, together we find $s > 0$ and by appropriate scaling we may take $s = 1$.

It is easy to see that the above conditions can be expressed alternatively as two optimization problems, as stated in the following proposition.

Proposition 1 (Zero-Order Necessary Conditions). *If* $\mathbf{x}^*$ *solves (7.14) under the stated convexity conditions, then there is a nonzero vector* $\lambda \in E^n$ *such that* $\mathbf{x}^*$ *is a solution to the two problems:*

$$\begin{aligned} &\text{minimize} && f(\mathbf{x}) + \lambda^T \mathbf{x} \\ &\text{subject to} && \mathbf{x} \in E^n \end{aligned} \tag{7.17}$$

and

$$\begin{aligned} &\text{maximize} && \lambda^T \mathbf{x} \\ &\text{subject to} && \mathbf{x} \in \Omega. \end{aligned} \tag{7.18}$$

Proof. Problem (7.17) follows from (7.15) (with $s = 1$) and the fact that $f(\mathbf{x}) \leqslant r$ for $r \geqslant f(\mathbf{x})$. The value c is attained from above at $(f^*, \mathbf{x}^*)$. Likewise (7.18) follows from (7.16) and the fact that $\mathbf{x}^*$ and the appropriate r attain c from below. ∎

Notice that problem (7.17) is completely unconstrained, since $\mathbf{x}$ may range over all of E^n. The second problem (7.18) is constrained by Ω but has a linear objective function. It is clear from Fig. 7.5 that the slope of the hyperplane is equal to the slope of the function f when f is continuously differentiable at the solution $\mathbf{x}^*$.

If the optimal solution $\mathbf{x}^*$ is in the interior of Ω, then the second problem (7.18) implies that $\lambda = \mathbf{0}$, for otherwise there would be a direction of movement from $\mathbf{x}^*$ that increases the product $\lambda^T \mathbf{x}$ above $\lambda^T \mathbf{x}^*$. The hyperplane is horizontal in that case.

The zeroth-order conditions provide no new information in this situation. However, when the solution is on a boundary point of Ω the conditions give very useful information.

Example 1 (Minimization Over an Interval). Consider a continuously differentiable function f of a single variable $x \in E^1$ defined on the unit interval [0,1] which plays the role of Ω here. The first problem (7.17) implies $f'(x^*) = -\lambda$. If the solution is at the left end of the interval (at $x = 0$) then the second problem (7.18) implies that $\lambda \leq 0$ which means that $f'(x^*) \geq 0$. The reverse holds if x^* is at the right end. These together are identical to the first-order conditions of Sect. 7.1.

Example 2. As a generalization of the above example, let $f \in C^1$ on E^n, and let f have a minimum with respect to Ω at $\mathbf{x}^*$. Let $\mathbf{d} \in E^n$ be a feasible direction at x^*. Then it follows again from (7.17) that $\nabla f(\mathbf{x}^*)\mathbf{d} \geq 0$.

Sufficient Conditions Theorem. The conditions of Proposition 1 are sufficient for $\mathbf{x}^*$ to be a minimum even without the convexity assumptions.

Proposition 2 (Zero-Order Sufficiency Conditions). *If there is a λ such that $\mathbf{x}^* \in \Omega$ solves the problems (7.17) and (7.18), then $\mathbf{x}^*$ solves (7.14).*

Proof. Suppose $\mathbf{x}_1$ is any other point in $\boldsymbol{\Omega}$. Then from (7.17)

$$f(\mathbf{x}_1) + \lambda^T \mathbf{x}_1 \geqslant f(\mathbf{x}^*) + \lambda^T \mathbf{x}^*.$$

This can be rewritten as

$$f(\mathbf{x}_1) - f(\mathbf{x}^*) \geqslant \lambda^T \mathbf{x}^* - \lambda^T \mathbf{x}_1.$$

By problem (7.18) the right hand side of this is greater than or equal to zero. Hence $f(\mathbf{x}_1) - f(\mathbf{x}^*) \geqslant 0$ which establishes the result. ∎

7.7 Global Convergence of Descent Algorithms

A good portion of the remainder of this book is devoted to presentation and analysis of various algorithms designed to solve nonlinear programming problems. Although these algorithms vary substantially in their motivation, application, and detailed analysis, ranging from the simple to the highly complex, they have the common heritage of all being iterative descent algorithms. By *iterative*, we mean, roughly, that the algorithm generates a series of points, each point being calculated on the basis of the points preceding it. By *descent*, we mean that as each new point is generated by the algorithm the corresponding value of some function (evaluated at the most recent point) decreases in value. Ideally, the sequence of points generated by the algorithm in this way converges in a finite or infinite number of steps to a solution of the original problem.

An iterative algorithm is initiated by specifying a starting point. If for arbitrary starting points the algorithm is guaranteed to generate a sequence of points converging to a solution, then the algorithm is said to be *globally convergent*. Quite definitely, not all algorithms have this obviously desirable property. Indeed, many of the most important algorithms for solving nonlinear programming problems are not globally convergent in their purest form and thus occasionally generate sequences that either do not converge at all or converge to points that are not solutions. It is often possible, however, to modify such algorithms, by appending special devices, so as to guarantee global convergence.

Fortunately, the subject of global convergence can be treated in a unified manner through the analysis of a general theory of algorithms developed mainly by Zangwill. From this analysis, which is presented in this section, we derive the Global Convergence Theorem that is applicable to the study of any iterative descent algorithm. Frequent reference to this important result is made in subsequent chapters.

Iterative Algorithms

We think of an algorithm as a mapping. Given a point $\mathbf{x}$ in some space X, the output of an algorithm applied to $\mathbf{x}$ is a new point. Operated iteratively, an algorithm is repeatedly reapplied to the new points it generates so as to produce a whole sequence of points. Thus, as a preliminary definition, we might formally define an algorithm A as a mapping taking points in a space X into (other) points in X. Operated iteratively, the algorithm A initiated at $\mathbf{x}_0 \in X$ would generate the sequence $\{\mathbf{x}_k\}$ defined by

$$\mathbf{x}_{k+1} = \mathbf{A}(\mathbf{x}_k).$$

In practice, the mapping $\mathbf{A}$ might be defined explicitly by a simple mathematical expression or it might be defined implicitly by, say, a lengthy complex computer program. Given an input vector, both define a corresponding output.

With this intuitive idea of an algorithm in mind, we now generalize the concept somewhat so as to provide greater flexibility in our analyses.

Definition. An *algorithm* $\mathbf{A}$ is a mapping defined on a space X that assigns to every point $\mathbf{x} \in X$ a subset of X.

In this definition the term "space" can be interpreted loosely. Usually X is the vector space E^n but it may be only a subset of E^n or even a more general metric space. The most important aspect of the definition, however, is that the mapping $\mathbf{A}$, rather than being a point-to-point mapping of X, is a *point-to-set mapping* of X.

An algorithm $\mathbf{A}$ generates a sequence of points in the following way. Given $\mathbf{x}_k \in X$ the algorithm yields $\mathbf{A}(\mathbf{x}_k)$ which is a subset of X. From this subset an arbitrary element $\mathbf{x}_{k+1}$ is selected. In this way, given an initial point $\mathbf{x}_0$, the algorithm generates sequences through the iteration

$$\mathbf{x}_{k+1} \in \mathbf{A}(\mathbf{x}_k).$$

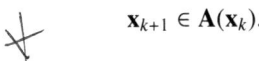

It is clear that, unlike the case where $\mathbf{A}$ is a point-to-point mapping, the sequence generated by the algorithm $\mathbf{A}$ cannot, in general, be predicted solely from knowledge of the initial point $\mathbf{x}_0$. This degree of uncertainty is designed to reflect uncertainty that we may have in practice as to specific details of an algorithm.

Example 1. Suppose for x on the real line we define

$$A(x) = [-|x|/2, \ |x|/2]$$

so that $A(x)$ is an interval of the real line. Starting at $x_0 = 100$, each of the sequences below might be generated from iterative application of this algorithm.

$$100, \ 50, \ 25, \ 12, \ -6, \ -2, \ 1, \ 1/2, \ldots$$
$$100, \ -40, \ 20, \ -5, \ -2, \ 1, \ 1/4, \ 1/8, \ldots$$
$$100, \ 10, \ -1, 1/16, 1/100, \ -1/1000, \ 1/10, \ 100, \ldots$$

The apparent ambiguity that is built into this definition of an algorithm is not meant to imply that actual algorithms are random in character. In actual implementation algorithms are not defined ambiguously. Indeed, a particular computer program executed twice from the same starting point will generate two copies of the same sequence. In other words, in practice algorithms are point-to-point mappings. The utility of the more general definition is that it allows one to analyze, in a single step, the convergence of an infinite family of similar algorithms. Thus, two computer programs, designed from the same basic idea, may differ slightly in some details, and therefore perhaps may not produce identical results when given the same starting point. Both programs may, however, be regarded as implementations of the same point-to-set mappings. In the example above, for instance, it is not necessary to know exactly how x_{k+1} is determined from x_k so long as it is known that its absolute value is no greater than one-half x_k's absolute value. The result will always tend toward zero. In this manner, the generalized concept of an algorithm sometimes leads to simpler analysis.

Descent

In order to describe the idea of a descent algorithm we first must agree on a subset Γ of the space X, referred to as the *solution set*. The basic idea of a *descent function*, which is defined below, is that for points outside the solution set, a single step of the algorithm yields a decrease in the value of the descent function.

Definition. Let $\Gamma \subset X$ be a given solution set and let $\mathbf{A}$ be an algorithm on X. A continuous real-valued function Z on X is said to be a *descent function* for Γ and $\mathbf{A}$ if it satisfies

i) if $\mathbf{x} \notin \Gamma$ and $\mathbf{y} \in \mathbf{A}(\mathbf{x})$, then $Z(\mathbf{y}) < Z(\mathbf{x})$
ii) if $\mathbf{x} \in \Gamma$ and $\mathbf{y} \in \mathbf{A}(\mathbf{x})$, then $Z(\mathbf{y}) \leqslant Z(\mathbf{x})$.

There are a number of ways a solution set, algorithm, and descent function can be defined. A natural set-up for the problem

$$\begin{aligned} &\text{minimize } f(\mathbf{x}) \\ &\text{subject to } \mathbf{x} \in \Omega \end{aligned} \tag{7.19}$$

is to let Γ be the set of minimizing points, and define an algorithm $\mathbf{A}$ on Ω in such a way that f decreases at each step and thereby serves as a descent function. Indeed, this is the procedure followed in a majority of cases. Another possibility for unconstrained problems is to let Γ be the set of points $\mathbf{x}$ satisfying $\nabla f(\mathbf{x}) = 0$. In this case we might design an algorithm for which $|\nabla f(\mathbf{x})|$ serves as a descent function or for which $f(\mathbf{x})$ serves as a descent function.

*Closed Mappings

An important property possessed by some algorithms is that they are closed. This property, which is a generalization for point-to-set mappings of the concept of continuity for point-to-point mappings, turns out to be the key to establishing a general global convergence theorem. In defining this property we allow the point-to-set mapping to map points in one space X into subsets of another space Y.

Definition. A point-to-set mapping $\mathbf{A}$ from X to Y is said to be *closed* at $\mathbf{x} \in X$ if the assumptions

i) $\mathbf{x}_k \to \mathbf{x}$, $\mathbf{x}_k \in X$,

ii) $\mathbf{y}_k \to \mathbf{y}$, $\mathbf{y}_k \in \mathbf{A}(\mathbf{x}_k)$

imply

iii) $\mathbf{y} \in \mathbf{A}(\mathbf{x})$.

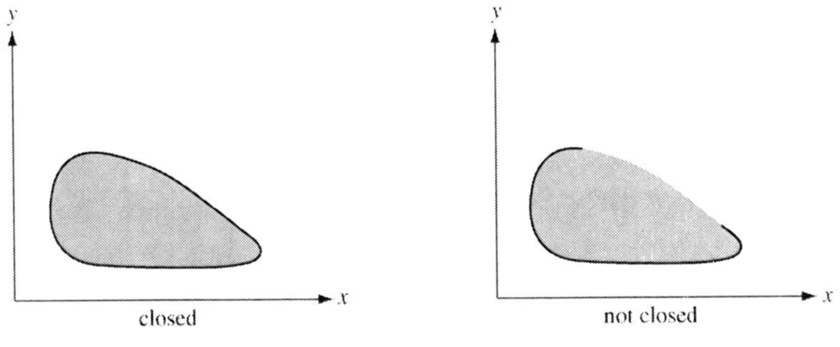

Fig. 7.6 Graphs of mappings

The point-to-set map $\mathbf{A}$ is said to be *closed* on X if it is closed at each point of X.

? Add to Pwpt ?

Example 2. As a special case, suppose that the mapping $\mathbf{A}$ is a point-to-point mapping; that is, for each $\mathbf{x} \in X$ the set $\mathbf{A}(\mathbf{x})$ consists of a single point in Y. Suppose also that $\mathbf{A}$ is continuous at $\mathbf{x} \in X$. This means that if $\mathbf{x}_k \to \mathbf{x}$ then $\mathbf{A}(\mathbf{x}_k) \to \mathbf{A}(\mathbf{x})$, and it follows that $\mathbf{A}$ is closed at $\mathbf{x}$. Thus for point-to-point mappings continuity implies closedness. The converse is, however, not true in general.

The definition of a closed mapping can be visualized in terms of the *graph* of the mapping, which is the set $\{(\mathbf{x}, \mathbf{y}) : \mathbf{x} \in X, \mathbf{y} \in \mathbf{A}(\mathbf{x})\}$. If X is closed, then $\mathbf{A}$ is closed throughout X if and only if this graph is a closed set. This is illustrated in Fig. 7.6. However, this equivalence is valid only when considering closedness everywhere. In general a mapping may be closed at some points and not at others.

Example 3. The reader should verify that the point-to-set mapping defined in Example 1 is closed.

Many complex algorithms that we analyze are most conveniently regarded as the composition of two or more simple point-to-set mappings. It is therefore natural to ask whether closedness of the individual maps implies closedness of the composite. The answer is a qualified "yes." The technical details of composition are described in the remainder of this subsection. They can safely be omitted at first reading while proceeding to the Global Convergence Theorem.

Definition. Let $\mathbf{A} : X \to Y$ and $\mathbf{B} : Y \to Z$ be point-to-set mappings. The composite mapping $\mathbf{C} = \mathbf{BA}$ is defined as the point-to-set mapping $\mathbf{C} : X \to Z$ with

$$\mathbf{C}(\mathbf{x}) = \bigcup_{\mathbf{y} \in \mathbf{A}(\mathbf{x})} \mathbf{B}(\mathbf{y}).$$

This definition is illustrated in Fig. 7.7.

Proposition. *Let $\mathbf{A} : X \to Y$ and $\mathbf{B} : Y \to Z$ be point-to-set mappings. Suppose $\mathbf{A}$ is closed at $\mathbf{x}$ and $\mathbf{B}$ is closed on $\mathbf{A}(\mathbf{x})$. Suppose also that if $\mathbf{x}_k \to \mathbf{x}$ and $\mathbf{y}_k \in \mathbf{A}(\mathbf{x}_k)$, there is a $\mathbf{y}$ such that, for some subsequence $\{\mathbf{y}_{k_i}\}$, $\mathbf{y}_{k_i} \to \mathbf{y}$. Then the composite mapping $\mathbf{C} = \mathbf{BA}$ is closed at $\mathbf{x}$.*

Proof. Let $\mathbf{x}_k \to \mathbf{x}$ and $\mathbf{z}_k \to \mathbf{z}$ with $\mathbf{z}_k \in \mathbf{C}(\mathbf{x}_k)$. It must be shown that $\mathbf{z} \in \mathbf{C}(\mathbf{x})$.

Select $\mathbf{y}_k \in \mathbf{A}(\mathbf{x}_k)$ such that $\mathbf{z}_k \in \mathbf{B}(\mathbf{y}_k)$ and according to the hypothesis let $\mathbf{y}$ and $\{\mathbf{y}_{k_i}\}$ be such that $\mathbf{y}_{k_i} \to \mathbf{y}$. Since $\mathbf{A}$ is closed at $\mathbf{x}$ it follows that $\mathbf{y} \in \mathbf{A}(\mathbf{x})$.

Likewise, since $\mathbf{y}_{k_i} \to \mathbf{y}$, $\mathbf{z}_{k_i} \to \mathbf{z}$ and $\mathbf{B}$ is closed at $\mathbf{y}$, it follows that $\mathbf{z} \in \mathbf{B}(\mathbf{y}) \subset \mathbf{BA}(\mathbf{x}) = \mathbf{C}(\mathbf{x})$. ∎

Two important corollaries follow immediately.

Corollary 1. *Let $\mathbf{A} : X \to Y$ and $\mathbf{B} : Y \to Z$ be point-to-set mappings. If $\mathbf{A}$ is closed at $\mathbf{x}$, $\mathbf{B}$ is closed on $\mathbf{A}(\mathbf{x})$ and Y is compact, then the composite map $\mathbf{C} = \mathbf{BA}$ is closed at $\mathbf{x}$.*

Corollary 2. *Let $\mathbf{A} : X \to Y$ be a point-to-point mapping and $\mathbf{B} : Y \to Z$ a point-to-set mapping. If $\mathbf{A}$ is continuous at $\mathbf{x}$ and $\mathbf{B}$ is closed at $\mathbf{A}(\mathbf{x})$, then the composite mapping $\mathbf{C} = \mathbf{BA}$ is closed at $\mathbf{x}$.*

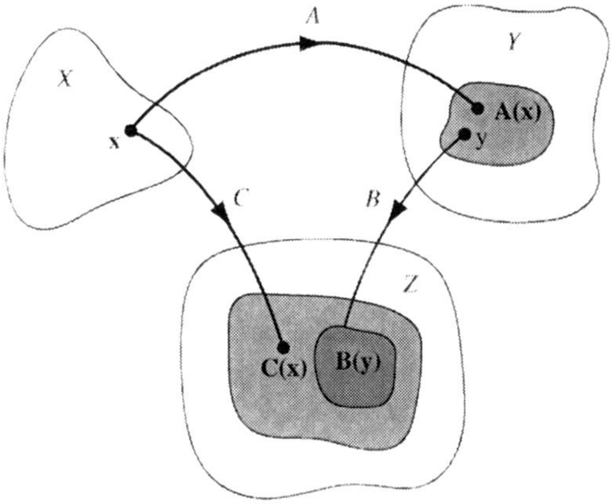

Fig. 7.7 Composition of mappings

Global Convergence Theorem

The Global Convergence Theorem is used to establish convergence for the following general situation. There is a solution set Γ. Points are generated according to the algorithm $\mathbf{x}_{k+1} \in \mathbf{A}(\mathbf{x}_k)$, and each new point always strictly decreases a descent function Z unless the solution set Γ is reached. For example, in nonlinear programming, the solution set may be the set of minimum points (perhaps only one point), and the descent function may be the objective function itself. A suitable algorithm is found that generates points such that each new point strictly reduces the value of the objective. Then, under appropriate conditions, it follows that the sequence converges to the solution set. The Global Convergence Theorem establishes technical conditions for which convergence is guaranteed.

***Global Convergence Theorem**. Let $\mathbf{A}$ be an algorithm on X, and suppose that, given $\mathbf{x}_0$ the sequence $\{\mathbf{x}_k\}_{k=0}^{\infty}$ is generated satisfying*

$$\mathbf{x}_{k+1} \in \mathbf{A}(\mathbf{x}_k).$$

Let a solution set $\Gamma \subset X$ be given, and suppose

i) all points $\mathbf{x}_k$ are contained in a compact set $S \subset X$
ii) there is a continuous function Z on X such that
  (a) if $\mathbf{x} \notin \Gamma$, then $Z(\mathbf{y}) < Z(\mathbf{x})$ for all $\mathbf{y} \in \mathbf{A}(\mathbf{x})$
  (b) if $\mathbf{x} \in \Gamma$, then $Z(\mathbf{y}) \leqslant Z(\mathbf{x})$ for all $\mathbf{y} \in \mathbf{A}(\mathbf{x})$
iii) the mapping $\mathbf{A}$ is closed at points outside Γ.

Then the limit of any convergent subsequence of $\{\mathbf{x}_k\}$ is a solution.

Proof. Suppose the convergent subsequence $\{\mathbf{x}_k\}$, $k \in \mathcal{K}$ converges to the limit $\mathbf{x}$. Since Z is continuous, it follows that for $k \in \mathcal{K}$, $Z(\mathbf{x}_k) \to Z(\mathbf{x})$. This means that Z is convergent with respect to the subsequence, and we shall show that it is convergent with respect to the entire sequence. By the monotonicity of Z on the sequence $\{\mathbf{x}_k\}$ we have $Z(\mathbf{x}_k) - Z(\mathbf{x}) \geqslant 0$ for all k. By the convergence of Z on the subsequence, there is, for a given $\varepsilon > 0$, a $K \in \mathcal{K}$ such that $Z(\mathbf{x}_k) - Z(\mathbf{x}) < \varepsilon$ for all $k > K$, $k \in \mathcal{K}$.

Thus for all $k > K$

$$Z(\mathbf{x}_k) - Z(\mathbf{x}) = Z(\mathbf{x}_k) - Z(\mathbf{x}_K) + Z(\mathbf{x}_K) - Z(\mathbf{x}) < \varepsilon,$$

which shows that $Z(\mathbf{x}_k) \to Z(\mathbf{x})$.

To complete the proof it is only necessary to show that $\mathbf{x}$ is a solution. Suppose $\mathbf{x}$ is not a solution. Consider the subsequence $\{\mathbf{x}_{k+1}\}_{\mathcal{K}}$. Since all members of this sequence are contained in a compact set, there is a $\bar{\mathcal{K}} \subset \mathcal{K}$ such that $\{\mathbf{x}_{k+1}\}_{\bar{\mathcal{K}}}$ converges to some limit $\bar{\mathbf{x}}$. We thus have $\mathbf{x}_k \to \mathbf{x}$, $k \in \bar{\mathcal{K}}$, and $\mathbf{x}_{k+1} \in \mathbf{A}(\mathbf{x}_k)$ with $\mathbf{x}_{k+1} \to \bar{\mathbf{x}}$, $k \in \bar{\mathcal{K}}$. Thus since $\mathbf{A}$ is closed at $\mathbf{x}$ it follows that $\bar{\mathbf{x}} \in \mathbf{A}(\mathbf{x})$. But from above, $Z(\bar{\mathbf{x}}) = Z(\mathbf{x})$ which contradicts the fact that Z is a descent function. ∎

Corollary. *If under the conditions of the Global Convergence Theorem Γ consists of a single point $\bar{\mathbf{x}}$, then the sequence $\{\mathbf{x}_k\}$ converges to $\bar{\mathbf{x}}$.*

Proof. Suppose to the contrary that there is a subsequence $\{\mathbf{x}_k\}_{\mathcal{K}}$ and an $\varepsilon > 0$ such that $|\mathbf{x}_k - \bar{\mathbf{x}}| > \varepsilon$ for all $k \in \mathcal{K}$. By compactness there must be $\mathcal{K}' \subset \mathcal{K}$ such that $\{\mathbf{x}_k\}_{\mathcal{K}'}$, converges, say to $\mathbf{x}'$. Clearly, $|\mathbf{x}' - \bar{\mathbf{x}}| \geqslant \varepsilon$, but by the Global Convergence Theorem $\mathbf{x}' \in \Gamma$, which is a contradiction. ∎

In later chapters the Global Convergence Theorem is used to establish the convergence of several standard algorithms. Here we consider some simple examples designed to illustrate the roles of the various conditions of the theorem.

Example 4. In many respects condition (iii) of the theorem, the closedness of $\mathbf{A}$ outside the solution set, is the most important condition. The failure of many popular algorithms can be traced to nonsatisfaction of this condition. On the real line consider the point-to-point algorithm

$$A(x) = \begin{cases} \frac{1}{2}(x-1)+1 & x > 1 \\ \frac{1}{2}x & x \leqslant 1 \end{cases}$$

and the solution set $\Gamma = \{0\}$. It is easily verified that a descent function for this solution set and this algorithm is $Z(x) = |x|$. However, starting from $x > 1$, the algorithm generates a sequence converging to $x = 1$ which is not a solution. The difficulty is that A is not closed at $x = 1$.

Example 5. On the real line X consider the solution set to be empty, the descent function $Z(x) = e^{-x}$, and the algorithm $A(x) = x + 1$. All conditions of the convergence theorem except (i) hold. The sequence generated from any starting condition diverges to infinity. This is not strictly a violation of the conclusion of the theorem but simply an example illustrating that if no compactness assumption is introduced, the generated sequence may have no convergent subsequence.

Example 6. Consider the point-to-set algorithm A defined by the graph in Fig. 7.8 and given explicitly on $X = [0,\ 1]$ by

$$A(x) = \begin{cases} [0,\ x) & 1 \geqslant x > 0 \\ 0 & x = 0, \end{cases}$$

where $[0,\ x)$ denotes a half-open interval (see Appendix A). Letting $\Gamma = \{0\}$, the function $Z(x) = x$ serves as a descent function, because for $x \neq 0$ all points in $A(x)$ are less than x.

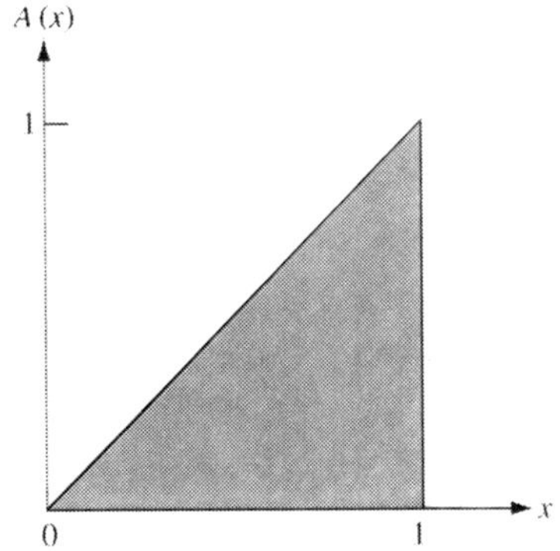

Fig. 7.8 Graph for Example 6

The sequence defined by

$$x_0 = 1$$
$$x_{k+1} = x_k - \frac{1}{2^{k+2}}$$

satisfies $x_{k+1} \in A(x_k)$ but it can easily be seen that $x_k \to \frac{1}{2} \notin \Gamma$. The difficulty here, of course, is that the algorithm A is not closed outside the solution set.

*Spacer Steps

In some of the more complex algorithms presented in later chapters, the rule used to determine a succeeding point in an iteration may depend on several previous points rather than just the current point, or it may depend on the iteration index k. Such features are generally introduced in order to obtain a rapid rate of convergence but they can grossly complicate the analysis of global convergence.

If in such a complex sequence of steps there is inserted, perhaps irregularly but infinitely often, a step of an algorithm such as steepest descent that is known to converge, then it is not difficult to insure that the entire complex process converges. The step which is repeated infinitely often and guarantees convergence is called a *spacer step*, since it separates disjoint portions of the complex sequence. Essentially the only requirement imposed on the other steps of the process is that they do not increase the value of the descent function.

This type of situation can be analyzed easily from the following viewpoint. Suppose **B** is an algorithm which together with the descent function Z and solution set Γ, satisfies all the requirements of the Global Convergence Theorem. Define the algorithm **C** by $\mathbf{C}(\mathbf{x}) = \{\mathbf{y} : Z(\mathbf{y}) \leqslant Z(\mathbf{x})\}$. In other words, **C** applied to **x** can give any point so long as it does not increase the value of Z. It is easy to verify that **C** is closed. We imagine that **B** represents the spacer step and the complex process between spacer steps is just some realization of C. Thus the overall process amounts merely to repeated applications of the composite algorithm **CB**. With this viewpoint we may state the Spacer Step Theorem.

Spacer Step Theorem. *Suppose* **B** *is an algorithm on X which is closed outside the solution set* Γ. *Let Z be a descent function corresponding to* **B** *and* Γ.
Suppose that the sequence $\{\mathbf{x}_k\}_{k=0}^{\infty}$ *is generated satisfying*

$$\mathbf{x}_{k+1} \in \mathbf{B}(\mathbf{x}_k)$$

for k in an infinite index set $\mathcal{K}$, *and that*

$$Z(\mathbf{x}_{k+1}) \leqslant Z(\mathbf{x}_k)$$

for all k. Suppose also that the set $S = \{\mathbf{x} : Z(\mathbf{x}) \leqslant Z(\mathbf{x}_0)\}$ *is compact. Then the limit of any convergent subsequence of* $\{\mathbf{x}_k\}_{\mathcal{K}}$ *is a solution.*

?

Proof. We first define for any $\mathbf{x} \in X$, $\bar{\mathbf{B}}(\mathbf{x}) = S \cap \mathbf{B}(\mathbf{x})$ and then observe that $\mathbf{A} = \mathbf{C}\bar{\mathbf{B}}$ is closed outside the solution set by Corollary 1. The Global Convergence Theorem can then be applied to A. Since S is compact, there is a subsequence of $\{\mathbf{x}_k\}_{k\in\mathcal{K}}$ converging to a limit **x**. In view of the above we conclude that $\mathbf{x} \in \Gamma$. ∎

7.8 Speed of Convergence

The study of speed of convergence is an important but sometimes complex subject. Nevertheless, there is a rich and yet elementary theory of convergence rates that enables one to predict with confidence the relative effectiveness of a wide class of algorithms. In this section we introduce various concepts designed to measure speed of convergence, and prepare for a study of this most important aspect of nonlinear programming.

Order of Convergence

Consider a sequence of real numbers $\{r_k\}_{k=0}^{\infty}$ converging to the limit r^*. We define several notions related to the speed of convergence of such a sequence.

Definition. Let the sequence $\{r_k\}$ converge to r^*. The *order* of convergence of $\{r_k\}$ is defined as the supremum of the nonnegative numbers p satisfying

$$0 \leqslant \overline{\lim_{k\to\infty}} \frac{|r_{k+1} - r^*|}{|r_k - r^*|^p} < \infty.$$

To ensure that the definition is applicable to any sequence, it is stated in terms of limit superior rather than just limit and 0/0 (which occurs if $r_k = r^*$ for all k) is regarded as finite. But these technicalities are rarely necessary in actual analysis, since the sequences generated by algorithms are generally quite well behaved.

It should be noted that the order of convergence, as with all other notions related to speed of convergence that are introduced, is determined only by the properties of the sequence that hold as $k \to \infty$. Somewhat loosely but picturesquely, we are therefore led to refer to the *tail* of a sequence—that part of the sequence that is arbitrarily far out. In this language we might say that the order of convergence is a measure of how good the worst part of the tail is. Larger values of the order p imply, in a sense, faster convergence, since the distance from the limit r^* is reduced, at least in the tail, by the pth power in a single step. Indeed, if the sequence has order p and (as is the usual case) the limit

$$\beta = \lim_{k\to\infty} \frac{|r_{k+1} - r^*|}{|r_k - r^*|^p}$$

exists, then asymptotically we have

$$|r_{k+1} - r^*| = \beta |r_k - r^*|^p.$$

Example 1. The sequence with $r_k = a^k$ where $0 < a < 1$ converges to zero with order unity, since $r_{k+1}/r_k = a$.

Example 2. The sequence with $r_k = a^{(2^k)}$ for $0 < a < 1$ converges to zero with order two, since $r_{k+1}/r_k^2 = 1$.

Linear Convergence

Most algorithms discussed in this book have an order of convergence equal to unity. It is therefore appropriate to consider this class in greater detail and distinguish certain cases within it.

Definition. If the sequence $\{r_k\}$ converges to r^* in such a way that

$$\lim_{k\to\infty} \frac{|r_{k+1} - r^*|}{|r_k - r^*|} = \beta < 1,$$

the sequence is said to converge *linearly* to r^* with *convergence ratio* (*or rate*) β.

Linear convergence is, for our purposes, without doubt the most important type of convergence behavior. A linearly convergent sequence, with convergence ratio β, can be said to have a tail that converges at least as fast as the geometric sequence $c\beta^k$ for some constant c. Thus linear convergence is sometimes referred to as *geometric convergence*, although in this book we reserve that phrase for the case when a sequence is exactly geometric.

As a rule, when comparing the relative effectiveness of two competing algorithms both of which produce linearly convergent sequences, the comparison is based on their corresponding convergence ratios—the smaller the ratio the faster the rate. The ultimate case where $\beta = 0$ is referred to as *superlinear convergence*. We note immediately that convergence of any order greater than unity is superlinear, but it is also possible for superlinear convergence to correspond to unity order.

Example 3. The sequence $r_k = (1/k)^k$ is of order unity, since $r_{k+1}/r_k^p \to \infty$ for $p > 1$. However, $r_{k+1}/r_k \to 0$ as $k \to \infty$ and hence this is superlinear convergence.

Arithmetic Convergence

Linear convergence is also called geometric convergence. There is another (slower) type of convergence:

Definition. If the sequence $\{r_k\}$ converges to r^* in such a way that

$$|r_k - r^*| \leq C\frac{|r_0 - r^*|}{k^p}, \quad k \geq 1, \ 0 < p < \infty$$

where C is a fixed positive number, the sequence is said to converge *arithmetically* to r^* with order p.

When $p = 1$, it is referred as arithmetic convergence. The greater of p the faster of the convergence.

Example 4. The sequence $r_k = 1/k$ converges to zero arithmetically. The convergence is of order one but it is not linear, since $\lim_{k\to\infty}(r_{k+1}/r_k) = 1$, that is, β is not strictly less than one.

**Average Rates*

All the definitions given above can be referred to as *step-wise* concepts of convergence, since they define bounds on the progress made by going a single step: from k to $k+1$. Another approach is to define concepts related to the average progress per step over a large number of steps. We briefly illustrate how this can be done.

Definition. Let the sequence $\{r_k\}$ converge to r^*. The *average order* of convergence is the infimum of the numbers $p > 1$ such that

$$\overline{\lim_{k\to\infty}} |r_k - r^*|^{1/p^k} = 1.$$

The order is infinity if the equality holds for no $p > 1$.

Example 5. For the sequence $r_k = a^{(2^k)}$, $0 < a < 1$, given in Example 2, we have

$$|r_k|^{1/2^k} = a,$$

while

$$|r_k|^{1/p^k} = a^{(2/p)^k} \to 1$$

for $p > 2$. Thus the average order is two.

Example 6. For $r_k = a^k$ with $0 < a < 1$ we have

$$(r_k)^{1/p^k} = a^{k(1/p)^k} \to 1$$

for any $p > 1$. Thus the average order is unity.

As before, the most important case is that of unity order, and in this case we define the *average convergence ratio* as $\overline{\lim_{k\to\infty}} |r_k - r^*|^{1/k}$. Thus for the geometric sequence $r_k = ca^k$, $0 < a < 1$, the average convergence ratio is a. Paralleling the earlier definitions, the reader can then in a similar manner define corresponding notions of average linear and average superlinear convergence.

Although the above array of definitions can be further embellished and expanded, it is quite adequate for our purposes. For the most part we work with the step-wise definitions, since in analyzing iterative algorithms it is natural to compare one step with the next. In most situations, moreover, when the sequences are well behaved and the limits exist in the definitions, then the step-wise and average concepts of convergence rates coincide.

*Convergence of Vectors

Suppose $\{\mathbf{x}_k\}_{k=0}^{\infty}$ is a sequence of vectors in E^n converging to a vector $\mathbf{x}^*$. The convergence properties of such a sequence are defined with respect to some particular function that converts the sequence of vectors into a sequence of numbers. Thus, if f is a given continuous function on E^n, the convergence properties of $\{\mathbf{x}_k\}$ can be defined with respect to f by analyzing the convergence of $f(\mathbf{x}_k)$ to $f(\mathbf{x}^*)$. The function f used in this way to measure convergence is called the *error function*.

In optimization theory it is common to choose the error function by which to measure convergence as the same function that defines the objective function of the original optimization problem. This means we measure convergence by how fast the objective converges to its minimum. alternatively, we sometimes use the function $|\mathbf{x} - \mathbf{x}^*|^2$ and thereby measure convergence by how fast the (squared) distance from the solution point decreases to zero.

Generally, the order of convergence of a sequence is insensitive to the particular error function used; but for step-wise linear convergence the associated convergence ratio is not. Nevertheless, the average convergence ratio is not too sensitive, as the following proposition demonstrates, and hence the particular error function used to measure convergence is not really very important.

Proposition. *Let f and g be two error functions satisfying $f(\mathbf{x}^*) = g(\mathbf{x}^*) = 0$ and, for all $\mathbf{x}$, a relation of the form*

$$0 \leqslant a_1 g(\mathbf{x}) \leqslant f(\mathbf{x}) \leqslant a_2 g(\mathbf{x})$$

for some fixed $a_1 > 0$, $a_2 > 0$. If the sequence $\{\mathbf{x}_k\}_{k=0}^{\infty}$ converges to $\mathbf{x}^*$ linearly with average ratio β with respect to one of these functions, it also does so with respect to the other.

Proof. The statement is easily seen to be symmetric in f and g. Thus we assume $\{\mathbf{x}_k\}$ is linearly convergent with average convergence ratio β with respect to f, and will prove that the same is true with respect to g. We have

$$\beta = \overline{\lim_{k\to\infty}}\, f(\mathbf{x}_k)^{1/k} \leqslant \overline{\lim_{k\to\infty}}\, a_2^{1/k} g(\mathbf{x}_k)^{1/k} = \overline{\lim_{k\to\infty}}\, g(\mathbf{x}_k)^{1/k}$$

and

$$\beta = \overline{\lim_{k\to\infty}}\, f(\mathbf{x}_k)^{1/k} \geqslant \overline{\lim_{k\to\infty}}\, a_1^{1/k} g(\mathbf{x}_k)^{1/k} = \overline{\lim_{k\to\infty}}\, g(\mathbf{x}_k)^{1/k}.$$

Thus

$$\beta = \overline{\lim_{k\to\infty}}\, g(\mathbf{x}_k)^{1/k}. \blacksquare$$

As an example of an application of the above proposition, consider the case where $g(\mathbf{x}) = |\mathbf{x} - \mathbf{x}^*|^2$ and $f(\mathbf{x}) = (\mathbf{x} - \mathbf{x}^*)^T \mathbf{Q}(\mathbf{x} - \mathbf{x}^*)$, where $\mathbf{Q}$ is a positive definite symmetric matrix. Then a_1 and a_2 correspond, respectively, to the smallest and largest eigenvalues of Q. Thus average linear convergence is identical with respect to any error function constructed from a positive definite quadratic form.

Complexity

Complexity theory as outlined in Sect. 5.1 is an important aspect of convergence theory. This theory can be used in conjunction with the theory of local convergence. If an algorithm converges according to any order greater than zero, then for a fixed problem, the sequence generated by the algorithm will converge in a time that is a function of the convergence order (and rate, if convergence is linear). For example, if the order is one with rate $0 < c < 1$ and the process begins with an error of R, a final error of r can be achieved by a number of steps n satisfying $c^n R \leqslant r$. Thus it requires approximately $n = \log(R/r)/\log(1/c)$ steps. In this form the number of steps is not affected by the size of the problem. However, problem size enters in two possible ways. First, the rate c may depend on the size-say going toward 1 as the size increases so that the speed is slower for large problems. The second way that size may enter, and this is the more important way, is that the time to execute a single step almost always increases with problem size. For instance if, for a problem seeking an optimal vector of dimension m, each step requires a Gaussian elimination inversion of an $m \times m$ matrix, the solution time will increase by a factor proportional to m^3. Overall the algorithm is therefore a polynomial time algorithm. Essentially all algorithms in this book employ steps, such as matrix multiplications or inversion or other algebraic operations, which are polynomial-time in character. Convergence analysis, therefore, focuses on whether an algorithm is globally convergent, on its local convergence properties, and also on the order of the algebraic operations required to execute the steps required. The last of these is usually easily deduced by listing the number and size of the required vector and matrix operations.

7.9 Summary

There are two different but complementary ways to characterize the solution to unconstrained optimization problems. In the local approach, one examines the relation of a given point to its neighbors. This leads to the conclusion that, at an unconstrained relative minimum point of a smooth function, the gradient of the function vanishes and the Hessian is positive semidefinite; and conversely, if at a point the gradient vanishes and the Hessian is positive definite, that point is a relative minimum point. This characterization has a natural extension to the global approach where convexity ensures that if the gradient vanishes at a point, that point is a global minimum point.

In considering iterative algorithms for finding either local or global minimum points, there are two distinct issues: global convergence properties and local convergence properties. The first is concerned with whether starting at an arbitrary point the sequence generated will converge to a solution. This is ensured if the algorithm is closed, has a descent function, and generates a bounded sequence. It is also explained that global convergence is guaranteed simply by the inclusion, in a complex algorithm, of spacer steps. This result is called upon frequently in what follows. Local convergence properties are a measure of the ultimate speed of convergence and generally determine the relative advantage of one algorithm to another.

7.10 Exercises

1. To approximate a function g over the interval $[0, 1]$ by a polynomial p of degree n (or less), we minimize the criterion

$$f(\mathbf{a}) = \int_0^1 [g(x) - p(x)]^2 dx,$$

where $p(x) = a_n x^n + a_{n-1} x^{n-1} + \ldots + a_0$. Find the equations satisfied by the optimal coefficients $\mathbf{a} = (a_0, a_1, \ldots, a_n)$.

2. In Example 4 of Sect. 7.2 show that if the solution has $x_1 > 0,\ x_1 + x_2 = 1$, then it is necessary that

$$b_1 - b_2 + (c_1 - c_2)h(x_1) = 0$$
$$b_2 + (c_2 - c_3)h(x_1 + x_2) \leqslant 0.$$

Hint: One way is to reformulate the problem in terms of the variables x_1 and $y = x_1 + x_2$.

3. (a) Using the first-order necessary conditions, find a minimum point of the function

$$f(x, y, z) = 2x^2 + xy + y^2 + yz + z^2 - 6x - 7y - 8z + 9.$$

(b) Verify that the point is a relative minimum point by verifying that the second-order sufficiency conditions hold.
(c) Prove that the point is a global minimum point.

4. In this exercise and the next we develop a method for determining whether a given symmetric matrix is positive definite. Given an $n \times n$ matrix $\mathbf{A}$ let $\mathbf{A}_k$ denote the principal submatrix made up of the first k rows and columns. Show (by induction) that if the first $n - 1$ principal submatrices are nonsingular, then there is a unique lower triangular matrix $\mathbf{L}$ with unit diagonal and a unique upper triangular matrix $\mathbf{U}$ such that $\mathbf{A} = \mathbf{LU}$. (See Appendix C.)
5. A symmetric matrix is positive definite if and only if the determinant of each of its principal submatrices is positive. Using this fact and the considerations of Exercise 4, show that an $n \times n$ symmetric matrix $\mathbf{A}$ is positive definite if and only if it has an $\mathbf{LU}$ decomposition (without interchange of rows) and the diagonal elements of $\mathbf{U}$ are all positive.
6. Using Exercise 5 show that an $n \times n$ matrix $\mathbf{A}$ is symmetric and positive definite if and only if it can be written as $\mathbf{A} = \mathbf{GG}^T$ where $\mathbf{G}$ is a lower triangular matrix with positive diagonal elements. This representation is known as the *Cholesky factorization* of $\mathbf{A}$.
7. Let $f_i,\ i \in I$ be a collection of convex functions defined on a convex set Ω. Show that the function f defined by $f(\mathbf{x}) = \sup\limits_{i \in I} f_i(\mathbf{x})$ is convex on the region where it is finite.
8. Let γ be a monotone nondecreasing function of a single variable (that is, $\gamma(r) \leqslant \gamma(r')$ for $r' > r$) which is also convex; and let f be a convex function defined on a convex set Ω. Show that the function $\gamma(f)$ defined by $\gamma(f)(\mathbf{x}) = \gamma[f(\mathbf{x})]$ is convex on Ω.
9. Let f be twice continuously differentiable on a region $\Omega \subset E^n$. Show that a sufficient condition for a point $\mathbf{x}^*$ in the interior of Ω to be a relative minimum point of f is that $\nabla f(\mathbf{x}^*) = \mathbf{0}$ and that f be locally convex at $\mathbf{x}^*$.

10. Define the point-to-set mapping on E^n by

$$\mathbf{A}(\mathbf{x}) = \{\mathbf{y} : \mathbf{y}^T\mathbf{x} \leqslant b\},$$

where b is a fixed constant. Is $\mathbf{A}$ closed?

11. Prove the two corollaries in Sect. 7.6 on the closedness of composite mappings.
12. Show that if $\mathbf{A}$ is a continuous point-to-point mapping, the Global Convergence Theorem is valid even without assumption (i). Compare with Example 2, Sect. 7.7.
13. Let $\{r_k\}_{k=0}^{\infty}$ and $\{c_k\}_{k=0}^{\infty}$ be sequences of real numbers. Suppose $r_k \to 0$ average linearly and that there are constants $c > 0$ and C such that $c \leqslant c_k \leqslant C$ for all k. Show that $c_k r_k \to 0$ average linearly.
14. Prove a proposition, similar to the one in Sect. 7.8, showing that the order of convergence is insensitive to the error function.
15. Show that if $r_k \to r^*$ (step-wise) linearly with convergence ratio β, then $r_k \to r^*$(average) linearly with average convergence ratio no greater than β.

References

7.1–7.5 For alternative discussions of the material in these sections, see Hadley [H2], Fiacco and McCormick [F4], Zangwill [Z2] and Luenberger [L8].

7.6 Although the general concepts of this section are well known, the formulation as zero-order conditions appears to be new.

7.7 The idea of using a descent function (usually the objective itself) in order to guarantee convergence of minimization algorithms is an old one that runs through most literature on optimization, and has long been used to establish global convergence. Formulation of the general Global Convergence Theorem, which captures the essence of many previously diverse arguments, and the idea of representing an algorithm as a point-to-set mapping are both due to Zangwill [Z2]. A version of the Spacer Step Theorem can be found in Zangwill [Z2] as well.

7.8 Most of the definitions given in this section have been standard for quite some time. A thorough discussion which contributes substantially to the unification of these concepts is contained in Ortega and Rheinboldt [O7].

Chapter 8
Basic Descent Methods

We turn now to a description of the basic techniques used for iteratively solving unconstrained minimization problems. These techniques are, of course, important for practical application since they often offer the simplest, most direct alternatives for obtaining solutions; but perhaps their greatest importance is that they establish certain reference plateaus with respect to difficulty of implementation and speed of convergence. Thus in later chapters as more efficient techniques and techniques capable of handling constraints are developed, reference is continually made to the basic techniques of this chapter both for guidance and as points of comparison.

There is a fundamental underlying structure for almost all the descent algorithms we discuss. One starts at an initial point; determines, according to a fixed rule, a direction of movement; and then moves in that direction to a (relative) minimum of the objective function on that line. At the new point a new direction is determined and the process is repeated. The primary differences between algorithms (steepest descent, Newton's method, etc.) rest with the rule by which successive directions of movement are selected. Once the selection is made, all algorithms call for movement to the minimum point on the corresponding line.

The process of determining the minimum point on a given line (one variable only) is called *line search*. For general nonlinear functions that cannot be minimized analytically, this process actually is accomplished by searching, in an intelligent manner, along the line for the minimum point. These line search techniques, which are really procedures for solving one-dimensional minimization problems, form the backbone of nonlinear programming algorithms, since higher dimensional problems are ultimately solved by executing a sequence of successive line searches. There are a number of different approaches to this important phase of minimization and the first half of this chapter is devoted to their, discussion.

The last sections of the chapter are devoted to a description and analysis of the basic descent algorithms for unconstrained problems; steepest descent, coordinate descent, and Newton's method. These algorithms serve as primary models for the development and analysis of all others discussed in the book.

D.G. Luenberger, Y. Ye, *Linear and Nonlinear Programming*, International Series in Operations Research & Management Science 228,
DOI 10.1007/978-3-319-18842-3_8

8.1 Line Search Algorithms

These algorithms are classified by the order of information of the objective functions $f(x)$ being evaluated.

0th-Order Method: Golden Section Search and Curve Fitting

A very popular method for resolving the line search problem is the Fibonacci search method described in this section. The method has a certain degree of theoretical elegance, which no doubt partially accounts for its popularity, but on the whole, as we shall see, there are other procedures which in most circumstances are superior.

The method determines the minimum value of a function f over a closed interval $[c_1, c_2]$. In applications, f may in fact be defined over a broader domain, but for this method a fixed interval of search must be specified. The only property that is assumed of f is that it is *unimodal*, that is, it has a single relative minimum (see Fig. 8.1). The minimum point of f is to be determined, at least approximately, by measuring the value of f at a certain number of points. It should be imagined, as is indeed the case in the setting of nonlinear programming, that each measurement of f is somewhat costly—of time if nothing more.

To develop an appropriate search strategy, that is, a strategy for selecting measurement points based on the previously obtained values, we pose the following problem: Find how to successively select N measurement points so that, without explicit knowledge of f, we can determine the smallest possible region of uncertainty in which the minimum must lie. In this problem the region of uncertainty is determined in any particular case by the relative values of the measured points in conjunction with our assumption that f is unimodal. Thus, after values are known at N points $x_1, x_2, \ldots, x_N$ with

$$c_1 \leqslant x_1 < x_2 \ldots < x_{N-1} < x_N \leqslant c_2,$$

the region of uncertainty is the interval $[x_{k-1}, x_{k+1}]$ where x_k is the minimum point among the N, and we define $x_0 = c_1$, $x_{N+1} = c_2$ for consistency. The minimum of f must lie somewhere in this interval.

The derivation of the optimal strategy for successively selecting measurement points to obtain the smallest region of uncertainty is fairly straight-forward but somewhat tedious. We simply state the result and give an example.

Let

$$d_1 = c_2 - c_1, \text{ the initial width of uncertainty}$$
$$d_k = \text{ width of uncertainty after } k \text{ measurements}$$

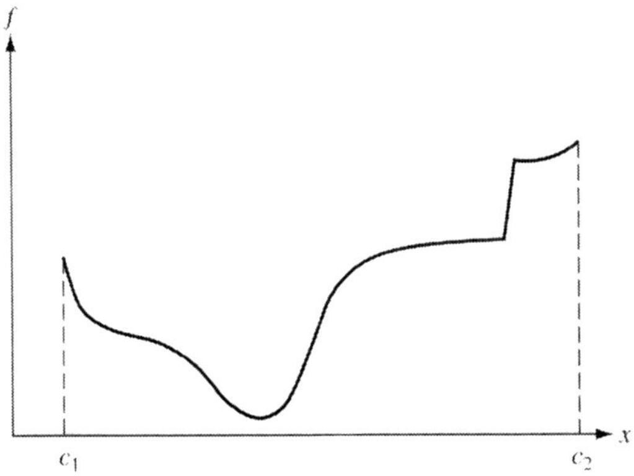

Fig. 8.1 A unimodal function

Then, if a total of N measurements are to be made, we have

$$d_k = \left(\frac{F_{N-k+1}}{F_N}\right) d_1, \tag{8.1}$$

where the integers F_k are members of the Fibonacci sequence generated by the recurrence relation

$$F_N = F_{N-1} + F_{N-2}, \qquad F_0 = F_1 = 1. \tag{8.2}$$

The resulting sequence is 1, 1, 2, 3, 5, 8, 13, ...

The procedure for reducing the width of uncertainty to d_N is this: The first two measurements are made symmetrically at a distance of $(F_{N-1}/F_N)d_1$ from the ends of the initial intervals; according to which of these is of lesser value, an uncertainty interval of width $d_2 = (F_{N-1}/F_N)d_1$ is determined. The third measurement point is placed symmetrically in this new interval of uncertainty with respect to the measurement already in the interval. The result of this third measurement gives an interval of uncertainty $d_3 = (F_{N-2}/F_N)d_1$. In general, each successive measurement point is placed in the current interval of uncertainty symmetrically with the point already existing in that interval.

Some examples are shown in Fig. 8.2. In these examples the sequence of measurement points is determined in accordance with the assumption that each measurement is of lower value than its predecessors. Note that the procedure always calls for the last two measurements to be made at the midpoint of the semifinal interval of uncertainty. We are to imagine that these two points are actually separated a small distance so that a comparison of their respective values will reduce the interval to nearly half. This terminal anomaly of the Fibonacci search process is, of course, of no great practical consequence.

Search by Golden Section

If the number N of allowed measurement points in a Fibonacci search is made to approach infinity, we obtain the golden section method. It can be argued, based on the optimal property of the finite Fibonacci method, that the corresponding infinite version yields a sequence of intervals of uncertainty whose widths tend to zero faster than that which would be obtained by other methods.

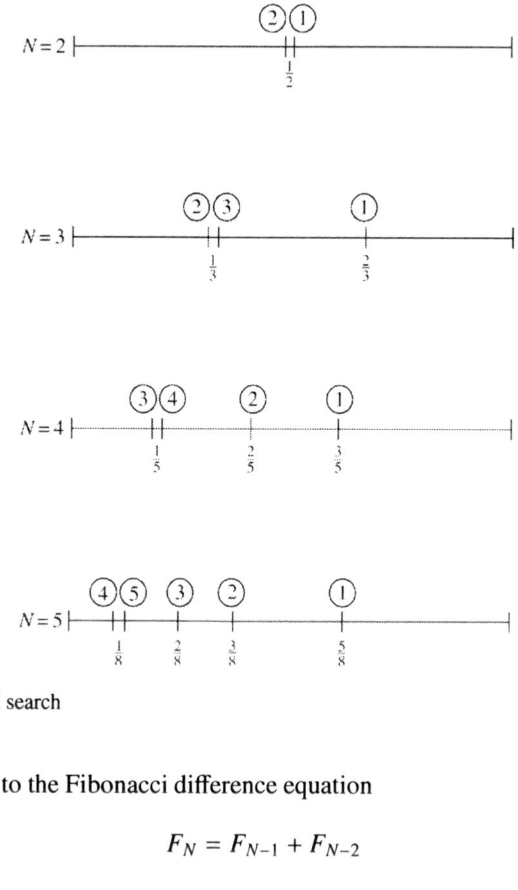

Fig. 8.2 Fibonacci search

The solution to the Fibonacci difference equation

$$F_N = F_{N-1} + F_{N-2} \tag{8.3}$$

is of the form

$$F_N = A\tau_1^N + B\tau_2^N, \tag{8.4}$$

where τ_1 and τ_2 are roots of the characteristic equation

$$\tau^2 = \tau + 1.$$

Explicitly,

$$\tau_1 = \frac{1+\sqrt{5}}{2},\ \tau_2 = \frac{1-\sqrt{5}}{2}.$$

(The number $\tau_1 \simeq 1.618$ is known as the *golden section* ratio and was considered by early Greeks to be the most aesthetic value for the ratio of two adjacent sides of a rectangle.) For large N the first term on the right side of (8.4) dominates the second, and hence

$$\lim_{N\to\infty} \frac{F_{N-1}}{F_N} = \frac{1}{\tau_1} \simeq 0.618.$$

It follows from (8.1) that the interval of uncertainty at any point in the process has width

$$d_k = \left(\frac{1}{\tau_1}\right)^{k-1} d_1, \tag{8.5}$$

and from this it follows that

$$\frac{d_{k+1}}{d_k} = \frac{1}{\tau_1} = 0.618. \tag{8.6}$$

Therefore, we conclude that, with respect to the width of the uncertainty interval, the search by golden section converges linearly (see Sect. 7.8) to the overall minimum of the function f with convergence ratio $1/\tau_1 = 0.618$.

The Fibonacci search method has a certain amount of theoretical appeal, since it assumes only that the function being searched is unimodal and with respect to this broad class of functions the method is, in some sense, optimal. In most problems, however, it can be safely assumed that the function being searched, as well as being unimodal, possesses a certain degree of smoothness, and one might, therefore, expect that more efficient search techniques exploiting this smoothness can be devised; and indeed they can. Techniques of this nature are usually based on curve fitting procedures where a smooth curve is passed through the previously measured points in order to determine an estimate of the minimum point. A variety of such techniques can be devised depending on whether or not derivatives of the function as well as the values can be measured, how many previous points are used to determine the fit, and the criterion used to determine the fit. In this section a number of possibilities are outlined and analyzed. All of them have orders of convergence greater than unity.

Quadratic Fit

The scheme that is often most useful in line searching is that of fitting a quadratic through three given points. This has the advantage of not requiring any derivative information. Given x_1, x_2, x_3 and corresponding values $f(x_1) = f_1$, $f(x_2) = f_2$, $f(x_3) = f_3$ we construct the quadratic passing through these points

$$q(x) = \sum_{i=1}^{3} f_i \frac{\prod_{j\neq i}(x - x_j)}{\prod_{j\neq i}(x_i - x_j)}, \tag{8.7}$$

and determine a new point x_4 as the point where the derivative of q vanishes. Thus

$$x_4 = \frac{1}{2} \frac{b_{23}f_1 + b_{31}f_2 + b_{12}f_3}{a_{23}f_1 + a_{31}f_2 + a_{12}f_3}, \tag{8.8}$$

where $a_{ij} = x_i - x_j$, $b_{ij} = x_i^2 - x_j^2$.

Define the errors $\varepsilon_i = x^* - x_i$, $i = 1, 2, 3, 4$. The expression for ε_4 must be a polynomial in ε_1, ε_2, ε_3. It must be second order (since it is a quadratic fit). It must go to zero if any two of the errors ε_1, ε_2, ε_3 is zero. (The reader should check this.) Finally, it must be symmetric (since the order of points is relevant). It follows that near a minimum point x^* of f, the errors are related approximately by

$$\varepsilon_4 = M(\varepsilon_1\varepsilon_2 + \varepsilon_2\varepsilon_3 + \varepsilon_1\varepsilon_3), \tag{8.9}$$

where M depends on the values of the second and third derivatives of f at x^*.

If we assume that $\varepsilon_k \to 0$ with an order greater than unity, then for large k the error is governed approximately by

$$\varepsilon_{k+2} = M\varepsilon_k\varepsilon_{k-1}.$$

Letting $y_k = \log M\varepsilon_k$ this becomes

$$y_{k+2} = y_k + y_{k-1}$$

with characteristic equation

$$\lambda^3 - \lambda - 1 = 0.$$

The largest root of this equation is $\lambda \simeq 1.3$ which thus determines the rate of growth of y_k and is the order of convergence of the quadratic fit method.

1*st-Order Method: Curve Fitting and Methods of False Position*

In this section a number fitting methods using the first derivative information are described. All of them have orders of convergence greater than unity.

Quadratic Fit: Method of False Position

Suppose that at two points x_k and x_{k-1} where measurements $f(x_k)$, $f'(x_k)$, $f'(x_{k-1})$ are available, it is possible to fit the quadratic

$$q(x) = f(x_k) + f'(x_k)(x - x_k) + \frac{f'(x_{k-1}) - f'(x_k)}{x_{k-1} - x_k} \cdot \frac{(x - x_k)^2}{2},$$

which has the same corresponding values. An estimate x_{k+1} can then be determined by finding the point where the derivative of q vanishes; thus

$$x_{k+1} = x_k - f'(x_k)\left[\frac{x_{k-1} - x_k}{f'(x_{k-1}) - f'(x_k)}\right]. \tag{8.10}$$

(See Fig. 8.3.) Comparing this formula with Newton's method, we see again that the value $f(x_k)$ does not enter; hence, our fit could have been passed through either $f(x_k)$ or $f(x_{k-1})$. Also the formula can be regarded as an approximation to Newton's method where the second derivative is replaced by the difference of two first derivatives.

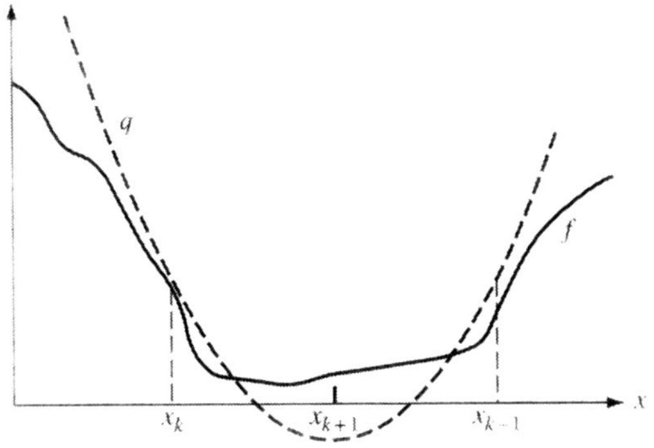

Fig. 8.3 False position for minimization

Again, since this method does not depend on values of f directly, it can be regarded as a method for solving $f'(x) \equiv g(x) = 0$. Viewed in this way the method, which is illustrated in Fig. 8.4, takes the form

$$x_{k+1} = x_k - g(x_k)\left[\frac{x_k - x_{k-1}}{g(x_k) - g(x_{k-1})}\right]. \tag{8.11}$$

We next investigate the order of convergence of the method of false position and discover that it is order $\tau_1 \simeq 1.618$, the golden mean.

Proposition. *Let g have a continuous second derivative and suppose x^* is such that $g(x^*) = 0$, $g'(x^*) \neq 0$. Then for x_0 sufficiently close to x^*, the sequence $\{x_k\}_{k=0}^{\infty}$ generated by the method of false position (8.11) converges to x^* with order $\tau_1 \simeq 1.618$.*

Proof. Introducing the notation

$$g[a, b] = \frac{g(b) - g(a)}{b - a}, \tag{8.12}$$

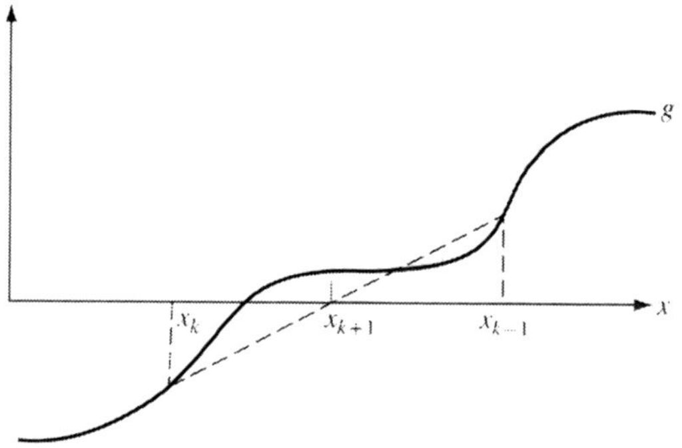

Fig. 8.4 False position for solving equations

we have

$$x_{k-1} - x^* = x_k - x^* - g(x_k)\left[\frac{x_k - x_{k-1}}{g(x_k) - g(x_{k-1})}\right]$$
$$= (x_k - x^*)\left\{\frac{g[x_{k-1}, x_k] - g[x_k, x^*]}{g[x_{k-1}, x_k]}\right\}. \tag{8.13}$$

Further, upon the introduction of the notation

$$g[a,\ b,\ c] = \frac{g[a, b] - g[b, c]}{a - c},$$

we may write (8.13) as

$$x_{k+1} - x^* = (x_k - x^*)(x_{k-1} - x^*)\left\{\frac{g[x_{k-1}, x_k, x^*]}{g[x_{k-1}, x_k]}\right\}.$$

Now, by the mean value theorem with remainder, we have (see Exercise 2)

$$g[x_{k-1},\ x_k] = g'(\xi_k) \tag{8.14}$$

and

$$g[x_{k-1},\ x_k,\ x^*] = \frac{1}{2}g''(\eta k), \tag{8.15}$$

where ξ_k and η_k are convex combinations of $x_k,\ x_{k-1}$ and $x_k,\ x_{k-1},\ x^*$, respectively. Thus

$$x_{k+1} - x^* = \frac{g''(\eta_k)}{2g'(\xi_k)}(x_k - x^*)(x_{k-1} - x^*). \tag{8.16}$$

It follows immediately that the process converges if it is started sufficiently close to x^*.

To determine the order of convergence, we note that for large k Eq. (8.16) becomes approximately

$$x_{k+1} - x^* = M(x_k - x^*)(x_{k-1} - x^*),$$

where

$$M = \frac{g''(x^*)}{2g'(x^*)}.$$

Thus defining $\varepsilon_k = (x_k - x^*)$ we have, in the limit,

$$\varepsilon_{k+1} = M\varepsilon_k\varepsilon_{k-1}. \tag{8.17}$$

Taking the logarithm of this equation we have, with $y_k = \log M\varepsilon_k$,

$$y_{k+1} = y_k + y_{k-1}, \tag{8.18}$$

which is the Fibonacci difference equation discussed in Sect. 7.1. A solution to this equation will satisfy

$$y_{k+1} - \tau_1 y_k \to 0.$$

Thus

$$\log M\varepsilon_{k+1} - \tau_1 \log M\varepsilon_k \to 0 \text{ or } \log \frac{M\varepsilon_{k+1}}{(M\varepsilon_k)^{\tau_1}} \to 0,$$

and hence

$$\frac{\varepsilon_{k+1}}{\varepsilon_k^{\tau_1}} \to M^{(\tau_1 - 1)}. \blacksquare$$

Having derived the error formula (8.17) by direct analysis, it is now appropriate to point out a short-cut technique, based on symmetry and other considerations, that can sometimes be used in even more complicated situations. The right side of error formula (8.17) must be a polynomial in ε_k and ε_{k-1}, since it is derived from approximations based on Taylor's theorem. Furthermore, it must be second order, since the method reduces to Newton's method when $x_k = x_{k-1}$. Also, it must go to zero if either ε_k or ε_{k-1} go to zero, since the method clearly yields $\varepsilon_{k+1} = 0$ in that case. Finally, it must be symmetric in ε_k and ε_{k-1}, since the order of points is irrelevant. The only formula satisfying these requirements is $\varepsilon_{k+1} = M\varepsilon_k\varepsilon_{k-1}$.

Cubic Fit

Given the points x_{k-1} and x_k together with the values $f(x_{k-1}), f'(x_{k-1}), f(x_k), f'(x_k)$, it is also possible to fit a cubic equation to the points having corresponding values. The next point x_{k+1} can then be determined as the relative minimum point of this cubic. This leads to

$$x_{k+1} = x_k - (x_k - x_{k-1})\left[\frac{f'(x_k) + u_2 - u_1}{f'(x_k) - f'(x_{k-1}) + 2u_2}\right], \tag{8.19}$$

where

$$u_1 = f'(x_{k-1}) + f'(x_k) - 3\frac{f(x_{k-1}) - f(x_k)}{x_{k-1} - x_k}$$

$$u_2 = [u_1^2 - f'(x_{k-1})f'(x_k)]^{1/2},$$

which is easily implementable for computations.

It can be shown (see Exercise 3) that the order of convergence of the cubic fit method is 2.0. Thus, although the method is exact for cubic functions indicating that its order might be three, its order is actually only two.

2nd-Order Method: Newton's Method

Suppose that the function f of a single variable x is to be minimized, and suppose that at a point x_k where a measurement is made it is possible to evaluate the three numbers $f(x_k)$, $f'(x_k)$, $f''(x_k)$. It is then possible to construct a quadratic function q which at x_k agrees with f up to second derivatives, that is

$$q(x) = f(x_k) + f'(x_k)(x - x_k) + \frac{1}{2}f''(x_k)(x - x_k)^2. \tag{8.20}$$

We may then calculate an estimate x_{k+1} of the minimum point of f by finding the point where the derivative of q vanishes. Thus setting

$$0 = q'(x_{k+1}) = f'(x_k) + f''(x_k)(x_{k+1} - x_k),$$

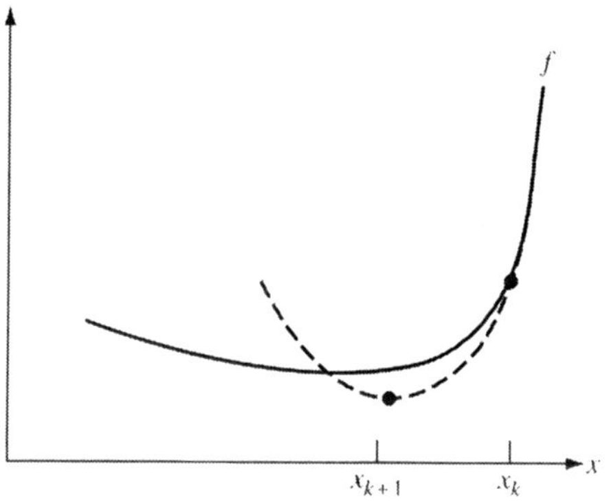

Fig. 8.5 Newton's method for minimization

we find

$$x_{k+1} = x_k - \frac{f'(x_k)}{f''(x_k)}. \tag{8.21}$$

This process, which is illustrated in Fig. 8.5, can then be repeated at x_{k+1}.

We note immediately that the new point x_{k+1} resulting from Newton's method does not depend on the value $f(x_k)$. The method can more simply be viewed as a technique for iteratively solving equations of the form

$$g(x) = 0,$$

where, when applied to minimization, we put $g(x) \equiv f'(x)$. In this notation Newton's method takes the form

$$x_{k+1} = x_k - \frac{g(x_k)}{g'(x_k)}. \tag{8.22}$$

This form is illustrated in Fig. 8.6.

We now show that Newton's method has order two convergence:

Proposition. *Let the function g have a continuous second derivative, and let x^* satisfy $g(x^*) = 0$, $g'(x^*) \neq 0$. Then, provided x_0 is sufficiently close to x^*, the sequence $\{x_k\}_{k=0}^{\infty}$ generated by Newton's method (8.22) converges to x^* with an order of convergence at least two.*

Proof. For points ξ in a region near x^* there is a k_1 such that $|g''(\xi)| < k_1$ and a k_2 such that $|g'(\xi)| > k_2$. Then since $g(x^*) = 0$ we can write

$$\begin{aligned} x_{k+1} - x^* &= x_k - x^* - \frac{g(x_k) - g(x^*)}{g'(x_k)} \\ &= -[g(x_k) - g(x^*) + g'(x_k)(x^* - x_k)]/g'(x_k). \end{aligned}$$

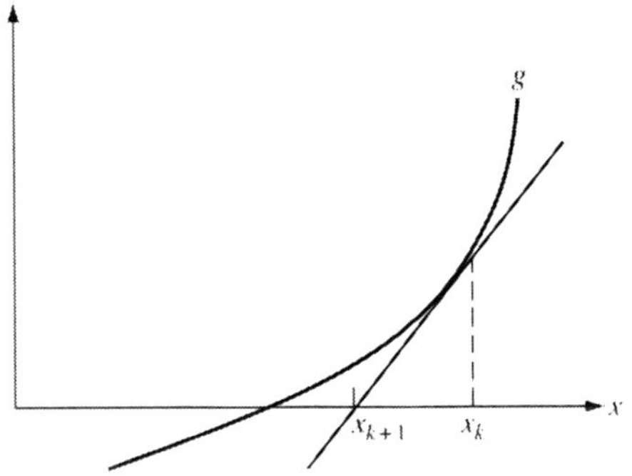

Fig. 8.6 Newton's method for solving equations

The term in brackets is, by Taylor's theorem, zero to first-order. In fact, using the remainder term in a Taylor series expansion about x_k, we obtain

$$x_{k+1} - x^* = \frac{1}{2}\frac{g''(\xi)}{g'(x_k)}(x_k - x^*)^2$$

for some ξ between x^* and x_k. Thus in the region near x^*,

$$|x_{k+1} - x^*| \leqslant \frac{k_1}{2k_2}|x_k - x^*|^2.$$

We see that if $|x_k - x^*|k_1/2k_2 < 1$, then $|x_{k+1} - x^*| < |x_k - x^*|$ and thus we conclude that if started close enough to the solution, the method will converge to x^* with an order of convergence at least two. ∎

Global Convergence of Curve Fitting

Above, we analyzed the convergence of various curve fitting procedures in the neighborhood of the solution point. If, however, any of these procedures were applied in pure form to search a line for a minimum, there is the danger—alas, the most likely possibility—that the process would diverge or wander about meaninglessly. In other words, the process may never get close enough to the solution for our detailed local convergence analysis to be applicable. It is therefore important to artfully combine our knowledge of the local behavior with conditions guaranteeing global convergence to yield a workable and effective procedure.

The key to guaranteeing global convergence is the Global Convergence Theorem of Chap. 7. Application of this theorem in turn hinges on the construction of a suitable descent function and minor modifications of a pure curve fitting algorithm. We offer below a particular blend of this kind of construction and analysis, taking as departure point the quadratic fit procedure discussed in Sect. 8.1 above.

Let us assume that the function f that we wish to minimize is strictly unimodal and has continuous second partial derivatives. We initiate our search procedure by searching along the line until we find three points x_1, x_2, x_3 with $x_1 < x_2 < x_3$ such that $f(x_1) \geqslant f(x_2) \leqslant f(x_3)$. In other words, the value at the middle of these three points is less than that at either end. Such a sequence of points can be determined in a number of ways—see Exercise 7.

The main reason for using points having this pattern is that a quadratic fit to these points will have a minimum (rather than a maximum) and the minimum point will lie in the interval $[x_1, x_3]$. See Fig. 8.7. We modify the pure quadratic fit algorithm so that it always works with points in this basic *three-point pattern*.

The point x_4 is calculated from the quadratic fit in the standard way and $f(x_4)$ is measured. Assuming (as in the figure) that $x_2 < x_4 < x_3$, and accounting for the unimodal nature of f, there are but two possibilities:

1. $f(x_4) \leqslant f(x_2)$
2. $f(x_2) < f(x_4) \leqslant f(x_3)$.

In either case a new three-point pattern, $\bar{x}_1, \bar{x}_2, \bar{x}_3$, involving x_4 and two of the old points, can be determined: In case (8.1) it is

$$(\bar{x}_1, \bar{x}_2, \bar{x}_3) = (x_2, x_4, x_3),$$

while in case (8.2) it is

$$(\bar{x}_1, \bar{x}_2, \bar{x}_3) = (x_1, x_2, x_4).$$

We then use this three-point pattern to fit another quadratic and continue. The pure quadratic fit procedure determines the next point from the current point and the previous two points. In the modification above, the next point is determined from the current point and the two out of three last points that form a three-point pattern with it. This simple modification leads to global convergence.

To prove convergence, we note that each three-point pattern can be thought of as defining a vector $\mathbf{x}$ in E^3. Corresponding to an $\mathbf{x} = (x_1, x_2, x_3)$ such that (x_1, x_2, x_3) form a three-point pattern with respect to f, we define $\mathbf{A}(\mathbf{x}) = (\bar{x}_1, \bar{x}_2, \bar{x}_3)$ as discussed above. For completeness we must consider the case where two or more of the x_i, $i = 1, 2, 3$ are equal, since this may occur. The appropriate definitions are simply limiting cases of the earlier ones. For example, if $x_1 = x_2$, then (x_1, x_2, x_3) form a three-point pattern if $f(x_2) \leqslant f(x_3)$ and $f'(x_2) < 0$ (which is the limiting case of $f(x_2) < f(x_1)$). A quadratic is fit in this case by using the values at the two distinct points and the derivative at the duplicated point. In case $x_1 = x_2 = x_3$, (x_1, x_2, x_3)forms a three-point pattern if $f'(x_2) = 0$ and $f''(x_2) \geqslant 0$.

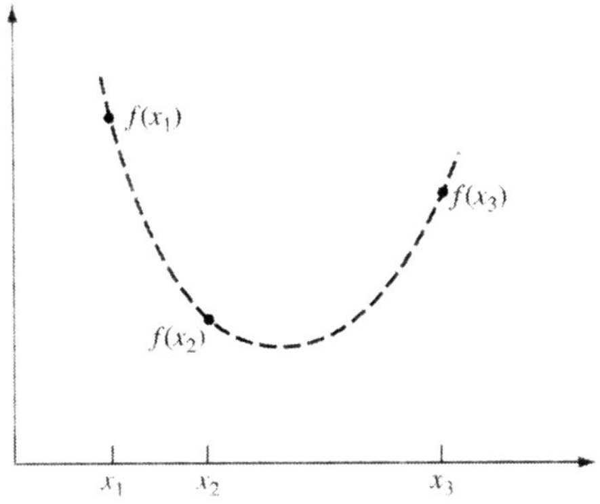

Fig. 8.7 Three-point pattern

With these definitions, the map $\mathbf{A}$ is well defined. It is also continuous, since curve fitting depends continuously on the data.

We next define the solution set $\Gamma \subset E^3$ as the points $\mathbf{x}^* = (x^*, x^*, x^*)$ where $f'(x^*) = 0$.

Finally, we let $Z(\mathbf{x}) = f(x_1) + f(x_2) + f(x_3)$. It is easy to see that Z is a descent function for $\mathbf{A}$. After application of $\mathbf{A}$ one of the values $f(x_1)$, $f(x_2)$, $f(x_3)$ will be replaced by $f(x_4)$, and by construction, and the assumption that f is unimodal, it will replace a strictly larger value. Of course, at $\mathbf{x}^* = (x^*, x^*, x^*)$ we have $\mathbf{A}(\mathbf{x}^*) = \mathbf{x}^*$ and hence $Z(\mathbf{A}(\mathbf{x}^*)) = Z(\mathbf{x}^*)$.

Since all points are contained in the initial interval, we have all the requirements for the Global Convergence Theorem. Thus the process converges to the solution. The order of convergence may not be destroyed by this modification, if near the solution the three-point pattern is always formed from the previous three points. In this case we would still have convergence of order 1.3. This cannot be guaranteed, however.

It has often been implicitly suggested, and accepted, that when using the quadratic fit technique one should require

$$f(x_{k+1}) < f(x_k)$$

so as to guarantee convergence. If the inequality is not satisfied at some cycle, then a special local search is used to find a better x_{k+1} that does satisfy it. This philosophy amounts to taking $Z(\mathbf{x}) = f(x_3)$ in our general framework and, unfortunately, this is not a descent function even for unimodal functions, and hence the special local search is likely to be necessary several times. It is true, of course, that a similar special local search may, occasionally, be required for the technique we suggest in regions of multiple minima, but it is never required in a unimodal region.

The above construction, based on the pure quadratic fit technique, can be emulated to produce effective procedures based on other curve fitting techniques. For application to smooth functions these techniques seem to be the best available in terms of flexibility to accommodate as much derivative information as is available, fast convergence, and a guarantee of global convergence.

*Closedness of Line Search Algorithms

Since searching along a line for a minimum point is a component part of most nonlinear programming algorithms, it is desirable to establish at once that this procedure is closed; that is, that the end product of the iterative procedures outlined above, when viewed as a single algorithmic step finding a minimum along a line, define closed algorithms. That is the objective of this section.

To initiate a line search with respect to a function f, two vectors must be specified: the initial point $\mathbf{x}$ and the direction $\mathbf{d}$ in which the search is to be made. The result of the search is a new point. Thus we define the search algorithm $\mathbf{S}$ as a mapping from E^{2n} to E^n.

We assume that the search is to be made over the semi-infinite line emanating from $\mathbf{x}$ in the direction $\mathbf{d}$. We also assume, for simplicity, that the search is not made in vain; that is, we assume that there is a minimum point along the line. This will be the case, for instance, if f is continuous and increases without bound as $\mathbf{x}$ tends toward infinity.

Definition. The mapping $\mathbf{S} : E^{2n} \to E^n$ is defined by

$$\mathbf{S}(\mathbf{x}, \mathbf{d}) = \{\mathbf{y} : \mathbf{y} = \mathbf{x} + \alpha\mathbf{d} \text{ for some } \alpha \geqslant 0,\ f(\mathbf{y}) = \min_{0 \leqslant \alpha \leqslant \infty} f(\mathbf{x} + \alpha\mathbf{d})\}. \tag{8.23}$$

In some cases there may be many vectors $\mathbf{y}$ yielding the minimum, so $\mathbf{S}$ is a set-valued mapping. We must verify that $\mathbf{S}$ is closed.

Theorem. *Let f be continuous on E^n. Then the mapping defined by (8.23) is closed at $(\mathbf{x}, \mathbf{d})$ if $\mathbf{d} \neq \mathbf{0}$.*

Proof. Suppose $\{\mathbf{x}_k\}$ and $\{\mathbf{d}_k\}$ are sequences with $\mathbf{x}_k \to \mathbf{x}$, $\mathbf{d}_k \to \mathbf{d} \neq \mathbf{0}$. Suppose also that $\mathbf{y}_k \in \mathbf{S}(\mathbf{x}_k, \mathbf{d}_k)$ and that $\mathbf{y}_k \to \mathbf{y}$. We must show that $\mathbf{y} \in \mathbf{S}(\mathbf{x}, \mathbf{d})$.

For each k we have $\mathbf{y}_k = \mathbf{x}_k + \alpha_k\mathbf{d}_k$ for some α_k. From this we may write

$$\alpha_k = \frac{|\mathbf{y}_k - \mathbf{x}_k|}{|\mathbf{d}_k|}.$$

Taking the limit of the right-hand side of the above, we see that

$$\alpha_k \to \overline{\alpha} \equiv \frac{|\mathbf{y} - \mathbf{x}|}{|\mathbf{d}|}.$$

It then follows that $\mathbf{y} = \mathbf{x} + \overline{\alpha}\mathbf{d}$. It still remains to be shown that $\mathbf{y} \in \mathbf{S}(\mathbf{x}, \mathbf{d})$.

For each k and each α, $0 \leqslant \alpha < \infty$,

$$f(\mathbf{y}_k) \leqslant f(\mathbf{x}_k + \alpha \mathbf{d}_k).$$

Letting $k \to \infty$ we obtain

$$f(\mathbf{y}) \leqslant f(\mathbf{x} + \alpha \mathbf{d}).$$

Thus

$$f(\mathbf{y}) \leqslant \min_{0 \leqslant \alpha < \infty} f(\mathbf{x} + \alpha \mathbf{d}),$$

and hence $\mathbf{y} \in \mathbf{S}(\mathbf{x}, \mathbf{d})$. ∎

The requirement that $\mathbf{d} \neq \mathbf{0}$ is natural both theoretically and practically. From a practical point of view this condition implies that, when constructing algorithms, the choice $\mathbf{d} = \mathbf{0}$ had better occur only in the solution set; but it is clear that if $\mathbf{d} = \mathbf{0}$, no search will be made. Theoretically, the map $\mathbf{S}$ can fail to be closed at $\mathbf{d} = \mathbf{0}$, as illustrated below.

Example. On E^1 define $f(x) = (x - 1)^2$. Then $S(x, d)$ is not closed at $x = 0$, $d = 0$. To see this we note that for any $d > 0$

$$\min_{0 \leqslant \alpha < \infty} f(\alpha d) = f(1),$$

and hence

$$S(0, d) = 1;$$

but

$$\min_{0 \leqslant \alpha < \infty} f(\alpha \cdot 0) = f(0)$$

so that

$$S(0, 0) = 0.$$

Thus as $d \to 0$, $S(0, d) \nrightarrow S(0, 0)$.

Inaccurate Line Search

In practice, of course, it is impossible to obtain the exact minimum point called for by the ideal line search algorithm $\mathbf{S}$ described above. As a matter of fact, it is often desirable to sacrifice accuracy in the line search routine in order to conserve overall computation time. Because of these factors we must, to be realistic, be certain, at every stage of development, that our theory does not crumble if inaccurate line searches are introduced.

Inaccuracy generally is introduced in a line search algorithm by simply terminating the search procedure before it has converged. The exact nature of the inaccuracy introduced may therefore depend on the particular search technique employed and the criterion used for terminating the search. We cannot develop a theory that simultaneously covers every important version of inaccuracy without seriously detracting from the underlying simplicity of the algorithms discussed later. For this reason our general approach, which is admittedly more free-wheeling in spirit than necessary but thereby more transparent and less encumbered than a detailed account of inaccuracy, will be to analyze algorithms as if an accurate line search were made at every step, and then point out in side remarks and exercises the effect of inaccuracy.

Armijo's Rule

A practical and popular criterion for terminating a line search is Armijo's rule. The essential idea is that the rule should first guarantee that the selected α is not too large, and next it should not be too small. Let us define the function

$$\phi(\alpha) = f(\mathbf{x}_k + \alpha \mathbf{d}_k).$$

Armijo's rule is implemented by consideration of the function $\phi(0) + \varepsilon\phi'(0)\alpha$ for fixed ε, $0 < \varepsilon < 1$. This function is shown in Fig. 8.8a as the dashed line. A value of α is considered to be not too large if the corresponding function value lies below the dashed line; that is, if

$$\phi(\alpha) \leqslant \phi(0) + \varepsilon\phi'(0)\alpha. \tag{8.24}$$

To insure that α is not too small, a value $\eta > 1$ is selected, and α is then considered to be not too small if

$$\phi(\eta\alpha) > \phi(0) + \varepsilon\phi'(0)\eta\alpha.$$

This means that if α is increased by the factor η, it will fail to meet the test (8.24). The acceptable region defined by the Armijo rule is shown in Fig. 8.8a when $\eta = 2$ (there are also other rules can be adapted).

Sometimes in practice, the Armijo test is used to define a simplified line search technique that does not employ curve fitting methods. One begins with an arbitrary α. If it satisfies (8.24), it is repeatedly increased by $\eta(\eta = 2$ or $\eta = 10$ and $\varepsilon = .2$ are often used) until (8.24) is not satisfied, and then the penultimate α is selected. If, on the other hand, the original α does not satisfy (8.24), it is repeatedly divided by η until the resulting α does satisfy (8.24).

8.2 The Method of Steepest Descent

One of the oldest and most widely known methods for minimizing a function of several variables is the method of steepest descent (often referred to as the gradient method). The method is extremely important from a theoretical viewpoint, since it is one of the simplest for which a satisfactory analysis exists. More advanced algorithms are often motivated by an attempt to modify the basic steepest descent technique in such a way that the new algorithm will have superior convergence properties. The method of steepest descent remains, therefore, not only the technique most often first tried on a new problem but also the standard of reference against which other techniques are measured. The principles used for its analysis will be used throughout this book.

The Method

Let f have continuous first partial derivatives on E^n. We will frequently have need for the gradient vector of f and therefore we introduce some simplifying notation. The gradient $\nabla f(\mathbf{x})$ is, according to our conventions, defined as a n-dimensional *row* vector. For convenience we define the n-dimensional *column* vector $\mathbf{g}(\mathbf{x}) = \nabla f(\mathbf{x})^T$. When there is no chance for ambiguity, we sometimes suppress the argument $\mathbf{x}$ and, for example, write $\mathbf{g}_k$ for $\mathbf{g}(\mathbf{x}_k) = \nabla f(\mathbf{x}_k)^T$.

The method of steepest descent is defined by the iterative algorithm

$$\mathbf{x}_{k+1} = \mathbf{x}_k - \alpha_k \mathbf{g}_k,$$

where stepsize α_k is a nonnegative scalar possibly minimizing $f(\mathbf{x}_k - \alpha\mathbf{g}_k)$. In words, from the point $\mathbf{x}_k$ we search along the direction of the negative gradient $-\mathbf{g}_k$ to a minimum point on this line; this minimum point is taken to be $\mathbf{x}_{k+1}$.

In formal terms, the overall algorithm $\mathbf{A} : E^n \to E^n$ which gives $\mathbf{x}_{k+1} \in \mathbf{A}(\mathbf{x}_k)$ can be decomposed in the form $\mathbf{A} = \mathbf{SG}$. Here $\mathbf{G} : E^n \to E^{2n}$ is defined by $\mathbf{G}(\mathbf{x}) = (\mathbf{x}, -\mathbf{g}(\mathbf{x}))$, giving the initial point and direction of a line search. This is followed by the line search $\mathbf{S} : E^{2n} \to E^n$ defined in Sect. 8.1.

Global Convergence and Convergence Speed

It was shown in Sect. 8.1 that $\mathbf{S}$ is closed if $\nabla f(\mathbf{x}) \neq \mathbf{0}$, and it is clear that $\mathbf{G}$ is continuous. Therefore, by Corollary 2 in Sect. 7.7 $\mathbf{A}$ is closed.

We define the solution set to be the points $\mathbf{x}$ where $\nabla f(\mathbf{x}) = \mathbf{0}$. Then $Z(\mathbf{x}) = f(\mathbf{x})$ is a descent function for $\mathbf{A}$, since for $\nabla f(\mathbf{x}) \neq \mathbf{0}$

$$\lim_{0 \leqslant \alpha < \infty} f(\mathbf{x} - \alpha\mathbf{g}(\mathbf{x})) < f(\mathbf{x}).$$

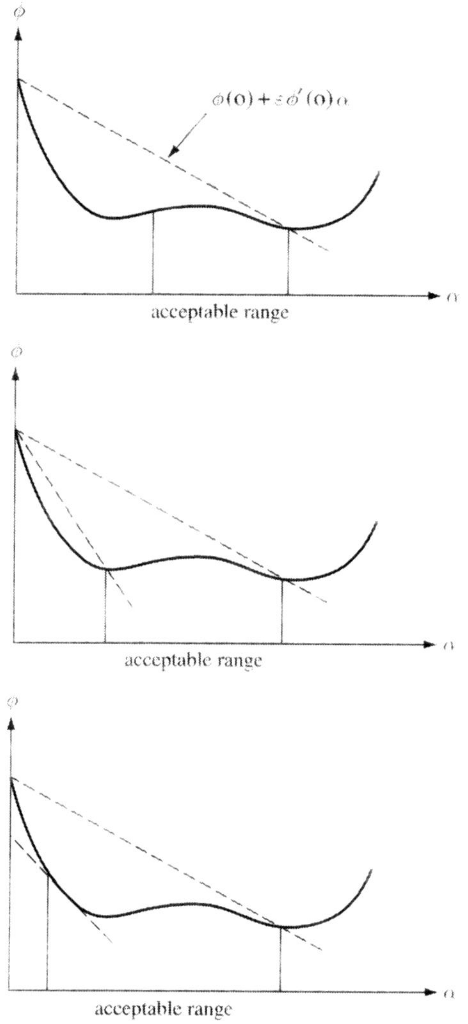

Fig. 8.8 Stopping rules. (**a**) Armijo rule. (**b**) Golden test. (**c**) Wolfe test

Thus by the Global Convergence Theorem, if the sequence $\{\mathbf{x}_k\}$ is bounded, it will have limit points and each of these is a solution. What about the convergence speed? Assume that $f(\mathbf{x})$ is convex and differentiable everywhere, admits a minimizer $\mathbf{x}^*$, and satisfies the (first-order) β-Lipschitz condition, that is, for any two points $\mathbf{x}$ and $\mathbf{y}$

$$|\nabla f(\mathbf{x}) - \nabla f(\mathbf{y})| \leq \beta|\mathbf{x} - \mathbf{y}|$$

for a positive real number β. Starting from any point $\mathbf{x}_0$, we consider the method of steepest descent with a fixed step size $\alpha_k = \frac{1}{\beta}$ for all k:

$$\mathbf{x}_{k+1} = \mathbf{x}_k - \frac{1}{\beta}\mathbf{g}_k = \mathbf{x}_k - \frac{1}{\beta}\nabla f(\mathbf{x}_k)^T. \quad (8.25)$$

We first prove a lemma.

Lemma 1. *Let $f(\boldsymbol{x})$ be differentiable everywhere and satisfy the (first-order) β-Lipschitz condition. Then, for any two points $\mathbf{x}$ and $\mathbf{y}$*

$$f(\boldsymbol{x}) - f(\boldsymbol{y}) - \nabla f(\boldsymbol{y})(\boldsymbol{x} - \boldsymbol{y}) \le \frac{\beta}{2}|\boldsymbol{x} - \boldsymbol{y}|^2.$$

Then we prove

Theorem 1 (Steepest Descent—Lipschitz Convex Case). *Let $f(\boldsymbol{x})$ be convex and differentiable everywhere, satisfy the (first-order) β-Lipschitz condition, and admit a minimizer $\mathbf{x}^*$. Then, the method of steepest descent (8.25) generates a sequence of solutions $\mathbf{x}_k$ such that*

$$|\nabla f(\boldsymbol{x}_k)| \le \frac{\beta^2}{\sqrt{k(k+1)}}|\boldsymbol{x}_0 - \boldsymbol{x}^*|,$$

and

$$f(\boldsymbol{x}_k) - f(\boldsymbol{x}^*) \le \frac{\beta}{2(k+1)}|\boldsymbol{x}_0 - \boldsymbol{x}^*|^2.$$

Proof. Consider the function $g_x(\mathbf{y}) = f(\mathbf{y}) - \nabla f(\mathbf{x})\mathbf{y}$ for any given $\mathbf{x}$. Note that g_x is also convex and satisfies the β-Lipschitz condition. Moreover, $\mathbf{x}$ is the minimizer of $g_x(\mathbf{y})$ and $\nabla g_x(\mathbf{y}) = \nabla f(\mathbf{y}) - \nabla f(\mathbf{x})$.

Applying Lemma 1 to g_x and noting the relations of g_x and $f(\mathbf{x})$, we have

$$\begin{aligned} f(\mathbf{x}) - f(\mathbf{y}) - \nabla f(\mathbf{x})(\mathbf{x} - \mathbf{y}) &= g_x(\mathbf{x}) - g_x(\mathbf{y}) \\ &\le g_x(\mathbf{y} - \tfrac{1}{\beta}\nabla g_x(\mathbf{y})) - g_x(\mathbf{y}) \\ &\le \nabla g_x(\mathbf{y})(-\tfrac{1}{\beta}\nabla g_x(\mathbf{y})^T) + \tfrac{\beta}{2}\tfrac{1}{\beta^2}|\nabla g_x(\mathbf{y})|^2 \\ &= -\tfrac{1}{2\beta}|\nabla g_x(\mathbf{y})|^2 \\ &= -\tfrac{1}{2\beta}|\nabla f(\mathbf{x}) - \nabla f(\mathbf{y})|^2. \end{aligned} \quad (8.26)$$

Similarly, we have

$$f(\mathbf{y}) - f(\mathbf{x}) - \nabla f(\mathbf{y})(\mathbf{y} - \mathbf{x}) \le -\frac{1}{2\beta}|\nabla f(\mathbf{x}) - \nabla f(\mathbf{y})|^2.$$

Adding the above two derived inequalities, we have for any $\mathbf{x}$ and $\mathbf{y}$:

$$(\nabla f(\mathbf{x}) - \nabla f(\mathbf{y}))(\mathbf{x} - \mathbf{y}) \ge \frac{1}{\beta}|\nabla f(\mathbf{x}) - \nabla f(\mathbf{y})|^2. \quad (8.27)$$

For simplification, in what follows let $\mathbf{d}_k = \mathbf{x}_k - \mathbf{x}^*$ and $\delta_k = [f(\mathbf{x}_k) - f(\mathbf{x}^*)] \ge 0$.

Now let $\mathbf{x} = \mathbf{x}_{k+1}$ and $\mathbf{y} = \mathbf{x}_k$ in (8.27). Then

$$-\frac{1}{\beta}(\mathbf{g}_k)^T(\mathbf{g}_{k+1} - \mathbf{g}_k) = (\mathbf{x}_{k+1} - \mathbf{x}_k)^T(\mathbf{g}_{k+1} - \mathbf{g}_k) \geq \frac{1}{\beta}|\mathbf{g}_{k+1} - \mathbf{g}_k|^2,$$

which leads to

$$|\mathbf{g}_{k+1}|^2 \leq (\mathbf{g}_{k+1})^T\mathbf{g}_k \leq |\mathbf{g}_{k+1}||\mathbf{g}_k|, \quad \text{that is} \quad |\mathbf{g}_{k+1}| \leq |\mathbf{g}_k|. \tag{8.28}$$

Inequality (8.28) implies that $|\mathbf{g}_k| = |\nabla f(\mathbf{x}_k)|$ is monotonically decreasing.

Applying inequality (8.26) for $\mathbf{x} = \mathbf{x}_k$ and $\mathbf{y} = \mathbf{x}^*$ and noting $\mathbf{g}^* = \mathbf{0}$ we have

$$\begin{aligned}\delta_k &\leq (\mathbf{g}_k)^T\mathbf{d}_k - \tfrac{1}{2\beta}|\mathbf{g}_k|^2\\ &= -\beta(\mathbf{x}_{k+1} - \mathbf{x}_k)\mathbf{d}_k - \tfrac{\beta}{2}|\mathbf{x}_{k+1} - \mathbf{x}_k|^2\\ &= -\tfrac{\beta}{2}(|\mathbf{x}_{k+1} - \mathbf{x}_k|^2 + 2(\mathbf{x}_{k+1} - \mathbf{x}_k)^T\mathbf{d}_k)\\ &= -\tfrac{\beta}{2}(|\mathbf{d}_{k+1} - \mathbf{d}_k|^2 + 2(\mathbf{d}_{k+1} - \mathbf{d}_k)^T\mathbf{d}_k)\\ &= \tfrac{\beta}{2}(|\mathbf{d}_k|^2 - |\mathbf{d}_{k+1}|^2).\end{aligned} \tag{8.29}$$

Summing up (8.29) from 0 to k, we have

$$\sum_{l=0}^{k} \delta_l \leq \frac{\beta}{2}(|\mathbf{d}_0|^2 - |\mathbf{d}_{k+1}|^2) \leq \frac{\beta}{2}|\mathbf{d}_0|^2. \tag{8.30}$$

Using (8.26) again for $\mathbf{x} = \mathbf{x}_{k+1}$ and $\mathbf{y} = \mathbf{x}_k$ and noting (8.25) we have

$$\begin{aligned}\delta_{k+1} - \delta_k &= f(\mathbf{x}_{k+1}) - f(\mathbf{x}_k)\\ &\leq \mathbf{g}_{k+1}^T(-\tfrac{1}{\beta}\mathbf{g}_k) - \tfrac{1}{2\beta}|\mathbf{g}_{k+1} - \mathbf{g}_k|^2\\ &= -\tfrac{1}{2\beta}(|\mathbf{g}_{k+1}|^2 + |\mathbf{g}_k|^2).\end{aligned} \tag{8.31}$$

Noting (8.31) holts for all k, we have

$$\begin{aligned}\textstyle\sum_{l=0}^{k}\delta_l &= \textstyle\sum_{l=0}^{k}\delta_l(l + 1 - l)\\ &= \textstyle\sum_{l=0}^{k}\delta_l(l + 1) - \sum_{l=0}^{k}\delta_l l\\ &= \textstyle\sum_{l=1}^{k+1}\delta_{l-1}l - \sum_{l=1}^{k}\delta_l l\\ &= \textstyle\delta_k(k + 1) + \sum_{l=1}^{k}(\delta_{l-1} - \delta_l)l\\ &\geq \textstyle\delta_k(k + 1) + \sum_{l=1}^{k}\frac{l}{2\beta}(|\mathbf{g}_l|^2 + |\mathbf{g}_{l-1}|^2)\\ &\geq \delta_k(k + 1) + \tfrac{k(k+1)}{2\beta}|\mathbf{g}_k|^2\end{aligned}$$

where the last inequality comes $|\mathbf{g}_k| = |\nabla f(\mathbf{x}_k)|$ is *monotonically* decreasing.

Using (8.30) we finally have

$$(k + 1)\delta_k + \frac{k(k + 1)}{2\beta}|\mathbf{g}_k|^2 \leq \frac{\beta}{2}|\mathbf{d}_0|^2. \tag{8.32}$$

Inequality (8.32), from $\delta_k = f(\mathbf{x}_k) - f(\mathbf{x}^*) \geq 0$ and $\mathbf{d}_0 = \mathbf{x}_0 - \mathbf{x}^*$, proves the desired bounds. ∎

Theorem 1 implies that the convergence speed of the steepest descent method is arithmetic.

The Quadratic Case

When $f(\mathbf{x})$ is strongly convex, the convergence speed can be increased from arithmetic to geometric or linear convergence. Since all of the important convergence characteristics of the method of steepest descent are revealed by an investigation of the method when applied to quadratic problems, we focus here on

$$f(\mathbf{x}) = \frac{1}{2}\mathbf{x}^T\mathbf{Q}\mathbf{x} - \mathbf{x}^T\mathbf{b}, \tag{8.33}$$

where $\mathbf{Q}$ is a positive definite symmetric $n \times n$ matrix. Since $\mathbf{Q}$ is positive definite, all of its eigenvalues are positive. We assume that these eigenvalues are ordered: $0 < a = \lambda_1 \leqslant \lambda_2 \ldots \leqslant \lambda_n = A$. With $\mathbf{Q}$ positive definite, it follows (from Proposition 5, Sect. 7.4) that f is strictly convex.

The unique minimum point of f can be found directly, by setting the gradient to zero, as the vector $\mathbf{x}^*$ satisfying

$$\mathbf{Q}\mathbf{x}^* = \mathbf{b}. \tag{8.34}$$

Moreover, introducing the function

$$E(\mathbf{x}) = \frac{1}{2}(\mathbf{x} - \mathbf{x}^*)^T\mathbf{Q}(\mathbf{x} - \mathbf{x}^*), \tag{8.35}$$

we have $E(\mathbf{x}) = f(\mathbf{x}) + (1/2)\mathbf{x}^{*T}\mathbf{Q}\mathbf{x}^*$, which shows that the function E differs from f only by a constant. For many purposes then, it will be convenient to consider that we are minimizing E rather than f.

The gradient (of both f and E) is given explicitly by

$$\mathbf{g}(\mathbf{x}) = \mathbf{Q}\mathbf{x} - \mathbf{b}. \tag{8.36}$$

Thus the method of steepest descent can be expressed as

$$\mathbf{x}_{k+1} = \mathbf{x}_k - \alpha_k\mathbf{g}_k, \tag{8.37}$$

where $\mathbf{g}_k = \mathbf{Q}\mathbf{x}_k - \mathbf{b}$ and where α_k minimizes $f(\mathbf{x}_k - \alpha\mathbf{g}_k)$. We can, however, in this special case, determine the value of α_k explicitly. We have, by definition (8.33),

$$f(\mathbf{x}_k - \alpha\mathbf{g}_k) = \frac{1}{2}(\mathbf{x}_k - \alpha\mathbf{g}_k)^T\mathbf{Q}(\mathbf{x}_k - \alpha\mathbf{g}_k) - (\mathbf{x}_k - \alpha\mathbf{g}_k)^T\mathbf{b},$$

which (as can be found by differentiating with respect to α) is minimized at

$$\alpha_k = \frac{\mathbf{g}_k^T\mathbf{g}_k}{\mathbf{g}_k^T\mathbf{Q}\mathbf{g}_k}. \tag{8.38}$$

Hence the method of steepest descent (8.37) takes the explicit form

$$\mathbf{x}_{k+1} = \mathbf{x}_k - \left(\frac{\mathbf{g}_k^T\mathbf{g}_k}{\mathbf{g}_k^T\mathbf{Q}\mathbf{g}_k}\right)\mathbf{g}_k, \tag{8.39}$$

where $\mathbf{g}_k = \mathbf{Q}\mathbf{x}_k - \mathbf{b}$.

The function f and the steepest descent process can be illustrated as in Fig. 8.9 by showing contours of constant values of f and a typical sequence developed by the process. The contours of f are n-dimensional ellipsoids with axes in the directions of the n-mutually orthogonal eigenvectors of $\mathbf{Q}$. The axis corresponding to the ith eigenvector has length proportional to $1/\lambda_i$. We now analyze this process and show that the rate of convergence depends on the ratio of the lengths of the axes of the elliptical contours of f, that is, on the eccentricity of the ellipsoids.

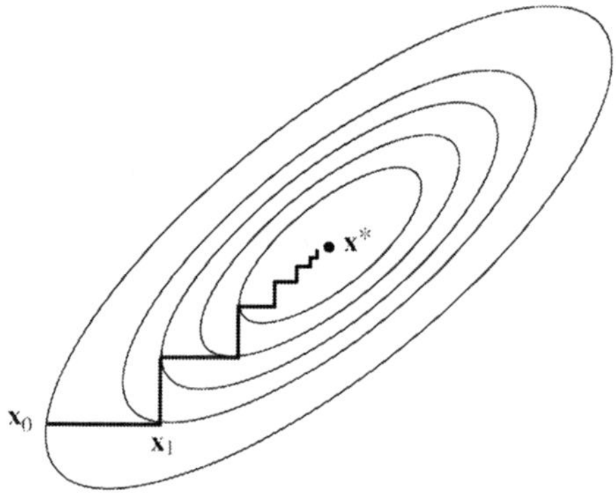

Fig. 8.9 Steepest descent

Lemma 2. *The iterative process (8.39) satisfies*

$$E(\boldsymbol{x}_{k+1}) = \left\{1 - \frac{(\mathbf{g}_k^T \mathbf{g}_k)^2}{(\mathbf{g}_k^T \mathbf{Q}\mathbf{g}_k)(\mathbf{g}_k^T \mathbf{Q}^{-1}\mathbf{g}_k)}\right\} E(\boldsymbol{x}_k). \tag{8.40}$$

Proof. The proof is by direct computation. We have, setting $\mathbf{y}_k = \mathbf{x}_k - \mathbf{x}^*$,

$$\frac{E(\mathbf{x}_k) - E(\mathbf{x}_{k+1})}{E(\mathbf{x}_k)} = \frac{2\alpha_k \mathbf{g}_k^T \mathbf{Q}\mathbf{y}_k - \alpha_k^2 \mathbf{g}_k^T \mathbf{Q}\mathbf{g}_k}{\mathbf{y}_k^T \mathbf{Q}\mathbf{y}_k}.$$

Using $\mathbf{g}_k = \mathbf{Q}\mathbf{y}_k$ we have

$$\begin{aligned}\frac{E(\mathbf{x}_k) - E(\mathbf{x}_{k+1})}{E(\mathbf{x}_k)} &= \frac{\frac{2(\mathbf{g}_k^T \mathbf{g}_k)^2}{(\mathbf{g}_k^T \mathbf{Q}\mathbf{g}_k)} - \frac{(\mathbf{g}_k^T \mathbf{g}_k)^2}{(\mathbf{g}_k^T \mathbf{Q}\mathbf{g}_k)}}{\mathbf{g}_k^T \mathbf{Q}^{-1}\mathbf{g}_k} \\ &= \frac{(\mathbf{g}_k^T \mathbf{g}_k)^2}{(\mathbf{g}_k^T \mathbf{Q}\mathbf{g}_k)(\mathbf{g}_k^T \mathbf{Q}^{-1}\mathbf{g}_k)}. \blacksquare\end{aligned}$$

In order to obtain a bound on the rate of convergence, we need a bound on the right-hand side of (8.40). The best bound is due to Kantorovich and his lemma, stated below, is a useful general tool in convergence analysis.

Kantorovich inequality: *Let* **Q** *be a positive definite symmetric* $n \times n$ *matrix. For any vector* **x** *there holds*

$$\frac{(\mathbf{x}^T\mathbf{x})^2}{(\mathbf{x}^T\mathbf{Q}\mathbf{x})(\mathbf{x}^T\mathbf{Q}^{-1}\mathbf{x})} \geqslant \frac{4aA}{(a+A)^2}, \tag{8.41}$$

where a and **A** *are, respectively, the smallest and largest eigenvalues of* **Q**.

Proof. Let the eigenvalues $\lambda_1, \lambda_2, \ldots, \lambda_n$ of **Q** satisfy

$$0 < a = \lambda_1 \leqslant \lambda_2 \ldots \leqslant \lambda_n = A.$$

By an appropriate change of coordinates the matrix **Q** becomes diagonal with diagonal $(\lambda_1, \lambda_2, \ldots, \lambda_n)$. In this coordinate system we have

$$\frac{(\mathbf{x}^T\mathbf{x})^2}{(\mathbf{x}^T\mathbf{Q}\mathbf{x})(\mathbf{x}^T\mathbf{Q}^{-1}\mathbf{x})} = \frac{(\sum_{i=1}^n x_i^2)^2}{(\sum_{i=1}^n \lambda_i x_i^2)(\sum_{i=1}^n (x_i^2/\lambda_i))},$$

which can be written as

$$\frac{(\mathbf{x}^T\mathbf{x})^2}{(\mathbf{x}^T\mathbf{Q}\mathbf{x})(\mathbf{x}^T\mathbf{Q}^{-1}\mathbf{x})} = \frac{1/\sum_{i=1}^n \xi_i\lambda_i}{\sum_{i=1}^n(\xi_i/\lambda_i)} \equiv \frac{\phi(\xi)}{\psi(\xi)},$$

where $\xi_i = x_i^2/\sum_{i=1}^n x_i^2$. We have converted the expression to the ratio of two functions involving convex combinations; one a combination of λ_i's; the other a combination of $1/\lambda_i$'s. The situation is shown pictorially in Fig. 8.10. The curve in the figure represents the function $1/\lambda$. Since $\sum_{i=1}^n \xi_i\lambda_i$ is a point between λ_1 and λ_n, the value of $\phi(\xi)$ is a point on the curve. On the other hand, the value of $\psi(\xi)$ is a convex combination of points on the curve and its value corresponds to a point in the shaded region. For the same vector ξ both functions are represented by points on the same vertical line. The minimum value of this ratio is achieved for some $\lambda = \xi_1\lambda_1 + \xi_n\lambda_n$, with $\xi_1 + \xi_n = 1$. Using the relation $\xi_1/\lambda_1 + \xi_n/\lambda_n = (\lambda_1 + \lambda_n - \xi_1\lambda_1 - \xi_n\lambda_n)/\lambda_1\lambda_n$, an appropriate bound is

$$\frac{\phi(\xi)}{\psi(\xi)} \geqslant \lim_{\lambda_1 \leqslant \lambda \leqslant \lambda_n} \frac{(1/\lambda)}{(\lambda_1 + \lambda_n - \lambda)/(\lambda_1\lambda_n)}.$$

The minimum is achieved at $\lambda = (\lambda_1 + \lambda_n)/2$, yielding

$$\frac{\phi(\xi)}{\psi(\xi)} \geqslant \frac{4\lambda_1\lambda_n}{(\lambda_1 + \lambda_n)^2}. \blacksquare$$

Combining the above two lemmas, we obtain the central result on the convergence of the method of steepest descent.

Theorem 2 (Steepest Descent—Quadratic Case). *For any* $\mathbf{x}_0 \in E^n$ *the method of steepest descent (8.39) converges to the unique minimum point* $\mathbf{x}^*$ *of f Furthermore, with* $E(\boldsymbol{x}) = \frac{1}{2}(\boldsymbol{x} - \boldsymbol{x}^*)^T\boldsymbol{Q}(\boldsymbol{x} - \boldsymbol{x}^*)$, *there holds at every step k*

$$E(\boldsymbol{x}_{k+1}) \leqslant \left(\frac{A-a}{A+a}\right)^2 E(\boldsymbol{x}_k). \tag{8.42}$$

Proof. By Lemma 2 and the Kantorovich inequality

$$E(\mathbf{x}_{k+1}) \leqslant \left\{1 - \frac{4aA}{(A+a)^2}\right\} E(\mathbf{x}_k) = \left(\frac{A-a}{A+a}\right)^2 E(\mathbf{x}_k).$$

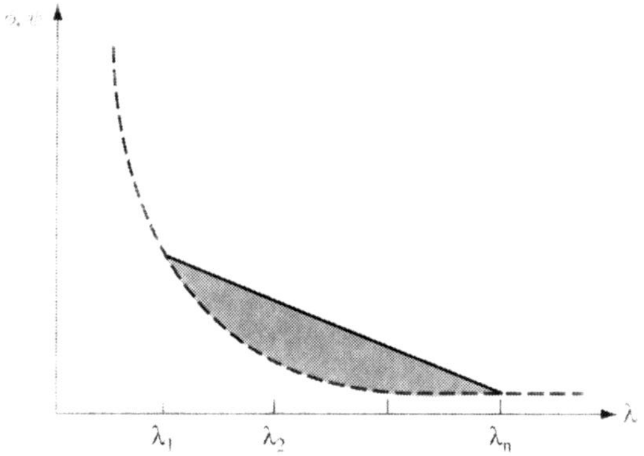

Fig. 8.10 Kantorovich inequality

It follows immediately that $E(\mathbf{x}_k) \to 0$ and hence, since $\mathbf{Q}$ is positive definite, that $\mathbf{x}_k \to \mathbf{x}^*$. ∎

Roughly speaking, the above theorem says that the convergence rate of steepest descent is slowed as the contours of f become more eccentric. If $a = A$, corresponding to circular contours, convergence occurs in a single step. Note, however, that even if $n - 1$ of the n eigenvalues are equal and the remaining one is a great distance from these, convergence will be slow, and hence a single abnormal eigenvalue can destroy the effectiveness of steepest descent.

In the terminology introduced in Sect. 7.8, the above theorem states that with respect to the error function E (or equivalently f) the method of steepest descent converges linearly with a ratio no greater than $[(A - a)/(A + a)]^2$. The actual rate depends on the initial point $\mathbf{x}_0$. However, for some initial points the bound is actually achieved. Furthermore, it has been shown by Akaike that, if the ratio is unfavorable, the process is very likely to converge at a rate close to the bound. Thus, somewhat loosely but with reasonable justification, we say that the convergence ratio of steepest descent *is* $[(A - a)/(A + a)]^2$.

It should be noted that the convergence rate actually depends only on the ratio $r = A/a$ of the largest to the smallest eigenvalue. Thus the convergence ratio is

$$\left(\frac{A-a}{A+a}\right)^2 = \left(\frac{r-1}{r+1}\right)^2,$$

which clearly shows that convergence is slowed as r increases. The ratio r, which is the single number associated with the matrix $\mathbf{Q}$ that characterizes convergence, is often called the *condition number* of the matrix.

Example. Let us take

$$\mathbf{Q} = \begin{bmatrix} 0.78 & -0.02 & -0.12 & -0.14 \\ -0.02 & 0.86 & -0.04 & 0.06 \\ -0.12 & -0.04 & 0.72 & -0.08 \\ -0.14 & 0.06 & -0.08 & 0.74 \end{bmatrix}$$

$$\mathbf{b} = (0.76, 0.08, 1.12, 0.68).$$

For this matrix it can be calculated that $a = 0.52$, $A = 0.94$ and hence $r = 1.8$. This is a very favorable condition number and leads to the convergence ratio $[(A - a)/(A + a)]^2 = 0.081$. Thus each iteration will reduce the error in the objective by more than a factor of ten; or, equivalently, each iteration will add about one more digit of accuracy. Indeed, starting from the origin the sequence of values obtained by steepest descent as shown in Table 8.1 is consistent with this estimate.

The Nonquadratic Case

For nonquadratic functions, we expect that steepest descent will also do reasonably well if the condition number is modest. Fortunately, we are able to establish estimates of the progress of the method when the Hessian matrix is always positive definite. Specifically, we assume that the Hessian matrix is bounded above and below as $a\mathbf{I} \leqslant \mathbf{F}(\bar{\mathbf{x}}) \leqslant A\mathbf{I}$. (Thus f is *strongly* convex.) We present three analyses:

Table 8.1 Solution to Example

Step k	$f(\mathbf{x}_k)$
0	0
1	-2.1563625
2	-2.1744062
3	-2.1746440
4	-2.1746585
5	-2.1746595
6	-2.1746595

Solution point $\mathbf{x}^* = (1.534965, 0.1220097, 1.975156, 1.412954)$

1. **Exact Line Search.** Given a point $\mathbf{x_k}$, we have for any α

$$f(\mathbf{x}_k - \alpha\mathbf{g}(\mathbf{x}_k)) \leqslant f(\mathbf{x}_k)) - \alpha\mathbf{g}(\mathbf{x}_k)^T\mathbf{g}(\mathbf{x}_k) + \frac{A\alpha^2}{2}\mathbf{g}(\mathbf{x}_k)^T\mathbf{g}(\mathbf{x}_k). \tag{8.43}$$

Minimizing both sides separately with respect to α the inequality will hold for the two minima. The minimum of the left hand side is $f(\mathbf{x}_{k+1})$. The minimum of the right hand side occurs at $\alpha = 1/A$, yielding the result

$$f(\mathbf{x}_{k+1}) \leqslant f(\mathbf{x}_k) - \frac{1}{2A}|\mathbf{g}(\mathbf{x}_k)|^2.$$

where $|\mathbf{g}(\mathbf{x_k})|^2 \equiv \mathbf{g}(\mathbf{x_k})^T\mathbf{g}(\mathbf{x_k})$. Subtracting the optimal value $f^* = f(\mathbf{x}^*)$ from both sides produces

$$f(\mathbf{x_{k+1}}) - \mathbf{f^*} \leqslant \mathbf{f}(\mathbf{x}_k) - \mathbf{f^*} - \frac{1}{2\mathbf{A}}|\mathbf{g}(\mathbf{x}_k)|^2. \tag{8.44}$$

In a similar way, for any $\mathbf{x}$ there holds

$$f(\mathbf{x}) \geqslant f(\mathbf{x_k}) + \mathbf{g}(\mathbf{x_k})^T(\mathbf{x} - \mathbf{x_k}) + \frac{\mathbf{a}}{2}|\mathbf{x} - \mathbf{x_k}|^2.$$

Again we can minimize both sides separately. The minimum of the left hand side is f^* the optimal solution value. Minimizing the right hand side leads to the quadratic optimization problem. The solution is $\overline{\mathbf{x}} = \mathbf{x}_k - \mathbf{g}(\mathbf{x}_k)/a$. Substituting this $\overline{\mathbf{x}}$ in the right hand side of the inequality gives

$$f^* \geqslant f(\mathbf{x}_k) - \frac{1}{2a}|\mathbf{g}(\mathbf{x}_k)|^2. \tag{8.45}$$

From (8.45) we have

$$-|\mathbf{g}(\mathbf{x}_k)|^2 \leqslant 2a[f^* - f(\mathbf{x}_k)]. \tag{8.46}$$

Substituting this in (8.44) gives

$$f(\mathbf{x}_{k+1}) - f^* \leqslant (1 - a/A)[f(\mathbf{x}_k) - f^*]. \tag{8.47}$$

This shows that the method of steepest descent makes progress even when it is not close to the solution.

2. **Other Stopping Criteria.** As an example of how other stopping criteria can be treated, we examine the rate of convergence when using Amijo's rule with $\varepsilon < 0.5$ and $\eta > 1$. Note first that the inequality $t \geqslant t^2$ for $0 \leqslant t \leqslant 1$ implies by a change of variable that

$$-\alpha + \frac{\alpha^2 A}{2} \leq -\alpha/2$$

for $0 \leqslant \alpha \leqslant 1/A$. Then using (8.43) we have that for $\alpha < 1/A$

$$\begin{aligned} f(\mathbf{x}_k - \alpha\mathbf{g}(\mathbf{x}_k)) &\leq f(\mathbf{x}_k) - \alpha|\mathbf{g}(\mathbf{x}_k)|^2 + 0.5\alpha^2 A|\mathbf{g}(\mathbf{x}_l)|^2 \\ &\leq f(\mathbf{x}_k) - 0.5\alpha|\mathbf{g}(\mathbf{x}_k)|^2 \\ &< f(\mathbf{x}_k) - \varepsilon\alpha|\mathbf{g}(\mathbf{x}_k)|^2 \end{aligned}$$

since $\varepsilon < 0.5$. This means that the first part of the stopping criterion is satisfied for $\alpha < 1/A$.

The second part of the stopping criterion states that $\eta\alpha$ does not satisfy the first criterion and thus the final α must satisfy $\alpha \geq 1/(\eta A)$. Therefore the inequality of the first part of the criterion implies

$$f(\mathbf{x}_{k+1}) \leq f(\mathbf{x}_k) - \frac{\varepsilon}{\eta A}|\mathbf{g}(\mathbf{x}_k)|^2.$$

Subtracting f^* from both sides,

$$f(\mathbf{x}_{k+1}) - f^* \leq f(\mathbf{x}_k) - f^* - \frac{\varepsilon}{\eta A}|\mathbf{g}(\mathbf{x}_k)|^2.$$

Finally, using (8.46) we obtain

$$f(\mathbf{x}_{k+1}) - f^* \leq [1 - (2\varepsilon a/\eta A)](f(\mathbf{x}_k) - f^*).$$

Clearly $2\varepsilon a/\eta A < 1$ and hence there is linear convergence. Notice if that in fact ε is chosen very close to 0.5 and η is chosen very close to 1, then the stopping condition demands that the α be restricted to a very small range, and the estimated rate of convergence is very close to the estimate obtained above for exact line search.

3. **Asymptotic Convergence.** We expect that as the points generated by steepest descent approach the solution point, the convergence characteristics will be close to those inherent for quadratic functions. This is indeed the case.

The general procedure for proving such a result, which is applicable to most methods having unity order of convergence, is to use the Hessian of the objective at the solution point as if it were the $\mathbf{Q}$ matrix of a quadratic problem. The particular theorem stated below is a special case of a theorem in Sect. 12.5 so we do not prove it here; but it illustrates the generalizability of an analysis of quadratic problems.

Theorem. *Suppose f is defined on E^n, has continuous second partial derivatives, and has a relative minimum at $\mathbf{x}^*$. Suppose further that the Hessian matrix of f, $\mathbf{F}(\mathbf{x}^*)$, has smallest eigenvalue $a > 0$ and largest eigenvalue $A > 0$. If $\{\mathbf{x}_k\}$ is a sequence generated by the method of steepest descent that converges to $\mathbf{x}^*$, then the sequence of objective values $\{f(\mathbf{x}_k)\}$ converges to $f(\mathbf{x}^*)$ linearly with a convergence ratio no greater than $[(A-a)/(A+a)]^2$.*

8.3 Applications of the Convergence Theory

Now that the basic convergence theory, as represented by the formula (8.42) for the rate of convergence, has been developed and demonstrated to actually characterize the behavior of steepest descent, it is appropriate to illustrate how the theory can be used. Generally, we do *not* suggest that one compute the numerical value of the formula—since it involves eigenvalues, or ratios of eigenvalues, that are not easily determined. Nevertheless, the formula itself is of immense practical importance, since it allows one to theoretically compare various situations. Without such a theory, one would be forced to rely completely on experimental comparisons.

Application 1 (Solution of Gradient Equation). One approach to the minimization of a function f is to consider solving the equations $\nabla f(\mathbf{x}) = \mathbf{0}$ that represent the necessary conditions. It has been proposed that these equations could be solved by applying steepest descent to the function $h(\mathbf{x}) = |\nabla f(\mathbf{x})|^2$. One advantage of this method is that the minimum value is known. We ask whether this method is likely to be faster or slower than the application of steepest descent to the original function f itself.

For simplicity we consider only the case where f is quadratic. Thus let $f(\mathbf{x}) = (1/2)\mathbf{x}^T\mathbf{Q}\mathbf{x} - \mathbf{b}^T\mathbf{x}$. Then the gradient of f is $\mathbf{g}(\mathbf{x}) = \mathbf{Q}\mathbf{x} - \mathbf{b}$, and $h(\mathbf{x}) = |\mathbf{g}(\mathbf{x})|^2 = \mathbf{x}^T\mathbf{Q}^2\mathbf{x} - 2\mathbf{x}^T\mathbf{Q}\mathbf{b} + \mathbf{b}^T\mathbf{b}$. Thus $h(\mathbf{x})$ is itself a quadratic function. The rate of convergence of steepest descent applied to h will be governed by the eigenvalues of the matrix $\mathbf{Q}^2$. In particular the rate will be

$$\left(\frac{\bar{r}-1}{\bar{r}+1}\right)^2,$$

where $\bar{r}$ is the condition number of the matrix $\mathbf{Q}^2$. However, the eigenvalues of $\mathbf{Q}^2$ are the squares of those of $\mathbf{Q}$ itself, so $\bar{r} = r^2$, where r is the condition number of $\mathbf{Q}$, and it is clear that the convergence rate for the proposed method will be worse than for steepest descent applied to the original function.

We can go further and actually estimate how much slower the proposed method is likely to be. If r is large, we have

$$\text{steepest descent rate} = \left(\frac{r-1}{r+1}\right)^2 \simeq (1 - 1/r)^4$$

$$\text{proposed method rate} = \left(\frac{r^2-1}{r^2+1}\right)^2 \simeq (1 - 1/r^2)^4.$$

Since $(1 - 1/r^2)^r \simeq 1 - 1/r$, it follows that it takes about r steps of the new method to equal one step of ordinary steepest descent. We conclude that if the original problem is difficult to solve with steepest descent, the proposed method will be quite a bit worse.

Application 2 (Penalty Methods). Let us briefly consider a problem with a single constraint:

$$\begin{aligned} &\text{minimize } f(\mathbf{x}) \\ &\text{subject to } h(\mathbf{x}) = 0. \end{aligned} \tag{8.48}$$

One method for approaching this problem is to convert it (at least approximately) to the unconstrained problem

$$\text{minimize } f(\mathbf{x}) + \frac{1}{2}\mu h(\mathbf{x})^2, \tag{8.49}$$

where μ is a (large) penalty coefficient. Because of the penalty, the solution to (8.49) will tend to have a small $h(\mathbf{x})$. Problem (8.49) can be solved as an unconstrained problem by the method of steepest descent. How will this behave?

For simplicity let us consider the case where f is quadratic and h is linear. Specifically, we consider the problem

$$\begin{aligned}&\text{minimize } \frac{1}{2}\mathbf{x}^T\mathbf{Q}\mathbf{x} - \mathbf{b}^T\mathbf{x} \\ &\text{subject to } \mathbf{c}^T\mathbf{x} = 0.\end{aligned} \tag{8.50}$$

The objective of the associated penalty problem is $(1/2)\{\mathbf{x}^T\mathbf{Q}\mathbf{x}+\mu\mathbf{x}^T\mathbf{c}\mathbf{c}^T\mathbf{x}\}-\mathbf{b}^T\mathbf{x}$. The quadratic form associated with this objective is defined by the matrix $\mathbf{Q}+\mu\mathbf{c}\mathbf{c}^T$ and, accordingly, the convergence rate of steepest descent will be governed by the condition number of this matrix. This matrix is the original matrix $\mathbf{Q}$ with a large rank-one matrix added. It should be fairly clear† that this addition will cause one eigenvalue of the matrix to be large (on the order of μ). Thus the condition number is roughly proportional to μ. Therefore, as one increases μ in order to get an accurate solution to the original constrained problem, the rate of convergence becomes extremely poor. We conclude that the penalty function method used in this simplistic way with steepest descent will not be very effective. (Penalty functions, and how to minimize them more rapidly, are considered in detail in Chap. 11.)

Scaling

The performance of the method of steepest descent is dependent on the particular choice of variables $\mathbf{x}$ used to define the problem. A new choice may substantially alter the convergence characteristics.

Suppose that $\mathbf{T}$ is an invertible $n \times n$ matrix. We can then represent points in E^n either by the standard vector $\mathbf{x}$ or by $\mathbf{y}$ where $\mathbf{T}\mathbf{y} = \mathbf{x}$. The problem of finding $\mathbf{x}$ to minimize $f(\mathbf{x})$ is equivalent to that of finding $\mathbf{y}$ to minimize $h(\mathbf{y}) = f(\mathbf{T}\mathbf{y})$. Using $\mathbf{y}$ as the underlying set of variables, we then have

$$\nabla h = \nabla f \mathbf{T}, \tag{8.51}$$

where ∇f is the gradient of f with respect to $\mathbf{x}$. Thus, using steepest descent, the direction of search will be

$$\nabla \mathbf{y} = -\mathbf{T}^T \nabla f^T, \tag{8.52}$$

which in the original variables is

$$\Delta \mathbf{x} = -\mathbf{T}\mathbf{T}^T \nabla f^T. \tag{8.53}$$

---

†See the Interlocking Eigenvalues Lemma in Sect. 10.6 for a proof that only one eigenvalue becomes large.

Thus we see that the change of variables changes the direction of search.

The rate of convergence of steepest descent with respect to $\mathbf{y}$ will be determined by the eigenvalues of the Hessian of the objective, taken with respect to $\mathbf{y}$. That Hessian is

$$\nabla^2 h(\mathbf{y}) \equiv \mathbf{H}(\mathbf{y}) = \mathbf{T}^T\mathbf{F}(\mathbf{T}\mathbf{y})\mathbf{T}.$$

Thus, if $\mathbf{x}^* = \mathbf{T}\mathbf{y}^*$ is the solution point, the rate of convergence is governed by the matrix

$$\mathbf{H}(\mathbf{y}^*) = \mathbf{T}^T\mathbf{F}(\mathbf{x}^*)\mathbf{T}. \tag{8.54}$$

Very little can be said in comparison of the convergence ratio associated with $\mathbf{H}$ and that of $\mathbf{F}$. If $\mathbf{T}$ is an orthonormal matrix, corresponding to $\mathbf{y}$ being defined from $\mathbf{x}$ by a simple rotation of coordinates, then $\mathbf{T}^T\mathbf{T} = \mathbf{I}$, and we see from (8.48) that the directions remain unchanged and the eigenvalues of $\mathbf{H}$ are the same as those of $\mathbf{F}$.

In general, before attacking a problem with steepest descent, it is desirable, if it is feasible, to introduce a change of variables that leads to a more favorable eigenvalue structure. Usually the only kind of transformation that is at all practical is one having $\mathbf{T}$ equal to a diagonal matrix, corresponding to the introduction of scale factors on each of the variables. One should strive, in doing this, to make the second derivatives with respect to each variable roughly the same. Although appropriate scaling can potentially lead to substantial payoff in terms of enhanced convergence rate, we largely ignore this possibility in our discussions of steepest descent. However, see the next application for a situation that frequently occurs.

Application 3 (Program Design). In applied work it is extremely rare that one solves just a single optimization problem of a given type. It is far more usual that once a problem is coded for computer solution, it will be solved repeatedly for various parameter values. Thus, for example, if one is seeking to find the optimal production plan (as in Example 2 of Sect. 7.2), the problem will be solved for the different values of the input prices. Similarly, other optimization problems will be solved under various assumptions and constraint values. It is for this reason that speed of convergence and convergence analysis is so important. One wants a program that can be used efficiently. In many such situations, the effort devoted to proper scaling repays itself, not with the first execution, but in the long run.

As a simple illustration consider the problem of minimizing the function

$$f(x) = x^2 - 5xy + y^4 - ax - by.$$

It is desirable to obtain solutions quickly for different values of the parameters a and b. We begin with the values $a = 25$, $b = 8$.

The result of steepest descent applied to this problem directly is shown in Table 8.2, column (a). It requires eighty iterations for convergence, which could be regarded as disappointing.

The reason for this poor performance is revealed by examining the Hessian matrix

$$\mathbf{F} = \begin{bmatrix} 2 & -5 \\ -5 & 12y^2 \end{bmatrix}$$

Using the results of our first experiment, we know that $y = 3$. Hence the diagonal elements of the Hessian, at the solution, differ by a factor of 54. (In fact, the condition number is about 61.) As a simple remedy we scale the problem by replacing the variable y by $z = ty$. The new lower right-corner term of the Hessian then becomes $12z^2/t^4$, which has magnitude $12 \times t^2 \times 3^2/t^4 = 108/t^2$. Thus we might put $t = 7$ in order to make the two diagonal terms approximately equal. The result of applying steepest descent to the problem scaled this way is shown in Table 8.2, column (b). (This superior performance is in accordance with our general theory, since the condition number of the scaled problem is about two.) For other nearby values of a and b, similar speeds will be attained.

Table 8.2 Solution to scaling application

Iteration no.	Value of f (a) Unscaled	(b) Scaled
0	0.0000	0.0000
1	−230.9958	−162.2000
2	−256.4042	−289.3124
4	−293.1705	−341.9802
6	−313.3619	−342.9865
8	−324.9978	−342.9998
9	−329.0408	−343.0000
15	−339.6124	
20	−341.9022	
25	−342.6004	
30	−342.8372	
35	−342.9275	
40	−342.9650	
45	−342.9825	
50	−342.9909	
55	−342.9951	
60	−342.9971	Solution
65	−342.9883	$x = 20.0$
70	−342.9990	$y = 3.0$
75	−342.9994	
80	−342.9997	

8.4 Accelerated Steepest Descent

There is an *accelerated* steepest descent method that works as follows:

$$\lambda^0 = 0,\ \lambda_{k+1} = \frac{1+\sqrt{1+4(\lambda_k)^2}}{2},\ \alpha_k = \frac{1-\lambda_k}{\lambda_{k+1}}, \tag{8.55}$$

$$\tilde{\mathbf{x}}_{k+1} = \mathbf{x}_k - \frac{1}{\beta}\nabla f(\mathbf{x}_k)^T, \quad \mathbf{x}_{k+1} = (1-\alpha_k)\tilde{\mathbf{x}}_{k+1} + \alpha_k\tilde{\mathbf{x}}_k. \tag{8.56}$$

Note that $(\lambda_k)^2 = \lambda_{k+1}(\lambda_{k+1} - 1)$, $\lambda_k > k/2$ and $\alpha_k \le 0$. One can prove:

Theorem (Accelerated Steepest Descent). *Let $f(\mathbf{x})$ be convex and differentiable everywhere, satisfies the (first-order) β-Lipschitz condition, and admits a minimizer $\mathbf{x}^*$. Then, the method of accelerated steepest descent generates a sequence of solutions such that*

$$f(\tilde{\mathbf{x}}_{k+1}) - f(\mathbf{x}^*) \le \frac{2\beta}{k^2}|\mathbf{x}^0 - \mathbf{x}^*|^2, \ \forall k \ge 1.$$

Proof. We now let $\mathbf{d}_k = \lambda_k \mathbf{x}_k - (\lambda_k - 1)\tilde{\mathbf{x}}_k - \mathbf{x}^*$, and $\delta_k = f(\tilde{\mathbf{x}}_k) - f(\mathbf{x}^*)(\ge 0)$.
Applying Lemma 1 for $\mathbf{x} = \tilde{\mathbf{x}}_{k+1}$ and $\mathbf{y} = \tilde{\mathbf{x}}_k$, convexity of f and (8.56), we have

$$\begin{aligned}\delta_{k+1} - \delta_k &= f(\tilde{\mathbf{x}}_{k+1}) - f(\mathbf{x}_k) + f(\mathbf{x}_k) - f(\tilde{\mathbf{x}}_k)\\ &\le -\tfrac{\beta}{2}|\tilde{\mathbf{x}}_{k+1} - \mathbf{x}_k|^2 + f(\mathbf{x}_k) - f(\tilde{\mathbf{x}}_k)\\ &\le -\tfrac{\beta}{2}|\tilde{\mathbf{x}}_{k+1} - \mathbf{x}_k|^2 + (\mathbf{g}_k)^T(\mathbf{x}_k - \tilde{\mathbf{x}}_k)\\ &= -\tfrac{\beta}{2}|\tilde{\mathbf{x}}_{k+1} - \mathbf{x}_k|^2 - \beta(\tilde{\mathbf{x}}_{k+1} - \mathbf{x}_k)^T(\mathbf{x}_k - \tilde{\mathbf{x}}_k).\end{aligned} \tag{8.57}$$

Applying Lemma 1 for $\mathbf{x} = \tilde{\mathbf{x}}_{k+1}$ and $\mathbf{y} = \mathbf{x}^*$, convexity of f and (8.56), we have

$$\begin{aligned}\delta_{k+1} &= f(\tilde{\mathbf{x}}_{k+1}) - f(\mathbf{x}_k) + f(\mathbf{x}_k) - f(\mathbf{x}^*)\\ &\le -\tfrac{\beta}{2}|\tilde{\mathbf{x}}_{k+1} - \mathbf{x}_k|^2 + f(\mathbf{x}_k) - f(\mathbf{x}^*)\\ &\le -\tfrac{\beta}{2}|\tilde{\mathbf{x}}_{k+1} - \mathbf{x}_k|^2 + (\mathbf{g}_k)^T(\mathbf{x}_k - \mathbf{x}^*)\\ &= -\tfrac{\beta}{2}|\tilde{\mathbf{x}}_{k+1} - \mathbf{x}_k|^2 - \beta(\tilde{\mathbf{x}}_{k+1} - \mathbf{x}_k)^T(\mathbf{x}_k - \mathbf{x}^*).\end{aligned} \tag{8.58}$$

Multiplying (8.57) by $\lambda_k(\lambda_k - 1)$ and (8.58) by λ_k respectively, and summing the two, we have

$$\begin{aligned}&(\lambda_k)^2\delta_{k+1} - (\lambda_{k-1})^2\delta_k\\ &\le -(\lambda_k)^2\tfrac{\beta}{2}|\tilde{\mathbf{x}}_{k+1} - \mathbf{x}_k|^2 - \lambda_k\beta(\tilde{\mathbf{x}}_{k+1} - \mathbf{x}_k)^T\mathbf{d}_k\\ &= -\tfrac{\beta}{2}((\lambda_k)^2|\tilde{\mathbf{x}}_{k+1} - \mathbf{x}_k|^2 + 2\lambda_k(\tilde{\mathbf{x}}_{k+1} - \mathbf{x}_k)^T\mathbf{d}_k)\\ &= -\tfrac{\beta}{2}(|\lambda_k\tilde{\mathbf{x}}_{k+1} - (\lambda_k - 1)\tilde{\mathbf{x}}_k - \mathbf{x}^*|^2 - |\mathbf{d}_k|^2)\\ &= \tfrac{\beta}{2}(|\mathbf{d}_k|^2 - |\lambda_k\tilde{\mathbf{x}}_{k+1} - (\lambda_k - 1)\tilde{\mathbf{x}}_k - \mathbf{x}^*|^2).\end{aligned}$$

Using (8.55) and (8.56) we derive

$$\lambda_k\tilde{\mathbf{x}}_{k+1} - (\lambda_k - 1)\tilde{\mathbf{x}}_k = \lambda_{k+1}\mathbf{x}_{k+1} - (\lambda_{k+1} - 1)\tilde{\mathbf{x}}_{k+1}.$$

Thus,

$$(\lambda_k)^2\delta_{k+1} - (\lambda_{k-1})^2\delta_k \le \frac{\beta}{2}(|\mathbf{d}_k|^2 - |\mathbf{d}_{k+1}|^2.) \tag{8.59}$$

Summing up (8.59) from 1 to k we have

$$\delta_{k+1} \le \frac{\beta}{2(\lambda_k)^2}|\mathbf{d}_1|^2 \le \frac{2\beta}{k^2}|\mathbf{d}_0|^2$$

where we used facts $\lambda_k \ge k/2$ and $|\mathbf{d}_1| \le |\mathbf{d}_0|$. ∎

The Method of False Position

Yet there is another steepest descent method, commonly called the BB method, that works as follows:

$$\Delta_k^x = \mathbf{x}_k - \mathbf{x}_{k-1} \quad \text{and} \quad \Delta_k^g = \nabla f(\mathbf{x}_k) - \nabla f(\mathbf{x}_{k-1}), \tag{8.60}$$

$$\alpha_k = \frac{(\Delta_k^x)^T \Delta_k^g}{(\Delta_k^g)^T \Delta_k^g} \quad \text{or} \quad \alpha_k = \frac{(\Delta_k^x)^T \Delta_k^x}{(\Delta_k^x)^T \Delta_k^g},$$

Then

$$\mathbf{x}_{k+1} = \mathbf{x}_k - \alpha_k \mathbf{g}_k = \mathbf{x}_k - \alpha_k \nabla f(\mathbf{x}_k)^T. \tag{8.61}$$

The step size of the BB method resembles the one used in quadratic curve fitting discussed for line search. There, the step size of (8.10) is given as $\frac{x_{k-1}-x_k}{f'(x_{k-1})-f'(x_k)}$. If we let $\delta_k^x = x_k - x_{k-1}$ and $\delta_k^g = f'(x_k) - f'(x_{k-1})$, this quantity can be written as $\frac{\delta_k^x \delta_k^g}{(\delta_k^g)^2}$ or $\frac{(\delta_k^x)^2}{\delta_k^x \delta_k^g}$. In the vector case, multiplication is replaced by inner product.

There was another explanation on the step size of the BB method. Consider convex quadratic minimization, and let the distinct positive eigenvalues of the Hessian $\mathbf{Q}$ be $\lambda_1, \lambda_2, \ldots \lambda_K$. Then, if we let the step size in the method of steepest descent be $\alpha_k = \frac{1}{\lambda_k}$, $k = 1, \ldots, K$, the method terminates in K iterations (which we leave as an exercise). In the BB method, α_k minimizes

$$|\Delta_k^x - \alpha \Delta_k^g| = |\Delta_k^x - \alpha \mathbf{Q} \Delta_k^x|.$$

If the error becomes 0 plus $|\Delta_k^x| \neq 0$, $\frac{1}{\alpha_k}$ will be a positive eigenvalue of $\mathbf{Q}$. Notice that the objective values of the iterates generated by the BB method is not monotonically decreasing; the method may overshoot in order to have a better position in the long run.

8.5 Newton's Method

The idea behind Newton's method is that the function f being minimized is approximated locally by a quadratic function, and this approximate function is minimized exactly. Thus near $\mathbf{x}_k$ we can approximate f by the truncated Taylor series

$$f(\mathbf{x}) \simeq f(\mathbf{x}_k) + \nabla f(\mathbf{x}_k)(\mathbf{x} - \mathbf{x}_k) + \frac{1}{2}(\mathbf{x} - \mathbf{x}_k)^T \mathbf{F}(\mathbf{x}_k)(\mathbf{x} - \mathbf{x}_k).$$

The right-hand side is minimized at

$$\mathbf{x}_{k+1} = \mathbf{x}_k - [\mathbf{F}(\mathbf{x}_k)]^{-1} \nabla f(\mathbf{x}_k)^T, \tag{8.62}$$

and this equation is the pure form of Newton's method.

In view of the second-order sufficiency conditions for a minimum point, we assume that at a relative minimum point, $\mathbf{x}^*$, the Hessian matrix, $\mathbf{F}(\mathbf{x}^*)$, is positive definite. We can then argue that if f has continuous second partial derivatives, $\mathbf{F}(\mathbf{x})$ is positive definite near $\mathbf{x}^*$ and hence the method is well defined near the solution.

Order Two Convergence

Newton's method has very desirable properties if started sufficiently close to the solution point. Its order of convergence is two.

> ***Theorem (Newton's Method).*** *Let $f \in C^3$ on E^n, and assume that at the local minimum point $\mathbf{x}^*$, the Hessian $\mathbf{F}(\mathbf{x}^*)$ is positive definite. Then if started sufficiently close to $\mathbf{x}^*$, the points generated by Newton's method converge to $\mathbf{x}^*$. The order of convergence is at least two.*

Proof. There are $\rho > 0$, $\beta_1 > 0$, $\beta_2 > 0$ such that for all $\mathbf{x}$ with $|\mathbf{x} - \mathbf{x}^*| < \rho$, there holds $|\mathbf{F}(\mathbf{x})^{-1}| < \beta_1$ (see Appendix A for the definition of the norm of a matrix) and $|\nabla f(\mathbf{x}^*)^T - \nabla f(\mathbf{x})^T - \mathbf{F}(\mathbf{x})(\mathbf{x}^* - \mathbf{x})| \leqslant \beta_2 |\mathbf{x} - \mathbf{x}^*|^2$. Now suppose $\mathbf{x}_k$ is selected with $\beta_1\beta_2|\mathbf{x}_k - \mathbf{x}^*| < 1$ and $|\mathbf{x}_k - \mathbf{x}^*| < \rho$. Then

$$\begin{aligned}
|\mathbf{x}_{k+1} - \mathbf{x}^*| &= |\mathbf{x}_k - \mathbf{x}^* - \mathbf{F}(\mathbf{x}_k)^{-1}\nabla f(\mathbf{x}_k)^T| \\
&= |\mathbf{F}(\mathbf{x}_k)^{-1}[\nabla f(\mathbf{x}^*)^T - \nabla f(\mathbf{x}_k)^T - \mathbf{F}(\mathbf{x}_k)(\mathbf{x}^* - \mathbf{x}_k)]| \\
&\leqslant |\mathbf{F}(\mathbf{x}_k)^{-1}|\beta_2|\mathbf{x}_k - \mathbf{x}^*|^2 \\
&\leqslant \beta_1\beta_2|\mathbf{x}_k - \mathbf{x}^*|^2 < |\mathbf{x}_k - \mathbf{x}^*|.
\end{aligned}$$

The final inequality shows that the new point is closer to $\mathbf{x}^*$ than the old point, and hence all conditions apply again to $\mathbf{x}_{k+1}$. The previous inequality establishes that convergence is second order. ∎

Modifications

Although Newton's method is very attractive in terms of its convergence properties near the solution, it requires modification before it can be used at points that are remote from the solution. The general nature of these modifications is discussed in the remainder of this section.

1. **Damping.** The first modification is that usually a search parameter α is introduced so that the method takes the form

$$\mathbf{x}_{k+1} = \mathbf{x}_k - \alpha_k[\mathbf{F}(\mathbf{x}_k)]^{-1}\nabla f(\mathbf{x}_k)^T,$$

where α_k is selected to minimize f. Near the solution we expect, on the basis of how Newton's method was derived, that $\alpha_k \simeq 1$. Introducing the parameter for general points, however, guards against the possibility that the objective might increase with $\alpha_k = 1$, due to nonquadratic terms in the objective function.

2. **Positive Definiteness.** A basic consideration for Newton's method can be seen most clearly by a brief examination of the general class of algorithms

$$\mathbf{x}_{k+1} = \mathbf{x}_k - \alpha \mathbf{M}_k \mathbf{g}_k, \tag{8.63}$$

where $\mathbf{M}_k$ is an $n \times n$ matrix, α is a positive search parameter, and $\mathbf{g}_k = \nabla f(\mathbf{x}_k)^T$. We note that both steepest descent ($\mathbf{M}_k = \mathbf{I}$) and Newton's method ($\mathbf{M}_k = [\mathbf{F}(\mathbf{x}_k)]^{-1}$) belong to this class. The direction vector $\mathbf{d}_k = -\mathbf{M}_k \mathbf{g}_k$ obtained in this way is a direction of descent if for small α the value of f decreases as α increases from zero. For small α we can say

$$f(\mathbf{x}_{k+1}) = f(\mathbf{x}_k) + \nabla f(\mathbf{x}_k)(\mathbf{x}_{k+1} - \mathbf{x}_k) + O(|\mathbf{x}_{k+1} - \mathbf{x}_k|^2).$$

Employing (8.51) this can be written as

$$f(\mathbf{x}_{k+1}) = f(\mathbf{x}_k) - \alpha \mathbf{g}_k^T \mathbf{M}_k \mathbf{g}_k + O(\alpha^2).$$

As $\alpha \to 0$, the second term on the right dominates the third. Hence if one is to guarantee a decrease in f for small α, we must have $\mathbf{g}_k^T \mathbf{M}_k \mathbf{g}_k > 0$. The simplest way to insure this is to require that $\mathbf{M}_k$ be positive definite.

The best circumstance is that where $\mathbf{F}(\mathbf{x})$ is itself positive definite throughout the search region. The objective function of many important optimization problems have this property, including for example interior-point approaches to linear programming using the logarithm as a barrier function. Indeed, it can be argued that convexity is an inherent property of the majority of well-formulated optimization problems.

Therefore, assume that the Hessian matrix $\mathbf{F}(\mathbf{x})$ is positive definite throughout the search region and that f has continuous third derivatives. At a given $\mathbf{x}_k$ define the symmetric matrix $\mathbf{T} = \mathbf{F}(\mathbf{x}_k)^{-1/2}$. As in Sect. 8.3 introduce the change of variable $\mathbf{Ty} = \mathbf{x}$. Then according to (8.48) a steepest descent direction with respect to $\mathbf{y}$ is equivalent to a direction with respect to $\mathbf{x}$ of $\mathbf{d} = -\mathbf{T}\mathbf{T}^T \mathbf{g}(\mathbf{x}_k)$, where $\mathbf{g}(\mathbf{x}_k)$ is the gradient of f with respect to $\mathbf{x}$ at $\mathbf{x}_k$. Thus, $\mathbf{d} = \mathbf{F}^{-1}\mathbf{g}(\mathbf{x}_k)$. In other words, a steepest descent direction in $\mathbf{y}$ is equivalent to a Newton direction in $\mathbf{x}$.

We can turn this relation around to analyze Newton steps in $\mathbf{x}$ as equivalent to gradient steps in $\mathbf{y}$. We know that convergence properties in $\mathbf{y}$ depend on the bounds on the Hessian matrix given by (8.49) as

$$\mathbf{H}(\mathbf{y}) = \mathbf{T}^T \mathbf{F}(\mathbf{x}) \mathbf{T} = \mathbf{F}^{-1/2} \mathbf{F}(\mathbf{x}) \mathbf{F}^{-1/2}. \tag{8.64}$$

Recall that $\mathbf{F} = \mathbf{F}(\mathbf{x}_k)$ which is fixed, whereas $\mathbf{F}(\mathbf{x})$ denotes the general Hessian matrix with respect to $\mathbf{x}$ near $\mathbf{x}_k$. The product (8.64) is the identity matrix at $\mathbf{y}_k$

but the rate of convergence of steepest descent in $\mathbf{y}$ depends on the bounds of the smallest and largest eigenvalues of $\mathbf{H}(\mathbf{y})$ in a region near $\mathbf{y}_k$.

These observations tell us that the damped method of Newton's method will converge at a linear rate at least as fast as $c = (1 - a/A)$ where a and A are lower and upper bounds on the eigenvalues of $\mathbf{F}(\mathbf{x}_0)^{-1/2}\mathbf{F}(\mathbf{x}^0)\mathbf{F}(\mathbf{x}_0)^{-1/2}$, where $\mathbf{x}_0$ and $\mathbf{x}^0$ are arbitrary points in the local search region. These bounds depend, in turn, on the bounds of the third-order derivatives of f. It is clear, however, by continuity of $\mathbf{F}(\mathbf{x})$ and its derivatives, that the rate becomes very fast near the solution, becoming superlinear, and in fact, as we know, quadratic.

3. **Backtracking.** The backtracking method of line search, using $\alpha = 1$ as the initial guess, is an attractive procedure for use with Newton's method. Using this method the overall progress of Newton's method divides naturally into two phases: first a damping phase where backtracking may require $\alpha < 1$, and second a quadratic phase where $\alpha = 1$ satisfies the backtracking criterion at every step. The damping phase was discussed above.

Let us now examine the situation when close to the solution. We assume that all derivatives of f through the third are continuous and uniformly bounded. We also assume that in the region close to the solution, $\mathbf{F}(\mathbf{x})$ is positive definite with $\overline{a} > 0$ and $\overline{A} > 0$ being, respectively, uniform lower and upper bounds on the eigenvalues of $\mathbf{F}(\mathbf{x})$. Using $\alpha = 1$ and $\varepsilon < 0.5$ we have for $\mathbf{d}_k = -\mathbf{F}(\mathbf{x}_k)^{-1}\mathbf{g}(\mathbf{x}_k)$

$$\begin{aligned} f(\mathbf{x}_k + \mathbf{d}_k) &= f(\mathbf{x}_k) - \mathbf{g}(\mathbf{x}_k)^T\mathbf{F}(\mathbf{x}_k)^{-1}\mathbf{g}(\mathbf{x}_k) + \frac{1}{2}\mathbf{g}(\mathbf{x}_k)^T\mathbf{F}(\mathbf{x}_k)^{-1}\mathbf{g}(\mathbf{x}_k) + o(|\mathbf{g}(\mathbf{x}_k)|^2) \\ &= f(\mathbf{x}_k) - \frac{1}{2}\mathbf{g}(\mathbf{x}_k)^T\mathbf{F}(\mathbf{x}_k)^{-1}\mathbf{g}(\mathbf{x}_k) + o(|\mathbf{g}(\mathbf{x}_k)|^2) \\ &< f(\mathbf{x}_k) - \varepsilon\mathbf{g}(\mathbf{x}_k)^T\mathbf{F}(\mathbf{x}_k)^{-1}\mathbf{g}(\mathbf{x}_k) + o(|\mathbf{g}(\mathbf{x}_k)|^2), \end{aligned}$$

where the o bound is uniform for all $\mathbf{x}_k$. Since $|\mathbf{g}(\mathbf{x}_k)| \to 0$ (uniformly) as $\mathbf{x}_k \to \mathbf{x}^*$, it follows that once $\mathbf{x}_k$ is sufficiently close to $\mathbf{x}^*$, then $f(\mathbf{x}_k+\mathbf{d}_k) < f(\mathbf{x}_k)-\varepsilon\mathbf{g}(\mathbf{x}_k)^T\mathbf{d}_k$ and hence the backtracking test (the first part of Amijo's rule) is satisfied. This means that $\alpha = 1$ will be used throughout the final phase.

4. **General Problems.** In practice, Newton's method must be modified to accommodate the possible nonpositive definiteness at regions remote from the solution.

A common approach is to take $\mathbf{M}_k = [\varepsilon_k\mathbf{I} + \mathbf{F}(\mathbf{x}_k)]^{-1}$ for some non-negative value of ε_k. This can be regarded as a kind of compromise between steepest descent (ε_k very large) and Newton's method ($\varepsilon_k = 0$). There is always an ε_k that makes $\mathbf{M}_k$ positive definite. We shall present one modification of this type.

Let $\mathbf{F}_k \equiv \mathbf{F}(\mathbf{x}_k)$. Fix a constant $\delta > 0$. Given $\mathbf{x}_k$, calculate the eigenvalues of $\mathbf{F}_k$ and let ε_k be the smallest nonnegative constant for which the matrix $\varepsilon_k\mathbf{I} + \mathbf{F}_k$ has eigenvalues greater than or equal to δ. Then define

$$\mathbf{d}_k = -(\varepsilon_k\mathbf{I} + \mathbf{F}_k)^{-1}\mathbf{g}_k \tag{8.65}$$

and iterate according to

$$\mathbf{x}_{k+1} = \mathbf{x}_k + \alpha_k \mathbf{d}_k, \tag{8.66}$$

where α_k minimizes $f(\mathbf{x}_k + \alpha \mathbf{d}_k)$, $\alpha \geqslant 0$.

This algorithm has the desired global and local properties. First, since the eigenvalues of a matrix depend continuously on its elements, ε_k is a continuous function of $\mathbf{x}_k$ and hence the mapping $\mathbf{D} : E^n \to E^{2n}$ defined by $\mathbf{D}(\mathbf{x}_k) = (\mathbf{x}_k, \mathbf{d}_k)$ is continuous. Thus the algorithm $\mathbf{A} = \mathbf{SD}$ is closed at points outside the solution set $\Omega = \{\mathbf{x} : \nabla f(\mathbf{x}) = \mathbf{0}\}$. Second, since $\varepsilon_k \mathbf{I} + \mathbf{F}_k$ is positive definite, $\mathbf{d}_k$ is a descent direction and thus $Z(\mathbf{x}) \equiv f(\mathbf{x})$ is a continuous descent function for A. Therefore, assuming the generated sequence is bounded, the Global Convergence Theorem applies. Furthermore, if $\delta > 0$ is smaller than the smallest eigenvalue of $\mathbf{F}(\mathbf{x}^*)$, then for $\mathbf{x}_k$ sufficiently close to $\mathbf{x}^*$ we will have $\varepsilon_k = 0$, and the method reduces to Newton's method. Thus this revised method also has order of convergence equal to two.

The selection of an appropriate δ is somewhat of an art. A small δ means that nearly singular matrices must be inverted, while a large δ means that the order two convergence may be lost. Experimentation and familiarity with a given class of problems are often required to find the best δ.

The utility of the above algorithm is hampered by the necessity to calculate the eigenvalues of $\mathbf{F}(\mathbf{x}_k)$, and in practice an alternate procedure is used. In one class of methods (Levenberg–Marquardt type methods), for a given value of ε_k, Cholesky factorization of the form $\varepsilon_\mathbf{k}\mathbf{I} + \mathbf{F}(\mathbf{x}_\mathbf{k}) = \mathbf{G}\mathbf{G}^T$ (see Exercise 6 of Chap. 7) is employed to check for positive definiteness. If the factorization breaks down, ε_k is increased. The factorization then also provides the direction vector through solution of the equations $\mathbf{G}\mathbf{G}^T\mathbf{d}_k = \mathbf{g}_k$, which are easily solved, since $\mathbf{G}$ is triangular. Then the value $f(\mathbf{x}_k + \mathbf{d}_k)$ is examined. If it is sufficiently below $f(\mathbf{x}_k)$, then $\mathbf{x}_{k+1}$ is accepted and a new ε_{k+1} is determined. Essentially, ε serves as a search parameter in these methods. It should be clear from this discussion that the simplicity that Newton's method first seemed to promise is not fully realized in practice.

Newton's Method and Logarithms

Interior point methods of linear and nonlinear programming use barrier functions, which usually are based on the logarithm. For linear programming especially, this means that the only nonlinear terms are logarithms. Newton's method enjoys some special properties in this case,

To illustrate, let us apply Newton's method to the one-dimensional problem

$$\min_x [tx - \ln x] \tag{8.67}$$

where t is a positive parameter. The derivative at x is

$$f'(x) = t - \frac{1}{x},$$

and of course the solution is $x^* = 1/t$, or equivalently $1 - tx^* = 0$. The second derivative is $f''(x) = 1/x^2$. Denoting by x^+ the result of one step of a pure Newton's method (with step length equal to 1) applied to the point x, we find

$$\begin{aligned} x^+ &= x - [f''(x)]^{-1} f'(x) = x - x^2\left(t - \frac{1}{x}\right) = x - tx^2 + x \\ &= 2x - tx^2. \end{aligned}$$

Thus

$$1 - tx^+ = 1 - 2tx + x^2t^2 = (1 - tx)^2 \tag{8.68}$$

Therefore, rather surprisingly, the quadratic nature of convergence of $(1 - tx) \to 0$ is directly evident and exact. Expression (8.68) represents a reduction in the error magnitude only if $|(1 - tx)| < 1$, or equivalently, $0 < x < 2/t$. If x is too large, then Newton's method must be used with damping until the region $0 < x < 2/t$ is reached. From then on, a step size of 1 will exhibit pure quadratic error reduction.

The situation is shown in Fig. 8.11. The graph is that of $f'(x) = t - 1/x$. The root-finding form of Newton's method (Sect. 8.1) is then applied to this function. At each point, the tangent line is followed to the x axis to find the new point. The starting value marked x_1 is far from the solution $1/t$ and hence following the tangent would lead to a new point that was negative. Damping must be applied at that starting point. Once a point x is reached with $0 < x < 1/t$, all further points will remain to the left of $1/t$ and move toward it quadratically.

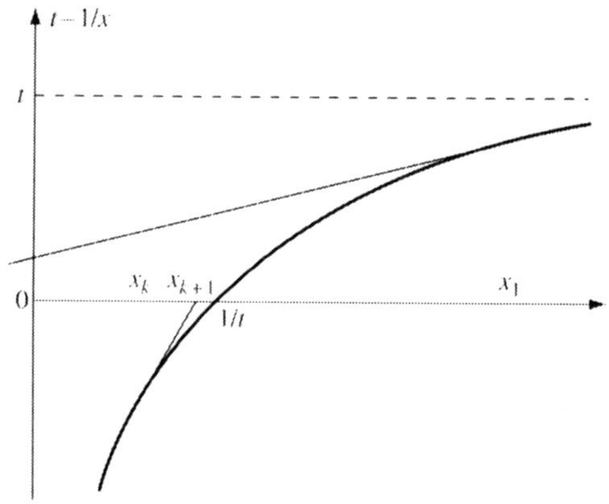

Fig. 8.11 Newton's method applied to minimization of $tx - \ln x$

In interior point methods for linear programming, a logarithmic barrier function is applied separately to the variables that must remain positive. The convergence analysis in these situations is an extension of that for the simple case given here, allowing for estimates of the rate of convergence that do not require knowledge of bounds of third-order derivatives.

Self-concordant Functions

The special properties exhibited above for the logarithm have been extended to the general class of *self-concordant functions* of which the logarithm is the primary example. A function f defined on the real line is self-concordant if it satisfies

$$|f'''(x)| \leq 2f''(x)^{3/2}, \tag{8.69}$$

throughout its domain. It is easily verified that $f(x) = -\ln x$ satisfies this inequality with equality for $x > 0$.

Self-concordancy is preserved by the addition of an affine term since such a term does not affect the second or third derivatives.

A function defined on E^n is said to be *self-concordant* if it is self-concordant in every direction: that is if $f(\mathbf{x} + \alpha\mathbf{d})$ is self-concordant with respect to α for every $\mathbf{d}$ throughout the domain of f.

Self-concordant functions can be combined by addition and even by composition with affine functions to yield other self-concordant functions. (See Exercise 29.) For example the function

$$f(\mathbf{x}) = -\sum_{i=1}^{m} \ln(b_j - \mathbf{a}_i^T \mathbf{x}),$$

often used in interior point methods for linear programming, is self-concordant.

When a self-concordant function is subjected to Newton's method, the quadratic convergence of final phase can be measured in terms of the function

$$\lambda(\mathbf{x}) = [\nabla f(\mathbf{x})\mathbf{F}(\mathbf{x})^{-1}\nabla f(\mathbf{x})^T]^{1/2},$$

where as usual $\mathbf{F}(\mathbf{x})$ is the Hessian matrix of f at $\mathbf{x}$. Then it can be shown that close to the solution

$$2\lambda(\mathbf{x}_{k+1}) \leq [2\lambda(\mathbf{x}_k)]^2. \tag{8.70}$$

Furthermore, in a backtracking procedure, estimates of both the stepwise progress in the damping phase and the point at which the quadratic phase begins can be expressed in terms of parameters that depend only on the backtracking parameters. Although, this knowledge does not generally influence practice, it is theoretically quite interesting.

Example 1 (The Logarithmic Case). Consider the earlier example of $f(x) = tx - \ln x$. There

$$\lambda(x) = [f'(x)^2/f''(x)]^{\frac{1}{2}} = |(t - 1/x)x| = |1 - tx|.$$

Then (8.70) gives

$$(1 - tx^+) \leq 2(1 - tx)^2.$$

Actually, for this example, as we found in (8.68), the factor of 2 is not required.

There is a relation between the analysis of self-concordant functions and our earlier convergence analysis.

Recall that one way to analyze Newton's method is to change variables from $\mathbf{x}$ to $\mathbf{y}$ according to $\tilde{\mathbf{y}} = [\mathbf{F}(\mathbf{x})]^{-(1/2)}\tilde{\mathbf{x}}$, where here $\mathbf{x}$ is a reference point and $\tilde{\mathbf{x}}$ is variable. The gradient with respect to $\mathbf{y}$ at $\tilde{\mathbf{y}}$ is then $\mathbf{F}(\mathbf{x})^{-(1/2)}\nabla f(\tilde{\mathbf{x}})$, and hence the norm of the gradient at $\mathbf{y}$ is $[\nabla f(\mathbf{x})\mathbf{F}(\mathbf{x})^{-1}\nabla f(\mathbf{x})^T]^{(1/2)} \equiv \lambda(\mathbf{x})$. Hence it is perhaps not surprising that $\lambda(\mathbf{x})$ plays a role analogous to the role played by the norm of the gradient in the analysis of steepest descent.

8.6 Coordinate Descent Methods

The algorithms discussed in this section are sometimes attractive because of their easy implementation. Generally, however, their convergence properties are poorer than steepest descent.

Let f be a function on E^n having continuous first partial derivatives. Given a point $\mathbf{x} = (x_1, x_2, \ldots, x_n)$, descent with respect to the coordinate x_i (i fixed) means that one solves

$$\underset{x_i}{\text{minimize}} \ f(x_1, x_2, \ldots, x_n).$$

Thus only changes in the single component x_i are allowed in seeking a new and better vector $\mathbf{x}$ (one can also consider $\mathbf{x}_i$ the ith block of variables, called the block-coordinate method). In our general terminology, each such descent can be regarded as a descent in the direction $\mathbf{e}_i$(or $-\mathbf{e}_i$) where $\mathbf{e}_i$ is the ith unit vector. By sequentially minimizing with respect to different components, a relative minimum of f might ultimately be determined.

There are a number of ways that this concept can be developed into a full algorithm. The *cyclic coordinate descent* algorithm minimizes f cyclically with respect to the coordinate variables. Thus x_1 is changed first, then x_2 and so forth through x_n. The process is then repeated starting with x_1 again. A variation of this is the *Aitken double sweep method*. In this procedure one searches over $x_1, x_2, \ldots, x_n$, in that order, and then comes back in the order $x_{n-1}, x_{n-2}, \ldots, x_1$. These cyclic methods have the advantage of not requiring any information about ∇f to determine the descent directions.

If the gradient of f is available, then it is possible to select the order of descent coordinates on the basis of the gradient. A popular technique is the *Gauss–Southwell Method* where at each stage the coordinate corresponding to the largest (in absolute value) component of the gradient vector is selected for descent. A *randomized strategy* can be also adapted in which one randomly chooses a coordinate to optimize in each step; see more discussions later.

Global Convergence

It is simple to prove global convergence for cyclic coordinate descent. The algorithmic map $\mathbf{A}$ is the composition of $2n$ maps

$$\mathbf{A} = \mathbf{SC}^n\mathbf{SC}^{n-1}\ldots\mathbf{SC}^1,$$

where $\mathbf{C}^i(\mathbf{x}) = (\mathbf{x}, \mathbf{e}_i)$ with $\mathbf{e}_i$ equal to the ith unit vector, and $\mathbf{S}$ is the usual line search algorithm but over the doubly infinite line rather than the semi-infinite line. The map $\mathbf{C}^i$ is obviously continuous and $\mathbf{S}$ is closed. If we assume that points are restricted to a compact set, then $\mathbf{A}$ is closed by Corollary 1, Sect. 7.7. We define the solution set $\Gamma = \{\mathbf{x} : \nabla f(\mathbf{x}) = \mathbf{0}\}$. If we impose the mild assumption on f that a search along any coordinate direction yields a unique minimum point, then the function $Z(\mathbf{x}) \equiv f(\mathbf{x})$ serves as a continuous descent function for $\mathbf{A}$ with respect to Γ. This is because a search along any coordinate direction either must yield a decrease or, by the uniqueness assumption, it cannot change position. Therefore, if at a point $\mathbf{x}$ we have $\nabla f(\mathbf{x}) \neq \mathbf{0}$, then at least one component of $\nabla f(\mathbf{x})$ does not vanish and a search along the corresponding coordinate direction must yield a decrease.

Local Convergence Rate

It is difficult to compare the rates of convergence of these algorithms with the rates of others that we analyze. This is partly because coordinate descent algorithms are from an entirely different general class of algorithms than, for example, steepest descent and Newton's method, since coordinate descent algorithms are unaffected by (diagonal) scale factor changes but are affected by rotation of coordinates—the opposite being true for steepest descent. Nevertheless, some comparison is possible.

It can be shown (see Exercise 20) that for the same quadratic problem as treated in Sect. 8.2, there holds for the Gauss–Southwell method

$$E(\mathbf{x}_{k+1}) \leqslant \left(1 - \frac{a}{A(n-1)}\right) E(\mathbf{x}_k), \tag{8.71}$$

where a, A are as in Sect. 8.2 and n is the dimension of the problem. Since

$$\left(\frac{A-a}{A+a}\right)^2 \leqslant \left(1 - \frac{a}{A}\right) \leqslant \left(1 - \frac{a}{A(n-1)}\right)^{n-1}, \tag{8.72}$$

we see that the bound we have for steepest descent is better than the bound we have for $n-1$ applications of the Gauss–Southwell scheme. Hence we might argue that it takes essentially $n-1$ coordinate searches to be as effective as a single gradient search. This is admittedly a crude guess, since (8.54) is generally not a tight bound, but the overall conclusion is consistent with the results of many experiments. Indeed, unless the variables of a problem are essentially uncoupled from each other (corresponding to a nearly diagonal Hessian matrix) coordinate descent methods seem to require about n line searches to equal the effect of one step of steepest descent.

The above discussion again illustrates the general objective that we seek in convergence analysis. By comparing the formula giving the rate of convergence for steepest descent with a bound for coordinate descent, we are able to draw some general conclusions on the relative performance of the two methods that are not dependent on specific values of a and A. Our analyses of local convergence properties, which usually involve specific formulae, are always guided by this objective of obtaining general qualitative comparisons.

Example. The quadratic problem considered in Sect. 8.2 with

$$\mathbf{Q} = \begin{bmatrix} 0.78 & -0.02 & -0.12 & -0.14 \\ -0.02 & 0.86 & -0.04 & 0.06 \\ -0.12 & -0.04 & 0.72 & -0.08 \\ -0.14 & 0.06 & -0.08 & 0.74 \end{bmatrix}$$

$$\mathbf{b} = (0.76, 0.08, 1.12, 0.68)$$

was solved by the various coordinate search methods. The corresponding values of the objective function are shown in Table 8.3. Observe that the convergence rates of the three coordinate search methods are approximately equal but that they all converge about three times slower than steepest descent. This is in accord with the estimate given above for the Gauss–Southwell method, since in this case $n - 1 = 3$.

Convergence Speed of a Randomized Coordinate Descent Method

We now describe a *randomized strategy* in selecting x_i in each step of the coordinate descent method for f that is differentiable and Lipschitz continuous; that is, there exist some constants $\beta_i > 0$, $i = 1, \ldots, n$, such that

$$|\nabla_i f(\mathbf{x} + h\mathbf{e}_i) - \nabla_i f(\mathbf{x})| \le \beta_i |h|, \quad \forall\, h \in E,\ \mathbf{x} \in E^n, \tag{8.73}$$

where $\nabla_i f(\mathbf{x})$ denotes the ith partial derivative of f at $\mathbf{x}$, and $\mathbf{e}_i$ is the ith unit vector with the ith entry equal 1 and everywhere else equal 0.

Randomized coordinate decent method. Given an initial point $\mathbf{x}_0$; repeat for $k = 0, 1, 2, \ldots$

1. Choose $i_k \in \{1, \ldots, n\}$ randomly with a uniform distribution.
2. Update $\mathbf{x}_{k+1} = \mathbf{x}_k - \frac{1}{\beta_{i_k}} \nabla_{i_k} f(\mathbf{x}_k)\mathbf{e}_{i_k}$.

Note that after k iterations, the randomized coordinate descent method generates a random sequence of $\mathbf{x}_k$, which depends on the observed realization of the random variable

$$\xi_{k-1} = \{i_0, i_1, \ldots, i_{k-1}\}.$$

Table 8.3 Solutions to Example

Iteration no.	Value of f for various methods: Gauss-Southwell	Cyclic	Double sweep
0	0.0	0.0	0.0
1	−0.871111	−0.370256	−0.370256
2	−1.445584	−0.376011	−0.376011
3	−2.087054	−1.446460	−1.446460
4	−2.130796	−2.052949	−2.052949
5	−2.163586	−2.149690	−2.060234
6	−2.170272	−2.149693	−2.060237
7	−2.172786	−2.167983	−2.165641
8	−2.174279	−2.173169	−2.165704
9	−2.174583	−2.174392	−2.168440
10	−2.174638	−2.174397	−2.173981
11	−2.174651	−2.174582	−2.174048
12	−2.174655	−2.174643	−2.174054
13	−2.174658	−2.174656	−2.174608
14	−2.174659	−2.174656	−2.174608
15	−2.174659	−2.174658	−2.174622
16		−2.174659	−2.174655
17		−2.174659	−2.174656
18			−2.174656
19			−2.174659
20			−2.174659

Theorem 3 (Randomized Coordinate Descent—Lipschitz Convex Case). *Let $f(x)$ be convex and differentiable everywhere, satisfy the Lipschitz condition (8.73), and admit a minimizer $\mathbf{x}^*$. Then, the randomized coordinate decent method generates a sequence of solutions $\mathbf{x}_k$ such that for any $k \geq 1$, the iterate $\mathbf{x}_k$ satisfies*

$$\mathrm{E}_{\xi_{k-1}}[f(\mathbf{x}_k)] - f(\mathbf{x}^*) \leq \frac{n}{n+k}\left(\frac{1}{2}|\mathbf{x}_0 - \mathbf{x}^*|_\beta^2 + f(\mathbf{x}_0) - f(\mathbf{x}^*)\right),$$

where $|\mathbf{x}|_\beta = \left(\sum_i \beta_i x_i^2\right)^{1/2}$ *for all* $\mathbf{x} \in E^n$.

Proof. Let $r_k^2 = |\mathbf{x}_k - \mathbf{x}^*|_\beta^2 = \sum_{i=1}^n \beta_i((\mathbf{x}_k)_i - x_i^*)^2$ for any $k \geq 0$. Since $\mathbf{x}_{k+1} = \mathbf{x}_k - \frac{1}{\beta_{i_k}}\nabla_{i_k} f(\mathbf{x}_k)\mathbf{e}_{i_k}$, we have

$$r_{k+1}^2 = r_k^2 - 2\nabla_{i_k} f(\mathbf{x}_k)((\mathbf{x}_k)_{i_k} - x_{i_k}^*) + \frac{1}{\beta_{i_k}}(\nabla_{i_k} f(\mathbf{x}_k))^2.$$

It follows from (8.73), Lemma 1, and $\mathbf{x}_{k+1} = \mathbf{x}_k - \frac{1}{\beta_{i_k}}\nabla_{i_k} f(\mathbf{x}_k)\mathbf{e}_{i_k}$ that

$$\begin{aligned} f(\mathbf{x}_{k+1}) &\leq f(\mathbf{x}_k) + \nabla_{i_k} f(\mathbf{x}_k)((\mathbf{x}_{k+1})_{i_k} - (\mathbf{x}_k)_{i_k}) + \frac{\beta_{i_k}}{2}((\mathbf{x}_{k+1})_{i_k} - (\mathbf{x}_k)_{i_k})^2 \\ &= f(\mathbf{x}_k) - \frac{1}{2\beta_{i_k}}(\nabla_{i_k} f(\mathbf{x}_k))^2. \end{aligned} \quad (8.74)$$

Combining the above two relations, one has

$$r_{k+1}^2 \le r_k^2 - 2\nabla_{i_k} f(\mathbf{x}_k)((\mathbf{x}_k)_{i_k} - \mathbf{x}_{i_k}^*) + 2(f(\mathbf{x}_k) - f(\mathbf{x}_{k+1})).$$

Multiplying both sides by $1/2$ and taking expectation with respect to i_k yields

$$\mathrm{E}_{i_k}\left[\frac{1}{2}r_{k+1}^2\right] \le \frac{1}{2}r_k^2 - \frac{1}{n}\nabla f(\mathbf{x}_k)(\mathbf{x}_k - \mathbf{x}^*) + f(\mathbf{x}_k) - \mathrm{E}_{i_k}\left[f(\mathbf{x}_{k+1})\right],$$

which together with the fact that $\nabla f(\mathbf{x}_k)(\mathbf{x}^* - \mathbf{x}_k) \le f(\mathbf{x}^*) - f(\mathbf{x}_k)$ yields

$$\mathrm{E}_{i_k}\left[\frac{1}{2}r_{k+1}^2\right] \le \frac{1}{2}r_k^2 + \frac{1}{n}f(\mathbf{x}^*) + \frac{n-1}{n}f(\mathbf{x}_k) - \mathrm{E}_{i_k}\left[f(\mathbf{x}_{k+1})\right].$$

By rearranging terms, we obtain that for each $k \ge 0$,

$$\mathrm{E}_{i_k}\left[\frac{1}{2}r_{k+1}^2 + f(\mathbf{x}_{k+1}) - f(\mathbf{x}^*)\right] \le \left(\frac{1}{2}r_k^2 + f(\mathbf{x}_k) - f(\mathbf{x}^*)\right) - \frac{1}{n}\left(f(\mathbf{x}_k) - f(\mathbf{x}^*)\right).$$

Let $f^* = f(\mathbf{x}^*)$. Then, taking expectation with respect to ξ_{k-1} on both sides of the above relation, we have

$$\mathrm{E}_{\xi_k}\left[\frac{1}{2}r_{k+1}^2 + f(\mathbf{x}_{k+1}) - f^*\right] \le \mathrm{E}_{\xi_{k-1}}\left[\frac{1}{2}r_k^2 + f(\mathbf{x}_k) - f^*\right] - \frac{\mathrm{E}_{\xi_{k-1}}\left[f(\mathbf{x}_k) - f^*\right]}{n}. \quad (8.75)$$

In addition, it follows from (8.74) that $\mathrm{E}_{\xi_j}[f(\mathbf{x}_{j+1})] \le \mathrm{E}_{\xi_{j-1}}[f(\mathbf{x}_j)]$ for all $j \ge 0$. Using this relation and applying the inequality (8.75) recursively, we further obtain that

$$\begin{aligned}
\mathrm{E}_{\xi_k}\left[f(\mathbf{x}_{k+1})\right] - f^* &\le \mathrm{E}_{\xi_k}\left[\frac{1}{2}r_{k+1}^2 + f(\mathbf{x}_{k+1}) - f^*\right] \\
&\le \frac{1}{2}r_0^2 + f(\mathbf{x}_0) - f^* - \frac{1}{n}\sum_{j=0}^{k}\left(\mathrm{E}_{\xi_{j-1}}\left[f(\mathbf{x}_j)\right] - f^*\right) \\
&\le \frac{1}{2}r_0^2 + f(\mathbf{x}_0) - f^* - \frac{k+1}{n}\left(\mathrm{E}_{\xi_k}\left[f(\mathbf{x}_{k+1})\right] - f^*\right).
\end{aligned}$$

This leads to the desired result by moving the last term on the right to the left side. ∎

If f is a strongly convex quadratic function, the randomized coordinate decent method would have an expected average convergence rate $(1 - \frac{a}{An})$. However, each step of the method does $\frac{1}{n}$ amount of work of the full steepest descent update; see an exercise.

8.7 Summary

Most iterative algorithms for minimization require a line search at every stage of the process. By employing any one of a variety of curve fitting techniques, however, the order of convergence of the line search process can be made greater than unity, which means that as compared to the linear convergence that accompanies most full descent algorithms (such as steepest descent) the individual line searches are rapid. Indeed, in common practice, only about three search points are required in any one line search. If the first derivatives are available, then two search points are required (method of false position); and if both first and second derivatives are available, then one search point is required (Newton's method).

It was also shown in Sect. 8.1 and the exercises that line search algorithms of varying degrees of accuracy are all closed. Thus line searching is not only rapid enough to be practical but also behaves in such a way as to make analysis of global convergence simple.

The most important results of this chapter are the arithmetic convergence of the method of steepest descent for solving convex minimization, the improved arithmetic convergence of the accelerated steepest descent method, and the geometric convergence of the method for solving strongly convex minimization. The fact that the method of steepest descent converges linearly with a convergence ratio equal to $[(A - a)/(A + a)]^2$, where a and A are, respectively, the smallest and largest eigenvalues of the Hessian of the objective function evaluated at the solution point. This formula, which arises frequently throughout the remainder of the book, serves as a fundamental reference point for other algorithms. It is, however, important to understand that it is the *formula* and not its *value* that serves as the reference. We rarely advocate that the formula be evaluated since it involves quantities (namely eigenvalues) that are generally not computable until after the optimal solution is known. The formula itself, however, even though its value is unknown, can be used to make significant comparisons of the effectiveness of steepest descent versus other algorithms.

Newton's method has order two convergence. However, it must be modified to insure global convergence, and evaluation of the Hessian at every point can be costly. Nevertheless, Newton's method provides another valuable reference point in the study of algorithms, and is frequently employed in interior point methods using a logarithmic barrier function.

As optimization problem sizes become bigger and bigger, various coordinate descent algorithms are extremely popular. They are valuable especially in situations where the variables are essentially uncoupled or there is special structure that makes searching in the coordinate directions particularly easy. Typically, steepest descent can be expected to be faster. Even if the gradient is not directly available, it would probably be better to evaluate a finite-difference approximation to the gradient, by taking a single step in each coordinate direction, and use this approximation in a steepest descent algorithm, rather than executing a full line search in each coordinate direction.

8.8 Exercises

1. Show that $g[a, b, c]$ defined by (8.14) is symmetric, that is, interchange of the arguments does not affect its value.
2. Prove (8.14) and (8.15).
   Hint: To prove (8.15) expand it, and subtract and add $g'(x_k)$ to the numerator.
3. Argue using symmetry that the error in the cubic fit method approximately satisfies an equation of the form

$$\varepsilon_{k+1} = M(\varepsilon_k^2\varepsilon_{k-1} + \varepsilon_k\varepsilon_{k-1}^2)$$

   and then find the order of convergence.
4. What conditions on the values and derivatives at two points guarantee that a cubic polynomial fit to this data will have a minimum between the two points? Use your answer to develop a search scheme, based on cubic fit, that is globally convergent for unimodal functions.
5. Using a symmetry argument, find the order of convergence for a line search method that fits a cubic to $x_{k-3}, x_{k-2}, x_{k-1}, x_k$ in order to find x_{k+1}.
6. Consider the iterative process

$$x_{k+1} = \frac{1}{2}\left(x_k + \frac{a}{x_k}\right),$$

   where $a > 0$. Assuming the process converges, to what does it converge? What is the order of convergence?
7. Suppose the continuous real-valued function f of a single variable satisfies

$$\min_{x\geqslant 0} f(x) < f(0).$$

   Starting at any $x > 0$ show that, through a series of halvings and doublings of x and evaluation of the corresponding $f(x)$'s, a three-point pattern can be determined.
8. For $\delta > 0$ define the map $\mathbf{S}^\delta$ by

$$\mathbf{S}^\delta(\mathbf{x}, \mathbf{d}) = \{\mathbf{y} : \mathbf{y} = \mathbf{x} + \alpha\mathbf{d},\ 0 \leqslant \alpha \leqslant \delta;\ f(\mathbf{y}) = \min_{0\leqslant\beta\leqslant\delta} f(\mathbf{x} + \beta\mathbf{d})\}.$$

   Thus $\mathbf{S}^\delta$ searches the interval $[0, \delta]$ for a minimum of $f(\mathbf{x} + \alpha\mathbf{d})$, representing a "limited range" line search. Show that if f is continuous, $\mathbf{S}^\delta$ is closed at all $(\mathbf{x}, \mathbf{d})$.
9. For $\varepsilon > 0$ define the map ${}^\varepsilon\mathbf{S}$ by

$${}^\varepsilon\mathbf{S}(\mathbf{x}, \mathbf{d}) = \{\mathbf{y} : \mathbf{y} = \mathbf{x} + \alpha\mathbf{d},\ \alpha \geqslant 0,\ f(\mathbf{y}) \leqslant \min_{0\leqslant\beta} f(\mathbf{x} + \beta\mathbf{d}) + \varepsilon\}.$$

   Show that if f is continuous, ${}^\varepsilon\mathbf{S}$ is closed at $(\mathbf{x}, \mathbf{d})$ if $\mathbf{d} \neq \mathbf{0}$. This map corresponds to an "inaccurate" line search.

10. Referring to the previous two exercises, define and prove a result for ${}^{\varepsilon}\mathbf{S}^{\delta}$.
11. Define $\bar{\mathbf{S}}$ as the line search algorithm that finds the first relative minimum of $f(\mathbf{x} + \alpha\mathbf{d})$ for $\alpha \geqslant 0$. If f is continuous and $\mathbf{d} \neq \mathbf{0}$, is $\bar{\mathbf{S}}$ closed?
12. Consider the problem

$$\text{minimize } 5x^2 + 5y^2 - xy - 11x + 11y + 11.$$

(a) Find a point satisfying the first-order necessary conditions for a solution.
(b) Show that this point is a global minimum.
(c) What would be the rate of convergence of steepest descent for this problem?
(d) Starting at $x = y = 0$, how many steepest descent iterations would it take (at most) to reduce the function value to 10^{-11}?

13. Define the search mapping $\mathbf{F}$ that determines the parameter α to within a given fraction c, $0 \leqslant c \leqslant 1$, by

$$\mathbf{F}(\mathbf{x}, \mathbf{d}) = \{\mathbf{y} : \mathbf{y} = \mathbf{x} + \alpha\mathbf{d},\ 0 \leqslant \alpha < \infty,\ |\overline{\alpha}| \leqslant c\overline{\alpha}, \text{ where } \frac{d}{d\alpha} f(\mathbf{x} + \overline{\alpha}\mathbf{d}) = 0\}.$$

Show that if $\mathbf{d} \neq \mathbf{0}$ and $(d/d\alpha)f(\mathbf{x}+\alpha\mathbf{d})$ is continuous, then $\mathbf{F}$ is closed at $(\mathbf{x}, \mathbf{d})$.
14. Let $\mathbf{e}_1, \mathbf{e}_2, \ldots, \mathbf{e}_n$ denote the eigenvectors of the symmetric positive definite $n \times n$ matrix $\mathbf{Q}$. For the quadratic problem considered in Sect. 8.2, suppose $\mathbf{x}_0$ is chosen so that $\mathbf{g}_0$ belongs to a subspace M spanned by a subset of the $\mathbf{e}_i$'s. Show that for the method of steepest descent $\mathbf{g}_k \in M$ for all k. Find the rate of convergence in this case.
15. Suppose we use the method of steepest descent to minimize the quadratic function $f(\mathbf{x}) = \frac{1}{2}(\mathbf{x} - \mathbf{x}^*)^T\mathbf{Q}(\mathbf{x} - \mathbf{x}^*)$ but we allow a tolerance $\pm\delta\alpha_k$ ($\delta \geqslant 0$) in the line search, that is $\mathbf{x}_{k+1} = \mathbf{x}_k - \alpha_k\mathbf{g}_k$, where

$$(1 - \delta)\overline{\alpha}_k \leqslant \alpha_k \leqslant (1 + \delta)\overline{\alpha}_k$$

and $\overline{\alpha}_k$ minimizes $f(\mathbf{x}_k - \alpha\mathbf{g}_k)$ over α.

(a) Find the convergence rate of the algorithm in terms of a and A, the smallest and largest eigenvalues of $\mathbf{Q}$, and the tolerance δ.
Hint: Assume the extreme case $\alpha_k = (1 + \delta)\overline{\alpha}_k$.
(b) What is the largest δ that guarantees convergence of the algorithm? Explain this result geometrically.
(c) Does the sign of δ make any difference?

16. Show that for a quadratic objective function the percentage test and the Goldstein test are equivalent.
17. Suppose in the method of steepest descent for the quadratic problem, the value of α_k is not determined to minimize $E(\mathbf{x}_{k+1})$ exactly but instead only satisfies

$$\frac{E(\mathbf{x}_k) - E(\mathbf{x}_{k+1})}{E(\mathbf{x}_k)} \geqslant \beta \frac{E(\mathbf{x}_k) - \overline{E}}{E(\mathbf{x}_k)}$$

for some β, $0 < \beta < 1$, where $\overline{E}$ is the value that corresponds to the best α_k. Find the best estimate for the rate of convergence in this case.

18. Suppose an iterative algorithm of the form $\mathbf{x}_{k+1} = \mathbf{x}_k + \alpha_k \mathbf{d}_k$ is applied to the quadratic problem with matrix $\mathbf{Q}$, where α_k as usual is chosen as the minimum point of the line search and where $\mathbf{d}_k$ is a vector satisfying $\mathbf{d}_k^T \mathbf{g}_k < 0$ and $(\mathbf{d}_k^T \mathbf{g}_k)^2 \geqslant \beta(\mathbf{d}_k^T \mathbf{Q} \mathbf{d}_k)(\mathbf{g}_k^T \mathbf{Q}^{-1} \mathbf{g}_k)$, where $0 < \beta \leqslant 1$. This corresponds to a steepest descent algorithm with "sloppy" choice of direction. Estimate the rate of convergence of this algorithm.
19. Repeat Exercise 18 with the condition on $(\mathbf{d}_k^T \mathbf{g}_k)^2$ replaced by

$$(\mathbf{d}_k^T \mathbf{g}_k)^2 \geqslant \beta(\mathbf{d}_k^T \mathbf{d}_k)(\mathbf{g}_k^T \mathbf{g}_k), \quad 0 < \beta \leqslant 1.$$

20. Use the result of Exercise 19 to derive (8.71) for the Gauss-Southwell method.
21. Let $f(x, y) = s^2 + y^2 + xy - 3x$.

    (a) Find an unconstrained local minimum point of f.
    (b) Why is the solution to (a) actually a global minimum point?
    (c) Find the minimum point of f subject to $x \geqslant 0$, $y \geqslant 0$.
    (d) If the method of steepest descent were applied to (a), what would be the rate of convergence of the objective function?

22. Find an estimate for the rate of convergence for the modified Newton method

$$\mathbf{x}_{k+1} = \mathbf{x}_k - \alpha_k(\varepsilon_k \mathbf{I} + \mathbf{F}_k)^{-1} \mathbf{g}_k$$

given by (8.65) and (8.66) when δ is larger than the smallest eigenvalue of $\mathbf{F}(\mathbf{x}^*)$.

23. Prove global convergence of the Gauss-Southwell method.
24. Consider a problem of the form

$$\begin{aligned} &\text{minimize } f(\mathbf{x}) \\ &\text{subject to } \mathbf{x} \geqslant \mathbf{0}, \end{aligned}$$

where $\mathbf{x} \in E^n$. A gradient-type procedure has been suggested for this kind of problem that accounts for the constraint. At a given point $\mathbf{x} = (x_1, x_2, \ldots, x_n)$, the direction $\mathbf{d} = (d_1, d_2, \ldots, d_n)$ is determined from the gradient $\nabla f(\mathbf{x})^T = \mathbf{g} = (g_1, g_2, \ldots, g_n)$ by

$$d_i = \begin{cases} -g_i \text{ if } x_i > 0 \text{ or } g_i < 0 \\ 0 \text{ if } x_i = 0 \text{ and } g_i \geqslant 0. \end{cases}$$

This direction is then used as a direction of search in the usual manner.

(a) What are the first-order necessary conditions for a minimum point of this problem?
(b) Show that $\mathbf{d}$, as determined by the algorithm, is zero only at a point satisfying the first-order conditions.
(c) Show that if $\mathbf{d} \neq \mathbf{0}$, it is possible to decrease the value of f by movement along $\mathbf{d}$.

(d) If restricted to a compact region, does the Global Convergence Theorem apply? Why?

25. Consider the quadratic problem and suppose $\mathbf{Q}$ has unity diagonal. Consider a coordinate descent procedure in which the coordinate to be searched is at every stage selected randomly, each coordinate being equally likely. Let $\boldsymbol{\varepsilon}_k = \mathbf{x}_k - \mathbf{x}^*$. Assuming $\boldsymbol{\varepsilon}_k$ is known, show that $\overline{\boldsymbol{\varepsilon}_{k+1}^T \mathbf{Q} \boldsymbol{\varepsilon}_{k+1}}$, the expected value of $\boldsymbol{\varepsilon}_{k+1}^T \mathbf{Q} \boldsymbol{\varepsilon}_{k+1}$, satisfies

$$\overline{\boldsymbol{\varepsilon}_{k+1}^T \mathbf{Q} \boldsymbol{\varepsilon}_{k+1}} = \left(1 - \frac{\boldsymbol{\varepsilon}_k^T \mathbf{Q}^2 \boldsymbol{\varepsilon}_k}{n \boldsymbol{\varepsilon}_k^T \mathbf{Q} \boldsymbol{\varepsilon}_k}\right) \boldsymbol{\varepsilon}_k^T \mathbf{Q} \boldsymbol{\varepsilon}_k \leqslant \left(1 - \frac{a^2}{nA}\right) \boldsymbol{\varepsilon}_k^T \mathbf{Q} \boldsymbol{\varepsilon}_k.$$

26. If the matrix $\mathbf{Q}$ has a condition number of 10, how many iterations of steepest descent would be required to get six place accuracy in the minimum value of the objective function of the corresponding quadratic problem?

27. *Stopping criterion.* A question that arises in using an algorithm such as steepest descent to minimize an objective function f is when to stop the iterative process, or, in other words, how can one tell when the current point is close to a solution. If, as with steepest descent, it is known that convergence is linear, this knowledge can be used to develop a stopping criterion. Let $\{f_k\}_{k=0}^{\infty}$ be the sequence of values obtained by the algorithm. We assume that $f_k \to f^*$ linearly, but both f^* and the convergence ratio β are unknown. However we know that, at least approximately,

$$f_{k+1} - f^* = \beta(f_k - f^*)$$

and

$$f_k - f^* = \beta(f_{k-1} - f^*).$$

These two equations can be solved for β and f^*.

(a) Show that

$$f^* = \frac{f_k^2 - f_{k-1} f_{k+1}}{2f_k - f_{k-1} - f_{k+1}}, \qquad \beta = \frac{f_{k+1} - f_k}{f_k - f_{k-1}}.$$

(b) Motivated by the above we form the sequence $\{f_k^*\}$ defined by

$$f_k^* = \frac{f_k^2 - f_{k-1} f_{k+1}}{2f_k - f_{k-1} - f_{k+1}}$$

as the original sequence is generated. (This procedure of generating $\{f_k^*\}$ from $\{f_k\}$ is called the Aitken δ^2-process.) If $|f_k - f^*| = \beta^k + o(\beta^k)$ show that $|f_k^* - f^*| = o(\beta^k)$ which means that $\{f_k^*\}$ converges to f^* faster than $\{f_k\}$ does. The iterative search for the minimum of f can then be terminated when $f_k - f_k^*$ is smaller than some prescribed tolerance.

28. Show that the concordant requirement (8.69) can be expressed as

$$\left| \frac{\mathrm{d}}{\mathrm{d}x} f''(x)^{-\frac{1}{2}} \right| \leq 1.$$

29. Assume $f(x)$ and $g(x)$ are self-concordant. Show that the following functions are also self-concordant.

    (a) $af(x)$ for $a > 1$
    (b) $ax + b + f(x)$
    (c) $f(ax + b)$
    (d) $f(x) + g(x)$

1. Prove Lemma 1
2. Consider convex quadratic minimization with matrix $\mathbf{Q}$, and let its distinct positive eigenvalues be $\lambda_1, \lambda_2, \ldots \lambda_K$. Then, if we let the step size in the method of steepest descent be $\alpha_k = \frac{1}{\lambda_k}$, $k = 1, \ldots, K$, the method terminates in K iterations.
3. Show that the *randomized coordinate descent method* has the expected average convergence rate $(1 - \frac{a}{An})$ for solving strongly convex quadratic programs where a and A are smallest and largest eigenvalues of the Hessian matrix.

References

8.1 For a detailed exposition of Fibonacci search techniques, see Wilde and Beightler [W1]. For an introductory discussion of difference equations, see Lanczos [L1]. Many of these techniques are standard among numerical analysts. See, for example, Kowalik and Osborne [K9], or Traub [T9]. Also see Tamir [T1] for an analysis of high-order fit methods. The use of symmetry arguments to shortcut the analysis is new. The closedness of line search algorithms was established by Zangwill [Z2]. For the line search stopping criteria, see Armijo [A8], Goldstein [G12], and Wolfe [W6].

8.2 For an alternate exposition of this well-known method, see Antosiewicz and Rheinboldt [A7] or Luenberger [L8]. For a proof that the estimate (8.42) is essentially exact, see Akaike [A2]. For early work on the nonquadratic case, see Curry [C10]. For recent work reports in this section see Boyd and Vandenberghe [B23]. The numerical problem considered in the example is a standard one. See Faddeev and Faddeeva [F1].

8.4 The accelerated method of steepest descent is due to Nesterov [190], also see Beck and Teboulle [23]. The BB method is due to Barzilai and Borwein [17], also see Dai and Fletcher [58].

8.5 For good reviews of modern Newton methods, see Fletcher [F9] and Gill, Murray, and Wright [G7]. The theory of self-concordant functions was developed by Nesterov and Nemirovskri, see [N2], [N4], there is a nice reformulation by Renegar [R2] and an introduction in Boyd and Vandenberghe [B23].

8.6 A detailed analysis of coordinate algorithms can be found in Fox [F17] and Isaacson and Keller [I1]. For a discussion of the Gauss-Southwell method, see Forsythe and Wasow [F16]. The proof of convergence speed of the randomized coordinate descent method is essentially due to Nesterov [188] and Lu and Lin [160].

Chapter 9
Conjugate Direction Methods

Conjugate direction methods can be regarded as being somewhat intermediate between the method of steepest descent and Newton's method. They are motivated by the desire to accelerate the typically slow convergence associated with steepest descent while avoiding the information requirements associated with the evaluation, storage, and inversion of the Hessian (or at least solution of a corresponding system of equations) as required by Newton's method.

Conjugate direction methods invariably are invented and analyzed for the purely quadratic problem

$$\text{minimize } \frac{1}{2}\mathbf{x}^T \mathbf{Q}\mathbf{x} - \mathbf{b}^T\mathbf{x},$$

where $\mathbf{Q}$ is an $n \times n$ symmetric positive definite matrix. The techniques once worked out for this problem are then extended, by approximation, to more general problems; it being argued that, since near the solution point every problem is approximately quadratic, convergence behavior is similar to that for the pure quadratic situation.

The area of conjugate direction algorithms has been one of great creativity in the nonlinear programming field, illustrating that detailed analysis of the pure quadratic problem can lead to significant practical advances. Indeed, conjugate direction methods, especially the method of conjugate gradients, have proved to be extremely effective in dealing with general objective functions and are considered among the best general purpose methods.

9.1 Conjugate Directions

Definition. Given a symmetric matrix $\mathbf{Q}$, two vectors $\mathbf{d}_1$ and $\mathbf{d}_2$ are said to be $\mathbf{Q}$-*orthogonal*, or *conjugate with respect to* $\mathbf{Q}$, if $\mathbf{d}_1^T\mathbf{Q}\mathbf{d}_2 = 0$.

D.G. Luenberger, Y. Ye, *Linear and Nonlinear Programming*, International Series in Operations Research & Management Science 228,
DOI 10.1007/978-3-319-18842-3_9

In the applications that we consider, the matrix $\mathbf{Q}$ will be positive definite but this is not inherent in the basic definition. Thus if $\mathbf{Q} = 0$, any two vectors are conjugate, while if $\mathbf{Q} = \mathbf{I}$, conjugacy is equivalent to the usual notion of orthogonality. A finite set of vectors $\mathbf{d}_0, \mathbf{d}_1, \ldots, \mathbf{d}_k$ is said to be a $\mathbf{Q}$-orthogonal set if $\mathbf{d}_i^T \mathbf{Q} \mathbf{d}_j = 0$ for all $i \neq j$.

Proposition. *If $\mathbf{Q}$ is positive definite and the set of nonzero vectors $\mathbf{d}_0, \mathbf{d}_1, \mathbf{d}_2, \ldots, \mathbf{d}_k$ are $\mathbf{Q}$-orthogonal, then these vectors are linearly independent.*

Proof. Suppose there are constants α_i, $i = 0, 1, 2, \ldots, k$ such that

$$\alpha_0 \mathbf{d}_0 + \cdots + \alpha_k \mathbf{d}_k = 0.$$

Multiplying by $\mathbf{Q}$ and taking the scalar product with $\mathbf{d}_i$ yields

$$\alpha_i \mathbf{d}_i^T \mathbf{Q} \mathbf{d}_i = 0.$$

Or, since $\mathbf{d}_i^T \mathbf{Q} \mathbf{d}_i > 0$ in view of the positive definiteness of $\mathbf{Q}$, we have $\alpha_i = 0$. ∎

Before discussing the general conjugate direction algorithm, let us investigate just why the notion of $\mathbf{Q}$-orthogonality is useful in the solution of the quadratic problem

$$\text{minimize } \frac{1}{2}\mathbf{x}^T \mathbf{Q} \mathbf{x} - \mathbf{b}^T \mathbf{x}, \tag{9.1}$$

when $\mathbf{Q}$ is positive definite. Recall that the unique solution to this problem is also the unique solution to the linear equation

$$\mathbf{Q}\mathbf{x} = \mathbf{b}, \tag{9.2}$$

and hence that the quadratic minimization problem is equivalent to a linear equation problem.

Corresponding to the $n \times n$ positive definite matrix $\mathbf{Q}$ let $\mathbf{d}_0, \mathbf{d}_1, \ldots, \mathbf{d}_{n-1}$ be n nonzero $\mathbf{Q}$-orthogonal vectors. By the above proposition they are linearly independent, which implies that the solution $\mathbf{x}^*$ of (9.1) or (9.2) can be expanded in terms of them as

$$\mathbf{x}^* = \alpha_0 \mathbf{d}_0 + \cdots + \alpha_{n-1} \mathbf{d}_{n-1} \tag{9.3}$$

for some set of α_i's. In fact, multiplying by $\mathbf{Q}$ and then taking the scalar product with $\mathbf{d}_i$ yields directly

$$\alpha_i = \frac{\mathbf{d}_i^T \mathbf{Q} \mathbf{x}^*}{\mathbf{d}_i^T \mathbf{Q} \mathbf{d}_i} = \frac{\mathbf{d}_i^T \mathbf{b}}{\mathbf{d}_i^T \mathbf{Q} \mathbf{d}_i}. \tag{9.4}$$

This shows that the α_i's and consequently the solution $\mathbf{x}^*$ can be found by evaluation of simple scalar products. The end result is

$$\mathbf{x}^* = \sum_{i=0}^{n-1} \frac{\mathbf{d}_i^T \mathbf{b}}{\mathbf{d}_i^T \mathbf{Q} \mathbf{d}_i} \mathbf{d}_i. \tag{9.5}$$

There are two basic ideas imbedded in (9.5). The first is the idea of selecting an orthogonal set of $\mathbf{d}_i$'s so that by taking an appropriate scalar product, all terms on the right side of (9.3), except the ***i*th**, vanish. This could, of course, have been accomplished by making the $\mathbf{d}_i$'s orthogonal in the ordinary sense instead of making them $\mathbf{Q}$-orthogonal. The second basic observation, however, is that by using $\mathbf{Q}$-orthogonality the resulting equation for α_i can be expressed in terms of the known vector $\mathbf{b}$ rather than the unknown vector $\mathbf{x}^*$; hence the coefficients can be evaluated without knowing $\mathbf{x}^*$.

The expansion for $\mathbf{x}^*$ can be considered to be the result of an iterative process of n steps where at the ith step $\alpha_i\mathbf{d}_i$ is added. Viewing the procedure this way, and allowing for an arbitrary initial point for the iteration, the basic conjugate direction method is obtained.

Conjugate Direction Theorem. *Let* $\{\mathbf{d}_i\}_{i=0}^{n-1}$ *be a set of nonzero* $\mathbf{Q}$*-orthogonal vectors. For any* $\mathbf{x}_0 \in E^n$ *the sequence* $\{\mathbf{x}_k\}$ *generated according to*

$$\mathbf{x}_{k+1} = \mathbf{x}_k + \alpha_k\mathbf{d}_k, \ k \geqslant 0 \tag{9.6}$$

with

$$\alpha_k = -\frac{\mathbf{g}_k^T\mathbf{d}_k}{\mathbf{d}_k^T\mathbf{Q}\mathbf{d}_k} \tag{9.7}$$

and

$$\mathbf{g}_k = \mathbf{Q}\mathbf{x}_k - \mathbf{b},$$

converges to the unique solution, $\mathbf{x}^*$, *of* $\mathbf{Q}\mathbf{x} = \mathbf{b}$ *after n steps, that is,* $\mathbf{x}_n = \mathbf{x}^*$.

Proof. Since the $\mathbf{d}_k$'s are linearly independent, we can write

$$\mathbf{x}^* - \mathbf{x}_0 = \alpha_0\mathbf{d}_0 + \alpha_1\mathbf{d}_1 + \cdots + \alpha_{n-1}\mathbf{d}_{n-1}$$

for some set of α_k's. As we did to get (9.4), we multiply by $\mathbf{Q}$ and take the scalar product with $\mathbf{d}_k$ to find

$$\alpha_k = \frac{\mathbf{d}_k^T\mathbf{Q}(\mathbf{x}^* - \mathbf{x}_0)}{\mathbf{d}_k^T\mathbf{Q}\mathbf{d}_k}. \tag{9.8}$$

Now following the iterative process (9.6) from $\mathbf{x}_0$ up to $\mathbf{x}_k$ gives

$$\mathbf{x}_k - \mathbf{x}_0 = \alpha_0\mathbf{d}_0 + \alpha_1\mathbf{d}_1 + \cdots + \alpha_{k-1}\mathbf{d}_{k-1}, \tag{9.9}$$

and hence by the $\mathbf{Q}$-orthogonality of the $\mathbf{d}_k$'s it follows that

$$\mathbf{d}_k^T\mathbf{Q}(\mathbf{x}_k - \mathbf{x}_0) = 0. \tag{9.10}$$

Substituting (9.10) into (9.8) produces

$$\alpha_k = \frac{\mathbf{d}_k^T\mathbf{Q}(\mathbf{x}^* - \mathbf{x}_k)}{\mathbf{d}_k^T\mathbf{Q}\mathbf{d}_k} = -\frac{\mathbf{g}_k^T\mathbf{d}_k}{\mathbf{d}_k^T\mathbf{Q}\mathbf{d}_k},$$

which is identical with (9.7). ∎

To this point the conjugate direction method has been derived essentially through the observation that solving (9.1) is equivalent to solving (9.2). The conjugate direction method has been viewed simply as a somewhat special, but nevertheless straightforward, orthogonal expansion for the solution to (9.2). This viewpoint, although important because of its underlying simplicity, ignores some of the most important aspects of the algorithm; especially those aspects that are important when extending the method to nonquadratic problems. These additional properties are discussed in the next section.

Also, methods for selecting or generating sequences of conjugate directions have not yet been presented. Some methods for doing this are discussed in the exercises; while the most important method, that of conjugate gradients, is discussed in Sect. 9.3.

9.2 Descent Properties of the Conjugate Direction Method

We define $\mathcal{B}_k$ as the subspace of E^n spanned by $\{\mathbf{d}_0, \mathbf{d}_1, \ldots, \mathbf{d}_{k-1}\}$. We shall show that as the method of conjugate directions progresses each $\mathbf{x}_k$ minimizes the objective over the k-dimensional linear variety $\mathbf{x}_0 + \mathcal{B}_k$.

Expanding Subspace Theorem. *Let $\{\mathbf{d}_i\}_{i=0}^{n-1}$ be a sequence of nonzero $\mathbf{Q}$-orthogonal vectors in E^n. Then for any $\mathbf{x}_0 \in E^n$ the sequence $\{\mathbf{x}_k\}$ generated according to*

$$\mathbf{x}_{k+1} = \mathbf{x}_k + \alpha_k \mathbf{d}_k \tag{9.11}$$

$$\alpha_k = -\frac{\mathbf{g}_k^T \mathbf{d}_k}{\mathbf{d}_k^T \mathbf{Q} \mathbf{d}_k} \tag{9.12}$$

has the property that $\mathbf{x}_k$ minimizes $f(\mathbf{x}) = \frac{1}{2}\mathbf{x}^T\mathbf{Q}\mathbf{x} - \mathbf{b}^T\mathbf{x}$ on the line $\mathbf{x} = \mathbf{x}_{k-1} + \alpha \mathbf{d}_{k-1}$, $-\infty < \alpha < \infty$, as well as on the linear variety $\mathbf{x}_0 + \mathcal{B}_k$.

Proof. It need only be shown that $\mathbf{x}_k$ minimizes f on the linear variety $\mathbf{x}_0 + \mathcal{B}_k$, since it contains the line $\mathbf{x} = \mathbf{x}_{k-1} + \alpha \mathbf{d}_{k-1}$. Since f is a strictly convex function, the conclusion will hold if it can be shown that $\mathbf{g}_k$ is orthogonal to $\mathcal{B}_k$ (that is, the gradient of f at $\mathbf{x}_k$ is orthogonal to the subspace $\mathcal{B}_k$). The situation is illustrated in Fig. 9.1. (Compare Theorem 2, Sect. 7.5.)

We prove $\mathbf{g}_k \perp \mathcal{B}_k$ by induction. Since $\mathcal{B}_0$ is empty that hypothesis is true for $k = 0$. Assuming that it is true for k, that is, assuming $\mathbf{g}_k \perp \mathcal{B}_k$, we show that $\mathbf{g}_{k+1} \perp \mathcal{B}_{k+1}$. We have

$$\mathbf{g}_{k+1} = \mathbf{g}_k + \alpha_k \mathbf{Q} \mathbf{d}_k, \tag{9.13}$$

and hence

$$\mathbf{d}_k^T \mathbf{g}_{k+1} = \mathbf{d}_k^T \mathbf{g}_k + \alpha_k \mathbf{d}_k^T \mathbf{Q} \mathbf{d}_k = 0 \tag{9.14}$$

by definition of α_k. Also for $i < k$

$$\mathbf{d}_i^T \mathbf{g}_{k+1} = \mathbf{d}_i^T \mathbf{g}_k + \alpha_k \mathbf{d}_i^T \mathbf{Q} \mathbf{d}_k. \tag{9.15}$$

The first term on the right-hand side of (9.15) vanishes because of the induction hypothesis, while the second vanishes by the **Q**-orthogonality of the $\mathbf{d}_i$'s. Thus $\mathbf{g}_{k+1} \perp \mathcal{B}_{k+1}$. ∎

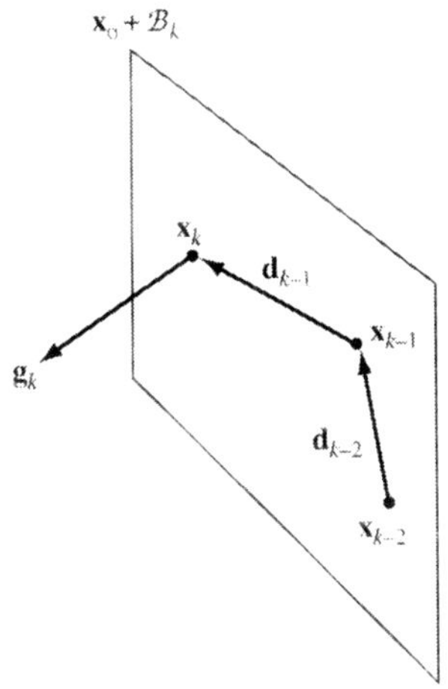

Fig. 9.1 Conjugate direction method

Corollary. *In the method of conjugate directions the gradients* $\mathbf{g}_k$, $k = 0, 1, \ldots, n$ *satisfy*

$$\mathbf{g}_k^T \mathbf{d}_i = 0 \text{ for } i < k.$$

The above theorem is referred to as the Expanding Subspace Theorem, since the $\mathcal{B}_k$'s form a sequence of subspaces with $\mathcal{B}_{k+1} \supset \mathcal{B}_k$. Since $\mathbf{x}_k$ minimizes f over $\mathbf{x}_0 + \mathcal{B}_k$, it is clear that $\mathbf{x}_n$ must be the overall minimum of f.

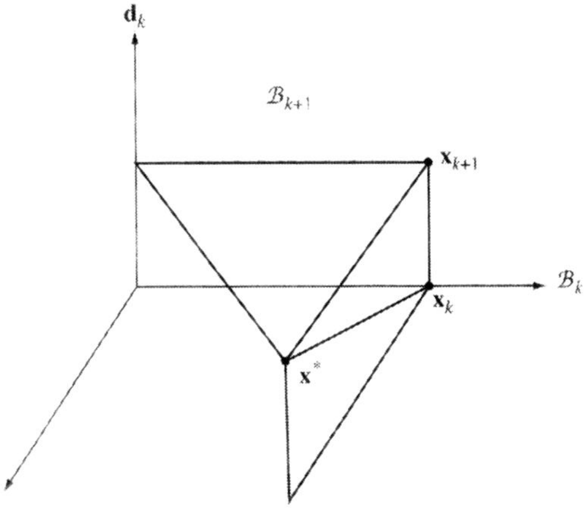

Fig. 9.2 Interpretation of expanding subspace theorem

To obtain another interpretation of this result we again introduce the function

$$E(\mathbf{x}) = \frac{1}{2}(\mathbf{x} - \mathbf{x}^*)^T \mathbf{Q}(\mathbf{x} - \mathbf{x}^*) \tag{9.16}$$

as a measure of how close the vector $\mathbf{x}$ is to the solution $\mathbf{x}^*$. Since $E(\mathbf{x}) = f(\mathbf{x}) + (1/2)\mathbf{x}^{*T}\mathbf{Q}\mathbf{x}^*$ the function E can be regarded as the objective that we seek to minimize.

By considering the minimization of E we can regard the original problem as one of minimizing a generalized distance from the point $\mathbf{x}^*$. Indeed, if we had $\mathbf{Q} = \mathbf{I}$, the generalized notion of distance would correspond (within a factor of two) to the usual Euclidean distance. For an arbitrary positive-definite $\mathbf{Q}$ we say E is a generalized Euclidean metric or distance function. Vectors $\mathbf{d}_i$, $i = 0, 1, \ldots, n-1$ that are $\mathbf{Q}$-orthogonal may be regarded as orthogonal in this generalized Euclidean space and this leads to the simple interpretation of the Expanding Subspace Theorem illustrated in Fig. 9.2. For simplicity we assume $\mathbf{x}_0 = \mathbf{0}$. In the figure $\mathbf{d}_k$ is shown as being orthogonal to $\mathcal{B}_k$ with respect to the generalized metric. The point $\mathbf{x}_k$ minimizes E over $\mathcal{B}_k$ while $\mathbf{x}_{k+1}$ minimizes E over $\mathcal{B}_{k+1}$. The basic property is that, since $\mathbf{d}_k$ is orthogonal to $\mathcal{B}_k$, the point $\mathbf{x}_{k+1}$ can be found by minimizing E along $\mathbf{d}_k$ and adding the result to $\mathbf{x}_k$.

9.3 The Conjugate Gradient Method

The conjugate gradient method is the conjugate direction method that is obtained by selecting the successive direction vectors as a conjugate version of the successive gradients obtained as the method progresses. Thus, the directions are not specified beforehand, but rather are determined sequentially at each step of the iteration. At step k one evaluates the current negative gradient vector and adds to it a linear combination of the previous direction vectors to obtain a new conjugate direction vector along which to move.

There are three primary advantages to this method of direction selection. First, unless the solution is attained in less than n steps, the gradient is always nonzero and linearly independent of all previous direction vectors. Indeed, the gradient $\mathbf{g}_k$ is orthogonal to the subspace $\mathcal{B}_k$ generated by $\mathbf{d}_0, \mathbf{d}_1, \ldots, \mathbf{d}_{k-1}$. If the solution is reached before n steps are taken, the gradient vanishes and the process terminates—it being unnecessary, in this case, to find additional directions.

Second, a more important advantage of the conjugate gradient method is the especially simple formula that is used to determine the new direction vector. This simplicity makes the method only slightly more complicated than steepest descent.

Third, because the directions are based on the gradients, the process makes good uniform progress toward the solution at every step. This is in contrast to the situation for arbitrary sequences of conjugate directions in which progress may be slight until the final few steps. Although for the pure quadratic problem uniform progress is of no great importance, it is important for generalizations to nonquadratic problems.

Conjugate Gradient Algorithm

Starting at any $\mathbf{x}_0 \in E^n$ define $\mathbf{d}_0 = -\mathbf{g}_0 = \mathbf{b} - \mathbf{Q}\mathbf{x}_0$ and

$$\mathbf{x}_{k+1} = \mathbf{x}_k + \alpha_k \mathbf{d}_k \tag{9.17}$$

$$\alpha_k = -\frac{\mathbf{g}_k^T \mathbf{d}_k}{\mathbf{d}_k^T \mathbf{Q} \mathbf{d}_k} \tag{9.18}$$

$$\mathbf{d}_{k+1} = -\mathbf{g}_{k+1} + \beta_k \mathbf{d}_k \tag{9.19}$$

$$\beta_k = \frac{\mathbf{g}_{k+1}^T \mathbf{Q} \mathbf{d}_k}{\mathbf{d}_k^T \mathbf{Q} \mathbf{d}_k}, \tag{9.20}$$

where $\mathbf{g}_k = \mathbf{Q}\mathbf{x}_k - \mathbf{b}$.

In the algorithm the first step is identical to a steepest descent step; each succeeding step moves in a direction that is a linear combination of the current gradient and the preceding direction vector. The attractive feature of the algorithm is the simple formulae, (9.19) and (9.20), for updating the direction vector. The method is only slightly more complicated to implement than the method of steepest descent but converges in a finite number of steps.

Verification of the Algorithm

To verify that the algorithm is a conjugate direction algorithm, it is necessary to verify that the vectors $\{\mathbf{d}_k\}$ are $\mathbf{Q}$-orthogonal. It is easiest to prove this by simultaneously proving a number of other properties of the algorithm. This is done in the theorem below where the notation $[\mathbf{d}_0, \mathbf{d}_1, \ldots, \mathbf{d}_k]$ is used to denote the subspace spanned by the vectors $\mathbf{d}_0, \mathbf{d}_1, \ldots, \mathbf{d}_k$.

Conjugate Gradient Theorem. *The conjugate gradient algorithm* (9.17)–(9.20) *is a conjugate direction method. If it does not terminate at* $\mathbf{x}_k$, *then*

a) $[\mathbf{g}_0, \mathbf{g}_1, \ldots, \mathbf{g}_k] = [\mathbf{g}_0, \mathbf{Q}\mathbf{g}_0, \ldots, \mathbf{Q}^k\mathbf{g}_0]$
b) $[\mathbf{d}_0, \mathbf{d}_1, \ldots, \mathbf{d}_k] = [\mathbf{g}_0, \mathbf{Q}\mathbf{g}_0, \ldots, \mathbf{Q}^k\mathbf{g}_0]$
c) $\mathbf{d}_k^T \mathbf{Q} \mathbf{d}_i = 0$ *for* $i \leqslant k-1$
d) $\alpha_k = \mathbf{g}_k^T \mathbf{g}_k / \mathbf{d}_k^T \mathbf{Q} \mathbf{d}_k$
e) $\beta_k = \mathbf{g}_{k+1}^T \mathbf{g}_{k+1} / \mathbf{g}_k^T \mathbf{g}_k$.

Proof. We first prove (a), (b) and (c) simultaneously by induction. Clearly, they are true for $k = 0$. Now suppose they are true for k, we show that they are true for $k+1$. We have

$$\mathbf{g}_{k+1} = \mathbf{g}_k + \alpha_k \mathbf{Q} \mathbf{d}_k.$$

By the induction hypothesis both $\mathbf{g}_k$ and $\mathbf{Q}\mathbf{d}_k$ belong to $[\mathbf{g}_0, \mathbf{Q}\mathbf{g}_0, \ldots, \mathbf{Q}^{k+1}\mathbf{g}_0]$, the first by (a) and the second by (b). Thus $\mathbf{g}_{k+1} \in [\mathbf{g}_0, \mathbf{Q}\mathbf{g}_0, \ldots, \mathbf{Q}^{k+1}\mathbf{g}_0]$. Furthermore $\mathbf{g}_{k+1} \notin [\mathbf{g}_0, \mathbf{Q}\mathbf{g}_0, \ldots, \mathbf{Q}^k\mathbf{g}_0] = [\mathbf{d}_0, \mathbf{d}_1, \ldots, \mathbf{d}_k]$ since otherwise $\mathbf{g}_{k+1} = 0$, because for any conjugate direction method $\mathbf{g}_{k+1}$ is orthogonal to $[\mathbf{d}_0, \mathbf{d}_1, \ldots, \mathbf{d}_k]$. (The induction hypothesis on (c) guarantees that the method is a conjugate direction method up to $\mathbf{x}_{k+1}$.) Thus, finally we conclude that

$$[\mathbf{g}_0, \mathbf{g}_1, \ldots, \mathbf{g}_{k+1}] = [\mathbf{g}_0, \mathbf{Q}\mathbf{g}_0, \ldots, \mathbf{Q}^{k+1}\mathbf{g}_0],$$

which proves (a).

To prove (b) we write

$$\mathbf{d}_{k+1} = -\mathbf{g}_{k+1} + \beta_k \mathbf{d}_k,$$

and (b) immediately follows from (a) and the induction hypothesis on (b).

Next, to prove (c) we have

$$\mathbf{d}_{k+1}^T \mathbf{Q}\mathbf{d}_i = -\mathbf{g}_{k+1}^T \mathbf{Q}\mathbf{d}_i + \beta_k \mathbf{d}_k^T \mathbf{Q}\mathbf{d}_i.$$

For $i = k$ the right side is zero by definition of β_k. For $i < k$ both terms vanish. The first term vanishes since $\mathbf{Q}\mathbf{d}_i \in [\mathbf{d}_1, \mathbf{d}_2, \ldots, \mathbf{d}_{i+1}]$, the induction hypothesis which guarantees the method is a conjugate direction method up to $\mathbf{x}_{k+1}$, and by the Expanding Subspace Theorem that guarantees that $\mathbf{g}_{k+1}$ is orthogonal to $[\mathbf{d}_0, \mathbf{d}_1, \ldots, \mathbf{d}_{i+1}]$. The second term vanishes by the induction hypothesis on (c). This proves (c), which also proves that the method is a conjugate direction method.

To prove (d) we have

$$-\mathbf{g}_k^T \mathbf{d}_k = \mathbf{g}_k^T \mathbf{g}_k - \beta_{k-1} \mathbf{g}_k^T \mathbf{d}_{k-1},$$

and the second term is zero by the Expanding Subspace Theorem.

Finally, to prove (e) we note that $\mathbf{g}_{k+1}^T \mathbf{g}_k = 0$, because $\mathbf{g}_k \in [\mathbf{d}_0, \ldots, \mathbf{d}_k]$ and $\mathbf{g}_{k+1}$ is orthogonal to $[\mathbf{d}_0, \ldots, \mathbf{d}_k]$. Thus since

$$\mathbf{Q}\mathbf{d}_k = \frac{1}{\alpha_k}(\mathbf{g}_{k+1} - \mathbf{g}_k),$$

we have

$$\mathbf{g}_{k+1}^T \mathbf{Q}\mathbf{d}_k = \frac{1}{\alpha_k} \mathbf{g}_{k+1}^T \mathbf{g}_{k+1}. \blacksquare$$

Parts (a) and (b) of this theorem are a formal statement of the interrelation between the direction vectors and the gradient vectors. Part (c) is the equation that verifies that the method is a conjugate direction method. Parts (d) and (e) are identities yielding alternative formulae for α_k and β_k that are often more convenient than the original ones.

9.4 The C–G Method as an Optimal Process

We turn now to the description of a special viewpoint that leads quickly to some very profound convergence results for the method of conjugate gradients. The basis of the viewpoint is part (b) of the Conjugate Gradient Theorem. This result tells us the spaces $\mathcal{B}_k$ over which we successively minimize are determined by the original gradient $\mathbf{g}_0$ and multiplications of it by $\mathbf{Q}$. Each step of the method brings into consideration an additional power of $\mathbf{Q}$ times $\mathbf{g}_0$. It is this observation we exploit.

Let us consider a new general approach for solving the quadratic minimization problem. Given an arbitrary starting point $\mathbf{x}_0$, let

$$\mathbf{x}_{k+1} = \mathbf{x}_0 + P_k(\mathbf{Q})\mathbf{g}_0, \tag{9.21}$$

where P_k is a polynomial of degree k. Selection of a set of coefficients for each of the polynomials P_k determines a sequence of $\mathbf{x}_k$'s. We have

$$\begin{aligned} \mathbf{x}_{k+1} - \mathbf{x}^* &= \mathbf{x}_0 - \mathbf{x}^* + P_k(\mathbf{Q})\mathbf{Q}(\mathbf{x}_0 - \mathbf{x}^*) \\ &= [\mathbf{I} + \mathbf{Q}P_k(\mathbf{Q})](\mathbf{x}_0 - \mathbf{x}^*), \end{aligned} \tag{9.22}$$

and hence

$$\begin{aligned} E(\mathbf{x}_{k+1}) &= \frac{1}{2}(\mathbf{x}_{k+1} - \mathbf{x}^*)^T\mathbf{Q}(\mathbf{x}_{k+1} - \mathbf{x}^*) \\ &= \frac{1}{2}(\mathbf{x}_0 - \mathbf{x}^*)^T\mathbf{Q}[\mathbf{I} + \mathbf{Q}P_k(\mathbf{Q})]^2(\mathbf{x}_0 - \mathbf{x}^*). \end{aligned} \tag{9.23}$$

We may now pose the problem of selecting the polynomial P_k in such a way as to minimize $E(\mathbf{x}_{k+1})$ with respect to all possible polynomials of degree k. Expanding (9.21), however, we obtain

$$\mathbf{x}_{k+1} = \mathbf{x}_0 + \gamma_0\mathbf{g}_0 + \gamma_1\mathbf{Q}\mathbf{g}_0 + \cdots + \gamma_k\mathbf{Q}^k\mathbf{g}_0, \tag{9.24}$$

where the γ_i's are the coefficients of P_k. In view of

$$\mathcal{B}_{k+1} = [\mathbf{d}_0, \mathbf{d}_1, \ldots, \mathbf{d}_k] = [\mathbf{g}_0, \mathbf{Q}\mathbf{g}_0, \ldots, \mathbf{Q}^k\mathbf{g}_0],$$

the vector $\mathbf{x}_{k+1} = \mathbf{x}_0 + \alpha_0\mathbf{d}_0 + \alpha_1\mathbf{d}_1 + \ldots + \alpha_k\mathbf{d}_k$ generated by the method of conjugate gradients has precisely this form; moreover, according to the Expanding Subspace Theorem, the coefficients γ_i determined by the conjugate gradient process are such as to minimize $E(\mathbf{x}_{k+1})$. Therefore, the problem posed of selecting the optimal P_k is solved by the conjugate gradient procedure.

The explicit relation between the optimal coefficients γ_i of P_k and the constants α_i, β_i associated with the conjugate gradient method is, of course, somewhat complicated, as is the relation between the coefficients of P_k and those of P_{k+1}. The power of the conjugate gradient method is that as it progresses it successively solves each of the optimal polynomial problems while updating only a small amount of information.

We summarize the above development by the following very useful theorem.

Theorem 1. *The point* $\mathbf{x}_{k+1}$ *generated by the conjugate gradient method satisfies*

$$E(\mathbf{x}_{k+1}) = \min_{P_k} \frac{1}{2}(\mathbf{x}_0 - \mathbf{x}^*)^T \mathbf{Q}[\mathbf{I} + \mathbf{Q}P_k(\mathbf{Q})]^2(\mathbf{x}_0 - \mathbf{x}^*), \tag{9.25}$$

where the minimum is taken with respect to all polynomials P_k *of degree* k.

Bounds on Convergence

To use Theorem 1 most effectively it is convenient to recast it in terms of eigenvectors and eigenvalues of the matrix $\mathbf{Q}$. Suppose that the vector $\mathbf{x}_0 - \mathbf{x}^*$ is written in the eigenvector expansion

$$\mathbf{x}_0 - \mathbf{x}^* = \xi_1\mathbf{e}_1 + \xi_2\mathbf{e}_2 + \cdots + \xi_n\mathbf{e}_n,$$

where the $\mathbf{e}_i$'s are normalized eigenvectors of $\mathbf{Q}$. Then since $\mathbf{Q}(\mathbf{x}_0 - \mathbf{x}^*) = \lambda_1\xi_1\mathbf{e}_1 + \lambda_2\xi_2\mathbf{e}_2 + \ldots + \lambda_n\xi_n\mathbf{e}_n$ and since the eigenvectors are mutually orthogonal, we have

$$E(\mathbf{x}_0) = \frac{1}{2}(\mathbf{x}_0 - \mathbf{x}^*)^T\mathbf{Q}(\mathbf{x}_0 - \mathbf{x}^*) = \frac{1}{2}\sum_{i=1}^{n}\lambda_i\xi_i^2, \tag{9.26}$$

where the λ_i's are the corresponding eigenvalues of $\mathbf{Q}$. Applying the same manipulations to (9.25), we find that for *any* polynomial P_k of degree k there holds

$$E(\mathbf{x}_{k+1}) \leqslant \frac{1}{2}\sum_{i=1}^{n}[1 + \lambda_i P_k(\lambda_i)]^2\lambda_i\xi_i^2.$$

It then follows that

$$E(\mathbf{x}_{k+1}) \leqslant \max_{\lambda_i}[1 + \lambda_i P_k(\lambda_i)]^2 \frac{1}{2}\sum_{i=1'}^{n}\lambda_i\xi_i^2,$$

and hence finally

$$E(\mathbf{x}_{k+1}) \leqslant \max_{\lambda_i}[1 + \lambda_i P_k(\lambda_i)]^2 E(\mathbf{x}_0).$$

We summarize this result by the following theorem.

Theorem 2. *In the method of conjugate gradients we have*

$$E(\mathbf{x}_{k+1}) \leqslant \max_{\lambda_i}[1 + \lambda_i P_k(\lambda_i)]^2 E(\mathbf{x}_0) \tag{9.27}$$

for any polynomial P_k *of degree* k, *where the maximum is taken over all eigenvalues* λ_i *of* $\mathbf{Q}$.

This way of viewing the conjugate gradient method as an optimal process is exploited in the next section. We note here that it implies the far from obvious fact that every step of the conjugate gradient method is at least as good as a steepest descent step would be from the same point. To see this, suppose $\mathbf{x}_k$ has been computed by the conjugate gradient method. From (9.24) we know $\mathbf{x}_k$ has the form

$$\mathbf{x}_k = \mathbf{x}_0 + \bar{\gamma}_0\mathbf{g}_0 + \bar{\gamma}_1\mathbf{Q}\mathbf{g}_0 + \cdots + \bar{\gamma}_{k-1}\mathbf{Q}^{k-1}\mathbf{g}_0.$$

Now if $\mathbf{x}_{k+1}$ is computed from $\mathbf{x}_k$ by steepest descent, then $\mathbf{x}_{k+1} = \mathbf{x}_k - \alpha_k\mathbf{g}_k$ for some α_k. In view of part (a) of the Conjugate Gradient Theorem $\mathbf{x}_{k+1}$ will have the form (9.24). Since for the conjugate direction method $E(\mathbf{x}_{k+1})$ is lower than any other $\mathbf{x}_{k+1}$ of the form (9.24), we obtain the desired conclusion.

Typically when some information about the eigenvalue structure of $\mathbf{Q}$ is known, that information can be exploited by construction of a suitable polynomial P_k to use in (9.27). Suppose, for example, it were known that $\mathbf{Q}$ had only $m < n$ distinct eigenvalues. Then it is clear that by suitable choice of P_{m-1} it would be possible to make the mth degree polynomial $1+\lambda P_{m-1}(\lambda)$ have its m zeros at the m eigenvalues. Using that particular polynomial in (9.27) shows that $E(\mathbf{x}_m) = 0$. Thus the optimal solution will be obtained in at most m, rather than n, steps. More sophisticated examples of this type of reasoning are contained in the next section and in the exercises at the end of the chapter.

9.5 The Partial Conjugate Gradient Method

A collection of procedures that are natural to consider at this point are those in which the conjugate gradient procedure is carried out for $m + 1 < n$ steps and then, rather than continuing, the process is restarted from the current point and $m + 1$ more conjugate gradient steps are taken. The special case of $m = 0$ corresponds to the standard method of steepest descent, while $m = n - 1$ corresponds to the full conjugate gradient method. These *partial conjugate gradient* methods are of extreme theoretical and practical importance, and their analysis yields additional insight into the method of conjugate gradients. The development of the last section forms the basis of our analysis.

As before, given the problem

$$\text{minimize } \frac{1}{2}\mathbf{x}^T\mathbf{Q}\mathbf{x} - \mathbf{b}^T\mathbf{x}, \tag{9.28}$$

we define for any point $\mathbf{x}_k$ the gradient $\mathbf{g}_k = \mathbf{Q}\mathbf{x}_k - \mathbf{b}$. We consider an iteration scheme of the form

$$\mathbf{x}_{k+1} = \mathbf{x}_k + P^k(\mathbf{Q})\mathbf{g}_k, \tag{9.29}$$

where P^k is a polynomial of degree m. We select the coefficients of the polynomial P^k so as to minimize

$$E(\mathbf{x}_{k+1}) = \frac{1}{2}(\mathbf{x}_{k+1} - \mathbf{x}^*)^T\mathbf{Q}(\mathbf{x}_{k+1} - \mathbf{x}^*), \tag{9.30}$$

where $\mathbf{x}^*$ is the solution to (9.28). In view of the development of the last section, it is clear that $\mathbf{x}_{k+1}$ can be found by taking $m + 1$ conjugate gradient steps rather than explicitly determining the appropriate polynomial directly. (The sequence indexing is slightly different here than in the previous section, since now we do not give separate indices to the intermediate steps of this process. Going from $\mathbf{x}_k$ to $\mathbf{x}_{k+1}$ by the partial conjugate gradient method involves m other points.)

The results of the previous section provide a tool for convergence analysis of this method. In this case, however, we develop a result that is of particular interest for $\mathbf{Q}$'s having a special eigenvalue structure that occurs frequently in optimization problems, especially, as shown below and in Chap. 12, in the context of penalty function methods for solving problems with constraints. We imagine that the eigenvalues of $\mathbf{Q}$ are of two kinds: there are m large eigenvalues that may or may not be located near each other, and $n - m$ smaller eigenvalues located within an interval $[a, b]$. Such a distribution of eigenvalues is shown in Fig. 9.3.

As an example, consider as in Sect. 8.3 the problem on E^n

$$\begin{aligned} &\text{minimize } \frac{1}{2}\mathbf{x}^T\mathbf{Q}\mathbf{x} - \mathbf{b}^T\mathbf{x} \\ &\text{subject to } \mathbf{c}^T\mathbf{x} = 0, \end{aligned}$$

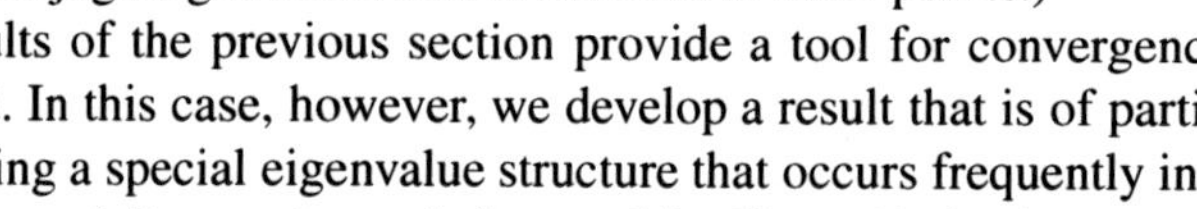

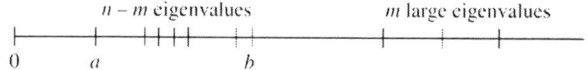

Fig. 9.3 Eigenvalue distribution

where $\mathbf{Q}$ is a symmetric positive definite matrix with eigenvalues in the interval $[a, A]$ and $\mathbf{b}$ and $\mathbf{c}$ are vectors in E^n. This is a constrained problem but it can be approximated by the unconstrained problem

$$\text{minimize } \frac{1}{2}\mathbf{x}^T\mathbf{Q}\mathbf{x} - \mathbf{b}^T\mathbf{x} + \frac{1}{2}\mu(\mathbf{c}^T\mathbf{x})^2,$$

where μ is a large positive constant. The last term in the objective function is called a *penalty term*; for large μ minimization with respect to $\mathbf{x}$ will tend to make $\mathbf{c}^T\mathbf{x}$ small.

The total quadratic term in the objective is $\frac{1}{2}\mathbf{x}^T(\mathbf{Q} + \mu\mathbf{c}\mathbf{c}^T)\mathbf{x}$, and thus it is appropriate to consider the eigenvalues of the matrix $\mathbf{Q} + \mu\mathbf{c}\mathbf{c}^T$. As μ tends to infinity it can be shown (see Chap. 13) that one eigenvalue of this matrix tends to infinity and the other $n - 1$ eigenvalues remain bounded within the original interval $[a, A]$.

As noted before, if steepest descent were applied to a problem with such a structure, convergence would be governed by the ratio of the smallest to largest eigenvalue, which in this case would be quite unfavorable. In the theorem below it is stated that by successively repeating $m + 1$ conjugate gradient steps the effects of the

m largest eigenvalues are eliminated and the rate of convergence is determined as if they were not present. A computational example of this phenomenon is presented in Sect. 13.5. The reader may find it interesting to read that section right after this one.

Theorem (Partial Conjugate Gradient Method). *Suppose the symmetric positive definite matrix* $\mathbf{Q}$ *has* $n-m$ *eigenvalues in the interval* $[a, b]$, $a > 0$ *and the remaining* m *eigenvalues are greater than* b. *Then the method of partial conjugate gradients, restarted every* $m+1$ *steps, satisfies*

$$E(\mathbf{x}_{k+1}) \leqslant \left(\frac{b-a}{b+a}\right)^2 E(\mathbf{x}_k). \tag{9.31}$$

(The point $\mathbf{x}_{k+1}$ *is found from* $\mathbf{x}_k$ *by taking* $m+1$ *conjugate gradient steps so that each increment in* k *is a composite of several simple steps.)*

Proof. Application of (9.27) yields

$$E(\mathbf{x}_{k+1} \leqslant \max_{\lambda_i}[1+\lambda_i P(\lambda_i)]^2 E(\mathbf{x}_k) \tag{9.32}$$

for any mth-order polynomial P, where the λ_i's are the eigenvalues of $\mathbf{Q}$. Let us select P so that the $(m+1)$th-degree polynomial $q(\lambda) = 1 + \lambda P(\lambda)$ vanishes at $(a+b)/2$ and at the m large eigenvalues of $\mathbf{Q}$. This is illustrated in Fig. 9.4. For this choice of P we may write (9.32) as

$$E(\mathbf{x}_{k+1}) \leqslant \max_{a \leqslant \lambda_i \leqslant b} [1+\lambda_i P(\lambda_i)]^2 E(\mathbf{x}_k).$$

Since the polynomial $q(\lambda) = 1 + \lambda P(\lambda)$ has $m+1$ real roots, $q'(\lambda)$ will have m real roots which alternate between the roots of $q(\lambda)$ on the real axis. Likewise, $q''(\lambda)$ will have $m-1$ real roots which alternate between the roots of $q'(\lambda)$. Thus, since $q(\lambda)$ has no root in the interval $(-\infty, (a+b)/2)$, we see that $q''(\lambda)$ does not change sign in that interval; and since it is easily verified that $q''(0) > 0$ it follows that $q(\lambda)$ is convex for $\lambda < (a+b)/2$. Therefore, on $[0, (a+b)/2]$, $q(\lambda)$ lies below the line $1 - [2\lambda/(a+b)]$. Thus we conclude that

$$q(\lambda) \leqslant 1 - \frac{2\lambda}{a+b}$$

on $[0, (a+b)/2]$ and that

$$q'\left(\frac{a+b}{2}\right) \geqslant -\frac{2}{a+b}.$$

We can see that on $[(a+b)/2, b]$

$$q(\lambda) \geqslant 1 - \frac{2\lambda}{a+b},$$

since for $q(\lambda)$ to cross first the line $1-[2\lambda/(a+b)]$ and then the λ-axis would require at least two changes in sign of $q''(\lambda)$, whereas, at most one root of $q''(\lambda)$ exists to the left of the second root of $q(\lambda)$. We see then that the inequality

$$|1 + \lambda P(\lambda)| \leqslant |1 - \frac{2\lambda}{a+b}|$$

is valid on the interval $[a, b]$. The final result (9.31) follows immediately. ∎

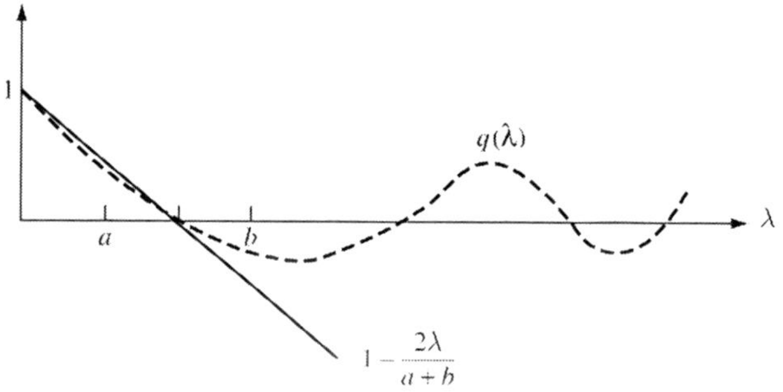

Fig. 9.4 Construction for proof

In view of this theorem, the method of partial conjugate gradients can be regarded as a generalization of steepest descent, not only in its philosophy and implementation, but also in its behavior. Its rate of convergence is bounded by exactly the same formula as that of steepest descent but with the largest eigenvalues removed from consideration. (It is worth noting that for $m = 0$ the above proof provides a simple derivation of the Steepest Descent Theorem.)

9.6 Extension to Nonquadratic Problems

The general unconstrained minimization problem on E^n

$$\text{minimize } f(\mathbf{x})$$

can be attacked by making suitable approximations to the conjugate gradient algorithm. There are a number of ways that this might be accomplished; the choice depends partially on what properties of f are easily computable. We look at three methods in this section and another in the following section.

Quadratic Approximation

In the quadratic approximation method we make the following associations at $\mathbf{x}_k$:

$$\mathbf{g}_k \leftrightarrow \nabla f(\mathbf{x}_k)^T, \qquad \mathbf{Q} \leftrightarrow \mathbf{F}(\mathbf{x}_k),$$

and using these associations, reevaluated at each step, all quantities necessary to implement the basic conjugate gradient algorithm can be evaluated. If f is quadratic, these associations are identities, so that the general algorithm obtained by using them is a generalization of the conjugate gradient scheme. This is similar to the philosophy underlying Newton's method where at each step the solution of a general problem is approximated by the solution of a purely quadratic problem through these same associations.

When applied to nonquadratic problems, conjugate gradient methods will not usually terminate within n steps. It is possible therefore simply to continue finding new directions according to the algorithm and terminate only when some termination criterion is met. Alternatively, the conjugate gradient process can be interrupted after n or $n+1$ steps and restarted with a pure gradient step. Since $\mathbf{Q}$-conjugacy of the direction vectors in the pure conjugate gradient algorithm is dependent on the initial direction being the negative gradient, the restarting procedure seems to be preferred. We always include this restarting procedure. The general conjugate gradient algorithm is then defined as below.

Step 1. Starting at $\mathbf{x}_0$ compute $\mathbf{g}_0 = \nabla f(\mathbf{x}_0)^T$ and set $\mathbf{d}_0 = -\mathbf{g}_0$.
Step 2. For $k = 0, 1, \ldots, n-1$:

(a) Set $\mathbf{x}_{k+1} = \mathbf{x}_k + \alpha_k \mathbf{d}_k$ where $\alpha_k = \frac{-\mathbf{g}_k^T \mathbf{d}_k}{\mathbf{d}_k^T \mathbf{F}(\mathbf{x}_k)\mathbf{d}_k}$.
(b) Compute $\mathbf{g}_{k+1} = \nabla f(\mathbf{x}_{k+1})^T$.
(c) Unless $k = n-1$, set $\mathbf{d}_{k+1} = -\mathbf{g}_{k+1} + \beta_k \mathbf{d}_k$ where

$$\beta_k = \frac{\mathbf{g}_{k+1}^T \mathbf{F}(\mathbf{x}_k)\mathbf{d}_k}{\mathbf{d}_k^T \mathbf{F}(\mathbf{x}_k)\mathbf{d}_k}$$

and repeat (a).

Step 3. Replace $\mathbf{x}_0$ by $\mathbf{x}_n$ and go back to Step 1.

An attractive feature of the algorithm is that, just as in the pure form of Newton's method, no line searching is required at any stage. Also, the algorithm converges in a finite number of steps for a quadratic problem. The undesirable features are that $\mathbf{F}(\mathbf{x}_k)$ must be evaluated at each point, which is often impractical, and that the algorithm is not, in this form, globally convergent.

Line Search Methods

It is possible to avoid the direct use of the association $\mathbf{Q} \leftrightarrow \mathbf{F}(\mathbf{x}_k)$. First, instead of using the formula for α_k in Step 2(a) above, α_k is found by a line search that minimizes the objective. This agrees with the formula in the quadratic case. Second, the formula for β_k in Step 2(c) is replaced by a different formula, which is, however, equivalent to the one in 2(c) in the quadratic case.

The first such method proposed was the *Fletcher–Reeves method*, in which Part (e) of the Conjugate Gradient Theorem is employed; that is,

$$\beta_k = \frac{\mathbf{g}_{k+1}^T \mathbf{g}_{k+1}}{\mathbf{g}_k^T \mathbf{g}_k}.$$

The complete algorithm (using restarts) is:

Step 1. Given $\mathbf{x}_0$ compute $\mathbf{g}_0 = \nabla f(\mathbf{x}_0)^T$ and set $\mathbf{d}_0 = -\mathbf{g}_0$.
Step 2. For $k = 0, 1, \ldots, n-1$:

(a) Set $\mathbf{x}_{k+1} = \mathbf{x}_k + \alpha_k \mathbf{d}_k$ where α_k minimizes $f(\mathbf{x}_k + \alpha \mathbf{d}_k)$.
(b) Compute $\mathbf{g}_{k+1} = \nabla f(\mathbf{x}_{k+1})^T$.
(c) Unless $k = n-1$, set $\mathbf{d}_{k+1} = -\mathbf{g}_{k+1} + \beta_k \mathbf{d}_k$ where

$$\beta_k = \frac{\mathbf{g}_{k+1}^T \mathbf{g}_{k+1}}{\mathbf{g}_k^T \mathbf{g}_k}.$$

Step 3. Replace $\mathbf{x}_0$ by $\mathbf{x}_n$ and go back to Step 1.

Another important method of this type is the *Polak–Ribiere method*, where

$$\beta_k = \frac{(\mathbf{g}_{k+1} - \mathbf{g}_k)^T \mathbf{g}_{k+1}}{\mathbf{g}_k^T \mathbf{g}_k}$$

is used to determine β_k. Again this leads to a value identical to the standard formula in the quadratic case. Experimental evidence seems to favor the Polak–Ribiere method over other methods of this general type.

Convergence

Global convergence of the line search methods is established by noting that a pure steepest descent step is taken every n steps and serves as a spacer step. Since the other steps do not increase the objective, and in fact hopefully they decrease it, global convergence is assured. Thus the restarting aspect of the algorithm is important for global convergence analysis, since in general one cannot guarantee that the directions $\mathbf{d}_k$ generated by the method are descent directions.

The local convergence properties of both of the above, and most other, nonquadratic extensions of the conjugate gradient method can be inferred from the quadratic analysis. Assuming that at the solution, $\mathbf{x}^*$, the matrix $\mathbf{F}(\mathbf{x}^*)$ is positive definite, we expect the asymptotic convergence rate per step to be at least as good as steepest descent, since this is true in the quadratic case. In addition to this bound on the single step rate we expect that the method is of order two with respect to each complete cycle of n steps. In other words, since one complete cycle solves a quadratic problem exactly just as Newton's method does in one step, we expect that for general nonquadratic problems there will hold $|\mathbf{x}_{k+n} - \mathbf{x}^*| \leqslant c|\mathbf{x}_k - \mathbf{x}^*|^2$ for some c and $k = 0, n, 2n, 3n, \ldots$. This can indeed be proved, and of course underlies the original motivation for the method. For problems with large n, however, a result of this type is in itself of little comfort, since we probably hope to terminate in fewer than n steps. Further discussion on this general topic is contained in Sect. 10.4.

Scaling and Partial Methods

Convergence of the partial conjugate gradient method, restarted every $m + 1$ steps, will in general be linear. The rate will be determined by the eigenvalue structure of the Hessian matrix $\mathbf{F}(\mathbf{x}^*)$, and it may be possible to obtain fast convergence by changing the eigenvalue structure through scaling procedures. If, for example, the eigenvalues can be arranged to occur in $m + 1$ bunches, the rate of the partial method will be relatively fast. Other structures can be analyzed by use of Theorem 2, Sect. 9.4, by using $\mathbf{F}(\mathbf{x}^*)$ rather than Q.

*9.7 *Parallel Tangents

In early experiments with the method of steepest descent the path of descent was noticed to be highly zig-zag in character, making slow indirect progress toward the solution. (This phenomenon is now quite well understood and is predicted by the convergence analysis of Sect. 8.2.) It was also noticed that in two dimensions the solution point often lies close to the line that connects the zig-zag points, as illustrated in Fig. 9.5. This observation motivated the *accelerated gradient method* in which a complete cycle consists of taking two steepest descent steps and then searching along the line connecting the initial point and the point obtained after the two gradient steps. The method of parallel tangents (PARTAN) was developed through an

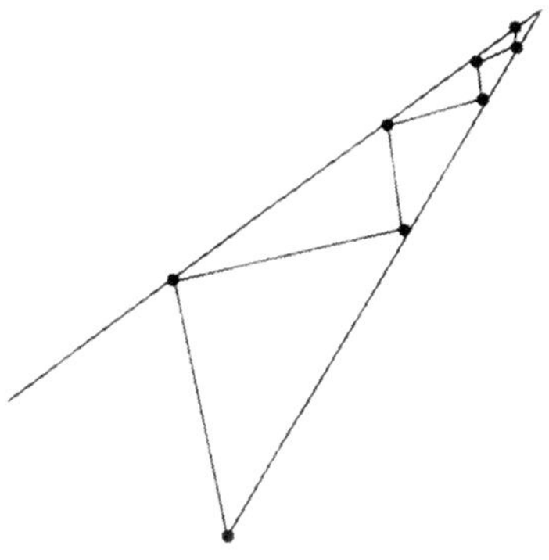

Fig. 9.5 Path of gradient method

attempt to extend this idea to an acceleration scheme involving all previous steps. The original development was based largely on a special geometric property of the tangents to the contours of a quadratic function, but the method is now recognized as a particular implementation of the method of conjugate gradients, and this is the context in which it is treated here.

The algorithm is defined by reference to Fig. 9.6. Starting at an arbitrary point $\mathbf{x}_0$ the point $\mathbf{x}_1$ is found by a standard steepest descent step. After that, from a point $\mathbf{x}_k$ the corresponding $\mathbf{y}_k$ is first found by a standard steepest descent step from $\mathbf{x}_k$, and then $\mathbf{x}_{k+1}$ is taken to be the minimum point on the line connecting $\mathbf{x}_{k-1}$ and $\mathbf{y}_k$. The process is continued for n steps and then restarted with a standard steepest descent step.

Notice that except for the first step, $\mathbf{x}_{k+1}$ is determined from $\mathbf{x}_k$, not by searching along a single line, but by searching along two lines. The direction $\mathbf{d}_k$ connecting two successive points (indicated as dotted lines in the figure) is thus determined only indirectly. We shall see, however, that, in the case where the objective function is quadratic, the $\mathbf{d}_k$'s are the same directions, and the $\mathbf{x}_k$'s are the same points, as would be generated by the method of conjugate gradients.

PARTAN Theorem. *For a quadratic function, PARTAN is equivalent to the method of conjugate gradients.*

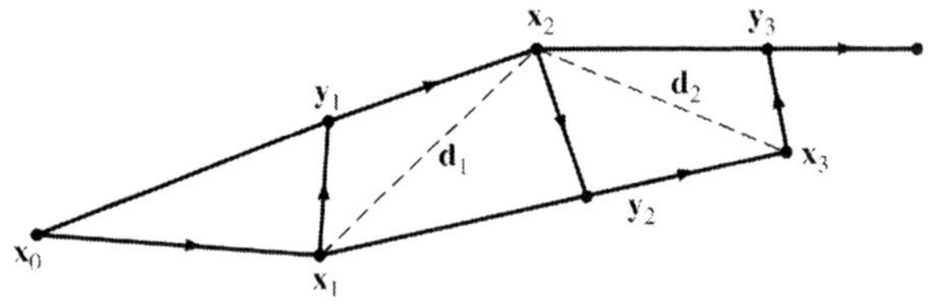

Fig. 9.6 PARTAN

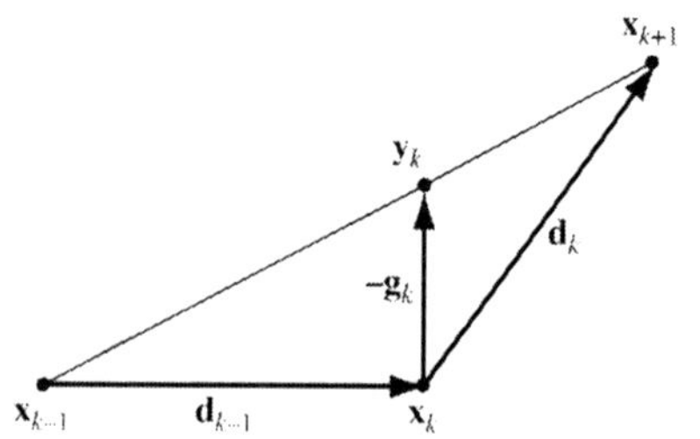

Fig. 9.7 One step of PARTAN

Proof. The proof is by induction. It is certainly true of the first step, since it is a steepest descent step. Suppose that $\mathbf{x}_0$, $\mathbf{x}_1$, ..., $\mathbf{x}_k$ have been generated by the conjugate gradient method and $\mathbf{x}_{k+1}$ is determined according to PARTAN. This single step is shown in Fig. 9.7. We want to show that $\mathbf{x}_{k+1}$ is the same point as would be generated by another step of the conjugate gradient method. For this to be true $\mathbf{x}_{k+1}$ must be that point which minimizes f over the plane defined by $\mathbf{d}_{k-1}$ and $\mathbf{g}_k = \nabla f(\mathbf{x}_k)^T$. From the theory of conjugate gradients, this point will also minimize f over the subspace determined by $\mathbf{g}_k$ and all previous $\mathbf{d}_i$' s. Equivalently, we must find the point $\mathbf{x}$ where $\nabla f(\mathbf{x})$ is orthogonal to both $\mathbf{g}_k$ and $\mathbf{d}_{k-1}$. Since $\mathbf{y}_k$ minimizes f along $\mathbf{g}_k$, we see that $\nabla f(\mathbf{y}_k)$ is orthogonal to $\mathbf{g}_k$. Since $\nabla f(\mathbf{x}_{k-1})$ is contained in the subspace $[\mathbf{d}_0, \mathbf{d}_1, \ldots, \mathbf{d}_{k-1}]$ and because $\mathbf{g}_k$ is orthogonal to this subspace by the Expanding Subspace Theorem, we see that $\nabla f(\mathbf{x}_{k-1})$ is also orthogonal to $\mathbf{g}_k$. Since $\nabla f(\mathbf{x})$ is linear in $\mathbf{x}$, it follows that at every point $\mathbf{x}$ on the line through $\mathbf{x}_{k-1}$ and $\mathbf{y}_k$ we have $\nabla f(\mathbf{x})$ orthogonal to $\mathbf{g}_k$. By minimizing f along this line, a point $\mathbf{x}_{k+1}$ is obtained where in addition $\nabla f(\mathbf{x}_{k+1})$ is orthogonal to the line. Thus $\nabla f(\mathbf{x}_{k+1})$ is orthogonal to both $\mathbf{g}_k$ and the line joining $\mathbf{x}_{k-1}$ and $\mathbf{y}_k$. It follows that $\nabla f(\mathbf{x}_{k+1})$ is orthogonal to the plane. ∎

There are advantages and disadvantages of PARTAN relative to other methods when applied to nonquadratic problems. One attractive feature of the algorithm is its simplicity and ease of implementation. Probably its most desirable property, however, is its strong global convergence characteristics. Each step of the process is at least as good as steepest descent; since going from $\mathbf{x}_k$ to $\mathbf{y}_k$ is exactly steepest descent, and the additional move to $\mathbf{x}_{k+1}$ provides further decrease of the objective function. Thus global convergence is not tied to the fact that the process is restarted every n steps. It is suggested, however, that PARTAN should be restarted every n steps (or $n + 1$ steps) so that it will behave like the conjugate gradient method near the solution.

An undesirable feature of the algorithm is that two line searches are required at each step, except the first, rather than one as is required by, say, the Fletcher–Reeves method. This is at least partially compensated by the fact that searches need not be as accurate for PARTAN, for while inaccurate searches in the Fletcher–Reeves method may yield nonsensical successive search directions, PARTAN will at least do as well as steepest descent.

9.8 Exercises

1. Let $\mathbf{Q}$ be a positive definite symmetric matrix and suppose $\mathbf{p}_0, \mathbf{p}_1, \ldots, \mathbf{p}_{n-1}$ are linearly independent vectors in E^n. Show that a Gram–Schmidt procedure can be used to generate a sequence of $\mathbf{Q}$-conjugate directions from the $\mathbf{p}_i$'s. Specifically, show that $\mathbf{d}_0, \mathbf{d}_1, \ldots, \mathbf{d}_{n-1}$ defined recursively by

$$\mathbf{d}_0 = \mathbf{p}_0$$
$$\mathbf{d}_{k+1} = \mathbf{p}_{k+1} - \sum_{i=0}^{k} \frac{\mathbf{p}_{k+1}^T \mathbf{Q} \mathbf{d}_i}{\mathbf{d}_i^T \mathbf{Q} \mathbf{d}_i} \mathbf{d}_i$$

form's a $\mathbf{Q}$-conjugate set.

2. Suppose the $\mathbf{p}_i$'s in Exercise 1 are generated as *moments* of $\mathbf{Q}$, that is, suppose $\mathbf{p}_k = \mathbf{Q}^k\mathbf{p}_0$, $k = 1, 2, \ldots, n-1$. Show that the corresponding $\mathbf{d}_k$'s can then be generated by a (three-term) recursion formula where $\mathbf{d}_{k+1}$ is defined only in terms of $\mathbf{Qd}_k$, $\mathbf{d}_k$ and $\mathbf{d}_{k-1}$.
3. Suppose the $\mathbf{p}_k$'s in Exercise 1 are taken as $\mathbf{p}_k = \mathbf{e}_k$ where $\mathbf{e}_k$ is the kth unit coordinate vector and the $\mathbf{d}_k$'s are constructed accordingly. Show that using $\mathbf{d}_k$'s in a conjugate direction method to minimize $(1/2)\mathbf{x}^T\mathbf{Qx} - \mathbf{b}^T\mathbf{x}$ is equivalent to the application of Gaussian elimination to solve $\mathbf{Qx} = \mathbf{b}$.
4. Let $f(\mathbf{x}) = (1/2)\mathbf{x}^T\mathbf{Qx} - \mathbf{b}^T\mathbf{x}$ be defined on E^n with $\mathbf{Q}$ positive definite. Let $\mathbf{x}_1$ be a minimum point of f over a subspace of E^n containing the vector $\mathbf{d}$ and let $\mathbf{x}_2$ be the minimum of f over another subspace containing $\mathbf{d}$. Suppose $f(\mathbf{x}_1) < f(\mathbf{x}_2)$. Show that $\mathbf{x}_1 - \mathbf{x}_2$ is $\mathbf{Q}$-conjugate to $\mathbf{d}$.
5. Let $\mathbf{Q}$ be a symmetric matrix. Show that any two eigenvectors of $\mathbf{Q}$, corresponding to distinct eigenvalues, are $\mathbf{Q}$-conjugate.
6. Let $\mathbf{Q}$ be an $n \times n$ symmetric matrix and let $\mathbf{d}_0, \mathbf{d}_1, \ldots, \mathbf{d}_{n-1}$ be $\mathbf{Q}$-conjugate. Show how to find an $\mathbf{E}$ such that $\mathbf{E}^T\mathbf{QE}$ is diagonal.
7. Show that in the conjugate gradient method $\mathbf{Qd}_{k-1} \in \mathcal{B}_{k+1}$.
8. Derive the rate of convergence of the method of steepest descent by viewing it as a one-step optimal process.
9. Let $P^k(\mathbf{Q}) = c_0 + c_1\mathbf{Q} + c_2\mathbf{Q}^2 + \cdots + c_m\mathbf{Q}^m$ be the optimal polynomial in (9.29) minimizing (9.30). Show that the c_i's can be found explicitly by solving the vector equation

$$-\begin{bmatrix} \mathbf{g}_k^T\mathbf{Qg}_k & \mathbf{g}_k^T\mathbf{Q}^2\mathbf{g}_k & \cdots & \mathbf{g}_k^T\mathbf{Q}^{m+1}\mathbf{g}_k \\ \mathbf{g}_k^T\mathbf{Q}^2 g_k & \mathbf{g}_k^T\mathbf{Q}^3\mathbf{g}_k & \cdots & \mathbf{g}_k^T\mathbf{Q}^{m+2}\mathbf{g}_k \\ \vdots & & & \\ \mathbf{g}_k^T\mathbf{Q}^{m+1}\mathbf{g}_k & & \cdots & \mathbf{g}_k^T\mathbf{Q}^{2m+1}\mathbf{g}_k \end{bmatrix} \begin{bmatrix} c_0 \\ c_1 \\ \vdots \\ c_m \end{bmatrix} = \begin{bmatrix} \mathbf{g}_k^T\mathbf{g}_k \\ \mathbf{g}_k^T\mathbf{Qg}_k \\ \vdots \\ \mathbf{g}_k^T\mathbf{Q}^m\mathbf{g}_k \end{bmatrix}$$

Show that this reduces to steepest descent when $m = 0$.
10. Show that for the method of conjugate directions there holds

$$E(\mathbf{x}_k) \leqslant 4\left(\frac{1-\sqrt{\gamma}}{1+\sqrt{\gamma}}\right)^{2k} E(\mathbf{x}_0),$$

where $\gamma = a/A$ and a and A are the smallest and largest eigenvalues of $\mathbf{Q}$. *Hint*: In (9.27) select $P_{k-1}(\lambda)$ so that

$$1 + \lambda P_{k-1}(\lambda) = \frac{T_k\left(\frac{A+a-2\lambda}{A-a}\right)}{T_k\left(\frac{A+a}{A-a}\right)},$$

where $T_k(\lambda) = \cos(k \text{ arc } \cos \lambda)$ is the kth Chebyshev polynomial. This choice gives the minimum maximum magnitude on $[a, A]$. Verify and use the inequality

$$\frac{(1-\gamma)^k}{(1+\sqrt{\gamma})^{2k} + (1-\sqrt{\gamma})^{2k}} \leqslant \left(\frac{1-\sqrt{\gamma}}{1+\sqrt{\gamma}}\right)^k.$$

11. Suppose it is known that each eigenvalue of $\mathbf{Q}$ lies either in the interval $[a, A]$ or in the interval $[a+\Delta, A+\Delta]$ where a, A, and Δ are all positive. Show that the partial conjugate gradient method restarted every two steps will converge with a ratio no greater than $[(A-a)/(A+a)]^2$ no matter how large Δ is.
12. Modify the first method given in Sect. 9.6 so that it is globally convergent.
13. Show that in the purely quadratic form of the conjugate gradient method $\mathbf{d}_k^T\mathbf{Q}\mathbf{d}_k = -\mathbf{d}_k^T\mathbf{Q}\mathbf{g}_k$. Using this show that to obtain $\mathbf{x}_{k+1}$ from $\mathbf{x}_k$ it is necessary to use $\mathbf{Q}$ only to evaluate $\mathbf{g}_k$ and $\mathbf{Q}\mathbf{g}_k$.
14. Show that in the quadratic problem $\mathbf{Q}\mathbf{g}_k$ can be evaluated by taking a unit step from $\mathbf{x}_k$ in the direction of the negative gradient and evaluating the gradient there. Specifically, if $\mathbf{y}_k = \mathbf{x}_k - \mathbf{g}_k$ and $\mathbf{p}_k = \nabla f(\mathbf{y}_k)^T$, then $\mathbf{Q}\mathbf{g}_k = \mathbf{g}_k - \mathbf{p}_k$.
15. Combine the results of Exercises 13 and 14 to derive a conjugate gradient method for general problems much in the spirit of the first method of Sect. 9.6 but which does not require knowledge of $\mathbf{F}(\mathbf{x}_k)$ or a line search.

References

9.1–9.3 For the original development of conjugate direction methods, see Hestenes and Stiefel [H10] and Hestenes [H7], [H9]. For another introductory treatment see Beckman [B8]. The method was extended to the case where Q is not positive definite, which arises in constrained problems, by Luenberger [L9], [L11].

9.4 The idea of viewing the conjugate gradient method as an optimal process was originated by Stiefel [S10]. Also see Daniel [D1] and Faddeev and Faddeeva [F1].

9.5 The partial conjugate gradient method presented here is identical to the so-called s-step gradient method. See Faddeev and Faddeeva [F1] and Forsythe [F14]. The bound on the rate of convergence given in this section in terms of the interval containing the $n-m$ smallest eigenvalues was first given in Luenberger [L13]. Although this bound cannot be expected to be tight, it is a reasonable conjecture that it becomes tight as the m largest eigenvalues tend to infinity with arbitrarily large separation.

9.6 For the first approximate method, see Daniel [D1]. For the line search methods, see Fletcher and Reeves [F12], Polak and Ribiere [P5], and Polak [P4]. For proof of the n-step, order two convergence, see Cohen [C4]. For a survey of computational experience of these methods, see Fletcher [F9].

9.7 PARTAN is due to Shah, Buehler, and Kempthome [S2]. Also see Wolfe [W5].

9.8 The approach indicated in Exercises 1 and 2 can be used as a foundation for the development of conjugate gradients; see Antosiewicz and Rheinboldt [A7], Vorobyev [V6], Faddeev and Faddeeva [F1], and Luenberger [L8]. The result stated in Exercise 3 is due to Hestenes and Stiefel [H10]. Exercise 4 is due to Powell [P6]. For the solution to Exercise 10, see Faddeev and Faddeeva [F1] or Daniel [D1].

Chapter 10
Quasi-Newton Methods

In this chapter we take another approach toward the development of methods lying somewhere intermediate to steepest descent and Newton's method. Again working under the assumption that evaluation and use of the Hessian matrix is impractical or costly, the idea underlying quasi-Newton methods is to use an approximation to the inverse Hessian in place of the true inverse that is required in Newton's method. The form of the approximation varies among different methods—ranging from the simplest where it remains fixed throughout the iterative process, to the more advanced where improved approximations are built up on the basis of information gathered during the descent process.

The quasi-Newton methods that build up an approximation to the inverse Hessian are analytically the most sophisticated methods discussed in this book for solving unconstrained problems and represent the culmination of the development of algorithms through detailed analysis of the quadratic problem. As might be expected, the convergence properties of these methods are somewhat more difficult to discover than those of simpler methods. Nevertheless, we are able, by continuing with the same basic techniques as before, to illuminate their most important features.

In the course of our analysis we develop two important generalizations of the method of steepest descent and its corresponding convergence rate theorem. The first, discussed in Sect. 10.1, modifies steepest descent by taking as the direction vector a positive definite transformation of the negative gradient. The second, discussed in Sect. 10.8, is a combination of steepest descent and Newton's method. Both of these fundamental methods have convergence properties analogous to those of steepest descent.

D.G. Luenberger, Y. Ye, *Linear and Nonlinear Programming*, International Series in Operations Research & Management Science 228,
DOI 10.1007/978-3-319-18842-3_10

10.1 Modified Newton Method

A very basic iterative process for solving the problem

$$\text{minimize} \quad f(\mathbf{x})$$

which includes as special cases most of our earlier ones is

$$\mathbf{x}_{k+1} = \mathbf{x}_k - \alpha_k \mathbf{S}_k \nabla f(\mathbf{x}_k)^T \tag{10.1}$$

where $\mathbf{S}_k$ is a symmetric $n \times n$ matrix and where, as usual, α_k is chosen to minimize $f(\mathbf{x}_{k+1})$. If $\mathbf{S}_k$ is the inverse of the Hessian of f, we obtain Newton's method, while if $\mathbf{S}_k = \mathbf{I}$ we have steepest descent. It would seem to be a good idea, in general, to select $\mathbf{S}_k$ as an approximation to the inverse of the Hessian. We examine that philosophy in this section.

First, we note, as in Sect. 8.5, that in order that the process (10.1) be guaranteed to be a descent method for small values of α, it is necessary in general to require that $\mathbf{S}_k$ be positive definite. We shall therefore always impose this as a requirement.

Because of the similarity of the algorithm (10.1) with steepest descent† it should not be surprising that its convergence properties are similar in character to our earlier results. We derive the actual rate of convergence by considering, as usual, the standard quadratic problem with

$$f(\mathbf{x}) = \frac{1}{2}\mathbf{x}^T \mathbf{Q}\mathbf{x} - \mathbf{b}^T \mathbf{x}, \tag{10.2}$$

where $\mathbf{Q}$ is symmetric and positive definite. For this case we can find an explicit expression for α_k in (10.1). The algorithm becomes

$$\mathbf{x}_{k+1} = \mathbf{x}_k - \alpha_k \mathbf{S}_k \mathbf{g}_k, \tag{10.3a}$$

where

$$\mathbf{g}_k = \mathbf{Q}\mathbf{x}_k - \mathbf{b} \tag{10.3b}$$

$$\alpha_k = \frac{\mathbf{g}_k^T \mathbf{S}_k \mathbf{g}_k}{\mathbf{g}_k^T \mathbf{S}_k \mathbf{Q} \mathbf{S}_k \mathbf{g}_k}. \tag{10.3c}$$

We may then derive the convergence rate of this algorithm by slightly extending the analysis carried out for the method of steepest descent.

Modified Newton Method Theorem (Quadratic Case). *Let* $\mathbf{x}^*$ *be the uniqueminimum point of* f, *and define* $E(\mathbf{x}) = \frac{1}{2}(\mathbf{x} - \mathbf{x}^*)^T \mathbf{Q}(\mathbf{x} - \mathbf{x}^*)$.

---

† The algorithm (10.1) is sometimes referred to as the *method of deflected gradients*, since the direction vector can be thought of as being determined by deflecting the gradient through multiplication by $\mathbf{S}_k$.

Then for the algorithm (10.3) *there holds at every step k*

$$E(\mathbf{x}_{k+1}) \leqslant \left(\frac{B_k - b_k}{B_k + b_k}\right)^2 E(\mathbf{x}_k), \tag{10.4}$$

where b_k and B_k are, respectively, the smallest and largest eigenvalues of thematrix $\mathbf{S}_k\mathbf{Q}$

Proof. We have by direct substitution

$$\frac{E(\mathbf{x}_k) - E(\mathbf{x}_{k+1})}{E(\mathbf{x}_k)} = \frac{(\mathbf{g}_k^T\mathbf{S}_k\mathbf{g}_k)^2}{(\mathbf{g}_k^T\mathbf{S}_k\mathbf{Q}\mathbf{S}_k\mathbf{g}_k)(\mathbf{g}_k^T\mathbf{Q}^{-1}\mathbf{g}_k)}.$$

Letting $\mathbf{T}_k = \mathbf{S}_k^{1/2}\mathbf{Q}\mathbf{S}_k^{1/2}$ and $\mathbf{p}_k = \mathbf{S}_k^{1/2}\mathbf{g}_k$ we obtain

$$\frac{E(\mathbf{x}_k) - E(\mathbf{x}_{k+1})}{E(\mathbf{x}_k)} = \frac{(\mathbf{p}_k^T\mathbf{P}_k)^2}{(\mathbf{p}_k^T\mathbf{T}_k\mathbf{p}_k)(\mathbf{p}_k^T\mathbf{T}_k^{-1}\mathbf{p}_k)}.$$

From the Kantorovich inequality we obtain easily

$$E(\mathbf{x}_{k+1}) \leqslant \left(\frac{B_k - b_k}{B_k + b_k}\right)^2 E(\mathbf{x}_k),$$

where b_k and B_k are the smallest and largest eigenvalues of $\mathbf{T}_k$. Since $\mathbf{S}_k^{1/2}\mathbf{T}_k\mathbf{S}_k^{-1/2} = \mathbf{S}_k\mathbf{Q}$, we see that $\mathbf{S}_k\mathbf{Q}$ is similar to $\mathbf{T}_k$ and therefore has the same eigenvalues. ∎

This theorem supports the intuitive notion that for the quadratic problem one should strive to make $\mathbf{S}_k$ close to $\mathbf{Q}^{-1}$ since then both b_k and B_k would be close to unity and convergence would be rapid. For a nonquadratic objective function f the analog to $\mathbf{Q}$ is the Hessian $\mathbf{F}(\mathbf{x})$, and hence one should try to make $\mathbf{S}_k$ close to $\mathbf{F}(\mathbf{x}_k)^{-1}$.

Two remarks may help to put the above result in proper perspective. The first remark is that both the algorithm (10.1) and the theorem stated above are only simple, minor, and natural extensions of the work presented in Chap. 8 on steepest descent. As such the result of this section can be regarded, correspondingly, not as a new idea but as an extension of the basic result on steepest descent. The second remark is that this one simple result when properly applied can quickly characterize the convergence properties of some fairly complex algorithms. Thus, rather than an isolated result concerned with a specific form of algorithm, the theorem above should be regarded as a general tool for convergence analysis. It provides significant insight into various quasi-Newton methods discussed in this chapter.

A Classical Method

We conclude this section by mentioning the *classical modified Newton's method*, a standard method for approximating Newton's method without evaluating $\mathbf{F}(\mathbf{x}_k)^{-1}$ for each k. We set

$$\mathbf{x}_{k+1} = \mathbf{x}_k - \alpha_k[\mathbf{F}(\mathbf{x}_0)]^{-1}\nabla f(\mathbf{x}_k)^T. \tag{10.5}$$

In this method the Hessian at the initial point $\mathbf{x}_0$ is used throughout the process. The effectiveness of this procedure is governed largely by how fast the Hessian is changing—in other words, by the magnitude of the third derivatives of f.

10.2 Construction of the Inverse

The fundamental idea behind most quasi-Newton methods is to try to construct the inverse Hessian, or an approximation of it, using information gathered as the descent process progresses. The current approximation $\mathbf{H}_k$ is then used at each stage to define the next descent direction by setting $\mathbf{S}_k = \mathbf{H}_k$ in the modified Newton method. Ideally, the approximations converge to the inverse of the Hessian at the solution point and the overall method behaves somewhat like Newton's method. In this section we show how the inverse Hessian can be built up from gradient information obtained at various points.

Let f be a function on E^n that has continuous second partial derivatives. If for two points $\mathbf{x}_{k+1}$, $\mathbf{x}_k$ we define $\mathbf{g}_{k+1} = \nabla f(\mathbf{x}_{k+1})^T$, $\mathbf{g}_k = \nabla f(\mathbf{x}_k)^T$ and $\mathbf{p}_k = \mathbf{x}_{k+1} - \mathbf{x}_k$, then

$$\mathbf{g}_{k+1} - \mathbf{g}_k \cong \mathbf{F}(\mathbf{x}_k)\mathbf{p}_k. \tag{10.6}$$

If the Hessian, $\mathbf{F}$, is constant, then we have

$$\mathbf{q}_k \equiv \mathbf{g}_{k+1} - \mathbf{g}_k = \mathbf{F}\mathbf{p}_k, \tag{10.7}$$

and we see that evaluation of the gradient at two points gives information about $\mathbf{F}$. If n linearly independent directions $\mathbf{p}_0, \mathbf{p}_1, \mathbf{p}_2, \ldots, \mathbf{p}_{n-1}$ and the corresponding $\mathbf{q}_k$'s are known, then $\mathbf{F}$ is uniquely determined. Indeed, letting $\mathbf{P}$ and $\mathbf{Q}$ be the $n \times n$ matrices with columns $\mathbf{p}_k$ and $\mathbf{q}_k$ respectively, we have $\mathbf{F} = \mathbf{Q}\mathbf{P}^{-1}$.

It is natural to attempt to construct successive approximations $\mathbf{H}_k$ to $\mathbf{F}^{-1}$ based on data obtained from the first k steps of a descent process in such a way that if $\mathbf{F}$ were constant the approximation would be consistent with (10.7) for these steps. Specifically, if $\mathbf{F}$ were constant $\mathbf{H}_{k+1}$ would satisfy

$$\mathbf{H}_{k+1}\mathbf{q}_i = \mathbf{p}_i, \quad 0 \leqslant i \leqslant k. \tag{10.8}$$

After n linearly independent steps we would then have $\mathbf{H}_n = \mathbf{F}^{-1}$.

For any $k < n$ the problem of constructing a suitable $\mathbf{H}_k$, with in general serves as an approximation to the inverse Hessian and which in the case of constant $\mathbf{F}$ satisfies (10.8), admits an infinity of solutions, since there are more degrees of freedom than there are constraints. Thus a particular method can take into account additional considerations. We discuss below one of the simplest schemes that has been proposed.

Rank One Correction

Since $\mathbf{F}$ and $\mathbf{F}^{-1}$ are symmetric, it is natural to require that $\mathbf{H}_k$, the approximation to $\mathbf{F}^{-1}$, be symmetric. We investigate the possibility of defining a recursion of the form

$$\mathbf{H}_{k+1} = \mathbf{H}_k + a_k \mathbf{z}_k \mathbf{z}_k^T, \tag{10.9}$$

which preserves symmetry. The vector $\mathbf{z}_k$ and the constant a_k define a matrix of (at most) rank one, by which the approximation to the inverse is updated. We select them so that (10.8) is satisfied. Setting i equal to k in (10.8) and substituting (10.9) we obtain

$$\mathbf{p}_k = \mathbf{H}_{k+1}\mathbf{q}_k = \mathbf{H}_k\mathbf{q}_k + a_k \mathbf{z}_k \mathbf{z}_k^T \mathbf{q}_k. \tag{10.10}$$

Taking the inner product with $\mathbf{q}_k$ we have

$$\mathbf{q}_k^T \mathbf{p}_k - \mathbf{q}_k^T \mathbf{H}_k \mathbf{q}_k = a_k \left(\mathbf{z}_k^T \mathbf{q}_k\right)^2 \tag{10.11}$$

On the other hand, using (10.10) we may write (10.9) as

$$\mathbf{H}_{k+1} = \mathbf{H}_k + \frac{(\mathbf{p}_k - \mathbf{H}_k\mathbf{q}_k)(\mathbf{p}_k - \mathbf{H}_k\mathbf{q}_k)^T}{a_k \left(\mathbf{z}_k^T \mathbf{q}_k\right)^2},$$

which in view of (10.11) leads finally to

$$\mathbf{H}_{k+1} = \mathbf{H}_k + \frac{(\mathbf{p}_k - \mathbf{H}_k\mathbf{q}_k)(\mathbf{p}_k - \mathbf{H}_k\mathbf{q}_k)^T}{\mathbf{q}_k^T(\mathbf{p}_k - \mathbf{H}_k\mathbf{q}_k)}. \tag{10.12}$$

We have determined what a rank one correction must be if it is to satisfy (10.8) for $i = k$. It remains to be shown that, for the case where $\mathbf{F}$ is constant, (10.8) is also satisfied for $i < k$. This in turn will imply that the rank one recursion converges to $\mathbf{F}^{-1}$ after at most n steps.

Theorem. *Let* $\boldsymbol{F}$ *be a fixed symmetric matrix and suppose that* $\mathbf{p}_0, \boldsymbol{p}_1, \mathbf{p}_2, \ldots, \boldsymbol{p}_k$ *are given vectors. Define the vectors* $\mathbf{q}_i = \boldsymbol{F}\boldsymbol{p}_i$, $i = 0, 1, 2, \ldots, k$.

Starting with any initial symmetric matrix $\mathbf{H}_0$ *let*

$$\boldsymbol{H}_{i+1} = \boldsymbol{H}_i + \frac{(\boldsymbol{p}_i - \boldsymbol{H}_i\boldsymbol{q}_i)(\boldsymbol{p}_i - \boldsymbol{H}_i\boldsymbol{q}_i)^T}{\boldsymbol{q}_i^T(\boldsymbol{p}_i - \boldsymbol{H}_i\boldsymbol{q}_i)}. \tag{10.13}$$

Then

$$\boldsymbol{p}_i = \boldsymbol{H}_{k+1}\boldsymbol{q}_i \quad \textit{for} \quad i \leqslant k. \tag{10.14}$$

Proof. The proof is by induction. Suppose it is true for $\mathbf{H}_k$, and $i \leqslant k-1$. The relation was shown above to be true for $\mathbf{H}_{k+1}$ and $i = k$. For $i < k$

$$\mathbf{H}_{k+1}\mathbf{q}_i = \mathbf{H}_k\mathbf{q}_i + \mathbf{y}_k(\mathbf{p}_k^T\mathbf{q}_i - \mathbf{q}_k^T\mathbf{H}_k\mathbf{q}_i), \tag{10.15}$$

where

$$\mathbf{y}_k = \frac{(\mathbf{p}_k - \mathbf{H}_k\mathbf{q}_k)}{\mathbf{q}_k^T(\mathbf{p}_k - \mathbf{H}_k\mathbf{q}_k)}.$$

By the induction hypothesis, (10.15) becomes

$$\mathbf{H}_{k+1}\mathbf{q}_i = \mathbf{p}_i + \mathbf{y}_k\left(\mathbf{p}_k^T\mathbf{q}_i - \mathbf{q}_k^T\mathbf{p}_i\right).$$

From the calculation

$$\mathbf{q}_k^T\mathbf{p}_i = \mathbf{p}_k^T\mathbf{F}\mathbf{p}_i = \mathbf{p}_k^T\mathbf{q}_i,$$

it follows that the second term vanishes. ∎

To incorporate the approximate inverse Hessian in a descent procedure while simultaneously improving it, we calculate the direction $\mathbf{d}_k$ From

$$\mathbf{d}_k = -\mathbf{H}_k\mathbf{g}_k$$

and then minimize $f(\mathbf{x}_k + \alpha\mathbf{d}_k)$ with respect to $\alpha \geqslant 0$. This determines $\mathbf{x}_{k+1} = \mathbf{x}_k + \alpha_k\mathbf{d}_k$, $\mathbf{p}_k = \alpha_k\mathbf{d}_k$, and $\mathbf{g}_{k+1}$. Then $\mathbf{H}_{k+1}$ can be calculated according to (10.12).

There are some difficulties with this simple rank one procedure. First, the updating formula (10.12) preserves positive definiteness only if $\mathbf{q}_k^T(\mathbf{p}_k - \mathbf{H}_k\mathbf{q}_k) > 0$, which cannot be guaranteed (see Exercise 6). Also, even if $\mathbf{q}_k^T(\mathbf{p}_k - \mathbf{H}_k\mathbf{q}_k)$ is positive, it may be small, which can lead to numerical difficulties. Thus, although an excellent simple example of how information gathered during the descent process can in principle be used to update an approximation to the inverse Hessian, the rank one method possesses some limitations.

10.3 Davidon-Fletcher-Powell Method

The earliest, and certainly one of the most clever schemes for constructing the inverse Hessian, was originally proposed by Davidon and later developed by Fletcher and Powell. It has the fascinating and desirable property that, for a quadratic objective, it simultaneously generates the directions of the conjugate gradient method while constructing the inverse Hessian. At each step the inverse Hessian is updated by the sum of two symmetric rank one matrices, and this scheme is therefore often referred to as a *rank two correction procedure*. The method is also often referred to as the *variable metric method*, the name originally suggested by Davidon.

The procedure is this: Starting with any symmetric positive definite matrix $\mathbf{H}_0$, any point $\mathbf{x}_0$, and with $k = 0$,

Step 1. Set $\mathbf{d}_k = -\mathbf{H}_k\mathbf{g}_k$.

Step 2. Minimize $f(\mathbf{x}_k + \alpha\mathbf{d}_k)$ with respect to $\alpha \geqslant 0$ to obtain $\mathbf{x}_{k+1}$, $\mathbf{p}_k = \alpha_k\mathbf{d}_k$, and $\mathbf{g}_{k+1}$.

Step 3. Set $\mathbf{q}_k = \mathbf{g}_{k+1} - \mathbf{g}_k$ and

$$\mathbf{H}_{k+1} = \mathbf{H}_k + \frac{\mathbf{p}_k\mathbf{p}_k^T}{\mathbf{p}_k^T\mathbf{q}_k} - \frac{\mathbf{H}_k\mathbf{q}_k\mathbf{q}_k^T\mathbf{H}_k}{\mathbf{q}_k^T\mathbf{H}_k\mathbf{q}_k}. \tag{10.16}$$

Update k and return to Step 1.

Positive Definiteness

We first demonstrate that if $\mathbf{H}_k$ is positive definite, then so is $\mathbf{H}_{k+1}$. For any $\mathbf{x} \in E^n$ we have

$$\mathbf{x}^T\mathbf{H}_{k+1}\mathbf{x} = \mathbf{x}^T\mathbf{H}_k\mathbf{x} + \frac{(\mathbf{x}^T\mathbf{p}_k)^2}{\mathbf{p}_k^T\mathbf{q}_k} - \frac{(\mathbf{x}^T\mathbf{H}_k\mathbf{q}_k)^2}{\mathbf{q}_k^T\mathbf{H}_k\mathbf{q}_k}. \tag{10.17}$$

Defining $\mathbf{a} = \mathbf{H}_k^{1/2}\mathbf{x}$, $\mathbf{b} = \mathbf{H}_k^{1/2}\mathbf{q}_k$ we may rewrite (10.17) as

$$\mathbf{x}^T\mathbf{H}_{k+1}\mathbf{x} = \frac{(\mathbf{a}^T\mathbf{a})(\mathbf{b}^T\mathbf{b}) - (\mathbf{a}^T\mathbf{b})^2}{(\mathbf{b}^T\mathbf{b})} + \frac{(\mathbf{x}^T\mathbf{p}_k)^2}{\mathbf{p}_k^T\mathbf{q}_k}.$$

We also have

$$\mathbf{p}_k^T\mathbf{q}_k = \mathbf{p}_k^T\mathbf{g}_{k+1} - \mathbf{p}_k^T\mathbf{g}_k = -\mathbf{p}_k^T\mathbf{g}_k, \tag{10.18}$$

since

$$\mathbf{p}_k^T\mathbf{g}_{k+1} = 0, \tag{10.19}$$

because $\mathbf{x}_{k+1}$ is the minimum point of f along $\mathbf{p}_k$. Thus by definition of $\mathbf{p}_k$

$$\mathbf{p}_k^T\mathbf{q}_k = \alpha_k\mathbf{g}_k^T\mathbf{H}_k\mathbf{g}_k, \tag{10.20}$$

and hence

$$\mathbf{x}^T\mathbf{H}_{k+1}\mathbf{x} = \frac{(\mathbf{a}^T\mathbf{a})(\mathbf{b}^T\mathbf{b}) - (\mathbf{a}^T\mathbf{b})^2}{(\mathbf{b}^T\mathbf{b})} + \frac{(\mathbf{x}^T\mathbf{p}_k)^2}{\alpha_k\mathbf{g}_k^T\mathbf{H}_k\mathbf{g}_k}. \tag{10.21}$$

Both terms on the right of (10.21) are nonnegative—the first by the Cauchy–Schwarz inequality. We must only show they do not both vanish simultaneously. The first term vanishes only if a and $\mathbf{b}$ are proportional. This in turn implies that $\mathbf{x}$ and $\mathbf{q}_k$ are proportional, say $\mathbf{x} = \beta\mathbf{q}_k$. In that case, however,

$$\mathbf{p}_k^T\mathbf{x} = \beta\mathbf{p}_k^T\mathbf{q}_k = \beta\alpha_k\mathbf{g}_k^T\mathbf{H}_k\mathbf{g}_k \neq 0$$

from (10.20). Thus $\mathbf{x}^T\mathbf{H}_{k+1}\mathbf{x} > 0$ for all nonzero $\mathbf{x}$.

It is of interest to note that in the proof above the fact that α_k is chosen as the minimum point of the line search was used in (10.19), which led to the important conclusion $\mathbf{p}_k^T\mathbf{q}_k > 0$. Actually any α_k, whether the minimum point or not, that gives $\mathbf{p}_k^T\mathbf{q}_k > 0$ can be used in the algorithm, and $\mathbf{H}_{k+1}$ will be positive definite (see Exercises 8 and 9).

Finite Step Convergence

We assume now that f is quadratic with (constant) Hessian $\mathbf{F}$. We show in this case that the Davidon-Fletcher-Powell method produces direction vectors $\mathbf{p}_k$ that are $\mathbf{F}$-orthogonal and that if the method is carried n steps then $\mathbf{H}_n = \mathbf{F}^{-1}$.

Theorem. *If f is quadratic with positive definite Hessian $\mathbf{F}$, then for the Davidon-Fletcher-Powell method*

$$\boldsymbol{p}_i^T \boldsymbol{F} \boldsymbol{p}_j = 0, \qquad 0 \leqslant i < j \leqslant k \tag{10.22}$$

$$\boldsymbol{H}_{k+1} \boldsymbol{F} \boldsymbol{p}_i = \boldsymbol{p}_i \quad for \quad 0 \leqslant i \leqslant k. \tag{10.23}$$

Proof. We note that for the quadratic case

$$\mathbf{q}_k = \mathbf{g}_{k+1} - \mathbf{g}_k = \mathbf{F}\mathbf{x}_{k+1} - \mathbf{F}\mathbf{x}_k = \mathbf{F}\mathbf{p}_k. \tag{10.24}$$

Also

$$\mathbf{H}_{k+1}\mathbf{F}\mathbf{p}_k = \mathbf{H}_{k+1}\mathbf{q}_k = \mathbf{p}_k \tag{10.25}$$

from (10.16).

We now prove (10.22) and (10.23) by induction. From (10.25) we see that they are true for $k = 0$. Assuming they are true for $k - 1$, we prove they are true for k. We have

$$\mathbf{g}_k = \mathbf{g}_{i+1} + \mathbf{F}(\mathbf{p}_{i+1} + \cdots + \mathbf{p}_{k-1}).$$

Therefore from (10.22) and (10.19)

$$\mathbf{p}_i^T \mathbf{g}_k = \mathbf{p}_i^T \mathbf{g}_{i+1} = 0 \quad \text{for} \quad 0 \leqslant i < k. \tag{10.26}$$

Hence from (10.23)

$$\mathbf{p}_i^T \mathbf{F}\mathbf{H}_k\mathbf{g}_k = 0. \tag{10.27}$$

Thus since $\mathbf{p}_k = -\alpha_k \mathbf{H}_k \mathbf{g}_k$ and since $\alpha_k \neq 0$, we obtain

$$\mathbf{p}_i^T \mathbf{F}\mathbf{p}_k = 0 \quad \text{for} \quad i < k, \tag{10.28}$$

which proves (10.22) for k.

Now since from (10.23) for $k - 1$, (10.24) and (10.28)

$$\mathbf{q}_k^T \mathbf{H}_k \mathbf{F}\mathbf{p}_i = \mathbf{q}_k^T \mathbf{p}_i = \mathbf{p}_k^T \mathbf{F}\mathbf{p}_i = 0, \quad 0 \leqslant i < k$$

we have

$$\mathbf{H}_{k+1}\mathbf{F}\mathbf{p}_i = \mathbf{H}_k\mathbf{F}\mathbf{p}_i = \mathbf{p}_i, \quad 0 \leqslant i < k.$$

This together with (10.25) proves (10.23) for k. ∎

Since the $\mathbf{p}_k$'s are $\mathbf{F}$-orthogonal and since we minimize f successively in these directions, we see that the method is a conjugate direction method. Furthermore, if the initial approximation $\mathbf{H}_0$ is taken equal to the identity matrix, the method becomes the conjugate gradient method. In any case the process obtains the overall minimum point within n steps.

Finally, (10.23) shows that $\mathbf{p}_0, \mathbf{p}_1, \mathbf{p}_2, \ldots, \mathbf{p}_k$ are eigenvectors corresponding to unity eigenvalue for the matrix $\mathbf{H}_{k+1}\mathbf{F}$. These eigenvectors are linearly independent, since they are $\mathbf{F}$-orthogonal, and therefore $\mathbf{H}_n = \mathbf{F}^{-1}$.

10.4 The Broyden Family

The updating formulae for the inverse Hessian considered in the previous two sections are based on satisfying

$$\mathbf{H}_{k+1}\mathbf{q}_i = \mathbf{p}_i, \quad 0 \leqslant i \leqslant k, \tag{10.29}$$

which is derived from the relation

$$\mathbf{q}_i = \mathbf{F}\mathbf{p}_i, \qquad 0 \leqslant i \leqslant k, \tag{10.30}$$

which would hold in the purely quadratic case. It is also possible to update approximations to the Hessian $\mathbf{F}$ itself, rather than its inverse. Thus, denoting the kth approximation of $\mathbf{F}$ by $\mathbf{B}_k$, we would, analogously, seek to satisfy

$$\mathbf{q}_i = \mathbf{B}_{k+1}\mathbf{p}_i, \quad 0 \leqslant i \leqslant k. \tag{10.31}$$

Equation (10.31) has exactly the same form as (10.29) except that $\mathbf{q}_i$ and $\mathbf{p}_i$ are interchanged and $\mathbf{H}$ is replaced by B. It should be clear that this implies that any update formula for $\mathbf{H}$ derived to satisfy (10.29) can be transformed into a corresponding update formula for B. Specifically, given any update formula for $\mathbf{H}$, the *complementary* formula is found by interchanging the roles of $\mathbf{B}$ and $\mathbf{H}$ and of $\mathbf{q}$ and $\mathbf{p}$. Likewise, any updating formula for $\mathbf{B}$ that satisfies (10.31) can be converted by the same process to a complementary formula for updating H. It is easily seen that taking the complement of a complement restores the original formula.

To illustrate complementary formulae, consider the rank one update of Sect. 10.2, which is

$$\mathbf{H}_{k+1} = \mathbf{H}_k + \frac{(\mathbf{p}_k - \mathbf{H}_k\mathbf{q}_k)(\mathbf{p}_k - \mathbf{H}_k\mathbf{q}_k)^T}{\mathbf{q}_k^T(\mathbf{p}_k - \mathbf{H}_k\mathbf{q}_k)}. \tag{10.32}$$

The corresponding complementary formula is

$$\mathbf{B}_{k+1} = \mathbf{B}_k + \frac{(\mathbf{q}_k - \mathbf{B}_k\mathbf{p}_k)(\mathbf{q}_k - \mathbf{B}_k\mathbf{p}_k)^T}{\mathbf{p}_k^T(\mathbf{q}_k - \mathbf{B}_k\mathbf{p}_k)}. \tag{10.33}$$

Likewise, the Davidon-Fletcher-Powell (or simply DFP) formula is

$$\mathbf{H}_{k+1}^{\mathrm{DFP}} = \mathbf{H}_k + \frac{\mathbf{p}_k\mathbf{p}_k^T}{\mathbf{p}_k^T\mathbf{q}_k} - \frac{\mathbf{H}_k\mathbf{q}_k\mathbf{q}_k^T\mathbf{H}_k}{\mathbf{q}_k^T\mathbf{H}_k\mathbf{q}_k}, \tag{10.34}$$

and its complement is

$$\mathbf{B}_{k+1} = \mathbf{B}_k + \frac{\mathbf{q}_k\mathbf{q}_k^T}{\mathbf{q}_k^T\mathbf{p}_k} - \frac{\mathbf{B}_k\mathbf{p}_k\mathbf{p}_k^T\mathbf{B}_k}{\mathbf{p}_k^T\mathbf{B}_k\mathbf{p}_k}. \tag{10.35}$$

This last update is known as the Broyden-Fletcher-Goldfarb-Shanno update of $\mathbf{B}_k$, and it plays an important role in what follows.

Another way to convert an updating formula for $\mathbf{H}$ to one for $\mathbf{B}$ or vice versa is to take the inverse. Clearly, if

$$\mathbf{H}_{k+1}\mathbf{q}_i = \mathbf{p}_i, \quad 0 \leqslant i \leqslant k, \tag{10.36}$$

then

$$\mathbf{q}_i = \mathbf{H}_{k+1}^{-1}\mathbf{p}_i, \quad 0 \leqslant i \leqslant k, \tag{10.37}$$

which implies that $\mathbf{H}_{k+1}^{-1}$ satisfies (10.31), the criterion for an update of B. Also, most importantly, the inverse of a rank two formula is itself a rank two formula.

The new formula can be found explicitly by two applications of the general inversion identity (often referred to as the Sherman-Morrison formula)

$$\left[\mathbf{A} + \mathbf{a}\mathbf{b}^T\right]^{-1} = \mathbf{A}^{-1} - \frac{\mathbf{A}^{-1}\mathbf{a}\mathbf{b}^T\mathbf{A}^{-1}}{1 + \mathbf{b}^T\mathbf{A}^{-1}\mathbf{a}}, \tag{10.38}$$

where A is an $n \times n$ matrix, and a and $\mathbf{b}$ are n-vectors, which is valid provided the inverses exist. (This is easily verified by multiplying through by $\mathbf{A} + \mathbf{a}\mathbf{b}^T$.)

The Broyden-Fletcher-Goldfard-Shanno update for $\mathbf{B}$ produces, by taking the inverse, a corresponding update for $\mathbf{H}$ of the form

$$\mathbf{H}_{k+1}^{\mathrm{BFGS}} = \mathbf{H}_k + \left(\frac{1 + \mathbf{q}_k^T\mathbf{H}_k\mathbf{q}_k}{\mathbf{q}_k^T\mathbf{q}_k}\right)\frac{\mathbf{p}_k\mathbf{p}_k^T}{\mathbf{p}_k^T\mathbf{q}_k} - \frac{\mathbf{p}_k\mathbf{q}_k^T\mathbf{H}_k + \mathbf{H}_k\mathbf{q}_k\mathbf{q}_k^T}{\mathbf{q}_k^T\mathbf{p}_k}. \tag{10.39}$$

This is an important update formula that can be used exactly like the DFP formula. Numerical experiments have repeatedly indicated that its performance is superior to that of the DFP formula, and for this reason it is now generally preferred.

It can be noted that both the DFP and the BFGS updates have symmetric rank two corrections that are constructed from the vectors $\mathbf{p}_k$ and $\mathbf{H}_k\mathbf{q}_k$. Weighted combinations of these formulae will therefore also be of this same type (symmetric, rank two, and constructed from $\mathbf{p}_k$ and $\mathbf{H}_k\mathbf{q}_k$). This observation naturally leads to consideration of a whole collection of updates, known as the Broyden family, defined by

$$\mathbf{H}^{\phi} = (1 - \phi)\mathbf{H}^{\mathrm{DFP}} + \phi\mathbf{H}^{\mathrm{BFGS}}, \tag{10.40}$$

where ϕ is a parameter that may take any real value. Clearly $\phi = 0$ and $\phi = 1$ yield the DFP and BFGS updates, respectively. The Broyden family also includes the rank one update (see Exercise 12).

An explicit representation of the Broyden family can be found, after a fair amount of algebra, to be

$$\mathbf{H}^{\phi}_{k+1} = \mathbf{H}_k + \frac{\mathbf{p}_k\mathbf{p}_k^T}{\mathbf{p}_k^T\mathbf{q}_k} - \frac{\mathbf{H}_k\mathbf{q}_k\mathbf{q}_k^T\mathbf{H}_k}{\mathbf{q}_k^T\mathbf{H}_k\mathbf{q}_k} + \phi\mathbf{v}_k\mathbf{v}_k^T = \mathbf{H}^{\mathbf{DFP}}_{k+1} + \phi\mathbf{v}_k\mathbf{v}_k^T, \tag{10.41}$$

where

$$\mathbf{v}_k = (\mathbf{q}_k^T\mathbf{H}_k\mathbf{q}_k)^{1/2}\left(\frac{\mathbf{p}_k}{\mathbf{p}_k^T\mathbf{q}_k} - \frac{\mathbf{H}_k\mathbf{q}_k}{\mathbf{q}_k^T\mathbf{H}_k\mathbf{q}_k}\right).$$

This form will be useful in some later developments.

A *Broyden method* is defined as a quasi-Newton method in which at each iteration a member of the Broyden family is used as the updating formula. The parameter ϕ is, in general, allowed to vary from one iteration to another, so a particular Broyden method is defined by a sequence $\phi_1, \phi_2, \ldots,$ of parameter values. A *pure* Broyden method is one that uses a constant ϕ.

Since both $\mathbf{H}^{\mathrm{DFP}}$ and $\mathbf{H}^{\mathrm{BFGS}}$ satisfy the fundamental relation (10.29) for updates, this relation is also satisfied by all members of the Broyden family. Thus it can be expected that many properties that were found to hold for the DFP method will also hold for any Broyden method, and indeed this is so. The following is a direct extension of the theorem of Sect. 10.3.

Theorem. *If f is quadratic with positive definite Hessian* $\mathbf{F}$, *then for a Broydenmethod*

$$\boldsymbol{p}_i^T\boldsymbol{F}\boldsymbol{p}_j = 0, \qquad 0 \leqslant i < j \leqslant k$$
$$\boldsymbol{H}_{k+1}\boldsymbol{F}\boldsymbol{p}_i = \boldsymbol{p}_i \quad \textit{for } 0 \leqslant i \leqslant k.$$

Proof. The proof parallels that of Sect. 10.3, since the results depend only on the basic relation (10.29) and the orthogonality (10.19) because of exact line search. ∎

The Broyden family does not necessarily preserve positive definiteness of $\mathbf{H}^{\phi}$ for all values of ϕ. However, we know that the DFP method does preserve positive definiteness. Hence from (10.41) it follows that positive definiteness is preserved for any $\phi \geqslant 0$, since the sum of a positive definite matrix and a positive semidefinite matrix is positive definite. For $\phi < 0$ there is the possibility that $\mathbf{H}^{\phi}$ may become singular, and thus special precautions should be introduced. In practice $\phi \geqslant 0$ is usually imposed to avoid difficulties.

There has been considerable experimentation with Broyden methods to determine superior strategies for selecting the sequence of parameters ϕ_k.

The above theorem shows that the choice is irrelevant in the case of a quadratic objective and accurate line search. More surprisingly, it has been shown that even for the case of *nonquadratic* functions and accurate line searches, the points generated by all Broyden methods will coincide (provided singularities are avoided and multiple minima are resolved consistently). This means that differences in methods are important only with inaccurate line search.

For general nonquadratic functions of modest dimension, Broyden methods seem to offer a combination of advantages as attractive general procedures. First, they require only that first-order (that is, gradient) information be available. Second, the directions generated can always be guaranteed to be directions of descent by arranging for $\mathbf{H}_k$ to be positive definite throughout the process. Third, since for a quadratic problem the matrices $\mathbf{H}_k$ converge to the inverse Hessian in at most n steps, it might be argued that in the general case $\mathbf{H}_k$ will converge to the inverse Hessian at the solution, and hence convergence will be superlinear. Unfortunately, while the methods are certainly excellent, their convergence characteristics require more careful analysis, and this will lead us to an important additional modification.

Partial Quasi-Newton Methods

There is, of course, the option of restarting a Broyden method every $m + 1$ steps, where $m + 1 < n$. This would yield a *partial quasi-Newton method* that, for small values of m, would have modest storage requirements, since the approximate inverse Hessian could be stored implicitly by storing only the vectors $\mathbf{p}_i$ and $\mathbf{q}_i$, $i \leqslant m+1$. In the quadratic case this method exactly corresponds to the partial conjugate gradient method and hence it has similar convergence properties.

10.5 Convergence Properties

The various schemes for simultaneously generating and using an approximation to the inverse Hessian are difficult to analyze definitively. One must therefore, to some extent, resort to the use of analogy and approximate analyses to determine their effectiveness. Nevertheless, the machinery we developed earlier provides a basis for at least a preliminary analysis.

Global Convergence

In practice, quasi-Newton methods are usually executed in a continuing fashion, starting with an initial approximation and successively improving it throughout the iterative process. Under various and somewhat stringent conditions, it can be proved that this procedure is globally convergent. If, on the other hand, the quasi-Newton methods are restarted every n or $n + 1$ steps by resetting the approximate inverse Hessian to its initial value, then global convergence is guaranteed by the presence of the first descent step of each cycle (which acts as a spacer step).

Local Convergence

The local convergence properties of quasi-Newton methods in the pure form discussed so far are not as good as might first be thought. Let us focus on the local convergence properties of these methods when executed with the restarting feature. Specifically, consider a Broyden method and for simplicity assume that at the beginning of each cycle the approximate inverse Hessian is reset to the identity matrix. Each cycle, if at least n steps in duration, will then contain one complete cycle of an approximation to the conjugate gradient method. Asymptotically, in the tail of the generated sequence, this approximation becomes arbitrarily accurate, and hence we may conclude, as for any method that asymptotically approaches the conjugate gradient method, that the method converges superlinearly (at least if viewed at the end of each cycle). Although superlinear convergence is attractive, the fact that in this case it hinges on repeated cycles of n steps in duration can seriously detract from its practical significance for problems with large n, since we might hope to terminate the procedure before completing even a single full cycle of n steps.

To obtain insight into the defects of the method, let us consider a special situation. Suppose that f is quadratic and that the eigenvalues of the Hessian, $\mathbf{F}$, of f are close together but all very large. If, starting with the identity matrix, an approximation to the inverse Hessian is updated m times, the matrix $\mathbf{H}_m\mathbf{F}$ will have m eigenvalues equal to unity and the rest will still be large. Thus, the ratio of smallest to largest eigenvalue of $\mathbf{H}_m\mathbf{F}$, the condition number, will be worse than for $\mathbf{F}$ itself. Therefore, if the updating were discontinued and $\mathbf{H}_m$ were used as the approximation to $\mathbf{F}^{-1}$ in future iterations according to the procedure of Sect. 10.1, we see that convergence would be poorer than it would be for ordinary steepest descent. In other words, the approximations to $\mathbf{F}^{-1}$ generated by the updating formulas, although accurate over the subspace traveled, do not necessarily improve and, indeed, are likely to worsen the eigenvalue structure of the iteration process.

In practice a poor eigenvalue structure arising in this manner will play a dominating role whenever there are factors that tend to weaken its approximation to the conjugate gradient method. Common factors of this type are round-off errors, inaccurate line searches, and nonquadratic terms in the objective function. Indeed, it has been frequently observed, empirically, that performance of the DFP method is highly sensitive to the accuracy of the line search algorithm—to the point where superior step-wise convergence properties can only be obtained through excessive time expenditure in the line search phase.

Example. To illustrate some of these conclusions we consider the six-dimensional problem defined by

$$f(\mathbf{x}) = \frac{1}{2}\mathbf{x}^T\mathbf{Q}\mathbf{x},$$

where

$$\mathbf{Q} = \begin{bmatrix} 40 & 0 & 0 & 0 & 0 & 0 \\ 0 & 38 & 0 & 0 & 0 & 0 \\ 0 & 0 & 36 & 0 & 0 & 0 \\ 0 & 0 & 0 & 34 & 0 & 0 \\ 0 & 0 & 0 & 0 & 32 & 0 \\ 0 & 0 & 0 & 0 & 0 & 30 \end{bmatrix}.$$

This function was minimized iteratively (the solution is obviously $\mathbf{x}^* = 0$) starting at $\mathbf{x}_0 = (10, 10, 10, 10, 10, 10)$, with $f(\mathbf{x}_0) = 10{,}500$, by using, alternatively, the method of steepest descent, the DFP method, the DFP method restarted every six steps, and the self-scaling method described in the next section. For this quadratic problem the appropriate step size to take at any stage can be calculated by a simple formula. On different computer runs of a given method, different levels of error were deliberately introduced into the step size in order to observe the effect of line search accuracy. This error took the form of a fixed percentage increase over the optimal value. The results are presented below:

CASE 1. No error in step size α

Iteration	Function value			
	Steepest descent	DFP	DFP (with restart)	Self-scaling
1	96.29630	96.29630	96.29630	96.29630
2	1.560669	6.900839×10^{-1}	6.900839×10^{-1}	6.900839×10^{-1}
3	2.932559×10^{-2}	3.988497×10^{-3}	3.988497×10^{-3}	3.988497×10^{-3}
4	5.787315×10^{-4}	1.683310×10^{-5}	1.683310×10^{-5}	1.683310×10^{-5}
5	1.164595×10^{-5}	3.878639×10^{-8}	3.878639×10^{-8}	3.878639×10^{-8}
6	2.359563×10^{-7}			

CASE 2. 0.1 % error in step size α

Iteration	Function value			
	Steepest descent	DFP	DFP (with restart)	Self-scaling
1	96.30669	96.30669	96.30669	96.30669
2	1.564971	6.994023×10^{-1}	6.994023×10^{-1}	6.902072×10^{-1}
3	2.939804×10^{-2}	1.225501×10^{-2}	1.225501×10^{-2}	3.989507×10^{-3}
4	5.810123×10^{-4}	7.301088×10^{-3}	7.301088×10^{-3}	1.684263×10^{-5}
5	1.169205×10^{-5}	2.636716×10^{-3}	2.636716×10^{-3}	3.881674×10^{-8}
6	2.372385×10^{-7}	1.031086×10^{-5}	1.031086×10^{-5}	
7		3.633330×10^{-9}	2.399278×10^{-8}	

CASE 3. 1 % error in step size α

Iteration	Function value			
	Steepest descent	DFP	DFP (with restart)	Self-scaling
1	97.33665	97.33665	97.33665	97.33665
2	1.586251	1.621908	1.621908	0.7024872
3	2.989875×10^{-2}	8.268893×10^{-1}	8.268893×10^{-1}	4.090350×10^{-3}
4	5.908101×10^{-4}	4.302943×10^{-1}	4.302943×10^{-1}	1.779424×10^{-5}
5	1.194144×10^{-5}	4.449852×10^{-3}	4.449852×10^{-3}	4.195668×10^{-8}
6	2.422985×10^{-7}	5.337835×10^{-5}	5.337835×10^{-5}	
7		3.767830×10^{-5}	4.493397×10^{-7}	
8		3.768097×10^{-9}		

CASE 4. 10 % error in step size α

Iteration	Function value			
	Steepest descent	DFP	DFP (with restart)	Self-scaling
1	200.333	200.333	200.333	200.333
2	2.732789	93.65457	93.65457	2.811061
3	3.836899×10^{-2}	56.92999	56.92999	3.562769×10^{-2}
4	6.376461×10^{-4}	1.620688	1.620688	4.200600×10^{-4}
5	1.219515×10^{-5}	5.251115×10^{-1}	5.251115×10^{-1}	4.726918×10^{-6}
6	2.457944×10^{-7}	3.323745×10^{-1}	3.323745×10^{-1}	
7		6.150890×10^{-3}	8.102700×10^{-3}	
8		3.025393×10^{-3}	2.973021×10^{-3}	
9		3.025476×10^{-5}	1.950152×10^{-3}	
10		3.025476×10^{-7}	2.769299×10^{-5}	
11			1.760320×10^{-5}	
12			1.123844×10^{-6}	

We note first that the error introduced is reported as a percentage of the step size itself. In terms of the change in function value, the quantity that is most often monitored to determine when to terminate a line search, the fractional error is the square of that in the step size. Thus, a one percent error in step size is equivalent to a 0.01 % error in the change in function value.

Next we note that the method of steepest descent is not radically affected by an inaccurate line search while the DFP methods are. Thus for this example while DFP is superior to steepest descent in the case of perfect accuracy, it becomes inferior at an error of only 0.1 % in step size.

10.6 Scaling

There is a general viewpoint about what makes up a desirable descent method that underlies much of our earlier discussions and which we now summarize briefly in order to motivate the presentation of scaling. A method that converges to the exact solution after n steps when applied to a quadratic function on E^n has obvious appeal especially if, as is usually the case, it can be inferred that for nonquadratic problems repeated cycles of length n of the method will yield superlinear convergence. For problems having large n, however, a more sophisticated criterion of performance needs to be established, since for such problems one usually hopes to be able to terminate the descent process before completing even a single full cycle of length n. Thus, with these sorts of problems in mind, the finite-step convergence property serves at best only as a sign post indicating that the algorithm *might*, make rapid progress in its early stages. It is essential to insure that in fact it *will* make rapid progress at every stage. Furthermore, the rapid convergence at each step must not be tied to an assumption on conjugate directions, a property easily destroyed by inaccurate line search and nonquadratic objective functions. With this viewpoint it is natural to look for quasi-Newton methods that simultaneously possess favorable eigenvalue structure at each step (in the sense of Sect. 10.1) and reduce to the conjugate gradient method if the objective function happens to be quadratic. Such methods are developed in this section.

Improvement of Eigenvalue Ratio

Referring to the example presented in the last section where the Davidon-Fletcher-Powell method performed poorly, we can trace the difficulty to the simple observation that the eigenvalues of $\mathbf{H}_0\mathbf{Q}$ are all much larger than unity. The DFP algorithm, or any Broyden method, essentially moves these eigenvalues, one at a time, to unity thereby producing an unfavorable eigenvalue ratio in each $\mathbf{H}_k\mathbf{Q}$ for $1 \leqslant k < n$. This phenomenon can be attributed to the fact that the methods are sensitive to simple scale factors. In particular if $\mathbf{H}_0$ were multiplied by a constant, the whole process would be different. In the example of the last section, if $\mathbf{H}_0$ were scaled by, for instance, multiplying it by 1/35, the eigenvalues of $\mathbf{H}_0\mathbf{Q}$ would be spread above and below unity, and in that case one might suspect that the poor performance would not show up.

Motivated by the above considerations, we shall establish conditions under which the eigenvalue ratio of $\mathbf{H}_{k+1}\mathbf{F}$ is at least as favorable as that of $\mathbf{H}_k\mathbf{F}$ in a Broyden method. These conditions will then be used as a basis for introducing appropriate scale factors.

We use (but do not prove) the following matrix theoretic result due to Loewner.

Interlocking Eigenvalues Lemma. *Let the symmetric $n \times n$ matrix A have eigenvalues $\lambda_1 \leqslant \lambda_2 \leqslant \ldots \leqslant \lambda_n$. Let a be any vector in E^n and denote the eigenvalues of the matrix $\mathbf{A} + \mathbf{a}\mathbf{a}^T$ by $\mu_1 \leqslant \mu_2 \ldots \leqslant \mu_n$. Then $\lambda_1 \leqslant \mu_1 \leqslant \lambda_2 \leqslant \mu_2 \ldots \leqslant \lambda_n \leqslant \mu_n$.*

For convenience we introduce the following definitions:

$$\mathbf{R}_k = \mathbf{F}_k^{1/2}\mathbf{H}_k\mathbf{F}_k^{1/2}$$
$$\mathbf{r}_k = \mathbf{F}_k^{1/2}\mathbf{p}_k.$$

Then using $\mathbf{q}_k = \mathbf{F}_k^{1/2}\mathbf{r}_k$, it can be readily verified that (10.41) is equivalent to

$$\mathbf{R}_{k+1}^{\phi} = \mathbf{R}_k - \frac{\mathbf{R}_k\mathbf{r}_k\mathbf{r}_k^T\mathbf{R}_k}{\mathbf{r}_k^T\mathbf{R}_k\mathbf{r}_k} + \frac{\mathbf{r}_k\mathbf{r}_k^T}{\mathbf{r}_k^T\mathbf{r}_k} + \phi\mathbf{z}_k\mathbf{z}_k^T, \tag{10.42}$$

where

$$\mathbf{z}_k = \mathbf{F}^{1/2}\mathbf{V}_k = \sqrt{\mathbf{r}_k^T\mathbf{R}_k\mathbf{r}_k}\left(\frac{\mathbf{r}_k}{\mathbf{r}_k^T\mathbf{r}_k} - \frac{\mathbf{R}_k\mathbf{r}_k}{\mathbf{r}_k^T\mathbf{R}_k\mathbf{r}_k}\right).$$

Since $\mathbf{R}_k$ is similar to $\mathbf{H}_k\mathbf{F}$ (because $\mathbf{H}_k\mathbf{F} = \mathbf{F}^{1/2}\mathbf{R}_k\mathbf{F}^{1/2}$), both have the same eigenvalues. It is most convenient, however, in view of (10.42) to study $\mathbf{R}_k$, obtaining conclusions about $\mathbf{H}_k\mathbf{F}$ indirectly.

Before proving the general theorem we shall consider the case $\phi = 0$ corresponding to the DFP formula. Suppose the eigenvalues of $\mathbf{R}_k$ are $\lambda_1, \lambda_2, \ldots, \lambda_n$ with $0 < \lambda_1 \leqslant \lambda_2 \leqslant \ldots \leqslant \lambda_n$. Suppose also that $1 \in [\lambda_1, \lambda_n]$. We will show that the eigenvalues of $\mathbf{R}_{k+1}$ are all contained in the interval $[\lambda_1, \lambda_n]$, which of course implies that $\mathbf{R}_{k+1}$ is no worse than $\mathbf{R}_k$ in terms of its condition number. Let us first consider the matrix

$$\mathbf{P} = \mathbf{R}_k - \frac{\mathbf{R}_k\mathbf{r}_k\mathbf{r}_k^T\mathbf{R}_k}{\mathbf{r}_k^T\mathbf{R}_k\mathbf{r}_k}.$$

We see that $\mathbf{P}\mathbf{r}_k = 0$ so one eigenvalue of $\mathbf{P}$ is zero. If we denote the eigenvalues of $\mathbf{P}$ by $\mu_1 \leqslant \mu_2 \leqslant \ldots \leqslant \mu_n$, we have from the above observation and the lemma on interlocking eigenvalues that

$$0 = \mu_1 \leqslant \lambda_1 \leqslant \mu_2 \leqslant \ldots \leqslant \mu_n \leqslant \lambda_n.$$

Next we consider

$$\mathbf{R}_{k+1} = \mathbf{R}_k - \frac{\mathbf{R}_k\mathbf{r}_k\mathbf{r}_k^T\mathbf{R}_k}{\mathbf{r}_k^T\mathbf{R}_k\mathbf{r}_k} + \frac{\mathbf{r}_k\mathbf{r}_k^T}{\mathbf{r}_k^T\mathbf{r}_k} = \mathbf{P} + \frac{\mathbf{r}_k\mathbf{r}_k^T}{\mathbf{r}_k^T\mathbf{r}_k}. \tag{10.43}$$

Since $\mathbf{r}_k$ is an eigenvector of $\mathbf{P}$ and since, by symmetry, all other eigenvectors of $\mathbf{P}$ are therefore orthogonal to $\mathbf{r}_k$, it follows that the only eigenvalue different in $\mathbf{R}_{k+1}$ from in $\mathbf{P}$ is the one corresponding to $\mathbf{r}_k$—it now being unity. Thus $\mathbf{R}_{k+1}$ has eigenvalues $\mu_2, \mu_3, \ldots, \mu_n$ and unity. These are all contained in the interval $[\lambda_1, \lambda_n]$. Thus updating does not worsen the eigenvalue ratio. It should be noted that this result in no way depends on α_k being selected to minimize f.

We now extend the above to the Broyden class with $0 \leqslant \phi \leqslant 1$.

Theorem. *Let the n eigenvalues of* $\mathbf{H}_k\mathbf{F}$ *be* $\lambda_1, \lambda_2, \ldots, \lambda_n$ *with* $0 < \lambda_1 \leqslant \lambda_2 \leqslant \ldots \leqslant \lambda_n$. *Suppose that* $1 \in [\lambda_1, \lambda_n]$. *Then for any* ϕ, $0 \leqslant \phi \leqslant 1$, *the eigenvalues of* $\mathbf{H}_{k+1}^{\phi}\mathbf{F}$, *where* $\mathbf{H}_{k+1}^{\phi}$ *is defined by (10.41), are all contained in* $[\lambda_1, \lambda_n]$.

Proof. The result shown above corresponds to $\phi = 0$. Let us now consider $\phi = 1$, corresponding to the BFGS formula. By our original definition of the BFGS update, $\mathbf{H}^{-1}$ is defined by the formula that is complementary to the DFP formula. Thus

$$\mathbf{H}_{k+1}^{-1} = \mathbf{H}_k^{-1} + \frac{\mathbf{q}_k\mathbf{q}_k^T}{\mathbf{q}_k^T\mathbf{p}_k} - \frac{\mathbf{H}_{k+1}^{-1}\mathbf{p}_k\mathbf{p}_k^T\mathbf{H}_k^{-1}}{\mathbf{p}_k^T\mathbf{H}_k^{-1}\mathbf{p}_k}.$$

This is equivalent to

$$\mathbf{R}_{k+1}^{-1} = \mathbf{R}_k^{-1} - \frac{\mathbf{R}_k^{-1}\mathbf{r}_k\mathbf{r}_k^T\mathbf{R}_k^{-1}}{\mathbf{r}_k^T\mathbf{R}_k^{-1}\mathbf{r}_k} + \frac{\mathbf{r}_k\mathbf{r}_k^T}{\mathbf{r}_k^T\mathbf{r}_k}, \tag{10.44}$$

which is identical to (10.43) except that $\mathbf{R}_k$ is replaced by $\mathbf{R}_k^{-1}$.

The eigenvalues of $\mathbf{R}_k^{-1}$ are $1/\lambda_n \leqslant 1/\lambda_{n-1} \leqslant \ldots \leqslant 1/\lambda_1$. Clearly, $1 \in [1/\lambda_n, 1/\lambda_1]$. Thus by the preliminary result, if the eigenvalues of $\mathbf{R}_{k+1}^{-1}$ are denoted $1/\mu_n < 1/\mu_{n-1} < \ldots < 1/\mu_1$, it follows that they are contained in the interval $[1/\lambda_n, 1/\lambda_1]$. Thus $1/\lambda_n < 1/\mu_n$ and $1/\lambda_1 > 1/\mu_1$. When inverted this yields $\mu_1 > \lambda_1$ and $\mu_n < \lambda_n$, which shows that the eigenvalues of $\mathbf{R}_{k+1}$ are contained in $[\lambda_1, \lambda_n]$. This establishes the result for $\phi = 1$.

For general ϕ the matrix $\mathbf{R}_{k+1}^{\phi}$ defined by (10.42) has eigenvalues that are all monotonically increasing with ϕ (as can be seen from the interlocking eigenvalues lemma). However, from above it is known that these eigenvalues are contained in $[\lambda_1, \lambda_n]$ for $\phi = 0$ and $\phi = 1$. Hence, they must be contained in $[\lambda_1, \lambda_n]$ for all ϕ, $0 \leqslant \phi \leqslant 1$. ∎

Scale Factors

In view of the result derived above, it is clearly advantageous to scale the matrix $\mathbf{H}_k$ so that the eigenvalues of $\mathbf{H}_k\mathbf{F}$ are spread both below and above unity. Of course in the ideal case of a quadratic problem with perfect line search this is strictly only necessary for $\mathbf{H}_0$, since unity is an eigenvalue of $\mathbf{H}_k\mathbf{F}$ for $k > 0$. But because of the inescapable deviations from the ideal, it is useful to consider the possibility of scaling every $\mathbf{H}_k$.

A scale factor can be incorporated directly into the updating formula. We first multiply $\mathbf{H}_k$ by the scale factor γ_k and then apply the usual updating formula. This is equivalent to replacing $\mathbf{H}_k$ by $\gamma_k\mathbf{H}_k$ in (10.42) and leads to

$$\mathbf{H}_{k+1} = \left(\mathbf{H}_k - \frac{\mathbf{H}_k\mathbf{q}_k\mathbf{q}_k^T\mathbf{H}_k}{\mathbf{q}_k^T\mathbf{H}_k\mathbf{q}_k} + \phi_k\mathbf{v}_k\mathbf{v}_k^T\right)\gamma_k + \frac{\mathbf{p}_k\mathbf{p}_k^T}{\mathbf{p}_k^T\mathbf{q}_k}. \tag{10.45}$$

This defines a two-parameter family of updates that reduces to the Broyden family for $\gamma_k = 1$.

Using $\gamma_0, \gamma_1, \ldots$ as arbitrary positive scale factors, we consider the algorithm: Start with any symmetric positive definite matrix $\mathbf{H}_0$ and any point $\mathbf{x}_0$, then starting with $k = 0$,

Step 1. Set $\mathbf{d}_k = -\mathbf{H}_k\mathbf{g}_k$.
Step 2. Minimize $f(\mathbf{x}_k + \alpha\mathbf{d}_k)$ with respect to $\alpha \geqslant 0$ to obtain $\mathbf{x}_{k+1}$, $\mathbf{P}_k = \alpha_k\mathbf{d}_k$, and $\mathbf{g}_{k+1}$.
Step 3. Set $\mathbf{q}_k = \mathbf{g}_{k+1} - \mathbf{g}_k$ and

$$\mathbf{H}_{k+1} = \left(\mathbf{H}_k - \frac{\mathbf{H}_k\mathbf{q}_k\mathbf{q}_k^T\mathbf{H}_k}{\mathbf{q}_k^T\mathbf{H}_k\mathbf{q}_k} + \phi_k\mathbf{v}_k\mathbf{v}_k^T\right)\gamma_k + \frac{\mathbf{p}_k\mathbf{p}_k^T}{\mathbf{p}_k^T\mathbf{q}_k}$$
$$\mathbf{v}_k = (\mathbf{q}_k^T\mathbf{H}\mathbf{q}_k)^{1/2}\left(\frac{\mathbf{p}_k}{\mathbf{p}_k^T\mathbf{q}_k} - \frac{\mathbf{H}_k\mathbf{q}_k}{\mathbf{q}_k^T\mathbf{H}_k\mathbf{q}_k}\right). \tag{10.46}$$

The use of scale factors does destroy the property $\mathbf{H}_n = \mathbf{F}^{-1}$ in the quadratic case, but it does not destroy the conjugate direction property. The following properties of this method can be proved as simple extensions of the results given in Sect. 10.3.

1. If $\mathbf{H}_k$ is positive definite and $\mathbf{p}_k^T\mathbf{q}_k > 0$, (10.46) yields an $\mathbf{H}_{k+1}$ that is positive definite.
2. If f is quadratic with Hessian $\mathbf{F}$, then the vectors $\mathbf{P}_0, \mathbf{p}_1, \ldots, \mathbf{p}_{n-1}$ are mutually $\mathbf{F}$-orthogonal, and, for each k, the vectors $\mathbf{P}_0, \mathbf{p}_1, \ldots, \mathbf{p}_k$ are eigenvectors of $\mathbf{H}_{k+1}\mathbf{F}$.

We can conclude that scale factors do not destroy the underlying conjugate behavior of the algorithm. Hence we can use scaling to ensure good single-step convergence properties.

A Self-scaling Quasi-Newton Algorithm

The question that arises next is how to select appropriate scale factors. If $\lambda_1 \leqslant \lambda_2 \leqslant \ldots \leqslant \lambda_n$ are the eigenvalues of $\mathbf{H}_k\mathbf{F}$, we want to multiply $\mathbf{H}_k$ by γ_k where $\lambda_1 \leqslant 1/\gamma_k \leqslant \lambda_n$. This will ensure that the new eigenvalues contain unity in the interval they span.

Note that in terms of our earlier notation

$$\frac{\mathbf{q}_k^T\mathbf{H}_k\mathbf{q}_k}{\mathbf{p}_k^T\mathbf{q}_k} = \frac{\mathbf{r}_k^T\mathbf{R}_k\mathbf{r}_k}{\mathbf{r}_k^T\mathbf{r}_k}.$$

Recalling that $\mathbf{R}_k$ has the same eigenvalues as $\mathbf{H}_k\mathbf{F}$ and noting that for any $\mathbf{r}_k$

$$\lambda_1 \leqslant \frac{\mathbf{r}_k^T\mathbf{R}_k\mathbf{r}_k}{\mathbf{r}_k^T\mathbf{r}_k} \leqslant \lambda_n,$$

we see that

$$\gamma_k = \frac{\mathbf{p}_k^T\mathbf{q}_k}{\mathbf{q}_k^T\mathbf{H}_k\mathbf{q}_k} \tag{10.47}$$

serves as a suitable scale factor.

We now state a complete self-scaling, restarting, quasi-Newton method based on the ideas above. For simplicity we take $\phi = 0$ and thus obtain a modification of the DFP method. Start at any point $\mathbf{x}_0$, $k = 0$.

Step 1. Set $\mathbf{H}_k = \mathbf{I}$.
Step 2. Set $\mathbf{d}_k = -\mathbf{H}_k\mathbf{g}_k$.
Step 3. Minimize $f(\mathbf{x}_k + \alpha\mathbf{d}_k)$ with respect to $\alpha \geqslant 0$ to obtain α_k, $\mathbf{x}_{k+1}$, $\mathbf{p}_k = \alpha_k\mathbf{d}_k$, $\mathbf{g}_{k+1}$ and $\mathbf{q}_k = \mathbf{g}_{k+1} - \mathbf{g}_k$. (Select α_k accurately enough to ensure $\mathbf{p}_k^T\mathbf{q}_k > 0$.)
Step 4. If k is not an integer multiple of n, set

$$\mathbf{H}_{k+1} = \left(\mathbf{H}_k - \frac{\mathbf{H}_k\mathbf{q}_k\mathbf{q}_k^T\mathbf{H}_k}{\mathbf{q}_k^T\mathbf{H}_k\mathbf{q}_k}\right)\frac{\mathbf{p}_k^T\mathbf{q}_k}{\mathbf{q}_k^T\mathbf{H}_k\mathbf{q}_k} + \frac{\mathbf{p}_k\mathbf{p}_k^T}{\mathbf{p}_k^T\mathbf{q}_k}. \tag{10.48}$$

Add one to k and return to Step 2. If k is an integer multiple of n, return to Step 1.

This algorithm was run, with various amounts of inaccuracy introduced in the line search, on the quadratic problem presented in Sect. 10.4. The results are presented in that section.

10.7 Memoryless Quasi-Newton Methods

The preceding development of quasi-Newton methods can be used as a basis for reconsideration of conjugate gradient methods. The result is an attractive class of new procedures.

Consider a simplification of the BFGS quasi-Newton method where $\mathbf{H}_{k+1}$ is defined by a BFGS update applied to $\mathbf{H} = \mathbf{I}$, rather than to $\mathbf{H}_k$. Thus $\mathbf{H}_{k+1}$ is determined without reference to the previous $\mathbf{H}_k$, and hence the update procedure is *memoryless*. This update procedure leads to the following algorithm: Start at any point $\mathbf{x}_0$, $k = 0$.

Step 1.

$$\text{Set } \mathbf{H}_k = \mathbf{I}. \tag{10.49}$$

Step 2.

$$\text{Set } \mathbf{d}_k = -\mathbf{H}_k\mathbf{g}_k. \tag{10.50}$$

Step 3. Minimize $f(\mathbf{x}_k + \alpha\mathbf{d}_k)$ with respect to $\alpha \geqslant 0$ to obtain α_k, $\mathbf{x}_{k+1}$, $\mathbf{p}_k = \alpha_k\mathbf{d}_k$, $\mathbf{g}_{k+1}$, and $\mathbf{q}_k = \mathbf{g}_{k+1} - \mathbf{g}_k$. (Select α_k accurately enough to ensure $\mathbf{p}_k^T\mathbf{q}_k > 0$.)

Step 4. If k is not an integer multiple of n, set

$$\mathbf{H}_{k+1} = \mathbf{I} - \frac{\mathbf{q}_k\mathbf{p}_k^T + \mathbf{p}_k\mathbf{q}_k^T}{\mathbf{p}_k^T\mathbf{q}_k} + \left(1 + \frac{\mathbf{q}_k^T\mathbf{q}_k}{\mathbf{p}_k^T\mathbf{q}_k}\right)\frac{\mathbf{p}_k\mathbf{p}_k^T}{\mathbf{p}_k^T\mathbf{q}_k}. \tag{10.51}$$

Add 1 to k and return to Step 2. If k is an integer multiple of n, return to Step 1.

Combining (10.50) and (10.51), it is easily seen that

$$\mathbf{d}_{k+1} = -\mathbf{g}_{k+1} + \frac{\mathbf{q}_k\mathbf{p}_k^T\mathbf{g}_{k+1} + \mathbf{p}_k\mathbf{q}_k^T\mathbf{g}_{k+1}}{\mathbf{p}_k^T\mathbf{q}_k} - \left(1 + \frac{\mathbf{q}_k^T\mathbf{q}_k}{\mathbf{p}_k^T\mathbf{q}_k}\right)\frac{\mathbf{p}_k\mathbf{p}_k^T\mathbf{g}_{k-1}}{\mathbf{p}_k^T\mathbf{q}_k}. \tag{10.52}$$

If the line search is exact, then $\mathbf{p}_k^T\mathbf{g}_{k+1} = 0$ and hence $\mathbf{p}_k^T\mathbf{q}_k = -\mathbf{p}_k^T\mathbf{g}_k$. In this case (10.52) is equivalent to

$$\mathbf{d}_{k+1} = -\mathbf{g}_{k+1} + \frac{\mathbf{q}_k^T\mathbf{g}_{k+1}}{\mathbf{p}_k^T\mathbf{q}_k}\mathbf{p}_k = -\mathbf{g}_{k+1} + \beta_k\mathbf{d}_k, \tag{10.53}$$

where

$$\beta_k = \frac{\mathbf{q}_k\mathbf{q}_{k+1}^T}{\mathbf{g}_k^T\mathbf{q}_k}.$$

This coincides exactly with the Polak-Ribiere form of the conjugate gradient method. Thus use of the BFGS update in this way yields an algorithm that is of the modified Newton type with positive definite coefficient matrix and which is equivalent to a standard implementation of the conjugate gradient method when the line search is exact.

The algorithm can be used without exact line search in a form that is similar to that of the conjugate gradient method by using (10.52). This requires storage of only the same vectors that are required of the conjugate gradient method. In light of the theory of quasi-Newton methods, however, the new form can be expected to be superior when inexact line searches are employed, and indeed experiments confirm this.

The above idea can be easily extended to produce a memoryless quasi-Newton method corresponding to any member of the Broyden family. The update formula (10.51) would simply use the general Broyden update (10.41) with $\mathbf{H}_k$ set equal to I. In the case of exact line search (with $\mathbf{p}_k^T\mathbf{g}_{k+1} = 0$), the resulting formula for $\mathbf{d}_{k+1}$ reduces to

$$\mathbf{d}_{k+1} = -\mathbf{g}_{k+1} + (1-\phi)\frac{\mathbf{q}_k^T\mathbf{g}_{k+1}}{\mathbf{q}_k^T\mathbf{q}_k}\mathbf{q}_k + \phi\frac{\mathbf{q}_k^T\mathbf{g}_{k+1}}{\mathbf{p}_k^T\mathbf{q}_k}\mathbf{p}_k. \tag{10.54}$$

We note that (10.54) is equivalent to the conjugate gradient direction (10.53) only for $\phi = 1$, corresponding to the BFGS update. For this reason the choice $\phi = 1$ is generally preferred for this type of method.

Scaling and Preconditioning

Since the conjugate gradient method implemented as a memoryless quasi-Newton method is a modified Newton method, the fundamental convergence theory based on condition number emphasized throughout this part of the book is applicable, as are the procedures for improving convergence. It is clear that the function scaling procedures discussed in the previous section can be incorporated.

According to the general theory of modified Newton methods, it is the eigenvalues of $\mathbf{H}_k\mathbf{F}(\mathbf{x}_k)$ that influence the convergence properties of these algorithms. From the analysis of the last section, the memoryless BFGS update procedure will, in the pure quadratic case, yield a matrix $\mathbf{H}_k\mathbf{F}$ that has a more favorable eigenvalue ratio than $\mathbf{F}$ itself only if the function f is scaled so that unity is contained in the interval spanned by the eigenvalues of F. Experimental evidence verifies that at least an initial scaling of the function in this way can lead to significant improvement. Scaling can be introduced at every step as well, and complete self-scaling can be effective in some situations.

It is possible to extend the scaling procedure to a more general *preconditioning* procedure. In this procedure the matrix governing convergence is changed from $\mathbf{F}(\mathbf{x}_k)$ to $\mathbf{HF}(\mathbf{x}_k)$ for some $\mathbf{H}$. If $\mathbf{HF}(\mathbf{x}_k)$ has its eigenvalues all close to unity, then the memoryless quasi-Newton method can be expected to perform exceedingly well, since it possesses simultaneously the advantages of being a conjugate gradient method and being a well-conditioned modified Newton method.

Preconditioning can be conveniently expressed in the basic algorithm by simply replacing $\mathbf{H}_k$ in the BFGS update formula by $\mathbf{H}$ instead of I and replacing I by $\mathbf{H}$ in Step 1. Thus (10.51) becomes

$$\mathbf{H}_{k+1} = \mathbf{H} - \frac{\mathbf{H}\mathbf{q}_k\mathbf{p}_k^T + \mathbf{p}_k\mathbf{q}_k^T\mathbf{H}}{\mathbf{q}_k^T\mathbf{q}_k} + \left(1 + \frac{\mathbf{q}_k^T\mathbf{H}\mathbf{q}_k}{\mathbf{p}_k^T\mathbf{q}_k}\right)\frac{\mathbf{p}_k\mathbf{p}_k^T}{\mathbf{p}_k^T\mathbf{p}_k}, \tag{10.55}$$

and the explicit conjugate gradient version (10.52) is also modified accordingly.

Preconditioning can also be used in conjunction with an $(m + 1)$-cycle partial conjugate gradient version of the memoryless quasi-Newton method. This is highly effective if a simple $\mathbf{H}$ can be found (as it sometimes can in problems with structure) so that the eigenvalues of $\mathbf{HF}(\mathbf{x}_k)$ are such that either all but m are equal to unity or they are in m bunches. For large-scale problems, methods of this type seem to be quite promising.

*10.8 *Combination of Steepest Descent and Newton's Method

In this section we digress from the study of quasi-Newton methods, and again expand our collection of basic principles. We present a combination of steepest descent and Newton's method which includes them both as special cases. The resulting combined method can be used to develop algorithms for problems having special structure, as illustrated in Chap. 13. This method and its analysis comprises a fundamental element of the modern theory of algorithms.

The method itself is quite simple. Suppose there is a subspace N of E^n on which the inverse Hessian of the objective function f is known (we shall make this statement more precise later). Then, in the quadratic case, the minimum of f over any linear variety parallel to N (that is, any translation of N) can be found in a single step. To minimize f over the whole space starting at any point $\mathbf{x}_k$, we could minimize f over the linear variety parallel to N and containing $\mathbf{x}_k$ to obtain $\mathbf{z}_k$; and then take a steepest descent step from there. This procedure is illustrated in Fig. 10.1. Since $\mathbf{z}_k$ is the minimum point of f over a linear variety parallel to N, the gradient at $\mathbf{z}_k$ will be orthogonal to N, and hence the gradient step is orthogonal to N. If f is not quadratic we can, knowing the Hessian of f on N, approximate the minimum point of f over a linear variety parallel to N by one step of Newton's method. To implement this scheme, that we described in a geometric sense, it is necessary to agree on a method for defining the subspace N and to determine what information about the inverse Hessian is required so as to implement a Newton step over N. We now turn to these questions.

Often, the most convenient way to describe a subspace, and the one we follow in this development, is in terms of a set of vectors that generate it. Thus, if $\mathbf{B}$ is an $n \times m$ matrix consisting of m column vectors that generate N, we may write N as the set of all vectors of the form $\mathbf{Bu}$ where $\mathbf{u} \in E^m$. For simplicity we always assume that the columns of $\mathbf{B}$ are linearly independent.

To see what information about the inverse Hessian is required, imagine that we are at a point $\mathbf{x}_k$ and wish to find the approximate minimum point $\mathbf{z}_k$ of f with respect to movement in N. Thus, we seek $\mathbf{u}_k$ so that

$$\mathbf{z}_k = \mathbf{x}_k + \mathbf{B}\mathbf{u}_k$$

approximately minimizes f. By "approximately minimizes" we mean that $\mathbf{z}_k$ should be the Newton approximation to the minimum over this subspace. We write

$$f(\mathbf{z}_k) \cong f(\mathbf{x}_k) + \nabla f(\mathbf{x}_k)\mathbf{B}\mathbf{u}_k + \frac{1}{2}\mathbf{u}_k^T\mathbf{B}^T\mathbf{F}(\mathbf{x}_k)\mathbf{B}\mathbf{u}_k$$

and solve for $\mathbf{u}_k$ to obtain the Newton approximation. We find

$$\begin{aligned}\mathbf{u}_k &= -(\mathbf{B}^T\mathbf{F}(\mathbf{x}_k)\mathbf{B})^{-1}\mathbf{B}^T\nabla f(\mathbf{x}_k)^T\\ \mathbf{z}_k &= \mathbf{x}_k - \mathbf{B}(\mathbf{B}^T\mathbf{F}(\mathbf{x}_k)\mathbf{B})^{-1}\mathbf{B}^T\nabla f(\mathbf{x}_k)^T.\end{aligned}$$

We see by analogy with the formula for Newton's method that the expression $\mathbf{B}(\mathbf{B}^T\mathbf{F}(\mathbf{x}_k)\mathbf{B})^{-1}\mathbf{B}^T$ can be interpreted as the inverse of $\mathbf{F}(\mathbf{x}_k)$ restricted to the subspace N.

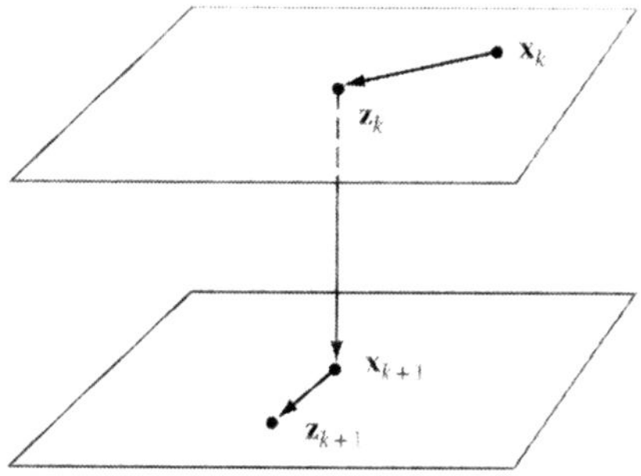

Fig. 10.1 Combined method

Example. Suppose

$$\mathbf{B} = \begin{bmatrix} \mathbf{I} \\ \mathbf{0} \end{bmatrix},$$

where $\mathbf{I}$ is an $m \times m$ identity matrix. This corresponds to the case where N is the subspace generated by the first m unit basis elements of E^n. Let us partition $\mathbf{F} = \nabla^2 f(\mathbf{x}_k)$ as

$$\mathbf{F} = \begin{bmatrix} \mathbf{F}_{11} & \mathbf{F}_{12} \\ \mathbf{F}_{21} & \mathbf{F}_{22} \end{bmatrix},$$

where $\mathbf{F}_{11}$ is $m \times m$. Then, in this case

$$(\mathbf{B}^T\mathbf{F}\mathbf{B})^{-1} = \mathbf{F}_{11}^{-1},$$

and

$$\mathbf{B}(\mathbf{B}^T\mathbf{F}\mathbf{B})^{-1}\mathbf{B}^T = \begin{bmatrix} \mathbf{F}_{11}^{-1} & \mathbf{0} \\ \mathbf{0} & \mathbf{0} \end{bmatrix},$$

which shows explicitly that it is the inverse of $\mathbf{F}$ on N that is required. The general case can be regarded as being obtained through partitioning in some skew coordinate system.

Now that the Newton approximation over N has been derived, it is possible to formalize the details of the algorithm suggested by Fig. 10.1. At a given point $\mathbf{x}_k$, the point $\mathbf{x}_{k+1}$ is determined through

a) Set $\mathbf{d}_k = -\mathbf{B}(\mathbf{B}^T\mathbf{F}(\mathbf{x}_k)\mathbf{B})^{-1}\mathbf{B}^T\nabla\mathbf{f}(\mathbf{x}_k)^T$.
b) $\mathbf{z}_k = \mathbf{x}_k + \beta_k\mathbf{d}_k$, where β_k minimizes $\mathbf{f}(\mathbf{x}_k + \beta\mathbf{d}_k)$. (10.56)
c) Set $\mathbf{p}_k = -\nabla\mathbf{f}(\mathbf{z}_k)^T$.
d) $\mathbf{x}_{k+1} = \mathbf{z}_k + \alpha_k\mathbf{p}_k$, where α_k minimizes $f(\mathbf{z}_k + \alpha\mathbf{p}_k)$.

The scalar search parameter β_k is introduced in the Newton part of the algorithm simply to assure that the descent conditions required for global convergence are met. Normally β_k will be approximately equal to unity. (See Sect. 8.5.)

Analysis of Quadratic Case

Since the method is not a full Newton method, we can conclude that it possesses only linear convergence and that the dominating aspects of convergence will be revealed by an analysis of the method as applied to a quadratic function. Furthermore, as might be intuitively anticipated, the associated rate of convergence is governed by the steepest descent part of algorithm (10.56), and that rate is governed by a Kantorovich-like ratio defined over the subspace orthogonal to N.

Theorem (Combined Method). *Let* $\mathbf{Q}$ *be an* $n \times n$ *symmetric positive definite matrix, and let* $\mathbf{x}^* \in E^n$. *Define the function*

$$E(\boldsymbol{x}) = \frac{1}{2}(\boldsymbol{x} - \boldsymbol{x}^*)^T \boldsymbol{Q}(\boldsymbol{x} - \boldsymbol{x}^*)$$

and let $\mathbf{b} = \boldsymbol{Q}\boldsymbol{x}^*$. *Let* $\mathbf{B}$ *be an* $n \times m$ *matrix of rank* m. *Starting at an arbitrary point* $\mathbf{x}_0$, *define the iterative process*

a) $\mathbf{u}_k = -(\boldsymbol{B}^T\boldsymbol{Q}\boldsymbol{B})^{-1}\boldsymbol{B}^T\boldsymbol{g}_k$, *where* $\mathbf{g}_k = \boldsymbol{Q}\boldsymbol{x}_k - \boldsymbol{b}$.

b) $\mathbf{z}_k = \boldsymbol{x}_k + \boldsymbol{B}\boldsymbol{u}_k$.

c) $\mathbf{p}_k = \boldsymbol{b} - \boldsymbol{Q}\boldsymbol{z}_k$.

d) $\mathbf{x}_{k+1} = z_k + \alpha_k \boldsymbol{p}_k$, *where* $\alpha_k = \frac{\boldsymbol{p}_k^T\boldsymbol{p}_k}{\boldsymbol{p}_k^T\boldsymbol{Q}\boldsymbol{p}_k}$.

This process converges to $\mathbf{x}^*$, *and satisfies*

$$E(\boldsymbol{x}_{k+1}) \leqslant (1 - \delta)E(\boldsymbol{x}_k) \tag{10.57}$$

where δ, $0 \leqslant 8 \leqslant 1$, *is the minimum of*

$$\frac{(\boldsymbol{p}^T\boldsymbol{p})^2}{(\boldsymbol{p}^T\boldsymbol{Q}\boldsymbol{p})(\boldsymbol{p}^T\boldsymbol{Q}^{-1}\boldsymbol{p})}$$

over all vectors $\mathbf{p}$ *in the nullspace of* $\mathbf{B}^T$.

Proof. The algorithm given in the theorem statement is exactly the general combined algorithm specialized to the quadratic situation. Next we note that

$$\begin{aligned}\mathbf{B}^T\mathbf{p}_k &= \mathbf{B}^T\mathbf{Q}(\mathbf{x}^* - \mathbf{z}_k) = \mathbf{B}^T\mathbf{Q}(\mathbf{x}^* - \mathbf{x}_k) - \mathbf{B}^T\mathbf{Q}\mathbf{B}\mathbf{u}_k \\ &= -\mathbf{B}^T\mathbf{g}_k + \mathbf{B}\mathbf{Q}\mathbf{B}^T(\mathbf{B}^T\mathbf{Q}\mathbf{B})^{-1}\mathbf{B}^T\mathbf{g}_k = \mathbf{0},\end{aligned} \tag{10.58}$$

which merely proves that the gradient at $\mathbf{z}_k$ is orthogonal to N. Next we calculate

$$\begin{aligned}2\{E(\mathbf{x}_k) - E(\mathbf{z}_k)\} &= (\mathbf{x}_k - \mathbf{x}^*)^T\mathbf{Q}(\mathbf{x}_k - \mathbf{x}^*) - (\mathbf{z}_k - \mathbf{x}^*)^T\mathbf{Q}(\mathbf{z}_k - \mathbf{x}^*) \\ &= -2\mathbf{u}_k^T\mathbf{B}^T\mathbf{Q}(\mathbf{x}_k - \mathbf{x}^*) - \mathbf{u}_k^T\mathbf{B}^T\mathbf{Q}\mathbf{B}\mathbf{u}_k \\ &= -2\mathbf{u}_k^T\mathbf{B}^T\mathbf{g}_k + \mathbf{u}_k^T\mathbf{B}^T\mathbf{Q}\mathbf{B}(\mathbf{B}^T\mathbf{Q}\mathbf{B})^{-1}\mathbf{B}^T\mathbf{g}_k \\ &= -\mathbf{u}_k^T\mathbf{B}^T\mathbf{g}_k = \mathbf{g}_k^T\mathbf{B}(\mathbf{B}^T\mathbf{Q}\mathbf{B})^{-1}\mathbf{B}^T\mathbf{g}_k.\end{aligned} \tag{10.59}$$

Then we compute

$$
\begin{aligned}
2\{E(\mathbf{z}_k) - E(\mathbf{x}_{k+1})\} &= (\mathbf{z}_k - \mathbf{x}^*)^T \mathbf{Q}(\mathbf{z}_k - \mathbf{x}^*) - (\mathbf{x}_{k+1} - \mathbf{x}^*)^T \mathbf{Q}(\mathbf{x}_{k+1} - \mathbf{x}^*) \\
&= -2\alpha_k \mathbf{p}_k^T \mathbf{Q}(\mathbf{z}_k - \mathbf{x}^*) - \alpha_k^2 \mathbf{p}_k^T \mathbf{Q}\mathbf{p}_k \\
&= 2\alpha_k \mathbf{p}_k^T \mathbf{p}_k - \alpha_k^2 \mathbf{p}_k^T \mathbf{Q}\mathbf{p}_k \qquad (10.60) \\
&= \alpha_k \mathbf{p}_k^T \mathbf{p}_k = \frac{(\mathbf{p}_k^T \mathbf{p}_k)^2}{\mathbf{p}_k^T \mathbf{Q}\mathbf{p}_k}.
\end{aligned}
$$

Now using (10.58) and $\mathbf{p}_k = -\mathbf{g}_k - \mathbf{QBu}_k$ we have

$$
\begin{aligned}
2E(\mathbf{x}_k) &= (\mathbf{x}_k - \mathbf{x}^*)^T \mathbf{Q}(\mathbf{x}_k - \mathbf{x}^*) = \mathbf{g}_k^T \mathbf{Q}^{-1} \mathbf{g}_k \\
&= (\mathbf{p}_k^T + \mathbf{u}_k^T \mathbf{B}^T \mathbf{Q})\mathbf{Q}^{-1}(\mathbf{p}_k + \mathbf{QBu}_k) \qquad (10.61) \\
&= \mathbf{p}_k^T \mathbf{Q}^{-1} \mathbf{p}_k + \mathbf{u}_k^T \mathbf{B}^T \mathbf{QBu}_k \\
&= \mathbf{p}_k^T \mathbf{Q}^{-1} \mathbf{p}_k + \mathbf{g}_k^T \mathbf{B}(\mathbf{B}^T \mathbf{QB})^{-1} \mathbf{B}^T \mathbf{g}_k.
\end{aligned}
$$

Adding (10.59) and (10.60) and dividing by (10.61) there results

$$
\begin{aligned}
\frac{E(\mathbf{x}_k) - E(\mathbf{x}_{k+1})}{E(\mathbf{x}_k)} &= \frac{\mathbf{g}_k^T \mathbf{B}(\mathbf{B}^T \mathbf{QB})^{-1} \mathbf{B}^T \mathbf{g}_k + (\mathbf{p}_k^T \mathbf{p}_k)^2 / \mathbf{p}_k^T \mathbf{Q}\mathbf{p}_k}{\mathbf{p}_k^T \mathbf{Q}^{-1} \mathbf{p}_k + \mathbf{g}_k^T \mathbf{B}(\mathbf{B}^T \mathbf{QB})^{-1} \mathbf{B}^T \mathbf{g}_k} \\
&= \frac{q + (\mathbf{p}_k^T \mathbf{p}_k)/(\mathbf{p}_k^T \mathbf{Q}\mathbf{p}_k)}{q + (\mathbf{p}_k^T \mathbf{Q}^{-1} \mathbf{p}_k)/(\mathbf{p}_k^T \mathbf{p}_k)},
\end{aligned}
$$

where $q \geqslant 0$. This has the form $(q + a)/(q + b)$ with

$$
a = \frac{\mathbf{p}_k^T \mathbf{p}_k}{\mathbf{p}_k^T \mathbf{Q}\mathbf{p}_k}, \qquad b = \frac{\mathbf{p}_k^T \mathbf{Q}^{-1} \mathbf{p}_k}{\mathbf{p}_k^T \mathbf{p}_k}.
$$

But for any $\mathbf{p}_k$, it follows that $a \leqslant b$. Hence

$$
\frac{q + a}{q + b} \geqslant \frac{a}{b},
$$

and thus

$$
\frac{E(\mathbf{x}_k) - E(\mathbf{x}_{k+1})}{E(\mathbf{x}_k)} \geqslant \frac{(\mathbf{p}_k^T \mathbf{p}_k)^2}{(\mathbf{p}_k^T \mathbf{Q}\mathbf{p}_k)(\mathbf{p}_k^T \mathbf{Q}^{-1} \mathbf{p}_k)}.
$$

Finally,

$$
E(\mathbf{x}_{k+1}) \leqslant E(\mathbf{x}_k) \left[1 - \frac{(\mathbf{p}_k^T \mathbf{p}_k)^2}{(\mathbf{p}_k^T \mathbf{Q}\mathbf{p}_k)(\mathbf{p}_k^T \mathbf{Q}^{-1} \mathbf{p}_k)} \right] \leqslant (1 - \delta) E(\mathbf{x}_k),
$$

since $\mathbf{B}^T \mathbf{p}_k = \mathbf{0}$. ∎

The value δ associated with the above theorem is related to the eigenvalue structure of $\mathbf{Q}$. If $\mathbf{p}$ were allowed to vary over the whole space, then the Kantorovich inequality

$$\frac{(\mathbf{p}^T\mathbf{p})^2}{(\mathbf{p}^T\mathbf{Q}\mathbf{p})(\mathbf{p}^T\mathbf{Q}^{-1}\mathbf{p})} \geqslant \frac{4aA}{(a+A)^2}, \tag{10.62}$$

where a and A are, respectively, the smallest and largest eigenvalues of $\mathbf{Q}$, gives explicitly

$$\delta = \frac{4aA}{(a+A)^2}.$$

When $\mathbf{p}$ is restricted to the nullspace of $\mathbf{B}^T$, the corresponding value of δ is larger. In some special cases it is possible to obtain a fairly explicit estimate of δ. Suppose, for example, that the subspace N were the subspace spanned by m eigenvectors of $\mathbf{Q}$. Then the subspace in which $\mathbf{p}$ is allowed to vary is the space orthogonal to N and is thus, in this case, the space generated by the other $n - m$ eigenvectors of $\mathbf{Q}$. In this case since for $\mathbf{p}$ in $N^\perp$ (the space orthogonal to N), both $\mathbf{Q}\mathbf{p}$ and $\mathbf{Q}^{-1}\mathbf{p}$ are also in $N^\perp$, the ratio δ satisfies

$$\delta = \frac{(\mathbf{p}^T\mathbf{p})^2}{(\mathbf{p}^T\mathbf{Q}\mathbf{p})(\mathbf{p}^T\mathbf{Q}^{-1}\mathbf{p})} \geqslant \frac{4aA}{(a+A)^2},$$

where now a and A are, respectively, the smallest and largest of the $n - m$ eigenvalues of $\mathbf{Q}$ corresponding to $N^\perp$. Thus the convergence ratio (10.57) reduces to the familiar form

$$E(\mathbf{x}_{k+1}) \leqslant \left(\frac{A-a}{A+a}\right)^2 E(\mathbf{x}_k),$$

where a and A are these special eigenvalues. Thus, if $\mathbf{B}$, or equivalently N, is chosen to include the eigenvectors corresponding to the most undesirable eigenvalues of $\mathbf{Q}$, the convergence rate of the combined method will be quite attractive.

Applications

The combination of steepest descent and Newton's method can be applied usefully in a number of important situations. Suppose, for example, we are faced with a problem of the form

$$\text{minimize} \quad f(\mathbf{x}, \mathbf{y}),$$

where $\mathbf{x} \in E^n$, $\mathbf{y} \in E^m$, and where the second partial derivatives with respect to $\mathbf{x}$ are easily computable but those with respect to $\mathbf{y}$ are not. We may then employ Newton steps with respect to $\mathbf{x}$ and steepest descent with respect to $\mathbf{y}$.

Another instance where this idea can be greatly effective is when there are a few vital variables in a problem which, being assigned high costs, tend to dominate the value of the objective function; in other words, the partial second derivatives with respect to these variables are large. The poor conditioning induced by these variables can to some extent be reduced by proper scaling of variables, but more effectively, by carrying out Newton's method with respect to them and steepest descent with respect to the others.

10.9 Summary

The basic motivation behind quasi-Newton methods is to try to obtain, at least on the average, the rapid convergence associated with Newton's method without explicitly evaluating the Hessian at every step. This can be accomplished by constructing approximations to the inverse Hessian based on information gathered during the descent process, and results in methods which viewed in blocks of n steps (where n is the dimension of the problem) generally possess superlinear convergence.

Good, or even superlinear, convergence measured in terms of large blocks, however, is not always indicative of rapid convergence measured in terms of individual steps. It is important, therefore, to design quasi-Newton methods so that their single step convergence is rapid and relatively insensitive to line search inaccuracies. We discussed two general principles for examining these aspects of descent algorithms. The first of these is the modified Newton method in which the direction of descent is taken as the result of multiplication of the negative gradient by a positive definite matrix $\mathbf{S}$. The single step convergence ratio of this method is determined by the usual steepest descent formula, but with the condition number of $\mathbf{SF}$ rather than just $\mathbf{F}$ used. This result was used to analyze some popular quasi-Newton methods, to develop the self-scaling method having good single step convergence properties, and to reexamine conjugate gradient methods.

The second principle method is the combined method in which Newton's method is executed over a subspace where the Hessian is known and steepest descent is executed elsewhere. This method converges at least as fast as steepest descent, and by incorporating the information gathered as the method progresses, the Newton portion can be executed over larger and larger subspaces.

At this point, it is perhaps valuable to summarize some of the main themes that have been developed throughout the four chapters comprising Part II. These chapters contain several important and popular algorithms that illustrate the range of possibilities available for minimizing a general nonlinear function. From a broad perspective, however, these individual algorithms can be considered simply as specific patterns on the analytical fabric that is woven through the chapters—the fabric that will support new algorithms and future developments.

One unifying element, that has reproved its value several times, is the Global Convergence Theorem. This result helped mold the final form of every algorithm presented in Part II and has effectively resolved the major questions concerning global convergence.

Another unifying element is the speed of convergence of an algorithm, which we have defined in terms of the asymptotic properties of the sequences an algorithm generates. Initially, it might have been argued that such measures, based on properties of the tail of the sequence, are perhaps not truly indicative of the actual time required to solve a problem—after all, a sequence generated in practice is a truncated version of the potentially infinite sequence, and asymptotic properties may not be representative of the finite version—a more complex measure of the speed of convergence may be required. It is fair to demand that the validity of the asymptotic measures we have proposed be judged in terms of how well they predict the performance of algorithms applied to specific examples. On this basis, as illustrated by the numerical examples presented in these chapters, and on others, the asymptotic rates are extremely reliable predictors of performance—provided that one carefully tempers one's analysis with common sense (by, for example, not concluding that superlinear convergence is necessarily superior to linear convergence when the superlinear convergence is based on repeated cycles of length n). A major conclusion, therefore, of the previous chapters is the essential validity of the asymptotic approach to convergence analysis. This conclusion is a major strand in the analytical fabric of nonlinear programming.

10.10 Exercises

1. Prove (10.4) directly for the modified Newton method by showing that each step of the modified Newton method is simply the ordinary method of steepest descent applied to a scaled version of the original problem.
2. Find the rate of convergence of the version of Newton's method defined by (10.50), (10.51) of Chap. 8. Show that convergence is only linear if δ is larger than the smallest eigenvalue of $\mathbf{F}(\mathbf{x}^*)$.
3. Consider the problem of minimizing a quadratic function

$$f(\mathbf{x}) = \frac{1}{2}\mathbf{x}^T\mathbf{Q}\mathbf{x} - \mathbf{x}^T\mathbf{b},$$

where $\mathbf{Q}$ is symmetric and sparse (that is, there are relatively few nonzero entries in Q). The matrix $\mathbf{Q}$ has the form

$$\mathbf{Q} = \mathbf{I} + \mathbf{V},$$

where $\mathbf{I}$ is the identity and $\mathbf{V}$ is a matrix with eigenvalues bounded by $e < 1$ in magnitude.

(a) With the given information, what is the best bound you can give for the rate of convergence of steepest descent applied to this problem?

(b) In general it is difficult to invert $\mathbf{Q}$ but the inverse can be approximated by $\mathbf{I}$–$\mathbf{V}$, which is easy to calculate. (The approximation is very good for small e.) We are thus led to consider the iterative process

$$\mathbf{x}_{k-l} = \mathbf{x}_k - \alpha_k[\mathbf{I} - \mathbf{V}]\mathbf{g}_k,$$

where $\mathbf{g}_k = \mathbf{Qx}_k - \mathbf{b}$ and α_k is chosen to minimize f in the usual way. With the information given, what is the best bound on the rate of convergence of this method?

(c) Show that for $e < (\sqrt{5} - 1)/2$ the method in part (b) is always superior to steepest descent.

4. This problem shows that the modified Newton's method is globally convergent under very weak assumptions.
Let $a > 0$ and $b \geqslant a$ be given constants. Consider the collection P of all $n \times n$ symmetric positive definite matrices $\mathbf{P}$ having all eigenvalues greater than or equal to a and all elements bounded in absolute value by b. Define the point-to-set mapping $\mathbf{B} : E^n \to E^{n+n^2}$ by $\mathbf{B}(\mathbf{x}) = \{(\mathbf{x}, \mathbf{P}) : \mathbf{P} \in P\}$. Show that $\mathbf{B}$ is a closed mapping.
Now given an objective function $f \in C^1$, consider the iterative algorithm

$$\mathbf{x}_{k+1} = \mathbf{x}_k - \alpha_k \mathbf{P}_k \mathbf{g}_k.$$

where $\mathbf{g}_k = \mathbf{g}(\mathbf{x}_k)$ is the gradient of f at $\mathbf{x}_k$, $\mathbf{P}_k$ is any matrix from P and α_k is chosen to minimize $f(\mathbf{x}_{k+1})$. This algorithm can be represented by A which can be decomposed as $\mathbf{A} = \mathbf{SCB}$ where $\mathbf{B}$ is defined above, $\mathbf{C}$ is defined by $\mathbf{C}(\mathbf{x}, \mathbf{P}) = (\mathbf{x}, -\mathbf{Pg}(\mathbf{x}))$, and $\mathbf{S}$ is the standard line search mapping. Show that if restricted to a compact set in $\mathbf{E}^n$, the mapping A is closed.
Assuming that a sequence $\{\mathbf{x}_k\}$ generated by this algorithm is bounded, show that the limit $\mathbf{x}^*$ of any convergent subsequence satisfies $\mathbf{g}(\mathbf{x}^*) = 0$.

5. The following algorithm has been proposed for minimizing unconstrained functions $f(\mathbf{x})$, $\mathbf{x} \in E^n$, without using gradients: Starting with some arbitrary point $\mathbf{x}_0$, obtain a direction of search $\mathbf{d}_k$ such that for each component of $\mathbf{d}_k$

$$f(\mathbf{x}_k = (\mathbf{d}_k)_i \mathbf{e}_i) = \min_{di} f(\mathbf{x}_k + d_i \mathbf{e}_i),$$

where $\mathbf{e}_j$ denotes the ith column of the identity matrix. In other words, the ith component of $\mathbf{d}_k$ is determined through a line search minimizing $f(\mathbf{x})$ along the ***i*th** component.
The next point $\mathbf{x}_{k+1}$ is then determined in the usual way through a line search along $\mathbf{d}_k$; that is,

$$\mathbf{x}_{k+1} = \mathbf{x}_k + \alpha_k \mathbf{d}_k,$$

where $\mathbf{d}_k$ minimizes $f(\mathbf{x}_{k+1})$.

(a) Obtain an explicit representation for the algorithm for the quadratic case where

$$f(\mathbf{x}) = \frac{1}{2}(\mathbf{x} - \mathbf{x}^*)^T \mathbf{Q}(\mathbf{x} - \mathbf{x}^*) + f(\mathbf{x}^*).$$

(b) What condition on $f(\mathbf{x})$ or its derivatives will guarantee descent of this algorithm for general $f(\mathbf{x})$?

(c) Derive the convergence rate of this algorithm (assuming a quadratic objective). Express your answer in terms of the condition number of some matrix.

6. Suppose that the rank one correction method of Sect. 10.2 is applied to the quadratic problem (10.2) and suppose that the matrix $\mathbf{R}_0 = \mathbf{F}^{1/2}\mathbf{H}_0\mathbf{F}^{1/2}$ has $m < n$ eigenvalues less than unity and $n - m$ eigenvalues greater than unity. Show that the condition $\mathbf{q}_k^T(\mathbf{p}_k - \mathbf{H}_k\mathbf{q}_k) > 0$ will be satisfied at most m times during the course of the method and hence, if updating is performed only when this condition holds, the sequence $\{\mathbf{H}_k\}$ will not converge to $\mathbf{F}^{-1}$. Infer from this that, in using the rank one correction method, $\mathbf{H}_0$ should be taken very small; but that, despite such a precaution, on nonquadratic problems the method is subject to difficulty.
7. Show that if $\mathbf{H}_0 = \mathbf{I}$ the Davidon-Fletcher-Powell method is the conjugate gradient method. What similar statement can be made when $\mathbf{H}_0$ is an arbitrary symmetric positive definite matrix?
8. In the text it is shown that for the Davidon-Fletcher-Powell method $\mathbf{H}_{k+1}$ is positive definite if $\mathbf{H}_k$ is. The proof assumed that α_k is chosen to exactly minimize $f(\mathbf{x}_k + \alpha\mathbf{d}_k)$. Show that any $\alpha_k > 0$ which leads to $\mathbf{p}_k^T\mathbf{q}_k > 0$ will guarantee the positive definiteness of $\mathbf{H}_{k+1}$. Show that for a quadratic problem any $\alpha_k \neq 0$ leads to a positive definite $\mathbf{H}_{k+1}$.
9. Suppose along the line $\mathbf{x}_k + \alpha\mathbf{d}_k$, $\alpha > 0$, the function $f(\mathbf{x}_k + \alpha\mathbf{d}_k)$ is unimodal and differentiable. Let $\overline{\alpha}_k$ be the minimizing value of α. Show that if any $\alpha_k > \overline{\alpha}_k$ is selected to define $\mathbf{x}_{k+1} = \mathbf{x}_k + \alpha_k\mathbf{d}_k$, then $\mathbf{p}_k^T\mathbf{q}_k > 0$. (Refer to Sect. 10.3.)
10. Let $\{\mathbf{H}_k\}$, $k = 0, 1, 2 \ldots$ be the sequence of matrices generated by the Davidon-Fletcher-Powell method applied, without restarting, to a function f having continuous second partial derivatives. Assuming that there is $a > 0$, $A > 0$ such that for all k we have $\mathbf{H}_k - a\mathbf{I}$ and $A\mathbf{I} - \mathbf{H}_k$ positive definite and the corresponding sequence of $\mathbf{x}_k$'s is bounded, show that the method is globally convergent.
11. Verify Eq. (10.41).
12.
(a) Show that starting with the rank one update formula for $\mathbf{H}$, forming the complementary formula, and then taking the inverse restores the original formula.
(b) What value of ϕ in the Broyden class corresponds to the rank one formula?
13. Explain how the partial Davidon method can be implemented for $m < n/2$, with less storage than required by the full method.
14. Prove statements (10.1) and (10.2) below Eq. (10.46) in Sect. 10.6.
15. Consider using

$$\gamma k = \frac{\mathbf{p}_k^T\mathbf{H}_k^{-1}\mathbf{p}_k}{\mathbf{p}_k^T\mathbf{q}_k}$$

instead of (10.47).
(a) Show that this also serves as a suitable scale factor for a self-scaling quasi-Newton method.
(b) Extend part (a) to

$$\gamma_k = (1-\phi)\frac{\mathbf{p}_k^T\mathbf{q}_k}{\mathbf{q}_k^T\mathbf{H}_k\mathbf{q}_k} + \phi\frac{\mathbf{p}_k^T\mathbf{H}_k^{-1}\mathbf{p}_k}{\mathbf{p}_k^T\mathbf{q}_k}$$

for $0 \leqslant \phi \leqslant 1$.

16. Prove global convergence of the combination of steepest descent and Newton's method.
17. Formulate a rate of convergence theorem for the application of the combination of steepest and Newton's method to nonquadratic problems.
18. Prove that if $\mathbf{Q}$ is positive definite

$$\frac{(\mathbf{p}^T\mathbf{p})}{\mathbf{p}^T\mathbf{Q}\mathbf{p}} \leqslant \frac{\mathbf{p}^T\mathbf{Q}^{-1}\mathbf{p}}{\mathbf{p}^T\mathbf{p}}$$

for any vector $\mathbf{p}$.
19. It is possible to combine Newton's method and the partial conjugate gradient method. Given a subspace $N \subset E^n$, $\mathbf{x}_{k+1}$ is generated from $\mathbf{x}_k$ by first finding $\mathbf{z}_k$ by taking a Newton step in the linear variety through $\mathbf{x}_k$ parallel to N, and then taking m conjugate gradient steps from $\mathbf{z}_k$. What is a bound on the rate of convergence of this method?
20. In this exercise we explore how the combined method of Sect. 10.7 can be updated as more information becomes available. Begin with $N_0 = \{0\}$. If N_k is represented by the corresponding matrix $\mathbf{B}_k$, define N_{k+1} by the corresponding $\mathbf{B}_{k+1} = [\mathbf{B}_k,\ \mathbf{p}_k]$, where $\mathbf{p}_k = \mathbf{x}_{k+1} - \mathbf{z}_k$.

(a) If $\mathbf{D}_k = \mathbf{B}_k(\mathbf{B}_k^T\mathbf{F}\mathbf{B}_k)^{-1}\mathbf{B}_k^T$ is known, show that

$$\mathbf{D}_{k+1} = \mathbf{D}_k = \frac{(\mathbf{p}_k - \mathbf{D}_k\mathbf{q}_k)(\mathbf{p}_k - \mathbf{D}_k\mathbf{q}_k)^T}{(\mathbf{p}_k - \mathbf{D}_k\mathbf{q}_k)^T\mathbf{q}_k},$$

where $\mathbf{q}_k = \mathbf{g}_{k+1} - \mathbf{g}_k$. (This is the rank one correction of Sect. 10.2.)

(b) Develop an algorithm that uses (a) in conjunction with the combined method of Sect. 10.8 and discuss its convergence properties.

References

10.1 An early analysis of this method was given by Crockett and Chenoff [C9].

10.2–10.3 The variable metric method was originally developed by Davidon [D12], and its relation to the conjugate gradient method was discovered by Fletcher and Powell [F11]. The rank one method was later developed by Davidon [D13] and Broyden [B24]. For an early general discussion of these methods, see Murtagh and Sargent [M10], and for an excellent recent review, see Dennis and Moré [D15].

10.4 The Broyden family was introduced in Broyden [B24]. The BFGS method was suggested independently by Broyden [B25], Fletcher [F6], Goldfarb [G9], and Shanno [S3]. The beautiful concept of complementarity, which leads easily to the BFGS update and definition of the Broyden class as presented in the text, is due to Fletcher. Another larger class was defined by Huang [H13]. A variational approach to deriving variable metric methods was introduced by Greenstadt [G15]. Also see Dennis and Schnabel [D16]. Originally there was considerable effort devoted to searching for a best sequence of ϕ_k's in a Broyden method, but Dixon [D17] showed that all methods are identical in the case of exact linear search. There are a number of numerical analysis and implementation issues that arise in connection with quasi-Newton updating methods. From this viewpoint Gill and Murray [G6] have suggested working directly with $\mathbf{B}_k$, an approximation to the Hessian itself, and updating a triangular factorization at each step.

10.5 Under various assumptions on the criterion function, it has been shown that quasi- Newton methods converge globally and superlinearly, provided that accurate exact line search is used. See Powell [P8] and Dennis and Moré [D15]. With inexact line search, restarting is generally required to establish global convergence.

10.6 The lemma on interlocking eigenvalues is due to Loewner [L6]. An analysis of the one-by-one shift of the eigenvalues to unity is contained in Fletcher [F6]. The scaling concept, including the self-scaling algorithm, is due to Oren and Luenberger [O5]. Also see Oren [O4]. The two-parameter class of updates defined by the scaling procedure can be shown to be equivalent to the symmetric Huang class. Oren and Spedicato [O6] developed a procedure for selecting the scaling parameter so as to optimize the condition number of the update.

10.7 The idea of expressing conjugate gradient methods as update formulae is due to Perry [P3]. The development of the form presented here is due to Shanno [S4]. Preconditioning for conjugate gradient methods was suggested by Bertsekas [B9].

10.8 The combined method appears in Luenberger [L10].

Part III
Constrained Minimization

Chapter 11
Constrained Minimization Conditions

We turn now, in this final part of the book, to the study of minimization problems having constraints. We begin by studying in this chapter the necessary and sufficient conditions satisfied at solution points. These conditions, aside from their intrinsic value in characterizing solutions, define Lagrange multipliers and a certain Hessian matrix which, taken together, form the foundation for both the development and analysis of algorithms presented in subsequent chapters.

The general method used in this chapter to derive necessary and sufficient conditions is a straightforward extension of that used in Chap. 7 for unconstrained problems. In the case of equality constraints, the feasible region is a curved surface embedded in E^n. Differential conditions satisfied at an optimal point are derived by considering the value of the objective function along curves on this surface passing through the optimal point. Thus the arguments run almost identically to those for the unconstrained case; families of curves on the constraint surface replacing the earlier artifice of considering feasible directions. There is also a theory of zero-order conditions that is presented in the final section of the chapter.

11.1 Constraints

We deal with general nonlinear programming problems of the form

$$\begin{aligned} &\text{minimize } f(\mathbf{x}) \\ &\text{subject to } h_1(\mathbf{x}) = 0, \quad g_1(\mathbf{x}) \leqslant 0 \\ &\qquad\qquad h_2(\mathbf{x}) = 0, \quad g_2(\mathbf{x}) \leqslant 0 \\ &\qquad\qquad \vdots \qquad\qquad\quad \vdots \\ &\qquad\qquad h_m(\mathbf{x}) = 0, \quad g_p(\mathbf{x}) \leqslant 0 \\ &\qquad\qquad \mathbf{x} \in \Omega \subset \mathrm{E}^n, \end{aligned} \tag{11.1}$$

D.G. Luenberger, Y. Ye, *Linear and Nonlinear Programming*, International Series in Operations Research & Management Science 228,
DOI 10.1007/978-3-319-18842-3_11

where $m \leqslant n$ and the functions $f,\ h_i, i = 1, 2, \dots, m$ and $g_j,\ j = 1, 2,\ \dots,\ p$ are continuous, and usually assumed to possess continuous second partial derivatives. For notational simplicity, we introduce the vector-valued functions $\mathbf{h} = (h_1,\ h_2,\ \dots,\ h_m)$ and $\mathbf{g} = (g_1,\ g_2,\ \dots,\ g_P)$ and rewrite (11.1) as

$$\begin{aligned} &\text{minimize } f(\mathbf{x}) \\ &\text{subject to } \mathbf{h}(\mathbf{x}) = \mathbf{0},\ \mathbf{g}(\mathbf{x}) \leqslant \mathbf{0} \\ &\qquad\qquad \mathbf{x} \in \Omega. \end{aligned} \tag{11.2}$$

The constraints $\mathbf{h}(\mathbf{x}) = \mathbf{0},\ \mathbf{g}(\mathbf{x}) \leqslant \mathbf{0}$ are referred to as *functional constraints*, while the constraint $\mathbf{x} \in \Omega$ is a *set constraint*. As before we continue to de-emphasize the set constraint, assuming in most cases that either Ω is the whole space E^n or that the solution to (11.2) is in the interior of Ω. A point $\mathbf{x} \in \Omega$ that satisfies all the functional constraints is said to be *feasible*.

A fundamental concept that provides a great deal of insight as well as simplifying the required theoretical development is that of an *active constraint*. An inequality constraint $g_i(\mathbf{x}) \leqslant 0$ is said to be *active* at a feasible point $\mathbf{x}$ if $g_i(\mathbf{x}) = 0$ and *inactive* at $\mathbf{x}$ if $g_i(\mathbf{x}) < 0$. By convention we refer to any equality constraint $h_i(\mathbf{x}) = 0$ as *active* at any feasible point. The constraints active at a feasible point $\mathbf{x}$ restrict the domain of feasibility in neighborhoods of $\mathbf{x}$, while the other, inactive constraints, have no influence in neighborhoods of $\mathbf{x}$. Therefore, in studying the properties of a local minimum point, it is clear that attention can be restricted to the active constraints. This is illustrated in Fig. 11.1 where local properties satisfied by the solution $\mathbf{x}^*$ obviously do not depend on the inactive constraints g_2 and g_3.

It is clear that, if it were known a priori which constraints were active at the solution to (11.1), the solution would be a local minimum point of the problem defined by ignoring the inactive constraints and treating all active constraints as equality constraints. Hence, with respect to local (or relative) solutions, the problem could be regarded as having equality constraints only. This observation suggests that the

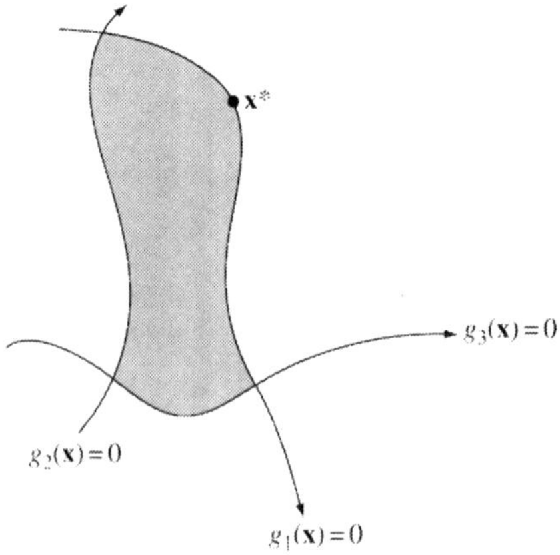

Fig. 11.1 Example of inactive constraints

majority of insight and theory applicable to (11.1) can be derived by consideration of equality constraints alone, later making additions to account for the selection of the active constraints. This is indeed so. Therefore, in the early portion of this chapter we consider problems having only equality constraints, thereby both economizing on notation and isolating the primary ideas associated with constrained problems. We then extend these results to the more general situation.

11.2 Tangent Plane

A set of equality constraints on E^n

$$\begin{aligned} h_1(\mathbf{x}) &= 0 \\ h_2(\mathbf{x}) &= 0 \\ &\vdots \\ h_m(\mathbf{x}) &= 0 \end{aligned} \tag{11.3}$$

defines a subset of E^n which is best viewed as a hypersurface. If the constraints are everywhere regular, in a sense to be described below, this hypersurface is of dimension $n - m$. If, as we assume in this section, the functions h_i, $i = 1, 2, \ldots, m$ belong to C^1, the surface defined by them is said to be *smooth*.

Associated with a point on a smooth surface is the *tangent plane* at that point, a term which in two or three dimensions has an obvious meaning. To formalize the general notion, we begin by defining curves on a surface. A *curve* on a surface S is a family of points $\mathbf{x}(t) \in S$ continuously parameterized by t for $a \leqslant t \leqslant b$. The curve is *differentiable* if $\dot{\mathbf{x}} \equiv (\mathrm{d}/\mathrm{d}t)\mathbf{x}(t)$ exists, and is *twice differentiable* if $\ddot{\mathbf{x}}(t)$ exists. A curve $\mathbf{x}(t)$ is said to pass through the point $\mathbf{x}^*$ if $\mathbf{x}^* = \mathbf{x}(t^*)$ for some t^*, $a \leqslant t^* \leqslant b$. The derivative of the curve at $\mathbf{x}^*$ is, of course, defined as $\dot{\mathbf{x}}(t^*)$. It is itself a vector in E^n.

Now consider all differentiable curves on S passing through a point $\mathbf{x}^*$. The *tangent plane* at $\mathbf{x}^*$ is defined as the collection of the derivatives at $\mathbf{x}^*$ of all these differentiable curves. The tangent plane is a subspace of E^n.

For surfaces defined through a set of constraint relations such as (11.3), the problem of obtaining an explicit representation for the tangent plane is a fundamental problem that we now address. Ideally, we would like to express this tangent plane in terms of derivatives of functions h_i that define the surface. We introduce the subspace

$$M = \{\mathbf{y} : \nabla \mathbf{h}(\mathbf{x}^*)\mathbf{y} = \mathbf{0}\}$$

and investigate under what conditions M is equal to the tangent plane at $\mathbf{x}^*$. The key concept for this purpose is that of a *regular point*. Figure 11.2 shows some examples where for visual clarity the tangent planes (which are sub-spaces) are translated to the point $\mathbf{x}^*$.

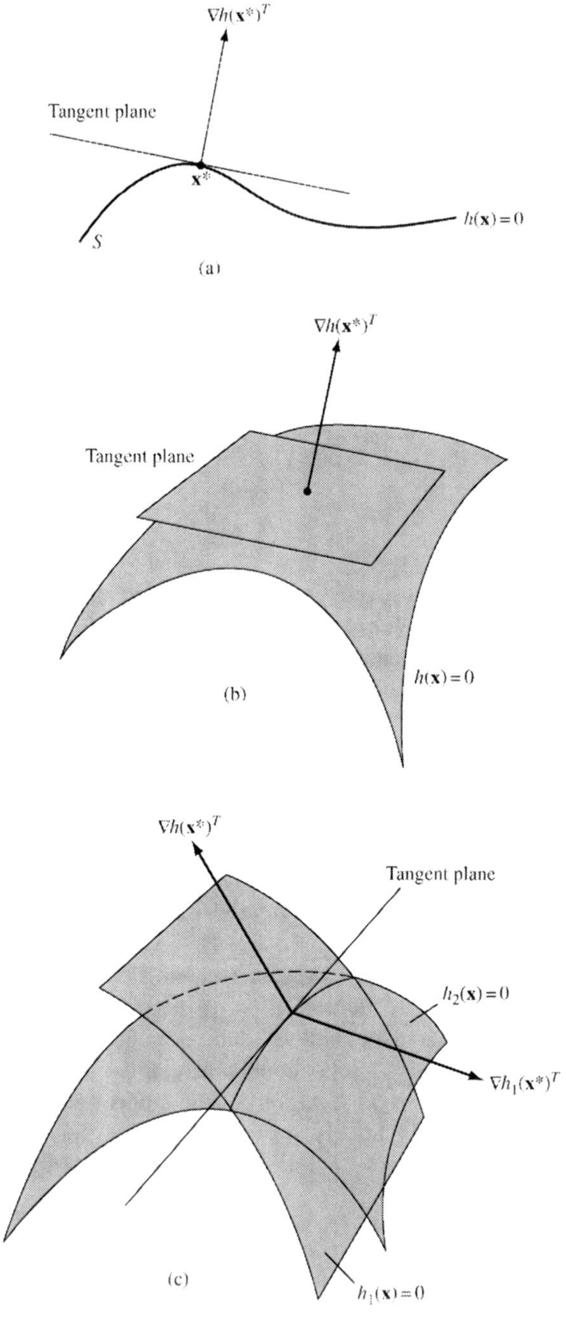

Fig. 11.2 Three examples of tangent planes (translated to $\mathbf{x}^*$)

Definition. A point $\mathbf{x}^*$ satisfying the constraint $\mathbf{h}(\mathbf{x}^*) = \mathbf{0}$ is said to be a *regular point* of the constraint if the gradient vectors $\nabla h_1(\mathbf{x}^*)$, $\nabla h_2(\mathbf{x}^*)$, ..., $\nabla h_m(\mathbf{x}^*)$ are linearly independent.

Note that if $\mathbf{h}$ is affine, $\mathbf{h}(\mathbf{x}) = \mathbf{A}\mathbf{x}+\mathbf{b}$, regularity is equivalent to A having rank equal to m, and this condition is independent of $\mathbf{x}$.

In general, at regular points it is possible to characterize the tangent plane in terms of the gradients of the constraint functions.

Theorem. *At a regular point* $\mathbf{x}^*$ *of the surface S defined by* $\mathbf{h}(x) = 0$ *the tangent plane is equal to*

$$M = \{y : \nabla \mathbf{h}(x^*)y = 0\}.$$

Proof. Let T be the tangent plane at $\mathbf{x}^*$. It is clear that $T \subset M$ whether $\mathbf{x}^*$ is regular or not, for any curve $\mathbf{x}(t)$ passing through $\mathbf{x}^*$ at $t = t^*$ having derivative $\dot{\mathbf{x}}(t^*)$ such that $\nabla\mathbf{h}(\mathbf{x}^*)\dot{\mathbf{x}}(t^*) \neq \mathbf{0}$ would not lie on S.

To prove that $M \subset T$ we must show that if $\mathbf{y} \in M$ then there is a curve on S passing through $\mathbf{x}^*$ with derivative $\mathbf{y}$. To construct such a curve we consider the equations

$$\mathbf{h}(\mathbf{x}^* + t\mathbf{y} + \nabla\mathbf{h}(\mathbf{x}^*)^T\mathbf{u}(t)) = \mathbf{0}, \tag{11.4}$$

where for fixed t we consider $\mathbf{u}(t) \in E^m$ to be the unknown. This is a nonlinear system of m equations and m unknowns, parameterized continuously, by t. At $t = 0$ there is a solution $\mathbf{u}(0) = \mathbf{0}$. The Jacobian matrix of the system with respect to $\mathbf{u}$ at $t = 0$ is the $m \times m$ matrix

$$\nabla\mathbf{h}(\mathbf{x}^*)\nabla\mathbf{h}(\mathbf{x}^*)^T,$$

which is nonsingular, since $\nabla\mathbf{h}(\mathbf{x}^*)$ is of full rank if $\mathbf{x}^*$ is a regular point. Thus, by the Implicit Function Theorem (see Appendix A) there is a continuously differentiable solution $\mathbf{u}(t)$ in some region $-a \leqslant t \leqslant a$.

The curve $\mathbf{x}(t) = \mathbf{x}^* + t\mathbf{y} + \nabla\mathbf{h}(\mathbf{x}^*)^T\mathbf{u}(t)$ is thus, by construction, a curve on S. By differentiating the system (11.4) with respect to t at $t = 0$ we obtain

$$\mathbf{0} = \frac{d}{dt}\mathbf{h}(\mathbf{x}(t))\bigg]_{t=0} = \nabla\mathbf{h}(\mathbf{x}^*)\mathbf{y} + \nabla\mathbf{h}(\mathbf{x}^*)\nabla\mathbf{h}(\mathbf{x}^*)^T\dot{\mathbf{u}}(0).$$

By definition of $\mathbf{y}$ we have $\nabla\mathbf{h}(\mathbf{x}^*)\mathbf{y} = \mathbf{0}$ and thus, again since $\nabla\mathbf{h}(\mathbf{x}^*)\nabla\mathbf{h}(\mathbf{x}^*)^T$ is nonsingular, we conclude that $\dot{\mathbf{x}}(0) = \mathbf{0}$. Therefore

$$\dot{\mathbf{x}}(0) = \mathbf{y} + \nabla\mathbf{h}(\mathbf{x}^*)^T\dot{\mathbf{x}}(0) = \mathbf{y},$$

and the constructed curve has derivative $\mathbf{y}$ at $\mathbf{x}^*$. ∎

It is important to recognize that the condition of being a regular point is not a condition on the constraint surface itself but on its representation in terms of an $\mathbf{h}$. The tangent plane is defined independently of the representation, while M is not.

Example. In E^2 let $h(x_1, x_2) = x_1$. Then $h(\mathbf{x}) = 0$ yields the x_2 axis, and every point on that axis is regular. If instead we put $h(x_1, x_2) = x_1^2$, again S is the x_2 axis but now no point on the axis is regular. Indeed in this case $M = E^2$, while the tangent plane is the x_2 axis.

11.3 First-Order Necessary Conditions (Equality Constraints)

The derivation of necessary and sufficient conditions for a point to be a local minimum point subject to equality constraints is fairly simple now that the representation of the tangent plane is known. We begin by deriving the first-order necessary conditions.

Lemma. *Let* $\mathbf{x}^*$ *be a regular point of the constraints* $\mathbf{h}(\boldsymbol{x}) = \boldsymbol{0}$ *and a local extremum point (a minimum or maximum) of* f *subject to these constraints.*

Then all $\mathbf{y} \in E^n$ *satisfying*

$$\nabla \mathbf{h}(\mathbf{x}^*)\mathbf{y} = \mathbf{0} \tag{11.5}$$

must also satisfy

$$\nabla f(\mathbf{x}^*)\mathbf{y} = 0. \tag{11.6}$$

Proof. Let $\mathbf{y}$ be any vector in the tangent plane at $\mathbf{x}^*$ and let $\mathbf{x}(t)$ be any smooth curve on the constraint surface passing through $\mathbf{x}^*$ with derivative $\mathbf{y}$ at $\mathbf{x}^*$; that is, $\mathbf{x}(0) = \mathbf{x}^*$, $\dot{\mathbf{x}}(0) = \mathbf{y}$, and $\mathbf{h}(\mathbf{x}(\mathbf{t})) = \mathbf{0}$ for $-a \leqslant t \leqslant a$ for some $a > 0$.

Since $\mathbf{x}^*$ is a regular point, the tangent plane is identical with the set of $\mathbf{y}$'s satisfying $\nabla \mathbf{h}(\mathbf{x}^*)\mathbf{y} = \mathbf{0}$. Then, since $\mathbf{x}^*$ is a constrained local extremum point of f, we have

$$\left.\frac{d}{dt} f(\mathbf{x}(t))\right|_{t=0} = 0,$$

or equivalently,

$$\nabla f(\mathbf{x}^*)\mathbf{y} = 0. \blacksquare$$

The above Lemma says that $\nabla f(\mathbf{x}^*)$ is orthogonal to the tangent plane. Next we conclude that this implies that $\nabla f(\mathbf{x}^*)$ is a linear combination of the gradients of $\mathbf{h}$ at $\mathbf{x}^*$, a relation that leads to the introduction of Lagrange multipliers. As in much of nonlinear programming, the Lagrange multiplier vector is often labeled λ rather than $\mathbf{y}$ in linear programming, and this convention is followed here.

Theorem. *Let* $\mathbf{x}^*$ *be a local extremum point of* f *subject to the constraints* $\mathbf{h}(\boldsymbol{x}) = \boldsymbol{0}$. *Assume further that* $\mathbf{x}^*$ *is a regular point of these constraints. Then there is a* $\boldsymbol{\lambda} \in E^m$ *such that*

$$\nabla f(\boldsymbol{x}^*) + \lambda^T \nabla \boldsymbol{h}(\boldsymbol{x}^*) = \boldsymbol{0}. \tag{11.7}$$

Proof. From the Lemma we may conclude that the value of the linear program

$$\begin{aligned} &\text{maximize} \quad \nabla f(\mathbf{x}^*)\mathbf{y} \\ &\text{subject to} \quad \nabla \mathbf{h}(\mathbf{x}^*)\mathbf{y} = \mathbf{0} \end{aligned}$$

is zero. Thus, by the Duality Theorem of linear programming (Sect. 4.2) the dual problem is feasible. Specifically, there is $\boldsymbol{\lambda} \in E^m$ such that $\nabla f(\mathbf{x}^*) + \lambda^T \nabla \mathbf{h}(\mathbf{x}^*) = 0$. $\blacksquare$

It should be noted that the first-order necessary conditions

$$\nabla f(\mathbf{x}^*) + \boldsymbol{\lambda}^T \nabla \mathbf{h}(\mathbf{x}^*) = \mathbf{0}$$

together with the constraints

$$\mathbf{h}(\mathbf{x}^*) = \mathbf{0}$$

give a total of $n+m$ (generally nonlinear) equations in the $n+m$ variables comprising $\mathbf{x}^*$, $\boldsymbol{\lambda}$. Thus the necessary conditions are a complete set since, at least locally, they determine a unique solution.

It is convenient to introduce the *Lagrangian* associated with the constrained problem, defined as

$$l(\mathbf{x}, \boldsymbol{\lambda}) = f(\mathbf{x}) + \boldsymbol{\lambda}^T \mathbf{h}(\mathbf{x}). \tag{11.8}$$

The necessary conditions can then be expressed in the form

$$\nabla_{\mathbf{x}} l(\mathbf{x}, \boldsymbol{\lambda}) = \mathbf{0} \tag{11.9}$$

$$\nabla_{\boldsymbol{\lambda}} l(\mathbf{x}, \boldsymbol{\lambda}) = \mathbf{0}, \tag{11.10}$$

the second of these being simply a restatement of the constraints.

11.4 Examples

We digress briefly from our mathematical development to consider some examples of constrained optimization problems. We present five simple examples that can be treated explicitly in a short space and then briefly discuss a broader range of applications.

Example 1. Consider the problem

$$\begin{aligned} \text{minimize} \quad & x_1 x_2 + x_2 x_3 + x_1 x_3 \\ \text{subject to} \quad & x_1 + x_2 + x_3 = 3. \end{aligned}$$

The necessary conditions become

$$\begin{aligned} x_2 + x_3 + \lambda &= 0 \\ x_1 \qquad + x_3 + \lambda &= 0 \\ x_1 + x_2 \qquad + \lambda &= 0. \end{aligned}$$

These three equations together with the one constraint equation give four equations that can be solved for the four unknowns x_1, x_2, x_3, λ. Solution yields $x_1 = x_2 = x_3 = 1$, $\lambda = -2$.

Example 2 (Maximum Volume). Let us consider an example of the type that is now standard in textbooks and which has a structure similar to that of the example above.

We seek to construct a cardboard box of maximum volume, given a fixed area of cardboard.

Denoting the dimensions of the box by $x,\ y,\ z$, the problem can be expressed as

$$\begin{aligned} &\text{maximize} \quad xyz \\ &\text{subject to} \quad (xy + yz + xz) = \frac{c}{2}, \end{aligned} \tag{11.11}$$

where $c > 0$ is the given area of cardboard. Introducing a Lagrange multiplier, the first-order necessary conditions are easily found to be

$$\begin{aligned} yz + \lambda(y + z) &= 0 \\ xz + \lambda(x + z) &= 0 \\ xy + \lambda(x + y) &= 0 \end{aligned} \tag{11.12}$$

together with the constraint. Before solving these, let us note that the sum of these equations is $(xy + yz + xz) + 2\lambda(x + y + z) = 0$. Using the constraint this becomes $c/2 + 2\lambda(x + y + z) = 0$. From this it is clear that $\lambda \neq 0$. Now we can show that $x,\ y$, and z are nonzero. This follows because $x = 0$ implies $z = 0$ from the second equation and $y = 0$ from the third equation. In a similar way, it is seen that if either $x,\ y$, or z are zero, all must be zero, which is impossible.

To solve the equations, multiply the first by x and the second by y, and then subtract the two to obtain

$$\lambda(x - y)z = 0.$$

Operate similarly on the second and third to obtain

$$\lambda(y - z)x = 0.$$

Since no variables can be zero, it follows that $x = y = z = \sqrt{c/6}$ is the unique solution to the necessary conditions. The box must be a cube.

Example 3 (Entropy). optimization problems often describe natural phenomena. An example is the characterization of naturally occurring probability distributions as maximum entropy distributions.

As a specific example consider a discrete probability density corresponding to a measured value taking one of n values $x_1,\ x_2,\ \ldots,\ x_n$. The probability associated with x_i is p_i. The p_i's satisfy $p_i \geqslant 0$ and $\sum_{i=1}^{n} p_i = 1$.

The *entropy* of such a density is

$$\varepsilon = -\sum_{i=1}^{n} p_i \log(p_i).$$

The *mean value* of the density is $\sum_{i=1}^{n} x_i p_i$.

If the value of mean is known to be m (by the physical situation), the maximum entropy argument suggests that the density should be taken as that which solves the following problem:

$$\begin{aligned}
\text{maximize} \quad & -\sum_{i=1}^{n} p_i \log(p_i) \\
\text{subject to} \quad & \sum_{i=1}^{n} p_i = 1 \\
& \sum_{i=1}^{n} x_i p_i = m \\
& p_i \geq 0, \quad i = 1, 2, \ldots, n.
\end{aligned} \tag{11.13}$$

We begin by ignoring the nonnegativity constraints, believing that they may be inactive. Introducing two Lagrange multipliers, λ and μ, the Lagrangian is

$$l = \sum_{i=1}^{n} \{-p_i \log p_i + \lambda p_i + \mu x_i p_i\} - \lambda - \mu m.$$

The necessary conditions are immediately found to be

$$-\log p_i - 1 + \lambda + \mu x_i = 0, \quad i = 1, 2, \ldots, n.$$

This leads to

$$p_i = \exp\{(\lambda - 1) + \mu x_i\}, \quad i = 1, 2, \ldots, n. \tag{11.14}$$

We note that $p_i > 0$, so the nonnegativity constraints are indeed inactive. The result (11.14) is known as an exponential density. The Lagrange multipliers λ and μ are parameters that must be selected so that the two equality constraints are satisfied.

Example 4 (Hanging Chain). A chain is suspended from two thin hooks that are 16 ft apart on a horizontal line as shown in Fig. 11.3. The chain itself consists of 20 links of stiff steel. Each link is one foot in length (measured inside). We wish to formulate the problem to determine the equilibrium shape of the chain.

The solution can be found by minimizing the potential energy of the chain. Let us number the links consecutively from 1 to 20 starting with the left end. We let link i span an x distance of x_i and a y distance of y_i. Then $x_i^2 + y_i^2 = 1$. The potential energy of a link is its weight times its vertical height (from some reference). The potential energy of the chain is the sum of the potential energies of each link. We may take the top of the chain as reference and assume that the mass of each link is concentrated at its center. Assuming unit weight, the potential energy is then

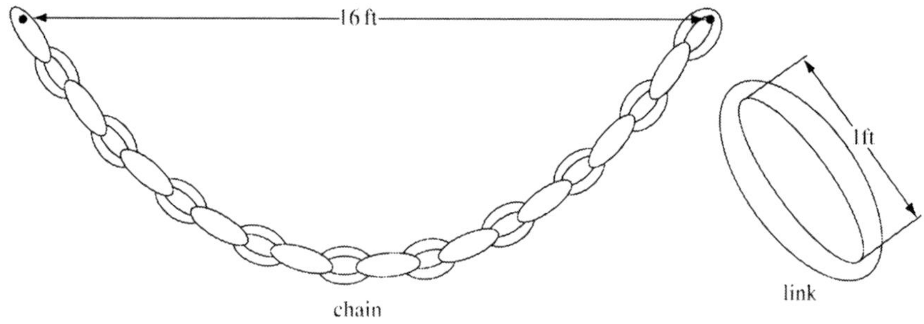

Fig. 11.3 A hanging chain

$$\frac{1}{2}y_1 + \left(y_1 + \frac{1}{2}y_2\right) + \left(y_1 + y_2 + \frac{1}{2}y_3\right) + \cdots$$
$$+ \left(y_1 + y_2 + \cdots + y_{n-1} + \frac{1}{2}y_n\right) = \sum_{i=1}^{n}\left(n - i + \frac{1}{2}\right)y_i,$$

where $n = 20$ in our example.

The chain is subject to two constraints: The total y displacement is zero, and the total x displacement is 16. Thus the equilibrium shape is the solution of

$$\begin{aligned} &\text{minimize} && \sum_{i=1}^{n}\left(n - i + \frac{1}{2}\right)y_i \\ &\text{subject to} && \sum_{i=1}^{n} y_i = 0 \\ & && \sum_{i=1}^{n}\sqrt{1 - y_i^2} = 16. \end{aligned} \tag{11.15}$$

The first-order necessary conditions are

$$\left(n - i + \frac{1}{2}\right) + \lambda - \frac{\mu y_i}{\sqrt{1 - y_i^2}} = 0 \tag{11.16}$$

for $i = 1, 2, \ldots, n$. This leads directly to

$$y_i = -\frac{n - i + \frac{1}{2} + \lambda}{\sqrt{\mu^2 + \left(n - i + \frac{1}{2} + \lambda\right)^2}}. \tag{11.17}$$

As in Example 2 the solution is determined once the Lagrange multipliers are known. They must be selected so that the solution satisfies the two constraints.

It is useful to point out that problems of this type may have local minimum points. The reader can examine this by considering a short chain of, say, four links and v and w configurations.

Example 5 (Portfolio Design). Suppose there are n securities indexed by $i = 1, 2, \ldots, n$. Each security i is characterized by its random rate of return r_i which has mean value $\bar{r}_i$. Its covariances with the rates of return of other securities are σ_{ij}, for $j = 1, 2, \ldots, n$. The portfolio problem is to allocate total available wealth among these n securities, allocating a fraction w_i of wealth to the security i.

The overall rate of return of a portfolio is $r = \sum_{i=1}^{n} w_i \bar{r}_i$ and variance $\sigma^2 = \sum_{i,j=1}^{n} w_i \sigma_{ij} w_j$.

Markowitz introduced the concept of devising *efficient* portfolios which for a given expected rate of return $\bar{r}$ have minimum possible variance. Such a portfolio is the solution to the problem

$$\begin{aligned} \min_{w_i, w_2, \ldots w_n} \ & \sum_{i,j=1}^{n} w_i \sigma_{ij} w_j \\ \text{subject to } & \sum_{i=1}^{n} w_i \bar{r}_i = \bar{r} \\ & \sum_{i=1}^{n} w_i = 1. \end{aligned}$$

The second constraint forces the sum of the weights to equal one. There may be the further restriction that each $w_i \geq 0$ which would imply that the securities must not be shorted (that is, sold short).

Introducing Lagrange multipliers λ and μ for the two constraints leads easily to the $n + 2$ linear equations

$$\sum_{j=1}^{n} \sigma_{ij} w_j + \lambda \bar{r}_i + \mu = 0 \quad \text{for } i = 1, 2, \ldots, n$$

$$\sum_{i=1}^{n} w_i \bar{r}_i = \bar{r}$$

$$\sum_{i=1}^{n} w_i = 1$$

in the $n + 2$ unknowns (the w_i's, λ and μ).

Large-Scale Applications

The problems that serve as the primary motivation for the methods described in this part of the book are actually somewhat different in character than the problems represented by the above examples, which by necessity are quite simple. Larger, more complex, nonlinear programming problems arise frequently in modern applied analysis in a wide variety of disciplines. Indeed, within the past few decades nonlinear programming has advanced from a relatively young and primarily analytic subject to a substantial general tool for problem solving.

Large nonlinear programming problems arise in problems of mechanical structures, such as determining optimal configurations for bridges, trusses, and so forth. Some mechanical designs and configurations that in the past were found by solving differential equations are now often found by solving suitable optimization problems. An example that is somewhat similar to the hanging chain problem is the determination of the shape of a stiff cable suspended between two points and supporting a load.

A wide assortment, of large-scale optimization problems arise in a similar way as methods for solving partial differential equations. In situations where the underlying continuous variables are defined over a two- or three-dimensional region, the continuous region is replaced by a grid consisting of perhaps several thousand discrete points. The corresponding discrete approximation to the partial differential equation is then solved indirectly by formulating an equivalent optimization problem. This approach is used in studies of plasticity, in heat equations, in the flow of fluids, in atomic physics, and indeed in almost all branches of physical science.

Problems of optimal control lead to large-scale nonlinear programming problems. In these problems a dynamic system, often described by an ordinary differential equation, relates control variables to a trajectory of the system state. This differential equation, or a discretized version of it, defines one set of constraints. The problem is to select the control variables so that the resulting trajectory satisfies various additional constraints and minimizes some criterion. An early example of such a problem that was solved numerically was the determination of the trajectory of a rocket to the moon that required the minimum fuel consumption.

There are many examples of nonlinear programming in industrial operations and business decision making. Many of these are nonlinear versions of the kinds of examples that were discussed in the linear programming part of the book. Nonlinearities can arise in production functions, cost curves, and, in fact, in almost all facets of problem formulation.

Portfolio analysis, in the context of both stock market investment and evaluation of a complex project within a firm, is an area where nonlinear programming is becoming increasingly useful. These problems can easily have thousands of variables.

In many areas of model building and analysis, optimization formulations are increasingly replacing the direct formulation of systems of equations. Thus large economic forecasting models often determine equilibrium prices by minimizing an objective termed *consumer surplus*. Physical models are often formulated as minimization of energy. Decision problems are formulated as maximizing expected utility. Data analysis procedures are based on minimizing an average error or maximizing a probability. As the methodology for solution of nonlinear programming improves, one can expect that this trend will continue.

11.5 Second-Order Conditions

By an argument analogous to that used for the unconstrained case, we can also derive the corresponding second-order conditions for constrained problems. Throughout this section it is assumed that f, $\mathbf{h} \in C^2$.

Second-Order Necessary Conditions. *Suppose that* $\mathbf{x}^*$ *is a local minimum of* f *subject to* $\mathbf{h}(\mathbf{x}) = 0$ *and that* $\mathbf{x}^*$ *is a regular point of these constraints. Then there is a* $\boldsymbol{\lambda} \in E^m$ *such that*

$$\nabla f(\mathbf{x}^*) + \boldsymbol{\lambda}^T \nabla \mathbf{h}(\mathbf{x}^*) = \mathbf{0}. \tag{11.18}$$

If we denote by M *the tangent plane* $M = \{\mathbf{y} : \nabla \mathbf{h}(\mathbf{x}^*)\mathbf{y} = \mathbf{0}\}$, *then the matrix*

$$\mathbf{L}(\mathbf{x}^*) = \mathbf{F}(\mathbf{x}^*) + \boldsymbol{\lambda}^T \mathbf{H}(\mathbf{x}^*) \tag{11.19}$$

is positive semidefinite on M, *that is,* $\mathbf{y}^T \mathbf{L}(\mathbf{x}^*)\mathbf{y} \geqslant \mathbf{0}$ *for all* $\mathbf{y} \in M$.

Proof. From elementary calculus it is clear that for every twice differentiable curve on the constraint surface S through $\mathbf{x}^*$ (with $\mathbf{x}(0) = \mathbf{x}^*$) we have

$$\left.\frac{d^2}{dt^2} f(\mathbf{x}(t))\right|_{t=0} \geqslant 0. \tag{11.20}$$

By definition

$$\left.\frac{d^2}{dt^2} f(\mathbf{x}(t))\right|_{t=0} = \dot{\mathbf{x}}(0)^T \mathbf{F}(\mathbf{x}^*)\dot{\mathbf{x}}(0) + \nabla f(\mathbf{x}^*)\ddot{\mathbf{x}}(0). \tag{11.21}$$

Furthermore, differentiating the relation $\boldsymbol{\lambda}^T \mathbf{h}(\mathbf{x}(t)) = 0$ twice, we obtain

$$\dot{\mathbf{x}}(0)^T \boldsymbol{\lambda}^T \mathbf{H}(\mathbf{x}^*)\dot{\mathbf{x}}(0) + \boldsymbol{\lambda}^T \nabla \mathbf{h}(\mathbf{x}^*)\ddot{\mathbf{x}}(\mathbf{0}) = 0. \tag{11.22}$$

Adding (11.22) to (11.21), while taking account of (11.20), yields the result

$$\left.\frac{d^2}{dt^2} f(\mathbf{x}(t))\right|_{t=0} = \dot{\mathbf{x}}(0)^T \mathbf{L}(\mathbf{x}^*)\dot{\mathbf{x}}(0) \geqslant 0.$$

Since $\dot{\mathbf{x}}(0)$ is arbitrary in M, we immediately have the stated conclusion. ∎

The above theorem is our first encounter with the matrix $\mathbf{L} = \mathbf{F} + \boldsymbol{\lambda}^T \mathbf{H}$ which is the matrix of second partial derivatives, with respect to $\mathbf{x}$, of the Lagrangian l. (See Appendix A, Sect. A.6, for a discussion of the notation $\boldsymbol{\lambda}^T \mathbf{H}$ used here.) This matrix is the backbone of the theory of algorithms for constrained problems, and it is encountered often in subsequent chapters.

We next state the corresponding set of sufficient conditions.

Second-Order Sufficiency Conditions. *Suppose there is a point* $\mathbf{x}^*$ *satisfying* $\mathbf{h}(\mathbf{x}^*) = \mathbf{0}$, *and a* $\boldsymbol{\lambda} \in E^m$ *such that*

$$\nabla f(\mathbf{x}^*) + \boldsymbol{\lambda}^T \nabla \mathbf{h}(\mathbf{x}^*) = \mathbf{0}. \tag{11.23}$$

Suppose also that the matrix $\mathbf{L}(\mathbf{x}^*) = \mathbf{F}(\mathbf{x}^*) + \boldsymbol{\lambda}^T\mathbf{H}(\mathbf{x}^*)$ *is positive definite on* $M = \{\mathbf{y} : \nabla\mathbf{h}(\mathbf{x}^*)\mathbf{y} = \mathbf{0}\}$, *that is, for* $\mathbf{y} \in M$, $\mathbf{y} \neq \mathbf{0}$ *there holds* $\mathbf{y}^T\mathbf{L}(\mathbf{x}^*)\mathbf{y} > 0$. *Then* $\mathbf{x}^*$ *is a strict local minimum of* f *subject to* $\mathbf{h}(\mathbf{x}) = \mathbf{0}$.

Proof. If $\mathbf{x}^*$ is not a strict relative minimum point, there exists a sequence of feasible points $\{\mathbf{y}_k\}$ converging to $\mathbf{x}^*$ such that for each k, $f(\mathbf{y}_k) \leqslant f(\mathbf{x}^*)$. Write each $\mathbf{y}_k$ in the form $\mathbf{y}_k = \mathbf{x}^* + \delta_k\mathbf{s}_k$ where $\mathbf{s}_k \in \mathbf{E}^n$, $|\mathbf{s}_k| = 1$, and $\delta_k > 0$ for each k. Clearly, $\delta_k \to 0$ and the sequence $\{\mathbf{s}_k\}$, being bounded, must have a convergent subsequence converging to some $\mathbf{s}^*$. For convenience of notation, we assume that the sequence $\{\mathbf{s}_k\}$ is itself convergent to $\mathbf{s}^*$. We also have $\mathbf{h}(\mathbf{y}_k) - \mathbf{h}(\mathbf{x}^*) = \mathbf{0}$, and dividing by δ_k and letting $k \to \infty$ we see that $\nabla\mathbf{h}(\mathbf{x}^*)\mathbf{s}^* = \mathbf{0}$.

Now by Taylor's theorem, we have for each j

$$0 = h_j(\mathbf{y}_k) = h_j(\mathbf{x}^*) + \delta_k\nabla h_j(\mathbf{x}^*)\mathbf{s}_k + \frac{\delta_k^2}{2}\mathbf{s}_k^T\nabla^2 h_j(\boldsymbol{\eta}_j)\mathbf{s}_k \tag{11.24}$$

and

$$0 \geqslant f(\mathbf{y}_k) - f(\mathbf{x}^*) = \delta_k\nabla f(\mathbf{x}^*)\mathbf{s}_k + \frac{\delta_k^2}{2}\mathbf{s}_k^T\nabla^2 f(\boldsymbol{\eta}_0)\mathbf{s}_k, \tag{11.25}$$

where each $\boldsymbol{\eta}_j$ is a point on the line segment joining $\mathbf{x}^*$ and $\mathbf{y}_k$. Multiplying (11.24) by $\boldsymbol{\lambda}_j$ and adding these to (11.25) we obtain, on accounting for (11.23),

$$0 \geqslant \frac{\delta_k^2}{2}\mathbf{s}_k^T\left\{\nabla^2 f(\boldsymbol{\eta}_0) + \sum_{i=1}^m \boldsymbol{\lambda}_i\nabla^2 h_i(\boldsymbol{\eta}_i)\right\}\mathbf{s}_k,$$

which yields a contradiction as $k \to \infty$. ∎

Example 1. Consider the problem

$$\begin{aligned} &\text{maximize} && x_1x_2 + x_2x_3 + x_1x_3 \\ &\text{subject to} && x_1 + x_2 + x_3 = 3. \end{aligned}$$

In Example 1 of Sect. 11.4 it was found that $x_1 = x_2 = x_3 = 1$, $\boldsymbol{\lambda} = -2$ satisfy the first-order conditions. The matrix $\mathbf{F} + \boldsymbol{\lambda}^T\mathbf{H}$ becomes in this case

$$L = \begin{bmatrix} 0 & 1 & 1 \\ 1 & 0 & 1 \\ 1 & 1 & 0 \end{bmatrix},$$

which itself is neither positive nor negative definite. On the subspace $M = \{\mathbf{y}\colon y_1 + y_2 + y_3 = 0\}$, however, we note that

$$\begin{aligned} \mathbf{y}^T\mathbf{L}\mathbf{y} &= y_1(y_2 + y_3) + y_2(y_1 + y_3) + y_3(y_1 + y_2) \\ &= -(y_1^2 + y_2^2 + y_3^2), \end{aligned}$$

and thus $\mathbf{L}$ is negative definite on M. Therefore, the solution we found is at least a local maximum.

11.6 Eigenvalues in Tangent Subspace

In the last section it was shown that the matrix $\mathbf{L}$ restricted to the subspace M that is tangent to the constraint surface plays a role in second-order conditions entirely analogous to that of the Hessian of the objective function in the unconstrained case. It is perhaps not surprising, in view of this, that the structure of $\mathbf{L}$ restricted to M also determines rates of convergence of algorithms designed for constrained problems in the same way that the structure of the Hessian of the objective function does for unconstrained algorithms. Indeed, we shall see that the eigenvalues of $\mathbf{L}$ restricted to M determine the natural rates of convergence for algorithms designed for constrained problems. It is important, therefore, to understand what these restricted eigenvalues represent. We first determine geometrically what we mean by the restriction of $\mathbf{L}$ to M which we denote by $\mathbf{L}_M$. Next we define the eigenvalues of the operator $\mathbf{L}_M$. Finally we indicate how these various quantities can be computed.

Given any vector $\mathbf{y} \in M$, the vector $\mathbf{Ly}$ is in E^n but not necessarily in M. We project $\mathbf{Ly}$ orthogonally back onto M, as shown in Fig. 11.4, and the result is said to be the restriction of $\mathbf{L}$ to M operating on $\mathbf{y}$. In this way we obtain a linear transformation from M to M. The transformation is determined somewhat implicitly, however, since we do not have an explicit matrix representation.

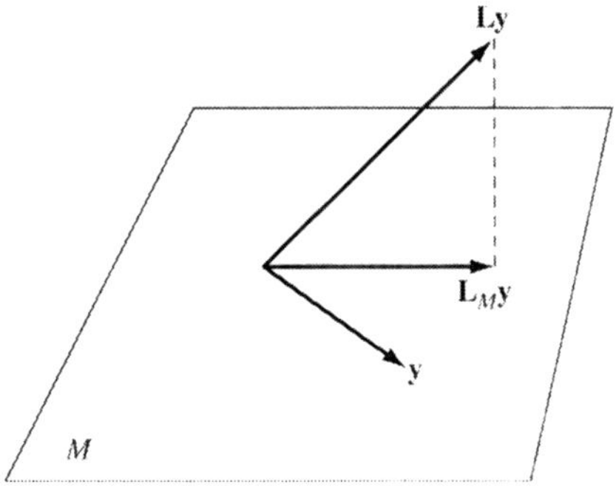

Fig. 11.4 Definition of $\mathbf{L}_M$

A vector $\mathbf{y} \in M$ is an *eigenvector* of $\mathbf{L}_M$ if there is a real number λ such that $\mathbf{L}_M\mathbf{y} = \lambda\mathbf{y}$; the corresponding λ is an *eigenvalue* of $\mathbf{L}_M$. This coincides with the standard definition. In terms of $\mathbf{L}$ we see that $\mathbf{y}$ is an eigenvector of $\mathbf{L}_M$ if $\mathbf{Ly}$ can be written as the sum of $\lambda\mathbf{y}$ and a vector orthogonal to M. See Fig. 11.5.

To obtain a matrix representation for $\mathbf{L}_M$ it is necessary to introduce a basis in the subspace M. For simplicity it is best to introduce an orthonormal basis, say $\mathbf{e}_1, \mathbf{e}_2, \ldots, \mathbf{e}_{n-m}$. Define the matrix $\mathbf{E}$ to be the $n \times (n-m)$ matrix whose columns consist of the vectors $\mathbf{e}_i$. Then any vector $\mathbf{y}$ in M can be written as $\mathbf{y} = \mathbf{Ez}$ for some $\mathbf{z} \in E^{n-m}$ and, of course, $\mathbf{LEz}$ represents the action of $\mathbf{L}$ on such a vector. To project this result back into M and express the result in terms of the basis $\mathbf{e}_1, \mathbf{e}_2, \ldots, \mathbf{e}_{n-m}$,

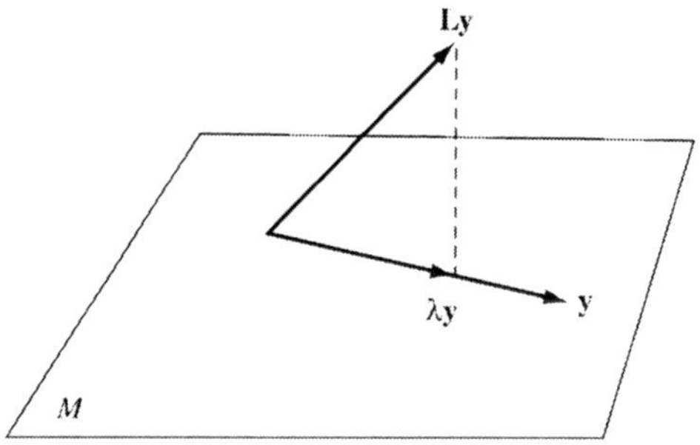

Fig. 11.5 Eigenvector of $\mathbf{L}_M$

we merely multiply by $\mathbf{E}^T$. Thus $\mathbf{E}^T\mathbf{LEz}$ is the vector whose components give the representation in terms of the basis; and, correspondingly, the $(n-m)\times(n-m)$ matrix $\mathbf{E}^T\mathbf{LE}$ is the matrix representation of $\mathbf{L}$ restricted to M.

The eigenvalues of $\mathbf{L}$ restricted to M can be found by determining the eigenvalues of $\mathbf{E}^T\mathbf{LE}$. These eigenvalues are independent of the particular orthonormal basis $\mathbf{E}$.

Example 1. In the last section we considered

$$L = \begin{bmatrix} 0 & 1 & 1 \\ 1 & 0 & 1 \\ 1 & 1 & 0 \end{bmatrix}$$

restricted to $M = \{\mathbf{y} : y_1 + y_2 + y_3 = 0\}$. To obtain an explicit matrix representation on M let us introduce the orthonormal basis:

$$\mathbf{e}_1 = \frac{1}{\sqrt{2}}(1, 0, -1)$$
$$\mathbf{e}_2 = \frac{1}{\sqrt{6}}(1, -2, 1).$$

This gives, upon expansion,

$$\mathbf{E}^T\mathbf{LE} = \begin{bmatrix} -1 & 0 \\ 0 & -1 \end{bmatrix},$$

and hence $\mathbf{L}$ restricted to M acts like the negative of the identity.

Example 2. Let us consider the problem

$$\begin{aligned} &\text{extremize} \quad x_1 + x_2^2 + x_2x_3 + 2x_3^2 \\ &\text{subject to} \quad \frac{1}{2}(x_1^2 + x_2^2 + x_3^2) = 1. \end{aligned}$$

The first-order necessary conditions are

$$\begin{aligned} 1 + \qquad \lambda x_1 &= 0 \\ 2x_2 + x_3 + \lambda x_2 &= 0 \\ x_2 + 4x_3 + \lambda x_3 &= 0. \end{aligned}$$

One solution to this set is easily seen to be $x_1 = 1,\ x_2 = 0,\ x_3 = 0,\ \lambda = -1$. Let us examine the second-order conditions at this solution point. The Lagrangian matrix there is

$$\mathbf{L} = \begin{bmatrix} -1 & 0 & 0 \\ 0 & 1 & 1 \\ 0 & 1 & 3 \end{bmatrix},$$

and the corresponding subspace M is

$$M = \{\mathbf{y} : y_1 = 0\}.$$

In this case M is the subspace spanned by the second two basis vectors in E^3 and hence the restriction of **L** to M can be found by taking the corresponding submatrix of L. Thus, in this case,

$$\mathbf{E}^T\mathbf{L}\mathbf{E} = \begin{bmatrix} 1 & 1 \\ 1 & 3 \end{bmatrix}.$$

The characteristic polynomial of this matrix is

$$\det \begin{bmatrix} 1-\lambda & 1 \\ 1 & 3-\lambda \end{bmatrix} = (1-\lambda)(3-\lambda) - 1 = \lambda^2 - 4\lambda + 2.$$

The eigenvalues of $\mathbf{L}_M$ are thus $\lambda = 2 \pm \sqrt{2}$, and $\mathbf{L}_M$ is positive definite.

Since the $\mathbf{L}_M$ matrix is positive definite, we conclude that the point found is a relative minimum point. This example illustrates that, in general, the restriction of **L** to M can be thought of as a submatrix of **L**, although it can be read directly from the original matrix only if the subspace M is spanned by a subset of the original basis vectors.

Projected Hessians

The above approach for determining the eigenvalues of **L** projected onto M is quite direct and relatively simple. There is another approach, however, that is useful in some theoretical arguments and convenient for simple applications. It is based on constructing matrices and determinants of order n rather than $n - m$, but there is no need to find the orthonormal basis **E**. For simplicity, let $\mathbf{A} = \nabla \mathbf{h}$ which has full row rank.

Any $\mathbf{x}$ satisfying $\mathbf{Ax} = \mathbf{0}$ can be expressed as

$$\mathbf{x} = (\mathbf{I} - \mathbf{A}^T(\mathbf{AA}^T)^{-1}\mathbf{A})\mathbf{z}$$

for some $\mathbf{z}$ (and the converse is also true), where $\mathbf{P}_A = (\mathbf{I} - \mathbf{A}^T(\mathbf{AA}^T)^{-1}\mathbf{A})$ is the so called projection matrix to the null space of $\mathbf{A}$ (that is, M). If $\mathbf{x}^T\mathbf{Lx} \geq 0$ for all $\mathbf{x}$ in this null space, then $\mathbf{z}^T\mathbf{P}_A\mathbf{LP}_A\mathbf{z} \geq 0$ for all $\mathbf{z} \in E^n$, or the n-dimensional symmetric matrix $\mathbf{P}_A\mathbf{LP}_A$ is positive semidefinite. Furthermore, if $\mathbf{P}_A\mathbf{LP}_A$ has rank $n - m$, then $\mathbf{L}_M$ is positive definite, which results the following test.

Projected Hessian Test. *The matrix* $\mathbf{L}$ *is positive definite on the subspace* $M = \{\mathbf{x} : \nabla\mathbf{hx} = \mathbf{0}\}$ *if and only if the projected Hessian matrix to the null space of* $\nabla\mathbf{h}$ *is positive semidefinite and has rank* $n - m$.

Example 3. Approaching Example 2 in this way and noting $\mathbf{A} = \nabla\mathbf{h} = (1, 0, 0)$ we have

$$\mathbf{P}_A = \mathbf{I} - \begin{bmatrix} 1 \\ 0 \\ 0 \end{bmatrix}\begin{bmatrix} 1 \\ 0 \\ 0 \end{bmatrix}^T = \begin{bmatrix} 0 & 0 & 0 \\ 0 & 1 & 0 \\ 0 & 0 & 1 \end{bmatrix}.$$

Then

$$\mathbf{P}_A\mathbf{LP}_A = \begin{bmatrix} 0 & 0 & 0 \\ 0 & 1 & 1 \\ 0 & 1 & 3 \end{bmatrix}$$

which is clearly positive semidefinite and has rank 2.

11.7 Sensitivity

The Lagrange multipliers associated with a constrained minimization problem have an interpretation as prices, similar to the prices associated with constraints in linear programming. In the nonlinear case the multipliers are associated with the particular solution point and correspond to incremental or marginal prices, *that* is, prices associated with small variations in the constraint requirements.

Suppose the problem

$$\begin{aligned} &\text{minimize} \quad f(\mathbf{x}) \\ &\text{subject to} \quad \mathbf{h}(\mathbf{x}) = \mathbf{0} \end{aligned} \tag{11.26}$$

has a solution at the point $\mathbf{x}^*$ which is a regular point of the constraints. Let $\boldsymbol{\lambda}$ be the corresponding Lagrange multiplier vector. Now consider the family of problems

$$\begin{aligned} &\text{minimize} \quad f(\mathbf{x}) \\ &\text{subject to} \quad \mathbf{h}(\mathbf{x}) = \mathbf{c}, \end{aligned} \tag{11.27}$$

where $\mathbf{c} \in E^m$. For a sufficiently small range of $\mathbf{c}$ near the zero vector, the problem will have a solution point $\mathbf{x}(\mathbf{c})$ near $\mathbf{x}(\mathbf{0}) \equiv \mathbf{x}^*$. For each of these solutions there is a corresponding value $f(\mathbf{x}(\mathbf{c}))$, and this value can be regarded as a function of $\mathbf{c}$, the right-hand side of the constraints. The components of the gradient of this function can be interpreted as the incremental rate of change in value per unit change in the constraint requirements. Thus, they are the incremental prices of the constraint requirements measured in units of the objective. We show below how these prices are related to the Lagrange multipliers of the problem having $\mathbf{c} = \mathbf{0}$.

Sensitivity Theorem. *Let f, $\mathbf{h} \in C^2$ and consider the family of problems*

$$\begin{aligned} &\text{minimize} \quad f(\mathbf{x}) \\ &\text{subject to} \quad \mathbf{h}(\mathbf{x}) = \mathbf{c}. \end{aligned} \tag{11.29}$$

Suppose for $\mathbf{c} = \mathbf{0}$ there is a local solution $\mathbf{x}^*$ that is a regular point and that, together with its associated Lagrange multiplier vector λ, satisfies the second-order sufficiency conditions for a strict local minimum. Then for every $\mathbf{c} \in E^m$ in a region containing $\mathbf{0}$ there is an $\mathbf{x}(\mathbf{c})$, depending continuously on $\mathbf{c}$, such that $\mathbf{x}(\mathbf{0}) = \mathbf{x}^*$ and such that $\mathbf{x}(\mathbf{c})$ is a local minimum of (11.27). Furthermore,

$$\nabla_{\mathbf{c}} f(\mathbf{x}(\mathbf{c}))\big]_{\mathbf{c}=0} = -\boldsymbol{\lambda}^T.$$

Proof. Consider the system of equations

$$\nabla f(\mathbf{x}) + \boldsymbol{\lambda}^T \nabla \mathbf{h}(\mathbf{x}) = \mathbf{0} \tag{11.30}$$

$$\mathbf{h}(\mathbf{x}) = \mathbf{c}. \tag{11.31}$$

By hypothesis, there is a solution $\mathbf{x}^*$, $\boldsymbol{\lambda}$ to this system when $\mathbf{c} = \mathbf{0}$. The Jacobian matrix of the system at this solution is

$$\begin{bmatrix} \mathbf{L}(\mathbf{x}^*) & \nabla \mathbf{h}(\mathbf{x}^*)^T \\ \nabla \mathbf{h}(\mathbf{x}^*) & \mathbf{0} \end{bmatrix}.$$

Because by assumption $\mathbf{x}^*$ is a regular point and $\mathbf{L}(\mathbf{x}^*)$ is positive definite on M, it follows that this matrix is nonsingular (see Exercise 11). Thus, by the Implicit Function Theorem, there is a solution $\mathbf{x}(\mathbf{c})$, $\boldsymbol{\lambda}(\mathbf{c})$ to the system which is in fact continuously differentiable.

By the chain rule we have

$$\nabla_{\mathbf{c}} f(\mathbf{x}(\mathbf{c}))\big]_{\mathbf{c}=0} = \nabla_{\mathbf{x}} f(\mathbf{x}^*) \nabla_{\mathbf{c}} \mathbf{x}(\mathbf{0}).$$

and

$$\nabla_{\mathbf{c}} \mathbf{h}(\mathbf{x}(\mathbf{c}))\big]_{\mathbf{c}=\mathbf{0}} = \nabla_{\mathbf{x}} \mathbf{h}(\mathbf{x}^*) \nabla_{\mathbf{c}} \mathbf{x}(\mathbf{0}).$$

In view of (11.31), the second of these is equal to the identity $\mathbf{I}$ on E^m, while this, in view of (11.30), implies that the first can be written

$$\nabla_{\mathbf{c}} f(\mathbf{x}(\mathbf{c}))\big]_{\mathbf{c}=\mathbf{0}} = -\boldsymbol{\lambda}^T. \blacksquare$$

11.8 Inequality Constraints

We consider now problems of the form

$$\begin{aligned}&\text{minimize} && f(\mathbf{x})\\ &\text{subject to} && \mathbf{h}(\mathbf{x}) = \mathbf{0}, \quad \mathbf{g}(\mathbf{x}) \leqslant \mathbf{0}.\end{aligned} \tag{11.32}$$

We assume that f and $\mathbf{h}$ are as before and that $\mathbf{g}$ is a p-dimensional function. Initially, we assume $f,\ \mathbf{h},\ \mathbf{g} \in C^1$.

There are a number of distinct theories concerning this problem, based on various regularity conditions or constraint qualifications, which are directed toward obtaining definitive general statements of necessary and sufficient conditions. One can by no means pretend that all such results can be obtained as minor extensions of the theory for problems having equality constraints only. To date, however, these alternative results concerning necessary conditions have been of isolated theoretical interest only—for they have not had an influence on the development of algorithms, and have not contributed to the theory of algorithms. Their use has been limited to small-scale programming problems of two or three variables. We therefore choose to emphasize the simplicity of incorporating inequalities rather than the possible complexities, not only for ease of presentation and insight, but also because it is this viewpoint that forms the basis for work beyond that of obtaining necessary conditions.

First-Order Necessary Conditions

With the following generalization of our previous definition it is possible to parallel the development of necessary conditions for equality constraints.

Definition. Let $\mathbf{x}^*$ be a point satisfying the constraints

$$\mathbf{h}(\mathbf{x}^*) = \mathbf{0},\ \mathbf{g}(\mathbf{x}^*) \leqslant \mathbf{0}, \tag{11.33}$$

and let J be the set of indices j for which $g_j(\mathbf{x}^*) = 0$. Then $\mathbf{x}^*$ is said to be a *regular point* of the constraints (11.33) if the gradient vectors $\nabla h_i(\mathbf{x}^*)$, $\nabla g_j(\mathbf{x}^*)$, $1 \leqslant i \leqslant m,\ j \in J$ are linearly independent.

We note that, following the definition of active constraints given in Sect. 11.1, a point $\mathbf{x}^*$ is a regular point if the gradients of the active constraints are linearly independent. Or, equivalently, $\mathbf{x}^*$ is regular for the constraints if it is regular in the sense of the earlier definition for equality constraints applied to the active constraints.

Karush-Kuhn-Tucker Conditions*. Let* $\mathbf{x}^*$ *be a relative minimum point for the problem*

$$\begin{aligned}&\text{minimize} && f(\mathbf{x})\\ &\text{subject to} && \mathbf{h}(\mathbf{x}) = \mathbf{0},\ \mathbf{g}(\mathbf{x}) \leqslant \mathbf{0},\end{aligned} \tag{11.34}$$

and suppose $\mathbf{x}^*$ *is a regular point for the constraints. Then there is a vector* $\boldsymbol{\lambda} \in E^m$ *and a vector* $\boldsymbol{\mu} \in E^p$ *with* $\boldsymbol{\mu} \geqslant \mathbf{0}$ *such that*

$$\nabla f(\mathbf{x}^*) + \boldsymbol{\lambda}^T \nabla \mathbf{h}(\mathbf{x}^*) + \boldsymbol{\mu}^T \nabla \mathbf{g}(\mathbf{x}^*) = \mathbf{0} \tag{11.35}$$

$$\boldsymbol{\mu}^T \mathbf{g}(\mathbf{x}^*) = 0. \tag{11.36}$$

Proof. We note first, since $\boldsymbol{\mu} \geqslant \mathbf{0}$ and $\mathbf{g}(\mathbf{x}^*) \leqslant \mathbf{0}$, (11.36) is equivalent to the statement that a component of $\boldsymbol{\mu}$ may be nonzero only if the corresponding constraint is active. This a *complementary slackness* condition, stating that $\mathbf{g}(\mathbf{x}^*)_i < 0$ implies $\mu_i = 0$ and $\mu_i > 0$ implies $\mathbf{g}(\mathbf{x}^*)_i = 0$.

Since $\mathbf{x}^*$ is a relative minimum point over the constraint set, it is also a relative minimum over the subset of that set defined by setting the active constraints to zero. Thus, for the resulting equality constrained problem defined in a neighborhood of $\mathbf{x}^*$, there are Lagrange multipliers. Therefore, we conclude that (11.35) holds with $\mu_j = 0$ if $g_j(\mathbf{x}^*) \neq 0$ (and hence (11.36) also holds).

It remains to be shown that $\boldsymbol{\mu} \geq \mathbf{0}$. Suppose $\mu_k < 0$ for some $k \in J$. Let S and M be the surface and tangent plane, respectively, defined by all other active constraints at $\mathbf{x}^*$. By the regularity assumption, there is a $\mathbf{y}$ such that $\mathbf{y} \in M$ and $\nabla g_k(\mathbf{x}^*)\mathbf{y} < 0$. Let $\mathbf{x}(t)$ be a curve on S passing through $\mathbf{x}^*$(**at** $t = 0$) with $\dot{\mathbf{x}}(0) = \mathbf{y}$. Then for small $t \geq 0$, $\mathbf{x}(t)$ is feasible, and

$$\left.\frac{df}{dt}(\mathbf{x}(t))\right]_{t=0} = \nabla f(\mathbf{x}^*)\mathbf{y} < \mathbf{0}$$

by (11.35), which contradicts the minimality of $\mathbf{x}^*$. ∎

Example. Consider the problem

$$\begin{aligned} \text{minimize} \quad & 2x_1^2 + 2x_1x_2 + x_2^2 - 10x_1 - 10x_2 \\ \text{subject to} \quad & x_1^2 + x_2^2 \leqslant 5 \\ & 3x_1 + x_2 \leqslant 6. \end{aligned}$$

The first-order necessary conditions, in addition to the constraints, are

$$\begin{gathered} 4x_1 + 2x_2 - 10 + 2\mu_1 x_1 + 3\mu_2 = 0 \\ 2x_1 + 2x_2 - 10 + 2\mu_1 x_2 + \mu_2 = 0 \\ \mu_1 \geq 0, \ \mu_2 \geq 0 \\ \mu_1(x_1^2 + x_2^2 - 5) = 0 \\ \mu_2(3x_1 + x_2 - 6) = 0. \end{gathered}$$

To find a solution we define various combinations of active constraints and check the signs of the resulting Lagrange multipliers. In this problem we can try setting none, one, or two constraints active. Assuming the first constraint is active and the second is inactive yields the equations

$$
\begin{aligned}
4x_1 + 2x_2 - 10 + 2\mu_1 x_1 &= 0 \\
2x_1 + 2x_2 - 10 + 2\mu_1 x_2 &= 0 \\
x_1^2 + x_2^2 &= 5,
\end{aligned}
$$

which has the solution

$$x_1 = 1,\ x_2 = 2,\ \mu_1 = 1.$$

This yields $3x_1 + x_2 = 5$ and hence the second constraint is satisfied. Thus, since $\mu_1 > 0$, we conclude that this solution satisfies the first-order necessary conditions.

Second-Order Conditions

The second-order conditions, both necessary and sufficient, for problems with inequality constraints, are derived essentially by consideration only of the equality constrained problem *that* is implied by the active constraints. The appropriate tangent plane for these problems is the plane tangent to the active constraints.

Second-Order Necessary Conditions. *Suppose the functions f, $\mathbf{g}$, $\mathbf{h} \in C^2$ and that $\mathbf{x}^*$ is a regular point of the constraints (11.33). If $\mathbf{x}^*$ is a relative minimumpoint for problem (11.32), then there is a $\boldsymbol{\lambda} \in E^m$, $\boldsymbol{\mu} \in E^p$, $\boldsymbol{\mu} \geq 0$ such that (11.35) and (36) hold and such that*

$$\mathbf{L}(\mathbf{x}^*) = \mathbf{F}(\mathbf{x}^*) + \boldsymbol{\lambda}^T\mathbf{H}(\mathbf{x}^*) + \boldsymbol{\mu}^T\mathbf{G}(\mathbf{x}^*) \tag{11.37}$$

is positive semidefinite on the tangent subspace of the active constraints at $\mathbf{x}^*$.

Proof. If $\mathbf{x}^*$ is a relative minimum point over the constraints (11.33), it is also a relative minimum point for the problem with the active constraints taken as equality constraints. ∎

Just as in the theory of unconstrained minimization, it is possible to formulate a converse to the Second-Order Necessary Condition Theorem and thereby obtain a Second-Order Sufficiency Condition Theorem. By analogy with the unconstrained situation, one can guess that the required hypothesis is that $\mathbf{L}(\mathbf{x}^*)$ be positive definite on the tangent plane M. This is indeed sufficient in most situations. However, if there are *degenerate inequality constraints* (that is, active inequality constraints having zero as associated Lagrange multiplier), we must require $\mathbf{L}(\mathbf{x}^*)$ to be positive definite on a subspace that is larger than M.

Second-Order Sufficiency Conditions. *Let f, $\mathbf{g}$, $\mathbf{h} \in C^2$. Sufficient conditions that a point $\mathbf{x}^*$ satisfying (33) be a strict relative minimum point of problem (11.32) is that there exist $\boldsymbol{\lambda} \in E^m$, $\boldsymbol{\mu} \in E^p$, such that*

$$\boldsymbol{\mu} \geq 0 \tag{11.38}$$

$$\boldsymbol{\mu}^T\mathbf{g}(\mathbf{x}^*) = 0 \tag{11.39}$$

$$\nabla f(\mathbf{x}^*) + \boldsymbol{\lambda}^T\nabla\mathbf{h}(\mathbf{x}^*) + \boldsymbol{\mu}^{T_1}\nabla\mathbf{g}(\mathbf{x}^*) = 0, \tag{11.40}$$

and the Hessian matrix

$$\mathbf{L}(\mathbf{x}^*) = \mathbf{F}(\mathbf{x}^*) + \boldsymbol{\lambda}^T\mathbf{H}(\mathbf{x}^*) + \mu^T\mathbf{G}(\mathbf{x}^*) \tag{11.41}$$

is positive definite on the subspace

$$M' = \left\{\mathbf{y} : \nabla\mathbf{h}(\mathbf{x}^*)\mathbf{y} = 0,\ \nabla g_j(\mathbf{x}^*)\mathbf{y} = 0 \text{ for all } j \in J\right\},$$

where $J = \left\{j : g_j(\mathbf{x}^*) = 0,\ \mu_j > 0\right\}$.

Proof. As in the proof of the corresponding theorem for equality constraints in Sect. 11.5, assume that $\mathbf{x}^*$ is not a strict relative minimum point; let $\{\mathbf{y}_k\}$ be a sequence of feasible points converging to $\mathbf{x}^*$ such that $f(\mathbf{y}_k) \leqslant f(\mathbf{x}^*)$, and write each $\mathbf{y}_k$ in the form $\mathbf{y}_k = \mathbf{x}^* + \delta_k\mathbf{s}_k$ with $|\mathbf{s}_k| = 1,\ \delta_k > 0$. We may assume that $\delta_k \to 0$ and $\mathbf{s}_k \to \mathbf{s}^*$. We have $0 \geq \nabla f(\mathbf{x}^*)\mathbf{s}^*$, and for each $i = 1, \ldots, m$ we have

$$\nabla h_i(\mathbf{x}^*)\mathbf{s}^* = 0.$$

Also for each active constraint g_j we have $g_j(\mathbf{y}_k) - g_j(\mathbf{x}^*) \leqslant 0$, and hence

$$\nabla g_j(\mathbf{x}^*)\mathbf{s}^* \leqslant 0.$$

If $\nabla g_j(\mathbf{x}^*)\mathbf{s}^* = 0$ for all $j \in J$, then the proof goes through just as in Sect. 11.5. If $\nabla g_j(\mathbf{x}^*)\mathbf{s}^* < 0$ for at least one $j \in J$, then

$$0 \geqslant \nabla f(\mathbf{x}^*)\mathbf{s}^* = -\boldsymbol{\lambda}^T\nabla\mathbf{h}(\mathbf{x}^*)\mathbf{s}^* - \boldsymbol{\mu}^T\nabla\mathbf{g}(\mathbf{x}^*)\mathbf{s}^* > 0,$$

which is a contradiction. ∎

We note in particular that if all active inequality constraints have strictly positive corresponding Lagrange multipliers (no degenerate inequalities), then the set J includes all of the active inequalities. In this case the sufficient condition is that the Lagrangian be positive definite on M, the tangent plane of active constraints.

Sensitivity

The sensitivity result for problems with inequalities is a simple restatement of the result for equalities. In this case, a nondegeneracy assumption is introduced so that the small variations produced in Lagrange multipliers when the constraints are varied will not violate the positivity requirement.

Sensitivity Theorem. *Let* $f,\ \mathbf{g},\ \mathbf{h} \in C^2$ *and consider the family of problems*

$$\begin{aligned} &\text{minimize} && f(\mathbf{x}) \\ &\text{subject to} && \mathbf{h}(\mathbf{x}) = \mathbf{c}, \quad \mathbf{g}(\mathbf{x}) \leqslant \mathbf{d}. \end{aligned} \tag{11.42}$$

Suppose that for $\mathbf{c} = \mathbf{0},\ \mathbf{d} = \mathbf{0}$, *there is a local solution* $\mathbf{x}^*$ *that is a regularpoint and that, together with the associated Lagrange multipliers,* $\boldsymbol{\lambda},\ \boldsymbol{\mu} \geq \mathbf{0}$,*satisfies the second-order* *sufficiency conditions for a strict local minimum.Assume further that no active inequality constraint is degenerate. Then for every* $(\mathbf{c}, \mathbf{d}) \in E^{m+p}$ *in a region containing* $(\mathbf{0}, \mathbf{0})$ *there is a solution* $\mathbf{x}(\mathbf{c}, \mathbf{d})$,*depending continuously on* $(\mathbf{c}, \mathbf{d})$, *such that* $\mathbf{x}(\mathbf{0}, \mathbf{0}) = \mathbf{x}^*$, *and such that* $\mathbf{x}(\mathbf{c}, \mathbf{d})$*is a relative minimum point of (11.42). Furthermore,*

$$\nabla_{\mathbf{c}} f(\mathbf{x}(\mathbf{c}, \mathbf{d}))\big]_{\mathbf{0},\mathbf{0}} = -\boldsymbol{\lambda}^T \tag{11.43}$$

$$\nabla_{\mathbf{d}} f(\mathbf{x}(\mathbf{c}, \mathbf{d}))\big]_{\mathbf{0},\mathbf{0}} = -\boldsymbol{\mu}^T. \tag{11.44}$$

11.9 Zero-Order Conditions and Lagrangian Relaxation

Zero-order conditions for functionally constrained problems express conditions in terms of Lagrange multipliers without the use of derivatives. This theory is not only of great practical value, but it also gives new insight into the meaning of Lagrange multipliers. Rather than regarding the Lagrange multipliers as separate scalars, they are identified as components of a single vector that has a strong geometric interpretation. As before, the basic constrained problem is

$$\begin{aligned} &\text{minimize } f(\mathbf{x}) \\ &\text{subject to } \mathbf{h}(\mathbf{x}) = \mathbf{0},\ \mathbf{g}(\mathbf{x}) \leq \mathbf{0} \\ &\qquad\qquad \mathbf{x} \in \Omega, \end{aligned} \tag{11.45}$$

where $\mathbf{x}$ is a vector in E^n, and $\mathbf{h}$ and $\mathbf{g}$ are m-dimensional and p-dimensional functions, respectively.

In purest form, zero-order conditions require that the functions that define the objective and the constraints are convex functions and sets. (See Appendix B.) The vector-valued function $\mathbf{g}$ consisting of p individual component functions $g_1, g_2, \ldots, g_p$ is said to be *convex* if each of the component functions is convex.

The programming problem (11.45) above is termed a *convex programming problem* if the functions f and $\mathbf{g}$ are convex, the function $\mathbf{h}$ is affine (that is, linear plus a constant and can be written as $\mathbf{Ax} - \mathbf{b}$), and the set $\Omega \subset \mathrm{E}^n$ is convex.

Notice that according to Proposition 3, Sect. 7.4, the set defined by each of the inequalities $g_j(\mathbf{x}) \leqslant 0$ is convex. This is true also of a set defined by $h_i(\mathbf{x}) = 0$. Since the overall constraint set is the intersection of these and Ω it follows from Proposition 1 of Appendix B that this overall constraint set is itself convex. Hence the problem can be regarded as minimize $f(\mathbf{x})$, $\mathbf{x} \in \Omega_1$ where Ω_1 is a convex subset of Ω.

With this view, one could apply the zero-order conditions of Sect. 7.6 to the problem with constraint set Ω_1. However, in the case of functional constraints it is common to keep the structure of the constraints explicit instead of folding them into an amorphous set.

Although it is possible to derive the zero-order conditions for (11.45) all at once, treating both equality and inequality constraints together, it is notationally cumbersome to do so and it may obscure the basic simplicity of the arguments. For this reason, we treat equality constraints first, then inequality constraints, and finally the combination of the two.

The equality problem is

$$\begin{aligned}&\text{minimize } f(\mathbf{x})\\&\text{subject to } \mathbf{h}(\mathbf{x}) = \mathbf{0}\\&\qquad\qquad \mathbf{x} \in \Omega.\end{aligned} \tag{11.46}$$

Letting $Y = E^m$, we have $\mathbf{h}(\mathbf{x}) \in Y$ for all $\mathbf{x}$. For this problem we require a regularity condition.

Definition. An affine function $\mathbf{h}$ is *regular* with respect to Ω if the set C in Y defined by $C = \{\mathbf{y} : \mathbf{h}(\mathbf{x}) = \mathbf{y} \text{ for some } \mathbf{x} \in \Omega\}$ contains an open sphere around $\mathbf{0}$; that is, C contains a set of the form $\{\mathbf{y} : |\mathbf{y}| < \varepsilon\}$ for some $\varepsilon > 0$.

This condition means that $\mathbf{h}(\mathbf{x})$ can attain $\mathbf{0}$ and can vary in arbitrary directions from $\mathbf{0}$. Notice that this condition is similar to the definition of a regular point in the context of first-order conditions. If $\mathbf{h}$ has continuous derivatives at a point $\mathbf{x}^*$ the earlier regularity condition implies that $\nabla\mathbf{h}(\mathbf{x}^*)$ is of full rank and the Implicit Function Theorem (of Appendix A) then guarantees that there is an $\varepsilon > 0$ such that for any $\mathbf{y}$ with $|\mathbf{y} - \mathbf{h}(\mathbf{x}^*)| < \varepsilon$ there is an $\mathbf{x}$ such that $\mathbf{h}(\mathbf{x}) = \mathbf{y}$. In other words, there is an open sphere around $\mathbf{y}^* = \mathbf{h}(\mathbf{x}^*)$ that is attainable. In the present situation we assume this attainability directly, at the point $\mathbf{0} \in \mathbf{Y}$.

Next we introduce the following important construction.

Definition. The *primal function* associated with problem (11.46) is

$$w(\mathbf{y}) = \inf\{f(\mathbf{x}) : \mathbf{h}(\mathbf{x}) = \mathbf{y},\ \mathbf{x} \in \Omega\},$$

defined for all $\mathbf{y} \in C$.

Notice that the primal function is defined by varying the right hand side of the constraint. The original problem (11.46) corresponds to $\omega(\mathbf{0})$. The primal function is illustrated in Fig. 11.6.

Proposition 1. *Suppose Ω is convex, the function f is convex, and $\mathbf{h}$ is affine. Then the primal function ω is convex.*

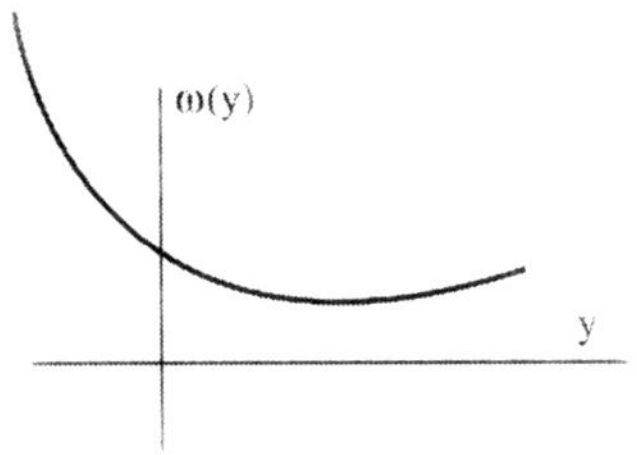

Fig. 11.6 The primal function

Proof. For simplicity of notation we assume that Ω is the entire space X. Then we observe

$$\begin{aligned}\Omega(\alpha\mathbf{y}_1 + (1-\alpha)\mathbf{y}_2) &= \inf\{f(\mathbf{x}) : \mathbf{h}(\mathbf{x}) = \alpha\mathbf{y}_1 + (1-\alpha)\mathbf{y}_2\} \\ &\leq \inf\{f(\mathbf{x}) : \mathbf{x} = \alpha\mathbf{x}_1 + (1-\alpha)\mathbf{x}_2,\ \mathbf{h}(\mathbf{x}_1) = \mathbf{y}_1,\ \mathbf{h}(\mathbf{x}_2) = \mathbf{y}_2\} \\ &\leq \alpha \inf\{f(\mathbf{x}_1) : \mathbf{h}(\mathbf{x}_1) = \mathbf{y}_1\} + (1-\alpha)\inf\{f\mathbf{x}_2) : \mathbf{h}(\mathbf{x}_2) = \mathbf{y}_2\} \\ &= \alpha\, \omega(\mathbf{y}_1) + (1-\alpha)\omega(\mathbf{y}_2). \blacksquare\end{aligned}$$

We now turn to the derivation of the Lagrange multiplier result for (11.46).

Proposition 2. *Assume that $\Omega \subset E^n$ is convex, f is a convex function on Ω and $\mathbf{h}$ is an m-dimensional affine function on Ω. Assume that $\mathbf{h}$ is regular with respect to Ω. If $\mathbf{x}^*$ solves (11.46), then there is $\boldsymbol{\lambda} \in E^m$ such that $\mathbf{x}^*$ solves the Lagrangian relaxation problem*

$$\begin{aligned}&\text{minimize} \quad f(\mathbf{x}) + \boldsymbol{\lambda}^T\mathbf{h}(\mathbf{x}) \\ &\text{subject to} \quad \mathbf{x} \in \Omega.\end{aligned}$$

Proof. Let $f^* = f(\mathbf{x}^*)$. Define the sets A and B in E^{m+1} as

$$A = \{(r, \mathbf{y}) : r \geq \omega(\mathbf{y}),\ \mathbf{y} \in C\} \text{ and } B = \{(r, \mathbf{y}) : r \leqslant f^*,\ \mathbf{y} = \mathbf{0}\}.$$

A is the epigraph of ω (see Sect. 7.6) and B is the vertical line extending below f^* and aligned with the origin. Both A and B are convex sets. Their only common point is at $(f^*, \mathbf{0})$. See Fig. 11.7.

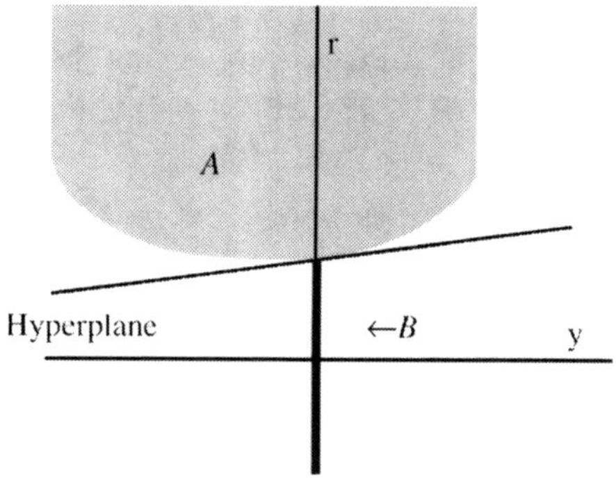

Fig. 11.7 The sets A and B and the separating hyperplane

According to the separating hyperplane theorem (Appendix B), there is a hyperplane separating A and B. This hyperplane can be represented by a nonzero vector in E^{m+1} of the form $(s, \boldsymbol{\lambda})$, with $\boldsymbol{\lambda} \in E^m$, and a separation constant c. The separation conditions are

$$sr + \boldsymbol{\lambda}^T\mathbf{y} \geq c \text{ for all } (r, \mathbf{y}) \in A \text{ and } sr + \boldsymbol{\lambda}^T\mathbf{y} \leqslant c \text{ for all } (r, \mathbf{y}) \in B.$$

It follows immediately that $s \geq 0$ for otherwise points $(r, \mathbf{0}) \in B$ with r very negative would violate the second inequality.

Geometrically, if $s = 0$ the hyperplane would be vertical. We wish to show that $s \neq 0$, and it is for this purpose that we make use of the regularity condition. Suppose $s = 0$. Then $\boldsymbol{\lambda} \neq \mathbf{0}$ since both s and $\boldsymbol{\lambda}$ cannot be zero. It follows from the second separation inequality that $c = 0$ because the hyperplane must include the point $(f^*, \mathbf{0})$. Now, as $\mathbf{y}$ ranges over a sphere centered at $\mathbf{0} \in C$, the left hand side of the first separation inequality ranges correspondingly over $\boldsymbol{\lambda}^T\mathbf{y}$ which is negative for some $\mathbf{y}$'s. This contradicts the first separation inequality. Thus $s \neq 0$ and thus in fact $s > 0$. Without loss of generality we may, by rescaling if necessary, assume that $s = 1$.

Finally, suppose $\mathbf{x} \in \Omega$. Then $(f(\mathbf{x}), \mathbf{h}(\mathbf{x})) \in A$ and $(f(\mathbf{x}^*), \mathbf{0}) \in B$. Thus, from the separation inequality (with $s = 1$) we have

$$f(\mathbf{x}) + \lambda^T\mathbf{h}(\mathbf{x}) \geq f(\mathbf{x}^*) = f(\mathbf{x}^*) + \lambda^T\mathbf{h}(\mathbf{x}^*).$$

Hence $\mathbf{x}^*$ solves the stated minimization problem. ∎

Example 1 (Best Rectangle). Consider the classic problem of finding the rectangle of maximum area while limiting the perimeter to a length of 4. This can be formulated as

$$\begin{aligned} &\text{minimize} && -x_1x_2 \\ &\text{subject to} && x_1 + x_2 - 2 = 0 \\ &&& x_1 \geq 0, \quad x_2 \geq 0. \end{aligned}$$

The regularity condition is met because it is possible to make the right hand side of the functional constraint slightly positive or slightly negative with nonnegative x_1 and x_2. We know the answer to the problem is $x_1 = x_2 = 1$. The Lagrange multiplier is $\lambda = 1$. The Lagrangian problem of Proposition 2 is

$$\begin{aligned} &\text{minimize} && -x_1x_2 + 1 \cdot (x_1 + x_2 - 2) \\ &\text{subject to} && x_1 \geq 0, \quad x_2 \geq 0. \end{aligned}$$

This can be solved by differentiation to obtain $x_1 = x_2 = 1$.

However the conclusion of the proposition is not satisfied! The value of the Lagrangian at the solution is $V = -1 + 1 + 1 - 2 = -1$. However, at $x_1 = x_2 = 0$ the value of the Lagrangian is $V' = -2$ which is less than V. The Lagrangian is *not* minimized at the solution. The proposition breaks down because the objective function $f(x_1, x_2) = -x_1x_2$ is not convex.

Example 2 (Best Diagonal). As an alternative problem, consider minimizing the length of the diagonal of a rectangle subject to the perimeter being of length 4. This problem can be formulated as

$$\begin{aligned} &\text{minimize} && \tfrac{1}{2}(x_1^2 + x_2^2) \\ &\text{subject to} && x_1 + x_2 - 2 = 0 \\ &&& x_1 \geq 0, \quad x_2 \geq 0. \end{aligned}$$

In this case the objective function is convex. The solution is $x_1 = x_2 = 1$ and the Lagrange multiplier is $\lambda = -1$. The Lagrangian problem is

$$\begin{aligned} \text{minimize} \quad & \frac{1}{2}(x_1^2 + x_2^2) - 1 \cdot (x_1 + x_2 - 2) \\ \text{subject to} \quad & x_1 \geq 0, \quad x_2 \geq 0. \end{aligned}$$

The value of the Lagrangian at the solution is $V = 1$ which in this case is a minimum as guaranteed by the proposition. (The value at $x_1 = x_2 = 0$ is $V' = 2$.)

Inequality Constraints

We outline the parallel results for the inequality constrained problem

$$\begin{aligned} \text{minimize} \quad & f(\mathbf{x}) \\ \text{subject to} \quad & \mathbf{g}(\mathbf{x}) \leq \mathbf{0} \\ & \mathbf{x} \in \Omega, \end{aligned} \tag{11.47}$$

where $\mathbf{g}$ is a p-dimensional function.

We let $Z = E^p$ and define $D \subset Z$ as $D = \{\mathbf{z} \in Z : \mathbf{g}(\mathbf{x}) \leqslant \mathbf{z} \text{ for some } \mathbf{x} \in \Omega\}$. The regularity condition (called the *Slater condition*) is that there is a $\mathbf{z}_1 \in D$ with $\mathbf{z}_1 < 0$.

As before we introduce the primal function.

Definition. The *primal function* associated with problem (11.47) is

$$w(\mathbf{z}) = \inf\{f(\mathbf{x}) : \mathbf{g}(\mathbf{x}) \leq \mathbf{z},\ \mathbf{x} \in \Omega\}.$$

The primal function is again defined by varying the right hand side of the constraint function, using the variable $\mathbf{z}$. Now the primal function in monotonically decreasing with $\mathbf{z}$, since an increase in $\mathbf{z}$ enlarges the constraint region.

Proposition 3. *Suppose $\Omega \subset E^n$ is convex and f and $\mathbf{g}$ are convex functions. Then the primal function ω is also convex.*

Proof. The proof parallels that of Proposition 1. One simply substitutes $\mathbf{g}(\mathbf{x}) \leq \mathbf{0}$ for $\mathbf{h}(\mathbf{x}) = \mathbf{y}$ throughout the series of inequalities. ∎

The zero-order necessary Lagrangian conditions are then given by the proposition below.

Proposition 4. *Assume Ω is a convex subset of E^n and that f and $\mathbf{g}$ are convex functions. Assume also that there is a point $\mathbf{x}_1 \in \Omega$ such that $\mathbf{g}(\boldsymbol{x}_1) < \boldsymbol{0}$. Then, if $\mathbf{x}^*$ solves (11.47), there is a vector $\boldsymbol{\mu} \in E^p$ with $\boldsymbol{\mu} \geq \boldsymbol{0}$ such that $\mathbf{x}^*$ solves the Lagrangian relaxation problem*

$$\begin{aligned} \text{minimize} \quad & f(\boldsymbol{x}^*) + \boldsymbol{\mu}^T \boldsymbol{g}(\boldsymbol{x}) \\ \text{subject to} \quad & \boldsymbol{x} \in \boldsymbol{\Omega}. \end{aligned} \tag{11.48}$$

Furthermore, $\boldsymbol{\mu}^T \boldsymbol{g}(\boldsymbol{x}^*) = 0$.

Proof. Here is the proof outline. Let $f^* = f(\mathbf{x}^*)$. In this case define in E^{p+1} the two sets

$$A = \{(r, \mathbf{0}) : r \geq f(\mathbf{x}),\ \mathbf{0} \geq \mathbf{g}(\mathbf{x}),\ \text{for some } \mathbf{x} \in \Omega\} \ \text{ and } \ B = \{(r, \mathbf{0}) : r \leq f^*,\ \mathbf{0} \leq \mathbf{0}\}.$$

A is the epigraph of the primal function ω. The set B is the rectangular region at or to the left of the vertical axis and at or lower than f^*. Both A and B are convex. See Fig. 11.8.

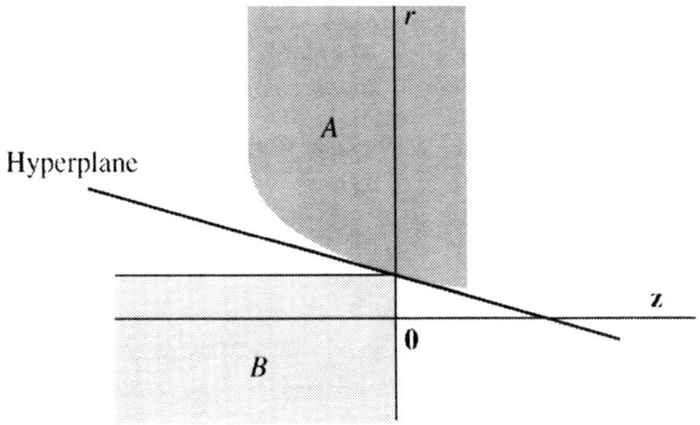

Fig. 11.8 The sets A and B and the separating hyperplane for inequalities

The proof is made by constructing a hyperplane separating A and B. The regularity condition guarantees that this hyperplane is not vertical. ∎

The condition $\boldsymbol{\mu}^T\mathbf{g}(\mathbf{x}^*) = 0$ is the complementary slackness condition that is characteristic of necessary conditions for problems with inequality constraints.

Example 4 (Quadratic Program). Consider the quadratic program

$$\begin{aligned} &\text{minimize } \mathbf{x}^T\mathbf{Q}\mathbf{x} + \mathbf{c}^T\mathbf{x} \\ &\text{subject to } \mathbf{a}^T\mathbf{x} \leq b \\ &\qquad\qquad\quad \mathbf{x} \geq \mathbf{0}. \end{aligned}$$

Let $\Omega = \{\mathbf{x} : \mathbf{x} \geq \mathbf{0}\}$ and $g(\mathbf{x}) = \mathbf{a}^T\mathbf{x} - b$. Assume that the $n \times n$ matrix $\mathbf{Q}$ is positive definite, in which case the objective function is convex. Assuming that $b > 0$, the Slater regularity condition is satisfied. Hence there is a Lagrange multiplier $\mu \geq 0$ (a scalar in this case) such that *the* solution $\mathbf{x}^*$ to the quadratic program is also a solution to

$$\begin{aligned} &\text{minimize} \quad \mathbf{x}^T\mathbf{Q}\mathbf{x} + \mathbf{c}^T\mathbf{x} + \mu(\mathbf{a}^T\mathbf{x} - b) \\ &\text{subject to} \quad \mathbf{x} \geq \mathbf{0} \quad \text{and} \quad \mu(\mathbf{a}^T\mathbf{x}^* - b) = 0. \end{aligned}$$

Mixed Constraints

The two previous results can be combined to obtain zero-order conditions for the problem

$$\begin{aligned} &\text{minimize } f(\mathbf{x}) \\ &\text{subject to } \mathbf{h}(\mathbf{x}) = \mathbf{0},\ \mathbf{g}(\mathbf{x}) \le \mathbf{0} \\ &\qquad \mathbf{x} \in \Omega. \end{aligned} \tag{11.49}$$

Zero-order Lagrange Theorem. Assume that $\Omega \subset E^n$ *is a convex set, f and* $\mathbf{g}$ *are convex functions of dimension 1 and p, respectively, and* $\mathbf{h}$ *is affine ofdimension m. Assume also that* $\mathbf{h}$ *satisfies the regularity condition with respect to Ω and that there is an* $\mathbf{x}_1 \in \Omega$ *with* $\mathbf{h}(\mathbf{x}_1) = \mathbf{0}$ *and* $\mathbf{g}(\mathbf{x}_1) < \mathbf{0}$. *Suppose* $\mathbf{x}^*$ *solves (11.49). Then there are vectors* $\boldsymbol{\lambda} \in E^m$ *and* $\boldsymbol{\mu} \in E^p$ *with* $\boldsymbol{\mu} \ge \mathbf{0}$ *such that* $\mathbf{x}^*$ *solves the Lagrangian relaxation problem*

$$\begin{aligned} &\text{minimize} \quad f(\mathbf{x}) + \boldsymbol{\lambda}^T\mathbf{h}(\mathbf{x}) + \boldsymbol{\mu}^T\mathbf{g}(\mathbf{x}) \\ &\text{subject to} \quad \mathbf{x} \in \Omega. \end{aligned} \tag{11.50}$$

Furthermore, $\boldsymbol{\mu}^T\mathbf{g}(\mathbf{x}^*) = 0$.

The convexity requirements of this result are satisfied in many practical problems. Indeed convex programming problems are both pervasive and relatively well treated by theory and numerical methods. The corresponding theory also motivates many approaches to general nonlinear programming problems. In fact, it will be apparent that many methods attempt to "convexify" a general nonlinear problem either by changing the formulation of the underlying application or by introducing devices that temporarily relax as the method progresses.

Zero-Order Sufficient Conditions

The sufficiency conditions are very strong and do not require convexity.

Proposition 5 (Sufficiency Conditions). *Suppose f is a real-valued function on a set $\Omega \subset E^n$. Suppose also that* $\mathbf{h}$ *and* $\mathbf{g}$ *are, respectively, m-dimensionaland p-dimensional functions on Ω. Finally, suppose there are vectors* $\mathrm{x}^* \in \Omega$, $\boldsymbol{\lambda} \in E^m$, *and* $\boldsymbol{\mu} \in E^p$ *with* $\boldsymbol{\mu} \ge \mathbf{0}$ *such that*

$$f(\mathrm{x}^*) + \boldsymbol{\lambda}^T\mathbf{h}(\mathrm{x}^*) + \boldsymbol{\mu}^T\mathbf{g}(\mathrm{x}^*) \le f(\mathrm{x}) + \boldsymbol{\lambda}^T\mathbf{h}(\mathrm{x}) + \boldsymbol{\mu}^T\mathbf{g}(\mathrm{x})$$

for all $\mathrm{x} \in \Omega$. *Then* x^* *solves*

$$\begin{aligned} &\text{minimize } f(\mathrm{x}) \\ &\text{subject to } \mathbf{h}(\mathrm{x}) = \mathbf{h}(\mathrm{x}^*),\ \mathbf{g}(\mathrm{x}) \le \mathbf{g}(\mathrm{x}^*) \\ &\qquad \mathrm{x} \in \Omega. \end{aligned}$$

Proof. Suppose there is $\mathbf{x}_1 \in \Omega$ with $f(\mathbf{x}_1) < f(\mathbf{x}^*)$, $\mathbf{h}(\mathbf{x}_1) = \mathbf{h}(\mathbf{x}^*)$, and $\mathbf{g}(\mathbf{x}_1) \le \mathbf{g}(\mathbf{x}^*)$. From $\boldsymbol{\mu} \ge \mathbf{0}$ it is clear that $\boldsymbol{\mu}^T\mathbf{g}(\mathbf{x}_1) \le \boldsymbol{\mu}^T\mathbf{g}(\mathbf{x}^*)$. It follows that $f(\mathbf{x}_1)+\boldsymbol{\lambda}^T\mathbf{h}(\mathbf{x}_1)+\boldsymbol{\mu}^T\mathbf{g}(\mathbf{x}_1) < f(\mathbf{x}^*) + \boldsymbol{\lambda}^T\mathbf{h}(\mathbf{x}^*) + \boldsymbol{\mu}^T\mathbf{g}(\mathbf{x}^*)$, which is a contradiction. ∎

Notice that the constraint of the Lagrangian relaxation problem is significantly simpler, and typically much easier to solve for given $\boldsymbol{\lambda}$ and $\boldsymbol{\mu}$. The result suggests that Lagrange multiplier values might be guessed and used to define an initial Lagrangian relaxation problem which is subsequently minimized. This will produce a solution of **x** and its constraint values. If these values meet the given right-hand side requirement, then **x** is optimal. Otherwise, one may adjust Lagrange multiplier values accordingly. Indeed, this approach, the Lagrangian relaxation method, will be characteristic of a duality method treated in Chap. 14.

The theory of this section has an inherent geometric simplicity captured clearly by Figs. 11.7 and 11.8. It raises ones' **s** level of understanding of Lagrange multipliers and sets the stage for the theory of convex duality presented in Chap. 14. It is certainly possible to jump ahead and read that now.

11.10 Summary

Given a minimization problem subject to equality constraints in which all functions are smooth, a necessary condition satisfied at a minimum point is that the gradient of the objective function is orthogonal to the tangent plane of the constraint surface. If the point is regular, then the tangent plane has a simple representation in terms of the gradients of the constraint functions, and the above condition can be expressed in terms of Lagrange multipliers.

If the functions have continuous second partial derivatives and Lagrange multipliers exist, then the Hessian of the Lagrangian restricted to the tangent plane plays a role in second-order conditions analogous to that played by the Hessian of the objective function in unconstrained problems. Specifically, the restricted Hessian must be positive semidefinite at a relative minimum point and, conversely, if it is positive definite at a point satisfying the first-order conditions, that point is a strict local minimum point.

Inequalities are treated by determining which of them are active at a solution. An active inequality then acts just like an equality, except that its associated Lagrange multiplier can never be negative because of the sensitivity interpretation of the multipliers.

The necessary conditions for convex problems can be expressed without derivatives, and these are according termed zero-order conditions. These conditions are highly geometric in character and explicitly treat the Lagrange multiplier as a vector in a space having dimension equal to that of the right-hand-side of the constraints. This Lagrange multiplier vector defines a hyperplane that separates the epigraph of the primal function from a set of unattainable objective and constraint value combinations.

The "zero-order" optimality condition developed in this chapter establishes a theoretical base of the Lagrangian relaxation method, which would be introduced later and is extremely popular for large-scale optimization.

11.11 Exercises

1. In E^2 consider the constraints

$$x_1 \geqslant 0, \quad x_2 \geqslant 0$$
$$x_2 - (x_1 - 1)^2 \leqslant 0.$$

   Show that the point $x_1 = 1$, $x_2 = 0$ is feasible but is not a regular point.
2. Find the rectangle of given perimeter that has greatest area by solving the first-order necessary conditions. Verify that the second-order sufficiency conditions are satisfied.
3. Verify the second-order conditions for the entropy example of Sect. 11.4.
4. A cardboard box for packing quantities of small foam balls is to be manufactured as shown in Fig. 11.9. The top, bottom, and front faces must be of double weight (i.e., two pieces of cardboard). A problem posed is to find the dimensions of such a box that maximize the volume for a given amount of cardboard, equal to 72 sq. ft.

   (a) What are the first-order necessary conditions?
   (b) Find x, y, z.
   (c) Verify the second-order conditions.

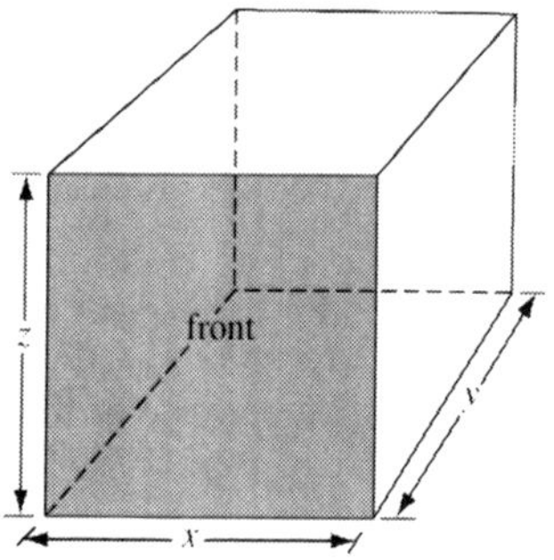

Fig. 11.9 Packing box

5. Define

$$\mathbf{L} = \begin{bmatrix} 4 & 3 & 2 \\ 3 & 1 & 1 \\ 2 & 1 & 1 \end{bmatrix}, \qquad \mathbf{h} = (1, 1, 0),$$

   and let M be the subspace consisting of those points $\mathbf{x} = (x_1, x_2, x_3)$ satisfying $\mathbf{h}^T\mathbf{x} = 0$.

   (a) Find $\mathbf{L}_M$.
   (b) Find the eigenvalues of $\mathbf{L}_M$.

(c) Find

$$p(\lambda) = \det \begin{bmatrix} 0 & \mathbf{h}^T \\ -\mathbf{h} & \mathbf{L} - \mathbf{I}\lambda \end{bmatrix}.$$

(d) Apply the projected Hessian test.

6. Show that $\mathbf{z}^T\mathbf{x} = 0$ for all $\mathbf{x}$ satisfying $\mathbf{A}\mathbf{x} = \mathbf{0}$ if and only if $\mathbf{z} = \mathbf{A}^T\mathbf{w}$ for some $\mathbf{w}$. (*Hint*: Use the Duality Theorem of Linear Programming.)
7. After a heavy military campaign a certain army requires many new shoes. The quartermaster can order three sizes of shoes. Although he does not know precisely how many of each size are required, he feels that the demand for the three sizes are independent and the demand for each size is uniformly distributed between zero and three thousand pairs. He wishes to allocate his shoe budget of \$4,000 among the three sizes so as to maximize the expected number of men properly shod. Small shoes cost \$1 per pair, medium shoes cost \$2 per pair, and large shoes cost \$4 per pair. How many pairs of each size should he order?
8. *Optimal control.* A one-dimensional dynamic process is governed by a difference equation

$$x(k+1) = \phi(x(k),\ u(k),\ k)$$

with initial condition $x(0) = x_0$. In this equation the value $x(k)$ is called the *state* at step k and $u(k)$ is the *control* at step k. Associated with this system there is an *objective function* of the form

$$J = \sum_{k=0}^{N} \psi(x(k),\ u(k),\ k).$$

In addition, there is a *terminal constraint* of the form

$$g(x(N+1)) = 0.$$

The problem is to find the sequence of controls $u(0),\ u(11.1),\ u(11.2),\ \ldots,\ u(N)$ and corresponding state values to minimize the objective function while satisfying the terminal constraint. Assuming all functions have continuous first partial derivatives and that the regularity condition is satisfied, show that associated with an optimal solution there is a sequence $\lambda(k)$, $k = 0, 1, \ldots, N$ and a μ such that

$$\begin{aligned} \lambda(k-1) &= \lambda(k)\phi_x(x(k),\ u(k),\ k) + \psi_x(x(k),\ u(k),\ k), \quad k = 1, 2, \ldots, N \\ \lambda(N) &= \mu g_x(x(N+1)) \\ \psi_u(x(k),\ u(k),\ k) &+ \lambda(k)\phi_u(x(k),\ u(k),\ k) = 0,\ k = 0, 1, 2, \ldots, N. \end{aligned}$$

9. Generalize Exercise 9 to include the case where the state $\mathbf{x}(k)$ is an n-dimensional vector and the control $\mathbf{u}(k)$ is an m-dimensional vector at each stage k.
10. An egocentric young man has just inherited a fortune F and is now planning how to spend it so as to maximize his total lifetime enjoyment. He deduces

that if $x(k)$ denotes his capital at the beginning of year k, his holdings will be approximately governed by the difference equation

$$x(k+1) = \alpha x(k) - u(k),\ x(0) = F,$$

where $\alpha \geqslant 1$ (with $\alpha - 1$ as the interest rate of investment) and where $u(k)$ is the amount spent in year k. He decides that the enjoyment achieved in year k can be expressed as $\psi(u(k))$ where ψ, his utility function, is a smooth function, and that his total lifetime enjoyment is

$$J = \sum_{k=0}^{N} \psi(u(k))\beta^k,$$

where the term $\beta^k (0 < \beta < 1)$ reflects the notion that future enjoyment is counted less today. The young man wishes to determine the sequence of expenditures that will maximize his total enjoyment subject to the condition $x(N+1) = 0$.

(a) Find the general optimality relationship for this problem.
(b) Find the solution for the special case $\psi(u) = u^{1/2}$.

11. Let $\mathbf{A}$ be an $m \times n$ matrix of rank m and let $\mathbf{L}$ be an $n \times n$ matrix that is symmetric and positive definite on the subspace $M = \{\mathbf{x} : \mathbf{A}\mathbf{x} = \mathbf{0}\}$. Show that the $(n+m) \times (n+m)$ matrix

$$\begin{bmatrix} \mathbf{L} & \mathbf{A}^T \\ \mathbf{A} & \mathbf{0} \end{bmatrix}$$

is nonsingular.
12. Consider the quadratic program

$$\begin{aligned} &\text{minimize} \quad \frac{1}{2}\mathbf{x}^T\mathbf{Q}\mathbf{x} - \mathbf{b}^T\mathbf{x} \\ &\text{subject to} \quad \mathbf{A}\mathbf{x} = \mathbf{c}. \end{aligned}$$

Prove that $\mathbf{x}^*$ is a local minimum point if and only if it is a global minimum point. (No convexity is assumed.)
13. Maximize $14\mathbf{x} - x^2 + 6y - y^2 + 7$ subject to $x + y \leqslant 2,\ x + 2y \leqslant 3$.
14. In the quadratic program example of Sect. 11.9, what are more general conditions on $\mathbf{a}$ and b that satisfy the Slater condition?
15. What are the general zero-order Lagrangian conditions for the problem (11.46) without the regularity condition? [The coefficient of f will be zero, so there is no real condition.]
16. Show that the problem of finding the rectangle of maximum area with a diagonal of unit length can be formulated as an unconstrained convex programming problem using trigonometric functions. [Hint: use variable θ over the range $0 \leq \theta \leq 45°$.]

References

11.1–11.5 For a classic treatment of Lagrange multipliers see Hancock [H4]. Also see Fiacco and McCormick [F4], Luenberger [L8], or McCormick [M2].

11.6 The simple formula for the characteristic polynomial of $\mathbf{L}_M$ as an $(n+m)$ th-order determinant is apparently due to Luenberger [L17].

11.8 The systematic treatment of inequality constraints was published by Kuhn and Tucker [K11]. Later it was found that the essential elements of the theory were contained in the 1939 unpublished M.Sci Disertation of W. Karush in the Department of Mathematics, University of Chicago. It is common to recognize this contribution by including his name to the conditions for optimality.

11.9 The theory of convex problems and the corresponding Lagrange multiplier theory was developed by Slater [S7]. For presentations similar to this section, see Hurwicz [H14] and Luenberger [L8].

Chapter 12
Primal Methods

In this chapter we initiate the presentation, analysis, and comparison of algorithms designed to solve constrained minimization problems. The four chapters that consider such problems roughly correspond to the following classification scheme. Consider a constrained minimization problem having n variables and m constraints. Methods can be devised for solving this problem that work in spaces of dimension $n - m$, n, m, or $n + m$. Each of the following chapters corresponds to methods in one of these spaces. Thus, the methods in the different chapters represent quite different approaches and are founded on different aspects of the theory. However, there are also strong interconnections between the methods of the various chapters, both in the final form of implementation and in their performance. Indeed, there soon emerges the theme that the rates of convergence of most practical algorithms are determined by the structure of the Hessian of the Lagrangian much like the structure of the Hessian of the objective function determines the rates of convergence for a wide assortment of methods for unconstrained problems. Thus, although the various algorithms of these chapters differ substantially in their motivation, they are ultimately found to be governed by a common set of principles.

12.1 Advantage of Primal Methods

We consider the question of solving the general nonlinear programming problem

$$\begin{aligned} &\text{minimize } f(\mathbf{x}) \\ &\text{subject to } \mathbf{h}(\mathbf{x}) = \mathbf{0}, \quad \mathbf{g}(\mathbf{x}) \leqslant \mathbf{0} \end{aligned} \tag{12.1}$$

where $\mathbf{x}$ is of dimension n, while f, $\mathbf{g}$, and $\mathbf{h}$ have dimensions 1, p, and m, respectively. It is assumed throughout the chapter that all of the functions have continuous partial derivatives of order three. Geometrically, we regard the problem as that of minimizing f over the region in E^n defined by the constraints.

D.G. Luenberger, Y. Ye, *Linear and Nonlinear Programming*, International Series in Operations Research & Management Science 228,
DOI 10.1007/978-3-319-18842-3_12

By a *primal method* of solution we mean a search method that works on the original problem directly by searching through the feasible region for the optimal solution. Each point in the process is feasible and the value of the objective function constantly decreases. For a problem with n variables and having m equality constraints only, primal methods work in the feasible space, which has dimension $n - m$.

Primal methods possess three significant advantages that recommend their use as general procedures applicable to almost all nonlinear programming problems. First, since each point generated in the search procedure is feasible, if the process is terminated before reaching the solution (as practicality almost always dictates for nonlinear problems), the terminating point is feasible. Thus this final point is a feasible and probably nearly optimal solution to the original problem and therefore may represent an acceptable solution to the practical problem that motivated the nonlinear program. A second attractive feature of primal methods is that, often, it can be guaranteed that if they generate a convergent sequence, the limit point of that sequence must be at least a local constrained minimum. Finally, a major advantage is that most primal methods do not rely on special problem structure, such as convexity, and hence these methods are applicable to general nonlinear programming problems.

Primal methods are not, however, without major disadvantages. They require a phase I procedure (see Sect. 3.5) to obtain an initial feasible point, and they are all plagued, particularly for problems with nonlinear constraints, with computational difficulties arising from the necessity to remain within the feasible region as the method progresses. Some methods can fail to converge for problems with inequality constraints unless elaborate precautions are taken.

The convergence rates of primal methods are competitive with those of other methods, and particularly for linear constraints, they are often among the most efficient. On balance their general applicability and simplicity place these methods in a role of central importance among nonlinear programming algorithms.

12.2 Feasible Direction Methods

The idea of feasible direction methods is to take steps through the feasible region of the form

$$\mathbf{x}_{k+1} = \mathbf{x}_k + \alpha_k \mathbf{d}_k, \tag{12.2}$$

where $\mathbf{d}_k$ is a direction vector and α_k is a nonnegative scalar. The scalar is chosen to minimize the objective function f with the restriction that the point $\mathbf{x}_{k+1}$ and the line segment joining $\mathbf{x}_k$ and $\mathbf{x}_{k+1}$ be feasible. Thus, in order that the process of minimizing with respect to α be nontrivial, an initial segment of the ray $\mathbf{x}_k + \alpha\mathbf{d}_k$, $\alpha > 0$ must be contained in the feasible region. This motivates the use of *feasible directions* for the directions of search. We recall from Sect. 7.1 that a vector $\mathbf{d}_k$ is a *feasible direction* (at $\mathbf{x}_k$) if there is an $\bar{\alpha} > 0$ such that $\mathbf{x}_k + \alpha\mathbf{d}_k$ is feasible for all α, $0 \leqslant \alpha \leqslant \bar{\alpha}$. A feasible direction method can be considered as a natural extension of our unconstrained descent methods. Each step is the composition of selecting a feasible direction and a constrained line search.

Let us consider the problem with linear inequality constraints

$$\begin{array}{ll} \text{minimize} & f(\mathbf{x}) \\ \text{subject to} & \mathbf{a}_i^T\mathbf{x} \leqslant b_i, \; i = 1, \ldots, m. \end{array} \tag{12.3}$$

Example 1 (Frank-Wolfe Method). One of the earliest proposals for a feasible direction method uses a sequential linear programming subproblem approach. Given a feasible point $\mathbf{x}_k$, the direction vector

$$\mathbf{d}_k = \mathbf{x}_k^* - \mathbf{x}_k$$

where $\mathbf{x}_k^*$ is a solution to the linear program

$$\begin{array}{ll} \text{minimize} & \nabla f(\mathbf{x}_k)\mathbf{x} \\ \text{subject to} & \mathbf{a}_i^T\mathbf{x} \leqslant b_i, \; i = 1, \ldots, m. \end{array}$$

Example 2 (Simplified Zoutendijk Method). Another proposal solves a sequence of linear subprograms as follows. Given a feasible point, $\mathbf{x}_k$, let I be the set of indices representing active constraints, that is, $\mathbf{a}_i^T\mathbf{x}_k = b_i$ for $i \in I$. The direction vector $\mathbf{d}_k$ is then chosen as a solution to the linear program

$$\begin{array}{ll} \text{minimize} & \nabla f(\mathbf{x}_k)\mathbf{d} \\ \text{subject to} & \mathbf{a}_i^T\mathbf{d} \leqslant 0, \quad i \in I \\ & \sum_{i=1}^{n} |d_i| = 1, \end{array} \tag{12.4}$$

where $\mathbf{d} = (d_1, d_2, \ldots, d_n)$. The last equation is a normalizing equation that ensures a bounded solution. (Even though it is written in terms of absolute values, the problem can be converted to a linear program; see Exercise 1.) The other constraints assure that vectors of the form $\mathbf{x}_k + \alpha\mathbf{d}_k$ will be feasible for sufficiently small $\alpha > 0$, and subject to these conditions, $\mathbf{d}$ is chosen to line up as closely as possible with the negative gradient of f. In some sense this will result in the locally best direction in which to proceed. The overall procedure progresses by generating feasible directions in this manner, and moving along them to decrease the objective.

There are two major shortcomings of feasible direction methods that require that they be modified in most cases. The first shortcoming is that for general problems there may not exist any feasible directions. If, for example, a problem had nonlinear equality constraints, we might find ourselves in the situation depicted by Fig. 12.1 where no straight line from $\mathbf{x}_k$ has a feasible segment. For such problems it is necessary either to relax our requirement of feasibility by allowing points to deviate slightly from the constraint surface or to introduce the concept of moving along curves rather than straight lines.

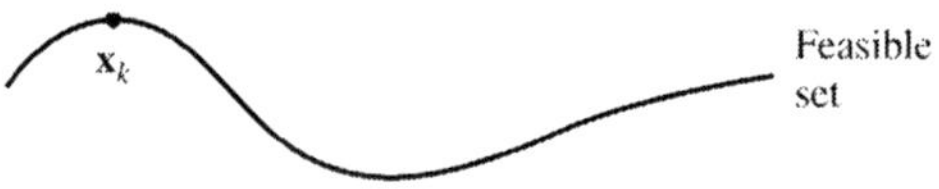

Fig. 12.1 No feasible direction

A second shortcoming is that in simplest form most feasible direction methods, such as the simplified Zoutendijk method, are not globally convergent. They are subject to *jamming* (sometimes referred to as *zigzagging*) where the sequence of points generated by the process converges to a point that is not even a constrained local minimum point. This phenomenon can be explained by the fact that the algorithmic map is not closed.

It is possible to develop feasible direction algorithms that are closed and hence not subject to jamming. Some procedures for doing so are discussed in Exercises 4–7. However, such methods can become somewhat complicated. A simpler approach for treating inequality constraints is to use an active set method, as discussed in the next section.

12.3 Active Set Methods

The idea underlying active set methods is to partition inequality constraints into two groups: those that are to be treated as active and those that are to be treated as inactive. The constraints treated as inactive are essentially ignored.

Consider the constrained problem

$$\begin{aligned} &\text{minimize} \quad f(\mathbf{x}) \\ &\text{subject to} \quad \mathbf{g}(\mathbf{x}) \leqslant \mathbf{0}, \end{aligned} \tag{12.5}$$

which for simplicity of the current discussion is taken to have inequality constraints only. The inclusion of equality constraints is straightforward, as will become clear.

The necessary conditions for this problem are

$$\begin{aligned} \nabla f(\mathbf{x}) + \boldsymbol{\lambda}^T \nabla \mathbf{g}(\mathbf{x}) &= \mathbf{0} \\ \mathbf{g}(\mathbf{x}) &\leqslant \mathbf{0} \\ \boldsymbol{\lambda}^T \mathbf{g}(\mathbf{x}) &= 0 \\ \boldsymbol{\lambda} &\geqslant \mathbf{0}. \end{aligned} \tag{12.6}$$

(See Sect. 11.8.) These conditions can be expressed in a somewhat simpler form in terms of the set of active constraints. Let A denote the index set of active constraints; that is, A is the set of i such that $g_i(\mathbf{x}^*) = 0$. Then the necessary conditions (12.6) become

$$
\begin{aligned}
\nabla f(\mathbf{x}) + \sum_{i \in A} \lambda_i \nabla g_i(\mathbf{x}) &= \mathbf{0} \\
g_i(\mathbf{x}) &= 0, \quad i \in A \\
g_i(\mathbf{x}) &< 0, \quad i \notin A \\
\lambda_i &\geqslant 0, \quad i \in A \\
\lambda_i &= 0, \quad i \notin A
\end{aligned}
\tag{12.7}
$$

The first two lines of these conditions correspond identically to the necessary conditions of the equality constrained problem obtained by requiring the active constraints to be zero. The next line guarantees that the inactive constraints are satisfied, and the sign requirement of the Lagrange multipliers guarantees that every constraint that is active *should* be active.

It is clear that if the active set were known, the original problem could be replaced by the corresponding problem having equality constraints only. Alternatively, suppose an active set was guessed and the corresponding equality constrained problem solved. Then if the other constraints were satisfied and the Lagrange multipliers turned out to be nonnegative, that solution would be correct.

The idea of active set methods is to define at each step, or at each phase, of an algorithm a set of constraints, termed the *working set*, that is to be treated as the active set. The working set is chosen to be a subset of the constraints that are actually active at the current point, and hence the current point is feasible for the working set. The algorithm then proceeds to move on the surface defined by the working set of constraints to an improved point. At this new point the working set may be changed. Overall, then, an active set method consists of the following components: (1) determination of a current working set that is a subset of the current active constraints, and (2) movement on the surface defined by the working set to an improved point.

There are several methods for determining the movement on the surface defined by the working set. (This surface will be called the *working surface*.) The most important of these methods are discussed in the following sections. The direction of movement is generally determined by first-order or second-order approximations of the functions at the current point in a manner similar to that for unconstrained problems. The asymptotic convergence properties of active set methods depend entirely on the procedure for moving on the working surface, since near the solution the working set is generally equal to the correct active set, and the process simply moves successively on the surface determined by those constraints.

Changes in Working Set

Suppose that for a given working set W the problem with equality constraints

$$
\begin{aligned}
&\text{minimize} \quad f(\mathbf{x}) \\
&\text{subject to} \quad g_i(\mathbf{x}) = 0, \quad i \in W
\end{aligned}
$$

is solved yielding the point $\mathbf{x}_W$ that satisfies $g_i(\mathbf{x}_W) < 0$, $i \notin W$. This point satisfies the necessary conditions

$$\nabla f(\mathbf{x}_W) + \sum_{i \in W} \lambda_i \nabla g_i(\mathbf{x}_W) = \mathbf{0}. \tag{12.8}$$

If $\lambda_i \geqslant 0$ for all $i \in W$, then the point $\mathbf{x}_W$ is a local solution to the original problem. If, on the other hand, there is an $i \in W$ such that $\lambda_i < 0$, then the objective can be decreased by relaxing constraint i. This follows directly from the sensitivity interpretation of Lagrange multipliers, since a small decrease in the constraint value from 0 to $-c$ would lead to a change in the objective function of $\lambda_i c$, which is negative. Thus, by dropping the constraint i from the working set, an improved solution can be obtained. The Lagrange multiplier of a problem thereby serves as an indication of which constraints should be dropped from the working set. This is illustrated in Fig. 12.2. In the figure, $\mathbf{x}$ is the minimum point of f on the surface (a curve in this case) defined by $g_1(x) = 0$. However, it is clear that the corresponding Lagrange multiplier λ_1 is negative, implying that g_1 should be dropped. Since ∇f points outside, it is clear that a movement toward the interior of the feasible region will indeed decrease f.

During the course of minimizing $f(\mathbf{x})$ over the working surface, it is necessary to monitor the values of the other constraints to be sure that they are not violated, since all points defined by the algorithm must be feasible. It often happens that while moving on the working surface a new constraint boundary is encountered. It is then convenient to add this constraint to the working set, proceeding on a surface of one lower dimension than before. This is illustrated in Fig. 12.3. In the figure the working constraint is just $g_1 = 0$ for $\mathbf{x}_1$, $\mathbf{x}_2$, $\mathbf{x}_3$. A boundary is encountered at the next step, and therefore $g_2 = 0$ is adjoined to the set of working constraints.

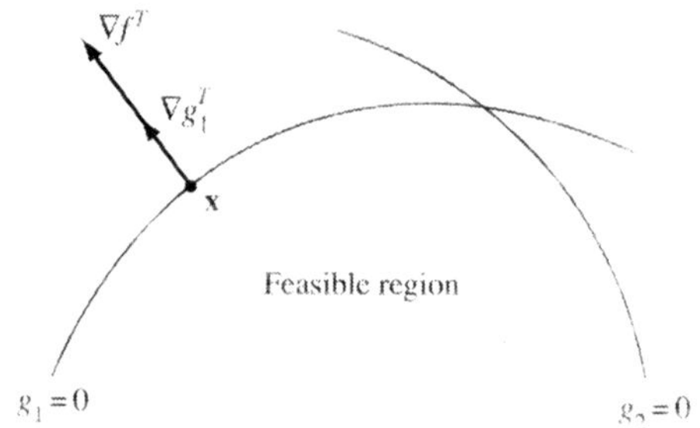

Fig. 12.2 Constraint to be dropped

A complete active set strategy for systematically dropping and adding constraints can be developed by combining the above two ideas. One starts with a given working set and begins minimizing over the corresponding working surface. If new constraint boundaries are encountered, they may be added to the working set, but no constraints are dropped from the working set. Finally, a point is obtained that minimizes f

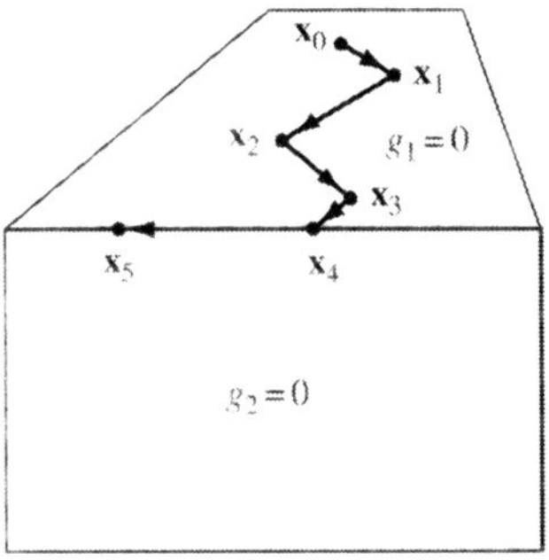

Fig. 12.3 Constraint added to working set

with respect to the current working set of constraints. The corresponding Lagrange multipliers are determined, and if they are all nonnegative the solution is optimal. Otherwise, one or more constraints with negative Lagrange multipliers are dropped from the working set. The procedure is reinitiated with this new working set, and f will strictly decrease on the next step.

An active set method built upon this basic active set strategy requires that a procedure be defined for minimization on a working surface that allows constraints to be added to the working set when they are encountered, and that, after dropping a constraint, insures that the objective is strictly decreased. Such a method is guaranteed to converge to the optimal solution, as shown below.

Active Set Theorem. *Suppose that for every subset W of the constraint indices, the constrained problem*

$$\begin{aligned} &\text{minimize} \quad f(\mathbf{x}) \\ &\text{subject to} \quad g_i(\mathbf{x}) = 0, \qquad i \in W \end{aligned} \tag{12.9}$$

is well-defined with a unique nondegenerate solution (that is, for all $i \in W$, $\lambda_i \neq 0$). Then the sequence of points generated by the basic active set strategy converges to the solution of the inequality constrained problem (12.6).

Proof. After the solution corresponding to one working set is found, a decrease in the objective is made, and hence it is not possible to return to that working set. Since there are only a finite number of working sets, the process must terminate. ∎

The difficulty with the above procedure is that several problems with incorrect active sets must be solved. Furthermore, the solutions to these intermediate problems must, in general, be exact global minimum points in order to determine the correct sign of the Lagrange multipliers and to assure that during the subsequent descent process the current working surface is not encountered again.

In practice one deviates from the ideal basic method outlined above by dropping constraints using various criteria before an exact minimum on the working surface is found. Convergence cannot be guaranteed for many of these methods, and indeed they are subject to zigzagging (or jamming) where the working set changes an infinite number of times. However, experience has shown that zigzagging is very rare for many algorithms, and in practice the active set strategy with various refinement is often very effective.

It is clear that a fundamental component of an active set method is the algorithm for solving a problem with equality constraints only, that is, for minimizing on the working surface. Such methods and their analyses are presented in the following sections.

12.4 The Gradient Projection Method

The gradient projection method is motivated by the ordinary method of steepest descent for unconstrained problems. The negative gradient is projected onto the working surface in order to define the direction of movement. We present it here in a simplified form that is based on a pure active set strategy.

Linear Constraints

Consider first problems of the form

$$\begin{aligned} &\text{minimize} \quad f(\mathbf{x}) \\ &\text{subject to} \quad \mathbf{a}_i^T\mathbf{x} \leqslant b_i, \quad i \in I_1 \\ &\qquad\qquad\quad \mathbf{a}_i^T\mathbf{x} = b_i, \quad i \in I_2 \end{aligned} \tag{12.10}$$

having linear equalities and inequalities.

A feasible solution to the constraints, if one exists, can be found by application of the phase I procedure of linear programming; so we shall always assume that our descent process is initiated at such a feasible point. At a given feasible point $\mathbf{x}$ there will be a certain number q of active constraints satisfying $\mathbf{a}_i^T\mathbf{x} = b_i$ and some inactive constraints $\mathbf{a}_i^T\mathbf{x} < b_i$. We initially take the working set $W(\mathbf{x})$ to be the set of active constraints.

At the feasible point $\mathbf{x}$ we seek a feasible direction vector $\mathbf{d}$ satisfying $\nabla f(\mathbf{x})\mathbf{d} < 0$, so that movement in the direction $\mathbf{d}$ will cause a decrease in the function f. Initially, we consider directions satisfying $\mathbf{a}_i^T\mathbf{d} = 0$, $i \in W(\mathbf{x})$ so that all working constraints remain active. This requirement amounts to requiring that the direction vector $\mathbf{d}$ lie in the tangent subspace $\mathbf{M}$ defined by the working set of constraints. The particular direction vector that we shall use is the projection of the negative gradient onto this subspace.

To compute this projection let $\mathbf{A}_q$ be defined as composed of the rows of working constraints. Assuming regularity of the constraints, as we shall always assume, $\mathbf{A}_q$ will be a $q \times n$ matrix of rank $q < n$. The tangent subspace M in which $\mathbf{d}$ must lie is the subspace of vectors satisfying $\mathbf{A}_q\mathbf{d} = \mathbf{0}$. This means that the subspace N

consisting of the vectors making up the rows of $\mathbf{A}_q$ (that is, all vectors of the form $\mathbf{A}_q^T\boldsymbol{\lambda}$ for $\boldsymbol{\lambda} \in E^q$) is orthogonal to M. Indeed, any vector can be written as the sum of vectors from each of these two complementary subspaces. In particular, the negative gradient vector $-\mathbf{g}_k$ can be written

$$-\mathbf{g}_k = \mathbf{d}_k + \mathbf{A}_q^T\boldsymbol{\lambda}_k \tag{12.11}$$

where $\mathbf{d}_k \in M$ and $\boldsymbol{\lambda}_k \in E^q$. We may solve for $\boldsymbol{\lambda}_k$ through the requirement that $\mathbf{A}_q\mathbf{d}_k = \mathbf{0}$. Thus

$$\mathbf{A}_q\mathbf{d}_k = -\mathbf{A}_q\mathbf{g}_k - (\mathbf{A}_q\mathbf{A}_q^T)\boldsymbol{\lambda}_k = \mathbf{0}, \tag{12.12}$$

which leads to

$$\boldsymbol{\lambda}_k = -(\mathbf{A}_q\mathbf{A}_q^T)^{-1}\mathbf{A}_q\mathbf{g}_k \tag{12.13}$$

and

$$\mathbf{d}_k = -[\mathbf{I} - \mathbf{A}_q^T(\mathbf{A}_q\mathbf{A}_q^T)^{-1}\mathbf{A}_q]\mathbf{g}_k = -\mathbf{P}_k\mathbf{g}_k. \tag{12.14}$$

The matrix

$$\mathbf{P}_k = [\mathbf{I} - \mathbf{A}_q^T(\mathbf{A}_q\mathbf{A}_q^T)^{-1}\mathbf{A}_q] \tag{12.15}$$

is called the projection matrix corresponding to the subspace M. Action by it on any vector yields the projection of that vector onto M. See Exercises 8 and 9 for other derivations of this result.

We easily check that if $\mathbf{d}_k \neq \mathbf{0}$, then it is a direction of descent. Since $\mathbf{g}_k + \mathbf{d}_k$ is orthogonal to $\mathbf{d}_k$, we have

$$\mathbf{g}_k^T\mathbf{d}_k = (\mathbf{g}_k^T + \mathbf{d}_k^T - \mathbf{d}_k^T)\mathbf{d}_k = -|\mathbf{d}_k|^2.$$

Thus if $\mathbf{d}_k$ as computed from (12.14) turns out to be nonzero, it is a feasible direction of descent on the working surface.

We next consider selection of the step size. As α is increased from zero, the point $\mathbf{x} + \alpha\mathbf{d}$ will initially remain feasible and the corresponding value of f will decrease. We find the length of the feasible segment of the line emanating from $\mathbf{x}$ and then minimize f over this segment. If the minimum occurs at the endpoint, a new constraint will become active and will be added to the working set.

Next, consider the possibility that the projected negative gradient is zero. We have in that case

$$\nabla f(\mathbf{x}_k) + \boldsymbol{\lambda}_k^T\mathbf{A}_q = \mathbf{0}, \tag{12.16}$$

and the point $\mathbf{x}_k$ satisfies the necessary conditions for a minimum on the working surface. If the components of $\boldsymbol{\lambda}_k$ corresponding to the active inequalities are all nonnegative, then this fact together with (12.16) implies that the Karush-Kuhn-Tucker conditions for the original problem are satisfied at $\mathbf{x}_k$ and the process terminates. In this case the $\boldsymbol{\lambda}_k$ found by projecting the negative gradient is essentially the Lagrange multiplier vector for the original problem (except that zero-valued multipliers must be appended for the inactive constraints).

If, however, at least one of those components of $\boldsymbol{\lambda}_k$ is negative, it is possible, by relaxing the corresponding inequality, to move in a new direction to an improved point. Suppose that λ_{jk}, the jth component of $\boldsymbol{\lambda}_k$, is negative and the indexing is arranged so that the corresponding constraint is the inequality $\mathbf{a}_j^T\mathbf{x} \leqslant b_j$. We determine the new direction vector by relaxing the jth constraint and projecting the negative gradient onto the subspace determined by the remaining $q-1$ active constraints. Let $\mathbf{A}_{\bar{q}}$ denote the matrix $\mathbf{A}_q$ with row $\mathbf{a}_j$ deleted. We have for some $\bar{\boldsymbol{\lambda}}_k$

$$-\mathbf{g}_k = \mathbf{A}_q^T\boldsymbol{\lambda}_k \tag{12.17}$$

$$-\mathbf{g}_k = \overline{\mathbf{d}}_k + \mathbf{A}_{\bar{q}}^T\overline{\boldsymbol{\lambda}}_k, \tag{12.18}$$

where $\overline{\mathbf{d}}_k$ is the projection of $-\mathbf{g}_k$ using $\mathbf{A}_{\bar{q}}$. It is immediately clear that $\overline{\mathbf{d}}_k \neq \mathbf{0}$, since otherwise (12.18) would be a special case of (12.17) with $\lambda_{jk} = 0$ which is impossible, since the rows of $\mathbf{A}_q$ are linearly independent. From our previous work we know that $\mathbf{g}_k^T\overline{\mathbf{d}}_k < 0$. Multiplying the transpose of (12.17) by $\overline{\mathbf{d}}_k$ and using $\mathbf{A}_{\bar{q}}\overline{\mathbf{d}}_k = \mathbf{0}$ we obtain

$$0 > \mathbf{g}_k^T\overline{\mathbf{d}}_k = -\lambda_{jk}\mathbf{a}_j^T\overline{\mathbf{d}}_k. \tag{12.19}$$

Since $\lambda_{jk} < 0$ we conclude that $\mathbf{a}_j^T\overline{\mathbf{d}}_k < 0$. Thus the vector $\overline{\mathbf{d}}_k$ is not only a direction of descent, but it is a feasible direction, since $\mathbf{a}_i^T\overline{\mathbf{d}}_k = 0$, $i \in W(\mathbf{x}_k)$, $i \neq j$, and $\mathbf{a}_j^T\overline{\mathbf{d}}_k < 0$. Hence j can be dropped from $W(\mathbf{x}_k)$.

In summary, one step of the algorithm is as follows: Given a feasible point $\mathbf{x}$

1. Find the subspace of active constraints M, and form $\mathbf{A}_q$, $W(\mathbf{x})$.
2. Calculate $\mathbf{P} = \mathbf{I} - \mathbf{A}_q^T(\mathbf{A}_q\mathbf{A}_q^T)^{-1}\mathbf{A}_q$ and $\mathbf{d} = -\mathbf{P}\nabla f(\mathrm{x})^T$.
3. If $\mathbf{d} \neq \mathbf{0}$, find α_1 and α_2 achieving, respectively,

$$\max\{\alpha : \mathbf{x} + \alpha\mathbf{d} \text{ is feasible}\}$$
$$\min\{f(\mathbf{x} + \alpha\mathbf{d}) : 0 \leqslant \alpha \leqslant \alpha_1\}.$$

   Set $\mathbf{x}$ to $\mathbf{x} + \alpha_2\mathbf{d}$ and return to (12.1).
4. If $\mathbf{d} = \mathbf{0}$, find $\boldsymbol{\lambda} = -(\mathbf{A}_q\mathbf{A}_q^T)^{-1}\mathbf{A}_q\nabla f(\mathbf{x})^T$.
   (a) If $\lambda_j \geqslant 0$, for all j corresponding to active inequalities, stop; $\mathbf{x}$ satisfies the Karush-Kuhn-Tucker conditions.
   (b) Otherwise, delete the row from $\mathbf{A}_q$ corresponding to the inequality with the most negative component of $\boldsymbol{\lambda}$ (and drop the corresponding constraint from $W(\mathbf{x})$) and return to (12.2).

The projection matrix need not be recomputed in its entirety at each new point. Since the set of active constraints in the working set changes by at most one constraint at a time, it is possible to calculate one required projection matrix from the previous one by an updating procedure. (See Exercise 11.) This is an important feature of the gradient projection method and greatly reduces the computation required at each step.

Example. Consider the problem

$$\begin{array}{ll} \text{minimize} & x_1^2 + x_2^2 + x_3^2 + x_4^2 - 2x_1 - 3x_4 \\ \text{subject to} & 2x_1 + x_2 + x_3 + 4x_4 = 7 \\ & x_1 + x_2 + 2x_3 + x_4 = 6 \\ & x_i \geqslant 0, \ i = 1, 2, 3, 4. \end{array} \tag{12.20}$$

Suppose that given the feasible point $\mathbf{x} = (2, 2, 1, 0)$ we wish to find the direction of the projected negative gradient. The active constraints are the two equalities and the inequality $x_4 \geqslant 0$. Thus

$$\mathbf{A}_q = \begin{bmatrix} 2 & 1 & 1 & 4 \\ 1 & 1 & 2 & 1 \\ 0 & 0 & 0 & 1 \end{bmatrix}, \tag{12.21}$$

and hence

$$\mathbf{A}_q \mathbf{A}_q^T = \begin{bmatrix} 22 & 9 & 4 \\ 9 & 7 & 1 \\ 4 & 1 & 1 \end{bmatrix}.$$

After considerable calculation we then find

$$(\mathbf{A}_q \mathbf{A}_q^T)^{-1} = \frac{1}{11} \begin{bmatrix} 6 & -5 & -19 \\ -5 & 6 & 14 \\ -19 & 14 & 73 \end{bmatrix}$$

and finally

$$\mathbf{P} = \frac{1}{11} \begin{bmatrix} 1 & -3 & 1 & 0 \\ -3 & 9 & -3 & 0 \\ 1 & -3 & 1 & 0 \\ 0 & 0 & 0 & 0 \end{bmatrix}. \tag{12.22}$$

The gradient at the point (2, 2, 1, 0) is $\mathbf{g} = (2, 4, 2, -3)$ and hence we find

$$\mathbf{d} = -\mathbf{P}\mathbf{g} = \frac{1}{11}(-8, 24, -8, 0),$$

or normalizing by 8/11

$$\mathbf{d} = (-1, 3, -1, 0). \tag{12.23}$$

It can be easily verified that movement in this direction does not violate the constraints.

Nonlinear Constraints

In extending the gradient projection method to problems of the form

$$\begin{array}{ll} \text{minimize} & f(\mathbf{x}) \\ \text{subject to} & \mathbf{h}(\mathbf{x}) = \mathbf{0}, \ \mathbf{g}(\mathbf{x}) \leqslant \mathbf{0}, \end{array} \tag{12.24}$$

the basic idea is that at a feasible point $\mathbf{x}_k$ one determines the active constraints and projects the negative gradient onto the subspace tangent to the surface determined by these constraints. This vector, if it is nonzero, determines the direction for the next step. The vector itself, however, is not in general a feasible direction, since the surface may be curved as illustrated in Fig. 12.4. It is therefore not always possible to move along this projected negative gradient to obtain the next point.

What is typically done in the face of this difficulty is essentially to search along a curve on the constraint surface, the direction of the curve being defined by the projected negative gradient. A new point is found in the following way: First, a move is made along the projected negative gradient to a point $\mathbf{y}$. Then a move is made in the direction perpendicular to the tangent plane at the original point to a nearby feasible point on the working surface, as illustrated in Fig. 12.4. Once this point is found the value of the objective is determined. This is repeated with various $\mathbf{y}$'s until a feasible point is found that satisfies one of the standard descent criteria for improvement relative to the original point.

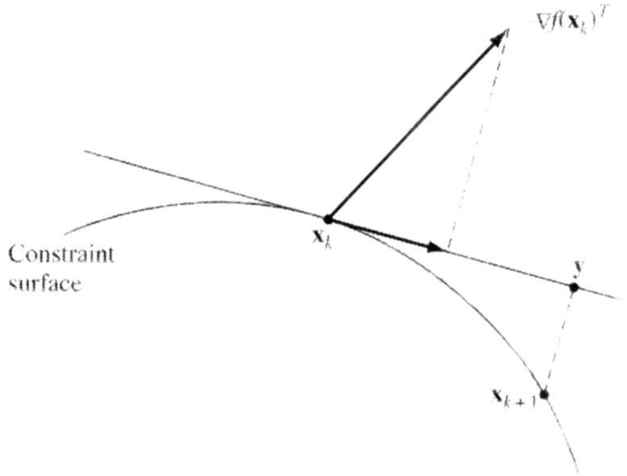

Fig. 12.4 Gradient projection method

This procedure of tentatively moving away from the feasible region and then coming back introduces a number of additional difficulties that require a series of interpolations and nonlinear equation solutions for their resolution. A satisfactory general routine implementing the gradient projection philosophy is therefore of necessity quite complex. It is not our purpose here to elaborate on these details but simply to point out the general nature of the difficulties and the basic devices for surmounting them.

One difficulty is illustrated in Fig. 12.5. If, after moving along the projected negative gradient to a point $\mathbf{y}$, one attempts to return to a point that satisfies the old active constraints, some inequalities that were originally satisfied may then be violated. One must in this circumstance use an interpolation scheme to find a new point $\bar{\mathbf{y}}$ along the negative gradient so that when returning to the active constraints no originally nonactive constraint is violated. Finding an appropriate $\bar{\mathbf{y}}$ is to some extent a trial and error process. Finally, the job of returning to the active constraints is itself a nonlinear problem which must be solved with an iterative technique. Such a technique is described below, but within a finite number of iterations, it cannot exactly reach the surface. Thus typically an error tolerance δ is introduced, and throughout the procedure the constraints are satisfied only to within δ.

Computation of the projections is also more difficult in the nonlinear case. Lumping, for notational convenience, the active inequalities together with the equalities into $\mathbf{h}(\mathbf{x}_k)$, the projection matrix at $\mathbf{x}_k$ is

$$\mathbf{P}_k = \mathbf{I} - \nabla\mathbf{h}(\mathbf{x}_k)^T[\nabla\mathbf{h}(\mathbf{x}_k)\nabla\mathbf{h}(\mathbf{x}_k)^T]^{-1}\nabla\mathbf{h}(\mathbf{x}_k). \tag{12.25}$$

At the point $\mathbf{x}_k$ this matrix can be updated to account for one more or one less constraint, just as in the linear case. When moving from $\mathbf{x}_k$ to $\mathbf{x}_{k+1}$, however, $\nabla\mathbf{h}$ will change and the new projection matrix cannot be found from the old, and hence this matrix must be recomputed at each step.

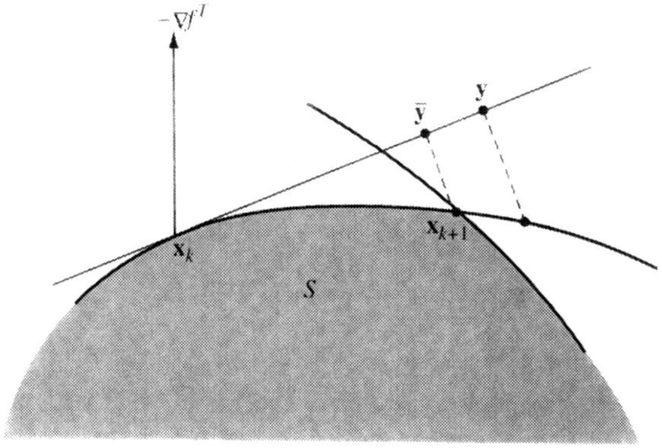

Fig. 12.5 Interpolation to obtain feasible point

The most important new feature of the method is the problem of returning to the feasible region from points outside this region. The type of iterative technique employed is a common one in nonlinear programming, including interior-point methods of linear programming, and we describe it here. The idea is, from any point near $\mathbf{x}_k$, to move back to the constraint surface in a direction orthogonal to the tangent plane at $\mathbf{x}_k$. Thus from a point $\mathbf{y}$ we seek a point of the form $\mathbf{y} + \nabla\mathbf{h}(\mathbf{x}_k)^T\boldsymbol{\alpha} = \mathbf{y}^*$ such that $\mathbf{h}(\mathbf{y}^*) = \mathbf{0}$. As shown in Fig. 12.6 such a solution may not always exist, but it does for $\mathbf{y}$ sufficiently close to $\mathbf{x}_k$.

To find a suitable first approximation to $\boldsymbol{\alpha}$, and hence to $\mathbf{y}^*$, we linearize the equation at $\mathbf{x}_k$ obtaining

$$\mathbf{h}(\mathbf{y} + \nabla\mathbf{h}(\mathbf{x}_k)^T\boldsymbol{\alpha}) \simeq \mathbf{h}(\mathbf{y}) + \nabla\mathbf{h}(\mathbf{x}_k)\nabla\mathbf{h}(\mathbf{x}_k)^T\boldsymbol{\alpha}, \tag{12.26}$$

the approximation being accurate for $|\boldsymbol{\alpha}|$ and $|\mathbf{y} - \mathbf{x}|$ small. This motivates the first approximation

$$\alpha_1 = -[\nabla \mathbf{h}(\mathbf{x}_k)\nabla \mathbf{h}(\mathbf{x}_k)^T]^{-1}\mathbf{h}(\mathbf{y}) \tag{12.27}$$

$$\mathbf{y}_1 = \mathbf{y} - \nabla \mathbf{h}(\mathbf{x}_k)^T[\nabla \mathbf{h}(\mathbf{x}_k)\nabla \mathbf{h}(\mathbf{x}_k)^T]^{-1}\mathbf{h}(\mathbf{y}). \tag{12.28}$$

Substituting $\mathbf{y}_1$ for $\mathbf{y}$ and successively repeating the process yields the sequence $\{\mathbf{y}_j\}$ generated by

$$\mathbf{y}_{j+1} = \mathbf{y}_j - \nabla \mathbf{h}(\mathbf{x}_k)^T[\nabla \mathbf{h}(\mathbf{x}_k)\nabla \mathbf{h}(\mathbf{x}_k)^T]^{-1}\mathbf{h}(\mathbf{y}_j), \tag{12.29}$$

which, started close enough to $\mathbf{x}_k$ and the constraint surface, will converge to a solution $\mathbf{y}^*$. We note that this process requires the same matrices as the projection operation.

The gradient projection method has been successfully implemented and has been found to be effective in solving general nonlinear programming problems. Successful implementation resolving the several difficulties introduced by the requirement of staying in the feasible region requires, as one would expect, some degree of skill. The true value of the method, however, can be determined only through an analysis of its rate of convergence.

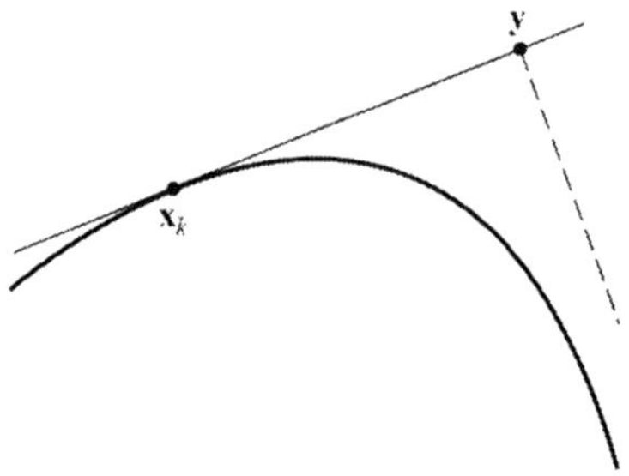

Fig. 12.6 Case in which it is impossible to return to surface

12.5 Convergence Rate of the Gradient Projection Method

An analysis that directly attacked the nonlinear version of the gradient projection method, with all of its iterative and interpolative devices, would quickly become monstrous. To obtain the asymptotic rate of convergence, however, it is not necessary to analyze this complex algorithm directly—instead it is sufficient to analyze an alternate simplified algorithm that asymptotically duplicates the gradient projection method near the solution. Through the introduction of this idealized algorithm we show that the rate of convergence of the gradient projection method is governed by the eigenvalue structure of the Hessian of the Lagrangian restricted to the constraint tangent subspace.

Geodesic Descent

For simplicity we consider first the problem having only equality constraints

$$\begin{array}{ll} \text{minimize} & f(\mathbf{x}) \\ \text{subject to} & \mathbf{h}(\mathbf{x}) = 0. \end{array} \tag{12.30}$$

The constraints define a continuous surface Ω in E^n.

In considering our own difficulties with this problem, owing to the fact that the surface is nonlinear thereby making directions of descent difficult to define, it is well to also consider the problem as it would be viewed by a small bug confined to the constraint surface who imagines it to be his total universe. To him the problem seems to be a simple one. It is unconstrained, with respect to his universe, and is only $(n - m)$-dimensional. He would characterize a solution point as a point where the gradient of f (as measured on the surface) vanishes and where the appropriate $(n - m)$-dimensional Hessian of f is positive semidefinite. If asked to develop a computational procedure for this problem, he would undoubtedly suggest, since he views the problem as unconstrained, the method of steepest descent. He would compute the gradient, as measured on his surface, and would move along what would appear to him to be straight lines.

Exactly what the bug would compute as the gradient and exactly what he would consider as straight lines would depend basically on how distance between two points on his surface were measured. If, as is most natural, we assume that he inherits his notion of distance from the one which we are using in E^n, then the path $\mathbf{x}(t)$ between two points $\mathbf{x}_1$ and $\mathbf{x}_2$ on his surface that minimizes $\int_{x_1}^{x_2} |\dot{\mathbf{x}}(t)| dt$ would be considered a straight line by him. Such a curve, having minimum arc length between two given points, is called a *geodesic*.

Returning to our own view of the problem, we note, as we have previously, that if we project the negative gradient onto the tangent plane of the constraint surface at a point $\mathbf{x}_k$, we cannot move along this projection itself and remain feasible. We might, however, consider moving along a curve which had the same initial heading as the projected negative gradient but which remained on the surface. Exactly which such curve to move along is somewhat arbitrary, but a natural choice, inspired perhaps by the considerations of the bug, is a geodesic. Specifically, at a given point on the surface, we would determine the geodesic curve passing through that point that had an initial heading identical to that of the projected negative gradient. We would then move along this geodesic to a new point on the surface having a lesser value of f.

The idealized procedure then, which the bug would use without a second thought, and which we would use if it were computationally feasible (which it definitely is not), would at a given feasible point $\mathbf{x}_k$ (see Fig. 12.7):

1. Calculate the projection $\mathbf{p}$ of $-\nabla f(\mathbf{x}_k)^T$ onto the tangent plane at $\mathbf{x}_k$.
2. Find the geodesic, $\mathbf{x}(t)$, $t \geqslant 0$, of the constraint surface having $\mathbf{x}(0) = \mathbf{x}_k$, $\dot{\mathbf{x}}(0) = \mathbf{p}$.
3. Minimize $f(\mathbf{x}(t))$ with respect to $t \geqslant 0$, obtaining t_k and $\mathbf{x}_{k+1} = \mathbf{x}(t_k)$.

At this point we emphasize that this technique (which we refer to as geodesic descent) is proposed essentially for theoretical purposes only. It does, however, capture the main philosophy of the gradient projection method. Furthermore, as the step size of the methods go to zero, as it does near the solution point, the distance between the point that would be determined by the gradient projection method and the point found by the idealized method goes to zero even faster. Thus the asymptotic rates of convergence for the two methods will be equal, and it is, therefore, appropriate to concentrate on the idealized method only.

Our bug confined to the surface would have no hesitation in estimating the rate of convergence of this method. He would simply express it in terms of the smallest and largest eigenvalues of the Hessian of f as measured on his surface. It should not be surprising, then, that we show that the asymptotic convergence ratio is

$$\left(\frac{A-a}{A+a}\right)^2, \tag{12.31}$$

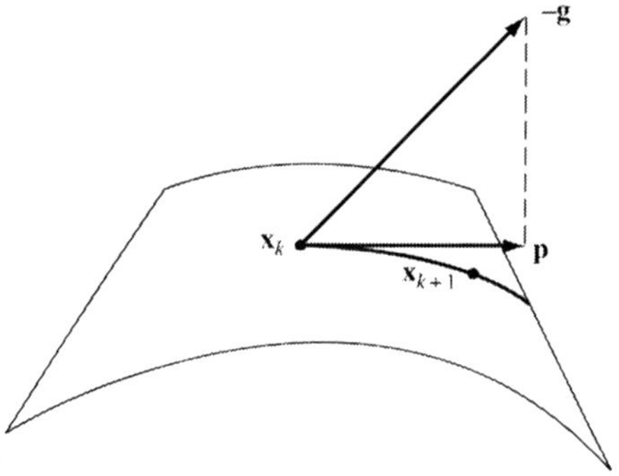

Fig. 12.7 Geodesic descent

where a and A are, respectively, the smallest and largest eigenvalues of $\mathbf{L}$, the Hessian of the Lagrangian, restricted to the tangent subspace M. This result parallels the convergence rate of the method of steepest descent, but with the eigenvalues determined from the same restricted Hessian matrix that is important in the general theory of necessary and sufficient conditions for constrained problems. This rate, which almost invariably arises when studying algorithms designed for constrained problems, will be referred to as the *canonical rate*.

We emphasize again that, since this convergence ratio governs the convergence of a large family of algorithms, it is the formula itself rather than its numerical value that is important. For any given problem we do not suggest that this ratio be evaluated, since this would be extremely difficult. Instead, the potency of the result derives from the fact that fairly comprehensive comparisons among algorithms can be made, on the basis of this formula, that apply to general classes of problems rather than simply to particular problems.

The remainder of this section is devoted to the analysis that is required to establish the convergence rate. Since this analysis is somewhat involved and not crucial for an understanding of remaining material, some readers may wish to simply read the theorem statement and proceed to the next section.

Geodesics

Given the surface $\mathbf{\Omega} = \{\mathbf{x} : \mathbf{h}(\mathbf{x}) = \mathbf{0}\} \subset E^n$, a smooth curve, $\mathbf{x}(t) \in \mathbf{\Omega}$, $0 \leqslant t \leqslant T$ starting at $\mathbf{x}(0)$ and terminating at $\mathbf{x}(T)$ that minimizes the total arc length

$$\int_0^T |\dot{\mathbf{x}}(t)|dt$$

with respect to all other such curves on $\mathbf{\Omega}$ is said to be a *geodesic* connecting $\mathbf{x}(0)$ and $\mathbf{x}(T)$.

It is common to parameterize a geodesic $\mathbf{x}(t)$, $0 \leqslant t \leqslant T$ so that $|\dot{\mathbf{x}}(t)| = 1$. The parameter t is then itself the arc length. If the parameter t is also regarded as time, then this parameterization corresponds to moving along the geodesic curve with unit velocity. Parameterized in this way, the geodesic is said to be *normalized.* On any linear subspace of E^n geodesics are straight lines. On a three-dimensional sphere, the geodesics are arcs of great circles.

It can be shown, using the calculus of variations, that any normalized geodesic on $\mathbf{\Omega}$ satisfies the condition

$$\ddot{\mathbf{x}}(t) = \nabla \mathbf{h}^T(\mathbf{x}(t))\boldsymbol{\omega}(t) \tag{12.32}$$

for some function $\boldsymbol{\omega}$ taking values in E^m. Geometrically, this condition says that if one moves along the geodesic curve with unit velocity, the acceleration at every point will be orthogonal to the surface. Indeed, this property can be regarded as the fundamental defining characteristic of a geodesic. To stay on the surface $\mathbf{\Omega}$, the geodesic must also satisfy the equation

$$\nabla \mathbf{h}(\mathbf{x}(t))\dot{\mathbf{x}}(t) = \mathbf{0}, \tag{12.33}$$

since the velocity vector at every point is tangent to $\mathbf{\Omega}$. At a regular point $\mathbf{x}_0$ these two differential equations, together with the initial conditions $\mathbf{x}(0) = \mathbf{x}_0, \dot{\mathbf{x}}(0)$ specified, and $|\dot{\mathbf{x}}(0)| = 1$, uniquely specify a curve $\mathbf{x}(t)$, $t \geqslant 0$ that can be continued as long as points on the curve are regular. Furthermore, $|\dot{\mathbf{x}}(t)| = 1$ for $t \geqslant 0$. Hence geodesic curves emanate in every direction from a regular point. Thus, for example, at any point on a sphere there is a unique great circle passing through the point in a given direction.

Lagrangian and Geodesics

Corresponding to any regular point $\mathbf{x} \in \boldsymbol{\Omega}$ we may define a corresponding Lagrange multiplier $\boldsymbol{\lambda}(\mathbf{x})$ by calculating the projection of the gradient of f onto the tangent subspace at $\mathbf{x}$, denoted $M(\mathbf{x})$. The matrix that, when operating on a vector, projects it onto $M(\mathbf{x})$ is

$$\mathbf{P}(\mathbf{x}) = \mathbf{I} - \nabla\mathbf{h}(\mathbf{x})^T[\nabla\mathbf{h}(\mathbf{x})\nabla\mathbf{h}(\mathbf{x})^T]^{-1}\nabla\mathbf{h}(\mathbf{x}),$$

and it follows immediately that the projection of $\nabla f(\mathbf{x})^T$ onto $M(\mathbf{x})$ has the form

$$\mathbf{y}(\mathbf{x}) = [\nabla f(\mathbf{x}) + \boldsymbol{\lambda}(\mathbf{x})^T\nabla\mathbf{h}(\mathbf{x})]^T, \tag{12.34}$$

where $\boldsymbol{\lambda}(\mathbf{x})$ is given explicitly as

$$\boldsymbol{\lambda}(\mathbf{x})^T = -\nabla f(\mathbf{x})\nabla\mathbf{h}(\mathbf{x})^T[\nabla\mathbf{h}(\mathbf{x})\nabla\mathbf{h}(\mathbf{x})^T]^{-1}. \tag{12.35}$$

Thus, in terms of the Lagrangian function $l(\mathbf{x}, \boldsymbol{\lambda}) = f(\mathbf{x}) + \boldsymbol{\lambda}^T\mathbf{h}(\mathbf{x})$, the projected gradient is

$$\mathbf{y}(\mathbf{x}) = l_{\mathbf{x}}(\mathbf{x}, \boldsymbol{\lambda}(\mathbf{x}))^T. \tag{12.36}$$

If a local solution to the original problem occurs at a regular point $\mathbf{x}^* \in \boldsymbol{\Omega}$, then as we know

$$l_{\mathbf{x}}(\mathbf{x}^*, \boldsymbol{\lambda}(\mathbf{x}^*)) = \mathbf{0}, \tag{12.37}$$

which states that the projected gradient must vanish at $\mathbf{x}^*$. Defining $\mathbf{L}(\mathbf{x}) = l_{\mathbf{xx}}(\mathbf{x}, \boldsymbol{\lambda}(\mathbf{x})) = \mathbf{F}(\mathbf{x}) + \boldsymbol{\lambda}(\mathbf{x})^T\mathbf{H}(\mathbf{x})$ we also know that at $\mathbf{x}^*$ we have the second-order necessary condition that $\mathbf{L}(\mathbf{x}^*)$ is positive semidefinite on $M(\mathbf{x}^*)$; that is, $\mathbf{z}^T\mathbf{L}(\mathbf{x}^*)\mathbf{z} \geqslant 0$ for all $\mathbf{z} \in M(\mathbf{x}^*)$. Equivalently, letting

$$\overline{\mathbf{L}}(\mathbf{x}) = \mathbf{P}(\mathbf{x})\mathbf{L}(\mathbf{x})\mathbf{P}(\mathbf{x}), \tag{12.38}$$

it follows that $\overline{\mathbf{L}}(\mathbf{x}^*)$ is positive semidefinite.

We then have the following fundamental and simple result, valid along a geodesic.

Proposition 1. *Let* $\mathbf{x}(t)$, $0 \leqslant t \leqslant T$, *be a geodesic on* $\boldsymbol{\Omega}$. *Then*

$$\frac{d}{dt}f(\mathbf{x}(t)) = l_{\mathbf{x}}(\mathbf{x}, \boldsymbol{\lambda}(\mathbf{x}))\dot{\mathbf{x}}(t) \tag{12.39}$$

$$\frac{d^2}{dt^2}f(\mathbf{x}(t)) = \dot{\mathbf{x}}(t)^T\mathbf{L}(\mathbf{x}(t))\dot{\mathbf{x}}(t). \tag{12.40}$$

Proof. We have

$$\frac{d}{dt}f(\mathbf{x}(t)) = \nabla f(\mathbf{x}(t))\dot{\mathbf{x}}(t) = l_{\mathbf{x}}(\mathbf{x}, \boldsymbol{\lambda}(\mathbf{x}))\dot{\mathbf{x}}(t),$$

the second equality following from the fact that $\dot{\mathbf{x}}(t) \in M(\mathbf{x})$. Next,

$$\frac{d^2}{dt^2}f(\mathbf{x}(t)) = \dot{\mathbf{x}}(t)^T\mathbf{F}(\mathbf{x}(t))\dot{\mathbf{x}}(t) + \nabla f(\mathbf{x}(t))\ddot{\mathbf{x}}(t). \tag{12.41}$$

But differentiating the relation $\boldsymbol{\lambda}^T\mathbf{h}(\mathbf{x}(t)) = 0$ twice, for fixed $\boldsymbol{\lambda}$, yields

$$\dot{\mathbf{x}}(t)^T\boldsymbol{\lambda}^T\mathbf{H}(\mathbf{x}(t))\dot{\mathbf{x}}(t) + \boldsymbol{\lambda}^T\nabla\mathbf{h}(\mathbf{x}(t))\ddot{\mathbf{x}}(t) = \mathbf{0}.$$

Adding this to (12.41), we have

$$\frac{d^2}{dt^2}f(\mathbf{x}(t)) = \dot{\mathbf{x}}(t)^T(\mathbf{F} + \boldsymbol{\lambda}^T\mathbf{H})\dot{\mathbf{x}}(t) + (\nabla f(\mathbf{x}) + \boldsymbol{\lambda}^T\nabla\mathbf{h}(\mathbf{x}))\ddot{\mathbf{x}}(t),$$

which is true for any fixed $\boldsymbol{\lambda}$. Setting $\boldsymbol{\lambda} = \boldsymbol{\lambda}(\mathbf{x})$ determined as above, $(\nabla\mathbf{f} + \boldsymbol{\lambda}^T\nabla\mathbf{h})^T$ is in $M(\mathbf{x})$ and hence orthogonal to $\ddot{\mathbf{x}}(t)$, since $\mathbf{x}(t)$ is a normalized geodesic. This gives (12.40). ∎

It should be noted that we proved a simplified version of this result in Chap. 11. There the result was given only for the optimal point $\mathbf{x}^*$, although it was valid for any curve. Here we have shown that essentially the same result is valid at any point provided that we move along a geodesic.

Rate of Convergence

We now prove the main theorem regarding the rate of convergence. We assume that all functions are three times continuously differentiable and that every point in a region near the solution $\mathbf{x}^*$ is regular. This theorem only establishes the rate of convergence and not convergence itself so for that reason the stated hypotheses assume that the method of geodesic descent generates a sequence $\{\mathbf{x}_k\}$ converging to $\mathbf{x}^*$.

Theorem. *Theorem. Let $\mathbf{x}^*$ be a local solution to the problem (12.30) and suppose that A and $a > 0$ are, respectively, the largest and smallest eigenvalues of $\mathbf{L}(\mathbf{x}^*)$ restricted to the tangent subspace $M(\mathbf{x}^*)$. If $\{\mathbf{x}_k\}$ is a sequence generated by the method of geodesic descent that converges to $\mathbf{x}^*$, then the sequence of objective values $\{f(\mathbf{x}_k)\}$ converges to $f(\mathbf{x}^*)$ linearly with a ratio no greater than $[(A-a)/(A+a)]^2$.*

Proof. Without loss of generality we may assume $f(\mathbf{x}^*) = 0$. Given a point $\mathbf{x}_k$ it will be convenient to define its distance from the solution point $\mathbf{x}^*$ as the arc length of the geodesic connecting $\mathbf{x}^*$ and $\mathbf{x}_k$. Thus if $\mathbf{x}(t)$ is a parameterized version of the geodesic with $\mathbf{x}(0) = \mathbf{x}^*$, $|\dot{\mathbf{x}}(t)| = 1$, $\mathbf{x}(T) = \mathbf{x}_k$, then T is the distance of $\mathbf{x}_k$ from $\mathbf{x}^*$. Associated with such a geodesic we also have the family $\mathbf{y}(t)$, $0 \leqslant t \leqslant T$, of corresponding projected gradients $\mathbf{y}(t) = l_{\mathbf{x}}(\mathbf{x}, \boldsymbol{\lambda}(\mathbf{x}))^T$, and Hessians $\mathbf{L}(t) = \mathbf{L}(\mathbf{x}(t))$. We write $\mathbf{y}_k = \mathbf{y}(\mathbf{x}_k)$, $\mathbf{L}_k = \mathbf{L}(\mathbf{x}_k)$.

We now derive an estimate for $f(\mathbf{x}_k)$. Using the geodesic discussed above we can write (setting $\dot{\mathbf{x}}_k = \dot{\mathbf{x}}(T)$)

$$f(\mathbf{x}^*) - f(\mathbf{x}_k) = -f(\mathbf{x}_k) = -\mathbf{y}_k^T\dot{\mathbf{x}}_k T + \frac{1}{2}T^2\dot{\mathbf{x}}_k^T\mathbf{L}_k\dot{\mathbf{x}}_k + o(T^2), \tag{12.42}$$

which follows from Proposition 1. We also have

$$\mathbf{y}_k = -\mathbf{y}(\mathbf{x}^*) + \mathbf{y}(\mathbf{x}_k) = \dot{\mathbf{y}}_k T + o(T). \tag{12.43}$$

But differentiating (12.34) we obtain

$$\dot{\mathbf{y}}_k = \mathbf{L}_k \dot{\mathbf{x}}_k + \nabla \mathbf{h}(\mathbf{x}_k)^T \dot{\boldsymbol{\lambda}}_k^T, \tag{12.44}$$

and hence if $\mathbf{P}_k$ is the projection matrix onto $M(\mathbf{x}_k) = M_k$, we have

$$\mathbf{P}_k \dot{\mathbf{y}}_k = \mathbf{P}_k \mathbf{L}_k \dot{\mathbf{x}}_k. \tag{12.45}$$

Multiplying (12.43) by $\mathbf{P}_k$ and accounting for $\mathbf{P}_k \mathbf{y}_k = \mathbf{y}_k$ we have

$$\mathbf{P}_k \dot{\mathbf{y}}_k T = \mathbf{y}_k + o(T). \tag{12.46}$$

Substituting (12.45) into this we obtain

$$\mathbf{P}_k \mathbf{L}_k \dot{\mathbf{x}}_k T = \mathbf{y}_k + o(T).$$

Since $\mathbf{P}_k \dot{\mathbf{x}}_k = \dot{\mathbf{x}}_k$ we have, defining $\overline{\mathbf{L}}_k = \mathbf{P}_k \mathbf{L}_k \mathbf{P}_k$,

$$\overline{\mathbf{L}}_k \dot{\mathbf{x}}_k T = \mathbf{y}_k + o(T). \tag{12.47}$$

The matrix $\overline{\mathbf{L}}_k$ is related to $\mathbf{L}_{M_k}$, the restriction of $\mathbf{L}_k$ to M_k, the only difference being that while $\mathbf{L}_{M_k}$ is defined only on M_k, the matrix $\overline{\mathbf{L}}_k$ is defined on all of E^n but in such a way that it agrees with $\mathbf{L}_{M_k}$ on M_k and is zero on $M_k^{\perp}$. The matrix $\overline{\mathbf{L}}_k$ is not invertible, but for $\mathbf{y}_k \in M_k$ there is a unique solution $\mathbf{z} \in M_k$ to the equation $\overline{\mathbf{L}}_k \mathbf{z} = \mathbf{y}_k$ which we denote† $\overline{\mathbf{L}}_k^{-1} \mathbf{y}_k$. With this notation we obtain from (12.47)

$$\dot{\mathbf{x}}_k T = \overline{\mathbf{L}}_k^{-1} \mathbf{y}_k + o(T). \tag{12.48}$$

Substituting this last result into (12.42) and accounting for $|\mathbf{y}_k| = O(T)$ (see (12.43)) we have

$$f(\mathbf{x}_k) = \frac{1}{2} \mathbf{y}_k^T \overline{\mathbf{L}}_k^{-1} \mathbf{y}_k + o(T^2), \tag{12.49}$$

which expresses the objective value at $\mathbf{x}_k$ in terms of the projected gradient.

Since $|\dot{\mathbf{x}}_k| = 1$ and since $\overline{\mathbf{L}}_k \to \overline{\mathbf{L}}^*$ as $\mathbf{x}_k \to \mathbf{x}^*$, we see from (12.47) that

$$o(T) + aT \leqslant |\mathbf{y}_k| \leqslant AT + o(T), \tag{12.50}$$

which means that not only do we have $|\mathbf{y}_k| = O(T)$, which was known before, but also $|\mathbf{y}_k| \neq o(T)$. We may therefore write our estimate (12.49) in the alternate form

† Actually a more standard procedure is to define the pseudoinverse $\overline{\mathbf{L}}_k^{\dagger}$, and then $\mathbf{z} = \overline{\mathbf{L}}_k^{\dagger} \mathbf{y}_k$.

$$f(\mathbf{x}_k) = \frac{1}{2}\mathbf{y}_k^T \overline{\mathbf{L}}_k^{-1} \mathbf{y}_k \left(1 + \frac{o(T^2)}{\mathbf{y}_k^T \overline{\mathbf{L}}_k^{-1} \mathbf{y}_k} \right), \tag{12.51}$$

and since $o(T^2) \neq \mathbf{y}_k^T \overline{\mathbf{L}}_k^{-1} \mathbf{y}_k = O(T^2)$, we have

$$f(\mathbf{x}_k) = \frac{1}{2}\mathbf{y}_k^T \overline{\mathbf{L}}_k^{-1} \mathbf{y}_k (1 + O(T)), \tag{12.52}$$

which is the desired estimate.

Next, we estimate $f(\mathbf{x}_{k+1})$ in terms of $f(\mathbf{x}_k)$. Given $\mathbf{x}_k$ now let $\mathbf{x}(t)$, $t \geqslant 0$, be the normalized geodesic emanating from $\mathbf{x}_k \equiv \mathbf{x}(0)$ in the direction of the negative projected gradient, that is,

$$\dot{\mathbf{x}}(0) \equiv \dot{\mathbf{x}}_k = -\mathbf{y}_k / |\mathbf{y}_k|.$$

Then

$$f(\mathbf{x}(t)) = f(\mathbf{x}_k) + t\mathbf{y}_k^T \dot{\mathbf{x}}_k + \frac{t^2}{2}\dot{\mathbf{x}}_k^T \mathbf{L}_k \dot{\mathbf{x}}_k + o(t^2). \tag{12.53}$$

This is minimized at

$$t_k = -\frac{\mathbf{y}_k^T \dot{\mathbf{x}}_k}{\dot{\mathbf{x}}_k^T \mathbf{L}_k \dot{\mathbf{x}}_k} + o(t_k). \tag{12.54}$$

In view of (12.50) this implies that $t_k = O(T)$, $t_k \neq o(T)$. Thus t_k goes to zero at essentially the same rate as T. Thus we have

$$f(\mathbf{x}_{k+1}) = f(\mathbf{x}_k) - \frac{1}{2}\frac{(\mathbf{y}_k^T \dot{\mathbf{x}}_k)^2}{\dot{\mathbf{x}}_k^T \mathbf{L}_k \dot{\mathbf{x}}_k} + o(T^2). \tag{12.55}$$

Using the same argument as before we can express this as

$$f(\mathbf{x}_k) - f(\mathbf{x}_{k+1}) = \frac{1}{2}\frac{(\mathbf{y}_k^T \mathbf{y}_k)^2}{\mathbf{y}_k^T \mathbf{L}_k \mathbf{y}_k}(1 + O(T)), \tag{12.56}$$

which is the other required estimate.

Finally, dividing (12.56) by (12.52) we find

$$\frac{f(\mathbf{x}_k) - f(\mathbf{x}_{k+1})}{f(\mathbf{x}_k)} = \frac{(\mathbf{y}_k^T \mathbf{y}_k)^2 (1 + O(T))}{(\mathbf{y}_k^T \mathbf{L}_k \mathbf{y}_k)(\mathbf{y}_k^T \overline{\mathbf{L}}_k^{-1} \mathbf{y}_k)}, \tag{12.57}$$

and thus

$$f(\mathbf{x}_{k+1}) = \left[1 - \frac{(\mathbf{y}_k^T \mathbf{y}_k)^2 (1 + O(T))}{(\mathbf{y}_k^T \mathbf{L}_k \mathbf{y}_k)(\mathbf{y}_k^T \overline{\mathbf{L}}_k^{-1} \mathbf{y}_k)} \right] f(\mathbf{x}_k). \tag{12.58}$$

Using the fact that $\mathbf{L}_k \to \mathbf{L}^*$ and applying the Kantorovich inequality leads to

$$f(\mathbf{x}_{k+1}) \leqslant \left[\left(\frac{A - a}{A + a} \right)^2 + O(T) \right] f(\mathbf{x}_k). \blacksquare \tag{12.59}$$

Problems with Inequalities

The idealized version of gradient projection could easily be extended to problems having nonlinear inequalities as well as equalities by following the pattern of Sect. 12.4. Such an extension, however, has no real value, since the idealized scheme cannot be implemented. The idealized procedure was devised only as a technique for analyzing the asymptotic rate of convergence of the analytically more complex, but more practical, gradient projection method.

The analysis of the idealized version of gradient projection given above, nevertheless, does apply to problems having inequality as well as equality constraints. If a computationally feasible procedure is employed that avoids jamming and does not bounce on and off constraint boundaries an infinite number of times, then near the solution the active constraints will remain fixed. This means that near the solution the method acts just as if it were solving a problem having the active constraints as equality constraints. Thus the asymptotic rate of convergence of the gradient projection method applied to a problem with inequalities is also given by (12.59) but with $\mathbf{L}(\mathbf{x}^*)$ and $M(\mathbf{x}^*)$ (and hence a and A) determined by the active constraints at the solution point $\mathbf{x}^*$. In every case, therefore, the rate of convergence is determined by the eigenvalues of the same restricted Hessian that arises in the necessary conditions.

12.6 The Reduced Gradient Method

From a computational viewpoint, the reduced gradient method, discussed in this section and the next, is closely related to the simplex method of linear programming in that the problem variables are partitioned into basic and nonbasic groups. From a theoretical viewpoint, the method can be shown to behave very much like the gradient projection method.

Linear Constraints

Consider the problem

$$\begin{aligned} &\text{minimize} \quad f(\mathbf{x}) \\ &\text{subject to} \quad \mathbf{A}\mathbf{x} = \mathbf{b}, \ \mathbf{x} \geqslant \mathbf{0}, \end{aligned} \tag{12.60}$$

where $\mathbf{x} \in E^n$, $\mathbf{b} \in E^m$, $\mathbf{A}$ is $m \times n$, and f is a function in C^2. The constraints are expressed in the format of the standard form of linear programming. For simplicity of notation it is assumed that each variable is required to be non-negative—if some variables were free, the procedure (but not the notation) would be somewhat simplified.

We invoke the *nondegeneracy assumptions* that every collection of m columns from $\mathbf{A}$ is linearly independent and every basic solution to the constraints has m

strictly positive variables. With these assumptions any feasible solution will have at most $n-m$ variables taking the value zero. Given a vector $\mathbf{x}$ satisfying the constraints, we partition the variables into two groups: $\mathbf{x} = (\mathbf{y}, \mathbf{z})$ where $\mathbf{y}$ has dimension m and $\mathbf{z}$ has dimension $n - m$. This partition is formed in such a way that all variables in $\mathbf{y}$ are strictly positive (for simplicity of notation we indicate the basic variables as being the first m components of $\mathbf{x}$ but, of course, in general this will not be so). With respect to the partition, the original problem can be expressed as

$$\text{minimize} \quad f(\mathbf{y}, \mathbf{z}) \tag{12.61a}$$

$$\text{subject to} \quad \mathbf{By} + \mathbf{Cz} = \mathbf{b} \tag{12.61b}$$

$$\mathbf{y} \geqslant \mathbf{0}, \ \mathbf{z} \geqslant \mathbf{0}, \tag{12.61c}$$

where, of course, $\mathbf{A} = [\mathbf{B}, \ \mathbf{C}]$. We can regard $\mathbf{z}$ as consisting of the independent variables and $\mathbf{y}$ the dependent variables, since if $\mathbf{z}$ is specified, (12.61b) can be uniquely solved for $\mathbf{y}$. Furthermore, a small change $\Delta\mathbf{z}$ from the original value that leaves $\mathbf{z}+\Delta\mathbf{z}$ nonnegative will, upon solution of (12.61b), yield another feasible solution, since $\mathbf{y}$ was originally taken to be strictly positive and thus $\mathbf{y} + \Delta\mathbf{y}$ will also be positive for small $\Delta\mathbf{y}$. We may therefore move from one feasible solution to another by selecting a $\Delta\mathbf{z}$ and moving $\mathbf{z}$ on the line $\mathbf{z} + \alpha\Delta\mathbf{z}$, $\alpha \geqslant 0$. Accordingly, $\mathbf{y}$ will move along a corresponding line $\mathbf{y} + \alpha\Delta\mathbf{y}$. If in moving this way some variable becomes zero, a new inequality constraint becomes active. If some independent variable becomes zero, a new direction $\Delta\mathbf{z}$ must be chosen. If a dependent (basic) variable becomes zero, the partition must be modified. The zero-valued basic variable is declared independent and one of the strictly positive independent variables is made dependent. Operationally, this interchange will be associated with a pivot operation.

The idea of the reduced gradient method is to consider, at each stage, the problem only in terms of the independent variables. Since the vector of dependent variables $\mathbf{y}$ is determined through the constraints (12.61b) from the vector of independent variables $\mathbf{z}$, the objective function can be considered to be a function of $\mathbf{z}$ only. Hence a simple modification of steepest descent, accounting for the constraints, can be executed. The gradient with respect to the independent variables $\mathbf{z}$ (the *reduced gradient*) is found by evaluating the gradient of $f(\mathbf{B}^{-1}\mathbf{b} - \mathbf{B}^{-1}\mathbf{C}\mathbf{z}\,\mathbf{z})$. It is equal to

$$\mathbf{r}^T = \nabla_{\mathbf{z}} f(\mathbf{y}, \mathbf{z}) - \nabla_{\mathbf{y}} f(\mathbf{y}, \mathbf{z})\mathbf{B}^{-1}\mathbf{C}. \tag{12.62}$$

It is easy to see that a point $(\mathbf{y}, \mathbf{z})$ satisfies the first-order necessary conditions for optimality if and only if

$$r_i = 0 \quad \text{for all} \quad z_i > 0$$
$$r_i \geqslant 0 \quad \text{for all} \quad z_i = 0.$$

In the active set form of the reduced gradient method the vector $\mathbf{z}$ is moved in the direction of the reduced gradient on the working surface. Thus at each step, a direction of the form

$$\Delta z_i = \begin{cases} -r_i, i \notin W(\mathbf{z}) \\ 0, \quad i \in W(\mathbf{z}) \end{cases}$$

is determined and a descent is made in this direction. The working set is augmented whenever a new variable reaches zero; if it is a basic variable, a new partition is also formed. If a point is found where $r_i = 0$ for all $i \notin W(\mathbf{z})$ (representing a vanishing reduced gradient on the working surface) but $r_j < 0$ for some $j \in W(\mathbf{z})$, then j is deleted from $W(\mathbf{z})$ as in the standard active set strategy.

It is possible to avoid the pure active set strategy by moving away from our active constraint whenever that would lead to an improvement, rather than waiting until an exact minimum on the working surface is found. Indeed, this type of procedure is often used in practice. One version progresses by moving the vector $\mathbf{z}$ in the direction of the overall negative reduced gradient, except that zero-valued components of $\mathbf{z}$ that would thereby become negative are held at zero. One step of the procedure is as follows:

1. Let $\mathbf{\Delta} z_i = \begin{cases} -r_i & \text{if } r_i < 0 \text{ or } z_i > 0 \\ 0 & \text{otherwise.} \end{cases}$
2. If $\mathbf{\Delta z}$ is zero, stop; the current point is a solution. Otherwise, find $\mathbf{\Delta y} = -\mathbf{B}^{-1}\mathbf{C}\mathbf{\Delta z}$.
3. Find α_1, α_2, α_3 achieving, respectively,

$$\max\{\alpha : \mathbf{y} + \alpha\mathbf{\Delta y} \geqslant 0\}$$
$$\max\{\alpha : \mathbf{z} + \alpha\mathbf{\Delta z} \geqslant 0\}$$
$$\min\{f(\mathbf{x} + \alpha\mathbf{\Delta x}) : 0 \leqslant \alpha \leqslant \alpha_1,\ 0 \leqslant \alpha \leqslant \alpha_2\}$$

   Let $\overline{\mathbf{x}} = \mathbf{x} + \alpha_3\mathbf{\Delta x}$.
4. If $\alpha_3 < \alpha_1$, return to (12.1). Otherwise, declare the vanishing variable in the dependent set independent and declare a strictly positive variable in the independent set dependent. Update $\mathbf{B}$ and $\mathbf{C}$.

Example. We consider the example presented in Sect. 12.4 where the projected negative gradient was computed:

$$\begin{aligned} \text{minimize } & x_1^2 + x_2^2 + x_3^2 + x_4^2 - 2x_1 - 3x_4 \\ \text{subject to } & 2x_1 + x_2 + x_3 + 4x_4 = 7 \\ & x_1 + x_2 + 2x_3 + x_4 = 6 \\ & x_i \geqslant 0,\ i = 1, 2, 3, 4. \end{aligned}$$

We are given the feasible point $\mathbf{x} = (2, 2, 1, 0)$. We may select any two of the strictly positive variables to be the basic variables. Suppose $\mathbf{y} = (x_1, x_2)$ is selected. In standard form the constraints are then

$$x_1 + 0 - x_3 + 3x_4 = 1$$
$$0 + x_2 + 3x_3 - 2x_4 = 5$$
$$x_i \geqslant 0, \quad i = 1, 2, 3, 4.$$

The gradient at the current point is $\mathbf{g} = (2, 4, 2, -3)$. The corresponding reduced gradient (with respect to $\mathbf{z} = (x_3, x_4)$) is then found by *pricing-out* in the usual manner. The situation at the current point can then be summarized by the tableau

Variable	x_1	x_2	x_3	x_4	
Constraints	1	0	−1	3	1
	0	1	3	−2	5
$\mathbf{r}^T$	0	0	−8	−1	
Current value	2	2	1	0	

Tableau for Example

In this solution x_3 and x_4 would be increased together in a ratio of eight to one. As they increase, x_1 and x_2 would follow in such a way as to keep the constraints satisfied. Overall, in E^4, the implied direction of movement is thus

$$\mathbf{d} = (5, -22, 8, 1).$$

If the reader carefully supplies the computational details not shown in the presentation of the example as worked here and in Sect. 12.4, he will undoubtedly develop a considerable appreciation for the relative simplicity of the reduced gradient method.

It should be clear that the reduced gradient method can, as illustrated in the example above, be executed with the aid of a tableau. At each step the tableau of constraints is arranged so that an identity matrix appears over the m dependent variables, and thus the dependent variables can be easily calculated from the values of the independent variables. The reduced gradient at any step is calculated by evaluating the n-dimensional gradient and "pricing out" the dependent variables just as the reduced cost vector is calculated in linear programming. And when the partition of basic and non-basic variables must be changed, a simple pivot operation is all that is required.

Global Convergence

The perceptive reader will note the direction finding algorithm that results from the second form of the reduced gradient method is not closed, since slight movement away from the boundary of an inequality constraint can cause a sudden change in the direction of search. Thus one might suspect, and correctly so, that this method is subject to jamming. However, a trivial modification will yield a closed mapping; and hence global convergence. This is discussed in Exercise 19.

Nonlinear Constraints

The *generalized reduced gradient method* solves nonlinear programming problems in the *standard form*

$$\begin{aligned} &\text{minimize} \quad f(\mathbf{x}) \\ &\text{subject to} \quad \mathbf{h}(\mathbf{x}) = \mathbf{0}, \ \mathbf{a} \leqslant \mathbf{x} \leqslant \mathbf{b}, \end{aligned}$$

where $\mathbf{h}(\mathbf{x})$ is of dimension m. A general nonlinear programming problem can always be expressed in this form by the introduction of slack variables, if required, and by allowing some components of $\mathbf{a}$ and $\mathbf{b}$ to take on the values $+\infty$ or $-\infty$, if necessary.

In a manner quite analogous to that of the case of linear constraints, we introduce a *nondegeneracy* assumption that, at each point $\mathbf{x}$, hypothesizes the existence of a partition of $\mathbf{x}$ into $\mathbf{x} = (\mathbf{y}, \mathbf{z})$ having the following properties:

(i) $\mathbf{y}$ is of dimension m, and $\mathbf{z}$ is of dimension $n - m$.
(ii) If $\mathbf{a} = (\mathbf{a}_\mathrm{y}, \mathbf{a}_\mathrm{z})$ and $\mathbf{b} = (\mathbf{b}_\mathrm{y}, \mathbf{b}_\mathrm{z})$ are the corresponding partitions of $\mathbf{a}$, $\mathbf{b}$, then $\mathbf{a}_\mathrm{y} < \mathbf{y} < \mathbf{b}_\mathrm{y}$.
(iii) The $m \times m$ matrix $\nabla_\mathrm{y}\mathbf{h}(\mathbf{y}, \mathbf{z})$ is nonsingular at $\mathbf{x} = (\mathbf{y}, \mathbf{z})$.

Again $\mathbf{y}$ and $\mathbf{z}$ are referred to as the vectors of *dependent* and *independent variables*, respectively.

The reduced gradient (with respect to $\mathbf{z}$) is in this case:

$$\mathbf{r}^T = \nabla_\mathrm{z} f(\mathbf{y}, \mathbf{z}) + \boldsymbol{\lambda}^T \nabla_\mathrm{z}\mathbf{h}(\mathbf{y}, \mathbf{z}),$$

where $\boldsymbol{\lambda}$ satisfies

$$\nabla_\mathrm{y} f(\mathbf{y}, \mathbf{z}) + \boldsymbol{\lambda}^T \nabla_\mathrm{y}\mathbf{h}(\mathbf{y}, \mathbf{z}) = \mathbf{0}.$$

Equivalently, we have

$$\mathbf{r}^T = \nabla_\mathrm{z} f(\mathbf{y}, \mathbf{z}) - \nabla_\mathrm{y} f(\mathbf{y}, \mathbf{z})[\nabla_\mathrm{y}\mathbf{h}(\mathbf{y}, \mathbf{z})]^{-1}\nabla_\mathrm{z}\mathbf{h}(\mathbf{y}, \mathbf{z}). \tag{12.63}$$

The actual procedure is roughly the same as for linear constraints in that moves are taken by changing $\mathbf{z}$ in the direction of the negative reduced gradient (with components of $\mathbf{z}$ on their boundary held fixed if the movement would violate the bound). The difference here is that although $\mathbf{z}$ moves along a straight line as before, the vector of dependent variables $\mathbf{y}$ must move nonlinearly to continuously satisfy the equality constraints. Computationally, this is accomplished by first moving linearly along the tangent to the surface defined by $\mathbf{z} \to \mathbf{z} + \Delta\mathbf{z}$, $\mathbf{y} \to \mathbf{y} + \Delta\mathbf{y}$ with $\Delta\mathbf{y} = -[\nabla_\mathrm{y}\mathbf{h}]^{-1}\nabla_\mathrm{z}\mathbf{h}\Delta\mathbf{z}$. Then a correction procedure, much like that employed in the gradient projection method, is used to return to the constraint surface and the magnitude bounds on the dependent variables are checked for feasibility. As with the gradient projection method, a feasibility tolerance must be introduced to acknowledge the impossibility of returning exactly to the constraint surface. An example corresponding to $n = 3$, $m = 1$, $a = 0$, $b = +\infty$ is shown in Fig. 12.8.

To return to the surface once a tentative move along the tangent is made, an iterative scheme is employed. If the point $\mathbf{x}_k$ was the point at the previous step, then from any point $\mathbf{x} = (\mathbf{v},\ \mathbf{w})$ near $\mathbf{x}_k$ one gets back to the constraint surface by solving the nonlinear equation

$$\mathbf{h}(\mathbf{y}, \mathbf{w}) = \mathbf{0} \tag{12.64}$$

for $\mathbf{y}$ (with $\mathbf{w}$ fixed). This is accomplished through the iterative process

$$\mathbf{y}_{j+1} = \mathbf{y}_j - [\nabla_y \mathbf{h}(\mathbf{x}_k)]^{-1} \mathbf{h}(\mathbf{y}_j,\ \mathbf{w}), \tag{12.65}$$

which, if started close enough to $\mathbf{x}_k$, will produce $\{\mathbf{y}_j\}$ with $\mathbf{y}_j \to \mathbf{y}$, solving (12.64).

The reduced gradient method suffers from the same basic difficulties as the gradient projection method, but as with the latter method, these difficulties can all be more or less successfully resolved. Computation is somewhat less complex in the case of the reduced gradient method, because rather than compute with $[\nabla \mathbf{h}(\mathbf{x}) \nabla \mathbf{h}(\mathbf{x})^T]^{-1}$ at each step, the matrix $[\nabla_y \mathbf{h}(\mathbf{y},\ \mathbf{z})]^{-1}$ is used.

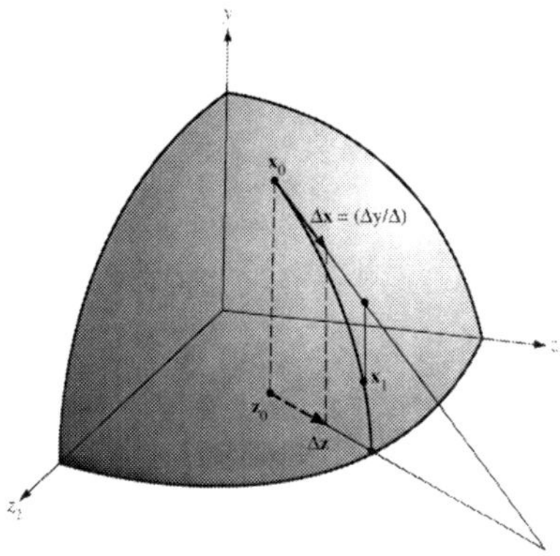

Fig. 12.8 Reduced gradient method

12.7 Convergence Rate of the Reduced Gradient Method

As argued before, for purposes of analyzing the rate of convergence, it is sufficient to consider the problem having only equality constraints

$$\begin{aligned} &\text{minimize} \ \ f(\mathbf{x}) \\ &\text{subject to} \ \ \mathbf{h}(\mathbf{x}) = \mathbf{0}. \end{aligned} \tag{12.66}$$

We then regard the problem as being defined over a surface $\boldsymbol{\Omega}$ of dimension $n - m$. At this point it is again timely to consider the view of our bug, who lives on this constraint surface. Invariably, he continues to regard the problem as extremely elementary, and indeed would have little appreciation for the complexity that seems to face us. To him the problem is an unconstrained problem in $n-m$ dimensions not, as we see it, a constrained problem in n dimensions. The bug will tenaciously hold to the method of steepest descent. We can emulate him provided that we know how he measures distance on his surface and thus how he calculates gradients and what he considers to be straight lines.

Rather than imagine that the measure of distance on his surface is the one that would be inherited from us in n dimensions, as we did when studying the gradient projection method, we, in this instance, follow the construction shown in Fig. 12.9. In our n-dimensional space, $n - m$ coordinates are selected as independent variables in such a way that, given their values, the values of the remaining (dependent) variables are determined by the surface. There is already a coordinate system in the space of independent variables, and it can be used on the surface by projecting it parallel to the space of the remaining dependent variables. Thus, an arc on the surface is considered to be straight if its projection onto the space of independent variables is a segment of a straight line. With this method for inducing a geometry on the surface, the bug's notion of steepest descent exactly coincides with an idealized version of the reduced gradient method.

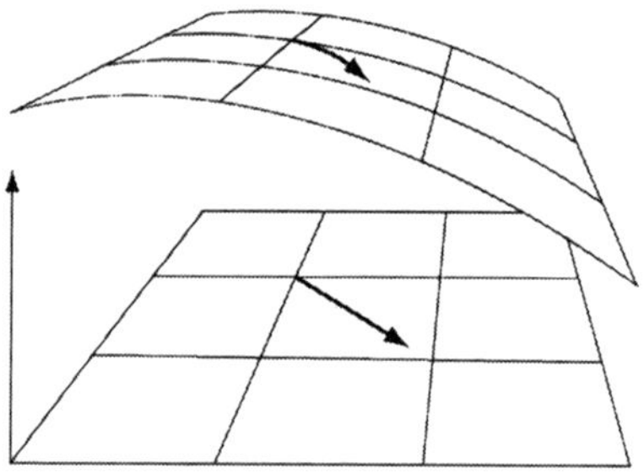

Fig. 12.9 Induced coordinate system

In the idealized version of the reduced gradient method for solving (12.66), the vector $\mathbf{x}$ is partitioned as $\mathbf{x} = (\mathbf{y},\ \mathbf{z})$ where $\mathbf{y} \in E^m$, $\mathbf{z} \in E^{n-m}$. It is assumed that the $m \times m$ matrix $\nabla_{\mathbf{y}}\mathbf{h}(\mathbf{y}, \mathbf{z})$ is nonsingular throughout a given region of interest. (With respect to the more general problem, this region is a small neighborhood around the solution where it is not necessary to change the partition.) The vector $\mathbf{y}$ is regarded as an implicit function of $\mathbf{z}$ through the equation

$$\mathbf{h}(\mathbf{y}(\mathbf{z}),\ \mathbf{z}) = \mathbf{0}. \tag{12.67}$$

The ordinary method of steepest descent is then applied to the function $q(\mathbf{z}) = f(\mathbf{y}(\mathbf{z}),\ \mathbf{z})$. We note that the gradient $\mathbf{r}^T$ of this function is given by (12.63).

Since the method is really just the ordinary method of steepest descent with respect to $\mathbf{z}$, the rate of convergence is determined by the eigenvalues of the Hessian of the function q at the solution. We therefore turn to the question of evaluating this Hessian.

Denote by $\mathbf{Y}(\mathbf{z})$ the first derivatives of the implicit function $\mathbf{y}(\mathbf{z})$, that is, $\mathbf{Y}(\mathbf{z}) \equiv \nabla_{\mathbf{z}}\mathbf{y}(\mathbf{z})$. Explicitly,

$$\mathbf{Y}(\mathbf{z}) = -[\nabla_{\mathbf{y}}\mathbf{h}(\mathbf{y}, \mathbf{z})]^{-1}\nabla_{\mathbf{z}}\mathbf{h}(\mathbf{y}, \mathbf{z}). \tag{12.68}$$

For any $\boldsymbol{\lambda} \in E^m$ we have

$$q(\mathbf{z}) = f(\mathbf{y}(\mathbf{z}), \mathbf{z}) = f(\mathbf{y}(\mathbf{z}), \mathbf{z}) + \boldsymbol{\lambda}^T\mathbf{h}(\mathbf{y}(\mathbf{z}), \mathbf{z}). \tag{12.69}$$

Thus

$$\nabla q(\mathbf{z}) = [\nabla_{\mathbf{y}} f(\mathbf{y}, \mathbf{z}) + \boldsymbol{\lambda}^T\nabla_{\mathbf{y}}\mathbf{h}(\mathbf{y}, \mathbf{z})]\mathbf{Y}(\mathbf{z}) + \nabla_{\mathbf{z}} f(\mathbf{y}, \mathbf{z}) + \boldsymbol{\lambda}^T\nabla_{\mathbf{z}}\mathbf{h}(\mathbf{y}, \mathbf{z}). \tag{12.70}$$

Now if at a given point $\mathbf{x}^* = (\mathbf{y}^*, \mathbf{z}^*) = (\mathbf{y}(\mathbf{z}^*), \mathbf{z}^*)$, we let $\boldsymbol{\lambda}$ satisfy

$$\nabla_{\mathbf{y}} f(\mathbf{y}^*, \mathbf{z}^*) + \boldsymbol{\lambda}^T\nabla_{\mathbf{y}}\mathbf{h}(\mathbf{y}^*, \mathbf{z}^*) = \mathbf{0}; \tag{12.71}$$

then introducing the Lagrangian $l(\mathbf{y}, \mathbf{z}, \boldsymbol{\lambda}) = f(\mathbf{y}, \mathbf{z}) + \boldsymbol{\lambda}^T\mathbf{h}(\mathbf{y}, \mathbf{z})$, we obtain by differentiating (12.70)

$$\begin{aligned}\nabla^2 q(\mathbf{z}^*) = \mathbf{Y}(\mathbf{z}^*)^T\nabla^2_{\mathbf{yy}} l(\mathbf{y}^*, \mathbf{z}^*)\mathbf{Y}(\mathbf{z}^*) + \nabla^2_{\mathbf{zy}} l(\mathbf{y}^*, \mathbf{z}^*)\mathbf{Y}(\mathbf{z}^*)\\ +\mathbf{Y}(\mathbf{z}^*)^T\nabla^2_{yz} l(\mathbf{y}^*, \mathbf{z}^*) + \nabla^2_{zz} l(\mathbf{y}^*, \mathbf{z}^*).\end{aligned} \tag{12.72}$$

Or defining the $n \times (n - m)$ matrix

$$\mathbf{C} = \left[\frac{\mathbf{Y}(\mathbf{z}^*)}{\mathbf{I}}\right], \tag{12.73}$$

where $\mathbf{I}$ is the $(n - m) \times (n - m)$ identity, we have

$$\mathbf{Q} \equiv \nabla^2 q(\mathbf{z}^*) = \mathbf{C}^T\mathbf{L}(\mathbf{x}^*)\mathbf{C}. \tag{12.74}$$

The matrix $\mathbf{L}(\mathbf{x}^*)$ is the $n \times n$ Hessian of the Lagrangian at $\mathbf{x}^*$, and $\nabla^2 q(\mathbf{z}^*)$ is an $(n - m) \times (n - m)$ matrix that is a restriction of $\mathbf{L}(\mathbf{x}^*)$ to the tangent subspace M, but it is not the usual restriction. We summarize our conclusion with the following theorem.

Theorem. *Let $\mathbf{x}^*$ be a local solution of problem (12.66). Suppose that the idealized reduced gradient method produces a sequence $\{\mathbf{x}_k\}$ converging to $\mathbf{x}^*$ and that the partition $\mathbf{x} = (\mathbf{y}, \mathbf{z})$ is used throughout the tail of the sequence. Let $\mathbf{L}$ be the Hessian of the Lagrangian at $\mathbf{x}^*$ and define the matrix $\mathbf{C}$ by (12.73) and (12.68). Then the sequence of objective values $\{f(\mathbf{x}_k)\}$ converges to $f(\mathbf{x}^*)$ linearly with a ratio no greater than $[(B - b)/(B + b)]^2$ where b and B are, respectively, the smallest and largest eigenvalues of the matrix $\mathbf{Q} = \mathbf{C}^T\mathbf{L}\mathbf{C}$.*

To compare the matrix $\mathbf{C}^T\mathbf{L}\mathbf{C}$ with the usual restriction of $\mathbf{L}$ to M that determines the convergence rate of most methods, we note that the $n \times (n - m)$ matrix $\mathbf{C}$ maps $\Delta\mathbf{z} \in E^{n-m}$ into $(\Delta\mathbf{y}, \Delta\mathbf{z}) \in E^n$ lying in the tangent subspace M; that is, $\nabla_{\mathbf{y}}\mathbf{h}\Delta\mathbf{y} + \nabla_{\mathbf{z}}\mathbf{h}\Delta\mathbf{z} = \mathbf{0}$. Thus the columns of $\mathbf{C}$ form a basis for the subspace M. Next note that the columns of the matrix

$$\mathbf{E} = \mathbf{C}(\mathbf{C}^T\mathbf{C})^{-1/2} \tag{12.75}$$

form an orthonormal basis for M, since each column of $\mathbf{E}$ is just a linear combination of columns of $\mathbf{C}$ and by direct calculation we see that $\mathbf{E}^T\mathbf{E} = \mathbf{I}$. Thus by the procedure described in Sect. 11.6 we see that a representation for the usual restriction of $\mathbf{L}$ to M is

$$\mathbf{L}_M = (\mathbf{C}^T\mathbf{C})^{-1/2}\mathbf{C}^T\mathbf{L}\mathbf{C}(\mathbf{C}^T\mathbf{C})^{-1/2}. \tag{12.76}$$

Comparing (12.76) with (12.74) we deduce that

$$\mathbf{Q} = (\mathbf{C}^T\mathbf{C})^{1/2}\mathbf{L}_M(\mathbf{C}^T\mathbf{C})^{1/2}. \tag{12.77}$$

This means that the Hessian matrix for the reduced gradient method is the restriction of $\mathbf{L}$ to M but pre- and post-multiplied by a positive definite symmetric matrix.

The eigenvalues of $\mathbf{Q}$ depend on the exact nature of $\mathbf{C}$ as well as $\mathbf{L}_M$. Thus, the rate of convergence of the reduced gradient method is not coordinate independent but depends strongly on just which variables are declared as independent at the final stage of the process. The convergence rate can be either faster or slower than that of the gradient projection method. In general, however, if $\mathbf{C}$ is well-behaved (that is, well-conditioned), the ratio of eigenvalues for the reduced gradient method can be expected to be the same order of magnitude as that of the gradient projection method. If, however, $\mathbf{C}$ should be ill-conditioned, as would arise in the case where the implicit equation $\mathbf{h}(\mathbf{y}, \mathbf{z}) = \mathbf{0}$ is itself ill-conditioned, then it can be shown that the eigenvalue ratio for the reduced gradient method will most likely be considerably worsened. This suggests that care should be taken to select a set of basic variables $\mathbf{y}$ that leads to a well-behaved $\mathbf{C}$ matrix.

Example (The Hanging Chain Problem). Consider again the hanging chain problem discussed in Sect. 11.4. This problem can be used to illustrate a wide assortment of theoretical principles and practical techniques. Indeed, a study of this example clearly reveals the predictive power that can be derived from an interplay of theory and physical intuition.

The problem is

$$\begin{aligned} \text{minimize} \quad & \sum_{i=1}^{n}(n-i+0.5)y_i \\ \text{subject to} \quad & \sum_{i=1}^{n} y_i = 0 \\ & \sum_{i=1}^{n} \sqrt{1-y_i^2} = 16, \end{aligned}$$

where in the original formulation $n = 20$.

This problem has been solved numerically by the reduced gradient method.* An initial feasible solution was the triangular shape shown in Fig. 12.10a with

* The exact solution is obviously symmetric about the center of the chain, and hence the problem could be reduced to having ten links and only one constraint. However, this symmetry disappears if the first constraint value is specified as nonzero. Therefore for generality we solve the full chain problem.

$$y_i = \begin{cases} -0.6, & 1 \leqslant i \leqslant 10 \\ 0.6, & 11 \leqslant i \leqslant 20. \end{cases}$$

Table 12.1 Results of original chain problem

Iteration	Value	Solution (1/2 of chain)
0	−60.00000	$y_1 = -0.8148260$
10	−66.47610	$y_2 = -0.7826505$
20	−66.52180	$y_3 = -0.7429208$
30	−66.53595	$y_4 = -0.6930959$
40	−66.54154	$y_5 = -0.6310976$
50	−66.54537	$y_6 = -0.5541078$
60	−66.54628	$y_7 = -0.4597160$
69	−66.54659	$y_8 = -0.3468334$
70	−66.54659	$y_9 = -0.2169879$
		$y_{10} = -0.07492541$

Lagrange multipliers −9.993817, −6.763148

The results obtained from a reduced gradient package are shown in Table 12.1. Note that convergence is obtained in approximately 70 iterations.

The Lagrange multipliers of the constraints are a by-product of the solution. These can be used to estimate the change in solution value if the constraint values are changed slightly. For example, suppose we wish to estimate, without resolving the problem, the change in potential energy (the objective function) that would result if the separation between the two supports were increased by, say, one inch. The change can be estimated by the formula $\Delta_u = -\lambda_2/12 = 0.0833 \times (6.76) = 0.563$. (When solved again numerically the change is found to be 0.568.)

Let us now pose some more challenging questions. Consider two variations of the original problem. In the first variation the chain is replaced by one having twice as many links, but each link is now half the size of the original links. The overall chain length is therefore the same as before. In the second variation the original chain is replaced by one having twice as many links, but each link is the same size as the original links. The chain length doubles in this case. If these problems are solved by the same method as the original problem, approximately how many iterations will be required—about the same number, many more, or substantially less?

These questions can be easily answered by using the theory of convergence rates developed in this chapter. The Hessian of the Lagrangian is

$$\mathbf{L} = \mathbf{F} + \lambda_1 \mathbf{H}_1 + \lambda_2 \mathbf{H}_2.$$

However, since the objective function and the first constraint are both linear, the only nonzero term in the above equation is $\lambda_2 \mathbf{H}_2$. Furthermore, since convergence rates depend only on eigenvalue ratios, the λ_2 can be ignored. Thus the eigenvalues of $\mathbf{H}_2$ determine the canonical convergence rate.

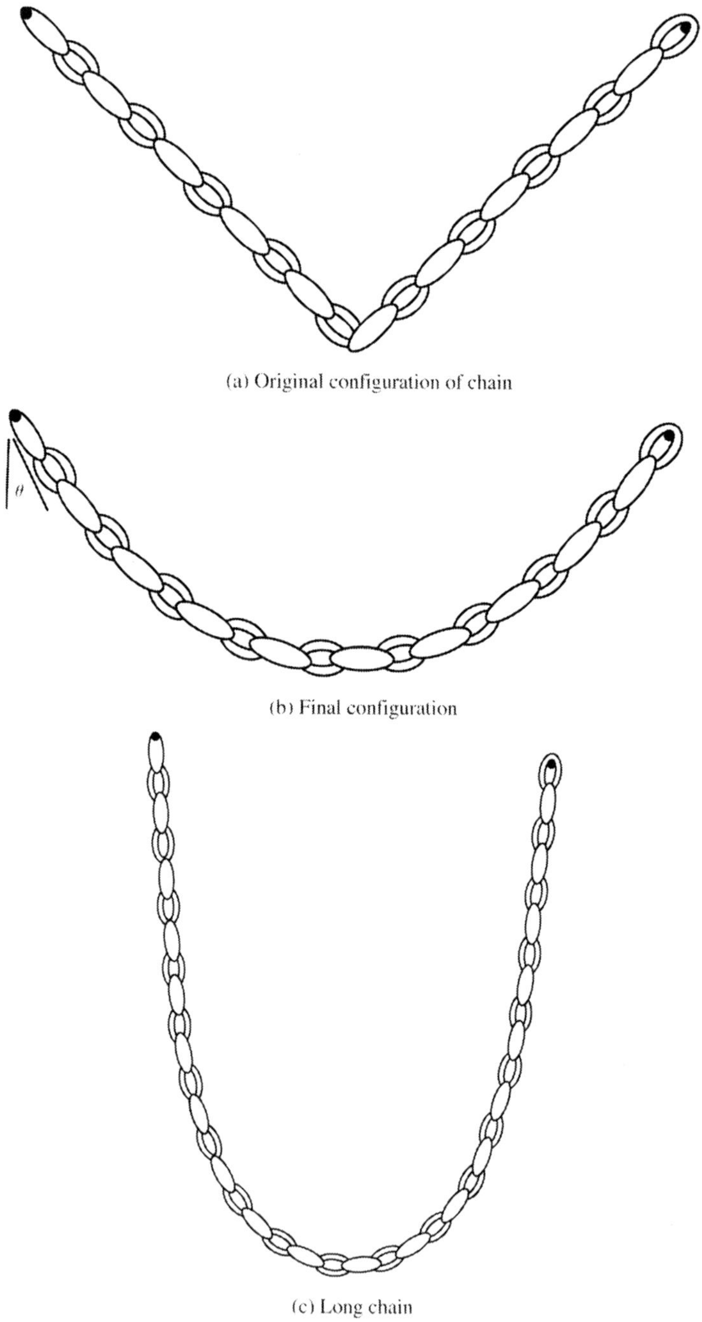

Fig. 12.10 The chain example. (**a**) Original configuration of chain. (**b**) Final configuration. (**c**) Long chain

It is easily seen that $\mathbf{H}_2$ is diagonal with ith diagonal term,

$$(\mathbf{H}_2)_{ii} = -(1 - y_i^2)^{-3/2},$$

and these values are the eigenvalues of $\mathbf{H}_2$. The canonical convergence rate is defined by the eigenvalues of $\mathbf{H}_{22}$ in the $(n-2)$-dimensional tangent subspace M. We cannot exactly determine these eigenvalues without a lot of work, but we can assume that they are close to the eigenvalues of $\mathbf{H}_{22}$. (Indeed, a version of the Interlocking Eigenvalues Lemma states that the $n-2$ eigenvalues are interlocked with the eigenvalues of $\mathbf{H}_{22}$.) Then the convergence rate of the gradient projection method will be governed by these eigenvalues. The reduced gradient method will most likely be somewhat slower.

The eigenvalue of smallest absolute value corresponds to the center links, where $y_i \simeq 0$. Conversely, the eigenvalue of largest absolute value corresponds to the first or last link, where y_i is largest in absolute value. Thus the relevant eigenvalue ratio is approximately

$$r = \frac{1}{(1 - y_1^2)^{3/2}} = \frac{1}{(\sin\theta)^{3/2}},$$

where θ is the angle shown in Fig. 12.10b.

For very little effort we have obtained a powerful understanding of the chain problem and its convergence properties. We can use this to answer the questions posed earlier. For the first variation, with twice as many links but each of half size, the angle θ will be about the same (perhaps a little smaller because of increased flexibility of the chain). Thus the number of iterations should be slightly larger because of the increase in θ and somewhat larger again because there are more variables (which tends to increase the condition number of $\mathbf{C}^T\mathbf{C}$). Note in Table 12.2 that about 122 iterations were required, which is consistent with this estimate.

For the second variation the chain will hang more vertically; hence y_1 will be larger, and therefore convergence will be fundamentally slower. To be more specific it is necessary to substitute a few numbers in our simple formula. For the original case we have $y_1 \simeq -.81$. This yields

$$r = (1 - .81^2)^{-3/2} = 4.9$$

and a convergence factor of

$$R = \left(\frac{r-1}{r+1}\right)^2 \simeq .44.$$

This is a modest value and quite consistent with the observed result of 70 iterations for a reduced gradient method. For the long chain we can estimate that $y_1 \simeq 98$. This yields

Table 12.2 Results of modified chain problems

Short links		Long chain	
Iteration	Value	Iteration	Value
0	−60.00000	0	−366.6061
10	−66.45499	10	−375.6423
20	−66.56377	20	−375.9123
40	−66.58443	50	−376.5128
60	−66.59191	100	−377.1625
80	−66.59514	200	−377.8983
100	−66.59656	500	−378.7989
120	−66.59825	1000	−379.3012
121	−66.59827	1500	−379.4994
122	−66.59827	2000	−379.5965
		2500	−379.6489
$y_1 = 0.4109519$		$y_1 = 0.9886223$	

$$r = (1 - .98^2)^{-3/2} \simeq 127$$

$$R = \left(\frac{r-1}{r+1}\right)^2 \simeq .969.$$

This last number represents extremely slow convergence. Indeed, since $(0.969)^{25} \simeq 0.44$, we expect that it may easily take 25 times as many iterations for the long chain problem to converge as the original problem (although quantitative estimates of this type are rough at best). This again is verified by the results shown in Table 12.2, where it is indicated that over 2,500 iterations were required by a version of the reduced gradient method.

*12.8 *Variations

It is possible to modify either the gradient projection method or the reduced gradient method so as to move in directions that are determined through additional considerations. For example, analogs of the conjugate gradient method, **PARTAN**, or any of the quasi-Newton methods can be applied to constrained problems by handling constraints through projection or reduction. The corresponding asymptotic rates of convergence for such methods are easily determined by applying the results for unconstrained problems on the $(n-m)$-dimensional surface of constraints, as illustrated in this chapter.

Although such generalizations can sometimes lead to substantial improvement in convergence rates, one must recognize that the detailed logic for a complicated generalization can become lengthy. If the method relies on the use of an approximate inverse Hessian restricted to the constraint surface, there must be an effective procedure for updating the approximation when the iterative process progresses from one set of active constraints to another. One would also like to insure that the poor eigenvalue structure sometimes associated with quasi-Newton methods does not dominate the short-term convergence characteristics of the extended method when the active constraint set changes. In other words, one would like to be able to achieve simultaneously both superlinear convergence and a guarantee of fast single step progress. There has been some work in this general area and it appears to be one of potential promise.

*Convex Simplex Method

A popular modification of the reduced gradient method, termed the *convex simplex method*, most closely parallels the highly effective simplex method for solving linear programs. The major difference between this method and the reduced gradient method is that instead of moving all (or several) of the independent variables in the direction of the negative reduced gradient, only one independent variable is changed at a time. The selection of the one independent variable to change is made much as in the ordinary simplex method.

At a given feasible point, let $\mathbf{x} = (\mathbf{y},\ \mathbf{z})$ be the partition of $\mathbf{x}$ into dependent and independent parts, and assume for simplicity that the bounds on $\mathbf{x}$ are $\mathbf{x} \geqslant \mathbf{0}$. Given the reduced gradient $\mathbf{r}^T$ at the current point, the component z_i to be changed is found from:

1. Let $r_{i1} = \min_i \{r_i\}$.
2. Let $r_{i2}z_{i2} = \max_i \{r_i z_i\}$

   If $r_{i1} = r_{i2}z_{i2} = 0$, terminate. Otherwise
   If $r_{i1} \leqslant -|r_{i2}z_{i2}|$, increase z_{i1}
   If $r_{i1} \geqslant -|r_{i2}z_{i2}|$, decrease z_{i2}.

The rule in Step 2 amounts to selecting the variable that yields the best potential decrease in the cost function. The rule accounts for the non-negativity constraint on the independent variables by weighting the cost coefficients of those variables that are candidates to be decreased by their distance from zero. This feature ensures global convergence of the method.

The remaining details of the method are identical to those of the reduced gradient method. Once a particular component of $\mathbf{z}$ is selected for change, according to the above criterion, the corresponding $\mathbf{y}$ vector is computed as a function of the change in that component so as to continuously satisfy the constraints. The component of $\mathbf{z}$ is continuously changed until either a local minimum with respect to that component is attained or the boundary of one nonnegativity constraint is reached.

Just as in the discussion of the reduced gradient method, it is convenient, for purposes of convergence analysis, to view the problem as unconstrained with respect to the independent variables. The convex simplex method is then seen to be a coordinate descent procedure in the space of these $n - m$ variables. Indeed, since the component selected is based on the magnitude of the components of the reduced gradient, the method is merely an adaptation of the Gauss-Southwell scheme discussed in Sect. 8.6 to the constrained situation. Hence, although it is difficult to pin down precisely, we expect that it would take approximately $n - m$ steps of this coordinate descent method to make the progress of a single reduced gradient step. To be competitive with the reduced gradient method; therefore, the difficulties associated with a single step—line searching and constraint evaluation—must be approximately $n - m$ times simpler when only a single component is varied than when all $n - m$ are varied simultaneously. This is indeed the case for linear programs and for some quadratic programs but not for nonlinear problems that require the full line search machinery. Hence, in general, the convex simplex method may not be a bargain.

12.9 Summary

The concept of feasible direction methods is a straightforward and logical extension of the methods used for unconstrained problems but leads to some subtle difficulties. These methods are susceptible to *jamming* (lack of global convergence) because many simple direction finding mappings and the usual line search mapping are not closed.

Problems with inequality constraints can be approached with an active set strategy. In this approach certain constraints are treated as active and the others are treated as inactive. By systematically adding and dropping constraints from the working set, the correct set of active constraints is determined during the search process. In general, however, an active set method may require that several constrained problems be solved exactly.

The most practical primal methods are the gradient projection methods and the reduced gradient method. Both of these basic methods can be regarded as the method of steepest descent applied on the surface defined by the active constraints. The rate of convergence for the two methods can be expected to be approximately equal and is determined by the eigenvalues of the Hessian of the Lagrangian restricted to the subspace tangent to the active constraints. Of the two methods, the reduced gradient method seems to be best. It can be easily modified to ensure against jamming and it requires fewer computations per iterative step and therefore, for most problems, will probably converge in less time than the gradient projection method.

12.10 Exercises

1. Show that the Frank-Wolfe method is globally convergent if the intersection of the feasible region and the objective level set $\{\mathbf{x} : f(\mathbf{x}) \leq f(\mathbf{x}^0)\}$ is bounded.
2. Sometimes a different normalizing term is used in (12.4). Show that the problem of finding $\mathbf{d} = (d_1, d_2, \ldots, d_n)$ to

$$\begin{aligned}&\text{minimize} \quad \mathbf{c}^T\mathbf{d}\\&\text{subject to} \quad \mathbf{Ad} \leqslant \mathbf{0}, \ (\sum_i |d_i|^p)^{1/p} = 1\end{aligned}$$

for $p = 1$ or $p = \infty$ can be converted to a linear program.

3. Perhaps the most natural normalizing term to use in (12.4) is one based on the Euclidean norm. This leads to the problem of finding $\mathbf{d} = (d_1, d_2, \ldots, d_n)$ to

$$\begin{aligned}&\text{minimize} \quad \mathbf{c}^T\mathbf{d}\\&\text{subject to} \quad \mathbf{Ad} \leqslant \mathbf{0}, \ \sum_{i=1}^{n} d_i^2 = 1.\end{aligned}$$

Find the Karush-Kuhn–Tucker necessary conditions for this problem and show how they can be solved by a modification of the simplex procedure.

4. Let $\Omega \subset E^n$ be a given feasible region. A set $\Gamma \subset E^{2n}$ consisting of pairs $(\mathbf{x}, \mathbf{d})$, with $\mathbf{x} \in \Omega$ and $\mathbf{d}$ a feasible direction at $\mathbf{x}$, is said to be a set of *uniformly feasible direction vectors* if there is a $\delta > 0$ such that $(\mathbf{x}, \mathbf{d}) \in \mathbf{\Gamma}$ implies that $\mathbf{x} + \alpha\mathbf{d}$ is feasible for all α, $0 \leqslant \alpha \leqslant \delta$. The number δ is referred to as the feasibility constant of the set $\mathbf{\Gamma}$.

   Let $\Gamma \subset E^{2n}$ be a set of uniformly feasible direction vectors for Ω, with feasibility constant δ. Define the mapping

$$\begin{aligned}\mathbf{M}_\delta(\mathbf{x}, \mathbf{d}) = \{\mathbf{y} : f(\mathbf{y}) \leqslant f(\mathbf{x} + \tau\mathbf{d}) \text{ for all } \tau, \ 0 \leqslant \tau \leqslant \delta; \mathbf{y} = \mathbf{x} + \alpha\mathbf{d},\\ \text{for some } \alpha, \ 0 \leqslant \alpha \leqslant \infty, \ \mathbf{y} \in \Omega\}.\end{aligned}$$

   Show that if $\mathbf{d} \neq \mathbf{0}$, the map $\mathbf{M}_\delta$ is closed at $(\mathbf{x}, \mathbf{d})$.

5. Let $\Gamma \subset E^{2n}$ be a set of uniformly feasible direction vectors for Ω with feasibility constant δ. For $\varepsilon > 0$ define the map ${}^{\varepsilon}\mathbf{M}_\delta$ or Γ by

$$\begin{aligned}{}^{\varepsilon}\mathbf{M}_\delta(\mathbf{x}, \mathbf{d}) = \{\mathbf{y} : f(\mathbf{y}) \leqslant f(\mathbf{x} + \tau\mathbf{d}) + \varepsilon \text{ for all } \tau, 0 \leqslant \tau \leqslant \delta; \mathbf{y} = \mathbf{x} + \alpha\mathbf{d},\\ \text{for some } \alpha, \ 0 \leqslant \alpha \leqslant \infty, \ \mathbf{y} \in \Omega\}.\end{aligned}$$

   The map ${}^{\varepsilon}\mathbf{M}_\delta$ corresponds to an "inaccurate" constrained line search. Show that this map is closed if $\mathbf{d} \neq \mathbf{0}$.

6. For the problem

$$\begin{aligned}&\text{minimize} \quad f(\mathbf{x})\\&\text{subject to} \quad \mathbf{a}_i^T\mathbf{x} \leqslant b_i, \ i = 1, 2, \ldots, m\end{aligned}$$

   consider selecting $\mathbf{d} = (d_1, d_2, \ldots, d_n)$ at a feasible point $\mathbf{x}$ by solving the problem

$$\begin{aligned}&\text{minimize} \quad \nabla f(\mathbf{x})\mathbf{d}\\&\text{subject to} \quad \mathbf{a}_i^T\mathbf{d} \leqslant (b_i - \mathbf{a}_i^T\mathbf{x})M, \ i = 1, 2, \ldots, m\\&\qquad\qquad\quad \sum_{i=1}^{n} |d_i| = 1,\end{aligned}$$

   where M is some given positive constant. For large M the ith inequality of this subsidiary problem will be active only if the corresponding inequality in the original problem is nearly active at $\mathbf{x}$ (indeed, note that $M \to \infty$ corresponds to Zoutendijk's method). Show that this direction finding mapping is closed and generates uniformly feasible directions with feasibility constant $1/M$.

7. Generalize the method of Exercise 6 so that it is applicable to nonlinear inequalities.
8. An alternate, but equivalent, definition of the projected gradient $\mathbf{p}$ is that it is the vector solving

$$\begin{aligned} &\text{minimize} \ |\mathbf{g} - \mathbf{p}|^2 \\ &\text{subject to} \ \mathbf{A}_q \mathbf{p} = \mathbf{0}. \end{aligned}$$

Using the Karush-Kuhn–Tucker necessary conditions, solve this problem and thereby derive the formula for the projected gradient.

9. Show that finding the $\mathbf{d}$ that solves

$$\begin{aligned} &\text{minimize} \ \mathbf{g}^T \mathbf{d} \\ &\text{subject to} \ \mathbf{A}_q \mathbf{d} = \mathbf{0}, \ |\mathbf{d}|^2 = 1 \end{aligned}$$

gives a vector $\mathbf{d}$ that has the same direction as the negative projected gradient.

10. Let $\mathbf{P}$ be a projection matrix. Show that $\mathbf{P}^T = \mathbf{P}$, $\mathbf{P}^2 = \mathbf{P}$.
11. Suppose $\mathbf{A}_q = [\mathbf{a}^T, \mathbf{A}_{\bar{q}}]$ so that $\mathbf{A}_q$ is the matrix $\mathbf{A}_{\bar{q}}$ with the row $\mathbf{a}^T$ adjoined. Show that $(\mathbf{A}_q \mathbf{A}_q^T)^{-1}$ can be found from $(\mathbf{A}_{\bar{q}} \mathbf{A}_{\bar{q}}^T)^{-1}$ from the formula

$$(\mathbf{A}_q \mathbf{A}_q^T)^{-1} = \begin{bmatrix} \varepsilon & -\varepsilon \mathbf{a}^T \mathbf{A}_{\bar{q}}^T (\mathbf{A}_{\bar{q}} \mathbf{A}_{\bar{q}}^T)^{-1} \\ -\varepsilon (\mathbf{A}_{\bar{q}} \mathbf{A}_{\bar{q}}^T)^{-1} \mathbf{A}_{\bar{q}} \mathbf{a} & (\mathbf{A}_{\bar{q}} \mathbf{A}_{\bar{q}}^T)^{-1} [\mathbf{I} + \mathbf{A}_{\bar{q}} \mathbf{a} \mathbf{a}^T \mathbf{A}_{\bar{q}}^T (\mathbf{A}_{\bar{q}} \mathbf{A}_{\bar{q}}^T)^{-1}] \end{bmatrix},$$

where

$$\varepsilon = \frac{1}{\mathbf{a}^T \mathbf{a} - \mathbf{a}^T \mathbf{A}_{\bar{q}}^T (\mathbf{A}_{\bar{q}} \mathbf{A}_{\bar{q}}^T)^{-1} \mathbf{A}_{\bar{q}} \mathbf{a}}.$$

Develop a similar formula for $(\mathbf{A}_{\bar{q}} \mathbf{A}_{\bar{q}})^{-1}$ in terms of $(\mathbf{A}_q \mathbf{A}_q)^{-1}$.

12. Show that the gradient projection method will solve a linear program in a finite number of steps.
13. Suppose that the projected negative gradient $\mathbf{d}$ is calculated satisfying

$$-\mathbf{g} = \mathbf{d} + \mathbf{A}_q^T \boldsymbol{\lambda}$$

and that some component λ_i of $\boldsymbol{\lambda}$, corresponding to an inequality, is negative. Show that if the ith inequality is dropped, the projection $\mathbf{d}_i$ of the negative gradient onto the remaining constraints is a feasible direction of descent.

14. Using the result of Exercise 13, it is possible to avoid the discontinuity at $\mathbf{d} = \mathbf{0}$ in the direction finding mapping of the simple gradient projection method. At a given point let $\gamma = -\min\{0, \lambda_i\}$, with the minimum taken with respect to the indices i corresponding the active inequalities. The direction to be taken at this point is $\mathbf{d} = -\mathbf{P}\mathbf{g}$ if $|\mathbf{P}\mathbf{g}| \geqslant \gamma$, or $\bar{\mathbf{d}}$, defined by dropping the inequality i for which $\lambda_i = -\gamma$, if $|\mathbf{P}\mathbf{g}| \leqslant \gamma$. (In case of equality either direction is selected.) Show that this direction finding map is closed over a region where the set of active inequalities does not change.

15. Consider the problem of maximizing entropy discussed in Example 3, Sect. 14.4. Suppose this problem were solved numerically with two constraints by the gradient projection method. Derive an estimate for the rate of convergence in terms of the optimal p_i's.
16. Find the geodesics of

    (a) a two-dimensional plane
    (b) a sphere.

17. Suppose that the problem

$$\begin{array}{ll}\text{minimize} & f(\mathbf{x})\\ \text{subject to} & \mathbf{h}(\mathbf{x}) = \mathbf{0}\end{array}$$

is such that every point is a regular point. And suppose that the sequence of points $\{\mathbf{x}_k\}_{k=0}^{\infty}$ generated by geodesic descent is bounded. Prove that every limit point of the sequence satisfies the first-order necessary conditions for a constrained minimum.
18. Show that, for linear constraints, if at some point in the reduced gradient method $\mathbf{\Delta z}$ is zero, that point satisfies the Karush-Kuhn–Tucker first-order necessary conditions for a constrained minimum.
19. Consider the problem

$$\begin{array}{ll}\text{minimize} & f(\mathbf{x})\\ \text{subject to} & \mathbf{A}\mathbf{x} = \mathbf{b},\ \mathbf{x} \geqslant \mathbf{0},\end{array}$$

where $\mathbf{A}$ is $m \times n$. Assume $f \in C^1$, that the feasible set is bounded, and that the nondegeneracy assumption holds. Suppose a "modified" reduced gradient algorithm is defined following the procedure in Sect. 12.6 but with two modifications: (1) the basic variables are, at the beginning of an iteration, always taken as the m largest variables (ties are broken arbitrarily); (2) the formula for $\mathbf{\Delta z}$ is replaced by

$$\Delta z_i = \begin{cases} -r_i & \text{if } r_i \leqslant 0\\ -x_i r_i & \text{if } r_i > 0\end{cases}$$

Establish the global convergence of this algorithm.
20. Find the exact solution to the example presented in Sect. 12.4.
21. Find the direction of movement that would be taken by the gradient projection method if in the example of Sect. 12.4 the constraint $x_4 = 0$ were relaxed. Show that if the term $-3x_4$ in the objective function were replaced by $-x_4$, then both the gradient projection method and the reduced gradient method would move in identical directions.
22. Show that in terms of convergence characteristics, the reduced gradient method behaves like the gradient projection method applied to a scaled version of the problem.

23. Let r be the condition number of $\mathbf{L}_M$ and s the condition number of $\mathbf{C}^T\mathbf{C}$. Show that the rate of convergence of the reduced gradient method is no worse than $[(sr-1)/(sr+1)]^2$.
24. Formulate the symmetric version of the hanging chain problem using a single constraint. Find an explicit expression for the condition number of the corresponding $\mathbf{C}^T\mathbf{C}$ matrix (assuming y_1 is basic). Use Exercise 23 to obtain an estimate of the convergence rate of the reduced gradient method applied to this problem, and compare it with the rate obtained in Table 12.1, Sect. 12.7. Repeat for the two-constraint formulation (assuming y_1 and y_n are basic).
25. Referring to Exercise 19 establish a global convergence result for the convex simplex method.

References

12.2 Feasible direction methods of various types were originally suggested and developed by Zoutendijk [Z4]. The systematic study of the global convergence properties of feasible direction methods was begun by Topkis and Veinott [T8] and by Zangwill [Z2]. The Frank-Wolfe method was initially proposed in [98].

12.3–12.4 The gradient projection method was proposed and developed (more completely than discussed here) by Rosen [R5, R6], who also introduced the notion of an active set strategy. See Gill, Murray, and Wright [G7] for a discussion of working sets and active set strategies.

12.5 This material is taken from Luenberger [L14].

12.6–12.7 The reduced gradient method was originally proposed by Wolfe [W5] for problems with linear constraints and generalized to nonlinear constraints by Abadie and Carpentier [A1]. Wolfe [W4] presents an example of jamming in the reduced gradient method. The convergence analysis given in this section is new.

12.8 The convex simplex method, for problems with linear constraints, together with a proof of its global convergence is due to Zangwill [Z2].

Chapter 13
Penalty and Barrier Methods

Penalty and barrier methods are procedures for approximating constrained optimization problems by unconstrained problems. The approximation is accomplished in the case of penalty methods by adding to the objective function a term that prescribes a high cost for violation of the constraints, and in the case of barrier methods by adding a term that favors points interior to the feasible region over those near the boundary. Associated with these methods is a parameter c or μ that determines the severity of the penalty or barrier and consequently the degree to which the unconstrained problem approximates the original constrained problem. For a problem with n variables and m constraints, penalty and barrier methods work directly in the n-dimensional space of variables, as compared to primal methods that work in $(n - m)$-dimensional space.

There are two fundamental issues associated with the methods of this chapter. The first has to do with how well the unconstrained problem approximates the constrained one. This is essential in examining whether, as the parameter c is increased toward infinity, the solution of the unconstrained problem converges to a solution of the constrained problem. The other issue, most important from a practical viewpoint, is the question of how to solve a given unconstrained problem when its objective function contains a penalty or barrier term. It turns out that as c is increased to yield a good approximating problem, the corresponding structure of the resulting unconstrained problem becomes increasingly unfavorable thereby slowing the convergence rate of many algorithms that might be applied. (Exact penalty functions also have a very unfavorable structure.) It is necessary, then, to devise acceleration procedures that circumvent this slow convergence phenomenon.

Penalty and barrier methods are of great interest to both the practitioner and the theorist. To the practitioner they offer a simple straightforward method for handling constrained problems that can be implemented without sophisticated computer programming and that possess much the same degree of generality as primal methods. The theorist, striving to make this approach practical by overcoming its inherently slow convergence, finds it appropriate to bring into play nearly all aspects

D.G. Luenberger, Y. Ye, *Linear and Nonlinear Programming*, International Series in Operations Research & Management Science 228,
DOI 10.1007/978-3-319-18842-3_13

of optimization theory; including Lagrange multipliers, necessary conditions, and many of the algorithms discussed earlier in this book. The canonical rate of convergence associated with the original constrained problem again asserts its fundamental role by essentially determining the natural *accelerated* rate of convergence for unconstrained penalty or barrier problems.

13.1 Penalty Methods

Consider the problem

$$\begin{array}{ll} \text{minimize} & f(\mathbf{x}) \\ \text{subject to} & \mathbf{x} \in S, \end{array} \tag{13.1}$$

where f is a continuous function on E^n and S is a constraint set in E^n. In most applications S is defined implicitly by a number of functional constraints, but in this section the more general description in (13.1) can be handled. The idea of a penalty function method is to replace problem (13.1) by an unconstrained problem of the form

$$\text{minimize} \quad f(\mathbf{x}) + cP(\mathbf{x}), \tag{13.2}$$

where c is a positive constant and P is a function on E^n satisfying: (i) P is continuous, (ii) $P(\mathbf{x}) \geqslant 0$ for all $\mathbf{x} \in E^n$, and (iii) $P(\mathbf{x}) = 0$ if and only if $\mathbf{x} \in S$.

Example 1. Suppose S is defined by a number of inequality constraints:

$$S = \{\mathbf{x} : g_i(\mathbf{x}) \leqslant 0, \quad i = 1, 2, \ldots, p\}.$$

A very useful penalty function in this case is

$$P(\mathbf{x}) = \frac{1}{2}\sum_{i=1}^{P}(\max[0, g_i(\mathbf{x})])^2.$$

The function $cP(\mathbf{x})$ is illustrated in Fig. 13.1 for the one-dimensional case with $g_1(x) = x - b$, $g_2(x) = a - x$.

For large c it is clear that the minimum point of problem (13.2) will be in a region where P is small. Thus, for increasing c it is expected that the corresponding solution points will approach the feasible region S and, subject to being close, will minimize f. Ideally then, as $c \to \infty$ the solution point of the penalty problem will converge to a solution of the constrained problem.

The Method

The procedure for solving problem (13.1) by the penalty function method is this: Let $\{c_k\}$, $k = 1, 2, \ldots$, be a sequence tending to infinity such that for each k, $c_k \geqslant 0$, $c_{k+1} > c_k$. Define the function

$$q(c, \mathbf{x}) = f(\mathbf{x}) + cP(\mathbf{x}). \tag{13.3}$$

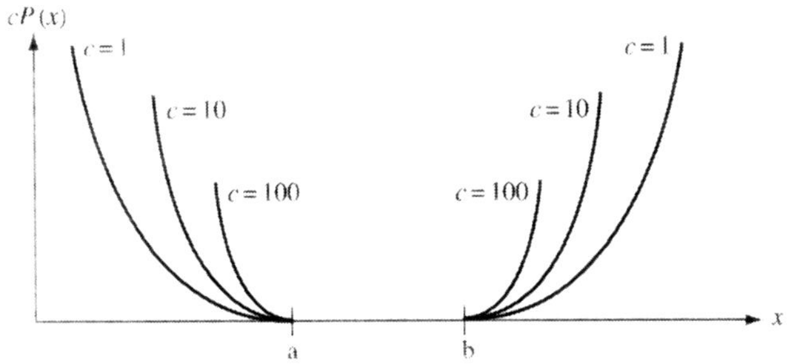

Fig. 13.1 Plot of $cP(x)$

For each k solve the problem

$$\text{minimize} \quad q(c_k, \mathbf{x}), \tag{13.4}$$

obtaining a solution point $\mathbf{x}_k$.

We assume here that, for each k, problem (13.4) has a solution. This will be true, for example, if $q(c, \mathbf{x})$ increases unboundedly as $|\mathbf{x}| \to \infty$. (Also see Exercise 2 to see that it is not necessary to obtain the minimum precisely.)

Convergence

The following lemma gives a set of inequalities that follow directly from the definition of $\mathbf{x}_k$ and the inequality $c_{k+1} > c_k$.

Lemma 1.

$$q(c_k, \mathbf{x}_k) \leqslant q(c_{k+1}, \mathbf{x}_{k+1}) \tag{13.5}$$

$$P(\mathbf{x}_k) \geqslant P(\mathbf{x}_{k+1}) \tag{13.6}$$

$$f(\mathbf{x}_k) \leqslant f(\mathbf{x}_{k+1}). \tag{13.7}$$

Proof.

$$\begin{aligned} q(c_{k+1}, \mathbf{x}_{k+1}) &= f(\mathbf{x}_{k+1}) + c_{k+1}P(\mathbf{x}_{k+1}) \geqslant f(\mathbf{x}_{k+1}) + c_kP(\mathbf{x}_{k+1}) \\ &\geqslant f(\mathbf{x}_k) + c_kP(\mathbf{x}_k) = q(c_k, \mathbf{x}_k), \end{aligned}$$

which proves (13.5).

We also have

$$f(\mathbf{x}_k) + c_kP(\mathbf{x}_k) \leqslant f(\mathbf{x}_{k+1}) + c_kP(\mathbf{x}_{k+1}) \tag{13.8}$$

$$f(\mathbf{x}_{k+1}) + c_{k+1}P(\mathbf{x}_{k+1}) \leqslant f(\mathbf{x}_k) + c_{k+1}P(\mathbf{x}_k). \tag{13.9}$$

Adding (13.8) and (13.9) yields

$$(c_{k+1} - c_k)P(\mathbf{x}_{k+1}) \leqslant (c_{k+1} - c_k)P(\mathbf{x}_k),$$

which proves (13.6).

Also

$$f(\mathbf{x}_{k+1}) + c_k P(\mathbf{x}_{k+1}) \geqslant f(\mathbf{x}_k) + c_k P(\mathbf{x}_k),$$

and hence using (13.6) we obtain (13.7). ∎

Lemma 2. *Let* $\mathbf{x}^*$ *be a solution to problem (13.1). Then for each k*

$$f(\mathbf{x}^*) \geqslant q(c_k, \mathbf{x}_k) \geqslant f(\mathbf{x}_k).$$

Proof.

$$f(\mathbf{x}^*) = f(\mathbf{x}^*) + c_k P(\mathbf{x}^*) \geqslant f(\mathbf{x}_k) + c_k P(\mathbf{x}_k) \geqslant f(\mathbf{x}_k). \blacksquare$$

Global convergence of the penalty method, or more precisely verification that any limit point of the sequence is a solution, follows easily from the two lemmas above.

Theorem. *Let* $\{x_k\}$ *be a sequence generated by the penalty method. Then, any limit point of the sequence is a solution to (13.1).*

Proof. Suppose the subsequence $\{\mathbf{x}_k\}$, $k \in \mathcal{K}$ is a convergent subsequence of $\{\mathbf{x}_k\}$ having limit $\overline{\mathbf{x}}$. Then by the continuity of f, we have

$$\operatorname*{limit}_{k\in\mathcal{K}} f(\mathbf{x}_k) = f(\overline{\mathbf{x}}). \tag{13.10}$$

Let f^* be the optimal value associated with problem (13.1). Then according to Lemmas 1 and 2, the sequence of values $q(c_k, \mathbf{x}_k)$ is nondecreasing and bounded above by f^*. Thus

$$\operatorname*{limit}_{k\in\mathcal{K}} q(c_k, \mathbf{x}_k) = q^* \leqslant f^*. \tag{13.11}$$

Subtracting (13.10) from (13.11) yields

$$\operatorname*{limit}_{k\in\mathcal{K}} c_k P(\mathbf{x}_k) = q^* - f(\overline{\mathbf{x}}). \tag{13.12}$$

Since $P(\mathbf{x}_k) \geqslant 0$ and $c_k \to \infty$, (13.12) implies

$$\operatorname*{limit}_{k\in\mathcal{K}} P(\mathbf{x}_k) = 0.$$

Using the continuity of P, this implies $P(\overline{\mathbf{x}}) = 0$. We therefore have shown that the limit point $\overline{\mathbf{x}}$ is feasible for (13.1).

To show that $\overline{\mathbf{x}}$ is optimal we note that from Lemma 2, $f(\mathbf{x}_k) \leqslant f^*$ and hence

$$f(\overline{\mathbf{x}}) = \text{limit}_{k\in\mathcal{K}} f(\mathbf{x}_k) \leqslant f^*. \blacksquare$$

13.2 Barrier Methods

Barrier methods are applicable to problems of the form

$$\begin{aligned} &\text{minimize} \ \ f(\mathbf{x}) \\ &\text{subject to} \ \ \mathbf{x} \in S, \end{aligned} \tag{13.13}$$

where the constraint set S has a nonempty interior that is arbitrarily close to any point of S. Intuitively, what this means is that the set has an interior and it is possible to get to any boundary point by approaching it from the interior. We shall refer to such a set as *robust*. Some examples of robust and nonrobust sets are shown in Fig. 13.2. This kind of set often arises in conjunction with inequality constraints, where S takes the form

$$S = \{\mathbf{x} : g_i(\mathbf{x}) \leqslant 0, \ i = 1, 2, \ \dots, \ p\}$$

Barrier methods are also termed *interior methods*. They work by establishing a barrier on the boundary of the feasible region that prevents a search procedure from leaving the region. A *barrier function* is a function B defined on the interior of S such that: (i) B is continuous, (ii) $B(\mathbf{x}) \geqslant 0$, (iii) $B(\mathbf{x}) \to \infty$ as $\mathbf{x}$ approaches the boundary of S.

Example 1. Let $g_i, \ i = 1, 2, \ \dots, \ p$ be continuous functions on E^n. Suppose

$$S = \{\mathbf{x} : g_i(\mathbf{x}) \leqslant 0, \ i = 1, 2, \ \dots, \ p\}.$$

is robust, and suppose the interior of S is the set of $\mathbf{x}$'s where $g_i(\mathbf{x}) < 0, \ i = 1, 2, \ \dots, \ p$. Then the function

$$B(\mathbf{x}) = -\sum_{i=1}^{p} \frac{1}{g_i(\mathbf{x})},$$

defined on the interior of S, is a barrier function. It is illustrated in one dimension for $g_1 = x - a, \ g_2 = x - b$ in Fig. 13.3.

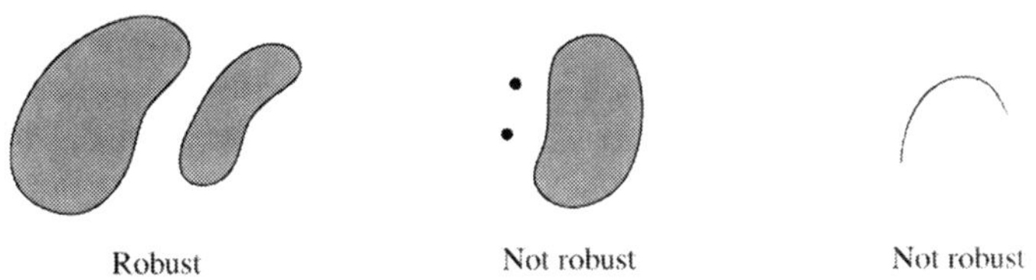

Fig. 13.2 Examples

Example 2. For the same situation as Example 1, we may use the logarithmic utility function

$$B(\mathbf{x}) = -\sum_{i=1}^{p} \log[-g_i(\mathbf{x})].$$

This is the barrier function commonly used in linear programming interior point methods, and it is frequently used more generally as well.

Corresponding to the problem (13.13), consider the approximate problem

$$\begin{aligned} &\text{minimize} \quad f(\mathbf{x}) + \tfrac{1}{c}B(\mathbf{x}) \\ &\text{subject to} \quad \mathbf{x} \in \text{interior of } S, \end{aligned} \tag{13.14}$$

where c is a positive constant.

Alternatively, it is common to formulate the barrier method as

$$\begin{aligned} &\text{minimize} \quad f(\mathbf{x}) + \mu B(\mathbf{x}) \\ &\text{subject to} \quad \mathbf{x} \in \text{interior of } S. \end{aligned} \tag{13.15}$$

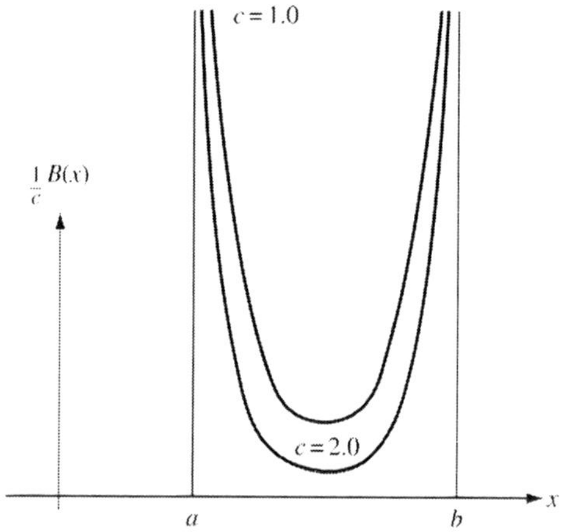

Fig. 13.3 Barrier function

When formulated with c we take c large (going to infinity); while when formulated with μ we take μ small (going to zero). Either way the result is a constrained problem, and indeed the constraint is somewhat more complicated than in the original problem (13.13). The advantage of this problem, however, is that it can be solved by using an unconstrained search technique. To find the solution one starts at an initial interior point and then searches from that point using steepest descent or some other iterative descent method applicable to unconstrained problems. Since the value of the objective function approaches infinity near the boundary of S, the search technique (if carefully implemented) will automatically remain within the interior of S, and the constraint need not be accounted for explicitly. Thus, although problem (13.14) or (13.15) is from a formal viewpoint a constrained problem, from a computational viewpoint it is unconstrained.

The Method

The barrier method is quite analogous to the penalty method. Let $\{c_k\}$ be a sequence tending to infinity such that for each k, $k = 1, 2, \ldots$, $c_k \geqslant 0$, $c_{k+1} > c_k$. Define the function

$$r(c, \mathbf{x}) = f(\mathbf{x}) + \frac{1}{c} B(\mathbf{x}).$$

For each k solve the problem

$$\begin{aligned} &\text{minimize} \quad r(c_k, \mathbf{x}) \\ &\text{subject to} \quad \mathbf{x} \in \text{interior of} S, \end{aligned}$$

obtaining the point $\mathbf{x}_k$.

Convergence

Virtually the same convergence properties hold for the barrier method as for the penalty method. We leave to the reader the proof of the following result.

Theorem. *Any limit point of a sequence $\{x_k\}$ generated by the barrier method is a solution to problem (13.13).*

13.3 Properties of Penalty and Barrier Functions

Penalty and barrier methods are applicable to nonlinear programming problems having a very general form of constraint set S. In most situations, however, this set is not given explicitly but is defined implicitly by a number of functional constraints. In these situations, the penalty or barrier function is invariably defined in terms of the constraint functions themselves; and although there are an unlimited number of ways in which this can be done, some important general implications follow from this kind of construction.

For economy of notation we consider problems of the form

$$\begin{aligned} &\text{minimize} \quad f(\mathbf{x}) \\ &\text{subject to} \quad g_i(\mathbf{x}) \leqslant 0, \quad i = 1, 2, \ldots, p. \end{aligned} \tag{13.16}$$

For our present purposes, equality constraints are suppressed, at least notationally, by writing each of them as two inequalities. If the problem is to be attacked with a barrier method, then, of course, equality constraints are not present even in an unsuppressed version.

Penalty Functions

A penalty function for a problem expressed in the form (13.16) will most naturally be expressed in terms of the auxiliary constraint functions

$$g_i^+(\mathbf{x}) \equiv \max[0,\ g_i(\mathbf{x})],\ i = 1, 2, \ldots, p. \tag{13.17}$$

This is because in the interior of the constraint region $P(\mathbf{x}) \equiv 0$ and hence P should be a function only of violated constraints. Denoting by $\mathbf{g}^+(\mathbf{x})$ the p-dimensional vector made up of the $g_i^+(\mathbf{x})$'s, we consider the general class of penalty functions

$$P(\mathbf{x}) = \gamma(\mathbf{g}^+(\mathbf{x})), \tag{13.18}$$

where γ is a continuous function from E^p to the real numbers, defined in such a way that P satisfies the requirements demanded of a penalty function.

Example 1. Set

$$P(\mathbf{x}) = \frac{1}{2}\sum_{i=1}^{p} g_i^+(\mathbf{x})^2 = \frac{1}{2}|\mathbf{g}^+(\mathbf{x})|^2,$$

which is without doubt the most popular penalty function. In this case γ is one-half times the identity quadratic form on E^p, that is, $\gamma(\mathbf{y}) = \frac{1}{2}|\mathbf{y}|^2$.

Example 2. By letting

$$\gamma(\mathbf{y}) = \mathbf{y}^T\mathbf{\Gamma}\mathbf{y},$$

where $\mathbf{\Gamma}$ is a symmetric positive definite $p \times p$ matrix, we obtain the penalty function

$$P(\mathbf{x}) = \mathbf{g}^+(\mathbf{x})^T\mathbf{\Gamma}\mathbf{g}^+(\mathbf{x}).$$

Example 3. A general class of penalty functions is

$$P(\mathbf{x}) = \sum_{i=1}^{p}(g_i^+(\mathbf{x}))^\varepsilon$$

for some $\varepsilon > 0$.

Lagrange Multipliers

In the penalty method we solve, for various c_k, the unconstrained problem

$$\text{minimize} \quad f(\mathbf{x}) + c_k P(\mathbf{x}). \tag{13.19}$$

Most algorithms require that the objective function has continuous first partial derivatives. Since we shall, as usual, assume that both f and $\mathbf{g} \in C^1$, it is natural to require, then, that the penalty function $P \in C^1$. We define

$$\nabla g_i^+(\mathbf{x}) = \begin{cases} \nabla g_i(\mathbf{x}) \text{ if } g_i(\mathbf{x}) \geqslant 0 \\ \mathbf{0} \qquad\quad \text{if } g_i(\mathbf{x}) < 0 \end{cases}, \tag{13.20}$$

and, of course, $\nabla \mathbf{g}^+(\mathbf{x})$ is the $m \times n$ matrix whose rows are the $\nabla \mathbf{g}_i^+$'s. Unfortunately, $\nabla \mathbf{g}^+$ is usually discontinuous at points where $g_i^+(\mathbf{x}) = 0$ for some $i = 1, 2, \ldots, p$, and thus some restrictions must be placed on γ in order to guarantee $P \in C^1$. We assume that $\gamma \in C^1$ and that if $\mathbf{y} = (y_1, y_2, \ldots, y_n)$, $\nabla\gamma(\mathbf{y}) = (\nabla\gamma_1, \nabla\gamma_2, \ldots, \nabla\gamma_n)$, then

$$y_i = 0 \quad \text{implies} \quad \nabla\gamma_i = 0. \tag{13.21}$$

(In Example 3 above, for instance, this condition is satisfied only for $\varepsilon > 1$.) With this assumption, the derivative of $\gamma(\mathbf{g}^+(\mathbf{x}))$ with respect to $\mathbf{x}$ is continuous and can be written as $\nabla\gamma(\mathbf{g}^+(\mathbf{x}))\nabla\mathbf{g}(\mathbf{x})$. In this result $\nabla\mathbf{g}(\mathbf{x})$ legitimately replaces the discontinuous $\nabla\mathbf{g}^+(\mathbf{x})$, because it is premultiplied by $\nabla\gamma(\mathbf{g}^+(\mathbf{x}))$. Of course, these considerations are necessary only for inequality constraints. If equality constraints are treated directly, the situation is far simpler.

In view of this assumption, problem (13.19) will have its solution at a point $\mathbf{x}_k$ satisfying

$$\nabla f(\mathbf{x}_k) + c_k \nabla\gamma(\mathbf{g}^+(\mathbf{x}_k))\nabla\mathbf{g}(\mathbf{x}_k) = \mathbf{0},$$

which can be written as

$$\nabla f(\mathbf{x}_k) + \boldsymbol{\lambda}_k^T \nabla\mathbf{g}(\mathbf{x}_k) = \mathbf{0}, \tag{13.22}$$

where

$$\boldsymbol{\lambda}_k^T \equiv c_k \nabla\gamma(\mathbf{g}^+(\mathbf{x}_k)). \tag{13.23}$$

Thus, associated with every c is a Lagrange multiplier vector that is determined after the unconstrained minimization is performed.

If a solution $\mathbf{x}^*$ to the original problem (13.16) is a regular point of the constraints, then there is a unique Lagrange multiplier vector $\boldsymbol{\lambda}^*$ associated with the solution. The result stated below says that $\boldsymbol{\lambda}_k \to \boldsymbol{\lambda}^*$.

Proposition. *Suppose that the penalty function method is applied to problem (13.16) using a penalty function of the form (13.18) with $\gamma \in C^1$ and satisfying (13.21). Corresponding to the sequence $\{\mathbf{x}_k\}$ generated by this method, define $\boldsymbol{\lambda}_k^T = c_k \nabla\gamma(\mathbf{g}^+(\mathbf{x}_k))$. If $\mathbf{x}_k \to \mathbf{x}^*$, a solution to (13.16), and this solution is a regular point, then $\boldsymbol{\lambda}_k \to \boldsymbol{\lambda}^*$, the Lagrange multiplier associated with problem (13.16).*

Proof left to the reader.

Example 4. For $P(\mathbf{x}) = \frac{1}{2}|\mathbf{g}^+(\mathbf{x})|^2$ we have $\boldsymbol{\lambda}_k = c_k\mathbf{g}^+(\mathbf{x}_k)$.

As a final observation we note that in general if $\mathbf{x}_k \to \mathbf{x}^*$, then since $\boldsymbol{\lambda}_k = c_k\nabla\gamma(\mathbf{g}^+(\mathbf{x}_k))^T \to \boldsymbol{\lambda}^*$, the sequence $\mathbf{x}_k$ approaches $\mathbf{x}^*$ from outside the constraint region. Indeed, as $\mathbf{x}_k$ approaches $\mathbf{x}^*$ all constraints that are active at $\mathbf{x}^*$ and have positive Lagrange multipliers will be violated at $\mathbf{x}_k$ because the corresponding components of $\nabla\gamma(\mathbf{g}^+(\mathbf{x}_k))$ are positive. Thus, if we assume that the active constraints are nondegenerate (all Lagrange multipliers are strictly positive), every active constraint will be approached from the outside.

The Hessian Matrix

Since the penalty function method must, for various (large) values of c, solve the unconstrained problem

$$\text{minimize } f(\mathbf{x}) + cP(\mathbf{x}), \tag{13.24}$$

it is important, in order to evaluate the difficulty of such a problem, to determine the eigenvalue structure of the Hessian of this modified objective function. We show here that the structure becomes increasingly unfavorable as c increases.

Although in this section we require that the function $P \in C^1$, we do not require that $P \in C^2$. In particular, the most popular penalty function $P(\mathbf{x}) = \frac{1}{2}|\mathbf{g}^+(\mathbf{x})|^2$, illustrated in Fig. 13.1 for one component, has a discontinuity in its second derivative at any point where a component of $\mathbf{g}$ is zero. At first this might appear to be a serious drawback, since it means the Hessian is discontinuous at the boundary of the constraint region—right where, in general, the solution is expected to lie. However, as pointed out above, the penalty method generates points that approach a boundary solution from outside the constraint region. Thus, except for some possible chance occurrences, the sequence will, as $\mathbf{x}_k \to \mathbf{x}^*$, be at points where the Hessian is well-defined. Furthermore, in iteratively solving the unconstrained problem (13.24) with a fixed c_k, a sequence will be generated that converges to $\mathbf{x}_k$ which is (for most values of k) a point where the Hessian is well-defined, and hence the standard type of analysis will be applicable to the tail of such a sequence.

Defining $q(c, \mathbf{x}) = f(\mathbf{x})+c\gamma(\mathbf{g}^+(\mathbf{x}))$ we have for the Hessian, $\mathbf{Q}$, of q (with respect to $\mathbf{x}$)

$$\mathbf{Q}(c, \mathbf{x}) = \mathbf{F}(\mathbf{x}) + c\nabla\gamma(\mathbf{g}^+(\mathbf{x}))\mathbf{G}(\mathbf{x}) + c\nabla\mathbf{g}^+(\mathbf{x})^T\boldsymbol{\Gamma}(\mathbf{g}^+(\mathbf{x}))\nabla\mathbf{g}^+(\mathbf{x}),$$

where $\mathbf{F}$, $\mathbf{G}$, and $\boldsymbol{\Gamma}$ are, respectively, the Hessians of f, $\mathbf{g}$, and γ. For a fixed c_k we use the definition of $\boldsymbol{\lambda}_k$ given by (13.23) and introduce the rather natural definition

$$\mathbf{L}_k(\mathbf{x}_k) = \mathbf{F}(\mathbf{x}_k) + \boldsymbol{\lambda}_k^T\mathbf{G}(\mathbf{x}_k), \tag{13.25}$$

which is the Hessian of the corresponding Lagrangian. Then we have

$$\mathbf{Q}(c_k, \mathbf{x}_k) = \mathbf{L}_k(\mathbf{x}_k) + c_k\nabla\mathbf{g}^+(\mathbf{x}_k)^T\boldsymbol{\Gamma}(\mathbf{g}^+(\mathbf{x}_k))\nabla\mathbf{g}^+(\mathbf{x}_k), \tag{13.26}$$

which is the desired expression.

The first term on the right side of (13.26) converges to the Hessian of the Lagrangian of the original constrained problem as $\mathbf{x}_k \to \mathbf{x}^*$, and hence has a limit that is independent of c_k. The second term is a matrix having rank equal to the rank of the active constraints and having a magnitude tending to infinity. (See Exercise 7.)

Example 5. For $P(\mathbf{x}) = \frac{1}{2}|\mathbf{g}^+(\mathbf{x})|^2$ we have

$$\boldsymbol{\Gamma}(\mathbf{g}^+(\mathbf{x}_k)) = \begin{bmatrix} e_1 & 0 & \cdots 0 \\ 0 & e_2 & 0 \\ 0 & \cdot & \cdot \\ \cdot & \cdot & \cdot \\ \cdot & \cdot & \cdot \\ 0 & \cdots 0 & e_p \end{bmatrix},$$

where

$$e_i = \begin{cases} 1 & \text{if } g_i(\mathbf{x}_k) > 0 \\ 0 & \text{if } g_i(\mathbf{x}_k) < 0 \\ \text{undefined} & \text{if } g_i(\mathbf{x}_k) = 0 \end{cases}$$

Thus

$$c_k \nabla \mathbf{g}^+(\mathbf{x}_k)^T (\mathbf{g}^+(\mathbf{x}_k)) \nabla \mathbf{g}^+(\mathbf{x}_k) = c_k \nabla \mathbf{g}^+(\mathbf{x}_k)^T \nabla \mathbf{g}^+(\mathbf{x}_k),$$

which is c_k times a matrix that approaches $\nabla \mathbf{g}^+(\mathbf{x}^*)^T \nabla \mathbf{g}^+(\mathbf{x}^*)$. This matrix has rank equal to the rank of the active constraints at $\mathbf{x}^*$ [refer to (13.20)].

Assuming that there are r active constraints at the solution $\mathbf{x}^*$, then for well-behaved γ, the Hessian matrix $\mathbf{Q}(c_k, \ \mathbf{x}_k)$ has r eigenvalues that tend to infinity as $c_k \to \infty$, arising from the second term on the right side of (13.26). There will be $n - r$ other eigenvalues that, although varying with c_k, tend to finite limits. These limits turn out to be, as is perhaps not too surprising at this point, the eigenvalues of $\mathbf{L}(\mathbf{x}^*)$ restricted to the tangent subspace M of the active constraints. The proof of this requires some further analysis.

Lemma 1. *Let* $\mathbf{A}(c)$ *be a symmetric matrix written in partitioned form*

$$\mathbf{A}(c) = \begin{bmatrix} \mathbf{A}_1(c) & \mathbf{A}_2(c) \\ \mathbf{A}_2^T(c) & \mathbf{A}_3(c) \end{bmatrix}, \tag{13.27}$$

where $\mathbf{A}_1(c)$ *tends to a positive definite matrix* $\mathbf{A}_1$, $\mathbf{A}_2(c)$ *tends to a finite matrix, and* $\mathbf{A}_3(c)$ *is a positive definite matrix tending to infinity with c (that is, for any* $s > 0$, $\mathbf{A}_3(c) - s\mathbf{I}$ *is positive definite for sufficiently large c). Then*

$$\mathbf{A}^{-1}(c) \to \begin{bmatrix} \mathbf{A}_1^{-1} & \mathbf{0} \\ \mathbf{0} & \mathbf{0} \end{bmatrix} \tag{13.28}$$

as $c \to \infty$.

Proof. We have the identity

$$\begin{bmatrix} \mathbf{A}_1 & \mathbf{A}_2 \\ \mathbf{A}_2^T & \mathbf{A}_3 \end{bmatrix}^{-1} = \begin{bmatrix} (\mathbf{A}_1 - \mathbf{A}_2\mathbf{A}_3^{-1}\mathbf{A}_2^T)^{-1} & -(\mathbf{A}_1 - \mathbf{A}_2\mathbf{A}_3^{-1}\mathbf{A}_2^T)\mathbf{A}_2\mathbf{A}_3^{-1} \\ -\mathbf{A}_3^{-1}\mathbf{A}_2^T(\mathbf{A}_1 - \mathbf{A}_2\mathbf{A}_3^{-1}\mathbf{A}_2^T)^{-1} & (\mathbf{A}_3 - \mathbf{A}_2^T\mathbf{A}_1^{-1}\mathbf{A}_2)^{-1} \end{bmatrix}. \tag{13.29}$$

Using the fact that $\mathbf{A}_3^{-1}(c) \to \mathbf{0}$ gives the result. ∎

To apply this result to the Hessian matrix (13.26) we associate $\mathbf{A}$ with $\mathbf{Q}(c_k, \mathbf{x}_k)$ and let the partition of $\mathbf{A}$ correspond to the partition of the space E^n into the subspace M and the subspace N that is orthogonal to M; that is, N is the subspace spanned by the gradients of the active constraints. In this partition, $\mathbf{L}_M$, the restriction of $\mathbf{L}$ to M, corresponds to the matrix $\mathbf{A}_1$.

We leave the details of the required continuity arguments to the reader. The important conclusion is that if $\mathbf{x}^*$ is a solution to (13.16), is a regular point, and has exactly r active constraints none of which are degenerate, then the Hessian matrices $\mathbf{Q}(c_k, \mathbf{x}_k)$ of a penalty function of form (13.18) have r eigenvalues tending to infinity as $c_k \to \infty$, and $n - r$ eigenvalues tending to the eigenvalues of $\mathbf{L}_M$.

This explicit characterization of the structure of penalty function Hessians is of great importance in the remainder of the chapter. The fundamental point is that virtually any choice of penalty function (within the class considered) leads both to an ill-conditioned Hessian and to consideration of the ubiquitous Hessian of the Lagrangian restricted to M.

Barrier Functions

Essentially the same story holds for barrier function. If we consider for Problem (13.16) barrier functions of the form

$$B(\mathbf{x}) = \eta(\mathbf{g}(\mathbf{x})), \tag{13.30}$$

then Lagrange multipliers and ill-conditioned Hessians are again inevitable. Rather than parallel the earlier analysis of penalty functions, we illustrate the conclusions with two examples.

Example 1. Define

$$B(\mathbf{x}) = \sum_{i=1}^{p} -\frac{1}{g_i(\mathbf{x})}. \tag{13.31}$$

The barrier objective

$$r(c_k, \mathbf{x}) = f(\mathbf{x}) - \frac{1}{c_k}\sum_{i=1}^{p} \frac{1}{g_i(\mathbf{x})}$$

has its minimum at a point $\mathbf{x}_k$ satisfying

$$\nabla f(\mathbf{x}_k) + \frac{1}{c_k}\sum_{i=1}^{p} \frac{1}{g_i(\mathbf{x}_k)^2}\nabla g_i(\mathbf{x}_k) = \mathbf{0}. \tag{13.32}$$

Thus, we define $\boldsymbol{\lambda}_k$ to be the vector having ith component $\frac{1}{c_k} \cdot \frac{1}{g_i(\mathbf{x}_k)^2}$. Then (13.32) can be written as

$$\nabla f(\mathbf{x}_k) + \boldsymbol{\lambda}_k^T \nabla \mathbf{g}(\mathbf{x}_k) = \mathbf{0}.$$

Again, assuming $\mathbf{x}_k \to \mathbf{x}^*$, the solution of (13.16), we can show that $\boldsymbol{\lambda}_k \to \boldsymbol{\lambda}^*$, the Lagrange multiplier vector associated with the solution. This implies that if g_i is an active constraint,

$$\frac{1}{c_k g_i(\mathbf{x}_k)^2} \to \lambda_i^* < \infty. \tag{13.33}$$

Next, evaluating the Hessian $\mathbf{R}(c_k,\ \mathbf{x}_k)$ of $r(c_k,\ \mathbf{x}_k)$, we have

$$\begin{aligned}\mathbf{R}(c_k,\ \mathbf{x}_k) &= \mathbf{F}(\mathbf{x}_k) + \frac{1}{c_k}\sum_{i=1}^{p}\frac{1}{g_i(\mathbf{x}_k)^2}\mathbf{G}_i(\mathbf{x}_k) - \frac{1}{c_k}\sum_{i=1}^{p}\frac{2}{g_i(\mathbf{x}_k)^3}\nabla g_i(\mathbf{x}_k)^T\nabla g_i(\mathbf{x}_k)\\ &= \mathbf{L}(\mathbf{x}_k) - \frac{1}{c_k}\sum_{i=1}^{p}\frac{2}{g_i(\mathbf{x}_k)^3}\nabla g_i(\mathbf{x}_k)^T\nabla g_i(\mathbf{x}_k).\end{aligned}$$

As $c_k \to \infty$ we have

$$\frac{-1}{c_k g_i(\mathbf{x}_k)^3} \to \begin{cases}\infty \text{ if } & g_i \text{ is active at } \mathbf{x}^*\\ 0 \text{ if } & g_i \text{ is inactive at } \mathbf{x}^*\end{cases}$$

so that we may write, from (13.33),

$$\mathbf{R}(c_k,\ \mathbf{x}_k) \to \mathbf{L}(\mathbf{x}_k) + \sum_{i\in I} -\frac{\lambda_{i,k}}{g_i(\mathbf{x}_k)}\nabla g_i(\mathbf{x}_k)^T\nabla g_i(\mathbf{x}_k),$$

where I is the set of indices corresponding to active constraints. Thus the Hessian of the barrier objective function has exactly the same structure as that of penalty objective functions.

Example 2. Let us use the logarithmic barrier function

$$B(\mathbf{x}) = -\sum_{i=1}^{p}\log[-g_i(\mathbf{x})]$$

In this case we will define the barrier objective in terms of μ as

$$r(\mu,\ \mathbf{x}) = f(\mathbf{x}) - \mu\sum_{i=1}^{p}\log[-g_i(\mathbf{x})]$$

The minimum point $\mathbf{x}_\mu$ satisfies

$$\mathbf{0} = \nabla f(\mathbf{x}_\mu) + \mu\sum_{i=1}^{p}\frac{-1}{g_i(\mathbf{x}_\mu)}\nabla g_i(\mathbf{x}_\mu). \tag{13.34}$$

Defining

$$\lambda_{\mu,i} = \mu\frac{-1}{g_i(\mathbf{x}_\mu)}$$

(13.34) can be written as

$$\nabla f(\mathbf{x}_\mu) + \boldsymbol{\lambda}_\mu^T \nabla \mathbf{g}(\mathbf{x}_\mu) = \mathbf{0}.$$

Further we expect that $\boldsymbol{\lambda}_\mu \to \boldsymbol{\lambda}^*$ as $\mu \to 0$.

The Hessian of $r(\mu, \mathbf{x})$ is

$$R(\mu, \mathbf{x}_\mu) = \mathbf{F}(\mathbf{x}_\mu) + \sum_{i=1}^{p} \lambda_{i,\mu} \mathbf{G}_i(\mathbf{x}_\mu) + \sum_{i=1}^{p} -\frac{\lambda_{i,\mu}}{g_i(\mathbf{x}_\mu)} \nabla g_i(\mathbf{x}_\mu)^T \nabla g_i(\mathbf{x}_\mu).$$

Hence, for small μ it has the same structure as that found in Example 1.

The Central Path

The definition of the central path associated with linear programs is easily extended to general nonlinear programs. For example, consider the problem

$$\begin{aligned} &\text{minimize} \quad f(\mathbf{x}) \\ &\text{subject to} \quad \mathbf{h}(\mathbf{x}) = \mathbf{0},\ \mathbf{g}(\mathbf{x}) \leq \mathbf{0}. \end{aligned}$$

We assume that $\mathring{\mathcal{F}} = \{\mathbf{x} : \mathbf{h}(\mathbf{x}) = \mathbf{0},\ \mathbf{g}(\mathbf{x}) < \mathbf{0}\} \neq \phi$. Then we use the logarithmic barrier function to define the problems

$$\begin{aligned} &\text{minimize} \quad f(\mathbf{x}) - \mu \textstyle\sum_{i=1}^{p} \log[-g_i(\mathbf{x})] \\ &\text{subject to} \quad \mathbf{h}(\mathbf{x}) = \mathbf{0}. \end{aligned}$$

The solution $\mathbf{x}_\mu$ parameterized by $\mu \to 0$ is called the central path; see Chap. 5.

The necessary conditions for the problem can be written as

$$\begin{aligned} \nabla f(\mathbf{x}_\mu) + \mathbf{s}^T \nabla \mathbf{g}(\mathbf{x}_\mu) + \mathbf{y}^T \nabla \mathbf{h}(\mathbf{x}_\mu) &= \mathbf{0} \\ \mathbf{h}(\mathbf{x}_\mu) &= \mathbf{0}. \\ s_i g_i(\mathbf{x}_\mu) &= -\mu;\ i = 1, 2, \ldots, p \end{aligned}$$

where $\mathbf{y}$ is the Lagrange multiplier vector for the constraint $\mathbf{h}(\mathbf{x}_\mu) = \mathbf{0}$. Then, the Newton method can be applied to solving the condition system as μ is gradually reduced to 0, that is, following the path.

Geometric Interpretation: The Primal Function

There is a geometric construction that provides a simple interpretation of penalty functions. The basis of the construction itself is also useful in other areas of optimization, especially duality theory, as explained in the next chapter.

Let us again consider the problem

$$\begin{aligned} \text{minimize } & f(\mathbf{x}) \\ \text{subject to } & \mathbf{h}(\mathbf{x}) = \mathbf{0}, \end{aligned} \tag{13.35}$$

where $\mathbf{h}(\mathbf{x}) \in E^m$. We assume that the solution point $\mathbf{x}^*$ of (13.35) is a regular point and that the second-order sufficiency conditions are satisfied. Corresponding to this problem we introduce the following definition:

Definition. Corresponding to the constrained minimization problem (13.35), the *primal function* ω is defined on E^m in a neighborhood of $\mathbf{0}$ to be

$$\omega(\mathbf{y}) = \min\{f(\mathbf{x}) : \mathbf{h}(\mathbf{x}) = \mathbf{y}\}. \tag{13.36}$$

The primal function gives the optimal value of the objective for various values of the right-hand side. In particular $\omega(0)$ gives the value of the original problem.

Strictly speaking the minimum in the definition (13.36) must be specified as a local minimum, in a neighborhood of $\mathbf{x}^*$. The existence of $\omega(\mathbf{y})$ then follows directly from the Sensitivity Theorem in Sect. 11.7. Furthermore, from that theorem it follows that $\nabla\omega(\mathbf{0}) = -\boldsymbol{\lambda}^{*T}$.

Now consider the penalty problem and note the following relations:

$$\begin{aligned} \min\{f(\mathbf{x}) + \frac{1}{2}c|\mathbf{h}(\mathbf{x})|^2\} &= \min_{\mathbf{x},\mathbf{y}}\{f(\mathbf{x}) + \frac{1}{2}c|\mathbf{y}|^2 : \mathbf{h}(\mathbf{x}) = \mathbf{y}\} \\ &= \min_{\mathbf{y}}\{\omega(\mathbf{y}) + \frac{1}{2}c|\mathbf{y}|^2\}. \end{aligned} \tag{13.37}$$

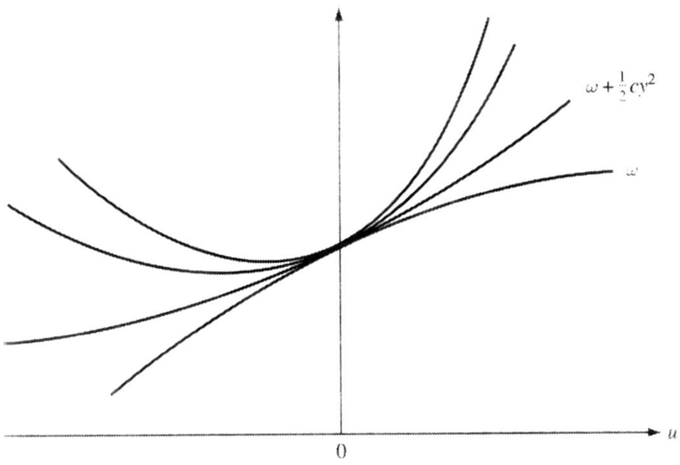

Fig. 13.4 The primal function

This is illustrated in Fig. 13.4 for the case where y is one-dimensional. The primal function is the lowest curve in the figure. Its value at $y = 0$ is the value of the original constrained problem. Above the primal function are the curves $\omega(y) + \frac{1}{2}cy^2$ for various values of c. The value of the penalty problem is shown by (13.37) to be the minimum point of this curve. For large values of c this curve becomes convex near 0 even if $\omega(y)$ is not convex. Viewed in this way, the penalty functions can be thought of as *convexifying* the primal.

Also, as c increases, the associated minimum point moves toward 0. However, it is never zero for finite c. Furthermore, in general, the criterion for $\mathbf{u}$ to be optimal for the penalty problem is that the gradient of $\omega(\mathbf{y}) + \frac{1}{2}c\mathbf{y}^2$ equals zero. This yields $\nabla\omega(\mathbf{y}) + c\mathbf{y}^T = \mathbf{0}$. Using $\nabla\omega(\mathbf{y}) = -\boldsymbol{\lambda}^T$ and $\mathbf{y} = \mathbf{h}(\mathbf{x}_c)$, where now $\mathbf{x}_c$ denotes the minimum point of the penalty problem, gives $\boldsymbol{\lambda} = c\mathbf{h}(\mathbf{x}_c)$, which is the same as (13.23).

13.4 Newton's Method and Penalty Functions

In the next few sections we address the problem of efficiently solving the unconstrained problems associated with a penalty or barrier method. The main difficulty is the extremely unfavorable eigenvalue structure that, as explained in Sect. 13.3, always accompanies unconstrained problems derived in this way. Certainly straightforward application of the method of steepest descent is out of the question!

One method for avoiding slow convergence for these problems is to apply Newton's method (or one of its variations), since the order two convergence of Newton's method is unaffected by the poor eigenvalue structure. In applying the method, however, special care must be devoted to the manner by which the Hessian is inverted, since it is ill-conditioned. Nevertheless, if second-order information is easily available, Newton's method offers an extremely attractive and effective method for solving modest size penalty or barrier optimization problems. When such information is not readily available, or if data handling and storage requirements of Newton's method are excessive, attention naturally focuses on first-order methods.

A simple modified Newton's method can often be quite effective for some penalty problems. For example, consider the problem having only equality constraints

$$\begin{aligned} &\text{minimize} \quad f(\mathbf{x}) \\ &\text{subject to} \quad \mathbf{h}(\mathbf{x}) = \mathbf{0} \end{aligned} \tag{13.38}$$

with $\mathbf{x} \in E^n$, $\mathbf{h}(\mathbf{x}) \in E^m$, $m < n$. Applying the standard quadratic penalty method we solve instead the unconstrained problem

$$\text{minimize } f(\mathbf{x}) + \tfrac{1}{2}c|\mathbf{h}(\mathbf{x})|^2 \tag{13.39}$$

for some large c. Calling the penalty objective function $q(\mathbf{x})$ we consider the iterative process

$$\mathbf{x}_{k+1} = \mathbf{x}_k - \alpha_k[\mathbf{I} + c\nabla\mathbf{h}(\mathbf{x}_k)^T\nabla\mathbf{h}(\mathbf{x}_k)]^{-1}\nabla q(\mathbf{x}_k)^T, \tag{13.40}$$

where α_k is chosen to minimize $q(\mathbf{x}_{k+1})$. The matrix $\mathbf{I} + c\nabla\mathbf{h}(\mathbf{x}_k)^T\nabla\mathbf{h}(\mathbf{x}_k)$ is positive definite and although quite ill-conditioned it can be inverted efficiently (see Exercise 11).

According to the Modified Newton Method Theorem (Sect. 10.1) the rate of convergence of this method is determined by the eigenvalues of the matrix

$$[\mathbf{I} + c\nabla\mathbf{h}(\mathbf{x}_k)^T\nabla\mathbf{h}(\mathbf{x}_k)]^{-1}\mathbf{Q}(\mathbf{x}_k), \tag{13.41}$$

where $\mathbf{Q}(\mathbf{x}_k)$ is the Hessian of q at $\mathbf{x}_k$. In view of (13.26), as $c \to \infty$ the matrix (13.41) will have m eigenvalues that approach unity, while the remaining $n - m$ eigenvalues approach the eigenvalues of $\mathbf{L}_M$ evaluated at the solution $\mathbf{x}^*$ of (13.38). Thus, if the smallest and largest eigenvalues of $\mathbf{L}_M$, a and A, are located such that the interval $[a, A]$ contains unity, the convergence ratio of this modified Newton's method will be equal (in the limit of $c \to \infty$) to the canonical ratio $[(A-a)/(A+a)]^2$ for problem (13.38).

If the eigenvalues of $\mathbf{L}_M$ are not spread below and above unity, the convergence rate will be slowed. If a point in the interval containing the eigenvalues of $\mathbf{L}_M$ is known, a scalar factor can be introduced so that the canonical rate is achieved, but such information is often not easily available.

Inequalities

If there are inequality as well as equality constraints in the problem, the analogous procedure can be applied to the associated penalty objective function. The unusual feature of this case is that corresponding to an inequality constraint $g_i(\mathbf{x}) \leqslant 0$, the term $\nabla g_i^+(\mathbf{x})^T\nabla g_i^+(\mathbf{x})$ used in the iteration matrix will suddenly appear if the constraint is violated. Thus the iteration matrix is discontinuous with respect to $\mathbf{x}$, and as the method progresses its nature changes according to which constraints are violated. This discontinuity does not, however, imply that the method is subject to jamming, since the result of Exercise 4, Chap. 10 is applicable to this method.

13.5 Conjugate Gradients and Penalty Methods

The partial conjugate gradient method proposed and analyzed in Sect. 9.5 is ideally suited to penalty or barrier problems having only a few active constraints. If there are m active constraints, then taking cycles of $m + 1$ conjugate gradient steps will yield a rate of convergence that is independent of the penalty constant c. For example, consider the problem having only equality constraints:

$$\begin{aligned} &\text{minimize} \;\; f(\mathbf{x}) \\ &\text{subject to} \;\; \mathbf{h}(\mathbf{x}) = \mathbf{0}, \end{aligned} \tag{13.42}$$

where $\mathbf{x} \in E^n$, $\mathbf{h}(\mathbf{x}) \in E^m$, $m < n$. Applying the standard quadratic penalty method, we solve instead the unconstrained problem

$$\text{minimize } f(\mathbf{x}) + \tfrac{1}{2}c|\mathbf{h}(\mathbf{x})|^2 \tag{13.43}$$

for some large c. The objective function of this problem has a Hessian matrix that has m eigenvalues that are of the order c in magnitude, while the remaining $n - m$ eigenvalues are close to the eigenvalues of the matrix $\mathbf{L}_M$, corresponding to problem (13.42). Thus, letting $\mathbf{x}_{k+1}$ be determined from $\mathbf{x}_k$ by taking $m + 1$ steps of a (non-quadratic) conjugate gradient method, and assuming $\mathbf{x}_k \to \overline{\mathbf{x}}$, a solution to (13.43), the sequence $\{f(\mathbf{x}_k)\}$ converges linearly to $f(\overline{\mathbf{x}})$ with a convergence ratio equal to approximately

$$\left(\frac{A - a}{A + a}\right)^2 \tag{13.44}$$

where a and A are, respectively, the smallest and largest eigenvalues of $\mathbf{L}_M(\overline{\mathbf{x}})$.

This is an extremely effective technique when m is relatively small. The programming logic required is only slightly greater than that of steepest descent, and the time per iteration is only about $m + 1$ times as great as for steepest descent. The method can be used for problems having inequality constraints as well but it is advisable to change the cycle length, depending on the number of constraints active at the end of the previous cycle.

Example 3.

$$\begin{aligned}
\text{minimize } & f(x_1, x_2, \ldots, x_{10}) = \sum_{k=1}^{10} kx_k^2 \\
\text{subject to } & 1.5x_1 + x_2 + x_3 + 0.5x_4 + 0.5x_5 = 5.5 \\
& 2.0x_6 - 0.5x_7 - 0.5x_8 + x_9 - x_{10} = 2.0 \\
& x_1 + x_3 + x_5 + x_7 + x_9 = 10 \\
& x_2 + x_4 + x_6 + x_8 + x_{10} = 15.
\end{aligned}$$

This problem was treated by the penalty function approach, and the resulting composite function was then solved for various values of c by using various cycle lengths of a conjugate gradient algorithm. In Table 13.1 p is the number of conjugate gradient steps in a cycle. Thus, $p = 1$ corresponds to ordinary steepest descent; $p = 5$ corresponds, by the theory of Sect. 9.5, to the smallest value of p for which the rate of convergence is independent of c; and $p = 10$ is the standard conjugate gradient method. Note that for $p < 5$ the convergence rate does indeed depend on c, while it is more or less constant for $p \geqslant 5$. The value of c's selected are not artificially large, since for $c = 200$ the constraints are satisfied only to within 0.5 % of their right-hand sides. For problems with nonlinear constraints the results will most likely be somewhat less favorable, since the predicted convergence rate would apply only to the tail of the sequence.

Table 13.1 Results for example 3

	p (steps per cycle)	Number of cycles to convergence	No. of steps	Value of modified objective
$c = 20$	1	90	90	388.565
	3	8	24	388.563
	5	3	15	388.563
	7	3	21	388.563
$c = 200$	1	230[a]	230	488.607
	3	21	63	487.446
	5	4	20	487.438
	7	2	14	487.433
$c = 2000$	1	260[a]	260	525.238
	3	45[a]	135	503.550
	5	3	15	500.910
	7	3	21	500.882

[a] Program not run to convergence due to excessive time

13.6 Normalization of Penalty Functions

There is a good deal of freedom in the selection of penalty or barrier functions that can be exploited to accelerate convergence. We propose here a simple normalization procedure that together with a two-step cycle of conjugate gradients yields the canonical rate of convergence. Again for simplicity we illustrate the technique for the penalty method applied to the problem

$$\begin{aligned} &\text{minimize } f(\mathbf{x}) \\ &\text{subject to } \mathbf{h}(\mathbf{x}) = \mathbf{0} \end{aligned} \tag{13.45}$$

as in Sects. 13.4 and 13.5, but the idea is easily extended to other penalty or barrier situations.

Corresponding to (13.45) we consider the family of quadratic penalty functions

$$P(\mathbf{x}) = \frac{1}{2}\mathbf{h}(\mathbf{x})^T \mathbf{\Gamma} \mathbf{h}(\mathbf{x}), \tag{13.46}$$

where $\mathbf{\Gamma}$ is a symmetric positive definite $m \times m$ matrix. We ask what the best choice of $\mathbf{\Gamma}$ might be.

Letting

$$q(c, \mathbf{x}) = f(\mathbf{x}) + cP(\mathbf{x}), \tag{13.47}$$

the Hessian of q turns out to be, using (13.26),

$$\mathbf{Q}(c, \mathbf{x}_k) = \mathbf{L}(\mathbf{x}_k) + c\nabla\mathbf{h}(\mathbf{x}_k)^T \mathbf{\Gamma} \nabla\mathbf{h}(\mathbf{x}_k). \tag{13.48}$$

The m large eigenvalues are due to the second term on the right. The observation we make is that although the m large eigenvalues are all proportional to c, they are not necessarily all equal. Indeed, for very large c these eigenvalues are determined almost exclusively by the second term, and are therefore c times the nonzero eigenvalues of the matrix $\nabla \mathbf{h}(\mathbf{x}_k)^T \boldsymbol{\Gamma} \nabla \mathbf{h}(\mathbf{x}_k)$. We would like to select $\boldsymbol{\Gamma}$ so that these eigenvalues are not spread out but are nearly equal to one another. An ideal choice for the kth iteration would be

$$\boldsymbol{\Gamma} = [\nabla \mathbf{h}(\mathbf{x}_k) \nabla \mathbf{h}(\mathbf{x}_k)^T]^{-1}, \tag{13.49}$$

since then all nonzero eigenvalues would be exactly equal. However, we do not allow to change at each step, and therefore compromise by setting

$$\boldsymbol{\Gamma} = [\nabla \mathbf{h}(\mathbf{x}_0) \nabla \mathbf{h}(\mathbf{x}_0)^T]^{-1}, \tag{13.50}$$

where $\mathbf{x}_0$ is the initial point of the iteration.

Using this penalty function, the corresponding eigenvalue structure will at any point look approximately like that shown in Fig. 13.5. The eigenvalues are bunched into two separate groups. As c is increased the smaller eigenvalues move into the interval $[a, A]$ where a and A are, as usual, the smallest and largest eigenvalues of $\mathbf{L}_M$ at the solution to (13.45). The larger eigenvalues move forward to the right and spread further apart.

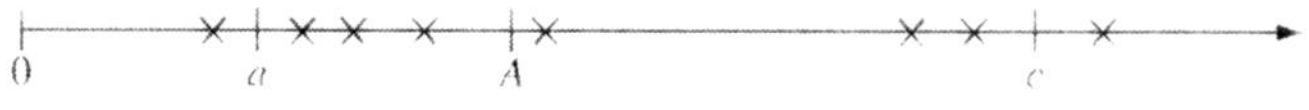

Fig. 13.5 Eigenvalue distributions

Using the result of Exercise 11, Chap. 9, we see that if $\mathbf{x}_{k+1}$ is determined from $\mathbf{x}_k$ by two conjugate gradient steps, the rate of convergence will be linear at a ratio determined by the widest of the two eigenvalue groups. If our normalization is sufficiently accurate, the large-valued group will have the lesser width. In that case convergence of this scheme is approximately that of the canonical rate for the original problem. Thus, by proper normalization it is possible to obtain the canonical rate of convergence for only about twice the time per iteration as required by steepest descent.

There are, of course, numerous variations of this method that can be used in practice. $\boldsymbol{\Gamma}$ can, for example, be allowed to vary at each step, or it can be occasionally updated.

Example. The example problem presented in the previous section was also solved by the normalization method presented above. The results for various values of c and for cycle lengths of one, two, and three are presented in Table 13.2. (All runs were initiated from the zero vector.)

Table 13.2 Results for example 3

	p (steps per cycle)	Number of cycles to convergence	No. of steps	Value of modified objective
$c = 10$	1	28	28	251.2657
	2	9	18	251.2657
	3	5	15	251.2657
$c = 100$	1	153	153	379.5955
	2	13	26	379.5955
	3	11	33	379.5955
$c = 1000$	1	261[a]	261	402.0903
	2	14	28	400.1687
	3	13	39	400.1687

[a] Program not run to convergence due to excessive time

13.7 Penalty Functions and Gradient Projection

The penalty function method can be combined with the idea of the gradient projection method to yield an attractive general purpose procedure for solving constrained optimization problems. The proposed combination method can be viewed either as a way of accelerating the rate of convergence of the penalty function method by eliminating the effect of the large eigenvalues, or as a technique for efficiently handling the delicate and usually cumbersome requirement in the gradient projection method that each point be feasible. The combined method converges at the canonical rate (the same as does the gradient projection method), is globally convergent (unlike the gradient projection method), and avoids much of the computational difficulty associated with staying feasible.

Underlying Concept

The basic theoretical result that motivates the development of this algorithm is the Combined Steepest Descent and Newton's Method Theorem of Sect. 10.7. The idea is to apply this combined method to a penalty problem. For simplicity we first consider the equality constrained problem

$$\begin{aligned} &\text{minimize} \quad f(\mathbf{x}) \\ &\text{subject to} \quad \mathbf{h}(\mathbf{x}) = \mathbf{0}, \end{aligned} \tag{13.51}$$

where $\mathbf{x} \in E^n$, $\mathbf{h}(\mathbf{x}) \in E^m$. The associated unconstrained penalty problem that we consider is

$$\text{minimize} \quad q(\mathbf{x}), \tag{13.52}$$

where

$$q(\mathbf{x}) = f(\mathbf{x}) + \frac{1}{2}c|\mathbf{h}(\mathbf{x})|^2.$$

At any point $\mathbf{x}_k$ let $M(\mathbf{x}_k)$ be the subspace tangent to the surface $S_k = \{\mathbf{x} : \mathbf{h}(\mathbf{x}) = \mathbf{h}(\mathbf{x}_k)\}$. This is a slight extension of the tangent subspaces that we have considered before, since $M(\mathbf{x}_k)$ is defined even for points that are not feasible. If the sequence $\{\mathbf{x}_k\}$ converges to a solution $\mathbf{x}_c$ of problem (13.52), then we expect that $M(\mathbf{x}_k)$ will in some sense converge to $M(\mathbf{x}_c)$. The orthogonal complement of $M(\mathbf{x}_k)$ is the space generated by the gradients of the constraint functions evaluated at $\mathbf{x}_k$. Let us denote this space by $N(\mathbf{x}_k)$. The idea of the algorithm is to take N as the subspace over which Newton's method is applied, and M as the space over which the gradient method is applied. A cycle of the algorithm would be as follows:

1. Given $\mathbf{x}_k$, apply one step of Newton's method over, the subspace $N(\mathbf{x}_k)$ to obtain a point $\mathbf{w}_k$ of the form

$$\mathbf{w}_k = \mathbf{x}_k + \nabla\mathbf{h}(\mathbf{x}_k)^T\mathbf{u}_k$$
$$\mathbf{u}_k \in E^m.$$

2. From $\mathbf{w}_k$, take an ordinary steepest descent step to obtain $\mathbf{x}_{k+1}$.

Of course, we must show how Step 1 can be easily executed, and this is done below, but first, without drawing out the details, let us examine the general structure of this algorithm.

The process is illustrated in Fig. 13.6. The first step is analogous to the step in the gradient projection method that returns to the feasible surface; except that here the criterion is reduction of the objective function rather than satisfaction of constraints. To interpret the second step, suppose for the moment that the original problem (13.51) has a quadratic objective and linear constraints; so that, consequently, the penalty problem (13.52) has a quadratic objective and $N(\mathbf{x})$, $M(\mathbf{x})$ and $\nabla\mathbf{h}(\mathbf{x})$ are independent of $\mathbf{x}$. In that case the first (Newton) step would exactly minimize q with respect to N, so that the gradient of q at $\mathbf{w}_k$ would be orthogonal to N; that is, the gradient would lie in the subspace M. Furthermore, since $\nabla q(\mathbf{w}_k) = \nabla f(\mathbf{w}_k) + c\mathbf{h}(\mathbf{w}_k)\nabla\mathbf{h}(\mathbf{w}_k)$, we see that $\nabla q(\mathbf{w}_k)$ would in that case be equal to the projection of the gradient of f onto M. Hence, the second step is, in the quadratic case exactly, and in the general case approximately, a move in the direction of the projected negative gradient of the original objective function.

The convergence properties of such a scheme are easily predicted from the theorem on the Combined Steepest Descent and Newton's Method, in Sect. 10.7, and our analysis of the structure of the Hessian of the penalty objective function given by (13.26). As $\mathbf{x}_k \to \mathbf{x}_c$ the rate will be determined by the ratio of largest to smallest eigenvalues of the Hessian restricted to $M(\mathbf{x}_c)$.

This leads, however, by what was shown in Sect. 12.3, to approximately the canonical rate for problem (13.51). Thus this combined method will yield again the canonical rate as $c \to \infty$.

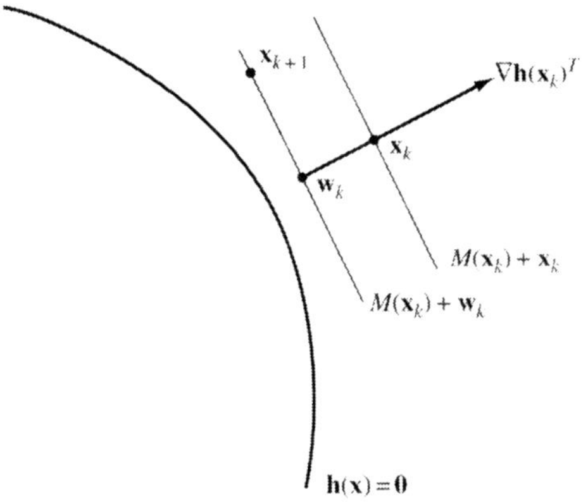

Fig. 13.6 Illustration of the method

Implementing the First Step

To implement the first step of the algorithm suggested above it is necessary to show how a Newton step can be taken in the subspace $N(\mathbf{x}_k)$. We show that, again for large values of c, this can be accomplished easily.

At the point $\mathbf{x}_k$ the function b, defined by

$$b(\mathbf{u}) = q(\mathbf{x}_k + \nabla \mathbf{h}(\mathbf{x}_k)^T \mathbf{u}) \tag{13.53}$$

for $\mathbf{u} \in E^m$, measures the variations in q with respect to displacements in $N(\mathbf{x}_k)$. We shall, for simplicity, assume that at each point, $\mathbf{x}_k$, $\nabla \mathbf{h}(\mathbf{x}_k)$ has rank m. We can immediately calculate the gradient with respect to $\mathbf{u}$,

$$\nabla b(\mathbf{u}) = \nabla q(\mathbf{x}_k + \nabla \mathbf{h}(\mathbf{x}_k)^T \mathbf{u}) \nabla \mathbf{h}(\mathbf{x}_k)^T, \tag{13.54}$$

and the $m \times n$ Hessian with respect to $\mathbf{u}$ at $\mathbf{u} = \mathbf{0}$,

$$\mathbf{B} = \nabla \mathbf{h}(\mathbf{x}_k) \mathbf{Q}(\mathbf{x}_k) \nabla \mathbf{h}(\mathbf{x}_k)^T. \tag{13.55}$$

where $\mathbf{Q}$ is the $n \times n$ Hessian of q with respect to $\mathbf{x}$. From (13.26) we have that at $\mathbf{x}_k$

$$\mathbf{Q}(\mathbf{x}_k) = \mathbf{L}_k(\mathbf{x}_k) + c \nabla \mathbf{h}(\mathbf{x}_k)^T \nabla \mathbf{h}(\mathbf{x}_k). \tag{13.56}$$

And given $\mathbf{B}$, the direction for the Newton step in N would be

$$\begin{aligned} \mathbf{d}_k &= -\nabla \mathbf{h}(\mathbf{x}_k)^T \mathbf{B}^{-1} \nabla c(\mathbf{0})^T \\ &= -\nabla \mathbf{h}(\mathbf{x}_k)^T \mathbf{B}^{-1} \nabla \mathbf{h}(\mathbf{x}_k) \nabla q(\mathbf{x}_k)^T. \end{aligned} \tag{13.57}$$

It is clear from (13.55) and (13.56) that exact evaluation of the Newton step requires knowledge of $\mathbf{L}(\mathbf{x}_k)$ which usually is costly to obtain. For large values of c, however, $\mathbf{B}$ can be approximated by

$$\mathbf{B} \simeq c[\nabla \mathbf{h}(\mathbf{x}_k) \nabla \mathbf{h}(\mathbf{x}_k)^T]^2, \tag{13.58}$$

and hence a good approximation to the Newton direction is

$$\mathbf{d}_k = -\frac{1}{c}\nabla \mathbf{h}(\mathbf{x}_k)^T[\nabla \mathbf{h}(\mathbf{x}_k)\nabla \mathbf{h}(\mathbf{x}_k)^T]^{-2}\nabla \mathbf{h}(\mathbf{x}_k)\nabla q(\mathbf{x}_k)^T. \tag{13.59}$$

Thus a suitable implementation of one cycle of the algorithm is:

1. Calculate

$$\mathbf{d}_k = -\frac{1}{c}\nabla \mathbf{h}(\mathbf{x}_k)^T[\nabla \mathbf{h}(\mathbf{x}_k)\nabla \mathbf{h}(\mathbf{x}_k)^T]^{-2}\nabla \mathbf{h}(\mathbf{x}_k)\nabla q(\mathbf{x}_k)^T.$$

2. Find β_k to minimize $q(\mathbf{x}_k + \beta\mathbf{d}_k)$ (using $\beta_k = 1$ as an initial search point), and set $\mathbf{w}_k = \mathbf{x}_k + \beta_k\mathbf{d}_k$.
3. Calculate $\mathbf{p}_k = -\nabla q(\mathbf{w}_k)^T$.
4. Find α_k to minimize $q(\mathbf{w}_k + \alpha\mathbf{p}_k)$, and set $\mathbf{x}_{k+1} = \mathbf{w}_k + \alpha_k\mathbf{p}_k$.

It is interesting to compare the Newton step of this version of the algorithm with the step for returning to the feasible region used in the ordinary gradient projection method. We have

$$\nabla q(\mathbf{x}_k)^T = \nabla f(\mathbf{x}_k)^T + c\nabla \mathbf{h}(\mathbf{x}_k)^T\mathbf{h}(\mathbf{x}_k). \tag{13.60}$$

If we neglect $\nabla f(\mathbf{x}_k)^T$ on the right (as would be valid if we are a long distance from the constraint boundary) then the vector $\mathbf{d}_k$ reduces to

$$\mathbf{d}_k = -\nabla \mathbf{h}(\mathbf{x}_k)^T[\nabla \mathbf{h}(\mathbf{x}_k)\nabla \mathbf{h}(\mathbf{x}_k)^T]^{-1}\mathbf{h}(\mathbf{x}_k),$$

which is precisely the first estimate used to return to the boundary in the gradient projection method. The scheme developed in this section can therefore be regarded as one which corrects this estimate by accounting for the variation in f.

An important advantage of the present method is that it is not necessary to carry out the search in detail. If $\beta = 1$ yields an improved value for the penalty objective, no further search is required. If not, one need search only until some improvement is obtained. At worst, if this search is poorly performed, the method degenerates to steepest descent. When one finally gets close to the solution, however, $\beta = 1$ is bound to yield an improvement and terminal convergence will progress at nearly the canonical rate.

Inequality Constraints

The procedure is conceptually the same for problems with inequality constraints. The only difference is that at the beginning of each cycle the subspace $M(\mathbf{x}_k)$ is calculated on the basis of those constraints that are either active or violated at $\mathbf{x}_k$, the others being ignored. The resulting technique is a descent algorithm in that the penalty objective function decreases at each cycle; it is globally convergent because of the pure gradient step taken at the end of each cycle; its rate of convergence approaches the canonical rate for the original constrained problem as $c \to \infty$; and there are no feasibility tolerances or subroutine iterations required.

*13.8 *Exact Penalty Functions

It is possible to construct penalty functions that are exact in the sense that the solution of the penalty problem yields the exact solution to the original problem for a finite value of the penalty parameter. With these functions it is not necessary to solve an infinite sequence of penalty problems to obtain the correct solution. However, a new difficulty introduced by these penalty functions is that they are non-differentiable.

For the general constrained problem

$$\begin{aligned} &\text{minimize} \;\; f(\mathbf{x}) \\ &\text{subject to} \;\; \mathbf{h}(\mathbf{x}) = \mathbf{0}, \;\; \mathbf{g}(\mathbf{x}) \leqslant \mathbf{0}, \end{aligned} \tag{13.61}$$

consider the absolute-value penalty function

$$P(\mathbf{x}) = \sum_{i=1}^{m} |h_i(\mathbf{x})| + \sum_{j=1}^{p} \max(0, \; g_j(\mathbf{x})). \tag{13.62}$$

The penalty problem is then, as usual,

$$\text{minimize } f(\mathbf{x}) + cP(\mathbf{x}) \tag{13.63}$$

for some positive constant c. We investigate the properties of the absolute-value penalty function through an example and then generalize the results.

Example 1. Consider the simple quadratic problem

$$\begin{aligned} &\text{minimize} \quad 2x^2 + 2xy + y^2 - 2y \\ &\text{subject to} \quad x = 0. \end{aligned} \tag{13.64}$$

It is easy to solve this problem directly by substituting $x = 0$ into the objective. This leads immediately to $x = 0, \; y = 1$.

If a standard quadratic penalty function is used, we minimize the objective

$$2x^2 + 2xy + y^2 - 2y + \frac{1}{2}cx^2 \tag{13.65}$$

for $c > 0$. The solution again can be easily found and is $x = -2/(2 + c), \; y = 1 - 2/(2 + c)$. This solution approaches the true solution as $c \to \infty$, as predicted by the general theory. However, for any finite c the solution is inexact.

Now let us use the absolute-value penalty function. We minimize the function

$$2x^2 + 2xy + y^2 - 2y + c|x|. \tag{13.66}$$

We rewrite (13.66) as

$$\begin{aligned} 2x^2 + 2xy + y^2 - 2y + c|x| \\ &= 2x^2 + 2xy + c|x| + (y-1)^2 - 1 \\ &= 2x^2 + 2x + c|x| + (y-1)^2 + 2x(y-1) - 1 \\ &= x^2 + (2x + c|x|) + (y - 1 + x)^2 - 1. \end{aligned} \tag{13.67}$$

All terms (except the $-l$) are nonnegative if $c > 2$. Therefore, the minimum value of this expression is -1, which is achieved (uniquely) by $x = 0,\ y = 1$. Therefore, for $c > 2$ the minimum point of the penalty problem is the correct solution to the original problem (13.64).

We let the reader verify that $\lambda = -2$ for this example. The fact that $c > |\lambda|$ is required for the solution to be exact is an illustration of a general result given by the following theorem.

Exact Penalty Theorem. *Suppose that the point* $\mathbf{x}^*$ *satisfies the second-order sufficiency conditions for a local minimum of the constrained problem* (13.61). Let $\boldsymbol{\lambda}$ and $\boldsymbol{\mu}$ be the corresponding Lagrange multipliers. Then for $c > \max\{|\lambda_i|,\ \mu_j : i = 1, 2, \ldots, m,\ j = 1, 2, \ldots, p\}$, $\mathbf{x}^*$ is also a local minimum of the absolute-value penalty objective (13.62).

Proof. For simplicity we assume that there are equality constraints only. Define the primal function

$$\omega(\mathbf{z}) = \min_{\mathbf{x}}\{f(\mathbf{x}) : h_i(\mathbf{x}) = z_i \text{ for } i = 1, 2, \ldots, m\}. \tag{13.68}$$

The primal function was introduced in Sect. 12.3. Under our assumption the function exists in a neighborhood of $\mathbf{x}^*$ and is continuously differentiable, with $\nabla\omega(\mathbf{0}) = -\boldsymbol{\lambda}^T$.

Now define

$$\omega_c(\mathbf{z}) = \omega(\mathbf{z}) + c\sum_{i=1}^m |z_i|.$$

Then we have

$$\begin{aligned} \min_{\mathbf{x}}\{f(\mathbf{x}) + c\sum_{i=1}^m |h_i(\mathbf{x})|\} &= \min_{\mathbf{x},\mathbf{u}}\{f(\mathbf{x}) + c\sum_{i=1}^m |z_i| : \mathbf{h}(\mathbf{x}) = \mathbf{z}\} \\ &= \min_{\mathbf{u}}\{p(\mathbf{z}) + c\sum_{i=1}^m |z_i|\} \\ &= \min_{\mathbf{u}} p_c(\mathbf{z}). \end{aligned}$$

By the Mean Value Theorem,

$$\omega(\mathbf{z}) = \omega(\mathbf{0}) + \nabla\omega(\alpha\mathbf{z})\mathbf{z}$$

for some α, $0 \leqslant \alpha \leqslant 1$. Therefore,

$$\omega_c(\mathbf{z}) = \omega(\mathbf{0}) + \nabla\omega(\alpha\mathbf{z})\mathbf{z} + c\sum_{i=1}^m |z_i|. \tag{13.69}$$

We know that $\nabla\omega(\mathbf{z})$ is continuous at $\mathbf{0}$, and thus given $\varepsilon > 0$ there is a neighborhood of $\mathbf{0}$ such that $|\nabla\omega(\mathbf{z})_i| < |\lambda_i| + \varepsilon$. Thus

$$\nabla\omega(\alpha\mathbf{z})\mathbf{z} = \sum_{i=1}^{m} \nabla\omega(\alpha\mathbf{z})_i z_i \geqslant -\{\max_i |\nabla\omega(\alpha\mathbf{z})_i|\} \sum_{i=1}^{m} |z_i|$$

$$\geqslant -\{\max_i(|\lambda_i| + \varepsilon)\} \sum_{i=1}^{m} |z_i|.$$

Using this in (13.69), we obtain

$$\omega_c(\mathbf{z}) \geqslant p(\mathbf{0}) + (c - \varepsilon - \max|\lambda_i|) \sum_{i=1}^{m} |z_i|.$$

For $c > \varepsilon + \max|\lambda_i|$ it follows that $\omega_c(\mathbf{z})$ is minimized at $\mathbf{z} = \mathbf{0}$. Since ε was arbitrary, the result holds for $c > \max|\lambda_i|$.

This result is easily extended to include inequality constraints. (See Exercise 16.)

∎

It is possible to develop a geometric interpretation of the absolute-value penalty function analogous to the interpretation for ordinary penalty functions given in Fig. 13.4. Figure 13.7 corresponds to a problem for a single constraint. The smooth curve represents the primal function of the problem. Its value at 0 is the value of the original problem, and its slope at 0 is $-\lambda$. The function $\omega_c(z)$ is obtained by adding $c|z|$ to the primal function, and this function has a discontinuous derivative at $z = 0$. It is clear that for $c > |\lambda|$, this composite function has a minimum at exactly $z = 0$, corresponding to the correct solution.

There are other exact penalty functions but, like the absolute-value penalty function, most are nondifferentiable at the solution. Such penalty functions are for this reason difficult to use directly; special descent algorithms for nondifferentiable objective functions have been developed, but they can be cumbersome. Furthermore, although these penalty functions are exact for a large enough c, it is not known at the outset what magnitude is sufficient. In practice a progression of c's must often be used. Because of these difficulties, the major use of exact penalty functions in nonlinear programming is as *merit functions–measuring* the progress of descent but not entering into the determination of the direction of movement. This idea is discussed in Chap. 15.

13.9 Summary

Penalty methods approximate a constrained problem by an unconstrained problem that assigns high cost to points that are far from the feasible region. As the approximation is made more exact (by letting the parameter c tend to infinity) the solution of the unconstrained penalty problem approaches the solution to the

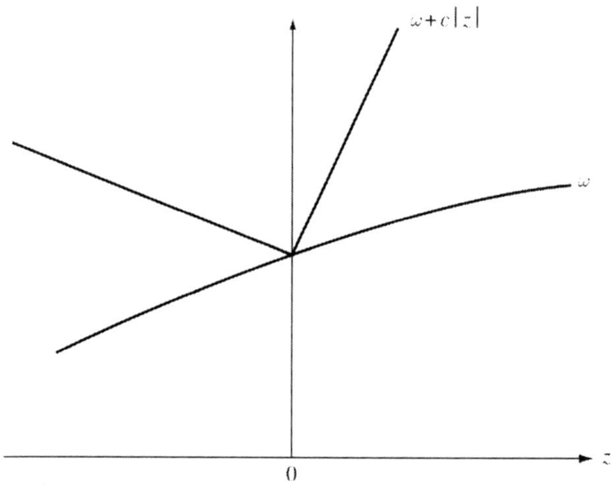

Fig. 13.7 Geometric interpretation of absolute-value penalty function

original constrained problem from outside the active constraints. Barrier methods, on the other hand, approximate a constrained problem by an (essentially) unconstrained problem that assigns high cost to being near the boundary of the feasible region, but unlike penalty methods, these methods are applicable only to problems having a robust feasible region. As the approximation is made more exact, the solution of the unconstrained barrier problem approaches the solution to the original constrained problem from inside the feasible region.

The objective functions of all penalty and barrier methods of the form $P(\mathbf{x}) = \gamma(h(\mathbf{x}))$, $B(\mathbf{x}) = \eta(g(\mathbf{x}))$ are ill-conditioned. If they are differentiable, then as $c \to \infty$ the Hessian (at the solution) is equal to the sum of $\mathbf{L}$, the Hessian of the Lagrangian associated with the original constrained problem, and a matrix of rank r that tends to infinity (where r is the number of active constraints). This is a fundamental property of these methods.

Effective exploitation of differentiable penalty and barrier functions requires that schemes be devised that eliminate the effect of the associated large eigenvalues. For this purpose the three general principles developed in earlier chapters, The Partial Conjugate Gradient Method, The Modified Newton Method, and The Combination of Steepest Descent and Newton's Method, when creatively applied, all yield methods that converge at approximately the canonical rate associated with the original constrained problem.

It is necessary to add a point of qualification with respect to some of the algorithms introduced in this chapter, lest it be inferred that they are offered as panaceas for the general programming problem. As has been repeatedly emphasized, the ideal study of convergence is a careful blend of analysis, good sense, and experimentation. The rate of convergence does not always tell the whole story, although it is often a major component of it. Although some of the algorithms presented in this chapter asymptotically achieve the canonical rate of convergence (at least approximately), for large c the points may have to be quite close to the solution before this rate characterizes the process. In other words, for large c the process may converge slowly in its initial phase, and, to obtain a truly representative analysis, one must look beyond the first-order convergence properties of these methods. For this reason many people find Newton's method attractive, although the work at each step can be substantial.

13.10 Exercises

1. Show that if $q(c, \mathbf{x})$ is continuous (with respect to $\mathbf{x}$) and $q(c, \mathbf{x}) \to \infty$ as $|\mathbf{x}| \to \infty$, then $q(c, \mathbf{x})$ has a minimum.
2. Suppose problem (13.1), with f continuous, is approximated by the penalty problem (13.2), and let $\{c_k\}$ be an increasing sequence of positive constants tending to infinity. Define $q(c, \mathbf{x}) = f(\mathbf{x}) + cP(\mathbf{x})$, and fix $\varepsilon > 0$. For each k let $\mathbf{x}_k$ be determined satisfying

$$q(c_k, \mathbf{x}_k) \leqslant [\min_{\mathbf{x}} q(c_k, \mathbf{x})] + \varepsilon.$$

Show that if $\mathbf{x}^*$ is a solution to (13.1), any limit point, $\overline{\mathbf{x}}$, of the sequence $\{\mathbf{x}_k\}$ is feasible and satisfies $f(\overline{\mathbf{x}}) \leqslant f(\mathbf{x}^*) + \varepsilon$.
3. Construct an example problem and a penalty function such that, as $c \to \infty$, the solution to the penalty problem diverges to infinity.
4. *Combined penalty and barrier method.* Consider a problem of the form

$$\begin{aligned} &\text{minimize} \;\; f(\mathbf{x}) \\ &\text{subject to} \;\; \mathbf{x} \in S \cap T \end{aligned}$$

and suppose P is a penalty function for S and B is a barrier function for T. Define

$$d(c, \mathbf{x}) = f(\mathbf{x}) + cP(\mathbf{x}) + \frac{1}{c} B(\mathbf{x}).$$

Let $\{c_k\}$ be a sequence $c_k \to \infty$, and for $k = 1, 2, \ldots$ let $\mathbf{x}_k$ be a solution to

$$\text{minimize } d(c_k, \mathbf{x})$$

subject to $\mathbf{x} \in$ interior of T. Assume all functions are continuous, T is compact (and robust), the original problem has a solution $\mathbf{x}^*$, and that $S \cap$ [interior of T] is not empty. Show that

(a) $\lim_{k \in \infty} d(c_k, \mathbf{x}_k) = f(\mathbf{x}^*)$.
(b) $\lim_{k \in \infty} c_k P(\mathbf{x}_k) = 0$.
(c) $\lim_{k \in \infty} \frac{1}{c_k} B(\mathbf{x}_k) = 0$.

5. Prove the Theorem at the end of Sect. 13.2.
6. Find the central path for the problem of minimizing x^2 subject to $x \geqslant 0$.
7. Consider a penalty function for the equality constraints

$$\mathbf{h}(\mathbf{x}) = \mathbf{0}, \ \mathbf{h}(\mathbf{x}) \in E^m,$$

having the form

$$P(\mathbf{x}) = \gamma(\mathbf{h}(\mathbf{x})) = \sum_{i=1}^{m} w(h_i(\mathbf{x})),$$

where w is a function whose derivative w' is analytic and has a zero of order $s \geqslant 1$ at zero.

(a) Show that corresponding to (13.26) we have

$$\mathbf{Q}(c_k, \ \mathbf{x}_k) = L_k(\mathbf{x}_k) + c_k \sum_{i=1}^{m} \{w''(h_i(\mathbf{x}_k))\} \nabla h_i(\mathbf{x}_k)^T \nabla h_i(\mathbf{x}_k).$$

(b) Show that as $c_k \to \infty$, m eigenvalues of $\mathbf{Q}(c_k, \ \mathbf{x}_k)$ have magnitude on the order of $(c_k)^{1/s}$.

8. Corresponding to the problem

$$\begin{aligned} &\text{minimize} \ \ f(\mathbf{x}) \\ &\text{subject to} \ \ g(\mathbf{x}) \leqslant 0, \end{aligned}$$

consider the sequence of unconstrained problems

$$\text{minimize} \ \ f(\mathbf{x}) + [g^+(\mathbf{x}) + 1]^k - 1,$$

and suppose $\mathbf{x}_k$ is the solution to the kth problem.

(a) Find an appropriate definition of a Lagrange multiplier λ_k to associate with $\mathbf{x}_k$.
(b) Find the limiting form of the Hessian of the associated objective function, and determine how fast the largest eigenvalues tend to infinity.

9. Repeat Exercise 8 for the sequence of unconstrained problems

$$\text{minimize} \ \ f(\mathbf{x}) + [(g(\mathbf{x}) + 1)^+]^k.$$

10. *Morrison 's method.* Suppose the problem

$$\begin{aligned} &\text{minimize} \ \ f(\mathbf{x}) \\ &\text{subject to} \ \ \mathbf{h}(\mathbf{x}) = \mathbf{0} \end{aligned} \tag{13.70}$$

has solution $\mathbf{x}^*$. Let M be an optimistic estimate of $f(\mathbf{x}^*)$, that is, $M \leqslant f(\mathbf{x}^*)$. Define $v(M, \ \mathbf{x}) = [f(\mathbf{x}) - M]^2 + |\mathbf{h}(\mathbf{x})|^2$ and define the unconstrained problem

$$\text{minimize} \ \ v(M, \ \mathbf{x}). \tag{13.71}$$

Given $M_k \leqslant f(\mathbf{x}^*)$, a solution $\mathbf{x}_{M_k}$ to the corresponding problem (13.71) is found, then M_k is updated through

$$M_{k+1} = M_k + [v(M_k, \mathbf{x}_{M_k})]^{1/2} \tag{13.72}$$

and the process repeated.

(a) Show that if $M = f(\mathbf{x}^*)$, a solution to (13.71) is a solution to (13.70).
(b) Show that if $\mathbf{x}_M$ is a solution to (13.71), then $f(\mathbf{x}_M) \leqslant f(\mathbf{x}^*)$.
(c) Show that if $M_k \leqslant f(\mathbf{x}^*)$ then M_{k+1} determined by (13.72) satisfies $M_{k+1} \leqslant f(\mathbf{x}^*)$.
(d) Show that $M_k \to f(\mathbf{x}^*)$.
(e) Find the Hessian of $v(M, \mathbf{x})$ (with respect to $\mathbf{x}^*$). Show that, to within a scale factor, it is identical to that associated with the standard penalty function method.

11. Let A be an $m \times n$ matrix of rank m. Prove the matrix identity

$$[\mathbf{I} + \mathbf{A}^T\mathbf{A}]^{-1} = \mathbf{I} - \mathbf{A}^T [\mathbf{I} + \mathbf{A}\mathbf{A}^T]^{-1}\mathbf{A}$$

and discuss how it can be used in conjunction with the method of Sect. 13.4.

12. Show that in the limit of large c, a single cycle of the normalization method of Sect. 13.6 is exactly the same as a single cycle of the combined penalty function and gradient projection method of Sect. 13.7.
13. Suppose that at some step k of the combined penalty function and gradient projection method, the $m \times n$ matrix $\nabla \mathbf{h}(\mathbf{x}_k)$ is not of rank m. Show how the method can be continued by temporarily executing the Newton step over a subspace of dimension less than m.
14. For a problem with equality constraints, show that in the combined penalty function and gradient projection method the second step (the steepest descent step) can be replaced by a step in the direction of the negative projected gradient (projected onto M_k) without destroying the global convergence property and without changing the rate of convergence.
15. Develop a method that is analogous to that of Sect. 13.7, but which is a combination of penalty functions and the reduced gradient method. Establish that the rate of convergence of the method is identical to that of the reduced gradient method.
16. Extend the result of the Exact Penalty Theorem of Sect. 13.8 to inequalities. Write $g_j(\mathbf{x}) \leqslant 0$ in the form of an equality as $g_j(\mathbf{x}) + y_j^2 = 0$ and show that the original theorem applies.
17. Develop a result analogous to that of the Exact Penalty Theorem of Sect. 13.8 for the penalty function

$$P(\mathbf{x}) = \max\{0, g_i(\mathbf{x}), g_2(\mathbf{x}), \ldots, g_p(\mathbf{x}), |h_i(\mathbf{x})|, |h_2(\mathbf{x})|, \ldots, |h_m(\mathbf{x})|\}.$$

18. Solve the problem

$$\begin{aligned} &\text{minimize} \quad x^2 + xy + y^2 - 2y \\ &\text{subject to} \quad x + y = 2 \end{aligned}$$

three ways analytically

(a) with the necessary conditions.
(b) with a quadratic penalty function.
(c) with an exact penalty function.

References

13.1 The penalty approach to constrained optimization is generally attributed to Courant [C8]. For more details than presented here, see Butler and Martin [B26] or Zangwill [Z1].

13.2 The barrier method is due to Carroll [C1], but was developed and popularized by Fiacco and McCormick [F4] who proved the general effectiveness of the method. Also see Frisch [F19].

13.3 It has long been known that penalty problems are solved slowly by steepest descent, and the difficulty has been traced to the ill-conditioning of the Hessian. The explicit characterization given here is a generalization of that in Luenberger [L10]. For the geometric interpretation, see Luenberger [L8]. The central path for nonlinear programming was analyzed by Nesterov and Nemirovskii [N2], Jarre [J2] and den Hertog [H6].

13.5 Most previous successful implementations of penalty or barrier methods have employed Newton's method to solve the unconstrained problems and thereby have largely avoided the effects of the ill-conditioned Hessian. See Fiacco and McCormick [F4] for some suggestions. The technique at the end of the section is new.

13.6 This method was first presented in Luenberger [L13].

13.8 See Luenberger [L10], for further analysis of this method.

13.9 The fact that the absolute-value penalty function is exact was discovered by Zangwill [Z1]. The fact that $c > |\lambda|$ is sufficient for exactness was pointed out by Luenberger [L12]. Line search methods have been developed for nonsmooth functions. See Lemarechal and Mifflin [L3].

13.10 For analysis along the lines of Exercise 7, see Lootsma [L7]. For the functions suggested in Exercises 8 and 9, see Levitin and Polyak [L5]. For the method of Exercise 10, see Morrison [M8].

Chapter 14
Duality and Dual Methods

We first derive the duality theory of for constrained optimization, which is based on our earlier *zero-order optimality conditions* and the Lagrangian relaxations. The variables of the dual are typically the Lagrange multipliers associated with the constraints in the primal problem—the original constrained optimization problem.

Thus, dual methods are based on the viewpoint that it is the Lagrange multipliers which are the fundamental unknowns associated with a constrained problem; once these multipliers are known determination of the solution point is simple (at least in some situations). Dual methods, therefore, do not attack the original constrained problem directly but instead attack an alternate problem, the dual problem, whose unknowns are the Lagrange multipliers of the first problem. For a problem with n variables and m equality constraints, dual methods thus work in the m-dimensional space of Lagrange multipliers. Because Lagrange multipliers measure sensitivities and hence often have meaningful intuitive interpretations as prices associated with constraint resources, searching for these multipliers, is often, in the context of a given practical problem, as appealing as searching for the values of the original problem variables.

The study of dual methods, and more particularly the introduction of the dual problem, precipitates some extensions of earlier concepts. One interesting feature of this chapter is the calculation of the Hessian of the dual problem and the discovery of a *dual canonical convergence ratio* associated with a constrained problem that governs the convergence of steepest ascent applied to the dual.

The convergence ratio theory lead to a popular method, the method of multipliers based on the augmented Lagrangian, in which the Hessian condition would be significantly improved to facilitate faster convergence.

The alternate direction method of multipliers is based on an idea resembling that in the coordinate descent method. Here, the gradient of the dual is calculated approximately in a block coordinate fashion using primal variables. This method is particularly effective for large-scale optimization.

D.G. Luenberger, Y. Ye, *Linear and Nonlinear Programming*, International Series in Operations Research & Management Science 228,
DOI 10.1007/978-3-319-18842-3_14

Cutting plane algorithms, exceedingly elementary in principle, develop a series of ever-improving approximating linear programs, whose solutions converge to the solution of the original problem. The methods differ only in the manner by which an improved approximating problem is constructed once a solution to the old approximation is known. The theory associated with these algorithms is, unfortunately, scant and their convergence properties are not particularly attractive. They are, however, often very easy to implement.

14.1 Global Duality

Duality in nonlinear programming takes its most elegant form when it is formulated globally in terms of sets and hyperplanes that touch those sets. This theory makes clear the role of Lagrange multipliers as defining hyperplanes which can be considered as dual to points in a vector space. The theory provides a symmetry between primal and dual problems and this symmetry can be considered as perfect for convex problems. For non-convex problems the "imperfection" is made clear by the duality gap which has a simple geometric interpretation. The global theory, which is presented in this section, serves as useful background when later we specialize to a local duality theory that can be used even without convexity and which is central to the understanding of the convergence of dual algorithms.

As a counterpoint to Sect. 11.9 where equality constraints were considered before inequality constraints, here we shall first consider a problem with inequality constraints. In particular, consider the problem

$$\begin{aligned} \text{minimize} \quad & f(\mathbf{x}) \\ \text{subject to} \quad & \mathbf{g}(\mathbf{x}) \leq \mathbf{0} \\ & \mathbf{x} \in \mathbf{\Omega}. \end{aligned} \tag{14.1}$$

$\mathbf{\Omega} \subset E^n$ is a convex set, and the functions f and $\mathbf{g}$ are defined on $\mathbf{\Omega}$. The function $\mathbf{g}$ is p-dimensional. The problem is not necessarily convex, but we assume that there is a feasible point. Recall that the primal function associated with (14.1) is defined for $\mathbf{z} \in E^p$ as

$$\omega(\mathbf{z}) = \inf\{f(\mathbf{x}) : \mathbf{g}(\mathbf{x}) \leq \mathbf{z},\ \mathbf{x} \in \mathbf{\Omega}\}, \tag{14.2}$$

defined by letting the right hand side of inequality constraint take on arbitrary values. It is understood that (14.2) is defined on the set $D = \{\mathbf{z} : \mathbf{g}(\mathbf{x}) \leq \mathbf{z}$, for some $\mathbf{x} \in \mathbf{\Omega}\}$.

If problem (14.1) has a solution $\mathbf{x}^*$ with value $f^* = f(\mathbf{x}^*)$, then f^* is the point on the vertical axis in E^{p+1} where the primal function passes through the axis. If (14.1) does not have a solution, then $f^* = \inf\{f(\mathbf{x}) : \mathbf{g}(\mathbf{x}) \leq \mathbf{0},\ \mathbf{x} \in \mathbf{\Omega}\}$ is the intersection point.

The duality principle is derived from consideration of all hyperplanes that lie below the primal function. As illustrated in Fig. 14.1 the intercept with the vertical axis of such a hyperplanes lies below (or at) the value f^*.

To express this property we define the *dual function* defined on the positive cone in E^p as

$$\phi(\boldsymbol{\mu}) = \inf\{f(\mathbf{x}) + \boldsymbol{\mu}^T\mathbf{g}(\mathbf{x}) : \mathbf{x} \in \boldsymbol{\Omega}\}. \tag{14.3}$$

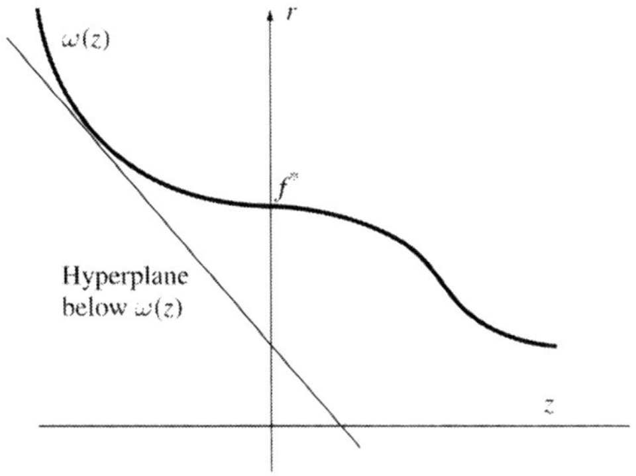

Fig. 14.1 Hyperplane below $\omega(\mathbf{z})$

In general, ϕ may not be finite throughout the positive orthant E^p_+ but the region where it is finite is convex.

Proposition 1. *The dual function is concave on the region where it is finite.*

Proof. Suppose $\boldsymbol{\mu}_1,\ \boldsymbol{\mu}_2$ are in the finite region, and let $0 \leq \alpha \leq 1$. Then

$$\begin{aligned}\phi(\alpha\boldsymbol{\mu}_1 + (1-\alpha\boldsymbol{\mu}_2)) &= \inf\{f(\mathbf{x}) + (\alpha\boldsymbol{\mu}_1 + (1-\alpha)\boldsymbol{\mu}_2)^T\mathbf{g}(\mathbf{x}) : \mathbf{x} \in \boldsymbol{\Omega}\}\\ &\geq \inf\{\alpha f(\mathbf{x}_1) + \alpha\boldsymbol{\mu}_1^T\mathbf{g}(\mathbf{x}_1) : \mathbf{x}_1 \in \boldsymbol{\Omega}\}\\ &+ \inf\{(1-\alpha)f(\mathbf{x}_2) + (1-\alpha)\boldsymbol{\mu}_2^T\mathbf{g}(\mathbf{x}_2) : \mathbf{x}_2 \in \boldsymbol{\Omega}\}\\ &= \alpha\phi(\boldsymbol{\mu}_1) + (1-\alpha)\phi(\boldsymbol{\mu}_2). \blacksquare\end{aligned}$$

We define $\phi^* = \sup\{\phi(\boldsymbol{\mu}) : \boldsymbol{\mu} \geq \mathbf{0}\}$ where it is understood that the supremum is taken over the region where ϕ is finite. We can now state the weak form of global duality.

Weak Duality Proposition. $\phi^* \leq f^*$.

Proof. For every $\boldsymbol{\mu} \geq \mathbf{0}$ we have

$$\begin{aligned}\phi(\boldsymbol{\mu}) &= \inf\{f(\mathbf{x}) + \boldsymbol{\mu}^T\mathbf{g}(\mathbf{x}) : \mathbf{x} \in \boldsymbol{\Omega}\}\\ &\leq \inf\{f(\mathbf{x}) + \boldsymbol{\mu}^T\mathbf{g}(\mathbf{x}) : \mathbf{g}(\mathbf{x}) \leq \mathbf{0},\ \mathbf{x} \in \boldsymbol{\Omega}\}\\ &\leq \inf\{f(\mathbf{x}) : \mathbf{g}(\mathbf{x}) \leq \mathbf{0},\ \mathbf{x} \in \boldsymbol{\Omega}\} = f^*.\end{aligned}$$

Taking the supremum over the left hand side gives $\phi^* \leq f^*$. $\blacksquare$

Hence the dual function gives lower bounds on the optimal value f^*.

This dual function has a strong geometric interpretation. Consider a $p+1$-dimensional vector $(1, \boldsymbol{\mu}) \in E^{p+1}$ with $\boldsymbol{\mu} \geq \mathbf{0}$ and a constant c. The set of vectors $(r, \mathbf{z})$ such that the inner product $(1, \boldsymbol{\mu})^T(r, \mathbf{z}) \equiv r + \boldsymbol{\mu}^T\mathbf{z} = c$ defines a hyperplane in E^{p+1}. Different values of c give different hyperplanes, all of which are parallel.

For a given $(1, \boldsymbol{\mu})$ we consider the lowest possible hyperplane of this form that just barely touches (supports) the region above the primal function of problem (14.1). Suppose $\mathbf{x}_1$ defines the touching point with values $r = f(\mathbf{x}_1)$ and $\mathbf{z} = \mathbf{g}(\mathbf{x}_1)$. Then $c = f(\mathbf{x}_1) + \boldsymbol{\mu}^T\mathbf{g}(\mathbf{x}_1) = \phi(\boldsymbol{\mu})$.

The hyperplane intersects the vertical axis at a point of the form $(r_0, \mathbf{0})$. This point also must satisfy $(1, \boldsymbol{\mu})^T(r_0, \mathbf{0}) = c = \phi(\boldsymbol{\mu})$. This gives $c = r_0$. Thus the intercept gives $\phi(\boldsymbol{\mu})$ directly. Thus the dual function at $\boldsymbol{\mu}$ is equal to the intercept of the hyperplane defined by $\boldsymbol{\mu}$ that just touches the epigraph of the primal function.

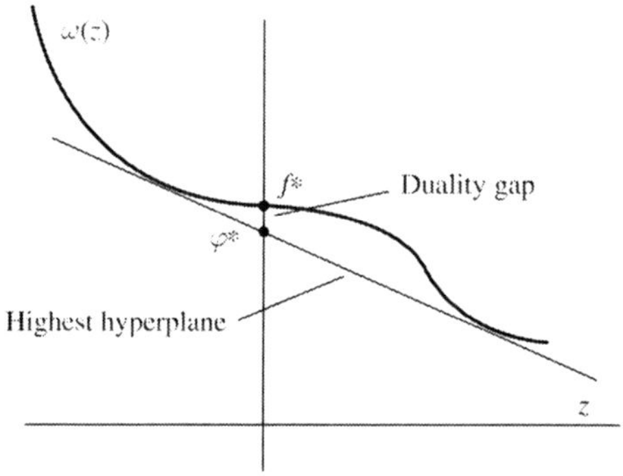

Fig. 14.2 The highest hyperplane

Furthermore, this intercept (and dual function value) is maximized by the Lagrange multiplier which corresponds to the largest possible intercept, at a point no higher than the optimal value f^*. See Fig. 14.2.

By introducing convexity assumptions, the foregoing analysis can be strengthened to give the strong duality theorem, with no duality gap when the intercept is at f^*. See Fig. 14.3.

We shall state the result for the more general problem that includes equality constraints of the form $\mathbf{h}(\mathbf{x}) = \mathbf{0}$, as in Sect. 11.9. Specifically, we consider the problem

$$\begin{aligned} &\text{maximize} \quad f(\mathbf{x}) \\ &\text{subject to} \quad \mathbf{h}(\mathbf{x}) = \mathbf{0}, \; \mathbf{g}(\mathbf{x}) \leq \mathbf{0} \\ &\qquad\qquad\quad \mathbf{x} \in \boldsymbol{\Omega} \end{aligned} \tag{14.4}$$

where $\mathbf{h}$ is affine of dimension m, $\mathbf{g}$ is convex of dimension p, and $\boldsymbol{\Omega}$ is a convex set.

In this case the dual function is

$$\phi(\boldsymbol{\lambda}, \boldsymbol{\mu}) = \inf\{f(\mathbf{x}) + \boldsymbol{\lambda}^T\mathbf{h}(\mathbf{x}) + \boldsymbol{\mu}^T\mathbf{g}(\mathbf{x}) : \mathbf{x} \in \boldsymbol{\Omega}\}.$$

And let

$$\phi^* = \sup\{\phi(\lambda, \mu) : \lambda \in E^m, \mu \in E^p, \mu \geq \mathbf{0}\}.$$

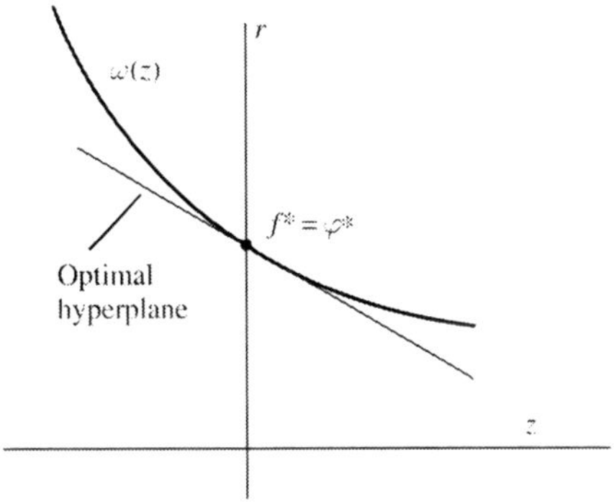

Fig. 14.3 The strong duality theorem. There is no duality gap

Strong Duality Theorem. *Suppose in the problem* (14.4), $\mathbf{h}$ *is affine and regular with respect to* $\boldsymbol{\Omega}$ *and there is a point* $\mathbf{x}_1 \in \boldsymbol{\Omega}$ *with that* $\mathbf{h}(\mathbf{x}) = \mathbf{0}$ *and* $\mathbf{g}(\mathbf{x}) < \mathbf{0}$.

Suppose the problem has solution $\mathbf{x}^*$ *with value* $f(\mathbf{x}^*) = f^*$. *Then for every* λ *and* $\mu \geq \mathbf{0}$ *there holds*

$$\phi^* \leq f^*.$$

Furthermore, there are λ, $\mu \geq \mathbf{0}$ *such that*

$$\phi(\lambda, \mu) = f^*$$

and hence $\phi^* = f^*$. *Moreover, the* λ *and* μ *above are Lagrange multipliers for the original problem.*

Proof. The proof follows almost immediately from the zero-order Lagrange theorem of Sect. 11.9. The Lagrange multipliers of that theorem give

$$\begin{aligned} f^* &= \max\{f(\mathbf{x}) + \lambda^T\mathbf{h}(\mathbf{x}) + \mu^T\mathbf{g}(\mathbf{x}) : \mathbf{x} \in \boldsymbol{\Omega}\} \\ &= \phi(\lambda, \mu) \leq \phi^* \leq f^*. \end{aligned}$$

Equality must hold across the inequalities, which establishes the results. ∎

As a nice summary we can place the primal and dual problems together for the problem with inequality constraints.

Primal	Dual
$f^* = \text{minimize } \omega(\mathbf{z})$	$\phi^* = \text{maximize } \phi(\mu)$
subject to $\mathbf{z} \leq \mathbf{0}$	subject to $\mu \geq \mathbf{0}$.

Example 1 (Quadratic Program). Consider the problem

$$\begin{aligned}&\text{minimize } \frac{1}{2}\mathbf{x}^T\mathbf{Q}\mathbf{x}\\&\text{subject to } \mathbf{B}\mathbf{x}-\mathbf{b}\le\mathbf{0}.\end{aligned}\tag{14.5}$$

The dual function is

$$\phi(\boldsymbol{\mu}) = \min_{\mathbf{x}} \frac{1}{2}\mathbf{x}^T\mathbf{Q}\mathbf{x} + \boldsymbol{\mu}^T(\mathbf{B}\mathbf{x}-\mathbf{b}).$$

This gives the necessary conditions

$$\mathbf{Q}\mathbf{x} + \mathbf{B}^T\boldsymbol{\mu} = \mathbf{0}$$

and hence $\mathbf{x} = -\mathbf{Q}^{-1}\mathbf{B}^T\boldsymbol{\mu}$. Substituting this into $\phi(\boldsymbol{\mu})$ gives

$$\phi(\mu) = -\frac{1}{2}\boldsymbol{\mu}^T\mathbf{B}\mathbf{Q}^{-1}\mathbf{B}^T\boldsymbol{\mu} - \boldsymbol{\mu}^T\mathbf{b}.$$

Hence the dual problem is

$$\begin{aligned}&\text{maximize } -\frac{1}{2}\boldsymbol{\mu}^T\mathbf{B}\mathbf{Q}^{-1}\mathbf{B}^T\boldsymbol{\mu} - \boldsymbol{\mu}^T\mathbf{b}\\&\text{subject to } \quad \boldsymbol{\mu}\ge\mathbf{0},\end{aligned}\tag{14.6}$$

which is also a quadratic programming problem. If this problem is solved for $\boldsymbol{\mu}$, that $\boldsymbol{\mu}$ will be the Lagrange multiplier for the primal problem (14.5).

Note that the first-order conditions for the dual problem (14.6) imply

$$\boldsymbol{\mu}^T[-\mathbf{B}\mathbf{Q}^{-1}\mathbf{B}^T\boldsymbol{\mu} - \mathbf{b}] = 0,$$

which by substituting the formula for $\mathbf{x}$ is equivalent to

$$\boldsymbol{\mu}^T[\mathbf{B}\mathbf{x}-\mathbf{b}] = 0.$$

This is the complementary slackness condition for the original (primal) problem (14.5).

Example 2 (Integer Solutions). Duality gaps may arise if the object function or the constraint functions are not convex. A gap may also arise if the underlying set is not convex. This is characteristic, for example, of problems in which the components of the solution vector are constrained to be integers. For instance, consider the problem

$$\begin{aligned}&\text{minimize} \quad x_1^2 + 2x_2^2\\&\text{subject to} \quad x_1 + x_2 \ge 1/2\\&\qquad\qquad\quad x_1,\ x_2 \text{ nonnegative integers}\end{aligned}$$

It is clear that the solution is $x_1 = 1$, $x_2 = 0$, with objective value $f^* = 1$. To put this problem in the standard form we have discussed, we write the constraint as

$$-x_1 - x_2 + 1/2 \leq z, \quad \text{where } z = 0.$$

The primal function $\omega(z)$ is equal to 0 for $z \geq 1/2$ since then $x_1 = x_2 = 0$ is feasible. The entire primal function has steps as z steps negatively integer by integer, as shown in Fig. 14.4.

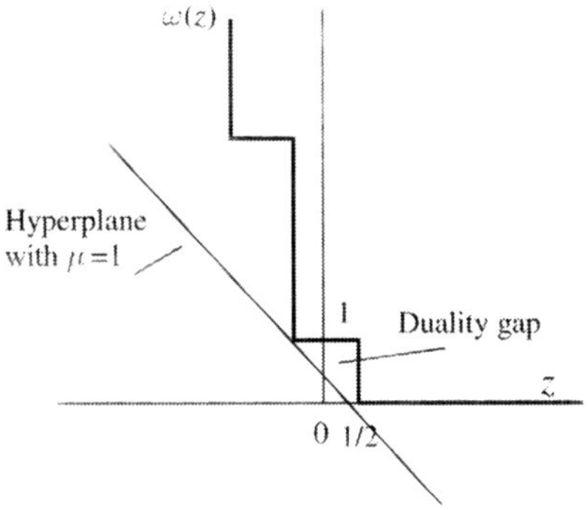

Fig. 14.4 Duality for an integer problem

The dual function is

$$\phi(\mu) = \max\{x_1^2 + x_2^2 - \lambda(x_1 + x_2 - 1/2)\}$$

where the maximum is taken with respect to the integer constraint. Analytically, the solution for small values of μ is

$$\begin{aligned} \phi(\mu) &= \mu/2 && \text{for } 0 \leq \mu \leq 1, \\ &= 1 - \mu/2 && \text{for } 1 \leq \mu \leq 2, \\ &\vdots && \text{and more} \end{aligned}$$

The maximum value of $\phi(\mu)$ is the maximum intercept of the corresponding hyperplanes (lines, in this case) with the vertical axis. This occurs for $\mu = 1$ with a corresponding value of $\phi^* = \phi(1) = 1/2$. We have $\phi^* < f^*$ and the difference $f^* - \phi^* = 1/2$ is the duality gap.

14.2 Local Duality

In practice the mechanics of duality are frequently carried out locally, by setting derivatives to zero, or moving in the direction of a gradient. For these operations the beautiful global theory can in large measure be replaced by a weaker but often more useful local theory. This theory requires a minimum of convexity assumptions defined locally. We present such a theory in this section, since it is in keeping with the spirit of the earlier chapters and is perhaps the simplest way to develop computationally useful duality results.

As often done before for convenience, we again consider nonlinear programming problems of the form

$$\begin{aligned} \text{minimize} \quad & f(\mathbf{x}) \\ \text{subject to} \quad & \mathbf{h}(\mathbf{x}) = \mathbf{0}, \end{aligned} \tag{14.7}$$

where $\mathbf{x} \in E^n$, $\mathbf{h}(\mathbf{x}) \in E^n$ and $f, \mathbf{h} \in C^2$. Global convexity is not assumed here. Everything we do can be easily extended to problems having inequality as well as equality constraints, for the price of a somewhat more involved notation.

We focus attention on a local solution $\mathbf{x}^*$ of (14.7). Assuming that $\mathbf{x}^*$ is a regular point of the constraints, then, as we know, there will be a corresponding Lagrange multiplier (row) vector λ^* such that

$$\nabla f(\mathbf{x}^*) + (\lambda^*)^T \nabla \mathbf{h}(\mathbf{x}^*) = \mathbf{0}, \tag{14.8}$$

and the Hessian of the Lagrangian

$$\mathbf{L}(\mathbf{x}^*) = \mathbf{F}(\mathbf{x}^*) + (\lambda^*)^T \mathbf{H}(\mathbf{x}^*) \tag{14.9}$$

must be positive semidefinite on the tangent subspace

$$M = \{\mathbf{x} : \nabla \mathbf{h}(\mathbf{x}^*)\mathbf{x} = \mathbf{0}\}.$$

At this point we introduce the special local convexity assumption necessary for the development of the local duality theory. Specifically, we assume that the Hessian $\mathbf{L}(\mathbf{x}^*)$ is positive definite. Of course, it should be emphasized that by this we mean $\mathbf{L}(\mathbf{x}^*)$ is positive definite on the whole space E^n, not just on the subspace M. The assumption guarantees that the Lagrangian $l(\mathbf{x}) = f(\mathbf{x}) + (\lambda^*)^T \mathbf{h}(\mathbf{x})$ is locally convex at $\mathbf{x}^*$.

With this assumption, the point $\mathbf{x}^*$ is not only a local solution to the constrained problem (14.7); it is also a local solution to the unconstrained problem

$$\text{minimize} \quad f(\mathbf{x}) + (\lambda^*)^T \mathbf{h}(\mathbf{x}), \tag{14.10}$$

since it satisfies the first- and second-order sufficiency conditions for a local minimum point. Furthermore, for any λ sufficiently close to λ^* the function $f(\mathbf{x}) + \lambda^T \mathbf{h}(\mathbf{x})$ will have a local minimum point at a point $\mathbf{x}$ near $\mathbf{x}^*$. This follows by noting that, by the Implicit Function Theorem, the equation

$$\nabla f(\mathbf{x}) + \lambda^T \nabla \mathbf{h}(\mathbf{x}) = \mathbf{0} \tag{14.11}$$

has a solution $\mathbf{x}$ near $\mathbf{x}^*$ when λ is near λ^*, because $\mathbf{L}^*$ is nonsingular; and by the fact that, at this solution $\mathbf{x}$, the Hessian $\mathbf{F}(\mathbf{x}) + \lambda^T \mathbf{H}(\mathbf{x})$ is positive definite. Thus locally there is a unique correspondence between λ and $\mathbf{x}$ through solution of the unconstrained problem

$$\text{minimize} \quad f(\mathbf{x}) + \lambda^T \mathbf{h}(\mathbf{x}). \tag{14.12}$$

Furthermore, this correspondence is continuously differentiable.

Near λ^* we define the *dual function* ϕ by the equation

$$\phi(\lambda) = \text{minimum}\,[f(\mathbf{x}) + \lambda^T \mathbf{h}(\mathbf{x})], \tag{14.13}$$

where here it is understood that the minimum is taken locally with respect to $\mathbf{x}$ near $\mathbf{x}^*$. We are then able to show (and will do so below) that locally the original constrained problem (14.7) is equivalent to unconstrained local maximization of the dual function ϕ with respect to λ. Hence we establish an equivalence between a constrained problem in $\mathbf{x}$ and an unconstrained problem in λ.

To establish the duality relation we must prove two important lemmas. In the statements below we denote by $\mathbf{x}(\lambda)$ the unique solution to (14.12) in the neighborhood of $\mathbf{x}^*$.

Lemma 1. *The dual function ϕ has gradient*

$$\nabla\phi(\lambda) = \mathbf{h}(\mathbf{x}(\lambda))^T \tag{14.14}$$

Proof. We have explicitly, from (14.13),

$$\phi(\lambda) = f(\mathbf{x}(\lambda)) + \lambda^T \mathbf{h}(\mathbf{x}(\lambda)).$$

Thus

$$\nabla\phi(\lambda) = [\nabla f(\mathbf{x}(\lambda)) + \lambda^T \nabla\mathbf{h}(\mathbf{x}(\lambda))]\nabla\mathbf{x}(\lambda) + \mathbf{h}(\mathbf{x}(\lambda))^T.$$

Since the first term on the right vanishes by definition of $\mathbf{x}(\lambda)$, we obtain (14.14). ∎

Lemma 1 is of extreme practical importance, since it shows that the gradient of the dual function is simple to calculate. Once the dual function itself is evaluated, by minimization with respect to $\mathbf{x}$, the corresponding $\mathbf{h}(\mathbf{x})^T$, which is the gradient, can be evaluated without further calculation.

The Hessian of the dual function can be expressed in terms of the Hessian of the Lagrangian. We use the notation $\mathbf{L}(\mathbf{x}, \lambda) = \mathbf{F}(\mathbf{x}) + \lambda^T \mathbf{H}(\mathbf{x})$, explicitly indicating the dependence on λ. (We continue to use $\mathbf{L}(\mathbf{x}^*)$ when $\lambda = \lambda^*$ is understood.) We then have the following lemma.

Lemma 2. *The Hessian of the dual function is*

$$\mathbf{\Phi}(\lambda) = -\nabla\mathbf{h}(\mathbf{x}(\lambda))\mathbf{L}^{-1}(\mathbf{x}(\lambda), \lambda)\nabla\mathbf{h}(\mathbf{x}(\lambda))^T. \tag{14.15}$$

Proof. The Hessian is the derivative of the gradient. Thus, by Lemma 1,

$$\Phi(\lambda) = \nabla\mathbf{h}(\mathbf{x}(\lambda))\nabla\mathbf{x}(\lambda). \tag{14.16}$$

By definition we have

$$\nabla f(\mathbf{x}(\lambda)) + \lambda^T \nabla \mathbf{h}(\mathbf{x}(\lambda)) = \mathbf{0},$$

and differentiating this with respect to λ we obtain

$$\mathbf{L}(\mathbf{x}(\lambda),\ \lambda)\nabla \mathbf{x}(\lambda) + \nabla \mathbf{h}(\mathbf{x}(\lambda))^T = \mathbf{0}.$$

Solving for $\nabla \mathbf{x}(\lambda)$ and substituting in (14.16) we obtain (14.15). ∎

Since $\mathbf{L}^{-1}(\mathbf{x}(\lambda))$ is positive definite, and since $\nabla \mathbf{h}(\mathbf{x}(\lambda))$ is of full rank near $\mathbf{x}^*$, we have as an immediate consequence of Lemma 2 that the $m \times m$ Hessian of ϕ is negative definite. As might be expected, this Hessian plays a dominant role in the analysis of dual methods.

Local Duality Theorem. Suppose that the problem

$$\begin{aligned} &\text{minimize} \quad f(\mathbf{x}) \\ &\text{subject to} \quad \mathbf{h}(\mathbf{x}) = \mathbf{0} \end{aligned} \tag{14.17}$$

has a local solution at $\mathbf{x}^*$ with *corresponding value* r^* and *Lagrange multiplier* λ^*. *Suppose also that* $\mathbf{x}^*$ *is a regular point of the constraints and that the corresponding Hessian of the Lagrangian* $\mathbf{L}^* = \mathbf{L}(\mathbf{x}^*)$ *is positive definite. Then the dual problem*

$$\text{maximize} \quad \phi(\lambda) \tag{14.18}$$

has a local solution at λ^* *with corresponding value* r^* *and* $\mathbf{x}^*$ *as the point corresponding to* λ^* *in the definition of* ϕ.

Proof. It is clear that $\mathbf{x}^*$ corresponds to λ^* in the definition of ϕ. Now at λ^* we have by Lemma 1

$$\nabla \phi(\lambda^*) = \mathbf{h}(\mathbf{x}^*)^T = \mathbf{0},$$

and by Lemma 2 the Hessian of ϕ is negative definite. Thus λ^* satisfies the first- and second-order sufficiency conditions for an unconstrained maximum point of ϕ. The corresponding value $\phi(\lambda^*)$ is found from the definition of ϕ to be r^*. ∎

Example 1. Consider the problem in two variables

$$\begin{aligned} &\text{minimize} \quad -xy \\ &\text{subject to} \quad (x-3)^2 + y^2 = 5. \end{aligned}$$

The first-order necessary conditions are

$$\begin{aligned} -y + (2x-6)\lambda &= 0 \\ -x + 2y\lambda &= 0 \end{aligned}$$

together with the constraint. These equations have a solution at

$$x = 4, \quad y = 2, \quad \lambda = 1.$$

The Hessian of the corresponding Lagrangian is

$$\mathbf{L} = \begin{bmatrix} 2 & -1 \\ -1 & 2 \end{bmatrix}.$$

Since this is positive definite, we conclude that the solution obtained is a local minimum. (It can be shown, in fact, that it is the global solution.)

Since $\mathbf{L}$ is positive definite, we can apply the local duality theory near this solution. We define

$$\phi(\lambda) = \min\{-xy + \lambda[(x-3)^2 + y^2 - 5]\},$$

which leads to

$$\phi(\lambda) = \frac{4\lambda + 4\lambda^3 - 80\lambda^5}{(4\lambda^2 - 1)^2}$$

valid for $\lambda > \frac{1}{2}$. It can be verified that ϕ has a local maximum at $\lambda = 1$.

Inequality Constraints

For problems having inequality constraints as well as equality constraints the above development requires only minor modification. Consider the problem

$$\begin{aligned} \text{minimize} \quad & f(\mathbf{x}) \\ \text{subject to} \quad & \mathbf{h}(\mathbf{x}) = \mathbf{0} \\ & \mathbf{g}(\mathbf{x}) \leq \mathbf{0}, \end{aligned} \tag{14.19}$$

where $\mathbf{g}(\mathbf{x}) \in E^p$, $\mathbf{g} \in C^2$ and everything else is as before. Suppose $\mathbf{x}^*$ is a local solution of (14.19) and is a regular point of the constraints. Then, as we know, there are Lagrange multipliers $\boldsymbol{\lambda}^*$ and $\boldsymbol{\mu}^* \geq \mathbf{0}$ such that

$$\nabla f(\mathbf{x}^*) + (\boldsymbol{\lambda}^*)^T \nabla \mathbf{h}(\mathbf{x}^*) + (\boldsymbol{\mu}^*)^T \nabla \mathbf{g}(\mathbf{x}^*) = \mathbf{0} \tag{14.20}$$

$$(\boldsymbol{\mu}^*)^T \mathbf{g}(\mathbf{x}^*) = 0. \tag{14.21}$$

We impose the local convexity assumptions that the Hessian of the Lagrangian

$$\mathbf{L}(\mathbf{x}^*) = \mathbf{F}(\mathbf{x}^*) + (\boldsymbol{\lambda}^*)^T \mathbf{H}(\mathbf{x}^*) + (\boldsymbol{\mu}^*)^T \mathbf{G}(\mathbf{x}^*) \tag{14.22}$$

is positive definite (on the whole space).

For $\boldsymbol{\lambda}$ and $\boldsymbol{\mu} \geq \mathbf{0}$ near $\boldsymbol{\lambda}^*$ and $\boldsymbol{\mu}^*$ we define the dual function

$$\phi(\boldsymbol{\lambda}, \boldsymbol{\mu}) = \min[f(\mathbf{x}) + \boldsymbol{\lambda}^T \mathbf{h}(\mathbf{x}) + \boldsymbol{\mu}^T \mathbf{g}(\mathbf{x})], \tag{14.23}$$

where the minimum is taken locally near $\mathbf{x}^*$. Then, it is easy to show, paralleling the development above for equality constraints, that ϕ achieves a local maximum with respect to $\boldsymbol{\lambda}$, $\boldsymbol{\mu} \geq \mathbf{0}$ at $\boldsymbol{\lambda}^*$, $\boldsymbol{\mu}^*$.

Partial Duality

It is not necessary to include the Lagrange multipliers of all the constraints of a problem in the definition of the dual function. In general, if the local convexity assumption holds, local duality can be defined with respect to any subset of functional constraints. Thus, for example, in the problem

$$\begin{aligned} \text{minimize} \quad & f(\mathbf{x}) \\ \text{subject to} \quad & \mathbf{h}(\mathbf{x}) = \mathbf{0} \\ & \mathbf{g}(\mathbf{x}) \leq \mathbf{0}, \end{aligned} \tag{14.24}$$

we might define the dual function with respect to only the equality constraints. In this case we would define

$$\phi(\boldsymbol{\lambda}) = \min_{\mathbf{g}(\mathbf{x}) \leq \mathbf{0}} \{f(\mathbf{x}) + \boldsymbol{\lambda}^T \mathbf{h}(\mathbf{x})\}, \tag{14.25}$$

where the minimum is taken locally near the solution $\mathbf{x}^*$ but constrained by the remaining constraints $\mathbf{g}(\mathbf{x}) \leq \mathbf{0}$. Again, the dual function defined in this way will achieve a local maximum at the optimal Lagrange multiplier $\boldsymbol{\lambda}^*$. The partial dual is especially useful when constraints $\mathbf{g}(\mathbf{x}) \leq \mathbf{0}$ are simple such as $\mathbf{x} \leq \mathbf{0}$ or in a box.

14.3 Canonical Convergence Rate of Dual Steepest Ascent

Constrained problems satisfying the local convexity assumption can be solved by solving the associated unconstrained dual problem, and any of the standard algorithms discussed in Chaps. 7 through 10 can be used for this purpose. Of course, the method that suggests itself immediately is the method of steepest ascent. It can be implemented by noting that, according to Lemma 1, Section 14.2, the gradient of ϕ is available almost without cost once ϕ itself is evaluated. Without some special properties, however, the method as a whole can be extremely costly to execute, since every evaluation of ϕ requires the solution of an unconstrained problem in the unknown $\mathbf{x}$. Nevertheless, as shown in the next section, many important problems do have a structure which is suited to this approach.

The method of steepest ascent, and other gradient-based algorithms, when applied to the dual problem will have a convergence rate governed by the eigenvalue structure of the Hessian of the dual function ϕ. At the Lagrange multiplier $\boldsymbol{\lambda}^*$ corresponding to a solution $\mathbf{x}^*$ this Hessian is (according to Lemma 2, Sect. 13.1)

$$\Phi = -\nabla \mathbf{h}(\mathbf{x}^*)(\mathbf{L}^*)^{-1} \nabla \mathbf{h}(\mathbf{x}^*)^T.$$

This expression shows that Φ is in some sense a restriction of the matrix $(\mathbf{L}^*)^{-1}$ to the subspace spanned by the gradients of the constraint functions, which is the orthogonal complement of the tangent subspace M. This restriction is not the orthogonal restriction of $(\mathbf{L}^*)^{-1}$ onto the complement of M since the particular representation of the constraints affects the structure of the Hessian. We see, however, that while the convergence of primal methods is governed by the restriction of $\mathbf{L}^*$ to M, the convergence of dual methods is governed by a restriction of $(\mathbf{L}^*)^{-1}$ to the orthogonal complement of M.

The *dual canonical convergence rate* associated with the original constrained problem, which is the rate of convergence of steepest ascent applied to the dual, is $(B-b)^2/(B+b)^2$ where b and B are, respectively, the smallest and largest eigenvalues of

$$-\Phi = \nabla \mathbf{h}(\mathbf{x}^*)(\mathbf{L}^*)^{-1}\nabla \mathbf{h}(\mathbf{x}^*)^T.$$

For locally convex programming problems, this rate is as important as the primal canonical rate.

Scaling

We conclude this section by pointing out a kind of complementarity that exists between the primal and dual rates. Suppose one calculates the primal and dual canonical rates associated with the locally convex constrained problem

$$\begin{aligned} &\text{minimize} \quad f(\mathbf{x}) \\ &\text{subject to} \quad \mathbf{h}(\mathbf{x}) = \mathbf{0}. \end{aligned}$$

If a change of primal variables $\mathbf{x}$ is introduced, the primal rate will in general change but the dual rate will not. On the other hand, if the constraints are transformed (by replacing them by $\mathbf{Th}(\mathbf{x}) = \mathbf{0}$ where $\mathbf{T}$ is a nonsingular $m \times m$ matrix), the dual rate will change but the primal rate will not.

14.4 Separable Problems and Their Duals

A structure that arises frequently in mathematical programming applications is that of the separable problem:

$$\text{minimize} \sum_{i=1}^{q} f_i(\mathbf{x}_i) \tag{14.26}$$

$$\text{subject to} \sum_{i=1}^{q} \mathbf{h}_i(\mathbf{x}_i) = \mathbf{0} \tag{14.27}$$

$$\sum_{i=1}^{q} \mathbf{g}_i(\mathbf{x}_i) \leq \mathbf{0}. \tag{14.28}$$

In this formulation the components of the n-vector $\mathbf{x}$ are partitioned into q disjoint groups, $\mathbf{x} = (\mathbf{x}_1, \mathbf{x}_2, \ldots, \mathbf{x}_q)$ where the groups may or may not have the same number of components. Both the objective function and the constraints separate into sums of functions of the individual groups. For each i, the functions f_i, $\mathbf{h}_i$, and $\mathbf{g}_i$ are twice continuously differentiable functions of dimensions 1, m, and p, respectively.

Example 1. Suppose that we have a fixed budget of, say, A dollars that may be allocated among n activities. If x_i dollars is allocated to the ith activity, then there will be a benefit (measured in some units) of $f_i(x_i)$. To obtain the maximum benefit within our budget, we solve the separable problem

$$\begin{aligned} \text{maximize} \quad & \sum_{i=1}^{n} f_i(x_i) \\ \text{subject to} \quad & \sum_{i=1}^{n} x_i \leqslant A \\ & x_i \geqslant 0. \end{aligned} \tag{14.29}$$

In the example $\mathbf{x}$ is partitioned into its individual components.

Example 2. Problems involving a series of decisions made at distinct times are often separable. For illustration, consider the problem of scheduling water release through a dam to produce as much electric power as possible over a given time interval while satisfying constraints on acceptable water levels. A discrete-time model of this problem is to

$$\begin{aligned} \text{maximize} \quad & \sum_{k=1}^{N} f(y(k),\ u(k)) \\ \text{subject to} \quad & y(k) = y(k-1) - u(k) + s(k),\ k = 1, \ldots, N \\ & c \leqslant y(k) \leqslant d,\ k = 1, \ldots, N \\ & 0 \leqslant u(k),\ k = 1, \ldots, N. \end{aligned}$$

Here $y(k)$ represents the water volume behind the dam at the end of period k, $u(k)$ represents the volume flow through the dam during period k, and $s(k)$ is the volume flowing into the lake behind the dam during period k from upper streams. The function f gives the power generation, and c and d are bounds on lake volume. The initial volume $y(0)$ is given.

In this example we consider $\mathbf{x}$ as the $2N$-dimensional vector of unknowns $y(k)$, $u(k)$, $k = 1, 2, \ldots, N$. This vector is partitioned into the pairs $\mathbf{x}_k = (y(k),\ u(k))$. The objective function is then clearly in separable form. The constraints can be viewed as being in the form (14.27) with $\mathbf{h}_k(\mathbf{x}_k)$ having dimension N and such that $\mathbf{h}_k(\mathbf{x}_k)$ is identically zero except in the k and $k+1$ components.

Decomposition

Separable problems are ideally suited to dual methods, because the required unconstrained minimization decomposes into small subproblems. To see this we recall that the generally most difficult aspect of a dual method is evaluation of the dual function. For a separable problem, if we associate λ with the equality constraints (14.27) and $\boldsymbol{\mu} \geqslant \mathbf{0}$ with the inequality constraints (14.28), the required dual function is

$$\phi(\lambda, \boldsymbol{\mu}) = \min \sum_{i=1}^{q} \left(f_i(\mathbf{x}_i) + \lambda^T \mathbf{h}_i(\mathbf{x}_i) + \boldsymbol{\mu}^T \mathbf{g}_i(\mathbf{x}_i) \right).$$

This minimization problem decomposes into the q separate problems

$$\min_{\mathbf{x}_i} \; f_i(\mathbf{x}_i) + \lambda^T \mathbf{h}_i(\mathbf{x}_i) + \boldsymbol{\mu}^T \mathbf{g}_i(\mathbf{x}_i).$$

The solution of these subproblems can usually be accomplished relatively efficiently, since they are of smaller dimension than the original problem.

Example 3. In Example 1 using duality with respect to the budget constraint, the ith subproblem becomes, for $\mu > 0$

$$\max_{x_i \geqslant 0} f_i(x_i) - \mu x_i,$$

which is only a one-dimensional problem. It can be interpreted as setting a benefit value μ for dollars and then maximizing total benefit from activity i, accounting for the dollar expenditure.

Example 4. In Example 1 using duality with respect to the equality constraints we denote the dual variables by $\lambda(k)$, $k = 1, 2, \ldots, N$. The kth subproblem becomes

$$\max_{\substack{c \leqslant y(k) \leqslant d \\ 0 \leqslant u(k)}} \{ f(y(k), u(k)) + [\lambda(k+1) - \lambda(k)] y(k) - \lambda(k)[u(k) - s(k)] \}$$

which is a two-dimensional optimization problem. Selection of $\lambda \in E^N$ decomposes the problem into separate problems for each time period. The variable $\lambda(k)$ can be regarded as a value, measured in units of power, for water at the beginning of period k. The kth subproblem can then be interpreted as that faced by an entrepreneur who leased the dam for one period. He can buy water for the dam at the beginning of the period at price $\lambda(k)$ and sell what he has left at the end of the period at price $\lambda(k+1)$. His problem is to determine $y(k)$ and $u(k)$ so that his net profit, accruing from sale of generated power and purchase and sale of water, is maximized.

Example 5 (The Hanging Chain). Consider again the problem of finding the equilibrium position of the hanging chain considered in Example 4, Sect. 11.3, and Example 1, Sect. 12.7. The problem is

$$\text{minimize} \sum_{i=1}^{n} c_i y_i$$
$$\text{subject to} \sum_{i=1}^{n} y_i = 0$$
$$\sum_{i=1}^{n} \sqrt{1 - y_i^2} = L,$$

where $c_i = n - i + \frac{1}{2}$, $L = 16$. This problem is locally convex, since as shown in Sect. 12.7 the Hessian of the Lagrangian is positive definite. The dual function is accordingly

$$\phi(\lambda, \mu) = \min \sum_{i=1}^{n} \left\{ c_i y_i + \lambda y_i + \mu \sqrt{1 - y_i^2} \right\} - L\mu.$$

Since the problem is separable, the minimization divides into a separate minimization for each y_i, yielding the equations

$$c_i + \lambda - \frac{\mu y_i}{\sqrt{1 - y_i^2}} = 0$$

or

$$(c_i + \lambda)^2 (1 - y_i^2) = \mu^2 y_i^2.$$

This yields

$$y_i = \frac{-(c_i + \lambda)}{[(c_i + \lambda)^2 + \mu^2]^{1/2}}. \tag{14.30}$$

The above represents a local minimum point provided $\mu < 0$; and the minus sign must be taken for consistency.

The dual function is then

$$\phi(\lambda, \mu) = \sum_{i=1}^{n} \left\{ \frac{-(c_i + \lambda)^2}{[(c_i + \lambda)^2 + \mu^2]^{1/2}} + \mu \left[\frac{\mu^2}{[(c_i + \lambda)^2 + \mu^2]} \right]^{1/2} \right\} - L\mu$$

or finally, using $\sqrt{\mu^2} = -\mu$ for $\mu < 0$,

$$\phi(\lambda, \mu) = -L\mu - \sum_{i=1}^{n} \sqrt{(c_i + \lambda)^2 + \mu^2}.$$

The correct values of λ and μ can be found by maximizing $\phi(\lambda, \mu)$. One way to do this is to use steepest ascent. The results of this calculation, starting at $\lambda = \mu = 0$, are shown in Table 14.1. The values of y_i can then be found from (14.30).

Table 14.1 Results of dual of chain problem

Iteration	Value	Final solution $\lambda = -10.00048$ $\mu = -6.761136$
0	−200.00000	$y_1 = -0.8147154$
1	−66.94638	$y_2 = -0.7825940$
2	−66.61959	$y_3 = -0.7427243$
3	−66.55867	$y_4 = -0.6930215$
4	−66.54845	$y_5 = -0.6310140$
5	−66.54683	$y_6 = -0.5540263$
6	−66.54658	$y_7 = -0.4596696$
7	−66.54654	$y_8 = -0.3467526$
8	−66.54653	$y_9 = -0.2165239$
9	−66.54653	$y_{10} = -0.0736802$

14.5 Augmented Lagrangian

One of the most effective general classes of nonlinear programming methods is the *augmented Lagrangian* methods, alternatively referred to as *methods of multiplier*. These methods can be viewed as a combination of penalty functions and local duality methods; the two concepts work together to eliminate many of the disadvantages associated with either method alone. The augmented Lagrangian for the equality constrained problem

$$\begin{aligned}&\text{minimize } \ f(\mathbf{x})\\&\text{subject to } \ \mathbf{h}(\mathbf{x}) = \mathbf{0}, \quad \mathbf{x} \in \Omega\end{aligned} \tag{14.31}$$

is the function

$$l_c(\mathbf{x}, \lambda) = f(\mathbf{x}) + \lambda^T \mathbf{h}(\mathbf{x}) + \frac{1}{2}c|\mathbf{h}(\mathbf{x})|^2$$

for some positive constant c. We shall briefly indicate how the augmented Lagrangian can be viewed as either a special penalty function or as the basis for a dual problem. These two viewpoints are then explored further in this and the next section.

From a penalty function viewpoint the augmented Lagrangian, for a fixed value of the vector λ, is simply the standard quadratic penalty function for the problem

$$\begin{aligned}&\text{minimize } \ f(\mathbf{x}) + \lambda^T \mathbf{h}(\mathbf{x})\\&\text{subject to } \ \mathbf{h}(\mathbf{x}) = \mathbf{0}, \quad \mathbf{x} \in \Omega\end{aligned} \tag{14.32}$$

This problem is clearly equivalent to the original problem (14.31), since combinations of the constraints adjoined to $f(\mathbf{x})$ do not affect the minimum point or the minimum value.

A typical step of an augmented Lagrangian method starts with a vector λ_k. Then $\mathbf{x}(\lambda_k)$ is found as the minimum point of

$$\text{minimize } f(\mathbf{x}) + \lambda_k^T \mathbf{h}(\mathbf{x}) + \frac{1}{2}c|\mathbf{h}(\mathbf{x})|^2 \quad \text{subject to} \quad \mathbf{x} \in \Omega. \tag{14.33}$$

Next λ_k is updated to λ_{k+1}. A standard method for the update is

$$\lambda_{k+1} = \lambda_k + c\mathbf{h}(\mathbf{x}(\lambda_k)).$$

To motivate the adjustment procedure, consider $\Omega = E^n$ and the constrained problem (14.32) with $\lambda = \lambda_k$. The Lagrange multiplier corresponding to this problem is $\lambda^* - \lambda_k$, where λ^* is the Lagrange multiplier of (14.31). On the other hand since (14.33) is the penalty function corresponding to (14.32), it follows from the results of Sect. 13.3 that $c\mathbf{h}(\mathbf{x}(\lambda_k))$ is approximately equal to the Lagrange multiplier of (14.32). Combining these two facts, we obtain $c\mathbf{h}(\mathbf{x}(\lambda_k)) \simeq \lambda^* - \lambda_k$. Therefore, a good approximation to the unknown λ^* is $\lambda_{k+1} = \lambda_k + c\mathbf{h}(\mathbf{x}(\lambda_k))$.

Although the main iteration in augmented Lagrangian methods is with respect to λ, the penalty parameter c may also be adjusted during the process. As in ordinary penalty function methods, the sequence of c's is usually preselected; c is either held fixed, is increased toward a finite value, or tends (slowly) toward infinity. Since in this method it is not necessary for c to go to infinity, and in fact it may remain of relatively modest value, the ill-conditioning usually associated with the penalty function approach is mediated.

From the viewpoint of duality theory, the augmented Lagrangian is simply the standard Lagrangian for the problem

$$\begin{aligned} &\text{minimize} && f(\mathbf{x}) + \frac{1}{2}c|\mathbf{h}(\mathbf{x})|^2 \\ &\text{subject to} && \mathbf{h}(\mathbf{x}) = \mathbf{0}, \quad \mathbf{x} \in \Omega. \end{aligned} \tag{14.34}$$

This problem is equivalent to the original problem (14.31), since the addition of the term $\frac{1}{2}c|\mathbf{h}(\mathbf{x})|^2$ to the objective does not change the optimal value, the optimum solution point, nor the Lagrange multiplier. However, whereas the original Lagrangian may not be convex near the solution, and hence the standard duality method cannot be applied, the term $\frac{1}{2}c|\mathbf{h}(\mathbf{x})|^2$ tends to "convexify" the Lagrangian. For sufficiently large c, the Lagrangian will indeed be locally convex. Thus the duality method can be employed, and the corresponding dual problem can be solved by an iterative process in λ. This viewpoint leads to the development of additional multiplier adjustment processes.

The Penalty Viewpoint

We begin our more detailed analysis of augmented Lagrangian methods by showing that if the penalty parameter c is sufficiently large, the augmented Lagrangian has a local minimum point near the true optimal point. This follows from the following simple lemma. (Again, we consider $\Omega = E^n$ for simplicity.)

Lemma. *Let* $\mathbf{A}$ *and* $\mathbf{B}$ *be* $n \times n$ *symmetric matrices. Suppose that* $\mathbf{B}$ *is positive semi-definite and that* $\mathbf{A}$ *is positive definite on the subspace* $\mathbf{Bx} = \mathbf{0}$. *Then there is a* c^* *such that for all* $c \geq c^*$ *the matrix* $\mathbf{A} + c\mathbf{B}$ *is positive definite.*

Proof. Suppose to the contrary that for every k there were an $\mathbf{x}_k$ with $|\mathbf{x}_k| = 1$ such that $\mathbf{x}_k^T(\mathbf{A} + k\mathbf{B})\mathbf{x}_k \leq 0$. The sequence $\{\mathbf{x}_k\}$ must have a convergent subsequence converging to a limit $\overline{\mathbf{x}}$. Now since $\mathbf{x}_k^T\mathbf{B}\mathbf{x}_k \geq 0$, it follows that $\overline{\mathbf{x}}^T\mathbf{B}\overline{\mathbf{x}} = 0$. It also follows that $\overline{\mathbf{x}}^T\mathbf{A}\overline{\mathbf{x}} \leq \mathbf{0}$. However, this contradicts the hypothesis of the lemma. ∎

This lemma applies directly to the Hessian of the augmented Lagrangian evaluated at the optimal solution pair $\mathbf{x}^*$, λ^*. We assume as usual that the second-order sufficiency conditions for a constrained minimum hold at $\mathbf{x}^*$, λ^*. The Hessian of the augmented Lagrangian evaluated at the optimal pair $\mathbf{x}^*$, λ^* is

$$\begin{aligned}\mathbf{L}_c(\mathbf{x}^*, \lambda^*) &= \mathbf{F}(\mathbf{x}^*) + (\lambda^*)^T\mathbf{H}(\mathbf{x}^*) + c\nabla\mathbf{h}(\mathbf{x}^*)^T\nabla\mathbf{h}(\mathbf{x}^*) \\ &= \mathbf{L}(\mathbf{x}^*) + c\nabla\mathbf{h}(\mathbf{x}^*)^T\nabla\mathbf{h}(\mathbf{x}^*).\end{aligned}$$

The first term, the Hessian of the normal Lagrangian, is positive definite on the subspace $\nabla\mathbf{h}(\mathbf{x}^*)\mathbf{x} = \mathbf{0}$. This corresponds to the matrix A in the lemma. The matrix $\nabla\mathbf{h}(\mathbf{x}^*)^T\nabla\mathbf{h}(\mathbf{x}^*)$ is positive semi-definite and corresponds to $\mathbf{B}$ in the lemma. It follows that there is a c^* such that for all $c > c^*$, $\mathbf{L}_c(\mathbf{x}^*, \lambda^*)$ is positive definite. This leads directly to the first basic result concerning augmented Lagrangian.

Proposition 1. *Assume that the second-order sufficiency conditions for a local minimum are satisfied at* $\mathbf{x}^*$, λ^*. *Then there is a* c^* *such that for all* $c \geq c^*$, *the augmented Lagrangian* $l_c(\mathbf{x}, \lambda^*)$ *has a local minimum point at* $\mathbf{x}^*$.

By a continuity argument the result of the above proposition can be extended to a neighborhood around $\mathbf{x}^*$, λ^*. That is, for any λ near λ^*, the augmented Lagrangian has a unique local minimum point near $\mathbf{x}^*$. This correspondence defines a continuous function. If a value of λ can be found such that $\mathbf{h}(\mathbf{x}(\lambda)) = \mathbf{0}$, then that λ must in fact be λ^*, since $\mathbf{x}(\lambda)$ satisfies the necessary conditions of the original problem. Therefore, the problem of determining the proper value of λ can be viewed as one of solving the equation $\mathbf{h}(\mathbf{x}(\lambda)) = \mathbf{0}$. For this purpose the iterative process

$$\lambda_{k+1} = \lambda_k + c\mathbf{h}(\mathbf{x}(\lambda_k)),$$

is a method of successive approximation. This process will converge linearly in a neighborhood around λ^*, although a rigorous proof is somewhat complex. We shall give more definite convergence results when we consider the duality viewpoint.

Example 1. Consider the simple quadratic problem studied in Sect. 13.8

$$\begin{aligned}&\text{minimize} \quad 2x^2 + 2xy + y^2 - 2y \\ &\text{subject to} \quad x = 0.\end{aligned}$$

The augmented Lagrangian for this problem is

$$l_c(x, y, \lambda) = 2x^2 + 2xy + y^2 - 2y + \lambda x + \frac{1}{2}cx^2.$$

The minimum of this can be found analytically to be $x = -(2+\lambda)/(2+c)$, $y = (4+c+\lambda)/(2+c)$. Since $h(x, y) = x$ in this example, it follows that the iterative process for λ_k is

$$\lambda_{k+1} = \lambda_k - \frac{c(2+\lambda_k)}{2+c}$$

or

$$\lambda_{k+1} = \left(\frac{2}{2+c}\right)\lambda_k - \frac{2c}{2+c}.$$

This converges to $\lambda = -2$ for any $c > 0$. The coefficient $2/(2+c)$ governs the rate of convergence, and clearly, as c is increased the rate improves.

Geometric Interpretation

The augmented Lagrangian method can be interpreted geometrically in terms of the primal function in a manner analogous to that in Sects. 13.3 and 13.8 for the ordinary quadratic penalty function and the absolute-value penalty function. Consider again the primal function $\omega(\mathbf{y})$ defined as

$$\omega(\mathbf{y}) = \min\{f(\mathbf{x}) : \mathbf{h}(\mathbf{x}) = \mathbf{y}\},$$

where the minimum is understood to be taken locally near $\mathbf{x}^*$. We remind the reader that $\omega(\mathbf{0}) = f(\mathbf{x}^*)$ and that $\nabla\omega(\mathbf{0})^T = -\boldsymbol{\lambda}^*$. The minimum of the augmented Lagrangian at step k can be expressed in terms of the primal function as follows:

$$\begin{aligned} \min l_c(\mathbf{x}, \boldsymbol{\lambda}_k) &= \min_{\mathbf{x}}\{f(\mathbf{x}) + \boldsymbol{\lambda}_k^T\mathbf{h}(\mathbf{x}) + \frac{1}{2}c|\mathbf{h}(\mathbf{x})|^2\} \\ &= \min_{\mathbf{x},\mathbf{u}}\{f(\mathbf{x}) + \boldsymbol{\lambda}_k^T\mathbf{y} + \frac{1}{2}c|\mathbf{y}|^2 : \mathbf{h}(\mathbf{x}) = \mathbf{y}\} \qquad (14.35) \\ &= \min_{\mathbf{u}}\{\omega(\mathbf{y}) + \boldsymbol{\lambda}_k^T\mathbf{y} + \frac{1}{2}c|\mathbf{y}|^2\}, \end{aligned}$$

where the minimization with respect to $\mathbf{y}$ is to be taken locally near $\mathbf{y} = \mathbf{0}$. This minimization is illustrated geometrically for the case of a single constraint in Fig. 14.5. The lower curve represents $\omega(\mathbf{y})$, and the upper curve represents $\omega(\mathbf{y}) + \frac{1}{2}c|\mathbf{y}|^2$. The minimum point $\mathbf{y}_k$ of (14.30) occurs at the point where this upper curve has slope equal to $-\lambda_k$. It is seen that for c sufficiently large this curve will be convex at $y = 0$. If λ_k is close to λ^*, it is clear that this minimum point will be close to 0; it will be exact if $\lambda_k = \lambda^*$.

The process for updating $\boldsymbol{\lambda}_k$ is also illustrated in Fig. 14.5. Note that in general, if $\mathbf{x}(\boldsymbol{\lambda}_k)$ minimizes $l_c(\mathbf{x}, \boldsymbol{\lambda}_k)$, then $\mathbf{y}_k = \mathbf{h}(\mathbf{x}(\boldsymbol{\lambda}_k))$ is the minimum point of $\omega(\mathbf{y}) + \boldsymbol{\lambda}_k^T\mathbf{y} + \frac{1}{2}c|\mathbf{y}|^2$. At that point we have as before

$$\nabla\omega(\mathbf{y}_k)^T + c\mathbf{y}_k = -\boldsymbol{\lambda}_k$$

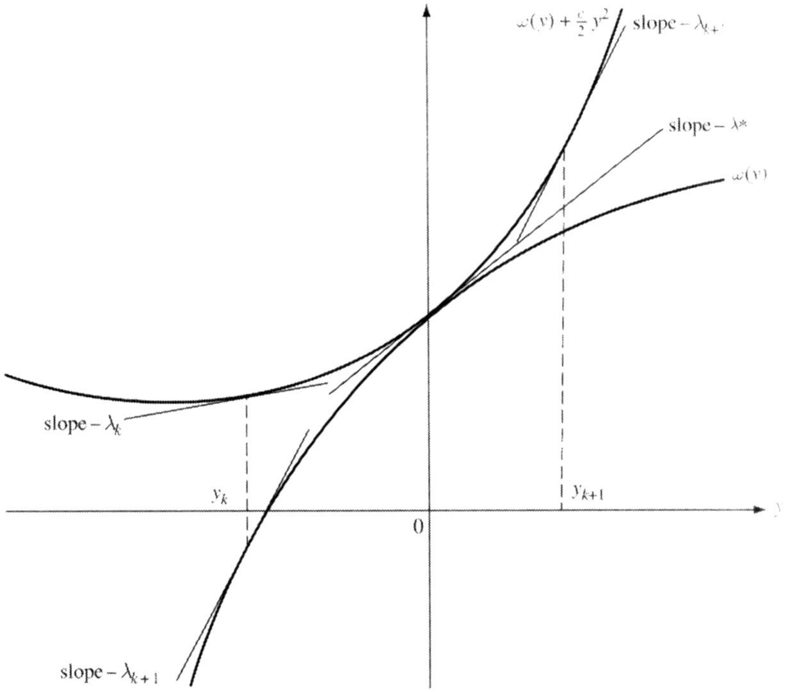

Fig. 14.5 Primal function and augmented Lagrangian

or equivalently,

$$\nabla\omega(\mathbf{y}_k)^T = -(\boldsymbol{\lambda}_k + c\mathbf{y}_k) = -(\boldsymbol{\lambda}_k + c\mathbf{h}(\mathbf{x}(\boldsymbol{\lambda}_k))).$$

It follows that for the next multiplier we have

$$\boldsymbol{\lambda}_{k+1} = \boldsymbol{\lambda}_k + c\mathbf{h}(\mathbf{x}(\boldsymbol{\lambda}_k)) = -\nabla\omega(\mathbf{y}_k)^T,$$

as shown in Fig. 14.5 for the one-dimensional case. In the figure the next point y_{k+1} is the point where $\omega(y) + \frac{1}{2}c|y|^2$ has slope $-\lambda_{k+1}$, which will yield a positive value of y_{k+1} in this case. It can be seen that if $\boldsymbol{\lambda}_k$ is sufficiently close to $\boldsymbol{\lambda}^*$, then $\boldsymbol{\lambda}_{k+1}$ will be even closer, and the iterative process will converge.

14.6 The Method of Multipliers

In the augmented Lagrangian method (the method of multipliers), the primary iteration is with respect to $\boldsymbol{\lambda}$, and therefore it is most natural to consider the method from the dual viewpoint. This is in fact the more powerful viewpoint and leads to improvements in the algorithm.

As we observed earlier, the constrained problem

$$\begin{aligned}&\text{minimize} \quad f(\mathbf{x})\\&\text{subject to} \quad \mathbf{h}(\mathbf{x}) = \mathbf{0}, \quad \mathbf{x} \in \Omega\end{aligned} \tag{14.36}$$

is *equivalent* to the problem

$$\begin{aligned}&\text{minimize} \quad f(\mathbf{x}) + \frac{1}{2}c|\mathbf{h}(\mathbf{x})|^2\\&\text{subject to} \quad \mathbf{h}(\mathbf{x}) = \mathbf{0}, \quad \mathbf{x} \in \Omega\end{aligned} \tag{14.37}$$

in the sense that the solution points, the optimal values, and the Lagrange multipliers are the same for both problems. However, as spelled out by Proposition 1 of the previous section, whereas problem (14.36) may not be locally convex, problem (14.37) is locally convex for sufficiently large c; specifically, the Hessian of the Lagrangian is positive definite at the solution pair $\mathbf{x}^*$, λ^*. Thus local duality theory is applicable to problem (14.37) for sufficiently large c.

To apply the dual method to (14.37), we define the dual function

$$\phi(\lambda) = \min\{f(\mathbf{x}) + \lambda^T\mathbf{h}(\mathbf{x}) + \frac{1}{2}c|\mathbf{h}(\mathbf{x})|^2\} \tag{14.38}$$

in a region near $\mathbf{x}^*$, λ^*. If $\mathbf{x}(\lambda)$ is the vector minimizing the right-hand side of (14.38), then as we have seen in Sect. 14.2, $\mathbf{h}(\mathbf{x}(\lambda))$ is the gradient of ϕ. Thus the iterative process

$$\lambda_{k+1} = \lambda_k + c\mathbf{h}(\mathbf{x}(\lambda_k))$$

used in the basic augmented Lagrangian method is seen to be *a steepest ascent iteration for maximizing the dual function* ϕ. It is a simple form of steepest ascent, using a constant stepsize c.

Although the stepsize c is a good choice (as will become even more evident later), it is clearly advantageous to apply the algorithmic principles of optimization developed previously by selecting the stepsize so that the new value of the dual function satisfies an ascent criterion. This can extend the range of convergence of the algorithm.

The rate of convergence of the optimal steepest ascent method (where the stepsize is selected to maximize ϕ in the gradient direction) is determined by the eigenvalues of the Hessian of ϕ. The Hessian of ϕ is found from (14.15) to be

$$\nabla\mathbf{h}(\mathbf{x}(\lambda))[\mathbf{L}(\mathbf{x}(\lambda), \lambda) + c\nabla\mathbf{h}(\mathbf{x}(\lambda))^T\nabla\mathbf{h}(\mathbf{x}(\lambda))]^{-1}\nabla\mathbf{h}(\mathbf{x})^T. \tag{14.39}$$

The eigenvalues of this matrix at the solution point $\mathbf{x}^*$, λ^* determine the convergence rate of the method of steepest ascent.

To analyze the eigenvalues we make use of the matrix identity

$$c\mathbf{B}(\mathbf{A} + c\mathbf{B}^T\mathbf{B})^{-1}\mathbf{B}^T = \mathbf{I} - (\mathbf{I} + c\mathbf{B}\mathbf{A}^{-1}\mathbf{B}^T)^{-1},$$

which is a generalization of the Sherman-Morrison formula. (See Sect. 10.4.) It is easily seen from the above identity that the matrices $\mathbf{B}(\mathbf{A} + c\mathbf{B}^T\mathbf{B})^{-1}\mathbf{B}^T$ and $(\mathbf{B}\mathbf{A}^{-1}\mathbf{B}^T)$ have identical eigenvectors. One way to see this is to multiply both sides of the identity by $(\mathbf{I} + c\mathbf{B}\mathbf{A}^{-1}\mathbf{B}^T)$ on the right to obtain

$$c\mathbf{B}(\mathbf{A} + c\mathbf{B}^T\mathbf{B})^{-1}\mathbf{B}^T(\mathbf{I} + c\mathbf{B}\mathbf{A}^{-1}\mathbf{B}^T) = c\mathbf{B}\mathbf{A}^{-1}\mathbf{B}^T.$$

Suppose both sides are applied to an eigenvector $\mathbf{e}$ of $\mathbf{B}\mathbf{A}^{-1}\mathbf{B}^T$ having eigenvalue w. Then we obtain

$$c\mathbf{B}(\mathbf{A} + c\mathbf{B}^T\mathbf{B})^{-1}\mathbf{B}^T(1 + cw)\mathbf{e} = cw\mathbf{e}.$$

It follows that $\mathbf{e}$ is also an eigenvector of $\mathbf{B}(\mathbf{A} + c\mathbf{B}^T\mathbf{B})^{-1}\mathbf{B}^T$, and if v is the corresponding eigenvalue, the relation

$$cu(1 + cw) = cw$$

must hold. Therefore, the eigenvalues are related by

$$u = \frac{w}{1 + cw}. \tag{14.40}$$

The above relations apply directly to the Hessian (14.39) through the associations $\mathbf{A} = \mathbf{L}(\mathbf{x}^*, \boldsymbol{\lambda}^*)$ and $\mathbf{B} = \nabla\mathbf{h}(\mathbf{x}^*)$. Note that the matrix $\nabla\mathbf{h}(\mathbf{x}^*)\mathbf{L}(\mathbf{x}^*, \boldsymbol{\lambda}^*)^{-1}\nabla\mathbf{h}(\mathbf{x}^*)^T$, corresponding to $\mathbf{B}\mathbf{A}^{-1}\mathbf{B}^T$ above, is the Hessian of the dual function of the original problem (14.36). As shown in Sect. 14.3 the eigenvalues of this matrix determine the rate of convergence for the ordinary dual method. Let w and W be the smallest and largest eigenvalues of this matrix. From (14.40) it follows that the ratio of smallest to largest eigenvalues of the Hessian of the dual for the augmented problem is

$$\frac{\frac{1}{w} + c}{\frac{1}{w} + c}.$$

This shows explicitly how the rate of convergence of the multiplier method depends on c. As c goes to infinity, the ratio of eigenvalues goes to unity, implying arbitrarily fast convergence.

Other unconstrained optimization techniques may be applied to the maximization of the dual function defined by the augmented Lagrangian; conjugate gradient methods, Newton's method, and quasi-Newton methods can all be used. The use of Newton's method requires evaluation of the Hessian matrix (14.39). For some problems this may be feasible, but for others some sort of approximation is desirable. One approximation is obtained by noting that for large values of c, the Hessian (14.39) is approximately equal to $(1/c)\mathbf{I}$. Using this value for the Hessian and $\mathbf{h}(\mathbf{x}(\boldsymbol{\lambda}))$ for the gradient, we are led to the iterative scheme

$$\boldsymbol{\lambda}_{k+1} = \boldsymbol{\lambda}_k + c\mathbf{h}(\mathbf{x}(\boldsymbol{\lambda}_k)),$$

which is exactly the simple method of multipliers originally proposed.

We might summarize the above observations by the following statement relating primal and dual convergence rates. If a penalty term is incorporated into a problem, the condition number of the primal problem becomes increasingly poor as $c \to \infty$ but the condition number of the dual becomes increasingly good. To apply the dual method, however, an unconstrained penalty problem of poor condition number must be solved at each step.

Inequality Constraints

The advantage of augmented Lagrangian methods is mostly in dealing with equalities. But certain inequality constraints can be easily incorporated. Let us consider the problem with p inequality constraints:

$$\begin{aligned} &\text{minimize} \;\; f(\mathbf{x}) \\ &\text{subject to} \;\; \mathbf{g}(\mathbf{x}) \leq \mathbf{0}. \end{aligned} \tag{14.41}$$

We assume that this problem has a well-defined solution $\mathbf{x}^*$, which is a regular point of the constraints and which satisfies the second-order sufficiency conditions for a local minimum as specified in Sect. 11.8. This problem can be written as an equivalent problem with equality constraints:

$$\begin{aligned} &\text{minimize} \;\; f(\mathbf{x}) \\ &\text{subject to} \;\; \mathbf{g}(\mathbf{x}) + \mathbf{u} = \mathbf{0}, \;\; \mathbf{u} \geq \mathbf{0}. \end{aligned} \tag{14.42}$$

Through this conversion we can hope to simply apply the theory for equality constraints to problems with inequalities.

In order to do so we must insure that (14.42) satisfies the second-order sufficiency conditions of Sect. 11.5. These conditions will not hold unless we impose a *strict complementarity* assumption that $g_j(\mathbf{x}^*) = 0$ implies $\mu_j^* > 0$ as well as the usual second-order sufficiency conditions for the original problem (14.41). (See Exercise 10.)

With these assumptions we define the (partial) dual function corresponding to the augmented Lagrangian method as

$$\phi(\boldsymbol{\mu}) = \min_{\mathbf{u} \geq \mathbf{0},\, \mathbf{x}} \quad f(\mathbf{x}) + \boldsymbol{\mu}^T[\mathbf{g}(\mathbf{x}) + \mathbf{u}] + \frac{1}{2}c|\mathbf{g}(\mathbf{x}) + \mathbf{u}|^2. \tag{14.43}$$

The minimization with respect to $\mathbf{u}$ in (14.43) can be carried out analytically, and this will lead to a definition of the dual function that only involves minimization with respect to $\mathbf{x}$. The variable u_j enters the objective of the dual function only through the univariate quadratic expression

$$P_j = \mu_j[g_j(\mathbf{x}) + u_j] + \frac{1}{2}c[g_j(\mathbf{x}) + u_j]^2. \tag{14.44}$$

It is this expression that we must minimize with respect to $u_j \geq 0$. This is easily accomplished by differentiation: If $u_j > 0$, the derivative must vanish; if $u_j = 0$, the derivative must be nonnegative. The derivative is zero at $z_j = -g_j(\mathbf{x}) - \boldsymbol{\mu}_j/c$. Thus we obtain the solution

$$u_j = \begin{cases} -g_j(\mathbf{x}) - \frac{\mu_j}{c}, & \text{if } -g_j(\mathbf{x}) - \frac{\mu_j}{c} \geq 0 \\ 0, & \text{otherwise} \end{cases}$$

or equivalently,

$$u_j = \max\left\{0, -g_j(\mathbf{x}) - \frac{\mu_j}{c}\right\}. \tag{14.45}$$

We now substitute this into (14.44) in order to obtain an explicit expression for the minimum of P_j.

For $u_j = 0$, we have

$$\begin{aligned} P_j &= \frac{1}{2c}\left(2\mu_j c g_j(\mathbf{x}) + c^2 g_j(\mathbf{x})^2\right) \\ &= \frac{1}{2c}\left([\mu_j + c g_j(\mathbf{x})]^2 - \mu_j^2\right). \end{aligned}$$

For $u_j = -g_j(\mathbf{x}) - \mu_j/c$ we have

$$P_j = -\mu_j^2/2c.$$

These can be combined into the formula

$$P_j = \frac{1}{2c}\left([\max\{0, \mu_j + c g_j(\mathbf{x})\}]^2 - \mu_j^2\right).$$

In view of the above, let us define the function of two scalar arguments t and μ:

$$P_c(t, \mu) = \frac{1}{2c}\left([\max\{0, \mu + ct\}]^2 - \mu^2\right). \tag{14.46}$$

For a fixed $\mu > 0$, this function is shown in Fig. 14.6. Note that it is a smooth function with derivative with respect to t equal to μ at $t = 0$.

The dual function for the inequality problem can now be written as

$$\phi(\mu) = \min_{\mathbf{x}}\left(f(\mathbf{x}) + \sum_{j=1}^{p} P_c(g_j(\mathbf{x}), \boldsymbol{\mu}_j)\right). \tag{14.47}$$

Thus inequality problems can be treated by adjoining to $f(\mathbf{x})$ a special penalty function (that depends on $\boldsymbol{\mu}$). The Lagrange multiplier $\boldsymbol{\mu}$ can then be adjusted to maximize ϕ, just as in the case of equality constraints.

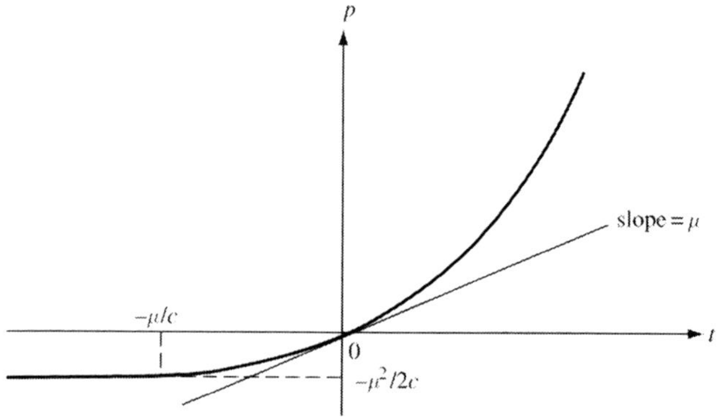

Fig. 14.6 Penalty function for inequality problem

14.7 The Alternating Direction Method of Multipliers

Consider the convex minimization model with linear constraints and an objective function which is the sum of two separable functions:

$$\begin{array}{ll} \text{minimize} & f_1(\mathbf{x}^1) + f_2(\mathbf{x}^2) \\ \text{subject to} & A_1\mathbf{x}^1 + A_2\mathbf{x}^2 = \mathbf{b}, \\ & \mathbf{x}^1 \in \Omega_1,\ \mathbf{x}^2 \in \Omega_2, \end{array} \tag{14.48}$$

where $A_i \in E^{m \times n_i}$ $(i = 1, 2)$, $\mathbf{b} \in E^m$, $\Omega_i \subset E^{n_i}$ $(i = 1, 2)$ are closed convex sets; and $f_i : E^{n_i} \to E$ $(i = 1, 2)$ are convex functions on Ω_i, respectively. Then, the augmented Lagrangian function for (14.48) would be

$$l_c(\mathbf{x}^1, \mathbf{x}^2, \boldsymbol{\lambda}) = f_1(\mathbf{x}^1) + f_2(\mathbf{x}^2) + \boldsymbol{\lambda}^T(A_1\mathbf{x}^1 + A_2\mathbf{x}^2 - \mathbf{b}) + \frac{c}{2}|A_1\mathbf{x}^1 + A_2\mathbf{x}^2 - \mathbf{b}|^2.$$

Throughout this section, we assume problem (14.48) has at least one optimal solution.

In contrast to the method of multipliers in the last section, the alternating direction method of multipliers (ADMM) is to (approximately) minimize $l_c(\mathbf{x}^1, \mathbf{x}^2, \boldsymbol{\lambda})$ in an alternative order:

$$\begin{aligned} \mathbf{x}^1_{k+1} &:= \arg\min\nolimits_{\mathbf{x}^1 \in \Omega_1} l_c(\mathbf{x}^1, \mathbf{x}^2_k, \boldsymbol{\lambda}_k), \\ \mathbf{x}^2_{k+1} &:= \arg\min\nolimits_{\mathbf{x}^2 \in \Omega_2} l_c(\mathbf{x}^1_{k+1}, \mathbf{x}^2, \boldsymbol{\lambda}_k), \\ \boldsymbol{\lambda}_{k+1} &:= \boldsymbol{\lambda}_k + c(A_1\mathbf{x}^1_{k+1} + A_2\mathbf{x}^2_{k+1} - \mathbf{b}). \end{aligned} \tag{14.49}$$

The idea is that each of the smaller minimization problems can be solved more efficiently or even in close forms for certain cases.

Convergence Speed Analysis

We present a convergence speed analysis of the ADMM. For simplicity, we shall let Ω_i be E^{n_i} and f_i be differentiable convex functions [the result is also valid for the ADMM applied to the aforementioned more general problem (14.48)]. Then, any optimal solution and multiplier $(\mathbf{x}_*^1, \mathbf{x}_*^2, \boldsymbol{\lambda}_*)$ satisfy

$$\nabla f_1(\mathbf{x}_*^1)^T + A_1^T\boldsymbol{\lambda}_* = \mathbf{0}, \quad \nabla f_2(\mathbf{x}_*^2)^T + A_2^T\boldsymbol{\lambda}_* = \mathbf{0}, \quad A_1\mathbf{x}_*^1 + A_2\mathbf{x}_*^2 - \mathbf{b} = \mathbf{0}, \quad (14.50)$$

and these conditions are also sufficient.

We first establish a key lemma.

Lemma 1. *Let* $\mathbf{d}_k^i = A_i(\mathbf{x}_k^i - \mathbf{x}_*^i)$, $i = 1, 2$, *and* $\mathbf{d}_k^\lambda = \boldsymbol{\lambda}_k - \boldsymbol{\lambda}_*$; *and* $\{\mathbf{x}_k^1, \mathbf{x}_k^2, \boldsymbol{\lambda}_k\}$ *be the sequence generated by ADMM* (14.49). *Then, it holds that*

$$c\left|A_2(\mathbf{x}_{k+1}^2 - \mathbf{x}_k^2)\right|^2 + \frac{1}{c}|\boldsymbol{\lambda}_{k+1} - \boldsymbol{\lambda}_k|^2 \le \left(c\left|A_2\mathbf{d}_k^2\right|^2 + \frac{1}{c}\left|\mathbf{d}_k^\lambda\right|^2\right) - \left(c\left|A_2\mathbf{d}_{k+1}^2\right|^2 + \frac{1}{c}\left|\mathbf{d}_{k+1}^\lambda\right|^2\right).$$

Proof. From the first-order optimality conditions of (14.49), we have

$$\begin{cases} \nabla f_1(\mathbf{x}_{k+1}^1)^T + A_1^T[\boldsymbol{\lambda}_k + c(A_1\mathbf{x}_{k+1}^1 + A_2\mathbf{x}_k^2 - \mathbf{b})] = \mathbf{0}, \\ \nabla f_2(\mathbf{x}_{k+1}^2)^T + A_2^T[\boldsymbol{\lambda}_k + c(A_1\mathbf{x}_{k+1}^1 + A_2\mathbf{x}_{k+1}^2 - \mathbf{b})] = \mathbf{0}, \\ \boldsymbol{\lambda}_{k+1} = \boldsymbol{\lambda}_k + c(A_1\mathbf{x}_{k+1}^1 + A_2\mathbf{x}_{k+1}^2 - \mathbf{b}). \end{cases} \quad (14.51)$$

Substituting the last equation into other equations in (14.51), we obtain

$$\begin{cases} \nabla f_1(\mathbf{x}_{k+1}^1)^T + A_1^T\boldsymbol{\lambda}_{k+1} = -cA_1^TA_2(\mathbf{x}_k^2 - \mathbf{x}_{k+1}^2), \\ \nabla f_2(\mathbf{x}_{k+1}^2)^T + A_2^T\boldsymbol{\lambda}_{k+1} = \mathbf{0}, \\ A_1\mathbf{x}_{k+1}^1 + A_2\mathbf{x}_{k+1}^2 - \mathbf{b} = \frac{1}{c}(\boldsymbol{\lambda}_{k+1} - \boldsymbol{\lambda}_k). \end{cases} \quad (14.52)$$

Moreover, the convexity of f_i, $i = 1, 2$, implies

$$(\nabla f_1(\mathbf{x}_{k+1}^1) - \nabla f_1(\mathbf{x}_*^1))(\mathbf{x}_{k+1}^1 - \mathbf{x}_*^1) \ge 0 \text{ and } (\nabla f_2(\mathbf{x}_{k+1}^2) - \nabla f_2(\mathbf{x}_*^2))(\mathbf{x}_{k+1}^2 - \mathbf{x}_*^2) \ge 0.$$

On the other hand, from (14.50) and (14.52),

$$\nabla f_1(\mathbf{x}_{k+1}^1)^T - \nabla f_1(\mathbf{x}_*^1)^T = \nabla f_1(\mathbf{x}_{k+1}^1) + A_1^T\boldsymbol{\lambda}_* = -A_1^T\mathbf{d}_{k+1}^\lambda - cA_1^TA_2(\mathbf{x}_k^2 - \mathbf{x}_{k+1}^2)$$

$$\nabla f_2(\mathbf{x}_{k+1}^2)^T - \nabla f_2(\mathbf{x}_*^2)^T = \nabla f_2(\mathbf{x}_{k+1}^2) + A_2^T\boldsymbol{\lambda}_* = -A_2^T\mathbf{d}_{k+1}^\lambda$$

and

$$\mathbf{0} = A_1\mathbf{x}_{k+1}^1 + A_2\mathbf{x}_{k+1}^2 - \mathbf{b} - \frac{1}{c}(\boldsymbol{\lambda}_{k+1} - \boldsymbol{\lambda}_k) = A_1\mathbf{d}_{k+1}^1 + A_2\mathbf{d}_2^{k+1} + \frac{1}{c}(\boldsymbol{\lambda}_k - \boldsymbol{\lambda}_{k+1}).$$

Thus,

$$
\begin{aligned}
0 &\le \begin{pmatrix} \mathbf{d}^1_{k+1} \\ \mathbf{d}_2^{k+1} \\ \mathbf{d}^\lambda_{k+1} \end{pmatrix}^T \begin{pmatrix} \nabla f_1(\mathbf{x}^1_{k+1})^T - \nabla f_1(\mathbf{x}^1_*)^T \\ \nabla f_2(\mathbf{x}^2_{k+1})^T - \nabla f_2(\mathbf{x}^2_*)^T \\ \mathbf{0} \end{pmatrix} \\
&= \begin{pmatrix} \mathbf{d}^1_{k+1} \\ \mathbf{d}_2^{k+1} \\ \mathbf{d}^\lambda_{k+1} \end{pmatrix}^T \begin{pmatrix} -A_1^T \mathbf{d}^\lambda_{k+1} - cA_1^T A_2(\mathbf{x}^2_k - \mathbf{x}^2_{k+1}) \\ -A_2^T \mathbf{d}^\lambda_{k+1} \\ A_1 \mathbf{d}^1_{k+1} + A_2 \mathbf{d}_2^{k+1} + \frac{1}{c}(\lambda_k - \lambda_{k+1}) \end{pmatrix} \\
&= \begin{pmatrix} \mathbf{d}^1_{k+1} \\ \mathbf{d}_2^{k+1} \\ \mathbf{d}^\lambda_{k+1} \end{pmatrix}^T \left(\begin{bmatrix} -A_1^T \mathbf{d}^\lambda_{k+1} \\ -A_2^T \mathbf{d}^\lambda_{k+1} \\ A_1 \mathbf{d}^1_{k+1} + A_2 \mathbf{d}_2^{k+1} \end{bmatrix} + \begin{bmatrix} -cA_1^T A_2(\mathbf{x}^2_k - \mathbf{x}^2_{k+1}) \\ \mathbf{0} \\ \frac{1}{c}(\lambda_k - \lambda_{k+1}) \end{bmatrix} \right) \\
&= \begin{pmatrix} \mathbf{d}^1_{k+1} \\ \mathbf{d}_2^{k+1} \\ \mathbf{d}^\lambda_{k+1} \end{pmatrix}^T \begin{pmatrix} -cA_1^T A_2(\mathbf{x}^2_k - \mathbf{x}^2_{k+1}) \\ \mathbf{0} \\ \frac{1}{c}(\lambda_k - \lambda_{k+1}) \end{pmatrix} = \begin{pmatrix} -A_1 \mathbf{d}^1_{k+1} \\ \mathbf{d}^\lambda_{k+1} \end{pmatrix}^T \begin{pmatrix} cA_2(\mathbf{x}^2_k - \mathbf{x}^2_{k+1}) \\ \frac{1}{c}(\lambda_k - \lambda_{k+1}) \end{pmatrix}
\end{aligned}
\tag{14.53}
$$

Again from $-A_1\mathbf{d}^1_{k+1} = \frac{1}{c}(\lambda_{k+1} - \lambda_k) + A_2\mathbf{d}^2_{k+1}$, inequality (14.53) implies

$$
\begin{aligned}
0 &\le \begin{pmatrix} \frac{1}{c}(\lambda_{k+1} - \lambda_k) + A_2\mathbf{d}^2_{k+1} \\ \mathbf{d}^\lambda_{k+1} \end{pmatrix}^T \begin{pmatrix} cA_2(\mathbf{x}^2_k - \mathbf{x}^2_{k+1}) \\ \frac{1}{c}(\lambda_k - \lambda_{k+1}) \end{pmatrix} \\
&= \begin{pmatrix} A_2\mathbf{d}^2_{k+1} \\ \mathbf{d}^\lambda_{k+1} \end{pmatrix}^T \begin{pmatrix} cA_2(\mathbf{x}^2_k - \mathbf{x}^2_{k+1}) \\ \frac{1}{c}(\lambda_k - \lambda_{k+1}) \end{pmatrix} + (\lambda_k - \lambda_{k+1})^T A_2(\mathbf{x}^2_k - \mathbf{x}^2_{k+1})
\end{aligned}
$$

Since $\nabla f_2(\mathbf{x}^2_k) = -\lambda_k^T A_2$ holds for every $k \ge 0$, it follows from the convexity of f_2 that

$$
(\lambda_k - \lambda_{k+1})^T A_2(\mathbf{x}^2_k - \mathbf{x}^2_{k+1}) = -(\nabla f_2(\mathbf{x}^2_k) - \nabla f_2(\mathbf{x}^2_{k+1}))(\mathbf{x}^2_k - \mathbf{x}^2_{k+1}) \le 0.
$$

Thus,

$$
\begin{pmatrix} A_2\mathbf{d}^2_{k+1} \\ \mathbf{d}^\lambda_{k+1} \end{pmatrix}^T \begin{pmatrix} cA_2(\mathbf{x}^2_k - \mathbf{x}^2_{k+1}) \\ \frac{1}{c}(\lambda_k - \lambda_{k+1}) \end{pmatrix} \ge 0 \quad \text{or} \quad \begin{pmatrix} \sqrt{c}A_2\mathbf{d}^2_{k+1} \\ \frac{1}{\sqrt{c}}\mathbf{d}^\lambda_{k+1} \end{pmatrix}^T \begin{pmatrix} \sqrt{c}A_2(\mathbf{x}^2_{k+1} - \mathbf{x}^2_k) \\ \frac{1}{\sqrt{c}}(\lambda_{k+1} - \lambda_k) \end{pmatrix} \le 0.
$$

Representing the left vector by $\mathbf{u}$ and the right one by $\mathbf{v}$ in the last inequality, we have

$$
0 \ge \mathbf{u}^T\mathbf{v} = \frac{1}{2}(|\mathbf{u}|^2 + |\mathbf{v}|^2 - |\mathbf{u} - \mathbf{v}|^2).
$$

Noting

$$
\mathbf{u} - \mathbf{v} = \begin{pmatrix} \sqrt{c}A_2\mathbf{d}^2_{k+1} \\ \frac{1}{\sqrt{c}}\mathbf{d}^\lambda_{k+1} \end{pmatrix} - \begin{pmatrix} \sqrt{c}A_2(\mathbf{x}^2_{k+1} - \mathbf{x}^2_k) \\ \frac{1}{\sqrt{c}}(\lambda_{k+1} - \lambda_k) \end{pmatrix} = \begin{pmatrix} \sqrt{c}A_2\mathbf{d}^2_k \\ \frac{1}{\sqrt{c}}\mathbf{d}^\lambda_k \end{pmatrix},
$$

we obtain the desired result in Lemma 1. ∎

For simplicity, let $c = 1$ in the following. Taking the sum from iterate 0 to iterate k for the inequality in Lemma 1, we obtain

$$
\sum_{t=0}^{k} \left(\left|A_2(\mathbf{x}^2_{t+1} - \mathbf{x}^2_t)\right|^2 + |\lambda_{t+1} - \lambda_t|^2 \right) \le \left|A_2\mathbf{x}^2_0 - A_2\mathbf{x}^2_*\right|^2 + |\lambda_0 - \lambda_*|^2 .
$$

Thus, we have

$$\min_{0\le t\le k}\left\{\left|A_2(\mathbf{x}^2_{t+1}-\mathbf{x}^2_t)\right|^2+|\boldsymbol{\lambda}_{t+1}-\boldsymbol{\lambda}_t|^2\right\}\le\frac{1}{k}\left(\left|A_2(\mathbf{x}^2_0-\mathbf{x}^2_*)\right|^2+|\boldsymbol{\lambda}_0-\boldsymbol{\lambda}_*|^2\right).$$

Therefore, from (14.52) we have

Theorem 1. *After k iterations of the ADMM method, there must be at least one iterate $0\le\bar{k}\le k$ such that*

$$\left\|\begin{pmatrix}\nabla f_1(\mathbf{x}^1_{\bar{k}+1})^T+A_1^T\boldsymbol{\lambda}_{\bar{k}+1}\\ \nabla f_2(\mathbf{x}^2_{\bar{k}+1})^T+A_2^T\boldsymbol{\lambda}_{\bar{k}+1}\\ A_1\mathbf{x}^1_{\bar{k}+1}+A_2\mathbf{x}^2_{\bar{k}+1}-\mathbf{b}\end{pmatrix}\right\|^2\le\frac{1+|A_1|}{k}\left(|A_2(\mathbf{x}^2_0-\mathbf{x}^2_*)|^2+|\boldsymbol{\lambda}_0-\boldsymbol{\lambda}_*|^2\right),$$

that is, $(\mathbf{x}^1_{\bar{k}+1},\mathbf{x}^2_{\bar{k}+1},\boldsymbol{\lambda}_{\bar{k}+1})$ has its optimality condition error square bounded by the quantity on the right-hand side that converges 0 arithmetically as $k\to\infty$.

The Three Block Extension

It is natural to consider the ADMM method for solving problems with more than two blocks:

$$\begin{array}{ll}\text{minimize} & f_1(\mathbf{x}^1)+f_2(\mathbf{x}^2)+f_3(\mathbf{x}^3)\\ \text{subject to} & A_1\mathbf{x}^1+A_2\mathbf{x}^2+A_3\mathbf{x}^3=\mathbf{b},\\ & \mathbf{x}^1\in\Omega_1,\ \mathbf{x}^2\in\Omega_2,\ \mathbf{x}^3\in\Omega_3,\end{array}\tag{14.54}$$

where $A_i\in E^{m\times n_i}$ $(i=1,2,3)$, $\mathbf{b}\in E^m$, $\Omega_i\subset E^{n_i}$ $(i=1,2,3)$ are closed convex sets; and $f_i:E^{n_i}\to E$ $(i=1,2,3)$ are convex functions on Ω_i, respectively. With the same philosophy as the ADMM to take advantage of the separable structure, one could consider the procedure

$$\begin{aligned}\mathbf{x}^1_{k+1}&:=\arg\min_{\mathbf{x}^1\in\Omega_1}l_c(\mathbf{x}^1,\mathbf{x}^2_k,\mathbf{x}^3_k,\boldsymbol{\lambda}_k),\\ \mathbf{x}^2_{k+1}&:=\arg\min_{\mathbf{x}^2\in\Omega_2}l_c(\mathbf{x}^1_{k+1},\mathbf{x}^2,\mathbf{x}^3_k,\boldsymbol{\lambda}_k),\\ \mathbf{x}^3_{k+1}&:=\arg\min_{\mathbf{x}^3\in\Omega_3}l_c(\mathbf{x}^1_{k+1},\mathbf{x}^2_{k+1},\mathbf{x}^3,\boldsymbol{\lambda}_k),\\ \boldsymbol{\lambda}_{k+1}&:=\boldsymbol{\lambda}_k+c(A_1\mathbf{x}^1_{k+1}+A_2\mathbf{x}^2_{k+1}+A_3\mathbf{x}^3_{k+1}-\mathbf{b}),\end{aligned}\tag{14.55}$$

where the augmented Lagrangian function

$$l_c(\mathbf{x}^1,\mathbf{x}^2,\mathbf{x}^3,\boldsymbol{\lambda})=\sum_{i=1}^3 f_i(\mathbf{x}^i)+\boldsymbol{\lambda}^T\Big(\sum_{i=1}^3A_i\mathbf{x}^i-\mathbf{b}\Big)+\frac{c}{2}\Big|\sum_{i=1}^3A_i\mathbf{x}^i-\mathbf{b}\Big|^2.$$

Unfortunately, unlike the convergence property for solving two-block problems, such a direct extension of ADMM not converge for problems with three blocks. Indeed, consider the following linear homogeneous equation with three variables

$$(A_1, A_2, A_3)\begin{pmatrix} x^1 \\ x^2 \\ x^3 \end{pmatrix} = \begin{pmatrix} 1 & 1 & 1 \\ 1 & 1 & 2 \\ 1 & 2 & 2 \end{pmatrix}\begin{pmatrix} x^1 \\ x^2 \\ x^3 \end{pmatrix} = \mathbf{0}.$$

Let $c = 1$ and each block contain one variable. Then, simple calculation will show that the direct extension of ADMM (14.55) is divergent from any point in a subspace of E^3. Note that the convergence of ADMM (14.55) applied to solving the linear equations with a null objective is independent of the selection of the penalty parameter c. We conclude:

Theorem 2. *For the three-block convex minimization problem (14.54), the direct extension of ADMM (14.55) may not converge for any penalty parameter $c > 0$ starting from any point in a subspace.*

*14.8 *Cutting Plane Methods

Cutting plane methods are applied to problems having the general form

$$\begin{aligned} &\text{minimize} \quad \mathbf{c}^T\mathbf{x} \\ &\text{subject to} \quad \mathbf{x} \in S, \end{aligned} \tag{14.56}$$

where $S \subset E^n$ is a closed convex set. Problems that involve minimization of a convex function over a convex set, such as the problem

$$\begin{aligned} &\text{minimize} \quad f(\mathbf{y}) \\ &\text{subject to} \quad \mathbf{y} \in R, \end{aligned} \tag{14.57}$$

where $R \subset E^{n-1}$ is a convex set and f is a convex function, can be easily converted to the form (14.56) by writing (14.57) equivalently as

$$\begin{aligned} &\text{minimize} \quad r \\ &\text{subject to} \quad f(\mathbf{y}) - r \leqslant 0, \ \mathbf{y} \in R \end{aligned} \tag{14.58}$$

which, with $\mathbf{x} = (r, \mathbf{y}) \in E^n$, is a special case of (14.56).

General Form of Algorithm

The general form of a cutting-plane algorithm for problem (14.56) is as follows: Given a polytope $P_k \supset S$

Step 1. Minimize $\mathbf{c}^T\mathbf{x}$ over P_k obtaining a point $\mathbf{x}_k$ in P_k. If $\mathbf{x}_k \in S$, stop; $\mathbf{x}_k$ is optimal. Otherwise,

Step 2. Find a hyperplane H_k separating the point $\mathbf{x}_k$ from S, that is, find $\mathbf{a}_k \in E^n$, $b_k \in E^1$ such that $S \subset \{\mathbf{x} : \mathbf{a}_k^T\mathbf{x} \leqslant b_k\}$, $\mathbf{x}_k \in \{\mathbf{x} : \mathbf{a}_k^T\mathbf{x} > b_k\}$. Update P_k to obtain P_{k+1} including as a constraint $\mathbf{a}_k^T\mathbf{x} \leqslant b_k$.

The process is illustrated in Fig. 14.7.

Specific algorithms differ mainly in the manner in which the hyperplane that separates the current point $\mathbf{x}_k$ from the constraint set S is selected. This selection is, of course, the most important aspect of the algorithm, since it is the deepness of the cut associated with the separating hyperplane, the distance of the hyperplane from the current point, that governs how much improvement there is in the approximation to the constraint set, and hence how fast the method converges.

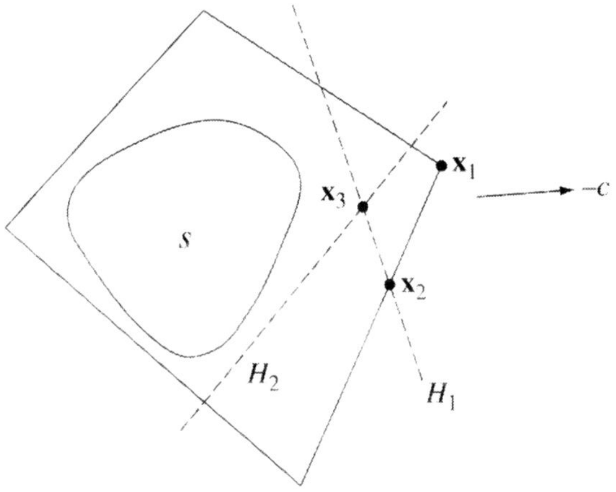

Fig. 14.7 Cutting plane method

Specific algorithms also differ somewhat with respect to the manner by which the polytope is updated once the new hyperplane is determined. The most straightforward procedure is to simply adjoin the linear inequality associated with that hyperplane to the ones determined previously. This yields the best possible updated approximation to the constraint set but tends to produce, after a large number of iterations, an unwieldy number of inequalities expressing the approximation. Thus, in some algorithms, older inequalities that are not binding at the current point are discarded from further consideration.

The general cutting plane algorithm can be regarded as an extended application of duality in linear programming, and although this viewpoint does not particularly aid in the analysis of the method, it reveals the basic interconnection between cutting plane and dual methods. The foundation of this viewpoint is the fact that S can be written as the intersection of all the half-spaces that contain it; thus

$$S = \{\mathbf{x} : \mathbf{a}_i^T\mathbf{x} \leq b_i,\ i \in I\},$$

where I is an (infinite) index set corresponding to all half-spaces containing S. With S viewed in this way problem (14.56) can be thought of as an (infinite) linear programming problem.

Corresponding to this linear program there is (at least formally) the dual problem

$$\begin{aligned} &\text{maximize} \ \sum_{i\in I} \lambda_i b_i \\ &\text{subject to} \ \sum_{i\in I} \lambda_i \mathbf{a}_i = \mathbf{c} \\ &\qquad\qquad \lambda_i \geqslant 0, \quad i \in I. \end{aligned} \tag{14.59}$$

Selecting a finite subset of I, say $\bar{I}$, and forming

$$P = \{\mathbf{x} : \mathbf{a}_i^T \mathbf{x} \leqslant b_i,\ i \in \bar{I}\}$$

gives a polytope that contains S. Minimizing $\mathbf{c}^T\mathbf{x}$ over this polytope yields a point and a corresponding subset of active constraints I_A. The dual problem with the additional restriction $\lambda_i = 0$ for $i \notin I_A$ will then have a feasible solution, but this solution will in general not be optimal. Thus, a solution to a polytope problem corresponds to a feasible but non-optimal solution to the dual. For this reason the cutting plane method can be regarded as working toward optimality of the (infinite dimensional) dual.

Kelley's Convex Cutting Plane Algorithm

The convex cutting plane method was developed to solve convex programming problems of the form

$$\begin{aligned} &\text{minimize} \ f(\mathbf{x}) \\ &\text{subject to} \ g_i(\mathbf{x}) \leqslant 0,\ i = 1, 2, \ldots, p, \end{aligned} \tag{14.60}$$

where $\mathbf{x} \in E^n$ and f and the g_i's are differentiable convex functions. As indicated in the last section, it is sufficient to consider the case where the objective function is linear; thus, we consider the problem

$$\begin{aligned} &\text{minimize} \ \mathbf{c}^T\mathbf{x} \\ &\text{subject to} \ \mathbf{g}(\mathbf{x}) \leqslant \mathbf{0} \end{aligned} \tag{14.61}$$

where $\mathbf{x} \in E^n$ and $\mathbf{g}(\mathbf{x}) \in E^p$ is convex and differentiable.

For $\mathbf{g}$ convex and differentiable we have the fundamental inequality

$$\mathbf{g}(\mathbf{x}) \geqslant \mathbf{g}(\mathbf{w}) + \nabla\mathbf{g}(\mathbf{w})(\mathbf{x} - \mathbf{w}) \tag{14.62}$$

for any $\mathbf{x}$, $\mathbf{w}$. We use this equation to determine the separating hyperplane. Specifically, the algorithm is as follows:

Let $S = \{\mathbf{x} : \mathbf{g}(\mathbf{x}) \leqslant \mathbf{0}\}$ and let P be an initial polytope containing S and such that $\mathbf{c}^T\mathbf{x}$ is bounded on P. Then

Step 1. Minimize $\mathbf{c}^T\mathbf{x}$ over P obtaining the point $\mathbf{x} = \mathbf{w}$. If $\mathbf{g}(\mathbf{w}) \leqslant \mathbf{0}$, stop; $\mathbf{w}$ is an optimal solution. Otherwise,

Step 2. Let i be an index maximizing $g_i(\mathbf{w})$. Clearly $g_i(\mathbf{w}) > 0$. Define the new approximating polytope to be the old one intersected with the half-space

$$\{\mathbf{x} : g_i(\mathbf{w}) + \nabla g_i(\mathbf{w})(\mathbf{x} - \mathbf{w}) \leqslant 0\}. \tag{14.63}$$

Return to Step 1.

The set defined by (14.63) is actually a half-space if $\nabla g_i(\mathbf{w}) \neq \mathbf{0}$. However, $\nabla g_i(\mathbf{w}) = \mathbf{0}$ would imply that $\mathbf{w}$ minimizes g_i which is impossible if S is nonempty. Furthermore, the half-space given by (14.63) contains S, since if $\mathbf{g}(\mathbf{x}) \leqslant \mathbf{0}$ then by (14.62) $g_i(\mathbf{w}) + \nabla g_i(\mathbf{w})(\mathbf{x} - \mathbf{w}) \leqslant g_i(\mathbf{x}) \leqslant \mathbf{0}$. The half-space does not contain the point $\mathbf{w}$ since $g_i(\mathbf{w}) > 0$. This method for selecting the separating hyperplane is illustrated in Fig. 14.8 for the one-dimensional case. Note that in one dimension, the procedure reduces to Newton's method.

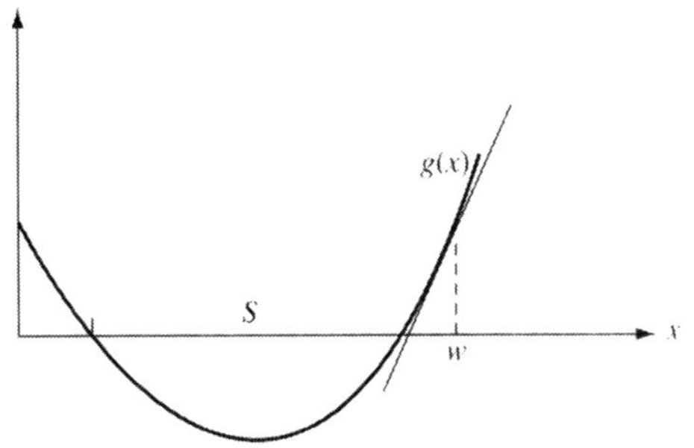

Fig. 14.8 Convex cutting plane

Calculation of the separating hyperplane is exceedingly simple in this algorithm, and hence the method really amounts to the solution of a series of linear programming problems. It should be noted that this algorithm, valid for any convex programming problem, does not involve any line searches. In that respect it is also similar to Newton's method applied to a convex function.

Convergence

Under fairly mild assumptions on the convex function, the convex cutting plane method is globally convergent. It is possible to apply the general convergence theorem to prove this, but somewhat easier, in this case, to prove it directly.

Theorem. *Let the convex functions g_i, $i = 1, 2, \ldots, p$ be continuously differentiable, and suppose the convex cutting plane algorithm generates the sequence of points $\{\mathbf{w}_k\}$. Any limit point of this sequence is a solution to problem (14.61).*

Proof. Suppose $\{\mathbf{w}_k\}$, $k \in \mathcal{K}$ is a subsequence of $\{\mathbf{w}_k\}$ converging to $\mathbf{w}$. By taking a further subsequence of this, if necessary, we may assume that the index i corresponding to Step 2 of the algorithm is fixed throughout the subsequence. Now if $k \in \mathcal{K}$, $k' \in \mathcal{K}$ and $k' > k$, then we must have

$$g_i(\mathbf{w}_k) + \nabla g_i(\mathbf{w}_k)(\mathbf{w}_{k'} - \mathbf{w}_k) \leqslant 0,$$

which implies that

$$g_i(\mathbf{w}_k) \leq |\nabla g_i(\mathbf{w}_k)||\mathbf{w}_{k'} - \mathbf{w}_k|. \tag{14.64}$$

Since $|\nabla g_i(\mathbf{w}_k)|$ is bounded with respect to $k \in \mathcal{K}$, the right-hand side of (14.64) goes to zero as k and k' go to infinity. The left-hand side goes to $g_i(\mathbf{w})$. Thus $g_i(\mathbf{w}) \leqslant 0$ and we see that $\mathbf{w}$ is feasible for problem (14.61).

If f^* is the optimal value of problem (14.61), we have $\mathbf{c}^T\mathbf{w}_k \leqslant f^*$ for each k since $\mathbf{w}_k$ is obtained by minimizing over a set containing S. Thus, by continuity, $\mathbf{c}^T\mathbf{w} \leqslant f^*$ and hence $\mathbf{w}$ is an optimal solution. ∎

As with most algorithms based on linear programming concepts, the rate of convergence of cutting plane algorithms has not yet been satisfactorily analyzed. Preliminary research shows that these algorithms converge arithmetically, that is, if $\mathbf{x}^*$ is optimal, then $|\mathbf{x}_k - \mathbf{x}^*|^2 \leqslant c/k$ for some constant c. This is an exceedingly poor type of convergence. This estimate, however, may not be the best possible and indeed there are indications that the convergence is actually geometric but with a ratio that goes to unity as the dimension of the problem increases.

Modifications

We now describe the supporting hyperplane algorithm (an alternative method for determining a cutting plane) and examine the possibility of dropping from consideration some old hyperplanes so that the linear programs do not grow too large. The convexity requirements are less severe for this algorithm. It is applicable to problems of the form

$$\begin{aligned} &\text{minimize} \quad \mathbf{c}^T\mathbf{x} \\ &\text{subject to} \quad \mathbf{g}(\mathbf{x}) \leqslant \mathbf{0}, \end{aligned}$$

where $\mathbf{x} \in E^n$, $\mathbf{g}(\mathbf{x}) \in E^p$, the g_i's are continuously differentiable, and the constraint region S defined by the inequalities is convex. Note that convexity of the functions themselves is not required. We also assume the existence of a point interior to the constraint region, that is, we assume the existence of a point $\mathbf{y}$ such that $\mathbf{g}(\mathbf{y}) < \mathbf{0}$, and we assume that on the constraint boundary $g_i(\mathbf{x}) = 0$ implies $\nabla g_i(\mathbf{x}) \neq \mathbf{0}$. The algorithm is as follows:

Start with an initial polytope P containing S and such that $\mathbf{c}^T\mathbf{x}$ is bounded below on S. Then

Step 1. Determine $\mathbf{w} = \mathbf{x}$ to minimize $\mathbf{c}^T\mathbf{x}$ over P. If $\mathbf{w} \in S$, stop. Otherwise,

Step 2. Find the point $\mathbf{u}$ on the line joining $\mathbf{y}$ and $\mathbf{w}$ that lies on the boundary of S. Let i be an index for which $g_i(\mathbf{u}) = 0$ and define the half-space $H = \{\mathbf{x}: \nabla g_i(\mathbf{u})(\mathbf{x} - \mathbf{u}) \leqslant 0\}$. Update P by intersecting with H. Return to Step 1.

The algorithm is illustrated in Fig. 14.9.

The price paid for the generality of this method over the convex cutting plane method is that an interpolation along the line joining $\mathbf{y}$ and $\mathbf{w}$ must be executed to find the point $\mathbf{u}$. This is analogous to the line search for a minimum point required by most programming algorithms.

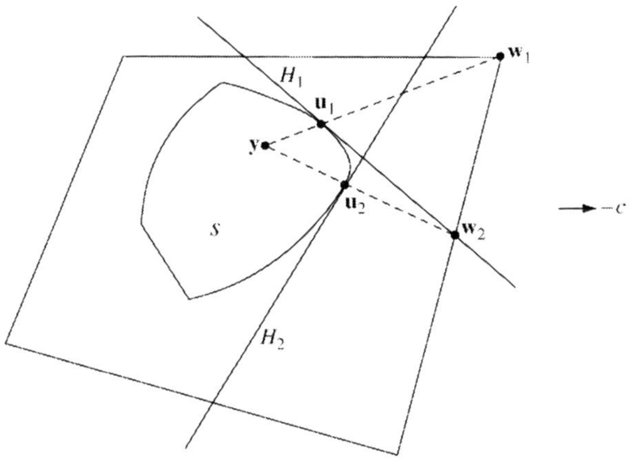

Fig. 14.9 Supporting hyperplane algorithm

Dropping Nonbinding Constraints

In all cutting plane algorithms nonbinding constraints can be dropped from the approximating set of linear inequalities so as to keep the complexity of the approximation manageable. Indeed, since n linearly independent hyperplanes determine a single point in E^n, the algorithm can be arranged, by discarding the nonbinding constraints at the end of each step, so that the polytope consists of exactly n linear inequalities at every stage.

Global convergence is not destroyed by this process, since the sequence of objective values will still be monotonically increasing. It is not known, however, what effect this has on the speed of convergence.

14.9 Exercises

1. (Linear programming) Use the global duality theorem to find the dual of the linear program

$$\begin{aligned}&\text{minimize} \quad \mathbf{c}^T\mathbf{x}\\&\text{subject to} \quad \mathbf{A}\mathbf{x} = \mathbf{b},\ \mathbf{x} \geq \mathbf{0}.\end{aligned}$$

Note that some of the regularity conditions may not be necessary for the linear case.

2. (Double dual) Show that the for a convex programming problem with a solution, the dual of the dual is in some sense the original problem.
3. (Non-convex?) Consider the problem

$$\begin{aligned}&\text{minimize} \quad xy\\&\text{subject to} \quad x + y - 4 \geq 0\\&\qquad\qquad 1 \leq x \leq 5,\ 1 \leq y \leq 5.\end{aligned}$$

Show that although the objective function is not convex, the primal function is convex. Find the optimal value and the Lagrange multiplier.

4. Find the global maximum of the dual function of Example 1, Sect. 14.2.
5. Show that the function ϕ defined for $\boldsymbol{\lambda}$, $\boldsymbol{\mu}$, ($\boldsymbol{\mu} \geqslant \mathbf{0}$), by $\phi(\boldsymbol{\lambda}, \boldsymbol{\mu}) = \min_{\mathbf{x}}[f(\mathbf{x}) + \boldsymbol{\lambda}^T\mathbf{h}(\mathbf{x}) + \boldsymbol{\mu}^T\mathbf{g}(\mathbf{x})]$ is concave over any convex region where it is finite.
6. Prove that the dual canonical rate of convergence is not affected by a change of variables in $\mathbf{x}$.
7. Corresponding to the dual function (14.23):
   (a) Find its gradient.
   (b) Find its Hessian.
   (c) Verify that it has a local maximum at $\boldsymbol{\lambda}^*$, $\boldsymbol{\mu}^*$.
8. Find the Hessian of the dual function for a separable problem.
9. Find an explicit formula for the dual function for the entropy problem (Example 3, Sect. 11.4).
10. Consider the problems

$$\begin{aligned}&\text{minimize} \quad f(\mathbf{x})\\&\text{subject to} \quad g_i(\mathbf{x}) \leqslant 0,\ j = 1, 2, \ldots, p\end{aligned} \tag{14.65}$$

and

$$\begin{aligned}&\text{minimize} \quad f(\mathbf{x})\\&\text{subject to} \quad g_i(\mathbf{x}) + z_j^2 = 0,\ j = 1, 2, \ldots, p.\end{aligned} \tag{14.66}$$

   (a) Let $\mathbf{x}^*$, μ_1^*, μ_2^*, …, μ_p^* be a point and set of Lagrange multipliers that satisfy the first-order necessary conditions for (14.65). For $\mathbf{x}^*$, $\boldsymbol{\mu}^*$, write the second-order sufficiency conditions for (14.66).

(b) Show that in general they are not satisfied unless, in addition to satisfying the sufficiency conditions of Sect. 11.8, $g_j(\mathbf{x}^*)$ implies $\mu_j^* > 0$.

11. Establish global convergence for the supporting hyperplane algorithm.
12. Establish global convergence for an imperfect version of the supporting hyperplane algorithm that in interpolating to find the boundary point $\mathbf{u}$ actually finds a point somewhere on the segment joining $\mathbf{u}$ and $\frac{1}{2}\mathbf{u} + \frac{1}{2}\mathbf{w}$ and establishes a hyperplane there.
13. Prove that the convex cutting plane method is still globally convergent if it is modified by discarding from the definition of the polytope at each stage hyperplanes corresponding to inactive linear inequalities.

References

14.1 Global duality was developed in conjunction with the theory of Sect. 11.9, by Hurwicz [H14] and Slater [S7]. The theory was presented in this form in Luenberger [L8].

14.2–14.3 An important early differential form of duality was developed by Wolfe [W3]. The convex theory can be traced to the Legendre transformation used in the calculus of variations but it owes its main heritage to Fenchel [F3]. This line was further developed by Karlin [K1] and Hurwicz [H14]. Also see Luenberger [L8].

14.4 The solution of separable problems by dual methods in this manner was pioneered by Everett [E2].

14.5–14.6 The method of multipliers was originally suggested by Hestenes [H8] and from a different viewpoint by Powell [P7]. The relation to duality was presented briefly in Luenberger [L15]. The method for treating inequality constraints was devised by Rockafellar [R3]. For an excellent survey of multiplier methods see Bertsekas [B12].

14.7 The alternating direction method of multipliers was due to Gabay and Mercier[109] and Glowinski and Marrocco [102]; also see Fortin and Glowinski[96], Eckstein and Bertsekas [78] and Boyd et al. [41]. The convergence speed analysis was initially done by He and Yuan [124] and Monteiro and Svaiter [180]. The non-convergence examples of three blocks were constructed by Chen et al. [50].

14.8 Cutting plane methods were first introduced by Kelley [K3] who developed the convex cutting plane method. The supporting hyperplane algorithm was suggested by Veinott [V5]. To see how global convergence of cutting plane algorithms can be established from the general convergence theorem see Zangwill [Z2]. For some results on the convergence rates of cutting plane algorithms consult Topkis [T7], Eaves and Zangwill [E1], and Wolfe [W7].

Chapter 15
Primal-Dual Methods

This chapter discusses methods that work simultaneously with primal and dual variables, in essence seeking to satisfy the first-order necessary conditions for optimality. The methods employ many of the concepts used in earlier chapters, including those related to active set methods, various first and second order methods, penalty methods, and barrier methods. Indeed, a study of this chapter is in a sense a review and extension of what has been presented earlier.

The first several sections of the chapter discuss methods for solving the standard nonlinear programming structure that has been treated in the Parts II and III of the text. These sections provide alternatives to the methods discussed earlier.

15.1 The Standard Problem

Consider again the standard nonlinear program

$$\begin{aligned} &\text{minimize} \;\; f(\mathbf{x}) \\ &\text{subject to} \;\; \mathbf{h}(\mathbf{x}) = \mathbf{0}, \;\; \mathbf{g}(\mathbf{x}) \leqslant \mathbf{0}. \end{aligned} \tag{15.1}$$

Together with the feasibility, the first-order necessary conditions for optimality are, as we know,

$$\begin{aligned} \nabla f(\mathbf{x}) + \boldsymbol{\lambda}^T \nabla \mathbf{h}(\mathbf{x}) + \boldsymbol{\mu}^T \nabla \mathbf{g}(\mathbf{x}) &= \mathbf{0} \\ \boldsymbol{\mu} &\geqslant \mathbf{0} \\ \boldsymbol{\mu}^T \mathbf{g}(\mathbf{x}) &= \mathbf{0} \end{aligned} \tag{15.2}$$

The last requirement is the complementary slackness condition. If it is known which of the inequality constraints is active at the solution, these active constraints can be

D.G. Luenberger, Y. Ye, *Linear and Nonlinear Programming*, International Series in Operations Research & Management Science 228,
DOI 10.1007/978-3-319-18842-3_15

rolled into the equality constraints $\mathbf{h}(\mathbf{x}) = \mathbf{0}$, and the inactive inequalities along with the complementary slackness condition dropped, to obtain a problem with equality constraints only. This indeed is the structure of the problem near the solution.

If in this structure the vector $\mathbf{x}$ is n-dimensional and $\mathbf{h}$ is m-dimensional, then $\boldsymbol{\lambda}$ will also be m-dimensional. The system (15.1) will, in this reduced form, consist of $n + m$ equations and $n + m$ unknowns, which is an indication that the system may be well defined, and hence that there is a solution for the pair $(\mathbf{x}, \boldsymbol{\lambda})$. In essence, primal-dual methods amount to solving this system of equations, and use additional strategies to account for inequality constraints.

In view of the above observation it is natural to consider whether in fact the system of necessary conditions is in fact well conditioned, possessing a unique solution $(\mathbf{x}, \boldsymbol{\lambda})$. We investigate this question by considering a linearized version of the conditions.

A useful and somewhat more generally useful approach is to consider the quadratic program

$$\begin{aligned} &\text{minimize} \quad \frac{1}{2}\mathbf{x}^T\mathbf{Q}\mathbf{x} + \mathbf{c}^T\mathbf{x} \\ &\text{subject to} \quad \mathbf{A}\mathbf{x} = \mathbf{b}, \end{aligned} \tag{15.3}$$

where $\mathbf{x}$ is n-dimensional and $\mathbf{b}$ is m-dimensional.

The first-order conditions for this problem are

$$\begin{aligned} \mathbf{Q}\mathbf{x} + \mathbf{A}^T\boldsymbol{\lambda} + \mathbf{c} &= \mathbf{0} \\ \mathbf{A}\mathbf{x} \qquad\quad - \mathbf{b} &= \mathbf{0}. \end{aligned} \tag{15.4}$$

These correspond to the necessary conditions (15.2) for equality constraints only. The following proposition gives conditions under which the system is nonsingular.

Proposition. *Let* $\mathbf{Q}$ *and* $\mathbf{A}$ *be* $n \times n$ *and* $m \times n$ *matrices, respectively. Suppose that* $\mathbf{A}$ *has rank* m *and that* $\mathbf{Q}$ *is positive definite on the subspace* $M = \{\mathbf{x} : \mathbf{A}\mathbf{x} = \mathbf{0}\}$. *Then the matrix*

$$\begin{bmatrix} \mathbf{Q} & \mathbf{A}^T \\ \mathbf{A} & \mathbf{0} \end{bmatrix} \tag{15.5}$$

is nonsingular.

Proof. Suppose $(\mathbf{x}, \mathbf{y}) \in E^{n+m}$ is such that

$$\begin{aligned} \mathbf{Q}\mathbf{x} + \mathbf{A}^T\mathbf{y} &= \mathbf{0} \\ \mathbf{A}\mathbf{x} \qquad &= \mathbf{0}. \end{aligned} \tag{15.6}$$

Multiplication of the first equation by $\mathbf{x}^T$ yields

$$\mathbf{x}^T\mathbf{Q}\mathbf{x} + \mathbf{x}^T\mathbf{A}^T\mathbf{y} = 0,$$

and substitution of $\mathbf{A}\mathbf{x} = \mathbf{0}$ yields $\mathbf{x}^T\mathbf{Q}\mathbf{x} = \mathbf{0}$. However, clearly $\mathbf{x} \in M$, and thus the hypothesis on $\mathbf{Q}$ together with $\mathbf{x}^T\mathbf{Q}\mathbf{x} = \mathbf{0}$ implies that $\mathbf{x} = \mathbf{0}$. It then follows from the first equation that $\mathbf{A}^T\mathbf{y} = \mathbf{0}$. The full-rank condition on $\mathbf{A}$ then implies that $\mathbf{y} = \mathbf{0}$. Thus the only solution to (15.6) is $\mathbf{x} = \mathbf{0}$, $\mathbf{y} = \mathbf{0}$. ∎

If, as is often the case, the matrix $\mathbf{Q}$ is actually positive definite (over the whole space), then an explicit formula for the solution of the system can be easily derived as follows: From the first equation in (15.4) we have

$$\mathbf{x} = -\mathbf{Q}^{-1}\mathbf{A}^T\boldsymbol{\lambda} - \mathbf{Q}^{-1}\mathbf{c}.$$

Substitution of this into the second equation then yields

$$-\mathbf{A}\mathbf{Q}^{-1}\mathbf{A}^T\boldsymbol{\lambda} - \mathbf{A}\mathbf{Q}^{-1}\mathbf{c} - \mathbf{b} = \mathbf{0},$$

from which we immediately obtain

$$\boldsymbol{\lambda} = -(\mathbf{A}\mathbf{Q}^{-1}\mathbf{A}^T)^{-1}[\mathbf{A}\mathbf{Q}^{-1}\mathbf{c} + \mathbf{b}] \tag{15.7}$$

and

$$\begin{aligned}\mathbf{x} &= \mathbf{Q}^{-1}\mathbf{A}^T(\mathbf{A}\mathbf{Q}^{-1}\mathbf{A}^T)^{-1}[\mathbf{A}\mathbf{Q}^{-1}\mathbf{c} + \mathbf{b}] - \mathbf{Q}^{-1}\mathbf{c} \\ &= -\mathbf{Q}^{-1}[\mathbf{I} - \mathbf{A}^T(\mathbf{A}\mathbf{Q}^{-1}\mathbf{A}^T)^{-1}\mathbf{A}\mathbf{Q}^{-1}]\mathbf{c} \\ &\quad + \mathbf{Q}^{-1}\mathbf{A}^T(\mathbf{A}\mathbf{Q}^{-1}\mathbf{A}^T)^{-1}\mathbf{b}.\end{aligned} \tag{15.8}$$

Strategies

There are some general strategies that guide the development of the primal-dual methods of this chapter.

1. **Descent Measures.** A fundamental concept that we have frequently used is that of assuring that progress is made at each step of an iterative algorithm. It is this that is used to guarantee global convergence. In primal methods this measure of descent is the objective function. Even the simplex method of linear programming is founded on this idea of making progress with respect to the objective function. For primal minimization methods, one typically arranges that the objective function decreases at each step.
   The objective function is not the only possible way to measure progress. We have, for example, when minimizing a function f, considered the quantity $(1/2)|\nabla f(\mathbf{x})|^2$, seeking to monotonically reduce it to zero.
   In general, a function used to measure progress is termed a *merit function*. Typically, it is defined so as to decrease as progress is made toward the solution of a minimization problem, but the sign may be reversed in some definitions. For primal-dual methods, the merit function may depend on both $\mathbf{x}$ and $\boldsymbol{\lambda}$. One especially useful merit function for equality constrained problems is

   $$m(\mathbf{x}, \boldsymbol{\lambda}) = \frac{1}{2}|\nabla f(\mathbf{x}) + \boldsymbol{\lambda}^T\nabla\mathbf{h}(\mathbf{x})|^2 + \frac{1}{2}|\mathbf{h}(\mathbf{x}))|^2.$$

   It is examined in the next section.

We shall examine other merit functions later in the chapter. With interior point methods or semidefinite programming, we shall use a potential function that serves as a merit function.

2. **Active Set Methods.** Inequality constraints can be treated using active set methods that treat the active constraints as equality constraints, at least for the current iteration. However, in primal-dual methods, both $\mathbf{x}$ and $\boldsymbol{\lambda}$ are changed. We shall consider variations of steepest descent, conjugate directions, and Newton's method where movement is made in the $(\mathbf{x}, \boldsymbol{\lambda})$ space.
3. **Penalty Functions.** In some primal-dual methods, a penalty function can serve as a merit function, even though the penalty function depends only on $\mathbf{x}$. This is particularly attractive for recursive quadratic programming methods where a quadratic program is solved at each stage to determine the direction of change in the pair $(\mathbf{x}, \boldsymbol{\lambda})$.
4. **Interior (Barrier) Methods.** Barrier methods lead to methods that move within the relative interior of the inequality constraints. This approach leads to the concept of the primal-dual central path. These methods are used for semidefinite programming since these problems are characterized as possessing a special form of inequality constraint.

15.2 A Simple Merit Function

It is very natural, when considering the system of necessary conditions (15.2), to form the function

$$m(\mathbf{x}, \boldsymbol{\lambda}) = \frac{1}{2}|\nabla f(\mathbf{x}) + \boldsymbol{\lambda}^T \nabla \mathbf{h}(\mathbf{x})|^2 + \frac{1}{2}|\mathbf{h}(\mathbf{x})|^2, \tag{15.9}$$

and use it as a measure of how close a point $(\mathbf{x}, \boldsymbol{\lambda})$ is to a solution.

It must be noted, however, that the function $m(\mathbf{x}, \boldsymbol{\lambda})$ is not always well-behaved; it may have local minima, and these are of no value in a search for a solution. The following theorem gives the conditions under which the function $m(\mathbf{x}, \boldsymbol{\lambda})$ can serve as a well-behaved merit function. Basically, the main requirement is that the Hessian of the Lagrangian be positive definite. As usual, we define $l(\mathbf{x}, \boldsymbol{\lambda}) = f(\mathbf{x}) + \boldsymbol{\lambda}^T \mathbf{h}(\mathbf{x})$.

Theorem. *Let f and $\mathbf{h}$ be twice continuously differentiable functions on E^n of dimension 1 and m, respectively. Suppose that $\mathbf{x}^*$ and $\boldsymbol{\lambda}^*$ satisfy the first-order necessary conditions for a local minimum of $m(\mathbf{x}, \boldsymbol{\lambda}) = \frac{1}{2}|\nabla f(\mathbf{x})+\boldsymbol{\lambda}^T \nabla \mathbf{h}(\mathbf{x})|^2 + \frac{1}{2}|\mathbf{h}(\mathbf{x})|^2$ with respect to $\mathbf{x}$ and $\boldsymbol{\lambda}$. Suppose also that at $\mathbf{x}^*$, $\boldsymbol{\lambda}^*$, (i) the rank of $\nabla \mathbf{h}(\mathbf{x}^*)$ is m and (ii) the Hessian matrix $\mathbf{L}(\mathbf{x}^*, \boldsymbol{\lambda}^*) = \mathbf{F}(\mathbf{x}^*)+\boldsymbol{\lambda}^{*T}\mathbf{H}(\mathbf{x}^*)$ is positive definite. Then, $\mathbf{x}^*$, $\boldsymbol{\lambda}^*$ is a (possibly nonunique) global minimum point of $m(\mathbf{x}, \boldsymbol{\lambda})$, with value $m(\mathbf{x}^*, \boldsymbol{\lambda}^*) = 0$.*

Proof. Since $\mathbf{x}^*$, $\boldsymbol{\lambda}^*$ satisfies the first-order conditions for a local minimum point of $m(\mathbf{x}, \boldsymbol{\lambda})$, we have

$$[\nabla f(\mathbf{x}^*) + \boldsymbol{\lambda}^{*T} \nabla \mathbf{h}(\mathbf{x}^*)]\mathbf{L}(\mathbf{x}^*, \boldsymbol{\lambda}^*) + \mathbf{h}(\mathbf{x}^*)^T \nabla \mathbf{h}(\mathbf{x}^*) = \mathbf{0} \tag{15.10}$$

$$[\nabla f(\mathbf{x}^*) + \boldsymbol{\lambda}^{*T} \nabla \mathbf{h}(\mathbf{x}^*)]\nabla \mathbf{h}(\mathbf{x}^*)^T = \mathbf{0}. \tag{15.11}$$

Multiplying (15.10) on the right by $[\nabla f(\mathbf{x}^*) + \boldsymbol{\lambda}^{*T}\nabla \mathbf{h}(\mathbf{x}^*)]^T$ and using (15.11) we obtain†

$$\nabla l(\mathbf{x}^*, \boldsymbol{\lambda}^*)\mathbf{L}(\mathbf{x}^*, \boldsymbol{\lambda}^*)\nabla l(\mathbf{x}^*, \boldsymbol{\lambda}^*)^T = 0.$$

Since $\mathbf{L}(\mathbf{x}^*, \boldsymbol{\lambda}^*)$ is positive definite, this implies that $\nabla l(\mathbf{x}^*, \boldsymbol{\lambda}^*) = \mathbf{0}$. Using this in (15.10), we find that $\mathbf{h}(\mathbf{x}^*)^T\nabla\mathbf{h}(\mathbf{x}^*) = \mathbf{0}$, which, since $\nabla\mathbf{h}(\mathbf{x}^*)$ is of rank m, implies that $\mathbf{h}(\mathbf{x}^*) = \mathbf{0}$. ∎

The requirement that the Hessian of the Lagrangian $\mathbf{L}(\mathbf{x}^*, \boldsymbol{\lambda}^*)$ be positive definite at a stationary point of the merit function m is actually not too restrictive. This condition will be satisfied in the case of a convex programming problem where f is strictly convex and $\mathbf{h}$ is linear. Furthermore, even in nonconvex problems one can often arrange for this condition to hold, at least near a solution to the original constrained minimization problem. If it is assumed that the second-order sufficiency conditions for a constrained minimum hold at $\mathbf{x}^*$, $\boldsymbol{\lambda}^*$, then $\mathbf{L}(\mathbf{x}^*, \boldsymbol{\lambda}^*)$ is positive definite on the subspace that defines the tangent to the constraints; that is, on the subspace defined by $\nabla\mathbf{h}(\mathbf{x}^*)\mathbf{x} = \mathbf{0}$. Now if the original problem is modified with a penalty term to the problem

$$\begin{aligned} &\text{minimize} \quad f(\mathbf{x}) + \frac{1}{2}c|\mathbf{h}(\mathbf{x})|^2 \\ &\text{subject to} \quad \mathbf{h}(\mathbf{x}) = \mathbf{0}, \end{aligned} \tag{15.12}$$

the solution point $\mathbf{x}^*$ will be unchanged. However, as discussed in Chap. 14, the Hessian of the Lagrangian of this new problem (15.12) at the solution point is $\mathbf{L}(\mathbf{x}^*, \boldsymbol{\lambda}^*) + c\nabla\mathbf{h}(\mathbf{x}^*)^T\nabla\mathbf{h}(\mathbf{x}^*)$. For sufficiently large c, this matrix will be positive definite. Thus a problem can be "convexified" (at least locally) before the merit function method is employed.

An extension to problems with inequality constraints can be defined by partitioning the constraints into the two groups *active* and *inactive*. However, at this point the simple merit function for problems with equality constraints is adequate for the purpose of illustrating the general idea.

15.3 Basic Primal-Dual Methods

Many primal-dual methods are patterned after some of the methods used in earlier chapters, except of course that the emphasis is on equation solving rather than explicit optimization.

† Unless explicitly indicated to the contrary, the notation $\nabla l(\mathbf{x}, \boldsymbol{\lambda})$ refers to the gradient of l with respect to $\mathbf{x}$, that is, $\nabla_{\mathbf{x}} l(\mathbf{x}, \boldsymbol{\lambda})$.

First-Order Method

We consider first a simple straightforward approach, which in a sense parallels the idea of steepest descent in that it uses only a first-order approximation to the primal–dual equations. It is defined by

$$\begin{aligned} \mathbf{x}_{k+1} &= \mathbf{x}_k - \alpha_k \nabla l(\mathbf{x}_k, \boldsymbol{\lambda}_k)^T \\ \boldsymbol{\lambda}_{k+1} &= \boldsymbol{\lambda}_k + \alpha_k \mathbf{h}(\mathbf{x}_k), \end{aligned} \tag{15.13}$$

where α_k is not yet determined. This is based on the error in satisfying (15.2). Assume that the Hessian of the Lagrangian $\mathbf{L}(\mathbf{x}, \boldsymbol{\lambda})$ is positive definite in some compact region of interest, and consider the simple merit function

$$m(\mathbf{x}, \boldsymbol{\lambda}) = \frac{1}{2}|\nabla l(\mathbf{x}, \boldsymbol{\lambda})|^2 + \frac{1}{2}|\mathbf{h}(\mathbf{x})|^2 \tag{15.14}$$

discussed above. We would like to determine whether the direction of change in (15.13) is a descent direction with respect to this merit function. The gradient of the merit function has components corresponding to $\mathbf{x}$ and $\boldsymbol{\lambda}$ of

$$\begin{aligned} &\nabla l(\mathbf{x}, \boldsymbol{\lambda})\mathbf{L}(\mathbf{x}, \boldsymbol{\lambda}) + \mathbf{h}(\mathbf{x})^T \nabla \mathbf{h}(\mathbf{x}) \\ &\nabla l(\mathbf{x}, \boldsymbol{\lambda}) \nabla \mathbf{h}(\mathbf{x})^T. \end{aligned} \tag{15.15}$$

Thus the inner product of this gradient with the direction vector having components $-\nabla l(\mathbf{x}, \boldsymbol{\lambda})^T$, $\mathbf{h}(\mathbf{x})$ is

$$\begin{aligned} -\nabla l(\mathbf{x}, \boldsymbol{\lambda})\mathbf{L}(\mathbf{x}, \boldsymbol{\lambda})\nabla l(\mathbf{x}, \boldsymbol{\lambda})^T - \mathbf{h}(\mathbf{x})^T \nabla \mathbf{h}(\mathbf{x}) \nabla l(\mathbf{x}, \boldsymbol{\lambda})^T + \nabla l(\mathbf{x}, \boldsymbol{\lambda}) \nabla \mathbf{h}(\mathbf{x})^T \mathbf{h}(\mathbf{x}) \\ = -\nabla l(\mathbf{x}, \boldsymbol{\lambda})\mathbf{L}(\mathbf{x}, \boldsymbol{\lambda})\nabla l(\mathbf{x}, \boldsymbol{\lambda})^T \leqslant 0. \end{aligned}$$

This shows that the search direction is in fact a descent direction for the merit function, unless $\nabla l(\mathbf{x}, \boldsymbol{\lambda}) = \mathbf{0}$. Thus by selecting α_k to minimize the merit function in the search direction at each step, the process will converge to a point where $\nabla l(\mathbf{x}, \boldsymbol{\lambda}) = \mathbf{0}$. However, there is no guarantee that $\mathbf{h}(\mathbf{x}) = \mathbf{0}$ at that point.

We can try to improve the method either by changing the way in which the direction is selected or by changing the merit function. In this case a slight modification of the merit function will work. Let

$$w(\mathbf{x}, \boldsymbol{\lambda}, \gamma) = m(\mathbf{x}, \boldsymbol{\lambda}) - \gamma[f(\mathbf{x}) + \boldsymbol{\lambda}^T \mathbf{h}(\mathbf{x})]$$

for some $\gamma > 0$. We then calculate that the gradient of w has the two components corresponding to $\mathbf{x}$ and $\boldsymbol{\lambda}$

$$\begin{aligned} &\nabla l(\mathbf{x}, \boldsymbol{\lambda})\mathbf{L}(\mathbf{x}, \boldsymbol{\lambda}) + \mathbf{h}(\mathbf{x})^T \nabla \mathbf{h}(\mathbf{x}) - \gamma \nabla l(\mathbf{x}, \boldsymbol{\lambda}) \\ &\nabla l(\mathbf{x}, \boldsymbol{\lambda}) \nabla \mathbf{h}(\mathbf{x})^T - \gamma \mathbf{h}(\mathbf{x})^T, \end{aligned}$$

and hence the inner product of the gradient with the direction $-\nabla l(\mathbf{x}, \boldsymbol{\lambda})^T$, $\mathbf{h}(\mathbf{x})$ is

$$-\nabla l(\mathbf{x}, \boldsymbol{\lambda})[\mathbf{L}(\mathbf{x}, \boldsymbol{\lambda}) - \gamma \mathbf{I}]\nabla l(\mathbf{x}, \boldsymbol{\lambda})^T - \gamma |\mathbf{h}(\mathbf{x})|^2.$$

Now since we are assuming that $\mathbf{L}(\mathbf{x}, \boldsymbol{\lambda})$ is positive definite in a compact region of interest, there is a $\gamma > 0$ such that $\mathbf{L}(\mathbf{x}, \boldsymbol{\lambda}) - \gamma\mathbf{I}$ is positive definite in this region. Then according to the above calculation, the direction $-\nabla l(\mathbf{x}, \boldsymbol{\lambda})^T$, $\mathbf{h}(\mathbf{x})$ is a descent direction, and the standard descent method will converge to a solution. This method will not converge very rapidly however. (See Exercise 2 for further analysis of this method.)

Conjugate Directions

One may also use the conjugate direction. Let us consider the quadratic program

$$\begin{aligned} &\text{minimize} \quad \frac{1}{2}\mathbf{x}^T\mathbf{Q}\mathbf{x} - \mathbf{b}^T\mathbf{x} \\ &\text{subject to} \quad \mathbf{A}\mathbf{x} = \mathbf{c}. \end{aligned} \tag{15.16}$$

The first-order necessary conditions for this problem are

$$\begin{aligned} \mathbf{Q}\mathbf{x} + \mathbf{A}^T\boldsymbol{\lambda} &= \mathbf{b} \\ \mathbf{A}\mathbf{x} \qquad &= \mathbf{c}. \end{aligned} \tag{15.17}$$

As discussed in the previous section, this problem is equivalent to solving a system of linear equations whose coefficient matrix is

$$\mathbf{M} = \begin{bmatrix} \mathbf{Q} & \mathbf{A}^T \\ \mathbf{A} & \mathbf{0} \end{bmatrix}. \tag{15.18}$$

This matrix is symmetric, but it is *not* positive definite (nor even semidefinite). However, it is possible to formally generalize the conjugate gradient method to systems of this type by just applying the conjugate-gradient formulae (15.17)–(15.20) of Sect. 9.3 with $\mathbf{Q}$ replaced by $\mathbf{M}$. A difficulty is that *singular directions* (defined as directions $\mathbf{p}$ such that $\mathbf{p}^T\mathbf{M}\mathbf{p} = 0$) may occur and cause the process to break down. Procedures for overcoming this difficulty have been developed, however. Also, as in the ordinary conjugate gradient method, the approach can be generalized to treat nonquadratic problems as well. Overall, however, the application of conjugate direction methods to the Lagrange system of equations, although very promising, is not currently considered practical.

Second-Order Method: Newton's Method

Newton's method for solving systems of equations can be easily applied to the Lagrange equations. In its most straightforward form, the method solves the system

$$\begin{aligned} \nabla l(\mathbf{x}, \boldsymbol{\lambda}) &= \mathbf{0} \\ \mathbf{h}(\mathbf{x}) &= \mathbf{0} \end{aligned} \tag{15.19}$$

by solving the linearized version recursively. That is, given $\mathbf{x}_k$, $\boldsymbol{\lambda}_k$ the new point $\mathbf{x}_{k+1}$, $\boldsymbol{\lambda}_{k+1}$ is determined from the equations

$$\begin{aligned} \nabla l(\mathbf{x}_k, \boldsymbol{\lambda}_k)^T + \mathbf{L}(\mathbf{x}_k, \boldsymbol{\lambda}_k)\mathbf{d}_k + \nabla \mathbf{h}(\mathbf{x}_k)^T \mathbf{y}_k &= \mathbf{0} \\ \mathbf{h}(\mathbf{x}_k) + \nabla \mathbf{h}(\mathbf{x}_k)\mathbf{d}_k &= \mathbf{0} \end{aligned} \tag{15.20}$$

by setting $\mathbf{x}_{k+1} = \mathbf{x}_k + \mathbf{d}_k$, $\boldsymbol{\lambda}_{k+1} = \boldsymbol{\lambda}_k + \mathbf{y}_k$. In matrix form the above Newton equations are

$$\begin{bmatrix} \mathbf{L}(\mathbf{x}_k, \boldsymbol{\lambda}_k) & \nabla \mathbf{h}(\mathbf{x}_k)^T \\ \nabla \mathbf{h}(\mathbf{x}_k) & \mathbf{0} \end{bmatrix} \begin{bmatrix} \mathbf{d}_k \\ \mathbf{y}_k \end{bmatrix} = \begin{bmatrix} -\nabla l(\mathbf{x}_k, \boldsymbol{\lambda}_k)^T \\ -\mathbf{h}(\mathbf{x}_k) \end{bmatrix}. \tag{15.21}$$

The Newton equations have some important structural properties. First, we observe that by adding $\nabla \mathbf{h}(\mathbf{x}_k)^T \boldsymbol{\lambda}_k$ to the top equation, the system can be transformed to the form

$$\begin{bmatrix} \mathbf{L}(\mathbf{x}_k, \boldsymbol{\lambda}_k) & \nabla \mathbf{h}(\mathbf{x}_k)^T \\ \nabla \mathbf{h}(\mathbf{x}_k) & \mathbf{0} \end{bmatrix} \begin{bmatrix} \mathbf{d}_k \\ \boldsymbol{\lambda}_{k+1} \end{bmatrix} = \begin{bmatrix} -\nabla f(\mathbf{x}_k)^T \\ -\mathbf{h}(\mathbf{x}_k) \end{bmatrix}, \tag{15.22}$$

where again $\boldsymbol{\lambda}_{k+1} = \boldsymbol{\lambda}_k + \mathbf{y}_k$. In this form $\boldsymbol{\lambda}_k$ appears only in the matrix $\mathbf{L}(\mathbf{x}_k, \boldsymbol{\lambda}_k)$. This conversion between (15.21) and (15.22) will be useful later.

Next we note that the structure of the coefficient matrix of (15.21) or (15.22) is identical to that of the Proposition of Sect. 15.1. The standard second-order sufficiency conditions imply that $\nabla \mathbf{h}(\mathbf{x}^*)$ is of full rank and that $\mathbf{L}(\mathbf{x}^*, \boldsymbol{\lambda}^*)$ is positive definite on $M = \{\mathbf{x} : \nabla \mathbf{h}(\mathbf{x}^*)\mathbf{x} = \mathbf{0}\}$ at the solution. By continuity these conditions can be assumed to hold in a region near the solution as well. Under these assumptions it follows from Proposition 1 that the Newton equation (15.21) has a unique solution.

It is again worthwhile to point out that, although the Hessian of the Lagrangian need be positive definite only on the tangent subspace in order for the system (15.21) to be nonsingular, it is possible to alter the original problem by incorporation of a quadratic penalty term so that the new Hessian of the Lagrangian is $\mathbf{L}(\mathbf{x}, \boldsymbol{\lambda}) + c\nabla \mathbf{h}(\mathbf{x})^T \nabla \mathbf{h}(\mathbf{x})$. For sufficiently large c, this new Hessian will be positive definite over the entire space.

If $\mathbf{L}(\mathbf{x}, \boldsymbol{\lambda})$ is positive definite (either originally or through the incorporation of a penalty term), it is possible to write an explicit expression for the solution of the system (15.21). Let us define $\mathbf{L}_k = \mathbf{L}(\mathbf{x}_k, \boldsymbol{\lambda}_k)$, $\mathbf{A}_k = \nabla \mathbf{h}(\mathbf{x}_k)$, $\mathbf{l}_k = \nabla l(\mathbf{x}_k, \boldsymbol{\lambda}_k)^T$, $\mathbf{h}_k = \mathbf{h}(\mathbf{x}_k)$. The system then takes the form

$$\begin{aligned} \mathbf{L}_k \mathbf{d}_k + \mathbf{A}_k^T \mathbf{y}_k &= -\mathbf{l}_k \\ \mathbf{A}_k \mathbf{d}_k \qquad &= -\mathbf{h}_k. \end{aligned} \tag{15.23}$$

The solution is readily found, as in (15.7) and (15.8) for quadratic programming, to be

$$\mathbf{y}_k = (\mathbf{A}_k \mathbf{L}_k^{-1} \mathbf{A}_k^T)^{-1} [\mathbf{h}_k - \mathbf{A}_k \mathbf{L}_k^{-1} \mathbf{l}_k] \tag{15.24}$$

$$\mathbf{d}_k = -\mathbf{L}_k^{-1} [\mathbf{I} - \mathbf{A}_k^T (\mathbf{A}_k \mathbf{L}_k^{-1} \mathbf{A}_k^T)^{-1} \mathbf{A}_k \mathbf{L}_k^{-1}] \mathbf{l}_k - \mathbf{L}_k^{-1} \mathbf{A}_k^T (\mathbf{A}_k \mathbf{L}_k^{-1} \mathbf{A}_k^T)^{-1} \mathbf{h}_k. \tag{15.25}$$

There are standard results concerning Newton's method applied to a system of nonlinear equations that are applicable to the system (15.19). These results state that if the linearized system is nonsingular at the solution (as is implied by our assumptions) and if the initial point is sufficiently close to the solution, the method will in fact converge to the solution and the convergence will be of order at least two. To guarantee convergence from remote initial points and hence be more broadly applicable, it is desirable to use the method as a descent process. Fortunately, we can show that the direction generated by Newton's method is a descent direction for the simple merit function

$$m(\mathbf{x}, \boldsymbol{\lambda}) = \frac{1}{2}|\nabla l(\mathbf{x}, \boldsymbol{\lambda})|^2 + \frac{1}{2}|\mathbf{h}(\mathbf{x})|^2.$$

Given $\mathbf{d}_k$, $\mathbf{y}_k$ satisfying (15.23), the inner product of this direction with the gradient of m at $\mathbf{x}_k$, $\boldsymbol{\lambda}_k$ is, referring to (15.15),

$$\begin{aligned}[\mathbf{L}_k\mathbf{l}_k + \mathbf{A}_k^T\mathbf{h}_k,\ \mathbf{A}_k\mathbf{l}_k]^T[\mathbf{d}_k,\ \mathbf{y}_k] &= \mathbf{l}_k^T\mathbf{L}_k\mathbf{d}_k + \mathbf{h}_k^T\mathbf{A}_k\mathbf{d}_k + \mathbf{l}_k^T\mathbf{A}_k^T\mathbf{y}_k \\ &= -|\mathbf{l}_k|^2 - |\mathbf{h}_k|^2.\end{aligned}$$

This is strictly negative unless both $\mathbf{l}_k = \mathbf{0}$ and $\mathbf{h}_k = \mathbf{0}$. Thus Newton's method has desirable global convergence properties when executed as a descent method with variable step size.

Note that the calculation above does not employ the explicit formulae (15.24) and (15.25), and hence it is not necessary that $\mathbf{L}(\mathbf{x}, \boldsymbol{\lambda})$ be positive definite, as long as the system (15.21) is invertible. We summarize the above discussion by the following theorem.

Theorem. *Define the Newton process by*

$$\begin{aligned}\mathbf{x}_{k+1} &= \mathbf{x}_k + \alpha_k\mathbf{d}_k \\ \boldsymbol{\lambda}_{k+1} &= \boldsymbol{\lambda}_k + \alpha_k\mathbf{y}_k,\end{aligned}$$

where $\mathbf{d}_k$, $\mathbf{y}_k$ are solutions to (15.24) and where α_k is selected to minimize the merit function

$$m(\mathbf{x}, \boldsymbol{\lambda}) = \frac{1}{2}|\nabla l(\mathbf{x}, \boldsymbol{\lambda})|^2 + \frac{1}{2}|\mathbf{h}(\mathbf{x})|^2.$$

Assume that $\mathbf{d}_k$, $\mathbf{y}_k$ exist and that the points generated lie in a compact set. Then any limit point of these points satisfies the first-order necessary conditions for a solution to the constrained minimization problem (15.1).

Proof. Most of this follows from the above observations and the Global Convergence Theorem. The one-dimensional search process is well-defined, since the merit function m is bounded below. ∎

In view of this result, it is worth pursuing Newton's method further. We would like to extend it to problems with inequality constraints. We would also like to avoid the necessity of evaluating $\mathbf{L}(\mathbf{x}_k, \boldsymbol{\lambda}_k)$ at each step and to consider alternative merit functions—perhaps those that might distinguish a local maximum from a local minimum, which the simple merit function does not do. These considerations guide the developments of the next several sections.

Relation to Sequential Quadratic Programming

It is clear from the development of the preceding discussion that Newton's method is closely related to quadratic programming with equality constraints. We explore this relationship more fully here, which will lead to a generalization of Newton's method to problems with inequality constraints.

Consider the problem

$$\begin{aligned} \text{minimize} \quad & \mathbf{l}_k^T \mathbf{d}_k + \frac{1}{2}\mathbf{d}_k^T \mathbf{L}_k \mathbf{d}_k \\ \text{subject to} \quad & \mathbf{A}_k \mathbf{d}_k + \mathbf{h}_k = \mathbf{0}. \end{aligned} \tag{15.26}$$

The first-order necessary conditions of this problem are exactly (15.21), or equivalently (15.23), where $\mathbf{y}_k$ corresponds to the Lagrange multiplier of (15.26). Thus, the solution of (15.26) produces a Newton step.

Alternatively, we may consider the quadratic program

$$\begin{aligned} \text{minimize} \quad & \nabla f(\mathbf{x}_k)\mathbf{d}_k + \frac{1}{2}\mathbf{d}_k^T \mathbf{L}_k \mathbf{d}_k \\ \text{subject to} \quad & \mathbf{A}_k \mathbf{d}_k + \mathbf{h}_k = \mathbf{0}. \end{aligned} \tag{15.27}$$

The necessary conditions of this problem are exactly (15.22), where $\boldsymbol{\lambda}_{k+1}$ now corresponds to the Lagrange multiplier of (15.27). The program (15.27) is obtained from (15.26) by merely subtracting $\boldsymbol{\lambda}_k^T \mathbf{A}_k \mathbf{d}_k$ from the objective function; and this change has no influence on $\mathbf{d}_k$, since $\mathbf{A}_k \mathbf{d}_k$ is fixed.

The connection with quadratic programming suggests a procedure for extending Newton's method to minimization problems with inequality constraints. Consider the problem

$$\begin{aligned} \text{minimize} \quad & f(\mathbf{x}) \\ \text{subject to} \quad & \mathbf{h}(\mathbf{x}) = \mathbf{0} \\ & \mathbf{g}(\mathbf{x}) \leqslant \mathbf{0}. \end{aligned}$$

Given an estimated solution point $\mathbf{x}_k$ and estimated Lagrange multipliers $\boldsymbol{\lambda}_k$, $\boldsymbol{\mu}_k$, one solves the quadratic program

$$\begin{aligned} \text{minimize} \quad & \nabla f(\mathbf{x}_k)\mathbf{d}_k + \tfrac{1}{2}\mathbf{d}_k^T \mathbf{L}_k \mathbf{d}_k \\ \text{subject to} \quad & \nabla \mathbf{h}(\mathbf{x}_k)\mathbf{d}_k + \mathbf{h}_k = \mathbf{0} \\ & \nabla \mathbf{g}(\mathbf{x}_k)\mathbf{d}_k + \mathbf{g}_k \leqslant \mathbf{0}, \end{aligned} \tag{15.28}$$

where $\mathbf{L}_k = \mathbf{F}(\mathbf{x}_k) + \boldsymbol{\lambda}_k^T \mathbf{H}(\mathbf{x}_k) + \boldsymbol{\mu}_k^T \mathbf{G}(\mathbf{x}_k)$, $\mathbf{h}_k = \mathbf{h}(\mathbf{x}_k)$, $\mathbf{g}_k = \mathbf{g}(\mathbf{x}_k)$. The new point is determined by $\mathbf{x}_{k+1} = \mathbf{x}_k + \mathbf{d}_k$, and the new Lagrange multipliers are the Lagrange multipliers of the quadratic program (15.28). This is the essence of an early method for nonlinear programming termed SOLVER. It is a very attractive procedure, since it applies directly to problems with inequality as well as equality constraints without the use of an active set strategy (although such a strategy might be used to solve the required quadratic program). Methods of this general type, where a quadratic program is solved at each step, are referred to as *recursive quadratic programming* methods, and several variations are considered in this chapter.

As presented here the recursive quadratic programming method extends Newton's method to problems with inequality constraints, but the method has limitations. The quadratic program may not always be well-defined, the method requires second-order derivative information, and the simple merit function is not a descent function for the case of inequalities. Of these, the most serious is the requirement of second-order information, and this is addressed in the next section.

15.4 Modified Newton Methods

A modified Newton method is based on replacing the actual linearized system by an approximation.

First, we concentrate on the equality constrained optimization problem

$$\begin{array}{ll} \text{minimize} & f(\mathbf{x}) \\ \text{subject to} & \mathbf{h}(\mathbf{x}) = \mathbf{0} \end{array} \tag{15.29}$$

in order to most clearly describe the relationships between the various approaches. Problems with inequality constraints can be treated within the equality constraint framework by an active set strategy or, in some cases, by recursive quadratic programming.

The basic equations for Newton's method can be written

$$\begin{bmatrix} \mathbf{x}_{k+1} \\ \boldsymbol{\lambda}_{k+1} \end{bmatrix} = \begin{bmatrix} \mathbf{x}_k \\ \boldsymbol{\lambda}_k \end{bmatrix} - \alpha_k \begin{bmatrix} \mathbf{L}_k & \mathbf{A}_k^T \\ \mathbf{A}_k & \mathbf{0} \end{bmatrix}^{-1} \begin{bmatrix} \mathbf{l}_k \\ \mathbf{h}_k \end{bmatrix},$$

where as before $\mathbf{L}_k$ is the Hessian of the Lagrangian, $\mathbf{A}_k = \nabla \mathbf{h}(\mathbf{x}_k)$, $\mathbf{l}_k = [\nabla f(\mathbf{x}_k) + \boldsymbol{\lambda}_k^T \nabla \mathbf{h}(\mathbf{x}_k)]^T$, $\mathbf{h}_k = \mathbf{h}(\mathbf{x}_k)$. A *structured modified Newton method* is a method of the form

$$\begin{bmatrix} \mathbf{x}_{k+1} \\ \boldsymbol{\lambda}_{k+1} \end{bmatrix} = \begin{bmatrix} \mathbf{x}_k \\ \boldsymbol{\lambda}_k \end{bmatrix} - \alpha_k \begin{bmatrix} \mathbf{B}_k & \mathbf{A}_k^T \\ \mathbf{A}_k & \mathbf{0} \end{bmatrix}^{-1} \begin{bmatrix} \mathbf{l}_k \\ \mathbf{h}_k \end{bmatrix}, \tag{15.30}$$

where $\mathbf{B}_k$ is an approximation to $\mathbf{L}_k$. The term "structured" derives from the fact that only second-order information in the original system of equations is approximated; the first-order information is kept intact.

Of course the method is implemented by solving the system

$$\begin{aligned} \mathbf{B}_k \mathbf{d}_k + \mathbf{A}_k^T \mathbf{y}_k &= -\mathbf{l}_k \\ \mathbf{A}_k \mathbf{d}_k \qquad &= -\mathbf{h}_k \end{aligned} \tag{15.31}$$

for $\mathbf{d}_k$ and $\mathbf{y}_k$ and then setting $\mathbf{x}_{k+1} = \mathbf{x}_k + \alpha_k \mathbf{d}_k$, $\boldsymbol{\lambda}_{k+1} = \boldsymbol{\lambda}_k + \alpha_k \mathbf{y}_k$ for some value of α_k. In this section we will not consider the procedure for selection of α_k, and thus for simplicity we take $\alpha_k = 1$. The simple transformation used earlier can be applied to write (15.31) in the form

$$\begin{aligned} \mathbf{B}_k\mathbf{d}_k + \mathbf{A}_k^T\boldsymbol{\lambda}_{k+1} &= -\nabla f(\mathbf{x}_k)^T \\ \mathbf{A}_k\mathbf{d}_k &= -\mathbf{h}_k. \end{aligned} \tag{15.32}$$

Then $\mathbf{x}_{k+1} = \mathbf{x}_k + \mathbf{d}_k$, and $\boldsymbol{\lambda}_{k+1}$ is found directly as a solution to system (15.32).

There are, of course, various ways to choose the approximation $\mathbf{B}_k$. One is to use a fixed, constant matrix throughout the iterative process. A second is to base $\mathbf{B}_k$ on some readily accessible information in $\mathbf{L}(\mathbf{x}_k, \boldsymbol{\lambda}_k)$, such as setting $\mathbf{B}_k$ equal to the diagonal of $\mathbf{L}(\mathbf{x}_k, \boldsymbol{\lambda}_k)$. Finally, a third possibility is to update $\mathbf{B}_k$ using one of the various quasi-Newton formulae.

One important advantage of the structured method is that $\mathbf{B}_k$ can be taken to be positive definite even though $\mathbf{L}_k$ is not. If this is done, we can write the explicit solution

$$\mathbf{y}_k = (\mathbf{A}_k\mathbf{B}_k^{-1}\mathbf{A}_k^T)^{-1}[\mathbf{h}_k - \mathbf{A}_k\mathbf{B}_k^{-1}\mathbf{l}_k] \tag{15.33}$$

$$\mathbf{d}_k = -\mathbf{B}_k^{-1}[\mathbf{I} - \mathbf{A}_k^T(\mathbf{A}_k\mathbf{B}_k^{-1}\mathbf{A}_k^T)^{-1}\mathbf{A}_k\mathbf{B}_k^{-1}]\mathbf{l}_k - \mathbf{B}_k^{-1}\mathbf{A}_k^T(\mathbf{A}_k\mathbf{B}_k^{-1}\mathbf{A}_k^T)^{-1}\mathbf{h}_k. \tag{15.34}$$

Consider the quadratic program

$$\begin{aligned} \text{minimize} \quad & \nabla f(\mathbf{x}_k)\mathbf{d}_k + \frac{1}{2}\mathbf{d}_k^T\mathbf{B}_k\mathbf{d}_k \\ \text{subject to} \quad & \mathbf{A}_k\mathbf{d}_k + \mathbf{h}(\mathbf{x}_k) = \mathbf{0}. \end{aligned} \tag{15.35}$$

The first-order necessary conditions for this problem are

$$\begin{aligned} \mathbf{B}_k\mathbf{d}_k + \mathbf{A}_k^T\boldsymbol{\lambda}_{k+1} &= -\nabla f(\mathbf{x}_k)^T \\ \mathbf{A}_k\mathbf{d}_k &= -\mathbf{h}(\mathbf{x}_k), \end{aligned} \tag{15.36}$$

which are again identical to the system of equations of the structured modified Newton method—in this case in the form (15.33). The Lagrange multiplier of the quadratic program is $\boldsymbol{\lambda}_{k+1}$. The equivalence of (15.35) and (15.36) leads to a recursive quadratic programming method, where at each $\mathbf{x}_k$ the quadratic program (15.35) is solved to determine the direction $\mathbf{d}_k$. In this case an arbitrary symmetric matrix $\mathbf{B}_k$ is used in place of the Hessian of the Lagrangian. Note that the problem (15.35) does not explicitly depend on $\boldsymbol{\lambda}_k$; but $\mathbf{B}_k$, often being chosen to approximate the Hessian of the Lagrangian, may depend on $\boldsymbol{\lambda}_k$.

As before, a principal advantage of the quadratic programming formulation is that there is an obvious extension to problems with inequality constraints: One simply employs a linearized version of the inequalities.

15.5 Descent Properties

In order to ensure convergence of the structured modified Newton methods of the previous section, it is necessary to find a suitable merit function—a merit function that is compatible with the direction-finding algorithm in the sense that it decreases along the direction generated. We must abandon the simple merit function at this point, since it is not compatible with these methods when $\mathbf{B}_k \neq \mathbf{L}_k$. However, two other penalty functions considered earlier, the absolute-value exact penalty function and the quadratic penalty function, *are* compatible with the modified Newton approach.

Absolute-Value Penalty Function

Let us consider the constrained minimization problem

$$\begin{aligned} &\text{minimize} \quad f(\mathbf{x}) \\ &\text{subject to} \quad \mathbf{g}(\mathbf{x}) \leqslant \mathbf{0}, \end{aligned} \tag{15.37}$$

where $\mathbf{g}(\mathbf{x})$ is r-dimensional. For notational simplicity we consider the case of inequality constraints only, since it is, in fact, the most difficult case. The extension to equality constraints is straightforward. In accordance with the recursive quadratic programming approach, given a current point $\mathbf{x}$, we select the direction of movement $\mathbf{d}$ by solving the quadratic programming problem

$$\begin{aligned} &\text{minimize} \quad \frac{1}{2}\mathbf{d}^T\mathbf{B}\mathbf{d} + \nabla f(\mathbf{x})\mathbf{d} \\ &\text{subject to} \quad \nabla \mathbf{g}(\mathbf{x})\mathbf{d} + \mathbf{g}(\mathbf{x}) \leqslant \mathbf{0}, \end{aligned} \tag{15.38}$$

where $\mathbf{B}$ is positive definite.

The first-order necessary conditions for a solution to this quadratic program are

$$\mathbf{B}\mathbf{d} + \nabla f(\mathbf{x})^T + \nabla \mathbf{g}(\mathbf{x})^T \boldsymbol{\mu} = \mathbf{0} \tag{15.39a}$$

$$\nabla \mathbf{g}(\mathbf{x})\mathbf{d} + \mathbf{g}(\mathbf{x}) \leqslant \mathbf{0} \tag{15.39b}$$

$$\boldsymbol{\mu}^T[\nabla \mathbf{g}(\mathbf{x})\mathbf{d} + \mathbf{g}(\mathbf{x})] = 0 \tag{15.39c}$$

$$\boldsymbol{\mu} \geqslant \mathbf{0}. \tag{15.39d}$$

Note that if the solution to the quadratic program has $\mathbf{d} = \mathbf{0}$, then the point $\mathbf{x}$, together with $\boldsymbol{\mu}$ from (15.39), satisfies the first-order necessary conditions for the original minimization problem (15.37). The following proposition is the fundamental result concerning the compatibility of the absolute-value penalty function and the quadratic programming method for determining the direction of movement.

Proposition 1. *Let* $\mathbf{d}$, $\boldsymbol{\mu}$ *(with* $\mathbf{d} \neq \mathbf{0}$*) be a solution of the quadratic program (15.38). Then if* $c \geqslant \max_j(\mu_j)$, *the vector* $\mathbf{d}$ *is a descent direction for the penalty function*

$$P(\mathbf{x}) = f(\mathbf{x}) + c\sum_{j=1}^{r} g_j(\mathbf{x})^+.$$

Proof. Let $J(\mathbf{x}) = \{j : g_j(\mathbf{x}) > 0\}$. Now for $\alpha > 0$,

$$\begin{aligned}
P(\mathbf{x} + \alpha\mathbf{d}) &= f(\mathbf{x} + \alpha\mathbf{d}) + c\sum_{j=1}^{r} g_j(\mathbf{x} + \alpha\mathbf{d})^+ \\
&= f(\mathbf{x}) + \alpha\nabla f(\mathbf{x})\mathbf{d} + c\sum_{j=1}^{r}[g_j(\mathbf{x}) + \alpha\nabla g_j(\mathbf{x})\mathbf{d}]^+ + o(\alpha) \\
&= f(\mathbf{x}) + \alpha\nabla f(\mathbf{x})\mathbf{d} + c\sum_{j=1}^{r} g_j(\mathbf{x})^+ + \alpha c\sum_{j\in J(\mathbf{x})}\nabla g_j(\mathbf{x})\mathbf{d} + o(\alpha) \\
&= P(\mathbf{x}) + \alpha\nabla f(\mathbf{x})\mathbf{d} + \alpha c\sum_{j\in J(\mathbf{x})}\nabla g_j(\mathbf{x})\mathbf{d} + o(\alpha).
\end{aligned} \tag{15.40}$$

Where (15.39b) was used in the third line to infer that $\nabla g_j(\mathbf{x}) \leq 0$ if $g_j(\mathbf{x}) = 0$. Again using (15.39b) we have

$$c\sum_{j\in J(\mathbf{x})}\nabla g_j(\mathbf{x})\mathbf{d} \leqslant c\sum_{j\in J(\mathbf{x})} -g_j(\mathbf{x}) = -c\sum_{j=1}^{r} g_j(\mathbf{x})^+. \tag{15.41}$$

Using (15.39a) we have

$$\nabla\mathbf{f}(\mathbf{x})\mathbf{d} = -\mathbf{d}^T\ \mathbf{B}\mathbf{d}\ - \sum_{j=1}^{r}\mu_j\nabla g_j(\mathbf{x})\mathbf{d},$$

which by using the complementary slackness condition (15.39c) leads to

$$\begin{aligned}
\nabla f(\mathbf{x})\mathbf{d} &= -\mathbf{d}^T\mathbf{B}\mathbf{d} + \sum_{j=1}^{r}\mu_j g_j(\mathbf{x}) \leqslant -\mathbf{d}^T\mathbf{B}\mathbf{d} + \sum_{j=1}^{r}\mu_j g_j(\mathbf{x})^+ \\
&\leq -\mathbf{d}^T\ \mathbf{B}\mathbf{d}\ + \max\ (\mu_j)\sum_{j=1}^{r} g_j(\mathbf{x})^+.
\end{aligned} \tag{15.42}$$

Finally, substituting (15.41) and (15.42) in (15.40), we find

$$P(\mathbf{x} + \alpha\mathbf{d}) \leqslant P(\mathbf{x}) + \alpha\{-\mathbf{d}^T\mathbf{B}\mathbf{d} - [c - \max(\mu_j)]\sum_{j=1}^{r} g_j(\mathbf{x})^+\} + o(\alpha),$$

Since $\mathbf{B}$ is positive definite and $c \geq \max(\mu_j)$, it follows that for α sufficiently small, $P(\mathbf{x} + \alpha\mathbf{d}) < P(\mathbf{x})$. ∎

The above proposition is exceedingly important, for it provides a basis for establishing the global convergence of modified Newton methods, including recursive quadratic programming. The following is a simple global convergence result based on the descent property.

Theorem. *Let* $\mathbf{B}$ *be positive definite and assume that throughout some compact region* $\subset E^n$*, the quadratic program (15.38) has a unique solution* $\mathbf{d}$*,* μ *such that at each point the Lagrange multipliers satisfy* $\max_j(\mu_j) \leqslant c$*. Let the sequence* $\{\mathbf{x}_k\}$ *be generated by*

$$\mathbf{x}_{k+1} = \mathbf{x}_k + \alpha_k \mathbf{d}_k,$$

where $\mathbf{d}_k$ *is the solution to (15.38) at* $\mathbf{x}_k$ *and where* α_k *minimizes* $P(\mathbf{x}_{k+1})$*. Assume that each* $\mathbf{x}_k \in \Omega$*. Then every limit point* $\overline{\mathbf{x}}$ *of* $\{\mathbf{x}_k\}$ *satisfies the first-order necessary conditions for the constrained minimization problem (15.37).*

Proof. The solution to a quadratic program depends continuously on the data, and hence the direction determined by the quadratic program (15.38) is a continuous function of $\mathbf{x}$. The function $P(\mathbf{x})$ is also continuous, and by Proposition 1, it follows that P is a descent function at every point that does not satisfy the first-order conditions. The result thus follows from the Global Convergence Theorem. ∎

In view of the above result, recursive quadratic programming in conjunction with the absolute-value penalty function is an attractive technique. There are, however, some difficulties to be kept in mind. First, the selection of the parameter α_k requires a one-dimensional search with respect to a nondifferentiable function. Thus the efficient curve-fitting search methods of Chap. 8 cannot be used without significant modification. Second, use of the absolute-value function requires an estimate of an upper bound for μ_j's, so that c can be selected properly. In some applications a suitable bound can be obtained from previous experience, but in general one must develop a method for revising the estimate upward when necessary.

Another potential difficulty with the quadratic programming approach above is that the quadratic program (15.38) may be infeasible at some point $\mathbf{x}_k$, even though the original problem (15.37) is feasible. If this happens, the method breaks down. However, see Exercise 8 for a method that avoids this problem.

The Quadratic Penalty Function

Another penalty function that is compatible with the modified Newton method approach is the standard quadratic penalty function. It has the added technical advantage that, since this penalty function is differentiable, it is possible to apply our earlier analytical principles to study the rate of convergence of the method. This leads to an analytical comparison of primal-dual methods with the methods of other chapters.

We shall restrict attention to the problem with equality constraints, since that is all that is required for a rate of convergence analysis. The method can be extended to problems with inequality constraints either directly or by an active set method. Thus we consider the problem

$$\begin{aligned} &\text{minimize} \quad f(\mathbf{x}) \\ &\text{subject to} \quad \mathbf{h}(\mathbf{x}) = \mathbf{0} \end{aligned} \tag{15.43}$$

and the standard quadratic penalty objective

$$P(\mathbf{x}) = f(\mathbf{x}) + \frac{1}{2}c|\mathbf{h}(\mathbf{x})|^2. \tag{15.44}$$

From the theory in Chap. 13, we know that minimization of the objective with a quadratic penalty function will *not* yield an exact solution to (15.43). In fact, the minimum of the penalty function (15.44) will have $c\mathbf{h}(\mathbf{x}) \simeq \boldsymbol{\lambda}$, where $\boldsymbol{\lambda}$ is the Lagrange multiplier of (15.43). Therefore, it seems appropriate in this case to consider the quadratic programming problem

$$\begin{aligned} &\text{minimize} \quad \frac{1}{2}\mathbf{d}^T\mathbf{B}\mathbf{d} + \nabla f(\mathbf{x})\mathbf{d} \\ &\text{subject to} \quad \nabla\mathbf{h}(\mathbf{x})\mathbf{d} + \mathbf{h}(\mathbf{x}) = \hat{\boldsymbol{\lambda}}/c, \end{aligned} \tag{15.45}$$

where $\hat{\boldsymbol{\lambda}}$ is an estimate of the Lagrange multiplier of the original problem. A particularly good choice is

$$\hat{\boldsymbol{\lambda}} = [(1/c)\mathbf{I} + \mathbf{Q}]^{-1}[\mathbf{h}(\mathbf{x}) - \mathbf{A}\mathbf{B}^{-1}\nabla f(\mathbf{x})^T], \tag{15.46}$$

where $\mathbf{A} = \nabla\mathbf{h}(\mathbf{x})$, $\mathbf{Q} = \mathbf{AB} - 1\mathbf{A}^T$ which is the Lagrange multiplier that would be obtained by the quadratic program with the penalty method. The proposed method requires that $\hat{\boldsymbol{\lambda}}$ be first estimated from (15.46) and then used in the quadratic programming problem (15.45).

The following proposition shows that this procedure produces a descent direction for the quadratic penalty objective.

Proposition 2. *For any $c > 0$, let $\mathbf{d}$, $\boldsymbol{\lambda}$ (with $\mathbf{d} \neq \mathbf{0}$) be a solution to the quadratic program (15.45). Then $\mathbf{d}$ is a descent direction of the function $P(\mathbf{x}) = f(\mathbf{x}) + (1/2)c|\mathbf{h}(\mathbf{x})|^2$.*

Proof. We have from the constraint equation

$$\mathbf{Ad} = (1/c)\hat{\boldsymbol{\lambda}} - \mathbf{h}(\mathbf{x}),$$

which yields

$$c\mathbf{A}^T\mathbf{Ad} = \mathbf{A}^T\hat{\boldsymbol{\lambda}} - \mathbf{c}\mathbf{A}^T\mathbf{h}(\mathbf{x}).$$

Solving the necessary conditions for (15.45) yields (see the top part of (15.9) for a similar expression with $\mathbf{Q} = \mathbf{B}$ there)

$$\mathbf{Bd} = \mathbf{A}^T\mathbf{Q}^{-1}[\mathbf{AB}^{-1}\nabla\mathbf{f}(\mathbf{x})^T + (1/c)\hat{\boldsymbol{\lambda}} - \mathbf{h}(\mathbf{x})] - \nabla f(\mathbf{x})^T.$$

Therefore,

$$\begin{aligned} (\mathbf{B} + c\mathbf{A}^T\mathbf{A})\mathbf{d} &= \mathbf{A}^T\mathbf{Q}^{-1}[\mathbf{AB}^{-1}\nabla f(\mathbf{x})^T - \mathbf{h}(\mathbf{x})] \\ &+ \mathbf{A}^T[(1/c)\mathbf{Q}^{-1} + \mathbf{I}]\hat{\boldsymbol{\lambda}} - \nabla f(\mathbf{x})^T - c\mathbf{A}^T\mathbf{h}(\mathbf{x}) \\ &= \mathbf{A}^T\mathbf{Q}^{-1}\{\mathbf{AB}^{-1}\nabla f(\mathbf{x})^T - \mathbf{h}(\mathbf{x}) + ((1/c)\mathbf{I} + \mathbf{Q})\hat{\boldsymbol{\lambda}}\} \\ &- \nabla f(\mathbf{x})^T - c\mathbf{A}^T\mathbf{h}(\mathbf{x}) \\ &= -\nabla f(\mathbf{x})^T - c\mathbf{A}^T\mathbf{h}(\mathbf{x}) = -\nabla P(\mathbf{x})^T. \end{aligned}$$

The matrix $(\mathbf{B}+c\mathbf{A}^T\mathbf{A})$ is positive definite for any $c \geqslant 0$. It follows that $\nabla P(\mathbf{x})\mathbf{d} < 0$. ∎

*15.6 *Rate of Convergence

It is now appropriate to apply the principles of convergence analysis that have been repeatedly emphasized in previous chapters to the recursive quadratic programming approach. We expect that, if this new approach is well founded, then the rate of convergence of the algorithm should be related to the familiar canonical rate, which we have learned is a fundamental measure of the complexity of the problem. If it is not so related, then some modification of the algorithm is probably required. Indeed, we shall find that a small but important modification *is* required.

From the proof of Proposition 2 of Sect. 15.5, we have the formula

$$(\mathbf{B} + c\mathbf{A}^T\mathbf{A})\mathbf{d} = -\nabla P(\mathbf{x})^T,$$

which can be written as

$$\mathbf{d} = -(\mathbf{B} + c\mathbf{A}^T\mathbf{A})^{-1}\nabla P(\mathbf{x})^T.$$

This shows that the method is a modified Newton method applied to the unconstrained minimization of $P(\mathbf{x})$. From the Modified Newton Method Theorem of Sect. 10.1, we see immediately that the rate of convergence is determined by the eigenvalues of the matrix that is the product of the coefficient matrix $(\mathbf{B} + c\mathbf{A}^T\mathbf{A})^{-1}$ and the Hessian of the function P at the solution point. The Hessian of P is $(\mathbf{L} + c\mathbf{A}^T\mathbf{A})$, where $\mathbf{L} = \mathbf{F}(\mathbf{x}) + c\mathbf{h}(\mathbf{x})^T\mathbf{H}(\mathbf{x})$. We know that the vector $c\mathbf{h}(\mathbf{x})$ at the solution of the penalty problem is equal to λ_c, where $\nabla f(\mathbf{x}) + \lambda_c^T\nabla\mathbf{h}(\mathbf{x}) = \mathbf{0}$. Therefore, the rate of convergence is determined by the eigenvalues of

$$(\mathbf{B} + c\mathbf{A}^T\mathbf{A})^{-1}(\mathbf{L} + c\mathbf{A}^T\mathbf{A}), \tag{15.47}$$

where all quantities are evaluated at the solution to the penalty problem and $\mathbf{L} = \mathbf{F} + \lambda_c^T\mathbf{H}$. For large values of c, all quantities are approximately equal to the values at the optimal solution to the constrained problem.

Now what we wish to show is that as $c \to \infty$, the matrix (15.47) looks like $\mathbf{B}_M^{-1}\mathbf{L}_M$ on the subspace, M, and like the identity matrix on $M^\perp$, the subspace orthogonal to M. To do this in detail, let $\mathbf{C}$ be an $n \times (n-m)$ matrix whose columns form an orthonormal basis for M, the tangent subspace $\{\mathbf{x} : \mathbf{A}\mathbf{x} = \mathbf{0}\}$. Let $\mathbf{D} = \mathbf{A}^T(\mathbf{A}\mathbf{A}^T)^{-1}$. Then $\mathbf{AC} = \mathbf{0}$, $\mathbf{AD} = \mathbf{I}$, $\mathbf{C}^T\mathbf{C} = \mathbf{I}$, $\mathbf{C}^T\mathbf{D} = \mathbf{0}$.

The eigenvalues of $(\mathbf{B} + c\mathbf{A}^T\mathbf{A})^{-1}(\mathbf{L} + c\mathbf{A}^T\mathbf{A})$ are equal to those of

$$\begin{aligned}
&[\mathbf{C},\ \mathbf{D}]^{-1}(\mathbf{B} + c\mathbf{A}^T\mathbf{A})^{-1}\{[\mathbf{C},\ \mathbf{D}]^T\}^{-1}[\mathbf{C},\ \mathbf{D}]^T(\mathbf{L} + c\mathbf{A}^T\mathbf{A})[\mathbf{C},\ \mathbf{D}]\\
&\quad = \begin{bmatrix} \mathbf{C}^T\mathbf{BC} & \mathbf{C}^T\mathbf{BD} \\ \mathbf{D}^T\mathbf{BC} & \mathbf{D}^T\mathbf{BC} + c\mathbf{I} \end{bmatrix}^{-1} \begin{bmatrix} \mathbf{C}^T\mathbf{LC} & \mathbf{C}^T\mathbf{LD} \\ \mathbf{D}^T\mathbf{LC} & \mathbf{D}^T\mathbf{LD} + c\mathbf{I} \end{bmatrix}.
\end{aligned}$$

Now as $c \to \infty$, the matrix above approaches

$$\begin{bmatrix} \mathbf{B}_M^{-1}\mathbf{L}_M & \mathbf{B}_M\mathbf{C}^T(\mathbf{L} - \mathbf{B})\mathbf{D} \\ \mathbf{0} & \mathbf{I} \end{bmatrix},$$

where $\mathbf{B}_M = \mathbf{C}^T\mathbf{B}\mathbf{C}$, $\mathbf{L}_M = \mathbf{C}^T\mathbf{L}\mathbf{C}$ (see Exercise 6). The eigenvalues of this matrix are those of $\mathbf{B}_M^{-1}\mathbf{L}_M$ together with those of $\mathbf{I}$. This analysis leads directly to the following conclusion:

Theorem. *Let a, A be the smallest and largest eigenvalues, respectively, of $\mathbf{B}_M^{-1}\mathbf{L}_M$ and assume that $a \leqslant 1 \leqslant A$. Then the structured modified Newton method with quadratic penalty function has a rate of convergence no greater than $[(A - a)/(A + a)]^2$ as $c \to \infty$.*

In the special case of $\mathbf{B} = \mathbf{I}$, the rate in the above proposition is precisely the canonical rate, defined by the eigenvalues of $\mathbf{L}$ restricted to the tangent plane. It is important to note, however, that in order for the rate of the theorem to be achieved, the eigenvalues of $\mathbf{B}_M^{-1}\mathbf{L}_M$ must be spread around unity; if not, the rate will be poorer. Thus, even if $\mathbf{L}_M$ is well-conditioned, but the eigenvalues differ greatly from unity, the choice $\mathbf{B} = \mathbf{I}$ may be poor. This is an instance where proper scaling is vital. (We also point out that the above analysis is closely related to that of Sect. 13.4, where a similar conclusion is obtained.)

There is a geometric explanation for the scaling property. Take $\mathbf{B} = \mathbf{I}$ for simplicity. Then the direction of movement $\mathbf{d}$ is $\mathbf{d} = -\nabla f(\mathbf{x})^T + \mathbf{A}^T\boldsymbol{\lambda}$ for some $\boldsymbol{\lambda}$. Using the fact that the projected gradient is $\mathbf{p} = \nabla f(\mathbf{x})^T + \mathbf{A}^T\boldsymbol{\mu}$ for some $\boldsymbol{\mu}$, we see that $\mathbf{d} = -\mathbf{p} + \mathbf{A}^T(\boldsymbol{\lambda} + \boldsymbol{\mu})$. Thus $\mathbf{d}$ can be decomposed into two components: one in the direction of the projected negative gradient, the other in a direction orthogonal to the tangent plane (see Fig. 15.1). Ideally, these two components should be in proper proportions so that the constraint surface is reached at the same point as would be reached by minimization in the direction of the projected negative gradient. If they are not, convergence will be poor.

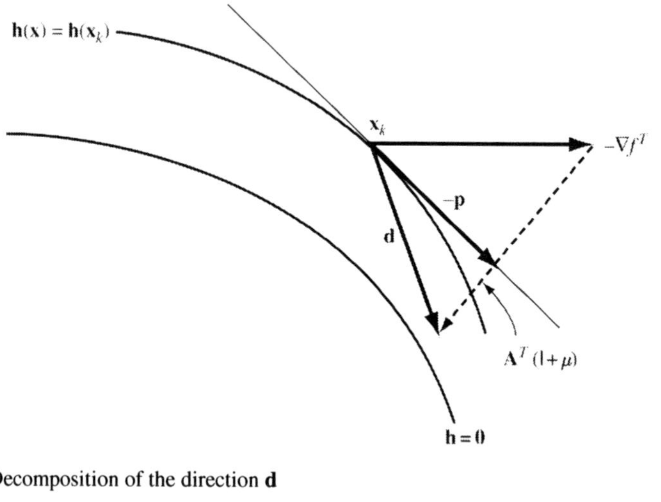

Fig. 15.1 Decomposition of the direction $\mathbf{d}$

15.7 Primal-Dual Interior Point Methods

The primal-dual interior-point methods discussed for linear programming in Chap. 5 are, as mentioned there, closely related to the barrier methods presented in Chap. 13 and the primal-dual methods of the current chapter. They can be naturally extended to solve nonlinear programming problems while maintaining both theoretical and practical efficiency.

Consider the inequality constrained problem

$$\begin{aligned} \text{minimize} \quad & f(\mathbf{x}) \\ \text{subject to} \quad & \mathbf{Ax} = \mathbf{b}, \\ & \mathbf{g}(\mathbf{x}) \leq \mathbf{0}, \end{aligned} \tag{15.48}$$

In general, a weakness of the active constraint method for such a problem is the combinatorial nature of determining which constraints should be active.

Logarithmic Barrier Function

A method that avoids the necessity to explicitly select a set of active constraints is based on the logarithmic barrier method, which solves a sequence of equality constrained minimization problems. Specifically,

$$\begin{aligned} \text{minimize} \quad & f(\mathbf{x}) - \mu \sum_{i=1}^{p} \log(-g_j(\mathbf{x})) \\ \text{subject to} \quad & \mathbf{Ax} = \mathbf{b}, \end{aligned} \tag{15.49}$$

where $\mu = \mu^k > 0$, $k = 1, \ldots,$, $\mu^k > \mu^{k+1}$, $\mu^k \to 0$. The μ^ks can be pre-determined. Typically, we have $\mu^{k+1} = \gamma\mu^k$ for some constant $0 < \gamma < 1$. Here, we also assume that the original problem has a feasible interior-point $\mathbf{x}^0$; that is,

$$\mathbf{Ax}^0 = \mathbf{b} \quad \text{and} \quad \mathbf{g}(\mathbf{x}^0) < \mathbf{0},$$

and $\mathbf{A}$ has full row rank.

For fixed μ, and using $s_i = \mu/g_i$, the first-order optimality conditions of the barrier problem (15.49) are:

$$\begin{aligned} -\mathbf{Sg}(\mathbf{x}) &= \mu\mathbf{1} \\ \mathbf{Ax} &= \mathbf{b} \\ -\mathbf{A}^T\mathbf{y} + \nabla f(\mathbf{x})^T + \nabla\mathbf{g}(\mathbf{x})^T\mathbf{s} &= \mathbf{0}, \end{aligned} \tag{15.50}$$

where $\mathbf{S} = \text{diag}(\mathbf{s})$; that is, a diagonal matrix whose diagonal entries are $\mathbf{s}$, and $\nabla\mathbf{g}(\mathbf{x})$ is the Jacobian matrix of $\mathbf{g}(\mathbf{x})$.

If $f(\mathbf{x})$ and $g_i(\mathbf{x})$ are convex functions for all i, $f(\mathbf{x}) - \mu \sum_i \log(-g_i(\mathbf{x}))$ is strictly convex in the interior of the feasible region, and the objective level set is bounded, then there is a unique minimizer for the barrier problem. Let $(\mathbf{x}(\mu) > \mathbf{0},\ \mathbf{y}(\mu),\ \mathbf{s}(\mu) > \mathbf{0})$ be the (unique) solution of (15.50). Then, these values form the *primal-dual central path* of (15.48):

$$C = \{(\mathbf{x}(\mu),\ \mathbf{y}(\mu),\ \mathbf{s}(\mu) > \mathbf{0}) : 0 < \mu < \infty\}.$$

This can be summarized in the following theorem.

Theorem 1. *Let* $(\mathbf{x}(\mu),\ \mathbf{y}(\mu),\ \mathbf{s}(\mu))$ *be on the central path.*

i) If $f(\mathbf{x})$ *and* $g_i(\mathbf{x})$ *are convex functions for all i, then* $\mathbf{s}(\mu)$ *is unique.*

ii) Furthermore, if $f(\mathbf{x}) - \mu \sum_i \log(-g_i(\mathbf{x}))$ *is strictly convex,*$(\mathbf{x}(\mu),\ \mathbf{y}(\mu),\ \mathbf{s}(\mu))$ *are unique, and they are bounded for* $0 < \mu \leqslant \mu^0$ *for any given* $\mu^0 > 0$.

iii) For $0 < \mu' < \mu$, $f(\mathbf{x}(\mu')) < f(\mathbf{x}(\mu))$ *if* $\mathbf{x}(\mu') \neq \mathbf{x}(\mu)$.

iv) $(\mathbf{x}(\mu),\ \mathbf{y}(\mu),\ \mathbf{s}(\mu))$ *converges to a point satisfying the first-order necessary conditions for a solution of (15.48) as* $\mu \to 0$.

Once we have an approximate solution point $(\mathbf{x},\ \mathbf{y},\ \mathbf{s}) = (\mathbf{x}_k,\ \mathbf{y}_k,\ \mathbf{s}_k)$ for (15.50) for $\mu = \mu^k > 0$, we can again use the primal-dual methods described for linear programming to generate a new approximate solution to (15.50) for $\mu = \mu^{k+1} < \mu^k$. The Newton direction vectors $(\mathbf{d}_\mathbf{x},\ \mathbf{d}_\mathbf{y},\ \mathbf{d}_\mathbf{s})$ is found from the system of linear equations:

$$\begin{aligned} -\mathbf{S}\nabla\mathbf{g}(\mathbf{x})\mathbf{d}_\mathbf{x} - \mathbf{G}(\mathbf{x})\mathbf{d}_\mathbf{s} &= \mu\mathbf{1} + \mathbf{S}\mathbf{g}(\mathbf{x}), \\ \mathbf{A}\mathbf{d}_\mathbf{x} &= \mathbf{b} - \mathbf{A}\mathbf{x}, \\ -\mathbf{A}^T\mathbf{d}_\mathbf{y} + \left(\nabla^2 f(\mathbf{x}) + \sum_i s_i \nabla^2 g_i(\mathbf{x})\right)\mathbf{d}_\mathbf{x} & \\ +\nabla\mathbf{g}(\mathbf{x})^T\mathbf{d}_\mathbf{s} &= \mathbf{A}^T\mathbf{y} - \nabla f(\mathbf{x})^T - \nabla\mathbf{g}(\mathbf{x})^T\mathbf{s}, \end{aligned} \tag{15.51}$$

where $\mathbf{G}(\mathbf{x}) = \text{diag}(\mathbf{g}(\mathbf{x}))$. Then, the new iterate is update to:

$$(\mathbf{x}_{k+1},\ \mathbf{y}_{k+1}, \mathbf{s}_{k+1}) = (\mathbf{x}_k,\ \mathbf{y}_k, \mathbf{s}_k) + \alpha_k(\mathbf{d}_\mathbf{x},\ \mathbf{d}_\mathbf{y}, \mathbf{d}_\mathbf{s})$$

for a stepsize α_k. Recently, this approach has also been used to find points satisfying the first-order conditions for problems when $f(\mathbf{x})$ and $g_i(\mathbf{x})$ are not generally convex functions.

Interior Point Method for Convex Quadratic Programming

Let $f(\mathbf{x}) = (1/2)\mathbf{x}^T\mathbf{Q}\mathbf{x} + \mathbf{c}^T\mathbf{x}$ and $g_i(\mathbf{x}) = -x_i$ for $i = 1, \ldots, n$, and consider the quadratic program

$$\begin{aligned} \text{minimize} \quad & \frac{1}{2}\mathbf{x}^T\mathbf{Q}\mathbf{x} + \mathbf{c}^T\mathbf{x} \\ \text{subject to} \quad & \mathbf{A}\mathbf{x} = \mathbf{b}, \\ & \mathbf{x} \geqslant \mathbf{0}, \end{aligned} \tag{15.52}$$

where the given matrix $\mathbf{Q} \in E^{n \times n}$ is positive semidefinite (that is, the objective is a convex function), $\mathbf{A} \in E^{n \times m}$, $\mathbf{c} \in E^n$ and $\mathbf{b} \in E^m$. The problem reduces to finding $\mathbf{x} \in E^n$, $\mathbf{y} \in E^m$ and $\mathbf{s} \in E^n$ satisfying the following optimality conditions:

$$\begin{aligned} \mathbf{Sx} &= \mathbf{0} \\ \mathbf{Ax} &= \mathbf{b} \\ -\mathbf{A}^T\mathbf{y} + \mathbf{Qx} - \mathbf{s} &= -\mathbf{c} \\ (\mathbf{x}, \mathbf{s}) &\geq \mathbf{0}. \end{aligned} \tag{15.53}$$

The optimality conditions with the logarithmic barrier function with parameter μ are be:

$$\begin{aligned} \mathbf{Sx} &= \mu\mathbf{1} \\ \mathbf{Ax} &= \mathbf{b} \\ -\mathbf{A}^T\mathbf{y} + \mathbf{Qx} - \mathbf{s} &= -\mathbf{c}. \end{aligned} \tag{15.54}$$

Note that the bottom two sets of constraints are linear equalities.

Thus, once we have an interior feasible point $(\mathbf{x}, \mathbf{y}, \mathbf{s})$ for (15.54), with $\boldsymbol{\mu} = \mathbf{x}^T\mathbf{s}/n$, we can apply Newton's method to compute a new (approximate) iterate $(\mathbf{x}^+, \mathbf{y}^+, \mathbf{s}^+)$ by solving for $(\mathbf{d_x}, \mathbf{d_y}, \mathbf{d_s})$ from the system of linear equations:

$$\begin{aligned} \mathbf{Sd_x} + \mathbf{Xd_s} &= \gamma\mu\mathbf{1} - \mathbf{Xs}, \\ \mathbf{Ad_x} &= \mathbf{0}, \\ -\mathbf{A}^T\mathbf{d_y} + \mathbf{Qd_x} - \mathbf{d_s} &= \mathbf{0}, \end{aligned} \tag{15.55}$$

where $\mathbf{X}$ and $\mathbf{S}$ are two diagonal matrices whose diagonal entries are $\mathbf{x} > \mathbf{0}$ and $\mathbf{s} > \mathbf{0}$, respectively. Here, γ is a fixed positive constant less than 1, which implies that our targeted μ is reduced by the factor γ at each step.

Potential Function as a Merit Function

For any interior feasible point $(\mathbf{x}, \mathbf{y}, \mathbf{s})$ of (15.52) and its dual, a suitable merit function is the potential function introduced in Chap. 5 for linear programming:

$$\psi_{n+\rho}(\mathbf{x}, \mathbf{s}) = (n+\rho)\log(\mathbf{x}^T\mathbf{s}) - \sum_{j=1}^{n} \log(x_j s_j).$$

The main result for this is stated in the following theorem.

Theorem 2. *In solving (15.55) for* $(\mathbf{d_x}, \mathbf{d_y}, \mathbf{d_s})$, *let* $\gamma = n/(n+\rho) < 1$ *for fixed* $\rho \geqslant \sqrt{n}$ *and assign* $\mathbf{x}^+ = \mathbf{x} + \alpha\mathbf{d_x}$, $\mathbf{y}^+ = \mathbf{y} + \alpha\mathbf{d_y}$, *and* $\mathbf{s}^+ = \mathbf{s} + \alpha\mathbf{d}_s$ *where*

$$\alpha = \frac{\overline{\alpha}\sqrt{\min(\mathbf{Xs})}}{|(\mathbf{XS})^{-1/2}(\frac{\mathbf{x}^T\mathbf{s}}{n+\rho}\mathbf{1} - \mathbf{Xs})|},$$

where $\overline{\alpha}$ is any positive constant less than 1. (Again **X** *and* **S** *are matrices with components on the diagonal being those of* **x** *and* **s**, *respectively.) Then,*

$$\psi_{n+\rho}(\mathbf{x}^+, \mathbf{s}^+) - \psi_{n+\rho}(\mathbf{x}, \mathbf{s}) \leqslant -\overline{\alpha}\sqrt{3/4} + \frac{\overline{\alpha}^2}{2(1-\overline{\alpha})}.$$

The proof of the theorem is also similar to that for linear programming; see Exercise 12. Notice that, since **Q** is positive semidefinite, we have

$$\mathbf{d_x}^T\mathbf{d_s} = (\mathbf{d_x}, \mathbf{d_y})^T(\mathbf{d_s}, \mathbf{0}) = \mathbf{d}_{\mathbf{x}}^T\mathbf{Q}\mathbf{d_x} \geqslant 0$$

while $\mathbf{d}_{\mathbf{x}}^T\mathbf{d_s} = 0$ in the linear programming case.

We outline the algorithm here:

Given any interior feasible $(\mathbf{x}_0, \mathbf{y}_0, \mathbf{s}_0)$ *of* (15.52) *and its dual. Set* $\rho \geqslant \sqrt{n}$ *and* $k = 0$.

1. *Set* $(\mathbf{x}, \mathbf{s}) = (\mathbf{x}_k, \mathbf{s}_k)$ *and* $\gamma = n/(n+\rho)$ *and compute* $(\mathbf{d_x}, \mathbf{d_y}, \mathbf{d_s})$ *from* (15.55).
2. *Let* $\mathbf{x}_{k+1} = \mathbf{x}_k + \overline{\alpha}\mathbf{d_x}$, $\mathbf{y}_{k+1} = \mathbf{y}_k + \overline{\alpha}\mathbf{d_y}$, and $\mathbf{s}_{k+1} = \mathbf{s}_k + \overline{\alpha}\mathbf{d_s}$ *where*

$$\overline{\alpha} = \arg\min_{\alpha \geqslant 0} \psi_{n+\rho}(\mathbf{x}_k + \alpha\mathbf{d_x}, \mathbf{s}_k + \alpha\mathbf{d_s}).$$

3. *Let* $k = k + 1$. If $\mathbf{s}_k^T\mathbf{x}_k/\mathbf{s}_0^T\mathbf{x}_0 \leq \varepsilon$, *stop. Otherwise, return to Step 1.*

This algorithm exhibits an iteration complexity bound that is identical to that of linear programming expressed in Theorem 1, Sect. 5.6.

15.8 Summary

A constrained optimization problem can be solved by directly solving the equations that represent the first-order necessary conditions for a solution. For a quadratic programming problem with linear constraints, these equations are linear and thus can be solved by standard linear procedures. Quadratic programs with inequality constraints can be solved by an active set method in which the direction of movement is toward the solution of the corresponding equality constrained problem. This method will solve a quadratic program in a finite number of steps.

For general nonlinear programming problems, many of the standard methods for solving systems of equations can be adapted to the corresponding necessary equations. One class consists of first-order methods that move in a direction related to the residual (that is, the error) in the equations. Another class of methods is based on extending the method of conjugate directions to nonpositive-definite systems. Finally, a third class is based on Newton's method for solving systems of nonlinear equations, and solving a linearized version of the system at each iteration. Under appropriate assumptions, Newton's method has excellent global as well as local convergence properties, since the simple merit function, $\frac{1}{2}|\nabla f(\mathbf{x})+\boldsymbol{\lambda}^T\nabla\mathbf{h}(\mathbf{x})|^2+\frac{1}{2}|\mathbf{h}(\mathbf{x})|^2$, decreases in the Newton direction. An individual step of Newton's method is equivalent to solving a quadratic programming problem, and thus Newton's method can be extended to problems with inequality constraints through recursive quadratic programming.

More effective methods are developed by accounting for the special structure of the linearized version of the necessary conditions and by introducing approximations to the second-order information. In order to assure global convergence of these methods, a penalty (or merit) function must be specified that is compatible with the method of direction selection, in the sense that the direction is a direction of descent for the merit function. The absolute-value penalty function and the standard quadratic penalty function are both compatible with some versions of recursive quadratic programming.

The best of the primal-dual methods take full account of special structure, and are based on direction-finding procedures that are closely related to methods described in earlier chapters. It is not surprising therefore that the convergence properties of these methods are also closely related to those of other chapters. Again we find that the canonical rate is fundamental for properly designed first-order methods.

Interior point methods in the primal-dual model are very effective for treating problems with inequality constraints, for they avoid (or at least minimize) the difficulties associated with determining which constraints will be active at the solution. Applied to general nonlinear programming problems, these methods closely parallel the interior point methods for linear programming. There is again a central path, and Newton's method is a good way to follow the path.

15.9 Exercises

1. Solve the quadratic program

$$\begin{aligned} \text{minimize } & x^2 - xy + y^2 - 3x \\ \text{subject to } & x \geqslant 0 \\ & y \geqslant 0 \\ & x + y \leqslant 4 \end{aligned}$$

by use of the active set method starting at $x = y = 0$.

2. Suppose $\mathbf{x}^*$, $\boldsymbol{\lambda}^*$ satisfy

$$\begin{aligned} \nabla f(\mathbf{x}^*) + \boldsymbol{\lambda}^{*T} \nabla \mathbf{h}(\mathbf{x}^*) &= \mathbf{0} \\ \mathbf{h}(\mathbf{x}^*) &= \mathbf{0}. \end{aligned}$$

Let

$$\mathbf{C} = \begin{bmatrix} \mathbf{L}(\mathbf{x}^*, \boldsymbol{\lambda}^*) & \nabla \mathbf{h}(\mathbf{x}^*)^T \\ \nabla \mathbf{h}(\mathbf{x}^*) & \mathbf{0} \end{bmatrix}.$$

Assume that $\mathbf{L}(\mathbf{x}^*, \boldsymbol{\lambda}^*)$ is positive definite and that $\nabla \mathbf{h}(\mathbf{x}^*)$ is of full rank.

(a) Show that the real part of each eigenvalue of $\mathbf{C}$ is positive.
(b) Using the result of Part (a), show that for some $\alpha > 0$ the iterative process

$$\mathbf{x}_{k+1} = \mathbf{x}_k - \alpha \nabla l(\mathbf{x}_k, \boldsymbol{\lambda}_k)^T \text{ and } \boldsymbol{\lambda}_{k+1} = \boldsymbol{\lambda}_k + \alpha \mathbf{h}(\mathbf{x}_k)$$

converges locally to $\mathbf{x}^*$, $\boldsymbol{\lambda}^*$. (That is, if started sufficiently close to $\mathbf{x}^*$, $\boldsymbol{\lambda}^*$, the process converges to $\mathbf{x}^*$, $\boldsymbol{\lambda}^*$.) *Hint*: Use Ostroski's Theorem: Let $\mathbf{A}(\mathbf{z})$ be a continuously differentiable mapping from E^p to E^p, assume $\mathbf{A}(\mathbf{z}^*) = \mathbf{0}$, and let $\nabla \mathbf{A}(\mathbf{z}^*)$ have all eigenvalues strictly inside the unit circle of the complex plane. Then $\mathbf{z}_{k+1} = \mathbf{z}_k + \mathbf{A}(\mathbf{z}_k)$ converges locally to $\mathbf{z}^*$.

3. Let $\mathbf{A}$ be a real symmetric matrix. A vector $\mathbf{x}$ is *singular* if $\mathbf{x}^T \mathbf{A}\mathbf{x} = 0$. A pair of vectors $\mathbf{x}$, $\mathbf{y}$ is a *hyperbolic pair* if both $\mathbf{x}$ and $\mathbf{y}$ are singular and $\mathbf{x}^T \mathbf{A}\mathbf{y} \neq 0$. Hyperbolic pairs can be used to generalize the conjugate gradient method to the nonpositive definite case.

(a) If $\mathbf{p}_k$ is singular, show that if $\mathbf{p}_{k+1}$ is defined as

$$\mathbf{p}_{k+1} = \mathbf{A}\mathbf{p}_k - \frac{(\mathbf{A}\mathbf{p}_k)^T \mathbf{A}^2 \mathbf{p}_k}{2|\mathbf{A}\mathbf{p}_k|^2} \mathbf{p}_k,$$

then $\mathbf{p}_k$, $\mathbf{p}_{k+1}$ is a hyperbolic pair.
(b) Consider a modification of the conjugate gradient process of Sect. 8.1, where if $\mathbf{p}_k$ is singular, $\mathbf{p}_{k+1}$ is generated as above, and then

$$\begin{aligned}
\mathbf{x}_{k+1} &= \mathbf{x}_k + \alpha_k \mathbf{p}_k \\
\mathbf{x}_{k+2} &= \mathbf{x}_{k+1} + \alpha_{k+1} \mathbf{p}_{k+1} \\
\alpha_k &= \frac{\mathbf{r}_k^T \mathbf{p}_{k+1}}{\mathbf{p}_k^T \mathbf{A}\mathbf{p}_{k+1}}, \quad \alpha_{k+1} = \frac{\mathbf{r}_k^T \mathbf{p}_k}{\mathbf{p}_k^T \mathbf{A}\mathbf{p}_{k+1}} \\
\mathbf{p}_{k+2} &= \mathbf{r}_{k+2} - \frac{\mathbf{r}_{k+2}^T \mathbf{A}\mathbf{p}_{k+1}}{\mathbf{p}_k \mathbf{A}\mathbf{p}_{k+1}} \mathbf{p}_k.
\end{aligned}$$

Show that if $\mathbf{p}_{k+1}$ is the second member of a hyperbolic pair and $\mathbf{r}_k \neq \mathbf{0}$, then $\mathbf{x}_{k+2} \neq \mathbf{x}_{k+1}$, which means the process does not get "stuck."

4. Another method for solving a system $\mathbf{A}\mathbf{x} = \mathbf{b}$ when A is nonsingular and symmetric is the *conjugate residual method*. In this method the direction vectors are constructed to be an $\mathbf{A}^2$-orthogonalized version of the residuals $\mathbf{r}_k = \mathbf{b} - \mathbf{A}\mathbf{x}_k$. The error function $E(\mathbf{x}) = |\mathbf{A}\mathbf{x} - \mathbf{b}|^2$ decreases monotonically in this process. Since the directions are based on $\mathbf{r}_k$ rather than the gradient of E, which is $2\mathbf{A}r_k$, the method extends the simplicity of the conjugate gradient method by implicit use of the fact that $\mathbf{A}^2$ is positive definite. The method is this: Set $\mathbf{p}_1 = \mathbf{r}_1 = \mathbf{b} - \mathbf{A}\mathbf{x}_1$ and repeat the following steps, omitting (a, b) on the first step.

If $\alpha_{k-1} \neq 0$,

$$\mathbf{p}_k = \mathbf{r}_k - \beta_k \mathbf{p}_{k-1}, \quad \beta_k = \frac{\mathbf{r}_k^T \mathbf{A}^2 \mathbf{p}_{k-1}}{\mathbf{p}_{k-1}^T \mathbf{A}^2 \mathbf{p}_{k-1}}. \tag{15.56a}$$

If $\alpha_{k-1} = 0$,

$$\mathbf{p}_k = \mathbf{A}\mathbf{r}_k - \gamma_k \mathbf{p}_{k-1} - \delta_k \mathbf{p}_{k-2}$$

$$\gamma_k = \frac{\mathbf{r}_k^T \mathbf{A}^3 \mathbf{p}_{k-1}}{\mathbf{p}_{k-1}^T \mathbf{A}^2 \mathbf{p}_{k-1}}, \qquad \delta_k = \frac{\mathbf{r}_k^T \mathbf{A}^3 \mathbf{p}_{k-2}}{\mathbf{p}_{k-2}^T \mathbf{A}^3 \mathbf{p}_{k-2}} \tag{15.56b}$$

$$\mathbf{x}_{k+1} = \mathbf{x}_k + \alpha_k \mathbf{p}_k, \qquad \alpha_k = \frac{\mathbf{r}_k^T \mathbf{A} \mathbf{p}_k}{\mathbf{p}_k^T \mathbf{A}^2 \mathbf{p}_k} \tag{15.56c}$$

$$\mathbf{r}_{k+1} = \mathbf{b} - \mathbf{A}\mathbf{x}_{k+1}. \tag{15.56d}$$

Show that the directions $\mathbf{p}_k$ are $\mathbf{A}^2$-orthogonal.

5. Consider the $(n+m)$-dimensional system of equations

$$\begin{bmatrix} \mathbf{L} & \mathbf{A}^T \\ \mathbf{A} & \mathbf{0} \end{bmatrix} \begin{bmatrix} \mathbf{x} \\ \boldsymbol{\lambda} \end{bmatrix} = \begin{bmatrix} \mathbf{a} \\ \mathbf{b} \end{bmatrix}.$$

Suppose that $\mathbf{A} = [\mathbf{B}, \ \mathbf{C}]$, where $\mathbf{B}$ is $m \times m$ and invertible. Let $\mathbf{x} = (\mathbf{x}_B, \ \mathbf{x}_C)$, where $\mathbf{x}_B$ is the first m components of $\mathbf{x}$. The system can then be written

$$\begin{bmatrix} \mathbf{L_{BB}} & \mathbf{L_{BC}} & \mathbf{B}^T \\ \mathbf{L_{CB}} & \mathbf{L_{cc}} & \mathbf{C}^T \\ \mathbf{B} & \mathbf{C} & \mathbf{0} \end{bmatrix} \begin{bmatrix} \mathbf{x_B} \\ \mathbf{x_C} \\ \boldsymbol{\lambda} \end{bmatrix} = \begin{bmatrix} \mathbf{a_B} \\ \mathbf{a_C} \\ \mathbf{b} \end{bmatrix}$$

(a) Assume that $\mathbf{L}$ is positive definite on the tangent space $\{\mathbf{x} : \mathbf{A}\mathbf{x} = \mathbf{0}\}$. Derive an explicit statement equivalent to this assumption in terms of the positive definiteness of some $(n-m) \times (n-m)$ matrix.

(b) Solve the system in terms of the submatrices of the partitioned form.

6. Consider the partitioned square matrix $\mathbf{M}$ of the form

$$\mathbf{M} = \begin{bmatrix} \mathbf{A} & \mathbf{B} \\ \mathbf{C} & \mathbf{D} \end{bmatrix}.$$

Show that

$$\mathbf{M}^{-1} = \begin{bmatrix} \mathbf{Q} & -\mathbf{Q}\mathbf{B}\mathbf{D}^{-1} \\ -\mathbf{D}^{-1}\mathbf{C}\mathbf{Q} & \mathbf{D}^{-1} + \mathbf{D}^{-1}\mathbf{C}\mathbf{Q}\mathbf{B}\mathbf{D}^{-1} \end{bmatrix},$$

where $\mathbf{Q} = (\mathbf{A} - \mathbf{B}\mathbf{D}^{-1}\mathbf{C})^{-1}$, provided that all indicated inverses exist. Use this result to verify the rate of convergence result in Sect. 15.6.

7. For the problem

$$\begin{aligned} &\text{minimize} \quad f(\mathbf{x}) \\ &\text{subject to} \quad \mathbf{g}(\mathbf{x}) \leqslant \mathbf{0}, \end{aligned}$$

where $\mathbf{g}(\mathbf{x})$ is r-dimensional, define the penalty function

$$p(\mathbf{x}) = f(\mathbf{x}) + c\ \max\{0,\ g_1(\mathbf{x}),\ g_2(\mathbf{x}),\ \ldots,\ g_r(\mathbf{x})\}.$$

Let $\mathbf{d}$, $(\mathbf{d} \neq \mathbf{0})$ be a solution to the quadratic program

$$\begin{aligned} &\text{minimize} \quad \frac{1}{2}\mathbf{d}^T\mathbf{B}\mathbf{d} + \nabla f(\mathbf{x})\mathbf{d} \\ &\text{subject to} \quad \mathbf{g}(\mathbf{x}) + \nabla\mathbf{g}(\mathbf{x})\mathbf{d} \leqslant \mathbf{0}, \end{aligned}$$

where $\mathbf{B}$ is positive definite. Show that $\mathbf{d}$ is a descent direction for p for sufficiently large c.

8. Suppose the quadratic program of Exercise 7 is not feasible. In that case one may solve

$$\begin{aligned} &\text{minimize} \quad \frac{1}{2}\mathbf{d}^T\mathbf{B}\mathbf{d} + \nabla f(\mathbf{x})\mathbf{d} + c\xi \\ &\text{subject to} \quad \mathbf{g}(\mathbf{x}) + \nabla\mathbf{g}(\mathbf{x})\mathbf{d} \leqslant \xi\mathbf{1} \\ &\qquad\qquad\qquad \xi \geqslant 0. \end{aligned}$$

(a) Show that if $\mathbf{d} \neq \mathbf{0}$ is a solution, then $\mathbf{d}$ is a descent direction for p.
(b) If $\mathbf{d} = \mathbf{0}$ is a solution, show that $\mathbf{x}$ is a critical point of p in the sense that for any $\mathbf{d} \neq \mathbf{0}$, $p(\mathbf{x} + \alpha\mathbf{d}) > p(\mathbf{x}) + o(\alpha)$.

9. For the equality constrained problem, consider the function

$$\phi(\mathbf{x}) = f(\mathbf{x}) + \boldsymbol{\lambda}(\mathbf{x})^T\mathbf{h}(\mathbf{x}) + c\mathbf{h}(\mathbf{x})^T\mathbf{C}(\mathbf{x})\mathbf{C}(\mathbf{x})^T\mathbf{h}(\mathbf{x}),$$

where

$$\mathbf{C}(\mathbf{x}) = [\nabla\mathbf{h}(\mathbf{x})\nabla\mathbf{h}(\mathbf{x})^T]^{-1}\nabla\mathbf{h}(\mathbf{x}) \quad \text{and} \quad \boldsymbol{\lambda}(\mathbf{x}) = \mathbf{C}(\mathbf{x})\nabla f(\mathbf{x})^T.$$

(a) Under standard assumptions on the original problem, show that for sufficiently large c, ϕ is (locally) an exact penalty function.
(b) Show that $\phi(\mathbf{x})$ can be expressed as

$$\phi(\mathbf{x}) = \mathbf{f}(\mathbf{x}) + \boldsymbol{\pi}(\mathbf{x})^T\mathbf{h}(\mathbf{x}),$$

where $\boldsymbol{\pi}(\mathbf{x})$ is the Lagrange multiplier of the problem

$$\begin{aligned} &\text{minimize} \quad \frac{1}{2}c\mathbf{d}^T\mathbf{d} + \nabla f(\mathbf{x})\mathbf{d} \\ &\text{subject to} \quad \nabla\mathbf{h}(\mathbf{x})\mathbf{d} + \mathbf{h}(\mathbf{x}) = \mathbf{0}. \end{aligned}$$

(c) Indicate how ϕ can be defined for problems with inequality constraints.

10. Let $\{\mathbf{B}_k\}$ be a sequence of positive definite symmetric matrices, and assume that there are constants $a > 0$, $b > 0$ such that $a|\mathbf{x}|^2 \leqslant \mathbf{x}^T\mathbf{B}_k\mathbf{x} \leqslant b|\mathbf{x}|^2$ for all $\mathbf{x}$. Suppose that $\mathbf{B}$ is replaced by $\mathbf{B}_k$ in the kth step of the recursive quadratic

programming procedure of the theorem in Sect. 15.4. Show that the conclusions of that theorem are still valid. *Hint*: Note that the set of allowable $\mathbf{B}_k$'s is closed.

11. (Central path theorem) Prove the central path theorem, Theorem 1 of Sect. 15.7, for convex optimization.
12. Prove the potential reduction theorem, Theorem 2 of Sect. 15.7, for convex quadratic programming. This theorem can be generalized to non-quadratic convex objective functions $f(\mathbf{x})$ satisfying the following condition: let

$$u : (0, 1) \rightarrow (1, \infty)$$

be a monotone increasing function; then

$$|\mathbf{X}(\nabla f(\mathbf{x} + \mathbf{d_x}) - \nabla f(\mathbf{x}) - \nabla^2 f(\mathbf{x})\mathbf{d_x})|_1 \leq u(\alpha)\mathbf{d_x}^T \nabla f(\mathbf{x})\mathbf{d_x}$$

whenever

$$\mathbf{x} > 0, \ |\mathbf{X}^{-1}\mathbf{d_x}|_\infty \leq \alpha < 1.$$

Such condition is called the scaled Lipschitz condition in $\{\mathbf{x} : \mathbf{x} > \mathbf{0}\}$.

References

15.1 An early method for solving quadratic programming problems is the principal pivoting method of Dantzig and Wolfe; see Dantzig [D6]. For a discussion of factorization methods applied to quadratic programming, see Gill, Murray, and Wright [G7].

15.2–15.4 Arrow and Hurwicz [A9] proposed a continuous process (represented as a system of differential equations) for solving the Lagrange equations. This early paper showed the value of the simple merit function in attacking the equations. A formal discussion of the properties of the simple merit function may be found in Luenberger [L17]. The first-order method was examined in detail by Polak [P4]. Also see Zangwill [Z2] for an early analysis of a method for inequality constraints. The conjugate direction method was first extended to nonpositive definite cases by the use of hyperbolic pairs and then by employing conjugate residuals. (See Exercises 3 and 4, and Luenberger [L9, L11].) Additional methods with somewhat better numerical properties were later developed by Paige and Saunders [P1] and by Fletcher [F8]. It is perhaps surprising that Newton's method was analyzed in this form only recently, well after the development of the SOLVER method discussed in Sect. 15.2. For a comprehensive account of Newton methods, see Bertsekas, Chap. 4 [B11]. The SOLVER method was proposed by Wilson [W2] for convex programming problems and was later interpreted by Beale [B7]. Garcia-Palomares and Mangasarian [G3] proposed a quadratic programming approach to the solution of the first-order equations. See Fletcher [F10] for a good overview discussion.

15.5–15.6 The discovery that the absolute-value penalty function is compatible with recursive quadratic programming was made by Pshenichny (see Pshenichny and Danilin [P10]) and later by Han [H3], who also suggested that the method be combined with a quasi-Newton update procedure.

The development of recursive quadratic programming for the standard quadratic penalty function is due to Biggs [B14, B15]. The convergence rate analysis of Sect. 15.6 first appeared in the second edition of this text.

15.7 Many researchers have applied interior-point algorithms to convex quadratic problems. These algorithms can be divided into three groups: the primal algorithm, the dual algorithm, and the primal-dual algorithm. Relations among these algorithms can be seen in den Hertog [H6], Anstreicher et al [A6], Sun and Qi [S12], Tseng [T12], and Ye [Y3]. For results similar to those of Exercises 2,7, and 8, see Bertsekas [B11]. For discussion of Exercise 9, see Fletcher [F10].

Appendix A
Mathematical Review

The purpose of this appendix is to set down for reference and review some basic definitions, notation, and relations that are used frequently in the text.

A.1 Sets

If x is a member of the set S, we write $x \in S$. We write $y \notin S$ if y is not a member of S.

A set S may be specified by listing its elements between braces; such as, for example, $S = \{1, 2, 3, 4\}$. Alternatively, a set can be specified in the form $S = \{x : P(x)\}$ as the set of elements satisfying property P; such as $S = \{x : 1 \leqslant x \leqslant 4,\ x \text{ integer}\}$

The *union* of two sets S and T is denoted $S \cup T$ and is the set consisting of the elements that belong to either S or T. The *intersection* of two sets S and T is denoted $S \cap T$ and is the set consisting of the elements that belong to both S and T. If S is a *subset* of T, that is, if every member of S is also a member of T, we write $S \subset T$ or $T \supset S$.

The empty set is denoted ϕ or $\emptyset$. There are two ways that operations such as minimization over a set are represented. Specifically we write either

$$\min_{x \in S} f(x) \text{ or } \min\{f(x) : x \in S\}$$

to denote the minimum value of f over the set S. The set of x's in S that achieve the minimum is denoted argmin $\{f(x) : x \in S\}$.

D.G. Luenberger, Y. Ye, *Linear and Nonlinear Programming*, International Series in Operations Research & Management Science 228,
DOI 10.1007/978-3-319-18842-3

Sets of Real Numbers

If a and b are real numbers, $[a, b]$ denotes the set of real numbers x satisfying $a \leqslant x \leqslant b$. A rounded, instead of square, bracket denotes strict inequality in the definition. Thus $(a, b]$ denotes all x satisfying $a < x \leqslant b$.

If S is a set of real numbers bounded above, then there is a smallest real number y such that $x \leqslant y$ for all $x \in S$. The number y is called the *least upper bound* or *supremum* of S and is denoted

$$\sup_{x \in S}(x) \text{ or } \sup\{x : x \in S\}.$$

Similarly, the *greatest lower bound* or *infimum* of a set S is denoted

$$\inf_{x \in S}(x) \text{ or } \inf\{x : x \in S\}.$$

A.2 Matrix Notation

A *matrix* is a rectangular array of numbers, called *elements*. The matrix itself is denoted by a boldface letter. When specific numbers are not used, the elements are denoted by italicized lower-case letters, having a double subscript. Thus we write

$$\mathbf{A} = \begin{bmatrix} a_{11} & a_{12} & \cdots & a_{1n} \\ a_{21} & a_{22} & \cdots & a_{2n} \\ \vdots & & & \\ a_{m1} & a_{m2} & \cdots & a_{mn} \end{bmatrix}$$

for a matrix $\mathbf{A}$ having m rows and n columns. Such a matrix is referred to as an $m \times n$ matrix. If we wish to specify a matrix by defining a general element, we use the notation $\mathbf{A} = [a_{ij}]$.

An $m \times n$ matrix all of whose elements are zero is called a *zero matrix* and denoted $\mathbf{0}$. A *square* matrix (a matrix with $m = n$) whose elements are $a_{ij} = 0$ for $i \neq j$, and $a_{ii} = 1$ for $i = 1, 2, \ldots, n$ is said to be an *identity matrix* and denoted $\mathbf{I}$.

The *sum* of two $m \times n$ matrices $\mathbf{A}$ and $\mathbf{B}$ is written $\mathbf{A} + \mathbf{B}$ and is the matrix whose elements are the sum of the corresponding elements in $\mathbf{A}$ and $\mathbf{B}$. The *product* of a matrix $\mathbf{A}$ and a scalar λ, written $\lambda\mathbf{A}$ or $\mathbf{A}\lambda$, is obtained by multiplying each element of $\mathbf{A}$ by λ. The *product* $\mathbf{AB}$ of an $m \times n$ matrix $\mathbf{A}$ and an $n \times p$ matrix $\mathbf{B}$ is the $m \times p$ matrix $\mathbf{C}$ with elements $c_{ij} = \sum_{k=1}^{n} a_{ik} b_{kj}$.

The *transpose* of an $m \times n$ matrix $\mathbf{A}$ is the $n \times m$ matrix $\mathbf{A}^T$ with elements $a_{ij}^T = a_{ji}$. A (square) matrix $\mathbf{A}$ is *symmetric* if $\mathbf{A}^T = \mathbf{A}$. A square matrix $\mathbf{A}$ is *nonsingular* if there is a matrix $\mathbf{A}^{-1}$, called the *inverse* of $\mathbf{A}$, such that $\mathbf{A}^{-1}\mathbf{A} = \mathbf{I} = \mathbf{A}\mathbf{A}^{-1}$. The *determinant* of a square matrix $\mathbf{A}$ is denoted by det ($\mathbf{A}$). The determinant is nonzero if and only if the matrix is nonsingular. Two square $n \times n$ matrices $\mathbf{A}$ and $\mathbf{B}$ are *similar* if there is a nonsingular matrix $\mathbf{S}$ such that $\mathbf{B} = \mathbf{S}^{-1}\mathbf{AS}$.

Matrices having a single row are referred to as *row vectors*; matrices having a single column are referred to as *column vectors*. *Vectors* of either type are usually denoted by lower-case boldface letters. To economize page space, row vectors are written $\mathbf{a} = [a_1, a_2, \ldots, a_n]$ and column vectors are written $\mathbf{a} = (a_1, a_2, \ldots, a_n)$. Since column vectors are used frequently, this notation avoids the necessity to display numerous columns. To further distinguish rows from columns, we write $\mathbf{a} \in E^n$ if $\mathbf{a}$ is a column vector with n components, and we write $\mathbf{b} \in E_n$ if $\mathbf{b}$ is a row vector with n components.

It is often convenient to partition a matrix into submatrices. This is indicated by drawing partitioning lines through the matrix, as for example,

$$\mathbf{A} = \left[\begin{array}{cc|cc} a_{11} & a_{12} & a_{13} & a_{14} \\ \hline a_{21} & a_{22} & a_{23} & a_{24} \\ a_{31} & a_{32} & a_{33} & a_{34} \end{array}\right] = \begin{bmatrix} \mathbf{A}_{11} & \mathbf{A}_{12} \\ \mathbf{A}_{21} & \mathbf{A}_{22} \end{bmatrix}.$$

The resulting submatrices are usually denoted $\mathbf{A}_{ij}$, as illustrated.

A matrix can be partitioned into either column or row vectors, in which case a special notation is convenient. Denoting the columns of an $m \times n$ matrix $\mathbf{A}$ by $\mathbf{a}_j$, $j = 1, 2, \ldots, n$, we write $\mathbf{A} = [\mathbf{a}_1, \mathbf{a}_2, \ldots, \mathbf{a}_n]$. Similarly, denoting the rows of $\mathbf{A}$ by $\mathbf{a}^i$, $i = 1, 2, \ldots, m$, we write $\mathbf{A} = (\mathbf{a}^1, \mathbf{a}^2, \ldots, \mathbf{a}^m)$. Following the same pattern, we often write $\mathbf{A} = [\mathbf{B}, \mathbf{C}]$ for the partitioned matrix $\mathbf{A} = [\mathbf{B}|\mathbf{C}]$.

A.3 Spaces

We consider the n-component vectors $\mathbf{x} = (x_1, x_2, \ldots, x_n)$ as elements of a vector space. The space itself, n-dimensional Euclidean space, is denoted E^n. Vectors in the space can be added or multiplied by a scalar, by performing the corresponding operations on the components. We write $\mathbf{x} \geqslant \mathbf{0}$ if each component of $\mathbf{x}$ is nonnegative.

The *line segment* connecting two vectors $\mathbf{x}$ and $\mathbf{y}$ is denoted $[\mathbf{x}, \mathbf{y}]$ and consists of all vectors of the form $\alpha\mathbf{x} + (1 - \alpha)\mathbf{y}$ with $0 \leqslant \alpha \leqslant 1$.

The *scalar product* of two vectors $\mathbf{x} = (x_1, x_2, \ldots, x_n)$ and $\mathbf{y} = (y_1, y_2, \ldots, y_n)$ is defined as $\mathbf{x}^T\mathbf{y} = \mathbf{y}^T\mathbf{x} = \sum_{i=1}^n x_i y_i$. The vectors $\mathbf{x}$ and $\mathbf{y}$ are said to be *orthogonal* if $\mathbf{x}^T\mathbf{y} = 0$. The *magnitude* or *norm* of a vector $\mathbf{x}$ is $|\mathbf{x}| = (\mathbf{x}^T\mathbf{x})^{1/2}$. For any two vectors $\mathbf{x}$ and $\mathbf{y}$ in E^n, the *Cauchy-Schwarz Inequality* holds: $|\mathbf{x}^T\mathbf{y}| \leqslant |\mathbf{x}| \cdot |\mathbf{y}|$.

A set of vectors $\mathbf{a}_1, \mathbf{a}_2, \ldots, \mathbf{a}_k$ is said to be *linearly dependent* if there are scalars $\lambda_1, \lambda_2, \ldots, \lambda_k$, not all zero, such that $\sum_{i=1}^k \lambda_i \mathbf{a}_i = \mathbf{0}$. If no such set of scalars exists, the vectors are said to be *linearly independent*. A *linear combination* of the vectors $\mathbf{a}_1, \mathbf{a}_2, \ldots, \mathbf{a}_k$ is a vector of the form $\sum_{i=1}^k \lambda_j \mathbf{a}_i$. The set of vectors that are linear combinations of $\mathbf{a}_1, \mathbf{a}_2, \ldots, \mathbf{a}_k$ is the set *spanned* by the vectors. A linearly independent set of vectors that span E^n is said to be a *basis* for E^n. Every basis for E^n contains exactly n vectors.

The *rank* of a matrix $\mathbf{A}$ is equal to the maximum number of linearly independent columns in $\mathbf{A}$. This number is also equal to the maximum number of linearly independent rows in $\mathbf{A}$. The $m \times n$ matrix $\mathbf{A}$ is said to be of *full rank* if the rank of $\mathbf{A}$ is equal to the minimum of m and n.

A *subspace* M of E^n is a subset that is closed under the operations of vector addition and scalar multiplication; that is, if $\mathbf{a}$ and $\mathbf{b}$ are vectors in M, then $\lambda\mathbf{a} + \mu\mathbf{b}$ is also in M for every pair of scalars λ, μ. The dimension of a subspace M is equal to the maximum number of linearly independent vectors in M. If M is a subspace of E^n, the *orthogonal complement* of M, denoted $M^\perp$, consists of all vectors that are orthogonal to every vector in M. The orthogonal complement of M is easily seen to be a subspace, and together M and $M^\perp$ span E^n in the sense that every vector $\mathbf{x} \in E^n$ can be written uniquely in the form $\mathbf{x} = \mathbf{a} + \mathbf{b}$ with $\mathbf{a} \in M$, $\mathbf{b} \in M^\perp$. In this case $\mathbf{a}$ and $\mathbf{b}$ are said to be the *orthogonal projections* of $\mathbf{x}$ onto the subspaces M and $M^\perp$, respectively.

A correspondence $\mathbf{A}$ that associates with each point in a space X a point in a space Y is said to be a *mapping from* X to Y. For convenience this situation is symbolized by $\mathbf{A} : X \to Y$. The mapping $\mathbf{A}$ may be either linear or nonlinear. The norm of linear mapping $\mathbf{A}$ is defined as $|\mathbf{A}| = \max_{|\mathbf{x}| \leq 1} |\mathbf{A}\mathbf{x}|$. It follows that for any $\mathbf{x}$, $|\mathbf{A}\mathbf{x}| \leq |\mathbf{A}| \cdot |\mathbf{x}|$.

A.4 Eigenvalues and Quadratic Forms

Corresponding to an $n \times n$ square matrix $\mathbf{A}$, a scalar λ and a nonzero vector $\mathbf{x}$ satisfying the equation $\mathbf{A}\mathbf{x} = \lambda\mathbf{x}$ are said to be, respectively, an eigenvalue and eigenvector of $\mathbf{A}$. In order that λ be an eigenvalue it is clear that it is necessary and sufficient for $\mathbf{A} - \lambda\mathbf{I}$ to be singular, and hence $\det(\mathbf{A} - \lambda\mathbf{I}) = 0$. This last result, when expanded, yields an nth-order polynomial equation which can be solved for n (possibly nondistinct) complex roots λ which are the eigenvalues of $\mathbf{A}$.

Now, for the remainder of this section, assume that $\mathbf{A}$ is symmetric. Then the following properties hold:

(i) The eigenvalues of $\mathbf{A}$ are real.
(ii) Eigenvectors associated with distinct eigenvalues are orthogonal.
(iii) There is an orthogonal basis for E^n, each element of which is an eigenvector of $\mathbf{A}$.

If the basis $\mathbf{u}_1, \mathbf{u}_2, \ldots, \mathbf{u}_n$ in (iii) is normalized so that each element has magnitude unity, then defining the matrix $\mathbf{Q} = [\mathbf{u}_1, \mathbf{u}_2, \ldots, \mathbf{u}_n]$ we note that $\mathbf{Q}^T\mathbf{Q} = \mathbf{I}$ and hence $\mathbf{Q}^T = \mathbf{Q}^{-1}$. A matrix with this property is said to be an *orthogonal* matrix. Also, we observe, in this case, that

$$\begin{aligned}\mathbf{Q}^{-1}\mathbf{A}\mathbf{Q} = \mathbf{Q}^T\mathbf{A}\mathbf{Q} &= \mathbf{Q}^T[\mathbf{A}\mathbf{u}_1, \mathbf{A}\mathbf{u}_2, \ldots, \mathbf{A}\mathbf{u}_n] \\ &= \mathbf{Q}^T[\lambda_1\mathbf{u}_1, \lambda_2\mathbf{u}_2, \ldots, \lambda_n\mathbf{u}_n].\end{aligned}$$

Thus

$$\mathbf{Q}^{-1}\mathbf{A}\mathbf{Q} = \begin{bmatrix} \lambda_1 & & & \\ & \lambda_2 & & \\ & & \ddots & \\ & & & \lambda_n \end{bmatrix},$$

and therefore **A** is similar to a diagonal matrix.

A symmetric matrix **A** is said to be *positive definite* if the *quadratic form* $\mathbf{x}^T\mathbf{A}\mathbf{x}$ is positive for all nonzero vectors **x**. Similarly, we define **A** to be *positive semidefinite, negative definite*, or *negative semidefinite* if $\mathbf{x}^T\mathbf{A}\mathbf{x} \geqslant 0, < 0$, or $\leqslant 0$ for all **x**. The matrix **A** is *indefinite* if $\mathbf{x}^T\mathbf{A}\mathbf{x}$ is positive for some **x** and negative for others.

It is easy to obtain a connection between definiteness and the eigenvalues of **A**. For any **x** let $\mathbf{y} = \mathbf{Q}^{-1}\mathbf{x}$ where **Q** is defined as above. Then $\mathbf{x}^T\mathbf{A}\mathbf{x} = \mathbf{y}^T\mathbf{Q}^T\mathbf{A}\mathbf{Q}\mathbf{y} = \sum_{i=1}^{n}\lambda_i y_i^2$. Since the y_i's are arbitrary (since **x** is), it is clear that **A** is positive definite (or positive semidefinite) if and only if all eigenvalues of **A** are positive (or nonnegative).

Through diagonalization we can also easily show that a positive semidefinite matrix **A** has a positive semidefinite (symmetric) square root $\mathbf{A}^{1/2}$ satisfying $\mathbf{A}^{1/2}\cdot\mathbf{A}^{1/2} = \mathbf{A}$. For this we use **Q** as above and define

$$\mathbf{A}^{1/2} = \mathbf{Q}\begin{bmatrix} \lambda_1^{1/2} & & & \\ & \lambda_2^{1/2} & & \\ & & \ddots & \\ & & & \lambda_n^{1/2} \end{bmatrix}\mathbf{Q}^T,$$

which is easily verified to have the desired properties.

A.5 Topological Concepts

A sequence of vectors $\mathbf{x}_0, \mathbf{x}_1, \ldots, \mathbf{x}_k, \ldots$, denoted by $\{\mathbf{x}_{k=0}\}_k^{\infty}$, or if the index set is understood, by simply $\{\mathbf{x}_k\}$, is said to *converge* to the limit **x** if $|\mathbf{x}_k - \mathbf{x}| \to 0$ as $k \to \infty$ (that is, if given $\varepsilon > 0$, there is a N such that $k \geqslant N$ implies $|\mathbf{x}_k - \mathbf{x}| < \varepsilon$). If $\{\mathbf{x}_k\}$ converges to **x**, we write $\mathbf{x}_k \to \mathbf{x}$ or $\lim \mathbf{x}_k = \mathbf{x}$.

A point **x** is a *limit point* of the sequence $\{\mathbf{x}_k\}$ if there is a subsequence of $\{\mathbf{x}_k\}$ convergent to **x**. Thus **x** is a limit point of $\{\mathbf{x}_k\}$ if there is a subset $\mathcal{K}$ of the positive integers such that $\{\mathbf{x}_k\}_{k\in\mathcal{K}}$ converges to **x**.

A *ball (sphere) around* **x** is a set of the form $\{\mathbf{y} : |\mathbf{y} - \mathbf{x}| < (=)\ \varepsilon\}$ for some $\varepsilon > 0$. Such a ball is also referred to as the *neighborhood* of **x** of radius ε.

A subset S of E^n is *open* if around every point in S there is a sphere that is contained in S. Equivalently, S is open if given $\mathbf{x} \in S$ there is an $\varepsilon > 0$ such that $|\mathbf{y} - \mathbf{x}| < \varepsilon$ implies $\mathbf{y} \in S$. Thus the sphere $\{\mathbf{x} : |\mathbf{x}| < 1\}$ is open. In general, open sets can be characterized as sets having no sharp boundaries. The *interior* of any set S in E^n is the set of points $\mathbf{x} \in S$ which are the center of some sphere contained in S.

It is denoted $\overset{\circ}{S}$. The interior of a set is always open; indeed it is the largest open set contained in S. The interior of the set $\{\mathbf{x} : |\mathbf{x}| \leqslant 1\}$ is the sphere $\{\mathbf{x} : |\mathbf{x}| < 1\}$.

A set P is *closed* if every point that is arbitrarily close to the set P is a member of P. Equivalently, P is closed if $\mathbf{x}_k \to \mathbf{x}$ with $\mathbf{x}_k \in P$ implies $\mathbf{x} \in P$. Thus the set $\{\mathbf{x} : |\mathbf{x}| \leqslant 1\}$ is closed. The *closure* of any set P in E^n is the smallest closed set containing P. It is denoted $\overline{S}$. The *boundary* of a set is that part of the closure that is not in the interior.

A set is *compact* if it is both closed and bounded (that is, if it is closed and is contained within some sphere of finite radius). An important result, due to Weierstrass, is that if S is a compact set and $\{\mathbf{x}_k\}$ is a sequence each member of which belongs to S, then $\{\mathbf{x}_k\}$ has a limit point in S (that is, there is subsequence converging to a point in S).

Corresponding to a bounded sequence $\{r_k\}_{k=0}^{\infty}$ of real numbers, if we let $s_k = \sup\{r_i : i \geqslant k\}$ then $\{s_k\}$ converges to some real number s_o. This number is called the *limit superior* of $\{r_k\}$ and is denoted $\overline{\lim_{k\to\infty}}(r_k)$.

A.6 Functions

A real-valued function f defined on a subset of E^n is said to be *continuous* at $\mathbf{x}$ if $\mathbf{x}_k \to \mathbf{x}$ implies $f(\mathbf{x}_k) \to f(\mathbf{x})$. Equivalently, f is continuous at $\mathbf{x}$ if given $\varepsilon > 0$ there is a $\delta > 0$ such that $|\mathbf{y}-\mathbf{x}| < \delta$ implies $|f(\mathbf{y})-f(\mathbf{x})| < \varepsilon$. An important result connected with continuous functions is a *theorem of Weierstrass*: A continuous function f defined on a compact set S has a minimum point in S; that is, there is an $\mathbf{x}^* \in S$ such that for all $\mathbf{x} \in S$, $f(\mathbf{x}) \geqslant f(\mathbf{x}^*)$.

A set of real-valued functions $f_1, f_2, \ldots, f_m$ on E^n can be regarded as a single vector function $\mathbf{f} = (f_1, f_2, \ldots, f_m)$. This function assigns a vector $\mathbf{f}(\mathbf{x}) = (f_1(\mathbf{x}), f_2(\mathbf{x}), \ldots, f_m(\mathbf{x}))$ in E^m to every vector $\mathbf{x} \in E^n$. Such a vector-valued function is said to be *continuous* if each of its component functions is continuous.

If each component of $\mathbf{f} = (f_1, f_2, \ldots, f_m)$ is continuous on some open set of E^n, then we write $\mathbf{f} \in C$. If in addition, each component function has first partial derivatives which are continuous on this set, we write $\mathbf{f} \in C^1$. In general, if the component functions have continuous partial derivatives of order p, we write $\mathbf{f} \in C^p$.

If $f \in C^1$ is a real-valued function on E^n, $f(\mathbf{x}) = f(x_1, x_2, \ldots, x_n)$, we define the *gradient* of f to be the vector

$$\nabla f(\mathbf{x}) = \left[\frac{\partial f(\mathbf{x})}{\partial x_1}, \frac{\partial f(\mathbf{x})}{\partial x_2}, \cdots, \frac{\partial f(\mathbf{x})}{\partial x_n}\right].$$

We sometimes use the alternative notation $f_{\mathbf{x}}(\mathbf{x})$ for $\nabla f(\mathbf{x})$. In matrix calculations the gradient is considered to be a row vector.

If $f \in C^2$ then we define the *Hessian* of f at $\mathbf{x}$ to be the $n \times n$ matrix denoted $\nabla^2 f(\mathbf{x})$ or $\mathbf{F}(\mathbf{x})$ as

$$\mathbf{F}(\mathbf{x}) = \left[\frac{\partial^2 f(\mathbf{x})}{\partial x_i \partial x_j}\right].$$

Since

$$\frac{\partial^2 f}{\partial x_i \partial x_j} = \frac{\partial^2 f}{\partial x_j \partial x_i},$$

it is easily seen that the Hessian is symmetric.

For a vector-valued function $\mathbf{f} = (f_1, f_2, \ldots, f_m)$ the situation is similar. If $\mathbf{f} \in C^1$, the first derivative is defined as the $m \times n$ matrix

$$\nabla \mathbf{f}(\mathbf{x}) = \left[\frac{\partial f_i(\mathbf{x})}{\partial x_j}\right].$$

If $\mathbf{f} \in C^2$ it is possible to define the m Hessians $\mathbf{F}_1(\mathbf{x}), \mathbf{F}_2(\mathbf{x}), \ldots, \mathbf{F}_m(\mathbf{x})$ corresponding to the m component functions. The second derivative itself, for a vector function, is a third-order tensor but we do not require its use explicitly. Given any $\boldsymbol{\lambda}^T = [\lambda_1, \lambda_2, \ldots, \lambda_m] \in E_m$, we note, however, that the real-valued function $\boldsymbol{\lambda}^T \mathbf{f}$ has gradient equal to $\boldsymbol{\lambda}^T \nabla \mathbf{f}(\mathbf{x})$ and Hessian, denoted $\boldsymbol{\lambda}^T \mathbf{F}(\mathbf{x})$, equal to

$$\boldsymbol{\lambda}^T \mathbf{F}(\mathbf{x}) = \sum_{i=1}^{m} \lambda_i \mathbf{F}_i(\mathbf{x}).$$

Also see Sect. 7.4 for a discussion of convex functions.

Taylor's Theorem

A group of results that are used frequently in analysis are referred to under the general heading of Taylor's Theorem or Mean Value Theorems. If $f \in C^1$ in a region containing the line segment $[\mathbf{x}_1, \mathbf{x}_2]$, then there is a θ, $0 \leqslant \theta \leqslant 1$ such that

$$f(\mathbf{x}_2) = f(\mathbf{x}_1) + \nabla f(\theta \mathbf{x}_1 + (1-\theta)\mathbf{x}_2)(\mathbf{x}_2 - \mathbf{x}_1).$$

Furthermore, if $f \in C^2$ then there is a θ, $0 \leqslant \theta \leqslant 1$ such that

$$\begin{aligned} f(\mathbf{x}_2) &= f(\mathbf{x}_1) + \nabla f(\mathbf{x}_1)(\mathbf{x}_2 - \mathbf{x}_1) \\ &\quad + \frac{1}{2}(\mathbf{x}_2 - \mathbf{x}_1)^T \mathbf{F}(\theta \mathbf{x}_1 + (1-\theta)\mathbf{x}_2)(\mathbf{x}_2 - \mathbf{x}_1), \end{aligned}$$

where $\mathbf{F}$ denotes the Hessian of f.

Implicit Function Theorem

Suppose we have a set of m equations in n variables

$$h_i(\mathbf{x}) = 0, \ i = 1, 2, \ \ldots, \ m.$$

The implicit function theorem addresses the question as to whether if $n - m$ of the variables are fixed, the equations can be solved for the remaining m variables. Thus selecting m variables, say $x_1, \ x_2, \ \ldots, \ x_m$, we wish to determine if these may be expressed in terms of the remaining variables in the form

$$x_i = \phi_i(x_{m+1}, \ x_{m+2}, \ \ldots, \ x_n), \qquad i = 1, 2, \ \ldots, \ m.$$

The functions ϕ_i, if they exist, are called *implicit* functions.

Theorem. *Let $\mathbf{x}^0 = (x_1^0, \ x_2^0, \ \ldots, \ x_n^0)$ be a point in E^n satisfying the properties:*

i) The functions $h_i \in C^p$, $i = 1, 2, \ \ldots, \ m$ in some neighborhood of $\mathbf{x}^0$, for some $p \geqslant 1$.
ii) $h_i(\mathbf{x}^0) = 0$, $i = 1, 2, \ \ldots, \ m$.
iii) The $m \times m$ Jacobian matrix

$$\mathbf{J} = \begin{bmatrix} \frac{\partial h_1(\mathbf{x}^0)}{\partial x_1} & \cdots & \frac{\partial h_1(\mathbf{x}^0)}{\partial x_m} \\ \vdots & & \vdots \\ \frac{\partial h_m(\mathbf{x}^0)}{\partial x_1} & \cdots & \frac{\partial h_m(\mathbf{x}^0)}{\partial x_m} \end{bmatrix}.$$

is nonsingular.

Then there is a neighborhood of $\hat{\mathbf{x}}^0 = (x_{m+1}^0, \ x_{m+2}^0, \ \ldots, \ x_n^0) \in E^{n-m}$ such that for $\hat{\mathbf{x}} = (x_{m+1}, \ x_{m+2}, \ \ldots, \ x_n)$ in this neighborhood there are functions $\phi_i(\hat{\mathbf{x}})$, $i = 1, 2, \ \ldots, \ m$ such that

i) $\phi_i \in C^p$.
ii) $x_i^0 = \phi_i(\hat{\mathbf{x}}^0), \qquad i = 1, 2, \ \ldots, \ m$.
iii) $h_i(\phi_1(\hat{\mathbf{x}}), \ \phi_2(\hat{\mathbf{x}}), \ \ldots, \ \phi_m(\hat{\mathbf{x}}), \hat{\mathbf{x}}) = 0$, $i = 1, 2, \ \ldots, \ m$.

Example 1. Consider the equation $x_1^2 + x_2 = 0$. A solution is $x_1 = 0, \ x_2 = 0$. However, in a neighborhood of this solution there is no function ϕ such that $x_1 = \phi(x_2)$. At this solution condition (iii) of the implicit function theorem is violated. At any other solution, however, such a ϕ exists.

Example 2. Let $\mathbf{A}$ be an $m \times n$ matrix $(m < n)$ and consider the system of linear equations $\mathbf{Ax} = \mathbf{b}$. If $\mathbf{A}$ is partitioned as $\mathbf{A} = [\mathbf{B}, \ \mathbf{C}]$ where $\mathbf{B}$ is $m \times m$ then condition (iii) is satisfied if and only if $\mathbf{B}$ is nonsingular. This condition corresponds, of course, exactly with what the theory of linear equations tells us. In view of this example, the implicit function can be regarded as a nonlinear generalization of the linear theory.

o, *O* Notation

If g is a real-valued function of a real variable, the notation $g(x) = O(x)$ means that $g(x)$ goes to zero at least as fast as x does. More precisely, it means that there is a $K \geqslant 0$ such that

$$\left|\frac{g(x)}{x}\right| \leqslant K \text{ as } x \to 0.$$

The notation $g(x) = o(x)$ means that $g(x)$ goes to zero faster than x does; or equivalently, that K above is zero.

Appendix B
Convex Sets

B.1 Basic Definitions

Concepts related to convex sets so dominate the theory of optimization that it is essential for a student of optimization to have knowledge of their most fundamental properties. In this appendix is compiled a brief summary of the most important of these properties.

Definition. A set C in E^n is said to be *convex* if for every $\mathbf{x}_1, \mathbf{x}_2 \in C$ and every real number α, $0 < \alpha < 1$, the point $\alpha\mathbf{x}_1 + (1-\alpha)\mathbf{x}_2 \in C$.

This definition can be interpreted geometrically as stating that a set is convex if, given two points in the set, every point on the line segment joining these two points is also a member of the set. This is illustrated in Fig. B.1.

The following proposition shows that certain familiar set operations preserve convexity.

Proposition 1. *Convex sets in E^n satisfy the following relations:*

i) If C is a convex set and β is a real number, the set

$$\beta C = \{\mathbf{x} : \mathbf{x} = \beta\mathbf{c},\ \mathbf{c} \in C\}$$

is convex.

ii) If C and D are convex sets, then the set

$$C + D = \{\mathbf{x} : \mathbf{x} = \mathbf{c} + \mathbf{d},\ \mathbf{c} \in C,\ \mathbf{d} \in D\}$$

is convex.

iii) The intersection of any collection of convex sets is convex.

The proofs of these three properties follow directly from the definition of a convex set and are left to the reader. The properties themselves are illustrated in Fig. B.2.

D.G. Luenberger, Y. Ye, *Linear and Nonlinear Programming*, International Series in Operations Research & Management Science 228,
DOI 10.1007/978-3-319-18842-3

Another important concept is that of forming the smallest convex set containing a given set.

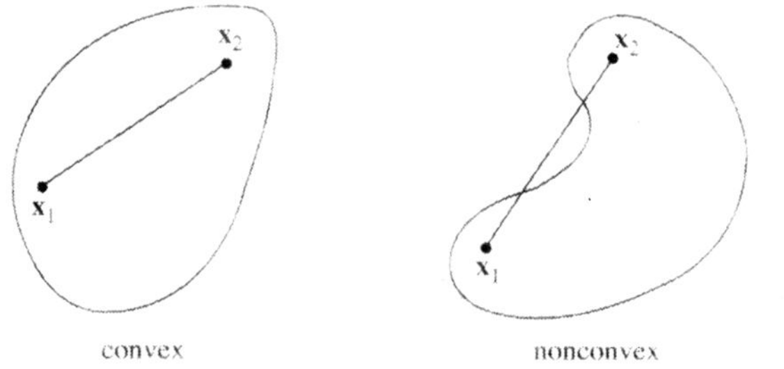

Fig. B.1 Convexity

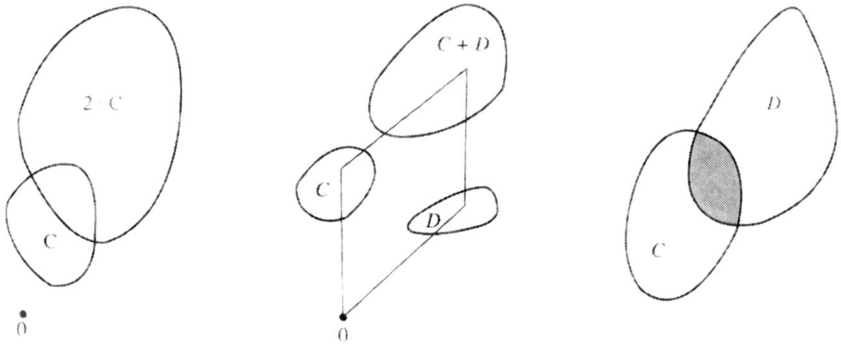

Fig. B.2 Properties of convex sets

Definition. Let S be a subset of E^n. The *convex hull* of S, denoted co(S), is the set which is the intersection of all convex sets containing S. The *closed convex hull* of S is defined as the closure of co(S).

Finally, we conclude this section by defining a *cone* and a *convex cone*. A convex cone is a special kind of convex set that arises quite frequently.

Definition. A set C is a *cone* if $\mathbf{x} \in C$ implies $\alpha\mathbf{x} \in C$ for all $\alpha > 0$. A cone that is also convex is a *convex cone*.

Some cones are shown in Fig. B.3. Their basic property is that if a point $\mathbf{x}$ belongs to a cone, then the entire half line from the origin through the point (but not the origin itself) also must belong to the cone.

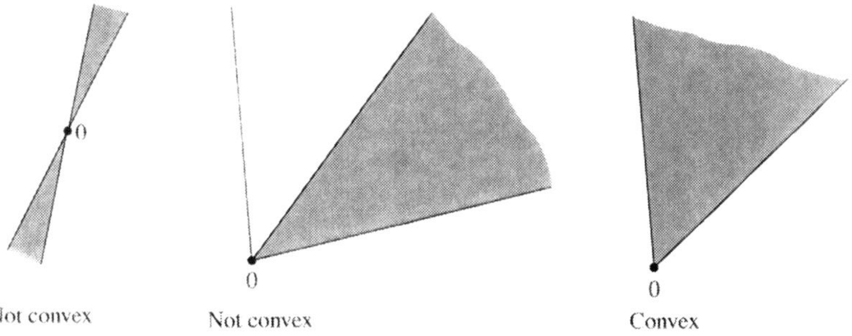

Fig. B.3 Cones

B.2 Hyperplanes and Polytopes

The most important type of convex set (aside from single points) is the hyperplane. Hyperplanes dominate the entire theory of optimization, appearing under the guise of Lagrange multipliers, duality theory, or gradient calculations.

The most natural definition of a hyperplane is the logical generalization of the geometric properties of a plane in three dimensions. We start by giving this geometric definition. For computations and for a concrete description of hyperplanes, however, there is an equivalent algebraic definition that is more useful. A major portion of this section is devoted to establishing this equivalence.

Definition. A set V in E^n is said to be a *linear variety*, if, given any $\mathbf{x}_1$, $\mathbf{x}_2 \in V$, we have $\lambda\mathbf{x}_1 + (1-\lambda)\mathbf{x}_2 \in V$ for all real numbers λ.

Note that the only difference between the definition of a linear variety and a convex set is that in a linear variety the entire line passing through any two points, rather than simply the line segment between them, must lie in the set. Thus in three dimensions the nonempty linear varieties are points, lines, two-dimensional planes, and the whole space. In general, it is clear that we may speak of the dimension of a linear variety. Thus, for example, a point is a linear variety of dimension zero and a line is a linear variety of dimension one. In the general case, the dimension of a linear variety in E^n can be found by translating it (moving it) so that it contains the origin and then determining the dimension of the resulting set, which is then a subspace of E^n.

Definition. A *hyperplane* in E^n is an $(n-1)$-dimensional linear variety.

We see that hyperplanes generalize the concept of a two-dimensional plane in three-dimensional space. They can be regarded as the largest linear varieties in a space, other than the entire space itself.

We now relate this abstract geometric definition to an algebraic one.

Proposition 2. *Let* **a** *be a nonzero n-dimensional column vector, and let c be a real number. The set*

$$H = \{\mathbf{x} \in E^n : \mathbf{a}^T\mathbf{x} = c\}$$

is a hyperplane in E^n.

Proof. It follows directly from the linearity of the equation $\mathbf{a}^T\mathbf{x} = c$ that H is a linear variety. Let $\mathbf{x}_1$ be any vector in H. Translating by $-\mathbf{x}_1$ we obtain the set $M = H - \mathbf{x}_1$ which is a linear subspace of E^n. This subspace consists of all vectors $\mathbf{x}$ satisfying $\mathbf{a}^T\mathbf{x} = 0$; in other words, all vectors orthogonal to $\mathbf{a}$. This is clearly an $(n-1)$-dimensional subspace. ∎

Proposition 3. *Let H be a hyperplane in* E^n. *Then there is a nonzero n- dimensional vector and a constant c such that*

$$H = \{\mathbf{x} \in E^n : \mathbf{a}^T\mathbf{x} = c\}.$$

Proof. Let $\mathbf{x}_1 \in H$ and translate by $-\mathbf{x}_1$ obtaining the set $M = H - \mathbf{x}_1$. Since H is a hyperplane, M is an $(n-1)$-dimensional subspace. Let $\mathbf{a}$ be any nonzero vector that is orthogonal to this subspace, that is, $\mathbf{a}$ belongs to the one-dimensional subspace $M^\perp$. Clearly $M = \{\mathbf{x} : \mathbf{a}^T\mathbf{x} = \mathbf{0}\}$. Letting $c = \mathbf{a}^T\mathbf{x}_1$ we see that if $\mathbf{x}_2 \in H$ we have $\mathbf{x}_2 - \mathbf{x}_1 \in M$ and thus $\mathbf{a}^T\mathbf{x}_2 - \mathbf{a}^T\mathbf{x}_1 = 0$ which implies $\mathbf{a}^T\mathbf{x}_2 = c$. Thus $H \subset \{\mathbf{x} : \mathbf{a}^T\mathbf{x} = c\}$. Since H is, by definition, of dimension $n-1$ and $\{\mathbf{x} : \mathbf{a}^T\mathbf{x} = c\}$ is of dimension $n-1$ by Proposition 2, these two sets must be equal. ∎

Combining Propositions 2 and 3, we see that a hyperplane is the set of solutions to a single linear equation. This is illustrated in Fig. B.4. We now use hyperplanes to build up other important classes of convex sets.

Definition. Let **a** be a nonzero vector in E^n and let c be a real number. Corresponding to the hyperplane $H = \{\mathbf{x} : \mathbf{a}^T\mathbf{x} = c\}$ are the *positive and negative closed half spaces*

$$H_+ = \{\mathbf{x} : \mathbf{a}^T\mathbf{x} \geqslant c\}$$
$$H_- = \{\mathbf{x} : \mathbf{a}^T\mathbf{x} \leqslant c\}$$

and the *positive* and *negative open half spaces*

$$\mathring{H}_+ = \{\mathbf{x} : \mathbf{a}^T\mathbf{x} > c\}$$
$$\mathring{H}_- = \{\mathbf{x} : \mathbf{a}^T\mathbf{x} < c\}.$$

It is easy to see that half spaces are convex sets and that the union of H_+ and H_- is the whole space.

Definition. A set which can be expressed as the intersection of a finite number of closed half spaces is said to be a *convex polytope*.

We see that convex polytopes are the sets obtained as the family of solutions to a set of linear inequalities of the form

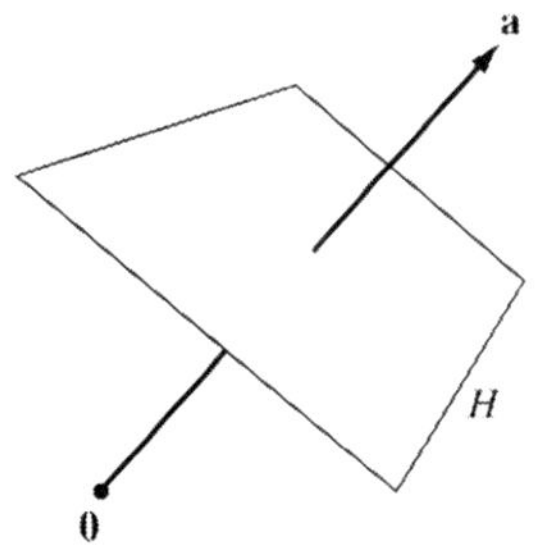

Fig. B.4 Hyperplane

Fig. B.5 Polytopes

$$\begin{aligned} \mathbf{a}_1^T\mathbf{x} &\leqslant b_1 \\ \mathbf{a}_2^T\mathbf{x} &\leqslant b_2 \\ \vdots \quad &\quad \vdots \\ \mathbf{a}_m^T\mathbf{x} &\leqslant b_m, \end{aligned}$$

since each individual inequality defines a half space and the solution family is the intersection of these half spaces. (If some $\mathbf{a}_i = \mathbf{0}$, the resulting set can still, as the reader may verify, be expressed as the intersection of a finite number of half spaces.)

Several polytopes are illustrated in Fig. B.5. We note that a polytope may be empty, bounded, or unbounded. The case of a nonempty bounded polytope is of special interest and we distinguish this case by the following.

Definition. A nonempty bounded polytope is called a *polyhedron*.

B.3 Separating and Supporting Hyperplanes

The two theorems in this section are perhaps the most important results related to convexity. Geometrically, the first states that given a point outside a convex set, a hyperplane can be passed through the point that does not touch the convex set. The second, which is a limiting case of the first, states that given a boundary point of a convex set, there is a hyperplane that contains the boundary point and contains the convex set on one side of it.

Theorem 1. *Let C be a convex set and let* $\mathbf{y}$ *be a point exterior to the closure of C. Then there is a vector* $\mathbf{a}$ *such that* $\mathbf{a}^T\mathbf{y} < \inf_{\mathbf{x}\in C} \mathbf{a}^T\mathbf{x}$.

Proof. Let

$$\delta = \inf_{\mathbf{x}\in C} |\mathbf{x} - \mathbf{y}| > 0.$$

There is an $\mathbf{x}_0$ on the boundary of C such that $|\mathbf{x}_0 - \mathbf{y}| = \delta$. This follows because the continuous function $f(\mathbf{x}) = |\mathbf{x} - \mathbf{y}|$ achieves its minimum over any closed and bounded set and it is clearly only necessary to consider $\mathbf{x}$ in the intersection of the closure of C and the sphere of radius 2δ centered at $\mathbf{y}$.

We shall show that setting $\mathbf{a} = \mathbf{x}_0 - \mathbf{y}$ satisfies the conditions of the theorem. Let $\mathbf{x} \in C$. For any α, $0 \leqslant \alpha \leqslant 1$, the point $\mathbf{x}_0 + \alpha(\mathbf{x} - \mathbf{x}_0) \in \overline{C}$ and thus

$$|\mathbf{x}_0 + \alpha(\mathbf{x} - \mathbf{x}_0) - \mathbf{y}|^2 \geqslant |\mathbf{x}_0 - \mathbf{y}|^2.$$

Expanding,

$$2\alpha(\mathbf{x}_0 - \mathbf{y})^T(\mathbf{x} - \mathbf{x}_0) + \alpha^2|\mathbf{x} - \mathbf{x}_0|^2 \geqslant 0.$$

Thus, considering this as $\alpha \to 0+$, we obtain

$$(\mathbf{x}_0 - \mathbf{y})^T(\mathbf{x} - \mathbf{x}_0) \geqslant 0$$

or,

$$\begin{aligned}(\mathbf{x}_0 - \mathbf{y})^T\mathbf{x} \geqslant (\mathbf{x}_0 - \mathbf{y})^T\mathbf{x}_0 &= (\mathbf{x}_0 - \mathbf{y})^T\mathbf{y} + (\mathbf{x}_0 - \mathbf{y})^T(\mathbf{x}_0 - \mathbf{y}) \\ &= (\mathbf{x}_0 - \mathbf{y})^T\mathbf{y} + \delta^2.\end{aligned}$$

Setting $\mathbf{a} = \mathbf{x}_0 - \mathbf{y}$ proves the theorem. ∎

The geometrical interpretation of Theorem 1 is that, given a convex set C and a point $\mathbf{y}$ exterior to the closure of C, there is a hyperplane containing $\mathbf{y}$ that contains C in one of its open half spaces. We can easily extend this theorem to include the case where $\mathbf{y}$ is a boundary point of C.

Theorem 2. *Let C be a convex set and let* $\mathbf{y}$ *be a boundary point of C. Then there is a hyperplane containing* $\mathbf{y}$ *and containing C in one of its closed half spaces.*

Proof. Let $\{\mathbf{y}_k\}$ be a sequence of vectors, exterior to the closure of C, converging to $\mathbf{y}$. Let $\{\mathbf{a}_k\}$ be the sequence of corresponding vectors constructed according to Theorem 1, normalized so that $|\mathbf{a}_k| = 1$, such that

$$\mathbf{a}_k^T\mathbf{y}_k < \inf_{\mathbf{x}\in C} \mathbf{a}_k^T\mathbf{x}.$$

Since $\{\mathbf{a}_k\}$ is a bounded sequence, it has a convergent subsequence $\{\mathbf{a}_k\}$, $k \in \mathcal{K}$ with limit $\mathbf{a}$. For this vector we have for any $\mathbf{x} \in C$.

$$\mathbf{a}^T\mathbf{y} = \lim_{k\in\mathcal{K}} \mathbf{a}_k^T\mathbf{y}_k \leqslant \lim_{k\in\mathcal{K}} \mathbf{a}_k^T\mathbf{x} = \mathbf{a}\mathbf{x}. \blacksquare$$

Definition. A hyperplane containing a convex set C in one of its closed half spaces and containing a boundary point of C is said to be a *supporting hyperplane* of C.

In terms of this definition, Theorem 2 says that, given a convex set C and a boundary point y of C, there is a hyperplane supporting C at $\mathbf{y}$.

It is useful in the study of convex sets to consider the *relative interior* of a convex set C defined as the largest subset of C that contains no boundary points of C.

Another variation of the theorems of this section is the one that follows, which is commonly known as the Separating Hyperplane Theorem.

Theorem 3. *Let B and C be convex sets with no common relative interior points. (That is the only common points are boundary points.) Then there is a hyperplane separating B and D. In particular, there is a nonzero vector* $\mathbf{a}$ *such that* $\sup_{\mathbf{b}\in B} \mathbf{a}^T\mathbf{b} \le \inf_{\mathbf{c}\in C} \mathbf{a}^T\mathbf{c}$.

Proof. Consider the set $G = C - B$. It is easily shown that G is convex and that $\mathbf{0}$ is not a relative interior point of G. Hence, Theorem 1 or Theorem 2 applies and gives the appropriate hyperplane. ∎

B.4 Extreme Points

Definition. A point $\mathbf{x}$ in a convex set C is said to be an *extreme point* of C if there are no two distinct points $\mathbf{x}_1$ and $\mathbf{x}_2$ in C such that $\mathbf{x} = \alpha\mathbf{x}_1 + (1-\alpha)\mathbf{x}_2$ for some α, $0 < \alpha < 1$.

For example, in E^2 the extreme points of a square are its four corners; the extreme points of a circular disk are all points on the boundary. Note that a linear variety consisting of more than one point has no extreme points.

Lemma 1. *Let C be a convex set, H a supporting hyperplane of C, and T the intersection of H and C. Every extreme point of T is an extreme point of C.*

Proof. Suppose $\mathbf{x}_0 \in T$ is not an extreme point of C. Then $\mathbf{x}_0 = \alpha\mathbf{x}_1 + (1-\alpha)\mathbf{x}_2$ for some $\mathbf{x}_1$, $\mathbf{x}_2 \in C$, $\mathbf{x}_1 \neq \mathbf{x}_2, 0 < \alpha < 1$. Let H be described as $H = \{\mathbf{x} : \mathbf{a}^T\mathbf{x} = c\}$ with C contained in its closed positive half space. Then

$$\mathbf{a}^T\mathbf{x}_1 \geqslant c, \quad \mathbf{a}^T\mathbf{x}_2 \geqslant c.$$

But, since $\mathbf{x}_0 \in H$,

$$c = \mathbf{a}^T\mathbf{x}_0 = \alpha\mathbf{a}^T\mathbf{x}_1 + (1-\alpha)\mathbf{a}^T\mathbf{x}_2,$$

and thus $\mathbf{x}_1$ and $\mathbf{x}_2 \in H$. Hence $\mathbf{x}_1$, $\mathbf{x}_2 \in T$ and $\mathbf{x}_0$ is not an extreme point of T. ∎

Theorem 4. *A closed bounded convex set in E^n is equal to the closed convex hull of its extreme points.*

Proof. The proof is by induction on the dimension of the space E^n. The statement is easily seen to be true for $n = 1$. Suppose that it is true for $n - 1$. Let C be a closed bounded convex set in E^n, and let K be the closed convex hull of the extreme points of C. We wish to show that $K = C$.

Assume there is $\mathbf{y} \in C$ $\mathbf{y} \notin K$. Then by Theorem 1, Sect. B.3, there is a hyperplane separating $\mathbf{y}$ and K; that is, there is $\mathbf{a} \neq \mathbf{0}$, such that $\mathbf{a}^T\mathbf{y} < \inf_{\mathbf{x}\in K} \mathbf{a}^T\mathbf{x}$. Let $c_0 = \inf_{\mathbf{x}\in C}(\mathbf{a}^T\mathbf{x})$. The number c_0 is finite and there is an $\mathbf{x}_0 \in C$ for which $\mathbf{a}^T\mathbf{x}_0 = c_0$, because by Weierstrass' Theorem, the continuous function $\mathbf{a}^T\mathbf{x}$ achieves its minimum over any closed bounded set. Thus the hyperplane $H = \{\mathbf{x} : \mathbf{a}^T\mathbf{x} = c_0\}$ is a supporting hyperplane to C. It is disjoint from K since $c_0 < \inf_{\mathbf{x}\in K}(\mathbf{a}^T\mathbf{x})$.

Let $T = H \cap C$. Then T is a bounded closed convex subset of H which can be regarded as a space of dimension $n - 1$. T is nonempty, since it contains $\mathbf{x}_0$. Thus, by the induction hypothesis, T contains extreme points; and by Lemma 1 these are also extreme points of C. Thus we have found extreme points of C not in K, which is a contradiction. ∎

Let us investigate the implications of this theorem for convex polyhedra. We recall that a convex polyhedron is a bounded polytope. Being the intersection of closed half spaces, a convex polyhedron is also closed. Thus any convex polyhedron is the closed convex hull of its extreme points. It can be shown (see Sect. 2.5) that any polytope has at most a finite number of extreme points and hence a convex polyhedron is equal to the convex hull of a finite number of points. The converse can also be established, yielding the following two equivalent characterizations.

Theorem 5. *A convex polyhedron can be described either as a bounded intersection of a finite number of closed half spaces, or as the convex hull of a finite number of points.*

Appendix C
Gaussian Elimination

This appendix describes the method for solving systems of linear equations that has proved to be, not only the most popular, but also the fastest and least susceptible to round-off error accumulation—the method of Gaussian elimination. Attention is directed toward explaining this classical elimination technique itself and its relation to the theory of **LU** decomposition of a non-singular square matrix.

We first note how easily triangular systems of equations can be solved. Thus the system

$$\begin{aligned} a_{11}x_1 & & & = b_1 \\ a_{21}x_1 + a_{22}x_2 & & & = b_2 \\ \vdots & & & \quad \vdots \\ a_{n1}x_1 + a_{n2}x_2 + \cdots + a_{nn}x_n & & & = b_n \end{aligned}$$

can be solved recursively as follows:

$$\begin{aligned} x_1 &= b_1/a_{11} \\ x_2 &= (b_2 - a_{21}x_1)/a_{22} \\ &\vdots \\ x_n &= (b_n - a_{n1}x_1 - a_{n2}x_2 \ldots - a_{nn-1}x_{n-1})/a_{nn}, \end{aligned}$$

provided that each of the diagonal terms a_{ii}, $i = 1, 2, \ldots, n$ is nonzero (as they must be if the system is nonsingular). This observation motivates us to attempt to reduce an arbitrary system of equations to a triangular one.

Definition. A square matrix $\mathbf{C} = [c_{ij}]$ is said to be *lower triangular* if $c_{ij} = 0$ for $i < j$. Similarly, $\mathbf{C}$ is said to be *upper triangular* if $c_{ij} = 0$ for $i > j$.

In matrix notation, the idea of Gaussian elimination is to somehow find a decomposition of a given $n \times n$ matrix $\mathbf{A}$ in the form $\mathbf{A} = \mathbf{LU}$ where $\mathbf{L}$ is a lower triangular and $\mathbf{U}$ an upper triangular matrix. The system

D.G. Luenberger, Y. Ye, *Linear and Nonlinear Programming*, International Series in Operations Research & Management Science 228,
DOI 10.1007/978-3-319-18842-3

$$\mathbf{Ax} = \mathbf{b} \tag{C.1}$$

can then be solved by solving the two triangular systems

$$\mathbf{Ly} = \mathbf{b}, \quad \mathbf{Ux} = \mathbf{y}. \tag{C.2}$$

The calculation of **L** and **U** together with solution of the first of these systems is usually referred to as *forward elimination*, while solution of the second triangular system is called *back substitution*.

Every nonsingular square matrix **A** has an **LU** decomposition, provided that interchanges of rows of **A** are introduced if necessary. This interchange of rows corresponds to a simple reordering of the system of equations, and hence amounts to no loss of generality in the method. For simplicity of notation, however, we assume that no such interchanges are required.

We turn now to the problem of explicitly determining **L** and **U**, by elimination, for a nonsingular matrix **A**. Given the system, we attempt to transform it so that zeros appear below the main diagonal. Assuming that $a_{11} \neq 0$ we subtract multiples of the first equation from each of the others in order to get zeros in the first column below a_{11}. If we define $m_{k1} = a_{k1}/a_{11}$ and let

$$\mathbf{M}_1 = \begin{bmatrix} 1 & & & \\ -m_{21} & 1 & & \\ -m_{31} & & 1 & \\ \bullet & & & \\ \bullet & & & \\ \bullet & & & \\ -m_{n1} & & & 1 \end{bmatrix},$$

the resulting new system of equations can be expressed as

$$\mathbf{A}^{(2)}\mathbf{x} = \mathbf{b}^{(2)}$$

with

$$\mathbf{A}^{(2)} = \mathbf{M}_1\mathbf{A}, \ \mathbf{b}^{(2)} = \mathbf{M}_1\mathbf{b}.$$

The matrix $\mathbf{A}^{(2)} = [a_{ij}^{(2)}]$ has $a_{k1}^{(2)} = 0, \ k > 1$.

Next, assuming $a_{22}^{(2)} \neq 0$, multiples of the second equation of the new system are subtracted from equations 3 through n to yield zeros below $a_{22}^{(2)}$ in the second column. This is equivalent in premultiplying $\mathbf{A}^{(2)}$ and $\mathbf{b}^{(2)}$ by

$$\mathbf{M}_2 = \begin{bmatrix} 1 & 0 & & \\ 0 & 1 & & \\ \bullet & -m_{32} & 1 & \\ \bullet & -m_{42} & & \\ \bullet & \bullet & & \\ & \bullet & & \\ & \bullet & & \\ & -m_{n2} & & 1 \end{bmatrix},$$

where $m_{k2} = a_{k2}^{(2)}/a_{22}^{(2)}$. This yields $\mathbf{A}^{(3)} = \mathbf{M}_2\mathbf{A}^{(2)}$ and $\mathbf{b}^{(3)} = \mathbf{M}_2\mathbf{A}^{(2)}$.

Proceeding in this way we obtain $\mathbf{A}^{(n)} = \mathbf{M}_{n-1}\mathbf{M}_{n-2}\ldots\mathbf{M}_1\mathbf{A}$, an upper triangular matrix which we denote by $\mathbf{U}$. The matrix $\mathbf{M} = \mathbf{M}_{n-1}\mathbf{M}_{n-2}\ldots\mathbf{M}_1$ is a lower triangular matrix, and since $\mathbf{MA} = \mathbf{U}$ we have $\mathbf{A} = \mathbf{M}^{-1}\mathbf{U}$. The matrix $\mathbf{L} = \mathbf{M}^{-1}$ is also lower triangular and becomes the $\mathbf{L}$ of the desired $\mathbf{LU}$ decomposition for $\mathbf{A}$.

The representation for $\mathbf{L}$ can be made more explicit by noting that $\mathbf{M}_k^{-1}$ is the same as $\mathbf{M}_k$ except that the off-diagonal terms have the opposite sign. Furthermore, we have $\mathbf{L} = \mathbf{M}^{-1} = \mathbf{M}_1^{-1}\mathbf{M}_2^{-1}\ldots\mathbf{M}_{n-1}^{-1}$ which is easily verified to be

$$\mathbf{L} = \begin{bmatrix} 1 & 0 & & & & \\ m_{21} & 1 & & & & \\ m_{31} & m_{32} & 1 & & & \\ \bullet & \bullet & & \bullet & & \\ \bullet & \bullet & & & \bullet & \\ \bullet & \bullet & & & & \bullet \\ m_{n1} & m_{n2} & \bullet\bullet\bullet & & & 1 \end{bmatrix}.$$

Hence $\mathbf{L}$ can be evaluated directly in terms of the calculations required by the elimination process. Of course, an explicit representation for $\mathbf{M} = \mathbf{L}^{-1}$ would actually be more useful but a simple representation for $\mathbf{M}$ does not exist. Thus we content ourselves with the explicit representation for $\mathbf{L}$ and use it in (C.2).

If the original system (C.1) is to be solved for a single $\mathbf{b}$ vector, the vector $\mathbf{y}$ satisfying $\mathbf{Ly} = \mathbf{b}$ is usually calculated simultaneously with $\mathbf{L}$ in the form $\mathbf{y} = \mathbf{b}^{(n)} = \mathbf{Mb}$. The final solution $\mathbf{x}$ is then found by a single back substitution, from $\mathbf{Ux} = \mathbf{y}$. Once the $\mathbf{LU}$ decomposition of $\mathbf{A}$ has been obtained, however, the solution corresponding to any right-hand side can be found by solving the two systems (C.2).

In practice, the diagonal element $a_{kk}^{(k)}$ of $\mathbf{A}^{(k)}$ may become zero or very close to zero. In this case it is important that the kth row be interchanged with a row that is below it. Indeed, for considerations of numerical accuracy, it is desirable to continuously introduce row interchanges of this type in such a way to insure $|m_{ij}| \leqslant 1$ for all i, j. If this is done, the Gaussian elimination procedure has exceptionally good stability properties.

Appendix D
Basic Network Concepts

This appendix describes some of the basic graph and network terminology and concepts necessary for the development of this alternative approach.

A *graph* consists of a finite collection of elements called *nodes* together with a subset of unordered pairs of the nodes called *arcs*. The nodes of a graph are usually numbered, say, $1, 2, 3, \ldots, n$. An arc between nodes i and j is then represented by the unordered pair $(i,\ j)$. A graph is typically represented as shown in Fig. D.1. The nodes are designated by circles, with the number inside each circle denoting the index of that node. The arcs are represented by the lines between the nodes.

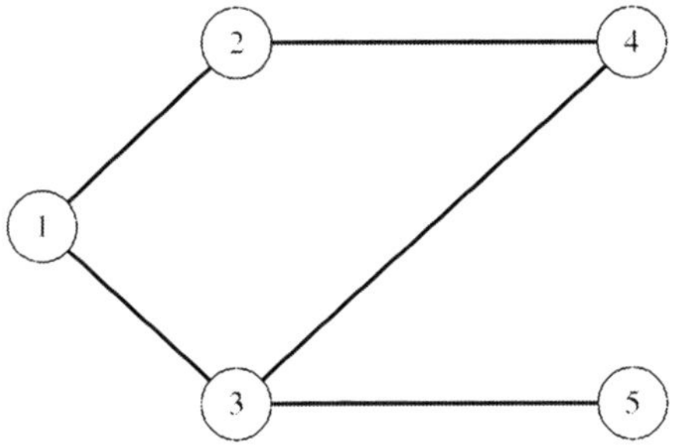

Fig. D.1 A graph

There are a number of other elementary definitions associated with graphs that are useful in describing their structure. A *chain* between nodes i and j is a sequence of arcs connecting them. The sequence must have the form $(i,\ k_1)$, $(k_1,\ k_2)$, $(k_2,\ k_3)$, $\ldots$, $(k_m,\ j)$. In Fig. D.1, (1, 2), (2, 4), (4, 3) is a chain between nodes 1 and 3. If a direction of movement along a chain is specified—say from node i to node j—it is then called a *path* from i to j. A *cycle* is a chain leading from node i back to node i. The chain (1, 2), (2, 4), (4, 3), (3, 1) is a cycle for the graph in Fig. D.1.

D.G. Luenberger, Y. Ye, *Linear and Nonlinear Programming*, International Series in Operations Research & Management Science 228,
DOI 10.1007/978-3-319-18842-3

A graph is *connected* if there is a chain between any two nodes. Thus, the graph of Fig. D.1 is connected. A graph is a *tree* if it is connected and has no cycles. Removal of any one of the arcs (1, 2), (1, 3), (2, 4), (3, 4) would transform the graph of Fig. D.1 into a tree. Sometimes we consider a tree within a graph G, which is just a tree made up of a subset of arcs from G. Such a tree is a *spanning* tree if it touches all nodes of G. It is easy to see that a graph is connected if and only if it contains a spanning tree.

In *directed graphs* a sense of orientation is given to each arc. In this case an arc is considered to be an *ordered pair* of nodes (i, j), and we say that the arc is from node i to node j. This is indicated on the graph by having an arrow on the arc pointing from i to j as shown in Fig. D.2. When working with directed graphs, some node pairs may have an arc in both directions between them. Rather than explicitly indicating both arcs in such a case, it is customary to indicate a single undirected arc. The notions of paths and cycles can be directly applied to directed graphs. In addition we say that node j is *reachable* from i if there is a path from node i to j.

In addition to the visual representation of a directed graph characterized by Fig. D.2, another common method of representation is in terms of a graph's node-arc incidence matrix. This is constructed by listing the nodes vertically and the arcs horizontally. Then in the column under arc (i, j), a $+1$ is placed in the position corresponding to node i and a -1 is placed in the position corresponding to node j. The incidence matrix for the graph of Fig. D.2 is shown in Table D.1.

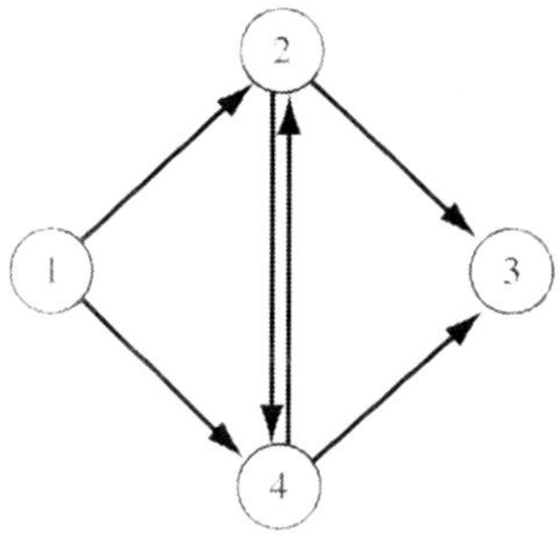

Fig. D.2 A directed graph

	(1,2)	(1,4)	(2,3)	(2,4)	(4,2)
1	1	1			
2	-1		1	1	-1
3			-1		
4		-1		-1	1

Table D.1 Incidence matrix for example

Clearly, all information about the structure of the graph is contained in the node-arc incidence matrix. This representation is often very useful for computational purposes, since it is easily stored in a computer.

D.1 Flows in Networks

A graph is an effective way to represent the communication structure between nodes. When there is the possibility of *flow* along the arcs, we refer to the directed graph as a *network*. In applications the network might represent a transportation system or a communication network, or it may simply be a representation used for mathematical purposes (such as in the assignment problem).

A flow in a given directed arc (i, j) is a number $x_{ij} \geqslant 0$. Flows in the arcs of the network must jointly satisfy a conservation criterion at each node. Specifically, unless the node is a *source* or *sink* as discussed below, flow cannot be created or lost at a node; the total flow into a node must equal the total flow out of the node. Thus at each such node i

$$\sum_{j=1}^{n} x_{ij} - \sum_{k=1}^{n} x_{ki} = 0.$$

The first sum is the total flow *from* i, and the second sum is the total flow *to* i. (Of course x_{ij} does not exist if there is no arc from i to j.) It should be clear that for nonzero flows to exist in a network without sources or sinks, the network must contain a cycle.

In many applications, some nodes are in fact designated as *sources* or *sinks* (or, alternatively, supply nodes or demand nodes). The net flow *out* of a source may be positive, and the level of this net flow may either be fixed or variable, depending on the application. Similarly, the net flow *into* a sink may be positive.

D.2 Tree Procedure

Recall that node j is *reachable* from node i in a directed graph if there is a path from node i to node j. For simple graphs, determination of reachability can be accomplished by inspection, but for large graphs it generally cannot. The problem can be solved systematically by a process of repeatedly labeling and scanning various nodes in the graph. This procedure is the backbone of a number of methods for solving more complex graph and network problems, as illustrated later. It can also be used to establish quickly some important theoretical results.

Assume that we wish to determine whether a path from node 1 to node m exists. At each step of the algorithm, each node is either unlabeled, labeled but unscanned, or labeled and scanned. The procedure consists of these steps:

Step 1. Label node 1 with any mark. All other nodes are unlabeled.

Step 2. For any labeled but unscanned node i, scan the node by finding all unlabeled nodes reachable from i by a single arc. Label these nodes with an i.

Step 3. If node m is labeled, stop; a *breakthrough* has been achieved—a path exists. If no unlabeled nodes can be labeled, stop; no connecting path exists. Otherwise, go to Step 2.

The process is illustrated in Fig. D.3, where a path between nodes 1 and 10 is sought. The nodes have been labeled and scanned in the order 1, 2, 3, 5, 6, 8, 4, 7, 9, 10. The labels are indicated close to the nodes. The arcs that were used in the scanning processes are indicated by heavy lines. Note that the collection of nodes and arcs selected by the process, regarded as an undirected graph, form a tree—a graph without cycles. This, of course, accounts for the name of the process, the tree procedure. If one is interested only in determining whether a connecting path exists and does not need to find the path itself, then the labels need only be simple check marks rather than node indices. However, if node indices are used as labels, then after successful completion of the algorithm, the actual connecting path can be found by tracing backward from node m by following the labels. In the example, one begins at 10 and moves to node 7 as indicated; then to 6, 3, and 1. The path follows the reverse of this sequence.

It is easy to prove that the algorithm does indeed resolve the issue of the existence of a connecting path. At each stage of the process, either a new node is labeled, it is impossible to continue, or node m is labeled and the process is successfully terminated. Clearly, the process can continue for at most $n - 1$ stages, where n is the number of nodes in the graph. Suppose at some stage it is impossible to continue. Let S be the set of labeled nodes at that stage and let $\bar{S}$ be the set of unlabeled nodes. Clearly, node 1 is contained in S, and node m is contained in $\bar{S}$. If there were a path connecting node 1 with node m, then there must be an arc in that path from a node k in S to a node in $\bar{S}$. However, this would imply that node k was not scanned, which is a contradiction. Conversely, if the algorithm does continue until reaching node m, then it is clear that a connecting path can be constructed backward as outlined above.

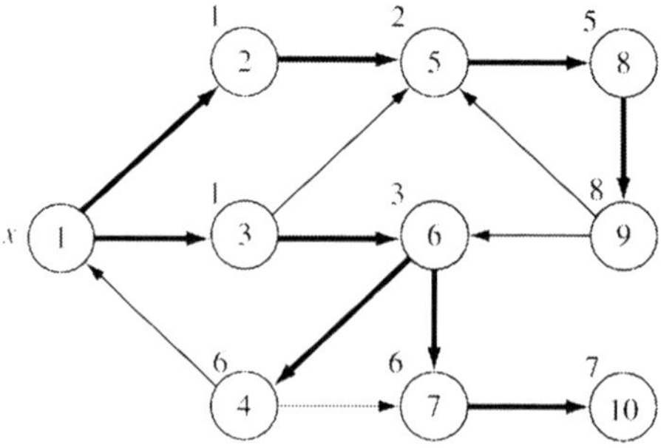

Fig. D.3 The scanning procedure

D.3 Capacitated Networks

In some network applications it is useful to assume that there are upper bounds on the allowable flow in various arcs. This motivates the concept of a capacitated network. A *capacitated network* is a network in which some arcs are assigned non-negative capacities, which define the maximum allowable flow in those arcs. The capacity of an arc (i, j) is denoted k_{ij}, and this capacity is indicated on the graph by placing the number k_{ij} adjacent to the arc. Figure 2.1 shows an example of a network with the capacities indicated. Thus the capacity from node 1 to node 2 is 12, while that from node 2 to node 1 is 6.

Bibliography

[A1] J. Abadie, J. Carpentier, Generalization of the Wolfe reduced gradient method to the case of nonlinear constraints, in *Optimization*, ed. by R. Fletcher (Academic, London, 1969), pp. 37–47

[A2] H. Akaike, On a successive transformation of probability distribution and its application to the analysis of the optimum gradient method. Ann. Inst. Stat. Math. **11**, 1–17 (1959)

[3] A.Y. Alfakih, A. Khandani, H. Wolkowicz, Solving euclidean distance matrix completion problems via semidefinite programming. Comput. Opt. Appl. **12**, 13–30 (1999)

[A3] F. Alizadeh, Combinatorial optimization with interior point methods and semi-definite matrices, Ph.D. thesis, University of Minnesota, Minneapolis, 1991

[A4] F. Alizadeh, Optimization over the positive semi-definite cone: interior-point methods and combinatorial applications, in *Advances in Optimization and Parallel Computing*, ed. by P.M. Pardalos (North Holland, Amsterdam, 1992), pp. 1–25

[6] E.D. Andersen, MOSEK: high performance software for large-scale LP, QP, SOCP, SDP and MIP, `http://www.mosek.com/` (1997)

[A5] E.D. Andersen, Y. Ye, On a homogeneous algorithm for the monotone complementarity problem. Math. Prog. **84**, 375–400 (1999)

[A6] K.M. Anstreicher, D. den Hertog, C. Roos, T. Terlaky, A long step barrier method for convex quadratic programming. Algorithmica **10**, 365–382 (1993)

[A7] H.A. Antosiewicz, W.C. Rheinboldt, Numerical analysis and functional analysis, in *Survey of Numerical Analysis*, ed. by J. Todd, Chap. 14 (McGraw-Hill, New York, 1962)

[A8] L. Armijo, Minimization of functions having lipschitz continuous first-partial derivatives. Pac. J. Math. **16**(1), 1–3 (1966)

D.G. Luenberger, Y. Ye, *Linear and Nonlinear Programming*, International Series in Operations Research & Management Science 228,
DOI 10.1007/978-3-319-18842-3

[A9] K.J. Arrow, L. Hurwicz, Gradient method for concave programming, I.: local results, in *Studies in Linear and Nonlinear Programming*, ed. by K.J. Arrow, L. Hurwicz, H. Uzawa (Stanford University Press, Stanford, CA, 1958)

[12] E. Balas, S. Ceria, G. Cornuejols, A lift-and-project cutting plane algorithm for mixed 0-1 programs. Math. Program. **58**, 295–324 (1993)

[B1] R.H. Bartels, A numerical investigation of the simplex method. Technical Report No. CS 104, Computer Science Department, Stanford University, Stanford, CA (31 July 1968)

[B2] R.H. Bartels, G.H Golub, The simplex method of linear programming using LU decomposition. Commun. ACM **12**(5), 266–268 (1969)

[15] A. Barvinok, A remark on the rank of positive semidefinite matrices subject to affine constraints. Discrete Comput. Geom. **25**, 23–31 (2001)

[16] A. Barvinok, *A Course in Convexity*. Graduate Studies in Mathematics, vol. 54 (American Mathematical Society, Providence, RI, 2002)

[17] J. Barzilai, J.M. Borwein, Two-point step size gradient methods. IMA J. Numer. Anal. **8**, 141–148 (2008)

[B3] D.A., Bayer, J.C. Lagarias, The nonlinear geometry of linear programming, part i: affine and projective scaling trajectories. Trans. Am. Math. Soc **314**(2), 499–526 (1989)

[B4] D.A. Bayer, J.C. Lagarias, The nonlinear geometry of linear programming, part ii: legendre transform coordinates. Trans. Am. Math. Soc. **314**(2), 527–581 (1989)

[B5] M.S. Bazaraa, J.J. Jarvis, *Linear Programming and Network Flows* (Wiley, New York, 1977)

[B6] M.S. Bazaraa, J.J. Jarvis, H.F. Sherali, Karmarkar's projective algorithm (Chap. 8.4), pp. 380–394; Analysis of Karmarkar's algorithm (Chap. 8.5), pp. 394–418, in *Linear Programming and Network Flows*, 2nd edn. (Wiley New York, 1990)

[B7] E.M.L. Beale, in *Numerical Methods, Nonlinear Programming*, ed. by J. Abadie (North-Holland, Amsterdam, 1967)

[23] A. Beck, M. Teboulle, A fast iterative shrinkage-thresholding algorithm for linear inverse problems. SIAM J. Imaging Sci. **2**, 183–202 (2009)

[B8] F.S. Beckman, The solution of linear equations by the conjugate gradient method, in *Mathematical Methods for Digital Computers*, ed. by A. Ralston, H.S. Wilf, vol. 1 (Wiley, New York, 1960)

[25] S.J. Benson, Y. Ye, X. Zhang, Solving large-scale sparse semidefinite programs for combinatorial optimization. SIAM J. Optim. **10**, 443–461 (2000)

[26] A. Ben-Tal, A. Nemirovski, Structural design via semidefinite programming, in *Handbook on Semidefinite Programming* (Kluwer, Boston, 2000), pp. 443–467

[B9] D.P. Bertsekas, Partial conjugate gradient methods for a class of optimal control problems. IEEE Trans. Autom. Control **19**, 209–217 (1973)

[B10] D.P. Bertsekas, Multiplier methods: a survey. Automatica **12**(2), 133–145 (1976)

[B11] D.P. Bertsekas, *Constrained Optimization and Lagrange Multiplier Methods* (Academic, New York, 1982)

[B12] D.P. Bertsekas, *Nonlinear Programming* (Athena Scientific, Belmont, 1995)

[31] D. Bertsimas, Y. Ye, Semidefinite relaxations, multivariate normal distributions, and order statistics, in *Handbook of Combinatorial Optimization* (Springer, New York, 1999), pp. 1473–1491

[B13] D.M. Bertsimas, J.N. Tsitsiklis, *Linear Optimization* (Athena Scientific, Belmont, 1997)

[B14] M.C. Biggs, Constrained minimization using recursive quadratic programming: some alternative sub-problem formulations, in *Towards Global Optimization*, ed. by L.C.W. Dixon, G.P. Szego (North-Holland, Amsterdam, 1975)

[B15] M.C. Biggs, On the convergence of some constrained minimization algorithms based on recursive quadratic programming. J. Inst. Math. Appl. **21**, 67–81 (1978)

[B16] G. Birkhoff, Three observations on linear algebra. Rev. Univ. Nac. Tucumán, Ser. A. **5**, 147–151 (1946)

[B17] P. Biswas, Y. Ye, Semidefinite programming for ad hoc wireless sensor network localization, in *Proceedings of the 3rd IPSN*, 2004, pp. 46–54

[B18] R.E. Bixby, Progress in linear programming. ORSA J. Comput. **6**(1), 15–22 (1994)

[B19] R.G. Bland, New finite pivoting rules for the simplex method. Math. Oper. Res. **2**(2), 103–107 (1977)

[B20] R.G. Bland, D. Goldfarb, M.J. Todd, The ellipsoidal method: a survey. Oper. Res. **29**, 1039–1091 (1981)

[B21] L. Blum, F. Cucker, M. Shub, S. Smale, *Complexity and Real Computation* (Springer, New York, 1996)

[41] S. Boyd, N. Parikh, E. Chu, B. Peleato, J. Eckstein, Distributed optimization and statistical learning via the alternating direction method of multipliers. Found. Trends Mach. Learn. **3**, 1–122 (2010)

[B22] S. Boyd, L.E. Ghaoui, E. Feron, V. Balakrishnan, *Linear Matrix Inequalities in System and Control Science* (SIAM, Philadelphia, 1994)

[B23] S. Boyd, L. Vandenberghe, *Convex Optimization* (Cambridge University Press, Cambridge, 2004)

[B24] C.G. Broyden, Quasi-Newton methods and their application to function minimization. Math. Comput. **21**, 368–381 (1967)

[B25] C.G. Broyden, The convergence of a class of double rank minimization algorithms: parts I and II. J. Inst. Math. Appl. **6**, 76–90, 222–231 (1970)

[B26] T. Butler, A.V. Martin, On a method of courant for minimizing functionals. J. Math. Phys. **41**, 291–299 (1962)

[C1] C.W. Carroll, The created response surface technique for optimizing nonlinear restrained systems. Oper. Res. **9**(12), 169–184 (1961)

[C2] A. Charnes, Optimality and degeneracy in linear programming. Econometrica **20**, 160–170 (1952)

[C3] A. Charnes, C.E. Lemke, The bounded variables problem. ONR Research Memorandum 10, Graduate School of Industrial Administration, Carnegie Institute of Technology, Pittsburgh (1954)

[50] C.H. Chen, B.S. He, Y.Y. Ye, X.M. Yuan, The direct extension of ADMM for multi-block convex minimization problems is not necessarily convergent. Math. Program. (2014). doi:10.1007/s10107-014-0826-5

[C4] A. Cohen, Rate of convergence for root finding and optimization algorithms. Ph.D. dissertation, University of California, Berkeley, 1970

[C5] S.A. Cook, The complexity of theorem-proving procedures, in *Proceedings of 3rd ACM Symposium on the Theory of Computing*, 1971, pp. 151–158

[C6] R.W. Cottle, *Linear Programming*. Lecture Notes for MS& E 310 (Stanford University, Stanford, 2002)

[C7] R. Cottle, J.S. Pang, R.E. Stone, Interior-Point Methods (Chap. 5.9), in *The Linear Complementarity Problem* (Academic, Boston, 1992), pp. 461–475

[C8] R. Courant, Calculus of variations and supplementary notes and exercises (mimeographed notes), supplementary notes by M. Kruskal and H. Rubin, revised and amended by J. Moser, New York University (1962)

[C9] J.B. Crockett, H. Chernoff, Gradient methods of maximization. Pac. J. Math. **5**, 33–50 (1955)

[C10] H. Curry, The method of steepest descent for nonlinear minimization problems. Q. Appl. Math. **2**, 258–261 (1944)

[58] Y.-H. Dai, R. Fletcher, Projected Barzilai-Borwein methods for large-scale box-constrained quadratic programming. Numer. Math. **100**, 21–47 (2005)

[D1] J.W. Daniel, The conjugate gradient method for linear and nonlinear operator equations. SIAM J. Numer. Anal. **4**(1), 10–26 (1967)

[D2] G.B. Dantzig, Maximization of a linear function of variables subject to linear inequalities (Chap. XXI), in *Activity Analysis of Production and Allocation*, ed. by T.C. Koopmans. Cowles Commission Monograph, vol. 13 (Wiley, New York, 1951)

[D3] G.B., Dantzig, Application of the simplex method to a transportation problem, in *Activity Analysis of Production and Allocation*, ed. by T.C. Koopmans (Wiley, New York, 1951), pp. 359–373

[D4] G.B. Dantzig, Computational algorithm of the revised simplex method. RAND Report RM-1266, The RAND Corporation, Santa Monica, CA (1953)

[D5] G.B. Dantzig, Variables with upper bounds in linear programming. RAND Report RM-1271, The RAND Corporation, Santa Monica, CA (1954)

[D6] G.B. Dantzig, *Linear Programming and Extensions* (Princeton University Press, Princeton, 1963)

[D7] G.B. Dantzig, L.R. Ford Jr., D.R. Fulkerson, A primal-dual algorithm, in *Linear Inequalities and Related Systems*. Annals of Mathematics Study, vol. 38 (Princeton University Press, Princeton, NJ, 1956), pp. 171–181

[D8] G.B. Dantzig, A. Orden, P. Wolfe, Generalized simplex method for minimizing a linear form under linear inequality restraints. RAND Report RM-1264, The RAND Corporation, Santa Monica, CA (1954)

[D9] G.B. Dantzig, M.N. Thapa, *Linear Programming 1: Introduction* (Springer, New York, 1997)

[D10] G.B. Dantzig, M.N. Thapa, *Linear Programming 2: Theory and Extensions* (Springer, New York, 2003)

[D11] G.B. Dantzig, P. Wolfe, Decomposition principle for linear programs. Oper. Res. **8**, 101–111 (1960)

[D12] W.C. Davidon, Variable metric method for minimization. Research and Development Report ANL-5990 (Ref.) U.S. Atomic Energy Commission, Argonne National Laboratories (1959)

[D13] W.C. Davidon, Variance algorithm for minimization. Comput. J. **10**, 406–410 (1968)

[72] E. de Klerk, C. Roos, T. Terlaky, Initialization in semidefinite programming via a self–dual skew–symmetric embedding. Oper. Res. Lett. **20**, 213–221 (1997)

[D14] R.S. Dembo, S.C. Eisenstat, T. Steinhaug, Inexact Newton methods. SIAM J. Numer. Anal. **19**(2), 400–408 (1982)

[D15] J.E. Dennis Jr., J.J. Moré, Quasi-Newton methods, motivation and theory. SIAM Rev. **19**, 46–89 (1977)

[D16] J.E. Dennis Jr., R.E. Schnabel, Least change secant updates for quasi-Newton methods. SIAM Rev. **21**, 443–469 (1979)

[D17] L.C.W. Dixon, Quasi-Newton algorithms generate identical points. Math. Prog. **2**, 383–387 (1972)

[E1] B.C. Eaves, W.I. Zangwill, Generalized cutting plane algorithms. Working Paper No. 274, Center for Research in Management Science, University of California, Berkeley (July 1969)

[78] J. Eckstein, D.P. Bertsekas, On the Douglas-Rachford splitting method and the proximal point algorithm for maximal monotone operators. Math. Program. **55**, 293–318 (1992)

[E2] H. Everett III, Generalized lagrange multiplier method for solving problems of optimum allocation of resources. Oper. Res. **11**, 399–417 (1963)

[F1] D.K. Faddeev, V.N. Faddeeva, *Computational Methods of Linear Algebra* (W. H. Freeman, San Francisco, CA, 1963)

[F2] S.C. Fang, S. Puthenpura, *Linear Optimization and Extensions* (Prentice-Hall, Englewood Cliffs, NJ, 1994)

[F3] W. Fenchel, *Convex Cones, Sets, and Functions.* Lecture Notes (Department of Mathematics, Princeton University, Princeton, NJ, 1953)

[F4] A.V. Fiacco, G.P. McCormick, *Nonlinear Programming: Sequential Unconstrained Minimization Techniques* (Wiley, New York, 1968)

[F5] A.V. Fiacco, G.P. McCormick, *Nonlinear Programming: Sequential Unconstrained Minimization Techniques* (Wiley, New York, 1968). Reprint: Volume 4 of *SIAM Classics in Applied Mathematics* (SIAM Publications, Philadelphia, PA, 1990)

[F6] R. Fletcher, A new approach to variable metric algorithms. Comput. J. **13**(13), 317–322 (1970)

[F7] R. Fletcher, An exact penalty function for nonlinear programming with inequalities. Math. Program. **5**, 129–150 (1973)

[F8] R. Fletcher, Conjugate gradient methods for indefinite systems. Numerical Analysis Report, 11. Department of Mathematics, University of Dundee, Scotland (September 1975)

[F9] R. Fletcher, *Practical Methods of Optimization 1: Unconstrained Optimization* (Wiley, Chichester, 1980)

[F10] R. Fletcher, *Practical Methods of Optimization 2: Constrained Optimization* (Wiley, Chichester, 1981)

[F11] R. Fletcher, M.J.D. Powell, A rapidly convergent descent method for minimization. Comput. J. **6**, 163–168 (1963)

[F12] R. Fletcher, C.M. Reeves, Function minimization by conjugate gradients. Comput. J. **7**, 149–154 (1964)

[F13] L.K. Ford Jr., D.K. Fulkerson, *Flows in Networks* (Princeton University Press, Princeton, NJ, 1962)

[F14] G.E. Forsythe, On the asymptotic directions of the s-dimensional optimum gradient method. Numer. Math. **11**, 57–76 (1968)

[F15] G.E. Forsythe, C.B. Moler, *Computer Solution of Linear Algebraic Systems* (Prentice-Hall, Englewood Cliffs, NJ, 1967)

[F16] G.E. Forsythe, W.R. Wasow, *Finite-Difference Methods for Partial Differential Equations* (Wiley, New York, 1960)

[96] M. Fortin, R. Glowinski, On decomposition-coordination methods using an augmented Lagrangian, in *Augmented Lagrangian Methods: Applications to the Solution of Boundary Problems*, ed. by M. Fortin, R. Glowinski (North-Holland, Amsterdam, 1983)

[F17] K. Fox, *An Introduction to Numerical Linear Algebra* (Clarendon Press, Oxford, 1964)

[98] M. Frank, P. Wolfe, An algorithm for quadratic programming. Naval Res. Logist. Q. **3**, 95–110 (1956)

[F18] R.M. Freund, Polynomial-time algorithms for linear programming based only on primal scaling and projected gradients of a potential function. Math. Program. **51**, 203–222 (1991)

[F19] K.R. Frisch, The logarithmic potential method for convex programming. Unpublished Manuscript, Institute of Economics, University of Oslo, Oslo (1955)

[G1] D. Gabay, Reduced quasi-Newton methods with feasibility improvement for nonlinear constrained optimization, in *Mathematical Programming Studies*, vol. 16 (North-Holland, Amsterdam, 1982), pp. 18–44

[102] D. Gabay, B. Mercier, A dual algorithm for the solution of nonlinear variational problems via finite element approximations. Comput. Math. Appl. **2**, 17–40 (1976)

[G2] D. Gale, *The Theory of Linear Economic Models* (McGraw-Hill, New York, 1960)

[G3] U.M. García-Palomares, O.L. Mangasarian, Superlinearly convergent quasi-Newton algorithms for nonlinearly constrained optimization problems. Math. Program. **11**, 1–13 (1976)

[G4] S.I. Gass, *Linear Programming*, 3rd edn. (McGraw-Hill, New York, 1969)

[G5] P.E. Gill, W. Murray, M.A. Saunders, J.A. Tomlin, M.H. Wright, On projected Newton barrier methods for linear programming and an equivalence to Karmarkar's projective method. Math. Program. **36**, 183–209 (1986)

[G6] P.E., Gill, W. Murray, Quasi-Newton methods for unconstrained optimization. J. Inst. Math. Appl. **9**, 91–108 (1972)

[G7] P.E. Gill, W. Murray, M.H. Wright, *Practical Optimization* (Academic, London, 1981)

[109] R. Glowinski, A. Marrocco, Approximation par éléments finis d'ordre un et résolution par pénalisation-dualité d'une classe de problèmes non linéaires. R.A.I.R.O. R2 **2**, 41–76 (1975)

[G8] M.X. Goemans, D.P. Williamson, Improved approximation algorithms for maximum cut and satisfiability problems using semidefinite programming. J. Assoc. Comput. Mach. **42**, 1115–1145 (1995)

[G9] D. Goldfarb, A family of variable metric methods derived by variational means. Math. Comput. **24**, 23–26 (1970)

[112] D. Goldfarb, G. Iyengar, Robust portfolio selection problems. Math. Oper. Res. **28**, 1–38 (2002)

[G10] D. Goldfarb, M.J. Todd, Linear programming, in *Optimization*, ed. by G.L. Nemhauser, A.H.G. Rinnooy Kan, M.J. Todd. Handbooks in Operations Research and Management Science, vol. 1 (North Holland, Amsterdam, 1989), pp. 141–170

[G11] D. Goldfarb, D. Xiao, A primal projective interior point method for linear programming. Math. Program. **51**, 17–43 (1991)

[G12] A.A. Goldstein, On steepest descent. SIAM J. Control **3**, 147–151 (1965)

[G13] C.C. Gonzaga, An algorithm for solving linear programming problems in $O(n^3L)$ operations, in *Progress in Mathematical Programming: Interior Point and Related Methods*, ed. by N. Megiddo (Springer, New York, 1989), pp. 1–28

[G14] C.C. Gonzaga, M.J. Todd, An $O(\sqrt{n}L)$-iteration large-step primal-dual affine algorithm for linear programming. SIAM J. Optim. **2**, 349–359 (1992)

[G15] J. Greenstadt, Variations on variable metric methods. Math. Comput. **24**, 1–22 (1970)

[H1] G. Hadley, *Linear Programming* (Addison-Wesley, Reading, MA, 1962)

[H2] G. Hadley, *Nonlinear and Dynamic Programming* (Addison-Wesley, Reading, MA, 1964)

[H3] S.P. Han, A globally convergent method for nonlinear programming. J. Optim. Theory Appl. **22**(3), 297–309 (1977)

[H4] H. Hancock, *Theory of Maxima and Minima* (Ginn, Boston, 1917)

[H5] J. Hartmanis, R.E. Stearns, On the computational complexity of algorithms. Trans. Am. Math. Soc. **117**, 285–306 (1965)

[124] B.S. He, X.M. Yuan, On the $O(1/n)$ convergence rate of the Douglas-Rachford alternating direction method. SIAM J. Numer. Anal. **50**, 700–709 (2012)

[H6] D. den Hertog, Interior point approach to linear, quadratic and convex programming, algorithms and complexity, Ph.D. thesis, Faculty of Mathematics and Informatics, TU Delft, BL Delft, 1992

[H7] M.R. Hestenes, The conjugate gradient method for solving linear systems, in *Proceeding of Symposium in Applied Mathematics*, vol. VI, Numerical Analysis (McGraw-Hill, New York 1956), pp. 83–102

[H8] M.R. Hestenes, Multiplier and gradient methods. J. Opt. Theory Appl. **4**(5), 303–320 (1969)

[H9] M.R. Hestenes, *Conjugate-Direction Methods in Optimization* (Springer, Berlin, 1980)

[H10] M.R. Hestenes, E.L. Stiefel, Methods of conjugate gradients for solving linear systems. J. Res. Natl. Bur. Stand. Sect. B **49**, 409–436 (1952)

[H11] F.L. Hitchcock, The distribution of a product from several sources to numerous localities. J. Math. Phys. **20**, 224–230 (1941)

[H12] P. Huard, Resolution of mathematical programming with nonlinear constraints by the method of centers, in *Nonlinear Programming*, ed. by J. Abadie (North Holland, Amsterdam, 1967), pp. 207–219

[H13] H.Y. Huang, Unified approach to quadratically convergent algorithms for function minimization. J. Optim. Theory Appl. **5**, 405–423 (1970)

[H14] L. Hurwicz, Programming in linear spaces, in *Studies in Linear and Nonlinear Programming*, ed. by K.J. Arrow, L. Hurwicz, H. Uzawa (Stanford University Press, Stanford, CA, 1958)

[I1] E. Isaacson, H.B. Keller, *Analysis of Numerical Methods* (Wiley, New York, 1966)

[J1] W. Jacobs, The caterer problem. Naval Res. Logist. Q. **1**, 154–165 (1954)

[J2] F. Jarre, Interior-point methods for convex programming. Appl. Math. Optim. **26**, 287–311 (1992)

[137] W.B. Johnson, J. Lindenstrauss, Extensions of lipshitz mapping into Hilbert space. Comtemp. Math. **26**, 189–206 (1984)

[K1] S. Karlin, *Mathematical Methods and Theory in Games, Programming, and Economics*, vol. I (Addison-Wesley, Reading, MA, 1959)

[K2] N.K. Karmarkar, A new polynomial-time algorithm for linear programming. Combinatorica **4**, 373–395 (1984)

[K3] J.E. Kelley, The cutting-plane method for solving convex programs. J. Soc. Ind. Appl. Math. **VIII**(4), 703–712 (1960)

[K4] L.G. Khachiyan, A polynomial algorithm for linear programming. Doklady Akad. Nauk USSR **244**, 1093–1096 (1979). Translated in Soviet Math. Doklady **20**, 191–194 (1979)

[K5] V. Klee, G.J. Minty, How good is the simplex method, in *Inequalities III*, ed. by O. Shisha (Academic, New York, 1972)

[K6] M. Kojima, S. Mizuno, A. Yoshise, A polynomial-time algorithm for a class of linear complementarity problems. Math. Program. **44**, 1–26 (1989)

[K7] M. Kojima, S. Mizuno, A. Yoshise, An $O(\sqrt{n}L)$ iteration potential reduction algorithm for linear complementarity problems. Math. Program. **50**, 331–342 (1991)

[K8] T.C. Koopmans, Optimum utilization of the transportation system, in *Proceedings of the International Statistical Conference*, Washington, DC, 1947

[K9] J. Kowalik, M.R. Osborne, *Methods for Unconstrained Optimization Problems* (Elsevier, New York, 1968)

[K10] H.W. Kuhn, The Hungarian method for the assignment problem. Naval Res. Logist. Q. **2**, 83–97 (1955)

[K11] H.W. Kuhn, A.W. Tucker, Nonlinear programming, in *Proceedings of the Second Berkeley Symposium on Mathematical Statistics and Probability*, ed. by J. Neyman (University of California Press, Berkeley/Los Angeles, CA, 1961), pp. 481–492

[L1] C. Lanczos, *Applied Analysis* (Prentice-Hall, Englewood Cliffs, NJ, 1956)

[150] J.B. Lasserre, Global optimization with polynomials and the problem of moments related. SIAM J. Optim. **11**, 796–817 (2001)

[151] M. Laurent, Matrix completion problems. Encycl. Optim. **3**, 221–229 (2001)

[L2] E. Lawler, *Combinatorial Optimization: Networks and Matroids* (Holt, Rinehart, and Winston, New York, 1976)

[L3] C. Lemarechal, R. Mifflin, Nonsmooth optimization, in *IIASA Proceedings III* (Pergamon Press, Oxford, 1978)

[L4] C.E. Lemke, The dual method of solving the linear programming problem. Naval Res. Logist. Q. **1**(1), 36–47 (1954)

[L5] E.S. Levitin, B.T. Polyak, Constrained minimization methods. Zh. vychisl. Math. Math. Fiz **6**(5), 787–823 (1966)

[156] M.S. Lobo, L. Vandenberghe, S. Boyd, Applications of second-order cone programming. Linear Algebra Appl. **284**, 193–228 (1998)

[L6] C. Loewner, Über monotone Matrixfunktionen. Math. Zeir. **38**, 177–216 (1934). Also see C. Loewner, Advanced matrix theory, mimeo notes, Stanford University, 1957

[L7] F.A. Lootsma, Boundary properties of penalty functions for constrained minimization, Doctoral dissertation, Technical University, Eindhoven, May 1970

[159] L. Lovász, A. Shrijver, Cones of matrices and setfunctions, and $0-1$ optimization. SIAM J. Optim. **1**, 166–190 (1990)

[160] Z. Lu, L. Xiao, On the complexity analysis of randomized block-coordinate descent methods. Math. Program. (2013). doi: 10.1007/s10107-014-0800-2

[L8] D.G. Luenberger, *Optimization by Vector Space Methods* (Wiley, New York, 1969)

[L9] D.G. Luenberger, Hyperbolic pairs in the method of conjugate gradients. SIAM J. Appl. Math. **17**(6), 1263–1267 (1969)

[L10] D.G. Luenberger, A combined penalty function and gradient projection method for nonlinear programming, Internal Memo, Department of Engineering-Economic Systems, Stanford University (June 1970)

[L11] D. G. Luenberger, The conjugate residual method for constrained minimization problems. SIAM J. Numer. Anal. **7**(3), 390–398 (1970)

[L12] D.G. Luenberger, Control problems with kinks. IEEE Trans. Autom. Control **AC-15**(5), 570–575 (1970)

[L13] D.G. Luenberger, Convergence rate of a penalty-function scheme. J. Optim. Theory Appl. **7**(1), 39–51 (1971)

[L14] D.G. Luenberger, The gradient projection method along geodesics. Manag. Sci. **18**(11), 620–631 (1972)

[L15] D.G. Luenberger, *Introduction to Linear and Nonlinear Programming*, 1st edn. (Addison-Wesley, Reading, MA, 1973)

[L16] D.G. Luenberger, *Linear and Nonlinear Programming*, 2nd edn. (Addison-Wesley, Reading, MA, 1984)

[L17] D.G. Luenberger, An approach to nonlinear programming. J. Optim. Theory Appl. **11**(3), 219–227 (1973)

[171] Z.Q. Luo, W. Ma, A.M. So, Y. Ye, S. Zhang, Semidefinite relaxation of quadratic optimization problems. IEEE Signal Process. Mag. **27**, 20–34 (2010)

[L18] Z.Q. Luo, J. Sturm, S. Zhang, Conic convex programming and self-dual embedding. Optim. Methods Softw. **14**, 169–218 (2000)

[L19] I.J. Lustig, R.E. Marsten, D.F. Shanno, On implementing mehrotra's predictor-corrector interior point method for linear programming. SIAM J. Optim. **2**, 435–449 (1992)

[M1] N. Maratos, Exact penalty function algorithms for finite dimensional and control optimization problems, Ph.D. thesis, Imperial College Sci. Tech., University of London, 1978

[M2] G.P. McCormick, Optimality criteria in nonlinear programming, in *Nonlinear Programming, SIAM-AMS Proceedings*, vol. IX, 1976, pp. 27–38

[M3] L. McLinden, The analogue of *Moreau's* proximation theorem, with applications to the nonlinear complementarity problem. Pac. J. Math. **88**, 101–161 (1980)

[M4] N. Megiddo, Pathways to the optimal set in linear programming, in *Progress in Mathematical Programming: Interior Point and Related Methods*, ed. by N. Megiddo (Springer, New York, 1989), pp. 131–158

[M5] S. Mehrotra, On the implementation of a primal-dual interior point method. SIAM J. Optim. **2**(4), 575–601 (1992)

[M6] S. Mizuno, M.J. Todd, Y. Ye, On adaptive step primal-dual interior point algorithms for linear programming. Math. Oper. Res. **18**, 964–981 (1993)

[180] R.D.C. Monteiro, B.F. Svaiter, Iteration-complexity of block-decomposition algorithms and the alternating direction method of multipliers. SIAM J. Optim. **23**, 475–507 (2013)

[M7] R.D.C. Monteiro, I. Adler, Interior path following primal-dual algorithms: part i: linear programming. Math. Program. **44**, 27–41 (1989)

[M8] D.D. Morrison, Optimization by least squares. SIAM J. Numer. Anal. **5**, 83–88 (1968)

[M9] B.A. Murtagh, *Advanced Linear Programming* (McGraw-Hill, New York, 1981)

[M10] B.A. Murtagh, R.W.H. Sargent, A constrained minimization method with quadratic convergence (Chap. 14), in *Optimization*, ed. by R. Fletcher (Academic, London, 1969)

[M11] K.G. Murty, *Linear and Combinatorial Programming* (Wiley, New York, 1976)

[M12] K.G. Murty, The Karmarkar's algorithm for linear programming (Chap. 11.4.1), in *Linear Complementarity, Linear and Nonlinear Programming*. Sigma Series in Applied Mathematics, vol. 3 (Heldermann Verlag, Berlin, 1988), pp. 469–494

[N1] S.G., Nash, A. Sofer *Linear and Nonlinear Programming* (McGraw-Hill Companies, New York, 1996)

[188] Y. Nesterov, Efficiency of coordinate descent methods on huge-scale optimization problems. SIAM J. Optim. **22**, 341–362 (2012)

[189] Y.E. Nesterov, Semidefinite relaxation and nonconvex quadratic optimization. Optim. Methods Softw. **9**, 141–160 (1998)

[190] Y. Nesterov, A method of solving a convex programming problem with convergence rate $O((1/k^2))$. Soviet Math. Doklady **27**(2), 372–376 (1983)

[191] Y. Nesterov, M.J. Todd, Y. Ye, Infeasible-start primal-dual methods and infeasibility detectors for nonlinear programming problems. Math. Program. **84**, 227–267 (1999)

[N2] Y. Nesterov, A. Nemirovskii, *Interior Point Polynomial Methods in Convex Programming: Theory and Algorithms* (SIAM Publications, Philadelphia, 1994)

[N3] Y. Nesterov, M.J. Todd, Self-scaled barriers and interior-point methods for convex programming. Math. Oper. Res. **22**(1) 1–42 (1997)

[N4] Y. Nesterov, *Introductory Lectures on Convex Optimization: A Basic Course* (Kluwer, Boston, 2004)

[O1] W. Orchard-Hays, Background development and extensions of the revised simplex method. RAND Report RM-1433, The RAND Corporation, Santa Monica, CA (1954)

[O2] A. Orden, Application of the simplex method to a variety of matrix problems, in *Directorate of Management Analysis: "Symposium on Linear Inequalities and Programming"*, ed. by A. Orden, L. Goldstein (DCS/Comptroller, Headquarters, U.S. Air Force, Washington, DC, 1952), pp. 28–50

[O3] A. Orden, The transshipment problem. Manag. Sci. **2**(3), 276–285 (1956)

[O4] S.S. Oren, Self-scaling variable metric (ssvm) algorithms ii: implementation and experiments. Manag. Sci. **20**, 863–874 (1974)

[O5] S.S. Oren, D.G. Luenberger, Self-scaling variable metric (ssvm) algorithms i: criteria and sufficient conditions for scaling a class of algorithms. Manag. Sci. **20**, 845–862 (1974)

[O6] S.S. Oren, E. Spedicato, Optimal conditioning of self-scaling variable metric algorithms. Math. Program. **10**, 70–90 (1976)

[O7] J.M. Ortega, W.C. Rheinboldt, *Iterative Solution of Nonlinear Equations in Several Variables* (Academic, New York, 1970)

[P1] C.C. Paige, M.A. Saunders, Solution of sparse indefinite systems of linear equations. SIAM J. Numer. Anal. **12**(4), 617–629 (1975)

[P2] C. Papadimitriou, K. Steiglitz, *Combinatorial Optimization Algorithms and Complexity* (Prentice-Hall, Englewood Cliffs, NJ, 1982)

[204] P. Parrilo, Semidefinite programming relaxations for semialgebraic problems. Math. Program. **96**, 293–320 (2003)

[205] G. Pataki, On the rank of extreme matrices in semidefinite programs and the multiplicity of optimal eigenvalues. Math. Oper. Res. **23**, 339–358 (1998)

[P3] A. Perry, A modified conjugate gradient algorithm, Discussion Paper No. 229, Center for Mathematical Studies in Economics and Management Science, North-Western University, Evanston, IL (1976)

[P4] E. Polak, *Computational Methods in Optimization: A Unified Approach* (Academic, New York, 1971)

[P5] E. Polak, G. Ribiere, Note sur la Convergence de Methods de Directions Conjugres. Revue Francaise Informat. Recherche Operationnelle **16**, 35–43 (1969)

[P6] M.J.D. Powell, An efficient method for finding the minimum of a function of several variables without calculating derivatives. Comput. J. **7**, 155–162 (1964)

[P7] M.J.D. Powell, A method for nonlinear constraints in minimization problems, in *Optimization*, ed. by R. Fletcher Powell (Academic, London, 1969), pp. 283–298

[P8] M.J.D. Powell, On the convergence of the variable metric algorithm. Mathematics Branch, Atomic Energy Research Establishment, Harwell, Berkshire, England, (October 1969)

[P9] M.J.D. Powell, Algorithms for nonlinear constraints that use lagrangian functions. Math. Program. **14**, 224–248 (1978)

[P10] B.N. Pshenichny, Y.M. Danilin, *Numerical Methods in Extremal Problems* (translated from Russian by V. Zhitomirsky) (MIR Publishers, Moscow, 1978)

[214] M. Ramana, An exact duality theory for semidefinite programming and its complexity implications. Math. Program. **77**, 129–162 (1997)

[215] M. Ramana, L. Tuncel H. Wolkowicz, Strong duality for semidefinite programming. SIAM J. Optim. **7**, 641–662 (1997)

[R1] J. Renegar, A polynomial-time algorithm, based on Newton's method, for linear programming. Math. Program. **40**, 59–93 (1988)

[R2] J. Renegar, *A Mathematical View of Interior-Point Methods in Convex Optimization* (Society for Industrial and Applied Mathematics, Philadelphia, 2001)

[R3] R.T. Rockafellar, The multiplier method of hestenes and powell applied to convex programming. J. Optim. Theory Appl. **12**, 555–562 (1973)

[219] R.T. Rockafellar, *Convex Analysis* (Princeton University Press, Princeton, NJ, 1970)

[R4] C. Roos, T. Terlaky, J.-Ph. Vial, *Theory and Algorithms for Linear Optimization: An Interior Point Approach* (Wiley, Chichester, 1997)

[R5] J. Rosen, The gradient projection method for nonlinear programming, I. Linear contraints. J. Soc. Ind. Appl. Math. **8**, 181–217 (1960)

[R6] J. Rosen, The gradient projection method for nonlinear programming, II. Non-linear constraints. J. Soc. Ind. Appl. Math. **9**, 514–532 (1961)

[S1] R. Saigal, *Linear Programming: Modern Integrated Analysis* (Kluwer Academic Publisher, Boston, 1995)

[S2] B. Shah, R. Buehler, O. Kempthorne, Some algorithms for minimizing a function of several variables. J. Soc. Ind. Appl. Math. **12**, 74–92 (1964)

[S3] D.F. Shanno, Conditioning of quasi-Newton methods for function minimization. Math. Comput. **24**, 647–656 (1970)

[S4] D.F. Shanno, Conjugate gradient methods with inexact line searches. Math. Oper. Res. **3**(3) 244–2560 (1978)

[S5] A. Shefi, Reduction of linear inequality constraints and determination of all feasible extreme points, Ph.D. dissertation, Department of Engineering-Economic Systems, Stanford University, Stanford, CA, October 1969

[228] W.F. Sheppard, On the calculation of the double integral expressing normal correlation. Trans. Camb. Philos. Soc. **19**, 23–66 (1900)

[S6] M. Simonnard, *Linear Programming*, translated by William S. Jewell (Prentice-Hall, Englewood Cliffs, NJ, 1966)

[S7] M. Slater, Lagrange multipliers revisited: a contribution to non-linear programming. Cowles Commission Discussion Paper, Math 403 (November 1950)

[231] A.M. So, Y. Ye, Theory of semidefinite programming for sensor network localization. Math. Program. **109**, 367–384 (2007)

[232] A.M. So, Y. Ye, J. Zhang, A unified theorem on SDP rank reduction. Math. Oper. Res. **33**, 910–920 (2008)

[S8] G. Sonnevend, An 'analytic center' for polyhedrons and new classes of global algorithms for linear (smooth, convex) programming, in *System Modelling and Optimization: Proceedings of the 12th IFIP-Conference held in Budapest, Hungary, September 1985*, ed. by A. Prekopa, J. Szelezsan, B. Strazicky. Lecture Notes in Control and Information Sciences, vol. 84 (Springer, Berlin, 1986), pp. 866–876

[S9] G.W. Stewart, A modification of Davidon's minimization method to accept difference approximations of derivatives. J. ACM **14**, 72–83 (1967)

[S10] E.L. Stiefel, Kernel polynomials in linear algebra and their numerical applications. Nat. Bur. Stand. Appl. Math. Ser. **49**, 1–22 (1958)

[S11] J.F. Sturm, Using SeDuMi 1.02, a MATLAB toolbox for optimization over symmetric cones. Optim. Methods Softw. **11&12**, 625–633 (1999)

[S12] J. Sun, L. Qi, An interior point algorithm of $O(\sqrt{n}|\ln(\epsilon)|)$ iterations for C^1-convex programming. Math. Program. **57**, 239–257 (1992)

[T1] A. Tamir, Line search techniques based on interpolating polynomials using function values only. Manag. Sci. **22**(5), 576–586 (1976)

[T2] K. Tanabe, Complementarity-enforced centered Newton method for mathematical programming, in *New Methods for Linear Programming*, ed. by K. Tone (The Institute of Statistical Mathematics, Tokyo, 1987), pp. 118–144

[T3] R.A. Tapia, Quasi-Newton methods for equality constrained optimization: equivalents of existing methods and new implementation, in *Symposium on Nonlinear Programming III*, ed. by O. Mangasarian, R. Meyer, S. Robinson (Academic, New York, 1978), pp. 125–164

[T4] M.J. Todd, A low complexity interior point algorithm for linear programming. SIAM J. Optim. **2**, 198–209 (1992)

[T5] M.J. Todd, Y. Ye, A centered projective algorithm for linear programming. Math. Oper. Res. **15**, 508–529 (1990)

[T6] K. Tone, Revisions of constraint approximations in the successive qp method for nonlinear programming problems. Math. Program. **26**(2), 144–152 (1983)

[T7] D.M. Topkis, A note on cutting-plane methods without nested constraint sets. ORC 69-36, Operations Research Center, College of Engineering, Berkeley, CA (December 1969)

[T8] D.M. Topkis, A.F. Veinott Jr., On the convergence of some feasible direction algorithms for nonlinear programming. J. SIAM Control **5**(2), 268–279 (1967)

[T9] J.F. Traub, *Iterative Methods for the Solution of Equations* (Prentice-Hall, Englewood Cliffs, NJ, 1964)

[T10] L. Tunçel, Constant potential primal-dual algorithms: a framework. Math. Prog. **66**, 145–159 (1994)

[T11] R. Tutuncu, An infeasible-interior-point potential-reduction algorithm for linear programming, Ph.D. thesis, School of Operations Research and Industrial Engineering, Cornell University, Ithaca, NY, 1995

[T12] P. Tseng, Complexity analysis of a linear complementarity algorithm based on a Lyapunov function. Math. Program. **53**, 297–306 (1992)

[V1] P.M. Vaidya, An algorithm for linear programming which requires $O((m + n)n^2 + (m + n)^{1.5}nL)$ arithmetic operations. Math. Prog. **47**, 175–201 (1990). Condensed version in: Proceedings of the 19th Annual ACM Symposium on Theory of Computing, 1987, pp. 29–38

[V2] L. Vandenberghe, S. Boyd, Semidefinite programming. SIAM Rev. **38**(1) 49–95 (1996)

[V3] R.J. Vanderbei, *Linear Programming: Foundations and Extensions* (Kluwer Academic Publishers, Boston, 1997)

[V4] S.A. Vavasis, *Nonlinear Optimization: Complexity Issues* (Oxford Science, New York, NY, 1991)

[V5] A.F. Veinott Jr., The supporting hyperplane method for unimodal programming. Oper. Res. XV **1**, 147–152 (1967)

[V6] Y.V. Vorobyev, *Methods of Moments in Applied Mathematics* (Gordon and Breach, New York, 1965)

[W1] D.J. Wilde, C.S. Beightler, *Foundations of Optimization* (Prentice-Hall, Englewood Cliffs, NJ, 1967)

[W2] R.B. Wilson, A simplicial algorithm for concave programming, Ph.D. dissertation, Harvard University Graduate School of Business Administration, 1963

[W3] P. Wolfe, A duality theorem for nonlinear programming. Q. Appl. Math. **19**, 239–244 (1961)

[W4] P. Wolfe, On the convergence of gradient methods under constraints. IBM Research Report RZ 204, Zurich (1966)

[W5] P. Wolfe, Methods of nonlinear programming (Chap. 6), in *Nonlinear Programming*, ed. by J. Abadie. Interscience (Wiley, New York, 1967), pp. 97–131

[W6] P. Wolfe, Convergence conditions for ascent methods. SIAM Rev. **11**, 226–235 (1969)

[W7] P. Wolfe, Convergence theory in nonlinear programming (Chap. 1), in *Integer and Nonlinear Programming*, ed. by J. Abadie (North-Holland Publishing Company, Amsterdam, 1970)

[W8] S.J. Wright, *Primal-Dual Interior-Point Methods* (SIAM, Philadelphia, 1996)

[264] G. Xue, Y. Ye, Efficient algorithms for minimizing a sum of Euclidean norms with applications. SIAM J. Optim. **7**, 1017–1036 (1997)

[265] Y. Ye, Approximating quadratic programming with bound and quadratic constraints. Math. Program. **84**, 219–226 (1999)

[Y1] Y. Ye, An $O(n^3L)$ potential reduction algorithm for linear programming. Math. Program. **50**, 239–258 (1991)

[Y2] Y. Ye, M.J. Todd, S. Mizuno, An $O(\sqrt{n}L)$-iteration homogeneous and self-dual linear programming algorithm. Math. Oper. Res. **19**, 53–67 (1994)

[Y3] Y. Ye, *Interior Point Algorithms* (Wiley, New York, 1997)

[269] Y. Ye, A new complexity result on solving the markov decision problem. Math. Oper. Res. **30**, 733–749 (2005)

[Z1] W.I. Zangwill, Nonlinear programming via penalty functions. Manag. Sci. **13**(5), 344–358 (1967)

[Z2] W.I. Zangwill, *Nonlinear Programming: A Unified Approach* (Prentice-Hall, Englewood Cliffs, NJ, 1969)

[Z3] Y. Zhang, D. Zhang, On polynomiality of the mehrotra-type predictor-corrector interior-point algorithms. Math. Program. **68**, 303–317 (1995)

[Z4] G. Zoutendijk, *Methods of Feasible Directions* (Elsevier, Amsterdam, 1960)

Index

D.G. Luenberger, Y. Ye, *Linear and Nonlinear Programming*, International Series in Operations Research & Management Science 228,
DOI 10.1007/978-3-319-18842-3

D

CPSIA information can be obtained at www.ICGtesting.com
Printed in the USA
LVOW10*1037051015

456930LV00010B/116/P